Ferromagnetism

IEEE Press
445 Hoes Lane, PO Box 1331
Piscataway, NJ 08855-1331

1993 Editorial Board
William Perkins, *Editor in Chief*

R. S. Blicq	R. F. Hoyt	J. M. F. Moura
M. Eden	J. D. Irwin	I. Peden
D. M. Etter	S. V. Kartalopoulos	L. Shaw
J. J. Farrell III	P. Laplante	M. Simaan
L. E. Frenzel	M. Lightner	D. J. Wells
G. F. Hoffnagle	E. K. Miller	

Dudley R. Kay, *Director of Book Publishing*
Carrie Briggs, *Administrative Assistant*

Valerie Zaborski, *Production Editor*

*Reissued in cooperation with the
IEEE Magnetics Society*

IEEE Magnetics Society Liaison to IEEE Press

Roger F. Hoyt
IBM Almaden Research Center

An IEEE Press Classic Reissue

FERROMAGNETISM

Richard M. Bozorth

IEEE PRESS

IEEE Magnetics Society, *Sponsor*

The Institute of Electrical and Electronics Engineers, Inc., New York

Copyright © 1951, D. Van Nostrand Company, Inc.
Copyright © 1978, American Telephone and Telegraph Company.

Printed in the United States of America
10 9 8 7 6 5 4 3 2

ISBN 0-7803-1032-2
IEEE Order number: PC03814

Dedicated to
L. H. B.

PREFACE

The material of this book is the background knowledge that has been useful to various members of the Bell Telephone Laboratories who have been actively interested in ferromagnetism. These staff members fall into three main categories: (1) physicists engaged in research on the nature of ferromagnetic phenomena, (2) those carrying on the development of new magnetic materials, and (3) engineers interested in the properties and behavior of magnetic materials suitable for use in apparatus. The knowledge has been drawn from the fields of metallurgy, chemistry, and electrical engineering, as well as from the field of physics.

The book is divided into two main parts, which describe respectively the properties of magnetic materials and the nature of magnetic phenomena. These two portions of ferromagnetism are brought together in one book in the belief that each part is necessary for the development of the other — the physicist needs the empirical data and the engineer should understand the principles governing the material that he uses. The approach is mainly descriptive and nonmathematical when discussing materials, and the emphasis is on the physical concepts when discussing theory. In many places only the outline and the results of analysis have been given, and the reader may regard these places as problems and fill in the missing steps, if he desires.

The author has believed for some time that expositions of subjects such as this can generally be improved by greater than customary use of figures and appropriate captions. In accordance with this idea the average number of figures per page has been increased to almost 0.9. For the preparation of these figures grateful acknowledgment is made to Miss M. Goertz, who plotted many of the data, and to Mr. S. Lund, who designed the final drawings. The drawing has been done under the able direction of Mr. H. P. Gridley. Mrs. E. M. Sparks has been responsible for the reproduction and collation of the text and figures of the manuscript.

The author has had the advantage of the reading and the criticism of the manuscript by a number of fellow physicists. L. W. McKeehan of Yale University has read the whole manuscript and suggested many valuable changes. In the chapters on materials, especially R. A. Chegwidden and J. A. Ashworth have supplied information and criticism. The second half of the book has been read by C. Kittel, and portions of it by K. K. Darrow and H. J. Williams, all of whom have given able criticism. Chapter 4 on Iron-Silicon Alloys has benefited much from the information obtained from G. H. Cole of the Armco Steel Corporation, from W. Morrill of the General Electric Company, and from C. W. Stoker of the Carnegie Illinois Steel Corporation, as well as from E. S. Greiner of the Bell Laboratories. E. M. Wise of the International Nickel Company graciously supplied information on the production, uses, and properties of nickel (Chapter 8).

Chapter 9 on Permanent Magnets has benefited from criticism by W. Ruder, A. H. Geisler, and D. L. Martin of the General Electric Company and by E. A. Nesbitt and R. A. Chegwidden of the Bell Laboratories. Comments on the preparation of magnetic material were received from J. H. Scaff, D. H. Wenny, and J. R. Townsend. J. E. Goldman of Carnegie Institute of Technology has read Chapter 12, and S. J. Barnett of the California Institute of Technology, portions of Chapter 10. C. D. Owens and V. E. Legg have commented on Chapter 17, and L. R. Maxwell and F. G. Brockman, on the description of the properties of ferrites. A number of people have kindly supplied me with information which has not previously been published; these are mentioned specifically in the text. Some figures are reproduced from articles published by the author in the *Physical Review*, *Journal of Applied Physics*, and *Encyclopaedia Britannica*, with the kind permission of the publishers.

Extensive use has been made of *Ferromagnetismus* by Becker and Döring (Springer, Berlin, 1939) and *Magnetism and Matter* by Stoner (Methuen, London, 1934), and of the useful compilations of von Auwers (1936), Hansen (1936), and others.

Finally, it is a pleasure to acknowledge my indebtedness to Dr. O. E. Buckley and Professor L. W. McKeehan for introducing me to the study of magnetism and directing my early efforts in this field.

In a book of this kind it is inevitable that there should be some errors, and some omissions of work important enough to be included. It is hoped that readers will be tolerant and notify the author of those that they find.

RICHARD M. BOZORTH

Murray Hill, N.J.
January 1951

CONTENTS

PART I. INTRODUCTION

CHAPTER	PAGE
1. CONCEPTS OF FERROMAGNETISM	1

    Magnetic Field 1
    Intensity of Magnetization and Magnetic Induction 2
    Magnetization and Permeability Curves 3
    Hysteresis Loop 4
    Ferromagnetism, Paramagnetism and Diamagnetism 5
    Kinds of Permeability 6
    Saturation 7
    Ideal Magnetization 8
    Demagnetization Curve 8
    Demagnetizing Field, Air Gaps 9
    Magnetostriction 11
    Magnetic Anisotropy 11
    Spontaneous Magnetization 12

2. FACTORS AFFECTING MAGNETIC QUALITY 14

    Phase Diagram 15
    Physical Factors 18

Production of Magnetic Materials   19

    Melting and Casting 20
    Fabrication 24
    Heat-Treatment 28

Effect of Composition   29

    Gross Chemical Composition 29
    Impurities 35

Some Important Physical Properties   38

    Properties Affected by Magnetization 46

PART II. MAGNETIC PROPERTIES OF MATERIALS

3. TECHNICAL AND PURE IRON 48

    Manufacture of Iron and Steel 48
    Chemical Composition 51
    Preparation of Pure Iron 51
    Physical Properties 53
    Magnetic Properties of Commercial Material 55
    Temperature 56
    Heat Treatment of Ordinary Iron 58

CHAPTER	PAGE
Hydrogen Treatment	60
Removal of Impurities	61
4. IRON-SILICON ALLOYS	**67**
History of Magnetic Alloys	67
Methods of Production	68
Phase Diagram	71
Physical Properties	74
Magnetic Saturation and Curie Point	76
Permeability and Losses of Hot-Rolled Sheet	79
Effect of Non-Metallic Impurities	83
Grain Size	86
Metallic Additions	87
Preferred Crystal Orientation	88
Properties at Low Inductions	92
Silicon Cast Iron	94
Structural Steels	95
Iron-Silicon-Aluminum Alloys	95
5. IRON-NICKEL ALLOYS	**102**
Binary Iron-Nickel Alloys	102
Structure	102
Density, Lattice Parameter and Thermal Expansion	103
Resistivity	104
Elastic and Mechanical Properties	106
History of Magnetic Investigations	106
Magnetic Saturation	111
Preparation and Annealing of Alloys	112
Dependence of Induction on Field Strength	113
Permeability, Coercive Force and Hysteresis Loss	115
Annealing Temperature and Time	115
Heat Treatment in a Magnetic Field	117
Heat Treatment in Hydrogen	121
Mechanical Treatment	124
Commercial Binary Alloys	128
78 Permalloy	128
45 Permalloy. Hipernik	129
Compressed Permalloy Powder	131
Isoperm	133
Permenorm 5000 Z	133
Thermoperm	133
Iron-Nickel-Molybdenum Alloys	134
Structure	134
Magnetic Properties	136
Commercial Alloys. Molybdenum Permalloy	139
Supermalloy	140
Compressed Molybdenum Permalloy Powder	144

CHAPTER	PAGE
Iron-Nickel-Chromium Alloys	146
Structure and Magnetic Properties	146
Commercial Alloys	152
Iron-Nickel-Copper Alloys	153
Structure and Physical Properties	153
Magnetic Properties	155
Commercial Alloys. Mumetal	159
1040 Alloy	160
Iron-Cobalt-Nickel Alloys	160
Structure and Physical Properties	161
Magnetic Properties	163
Heat Treatment in a Magnetic Field	171
Reversible Permeability	177
Commercial Alloys	179
Other Iron-Nickel Alloys	180
Iron-Nickel-Manganese Alloys	180
Iron-Nickel-Aluminum Alloys	183
Iron-Nickel-Silicon Alloys	184
Iron-Nickel Alloys Containing Ti, V, Ta or W	186
Alloys Containing Silver or Beryllium	189
6. IRON-COBALT ALLOYS	190
Structure and Physical Properties	190
Magnetic Properties	193
Iron-Cobalt Alloys with Additions	199
Iron-Cobalt-Vanadium Alloys	200
Other Iron-Cobalt Alloys	205
7. OTHER IRON ALLOYS OF HIGH PERMEABILITY	210
Iron-Aluminum Alloys	210
Iron-Antimony to Iron-Oxygen	220
Ferrites	244
Iron Alloyed with Palladium and Platinum Metals	249
Iron-Phosphorus to Iron-Zirconium	251
8. OTHER HIGH PERMEABILITY MATERIALS	261
Cobalt	261
Metallurgy of Cobalt	261
Physical Properties	262
Magnetic Properties	264
Nickel	267
Metallurgy of Nickel	268
Physical Properties	269
Magnetic Properties	269

CHAPTER	PAGE
Cobalt-Nickel Alloys	276
Structure and Physical Properties	276
Magnetic Properties	278
Ternary Alloys	280
Other Cobalt Alloys	281
Cobalt-Aluminum to Cobalt-Zirconium	283
Other Nickel Alloys	299
Nickel-Aluminum to Nickel-Zirconium	299
Heusler Alloys	328
Structure	328
Saturation and Curie Point	330
Permeability and Hysteresis	333
Other Manganese Alloys	334
Manganese-Antimony to Manganese-Tin	334
Curie Points of Arsenic Group	341
Other Alloys and Elements	341
Chromium-Antimony to Chromium-Tellurium	341
Silver-Fluorine	342
Potassium-Sulfur	342
Gadolinium	342
9. PERMANENT MAGNETS	344
Introduction	344
History	345
Energy Product	348
Demagnetization Curve	349
Reversible Permeability and Spring Back	353
Stability	354
Design of Permanent Magnets	359
Properties of Alloy Systems	364
Iron-Carbon Alloys. Constitution	364
Physical Properties	367
Magnetic Properties	367
Additions of Manganese	370
Tungsten Steels	371
Chrome Steels	374
Chrome-Tungsten Steels	378
Molybdenum Steel	378
Cobalt Steels	379
Other Steels	382
Iron-Cobalt-Molybdenum and Iron-Cobalt-Tungsten Alloys	382
Iron-Nickel-Aluminum Alloys	385
Additions of Cobalt and Copper	388
Heat Treatment in Magnetic Field	389

CHAPTER	PAGE
Alloys Containing Titanium	394
Iron-Nickel-Copper Alloys	396
Cobalt-Nickel-Copper Alloys	402
Iron-Cobalt-Vanadium Alloys	405
Alloys with Noble Metals. Iron-Platinum	410
Iron-Palladium	411
Iron with Other Noble Metals	412
Cobalt-Platinum	412
Cobalt-Palladium	414
Nickel with Noble Metals	414
Chromium-Platinum	415
Manganese-Silver-Aluminum Alloys	415
Other Binary Alloys	416
Other Ternary and Complex Alloys	416
Powder Metallurgy	416
Magnets from Electrodeposited Powder	417
Powder Magnets	419
Oxide Magnets	421

PART III. MAGNETIC PHENOMENA AND THEORIES

10. **MAGNETIC THEORY** . . . 423

Ferromagnetism 423

Ewing's Theory	423
Limitations of Ewing's Theory	425
The Weiss Theory	427
Quantum Theory	430
Ferromagnetics Above the Curie Point	432
Molecular Field Constant, N	433
Atomic Structure of Ferromagnetic Materials	434
Collective Electron Ferromagnetism	442
Interpretation of the Molecular Field	443
Thermal Expansion Near $T = \theta$	447
$T^{3/2}$ Law	448
Gyromagnetic Effect	449
Gyromagnetic Experiments	451
Experimental Values of g	453

Diamagnetism 455

Theory of Diamagnetism	458
Simple Salts and Their Solutions	459
Other Compounds	460

Paramagnetism 461

Rare Earths	463
Ions of the Iron Group	464
Paramagnetism at Low Temperatures	465
Paramagnetic Gases	466
Paramagnetism of Free Electrons	467

CHAPTER	PAGE
Molecular Beams	468
Nuclear Moments	468
Antiferromagnetism	470

11. The Magnetization Curve and the Domain Theory 476

Three Parts of Magnetization Curve	476
General Description of Domain Theory	477
Stability of Domain Orientation	481
Approach to Saturation	484
Initial Portion of Curve	489
Experimental Test of Rayleigh's Law	492
Rectangular Hysteresis Loops	494
Constricted Loops	498
Coercive Force and Residual Induction	499
Interpretation of Retentivity by Domain Theory	501
Hysteresis	507
Rotational Hysteresis	514
Distribution of Heat-Loss Over Magnetic Cycle	518
Barkhausen Effect	524
Powder Patterns	532
Incremental and Reversible Permeability	538
Reversible Permeability vs Induction	543
Incremental Permeability vs Amplitude	546
Hysteresis with Superposed Fields	549
Superposed Non-Parallel Fields	552

12. Magnetic Properties of Crystals 555

Permeability	558
Remanence and Coercive Force	561
Crystal Anisotropy Energy	563
Torque Curves	565
Values of Constants	567
Calculation of Magnetization Curves	576
Calculation of Torque Curves	584
Anisotropy in Polycrystalline Sheet	586
Rotational Hysteresis	590
Origin of Anistropy	592

13. Stress and Magnetostriction 595

Stress   595

Strain Beyond Elastic Limit	596
Stress Within the Elastic Limit	600
Domain Theory of Effect of Stress	610
Effect of Very Small Stresses	613
Internal Strains	619

Magnetostriction   627

Experimental Methods	628
Brief Survey	630

CHAPTER	PAGE
Domain Theory	634
Preferred Domain Orientations	636
Reversible Magnetostriction	639
Volume Magnetostriction	641
Magnetostriction of Single Crystals	645
Theory of Saturation in Single Crystals	649
Magnetostriction in Unsaturated Crystals	652
Origin of Magnetostriction	654

Magnetostriction Data — 655

Iron	655
Cobalt	657
Nickel	659
Iron-Cobalt Alloys	663
Iron-Nickel Alloys	664
Cobalt-Nickel Alloys	672
Other Iron-Cobalt-Nickel Alloys	674
Other Iron Alloys	676
Other Soft Materials	680
Permanent Magnet Materials	682

Change of Elastic Modulus — 684

Introduction	684
Experimental Methods	686
Some Experimental Results	687
Theory for Large Internal Strain	690
Small Internal Strains	692
Comparison with Data	693
Change of E Near Curie Point	697
Other Data	699

Magnetomechanical Damping — 699

Damping Constants	700
Damping in Some Materials	701
Macro Eddy Currents	703
Micro Eddy Currents	706
Magnetomechanical Hysteresis	707
Separation of Losses	709
Additional Data	712

14. **TEMPERATURE AND THE CURIE POINT** — 713

Effect of Phase Changes	715
Magnetization Near the Curie Point	716
Low Temperatures	719
Variation of Curie Point with Composition	720
Curie Point and Pressure	723
Changes in Properties Near Curie Point	728
Applications	728

15. **ENERGY, SPECIFIC HEAT, AND MAGNETOCALORIC EFFECT** — 729

Energy of Magnetization	729

CHAPTER	PAGE
Thermodynamic Expressions	731
Application to Magnetostriction	732
Heat Capacity	732
Comparison with Experiment	734
Low Temperatures	738
Magnetocaloric Effect	740
Temperature Change and Torque	743
16. Magnetism and Electrical Properties	**745**
Resistivity and Field Strength (Magnetoresistance)	745
Effect of Tension	748
Quantitative Aspects of Domain Theory. Polycrystalline Material	752
Results for Alloy Series	756
Effect of Temperature	760
Magnetoresistance in Single Crystals	764
Effect of Anisotropy of Magnetostriction	766
Effect of Phase Change	767
17. Change of Magnetization with Time	**769**
Effect of Eddy Currents	769
Eddy Currents in Sheets	770
Eddy Currents in Cylinders	775
Wire Carrying Current	776
Eddy Current Losses	778
Losses in Low Fields	780
Losses at Intermediate and High Inductions	781
Eddy Currents with Change in Field	783
Propagation of Magnetic Waves	788
Magnetic Lag (not due to eddy currents)	788
Lag Dependent on Impurities	789
Jordan Lag	796
Long Period Lag. Aging	797
Permeability at Very High Frequencies	798
Ferromagnetic Resonance	803
Experimental Values of g	808
Resonance in Single Crystals	809
18. Special Problems in Domain Theory	**811**
Kinds of Magnetic Energy	811
Introduction	811
Crystal Anistropy	811
Magnetic Strain Energy	812
Mutual Energy Between Magnetization and External Field	813
Magnetostatic Energy	814
Energy of Bloch Wall	814
Evaluation of Wall Energy and Thickness	816
Magnetic Processes	818
Rotational Processes	818
Boundary Displacement	820

CHAPTER	PAGE
Applications to Specific Problems	821
Initial Permeability	821
Coercive Force	823
Fine Particles	828
Domain Geometry	834

PART IV. MEASUREMENTS

19. MEASUREMENT OF MAGNETIC QUANTITIES 838

Basic Relations — 838

Field of a Magnet 838
Fields Produced by Currents 839
Force on Magnet in Field 841
Force on Current in Field 842
Electromotive Force and Magnetomotive Force 842

Common Methods — 843

Ballistic Method with Ring 843
Rod Specimens. Demagnetizing Factors 845
Yokes and Permeameters 849
Alternating Current Methods 852

Special Methods — 855

Production of High Fields 855
Magnetometers 857
Para- and Diamagnetic Materials 857
Liquids and Gases 859
Other Special Measurements 860

APPENDIX 1. SYMBOLS USED IN TEXT 862

APPENDIX 2. SOME PHYSICAL PROPERTIES OF THE ELEMENTS 864

APPENDIX 3. VALUES OF SOME CONSTANTS 867

APPENDIX 4. MAGNETIC PROPERTIES OF VARIOUS MATERIALS 868

BIBLIOGRAPHY 875

NAME INDEX 947

SUBJECT INDEX 959

CHAPTER 1

CONCEPTS OF FERROMAGNETISM

The quality of magnetism first apparent to the ancient world, and to us today, is the tractive force that exists between two bodies such as lodestone or iron. When a magnetized body is dipped into iron filings they cling to it, especially in certain places called *poles* usually located near the ends of the magnet (Fig. 1–1). The concept of poles is useful in defining the two quantities basic to magnetism, magnetic field strength and intensity of magnetization.

FIG. 1-1. A magnetized bar attracts iron filings most strongly at its *poles*.

This chapter is to recall to the reader these and various other concepts associated with ferromagnetism. It is not intended to comprise definitions. Some definitions have been formulated by the American Society for Testing Materials [49A2] and the American Standards Association [42A2]. In this book the cgs and practical systems are used.

Magnetic Field.—A magnet will attract a piece of iron even though the two are not in contact, and this action-at-a-distance is said to be caused by the magnetic field, or field of force. This field may be explored by sprinkling iron filings around a magnet, whereupon they form in lines that converge on the poles and indicate also the direction a small compass needle would take if placed at any point (Fig. 1–2).

Poles exert forces on each other: north and south poles attract each other and like poles repel with a force that varies inversely as the square of the distance between them. A *unit pole* is a convenient concept defined so that two unit poles of like kind, one centimeter apart in vacuum, would repel each other with a force of one dyne. The strength of the field of force, the *magnetic field strength*, or magnetizing force H, may be defined in terms of magnetic poles: one centimeter from a unit pole the field strength is one

2   CONCEPTS OF FERROMAGNETISM

oersted. A magnetic field may be produced by a current of electricity as well as by a magnet, and the unit of field strength can also be defined in terms of current. (In the rationalized MKS system, not used in this book, the unit of field strength is one ampere-turn/meter, and is $4\pi/1000$ or 0.01256 oersteds.)

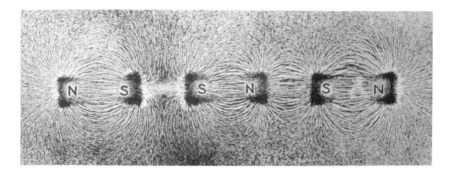

FIG. 1-2. Iron filings indicate the directions of the magnetic field near a group of magnets and show the lines of force emanating from S poles and converging on N poles.

A magnetic field has direction as well as strength; the direction is that in which a north pole, subjected to it, tends to move, or that indicated by the north-seeking end of a small compass needle placed at the point.

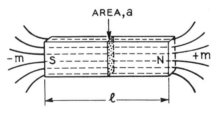

FIG. 1-3. A pole of strength m and area a corresponds to an intensity of magnetization $I = m/a$. Lines of induction are endless, pass into the magnet at S and leave at N.

Intensity of Magnetization and Magnetic Induction.—In order to describe the magnetic properties of materials, one must have a quantitative measure of magnetization. Such a measure is the *intensity of magnetization*, defined in terms of the number of unit poles in a piece of given cross-sectional area. Suppose that a uniformly magnetized bar, of length l and cross-sectional area a, has m unit north poles at one end and m unit south poles at the other (Fig. 1-3). The intensity of magnetization is then m/a and is represented by the symbol I.

It may be shown that I is also the *magnetic moment per unit volume;* for the magnetic moment is pole strength times interpolar distance ($M = ml$), and $I = M/v$, if v is the volume al.

Faraday showed that some of the properties of magnetism may be likened to a flow and conceived endless *lines of induction* that represent the direction and, by their concentration, the flow at any point. The lines pass from a magnetized material into the air at a north pole, enter again at a south pole, and pass through the material from the south pole back to the north to form a closed loop.

The total number of lines crossing a given area at right angles is the *flux* in that area. The flux per unit area is the flux density, or *magnetic induction*, and is represented by the symbol B. Both H and I contribute to the lines of induction, but in magnetic materials the contribution of I is generally the larger. The magnetic induction is defined by the relation:

$$B = H + 4\pi I,$$

the addition being vectorial when H and I differ in direction. The occurrence of the factor 4π is caused by the fact that a unit pole gives rise to a unit field everywhere on the surface of a sphere of unit radius enclosing the pole, and this sphere has an area of 4π. The lines of induction may be visualized with the aid of Fig. 1-2 showing the pattern obtained with iron filings. The cgs unit of induction is the *gauss*. (The rationalized MKS unit, the weber/(meter)^2, is equal to 10^4 gauss.)

Alternatively, B can be defined in terms of the electromotive force created by the relative movement of electric circuit elements and lines of induction, and I can be derived from the definitions of B and H.

Magnetization and Permeability Curves.—When a piece of unmagnetized iron is brought near a magnet or subjected to the magnetic field of an electric current, the magnetization induced in the iron by the field is described by a magnetization curve obtained by plotting the intensity of magnetization I or the magnetic induction B against the field strength H. Such curves are of fundamental importance for describing the magnetic properties of materials, and many of them are shown on the following pages. A magnetization curve for iron is shown as the solid line of Fig. 1-4.

The behavior of a material is also described by its permeability curve and hysteresis loop. The ratio B/H is called the permeability μ, and this represents the relative increase in flux caused by the presence of the magnetic material.* It is quite useful when B is considered to be due to H. The permeability curve is obtained by plotting the permeability μ against either H, B (Fig. 1-5), $B-H$, or I. In any case the curve rises from a point on the μ-axis above the origin (the initial permeability is non-zero) to a maximum (the maximum permeability) and falls off rapidly and then more slowly toward a value of one (not zero). The quantity $B-H = 4\pi I$ attains a ceiling value, known as *saturation induction* and represented by the symbol B_s; when it is used as abscissa, the axis of

---

* In the rationalized MKS system the permeability of free space is $4\pi \cdot 10^{-7}$ (weber/meter^2) / (ampere-turn/meter).

abscissa ends at a finite distance from the origin, and so does the curve of μ vs $B-H$.

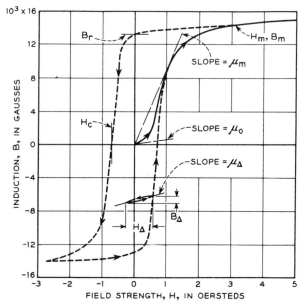

FIG. 1-4. Magnetization curve (solid) and hysteresis loop (dotted). Some important magnetic quantities are illustrated.

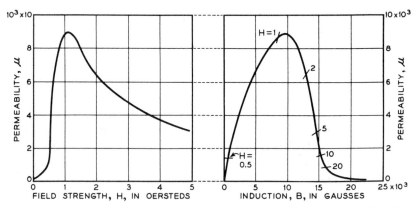

FIG. 1-5. Permeability curves of iron, with μ plotted against H and B. I and $B-H$ are also used as abscissae.

Hysteresis Loop.—If the field strength is first increased from zero to a high value and then decreased again, as indicated by the arrows of Fig. 1-4, it is observed that the original curve is not retraced; the induction "lags behind" the field and follows a characteristic curve, shown by the broken line. This phenomenon was named *hysteresis* by Ewing, and the characteristic curve is called a hysteresis loop. On a loop symmetrical about the origin, the value of H for which $B = 0$ is called the *coercive force* H_c, and this is often used as a measure of quality of the material. The value of induction for $H = 0$ is the *residual induction* B_r. When the field strength has been sufficient to magnetize the material practically to saturation, the coercive force and residual induction become the *coercivity* and *retentivity*. The values of H and B at the tips of a loop are usually called H_m and B_m.

Ferromagnetism, Paramagnetism, and Diamagnetism.—Materials which have magnetic properties similar to iron (e.g., nickel and cobalt and many alloys of these three elements) are ferromagnetic. In another class of materials, more numerous, the permeabilities are only slightly greater than one, usually between 1.000 and 1.001 (except near 0°K when they may be much larger). As a rule these materials do not show hysteresis, and their permeabilities are independent of field strength and are either independent of temperature or decrease with increasing temperature. Such materials are *paramagnetic*. Among the paramagnetic substances are many of the salts of the iron and the rare earth families and the platinum and palladium metals, the elements sodium, potassium and oxygen, and the ferromagnetic metals above the Curie points. They may be solids, liquids, or gases. In *diamagnetic* substances the magnetization is directed oppositely to the field, i.e., they have permeabilities somewhat less than one. They are, therefore, repelled from the poles of an electromagnet and tend to move toward a weaker field. Many of the metals and most of the nonmetals are diamagnetic.

Paramagnetic and diamagnetic substances are described more conveniently by their susceptibilities than by their permeabilities. The *susceptibility* is a measure of the increase in magnetic moment caused by the application of a field, and is defined as

$$\kappa = I/H,$$

or the equivalent,

$$\kappa = (\mu - 1)/4\pi.$$

For diamagnetic materials the susceptibility is negative, and for bismuth it has a value of -0.000013. For substances like iron the susceptibility may be 1000 or more; values as high as 10 000 have been recorded. Other kinds of susceptibility are referred to in Chap. 11.

It is sometimes difficult to draw the line separating ferromagnetic from paramagnetic substances. The important attributes of a ferromagnetic substance are dependence of permeability on the field strength and on the previous magnetic history (hysteresis), approach of the magnetization to a finite limit as the field strength is indefinitely increased (saturation), the presence of small, magnetized regions containing many atoms and having magnet moments comparable with the saturation moment of the material even when the material is unmagnetized (spontaneous magnetization), and disappearance of the characteristics already mentioned when the temperature is raised to a certain temperature, the *Curie point*. (See Figs. 2–5 and 7.) From a practical point of view one may say arbitrarily that a material is ferromagnetic if it has a permeability greater than 1.1. From an atomic point of view the atomic moments of a ferromagnetic material align themselves parallel to each other, against the forces of thermal agitation. In some materials the atomic forces align neighboring atoms antiparallel—this phenomenon is *antiferromagnetism*, and it is characterized by hysteresis and a Curie point. On account of their small permeabilities these materials are classed as paramagnetic.

In the broad sense of the word, as used by Faraday, paramagnetism includes ferromagnetism ($\mu > 1$). More often, and in this book, ferromagnetism is considered a separate classification. Ferromagnetic materials are usually designated "magnetic"; all materials are either ferromagnetic, paramagnetic, or diamagnetic.

Kinds of Permeability.—The *normal permeability*, often referred to simply as the permeability, is $\mu = B/H$, measured when the specimen is in the "cyclic magnetic state." Under these conditions the material responds equally when the field is applied in either of the two opposite directions. Before such a measurement the specimen is ordinarily demagnetized by applying an alternating field with amplitude high enough to cause the induction to approach saturation, then slowly reducing the amplitude to zero. The material may be demagnetized also by heating the material above the Curie point and cooling in zero field; in this case the magnetization curve is referred to as the "virgin" curve. In either case the curve represents the values of B (or $B-H$ or I) and H measured as H is increased from zero, or it is (preferably) the locus of the tips of hysteresis loops taken with increasing amplitudes of H, the field having been reversed several times at each amplitude. The latter is sometimes called the "commutation" curve.

The *initial permeability* μ_0 is the limit approached by the normal permeability as B and H are decreased toward zero. Some permeability curves in weak fields are shown in Fig. 1–6; extrapolation is linear when H is sufficiently small.

The *maximum permeability* μ_m is the largest value of normal permeability obtained by varying the amplitude of H (see Figs. 1–4 and 5).

The *incremental permeability* μ_Δ refers to the permeability measured with superposed fields. Let one ("biasing") field H_b be applied and held constant, and another field H_Δ be applied and alternated cyclically, causing an alternating induction of amplitude B_Δ. Then $\mu_\Delta = B_\Delta/H_\Delta$ (Fig. 1–4).

When H_Δ approaches zero, μ_Δ approaches a limiting value μ_r, the reversible permeability. When the material is demagnetized and $H_b = 0$, $\mu_r = \mu_0$. Both μ_Δ and μ_r are dependent on the value of H_b and on the previous

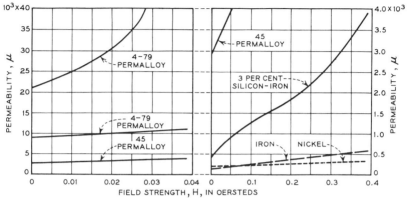

Fig. 1-6. Curves showing μ vs H for initial portions of magnetization curves of various materials. Lines are usually straight near axis of μ.

magnetic history of the specimen; μ_Δ is also dependent on the magnitude of H_Δ.

Occasionally the term "differential permeability" is used—it is simply the slope of the B vs H curve, dB/dH.

Saturation.—As H is increased indefinitely the intensity of magnetization I and the *intrinsic induction* $B - H$ of a ferromagnetic material approach finite limits, commonly referred to as "saturation." The induction B, on the contrary, increases indefinitely. This is shown for a permanent magnet material, Alnico 5, in Fig. 1-7. The limit of $B - H$ at saturation is designated simply by B_s, and the limit of I is I_s. Thus $B_s = 4\pi I_s$. The saturation at the absolute zero of temperature is represented by the symbol I_0.

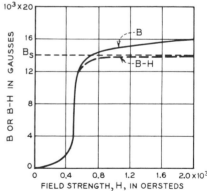

Fig. 1-7. In high fields $B-H$ (and I) approach asymptotically to limit B_s (and I_s), *saturation*. Material is permanent magnet, Alnico 5.

The magnetic moment per gram σ is equal to the intensity of magnetization divided by the density d:

$$\sigma = I/d.$$

The saturation per gram is $\sigma_s = I_s/d$. The magnetic moment per gram atom σ_A is

$$\sigma_A = AI/d,$$

A being the atomic weight.

There is an advantage in using σ when measurements of magnetic moment are made at various temperatures, for then one merely divides the magnetic moment by the known weight, whereas, to calculate I, the density at each temperature must also be known.

The magnetic moment per atom is

$$\mu_A = \frac{AI_0}{N_0 d} = \frac{A\sigma_0}{N_0},$$

N_0 being the number of atoms per gram atom (6.025×10^{23}). The value of μ_A is 2.06×10^{-20} for Fe. It is more usual to express the atomic moment in terms of the Bohr magneton β, the moment which arises from the motion of a single electron moving in its smallest orbit. The number of Bohr magnetons per atom is

$$n_0 = \mu_A/\beta,$$

β being 9.27×10^{-21} erg/gauss. The values of n_0 for Fe, Co, and Ni are 2.22, 1.71, and 0.60, respectively.

Ideal Magnetization.—This refers to the magnetization remaining after applying a constant field H, superposing on it a field varying continually from $+H_\Delta$ to $-H_\Delta$, large enough in amplitude to cause practical saturation in each direction, then reducing the amplitude of H_Δ slowly to zero. The resulting I or B is plotted against H, and the *ideal* or *anhysteretic* magnetization curve so obtained has the characteristic shape shown in Fig. 1-8, with no point of inflection. The *ideal permeability*, the ratio of the B so obtained to the constant H, has a very high finite maximum at $H = 0$, of about the same magnitude as the maximum value of dB/dH for the hysteresis loop. The course of the ideal curve is near the curve defined by the midpoints of horizontal chords in the maximum loop.

Demagnetization Curve.—This is the portion of the hysteresis loop that lies in the second quadrant, between the points marked B_r and H_c in Fig. 1-4, and is shown for an important permanent magnet material on the left side of Fig. 1-9. This curve is especially important in the evaluation of materials used for permanent magnets, for in the use of such materials they are subjected to fields that tend to reduce the magnetization they originally possessed. A more specific criterion is the so-called *energy product*, obtained by multiplying together the magnitudes of B and H for a given point on the demagnetization curve. The energy product for all points on the demagnetization curve of Fig. 1-9 is plotted against B in the right-hand side of the same figure. The maximum value $(BH)_m$ of this curve is the best single criterion for a material for use as a permanent magnet. More detailed consideration of the demagnetization curve and the energy product are given in Chap. 9.

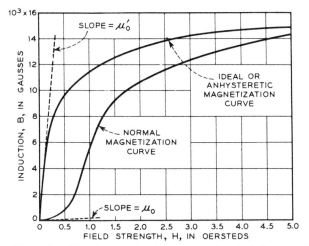

FIG. 1-8.—Normal and ideal magnetization curves of typical shape. Material is iron.

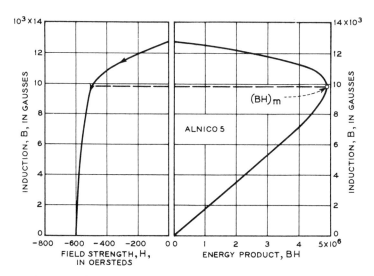

FIG. 1-9.—Demagnetization curve (left) and energy product curve (right) of Alnico 5.

Demagnetizing Field. Air Gaps.—When a rod is magnetized by an applied field H_a, its ends carry magnetic poles which themselves cause magnetic fields to be present in all parts of the rod. Normally these fields are directed in the opposite direction to the field H_a, as shown in Fig. 1–10, and are therefore called demagnetizing fields. The true field acting on a given section of the bar, e.g., its middle, is then the resultant of the applied field and the demagnetizing field, ΔH:

$$H = H_a - \Delta H.$$

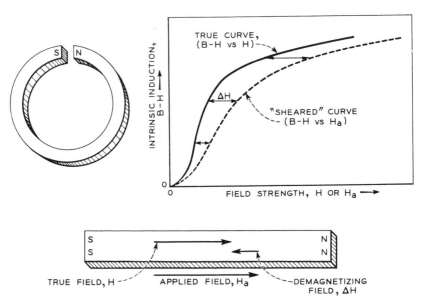

FIG. 1–10. Effect of demagnetizing field on the magnetization curve of a short bar, or ring with slot. The field strength in the material is the applied field minus the demagnetizing field, ΔH, which is proportional to $B-H$.

The demagnetizing field is approximately proportional to the intensity of magnetization

$$\Delta H = NI = \frac{N}{4\pi}(B - H).$$

The proportionality factor N is called the *demagnetizing factor* and depends primarily on the shape of the test body. Sometimes $N/4\pi$ is represented by the symbol D_B. It is zero in a ring, and in a rod when the ratio of length to diameter is very large. A more complete discussion is given in Chap. 19.

When $B-H$ is plotted against H_a, the curve is lower than the $B-H$ vs H curve, as illustrated in Fig. 1–10, and is said to be "sheared." The horizontal distance between them is proportional to $B-H$, as indicated.

Similarly a ring specimen containing an air gap has a sheared magnetization curve, and a demagnetizing factor may be associated with an air gap the length of which is a given fraction of the length of the magnetic circuit.

Magnetostriction.—When a body is magnetized, its dimensions are changed slightly, by not more than a few parts per million, and such changes are referred to as *magnetostriction*. The change in length in the direction parallel to the magnetization is that most often measured, and this change Δl divided by the original length l is the *Joule magnetostriction*.

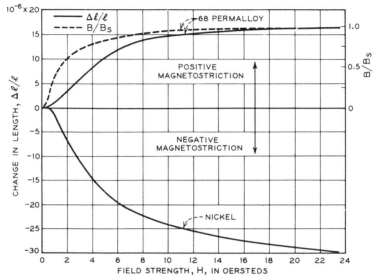

FIG. 1–11. The magnetostrictive change in length of 68 Permalloy and nickel.

The symbol λ is usually used for $\Delta l/l$. Unless otherwise specified, "magnetostriction" refers to this Joule magnetostriction. The magnetostriction of nickel and 68 Permalloy, as dependent on field strength, is shown in Fig. 1–11. Changes of volume are measurable, but the relative change in volume, $\Delta v/v$, is usually much smaller than the relative change in length, $\Delta l/l$.

Magnetic Anisotropy.—In single crystals of iron (and of other substances) the magnetic properties depend on the direction in which they are measured. Although some of these crystals are cubic and have some isotropic properties, such as their interaction with light, they are magnetically anisotropic in their response to magnetic fields of any considerable magnitude. Crystals of non-cubic symmetry are anisotropic with respect to light and to magnetic fields of any magnitude.

In some polycrystalline materials the various crystals are oriented more or less at random, and the properties in different directions are not greatly different. In many materials, however, as in rolled metal sheets, the process of fabrication produces some regularity in the distribution of orientations, and the magnetic properties are markedly anisotropic. The anisotropy usually persists even after the material is annealed.

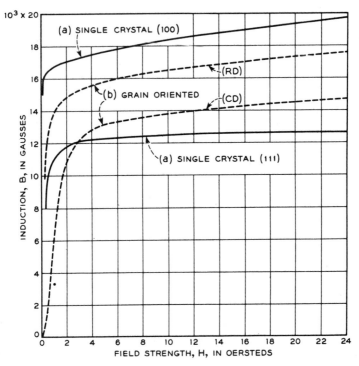

Fig. 1-12. Magnetization curves measured in different directions in a single crystal (a) and in anisotropic sheet (b) of silicon-iron. RD, measured parallel to the rolling direction; CD, measured parallel to the cross direction (90° to rolling direction, in plane of sheet).

Figure 1-12 shows a B vs H curve obtained by applying H, and measuring B, in the direction of a crystal axis ([100] direction) in iron containing 3% silicon, and another curve for which H and B are equally inclined to the three cubic axes ([111] direction). Curves are also shown for a sheet that has been rolled and then annealed, as measured in directions (1) parallel to the direction in which it was rolled (rolling direction, RD) and (2) at right angles to this direction (cross direction, CD), respectively.

Spontaneous Magnetization.—A ferromagnetic substance has long been regarded as an assemblage of small permanent magnets. When the material is unmagnetized, the magnets are arranged with haphazard orientations; when it is magnetized, they are lined up with their axes approximately parallel. The nature of this small magnet has been the subject of much consideration over a period of years. It is now known that all ferromagnetic materials are composed of many small magnets or *domains*, each of which consists of many atoms. Within a domain all of the atoms are aligned parallel and the domain is thus saturated, even when no field is applied. The material is therefore said to be "spontaneously magnetized." When the magnetization of the material is changed, the atoms turn together in groups (each atomic magnet about its own axis), the atoms in each group remaining parallel to each other so that they are aligned more nearly with the magnetic field applied to the material. The domain theory is discussed in more detail later.

CHAPTER 2

FACTORS AFFECTING MAGNETIC QUALITY

The properties of magnetic materials depend on chemical composition, fabrication, and heat treatment. Some properties, such as saturation magnetization, change only slowly with chemical composition and are usually unaffected by fabrication or heat treatment. However, permeability, coercive force, and hysteresis loss are highly sensitive and

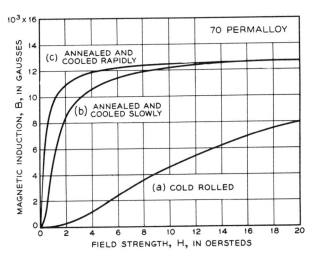

FIG. 2-1. Effect of mechanical and heat treatment of 70 Permalloy (70% Ni, 30% Fe) on its magnetization curve.

show changes which are extreme among all the physical properties, when changes are made in impurities or heat treatment. Properties may thus be divided into *structure-sensitive* and *structure-insensitive* groups. As an example, Fig. 2-1 shows magnetization curves of Permalloy after it has been (a) cold rolled, (b) annealed and cooled slowly, and (c) annealed and cooled rapidly. The maximum permeability varies with the treatment over a range of about twentyfold, while the saturation induction is the same within a few per cent. Structure-sensitive properties such as permeability depend on small irregularities in atomic arrangements, which have little effect on properties such as saturation induction.

Some of the more common sensitive and insensitive properties are listed in Table 1. The principal physical and chemical factors which affect these

TABLE 1. PROPERTIES COMMONLY SENSITIVE OR INSENSITIVE TO SMALL CHANGES IN STRUCTURE, AND SOME OF THE FACTORS WHICH EFFECT SUCH CHANGES

Structure-Insensitive Properties	Structure-Sensitive Properties	Factors Affecting the Properties
I_s, Saturation magnetization	μ, Permeability	Composition (gross)
θ, Curie point	H_c, Coercive force	Impurities
λ_s, Magnetostriction at saturation	W_h, Hysteresis loss	Strain
K, Crystal anisotropy constant		Temperature
		Crystal structure
		Crystal orientation

properties are listed in column 3. Their various effects will be briefly discussed and illustrated.

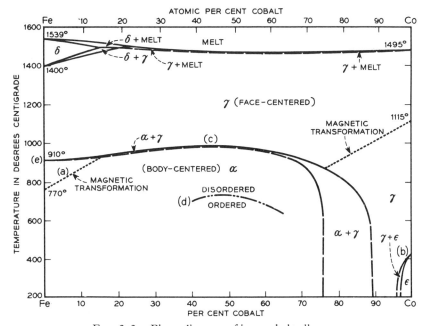

FIG. 2–2. Phase diagram of iron-cobalt alloys.

Phase Diagram.—Some of the most drastic changes in properties occur when the fabrication or heat treatment has brought about a change in structure of the material. For this reason the phase diagram or constitutional diagram is of the utmost importance in relation to the preparation and properties of magnetic materials. As an example consider the phase diagram of the binary iron-cobalt alloys of Fig. 2–2. Here the various areas show the phases, of different composition or structure, which are stable at the temperatures and compositions indicated. The α phase has the body-centered cubic crystal structure characteristic of iron. (The relative positions of the atoms are shown in Fig. 12-1.) At 900°C iron transforms into the face-centered phase γ, and at 1400°C into the δ phase, which has the same structure as the α phase. At about 400°C cobalt transforms, on heating, from the ε phase (hexagonal structure) into the γ phase.

The dotted lines indicate the Curie point, at which the material becomes non-magnetic.

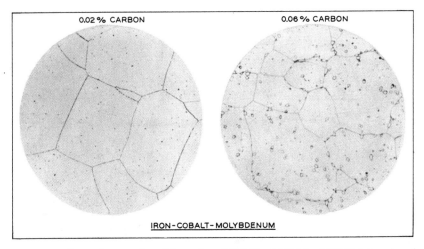

Fig. 2-3. Photomicrographs of Remalloy (12% Co, 17% Mo, 71% Fe) showing the precipitation of a second phase in the specimen containing an excess of carbon (0.06%). Courtesy of E. E. Thomas. Magnification: (a) 50 times, (b) 200 times.

In between the areas corresponding to the single phases α, γ, δ, and ε there are two-phase regions in which two crystal structures co-exist, some of the crystal grains having one structure and others the other. Such a two-phase structure is usually evident upon microscopic or X-ray examination. Photomicrographs of a single-phase alloy and a two-phase alloy of iron-cobalt-molybdenum are reproduced in Fig. 2-3(a) and (b).

The diagram of Fig. 2-2 shows several kinds of changes that affect the magnetic properties. At (a) the material becomes non-magnetic on heating, without change in phase. At (b) there is a change of phase, both phases being magnetic. Figure 2-4 shows the kind of change that occurs during this latter transition. At (c) of Fig. 2-2 there is a change from a ferromagnetic to a non-magnetic phase, and Fig. 2-5 shows the rapid change in magnetization that occurs when the temperature rises in this area. The line (d) shows that the α phase becomes ordered on cooling, i.e., the iron and cobalt atoms tend to distribute themselves regularly among the various atom positions so that each atom is surrounded by atoms

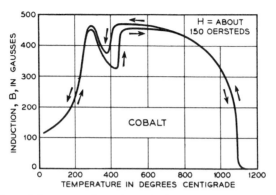

Fig. 2-4. Effect of phase transformation of cobalt on magnetization with a constant field of 150 oersteds. Both phases magnetic.

of the other kind. This phenomenon will be discussed especially in connection with the properties of iron-aluminum alloys.

The transition at (e) is entirely in the non-magnetic region, but it has

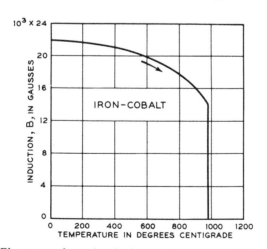

Fig. 2-5. Phase transformation in iron-cobalt (50% Co). High-temperature phase is non-magnetic.

its influence on the properties of iron at room temperature. If iron is cooled very slowly through (e), the internal strains caused by the change in structure will be relieved by diffusion of the metal atoms, but if the cooling is too rapid there will not be sufficient time for strain relief. Practically this means that to obtain high permeability in iron it must be annealed for some time below 900°C, or cooled slowly through this temperature so that diffusion will have time to occur. In most ferromagnetic materials diffusion occurs at a reasonably rapid rate only at temperatures above about 500–600°C. Other effects of phase change will be considered in a later section.

Physical Factors.—The effect of a homogeneous *strain* on the magnetization curve can be observed in a simple way by applying tension to an annealed wire and then measuring B and H. The effect of tension on some

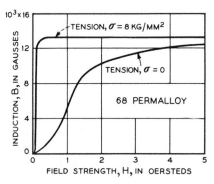
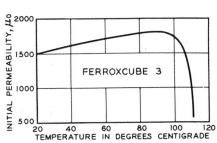

Fig. 2–6. Effect of tension on the magnetization curve of 68 Permalloy.

Fig. 2–7. Variation of initial permeability of Ferroxcube 3, showing maximum at temperature just below the Curie temperature.

materials is to increase the permeability, as shown in Fig. 2–6, and on other materials to decrease it. Compression usually causes a change in the sense opposite to that caused by tension.

The internal strains resulting from *plastic deformation* of the material, brought about by stressing beyond the elastic limit, as by pulling, rolling, or drawing, almost always reduce the permeability. The material is then under rather severe local strains similar to those present after phase change, and these are different in magnitude and direction in different places in the material and have quite different values at points close together. Strains of this kind can usually be relieved by annealing; therefore metal that has been fabricated by plastic deformation is customarily annealed to raise its permeability. Figure 2–1 shows the effect of annealing a Permalloy strip that has been cold rolled to 15% of its original thickness.

The *temperature* also is effective in changing permeability and other properties, even when no change in phase occurs. Figure 2–7 shows the rapidity with which the initial permeability decreases as the Curie point is approached. For this material, Ferroxcube III, a zinc manganese ferrite ($ZnMnFe_4O_8$), the Curie point is not far above room temperature.

The effect of *impurities* may be illustrated by the B vs H curves for iron containing various amounts of carbon. Curve (a) of Fig. 2-8 is for a mild steel having 0.2% carbon; (b) is for the iron commonly used in electromagnetic apparatus, containing about 0.02% carbon and annealed at about 900°C. When this same iron is purified by heating for several hours at 1400°C in hydrogen, the carbon is reduced to less than 0.001%, other impurities are removed, and curve (c) is obtained.

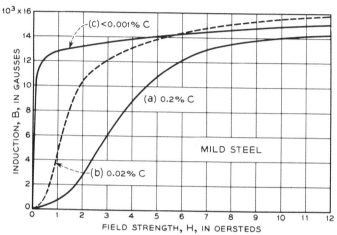

FIG. 2-8. Effect of impurities on magnetic properties of iron. Annealing at 1400°C in hydrogen reduces the carbon content from about 0.02% to less than 0.001%

Finally, Fig. 2-9 shows that large differences in permeability may be found by simply varying the *direction of measurement* of the magnetic properties in a single specimen. The material is a single crystal of iron containing about 4% silicon, and the directions in which the properties are measured are [100] (parallel to one of the crystal axes), and [111] (as far removed as possible from an axis). The magnetic properties in the two directions are different because different "views" of the atomic arrangement are obtained in the two directions.

PRODUCTION OF MAGNETIC MATERIALS

In the preparation of magnetic materials for either laboratory or commercial use there are many processes which influence the chemical and physical structure of the product. The selection of raw materials, the melting and casting, the fabrication and the heat treatment, all are important and must be carried out with a proper knowledge of the metallurgy of the material. A brief description of the common practices is now given.

Melting and Casting.—For experimental investigation of magnetic materials in the laboratory, the raw materials easily obtainable on the market are generally satisfactory. When high purity is desirable, specially prepared materials and crucibles must be used and the atmosphere in contact with the melt must be controlled. The impurities that have the greatest influence on the magnetic properties of high permeability materials

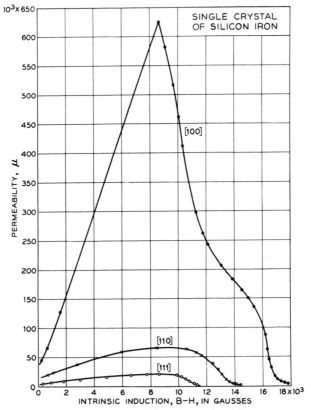

FIG. 2-9. Dependence of permeability on crystallographic direction in iron containing 3.8% silicon [37W5].

are the non-metallic elements, particularly oxygen, carbon, and sulfur, and the presence of these impurities is therefore watched carefully and their analyses are carried out with special accuracy. Impurities are likely to change in important respects during the melting and pouring because of reactions of the melt with the atmosphere, the slag or the crucible lining, or because of reactions taking place among the constituents of the metal.

Melting of small lots (10 lb) is best carried out in a high-frequency induction furnace. Figure 2-10 shows such a furnace designed for melting 10-50 lb, and casting by tilting the furnace, the whole operation being carried out in a controlled atmosphere. High-frequency currents (usually 1000-2000 cps but sometimes much higher) are passed through the water-cooled copper coils, and the alternating magnetic field thus produced heats the charge by inducing eddy currents in it. Crucibles are usually composed of alumina or magnesia.

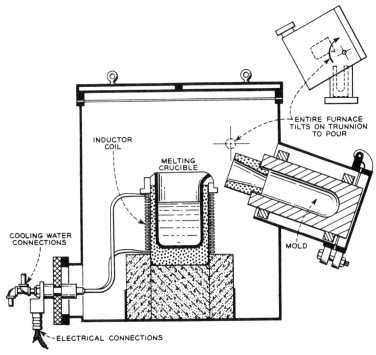

FIG. 2-10. Induction furnace designed for small melts in controlled atmosphere, as designed by J. H. Scaff and constructed by the Ajax Northrup Company.

On a commercial scale, melts of silicon-iron are usually made in the open hearth furnace (Chap. 3) in which pig iron and scrap are refined and ferro-silicon added. The furnace capacity may be as large as 100 tons. Sometimes silicon-iron, and usually iron-nickel, alloys are melted in the arc furnace, in amounts varying from a few tons to 50 tons. A photograph of such a furnace, in position for pouring, is shown in Fig. 2-11. The heat is produced in the arc drawn between large carbon electrodes immersed in the metal, the current sometimes rising to over 10 000 amperes. By tipping the furnace the melt is poured into a ladle, and from this it is poured into cast-iron molds through a valve-controlled hole in the bottom

Fig. 2-11. Arc furnace for large commercial melts. (*Courtesy of J. S. Marsh of the Bethlehem Steel Company.*)

of the ladle. Special-purpose alloys, including permanent magnets, are prepared commercially in high-frequency induction furnaces or in arc furnaces in quantities ranging from a fraction of a ton to several tons.

Slags are commonly used when melting in air, both to protect from oxidation and to reduce the amounts of undesirable impurities. Common protective coverings are mixtures of lime, magnesia, silica, fluorite, alumina, and borax in varying proportions. In commercial production different slags are used at different stages to refine the melt; e.g., iron oxide may be used to decarburize and basic oxides to desulfurize.

Melting in vacuum requires special technique that has been described in some detail by Yensen [15Y1]. Commercial use has been described by Rohn [33R5] and others [44H2]. Melting in hydrogen has been used on an experimental scale in both high-frequency and resistance-wound furnaces. In commercial furnaces Rohn has used hydrogen and vacuum alternately before pouring, for purification in the melt, in low-frequency induction furnaces having capacities of several tons.

Just before casting a melt of a high permeability alloy such as iron-nickel, a deoxidizer may be added, e.g., aluminum, magnesium, calcium, or silicon, in an amount averaging around 0.1%. The efficacy of a deoxidizer is measured by its heat of formation, and this is given for the common elements in Table 2, taken from Sachs and Van Horn [40S6].

TABLE 2. HEATS OF FORMATION AND OTHER PROPERTIES OF SOME OXIDES
(Sachs and Van Horn [40S6])

Oxide	Heat of Formation (Kilo-cal per gram atom of metal)	Melting Point (°C)	Density (g/cm³)
CaO	152	>2500	3.4
BeO	144	>2500	3.0
MgO	144	2800	3.65
Li_2O	141	>1700	2.0
Al_2O_3	127	2050	3.5
V_2O_2	116	1970	4.9
TiO_2	109	1640	4.3
Na_2O	101	*	2.3
SiO_2	95	1670	2.3
B_2O_3	94	580	1.8
MnO	91	1650	5.5
ZrO_2	89	2700	5.5
ZnO	85	*	5.5
P_2O_5	73	*	2.4
SnO_2	68	1130	6.95
FeO	66	1420	5.7
NiO	58	†	7.45

* Sublimes.   † Decomposes before melting.

Also several tenths of a per cent of manganese may be put in to counteract the sulfur so that the material may be more readily worked; the manganese sulfide so formed collects into small globular masses which do not interfere seriously with the magnetic or mechanical properties of most materials.

Ordinarily a quantity of gas is dissolved in molten metal, and this is likely to separate during solidification and cause unsound ingots. The solubilities of some gases in iron and nickel have been determined by

Sieverts [29S8] and others and are given in Fig. 2–12, adapted from the compilation by Dushman [49D1]. The characteristic decrease of solubility during freezing is apparent. Most of the gases given off by magnetic metals during heating are formed from the impurities carbon, oxygen, nitrogen, and sulfur; CO is usually given off in greatest amount from cast metal, and some N_2 and H_2 are also found. Refining of the melt is

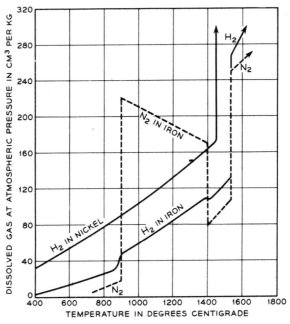

Fig. 2–12. Solubility of some gases in iron and nickel at various temperatures.

therefore of obvious advantage, and the furnace of Fig. 2–10 is especially useful for this purpose.

Small ingots are sometimes made by cooling in the crucible. Usually, however, ingots are poured into cast-iron molds for subsequent reduction by rolling, etc.; permanent magnet or other materials are often cast in sand in shapes which require only nominal amounts of machining or grinding for use in apparatus or in testing. Special techniques for specific materials will be discussed in the following chapters.

Other considerations important in the melting and pouring of ingots are proper mixing in the melt, the temperature of pouring, mold construction, avoidance of inclusions of slag, segregation, shrinkage, cracks, blow holes, etc.

Fabrication.—Magnetic materials require a wide variety of modes of fabrication, and these will be discussed in connection with the specific materials. The methods include hot and cold rolling, forging, swaging, drawing, pulverization, electrodeposition, and numerous operations such as punching, pressing, and spinning. In the commercial fabrication of ductile material it is common practice to start the reduction in a breakdown or blooming mill (Fig. 2–13) after heating the ingot to a high temperature (1200–1400°C). Large ingots, of several tons weight, are often led to the mill before they have cooled below the proper temperature. The reduction is continued as the metal cools, in a rod or flat rolling mill, depending on the desired form of the final product. When the thickness is decreased

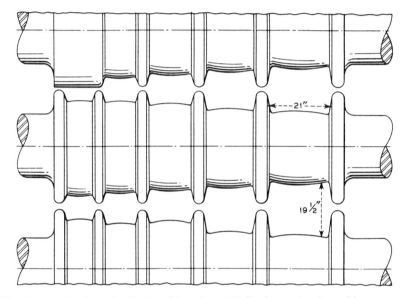

FIG. 2–13. Design of rolls in a blooming mill for hot reduction of ingots to rod. (*Courtesy of Carnegie Illinois Steel Corp.*)

to 0.2–0.5 in., the material has usually cooled below the recrystallization temperature. Because of the difficulty in handling hot sheets or rod of small thickness, they are rolled at or near room temperature, with intermediate annealings if necessary to soften or to develop the proper structure. In experimental work, rod is often swaged instead of rolled.

In recent years the outstanding trends in methods of fabricating materials have been toward the construction of the multiple-roll rolling mill for rolling thin strip, and the continuous strip mill for high-speed production on a large scale. Figure 2–14 shows the principle of construction of a typical 4-high mill, (a) and (b), and of two special mills, (c) and (d). In the 20-high Rohn mill [33R4] and 12-high Sendzimir mill [46S3] the two working rolls are quite small (0.2–1 in. in diameter). These are each backed by two larger rolls and these in turn by others as indicated. In the Rohn mill (c), power is supplied to the two smallest rolls and the final bearing surfaces are at the ends of the largest rolls. In the Sendzimir mill (d) the power is supplied to the rolls of intermediate size, and the bearing surfaces are distributed along the whole length of the largest rolls so that

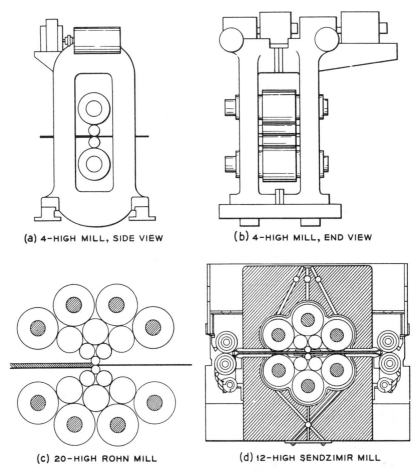

Fig. 2-14. Arrangement of rolls in mills used for reduction of thin sheet: (a) and (b) conventional 4-high mill; (c) Rohn 20-high; (d) Sendzimir 12-high.

no appreciable bending of the rolls occurs. The small rolls reduce the thickness of thin stock with great efficiency, and the idling rolls permit the application of high pressure. In the Steckel mill power is used to pull the sheet through the rolls, which are usually four high with small working rolls.

FIG. 2-15. Continuous strip mill designed for large output, having 6 individual mills in tandem. (*Courtesy of C. W. Stoker of Carnegie Illinois Steel Corp.*)

The continuous strip mill is an arrangement of individual mills such that the strip is fed continuously from one to another and may be undergoing reduction in thickness in several mills simultaneously. Figure 2–15 shows a mill of this kind, used for cold reduction, with six individual mills in tandem.

For magnetic testing numerous forms of specimens are required for various kinds of tests; these include strips for standard tests for transformer sheet, rings or parallelograms for conventional ballistic tests, "pancakes" of thin tape spirally wound for measurement by alternating current, ellipsoids for high field measurements, and many others. The various forms are required to study or eliminate the effects of eddy currents, demagnetizing fields and directional effects and to simulate the use of material in apparatus. Most of the sizes needed for commerce and for experimental investigation are filled by strips or sheets of thicknesses 0.002–0.1 in., from which coils can be wound or parts cut, by rods from which relay cores or other forms can be made, by powdered material used for pressing into cores for coils for inductive loading, and by castings for permanent magnets or other objects which may be machined or ground to final shape.

Heat Treatment.—High permeability materials are annealed primarily to relieve the internal strains introduced during fabrication. However, permanent magnet materials are heat-treated to *introduce* strains by precipitating a second phase. Heat treatments are decidedly characteristic of the materials and their intended uses and will be discussed in detail in connection with them. Figure 2–16 shows some of the commonest treatments in the form of temperature-time curves. The purposes of these various heating and cooling cycles, and typical materials subjected to them, may be listed as follows:

(1) Relief of internal strains due to fabrication or phase changes. Magnetic iron.
(2) Increase of internal strains by precipitation hardening. Alnico type of permanent magnets.
(3) Purification by contact with hydrogen or other gases. Silicon-iron (cold rolled), hydrogen-treated iron.

There are also special treatments, such as those used for "double-treated" Permalloy, "magnetically annealed" Permalloy, and Perminvar.

Occasionally it is necessary to homogenize a material by maintaining the temperature just below the freezing point for many hours. Heat treatments also may affect grain size, crystal orientation, or atomic ordering.

Furnaces for heat treating have various designs that will not be considered here. A modern improvement has been the use of globar (silicon carbide) heating elements that permit treatment at 1300–1350°C in an atmosphere of hydrogen or air.

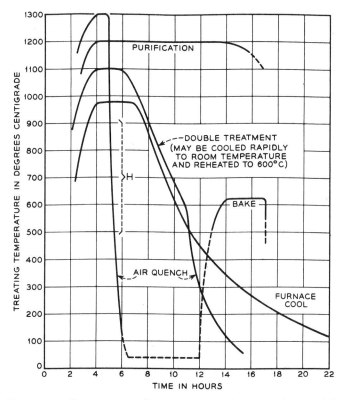

FIG. 2-16. Some common heat treatments for magnetic materials.

EFFECT OF COMPOSITION

Gross Chemical Composition.—The effect of composition on magnetic properties will now be considered, using as examples the more important binary alloys of iron-silicon, iron-nickel, and iron-cobalt, on which are based the most useful and interesting materials. The iron-silicon alloys are used commercially without additions, the iron-nickel and iron-cobalt alloys are most useful in the ternary form, and many special alloys, e.g., material for permanent magnets, contain four or five components.

Figure 2-17 shows four important properties of the iron-silicon alloys of low silicon content, after they have been hot rolled and annealed. The commercial alloys (3–5% silicon) are the most useful because they have the best combination of properties of various kinds. The properties shown in the figure are important for the best commercial balance; the maximum permeability μ_m only indirectly (it is a good measure of hysteresis loss and maximum field necessary in use), and the Curie point θ only in a minor role. The saturation I_s, permeability, and resistivity ρ, should all be as high as possible. I_s, θ, and ρ are structure-insensitive and vary with com-

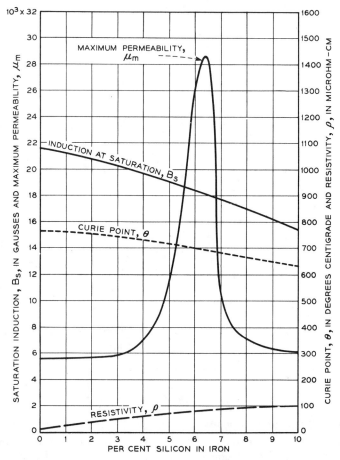

Fig. 2-17. Variation of some properties of iron-silicon alloys with composition: B_s, saturation intrinsic induction; θ, magnetic transformation point; ρ, electrical resistivity; μ_m, maximum permeability.

position in a characteristically smooth way, practically independent of heat treatment; μ_m depends on heat treatment (strain), impurities, and crystal orientation. There are no phase changes to give sudden changes with composition of properties measured at room temperature.

Some of the properties of the iron-nickel alloys are given in Figs. 2-18, 19, and 20. The change in phase from α to γ at about 30% nickel is respon-

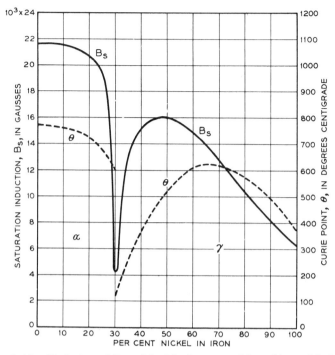

Fig. 2-18. Variation of B_s and θ with the composition of iron-nickel alloys.

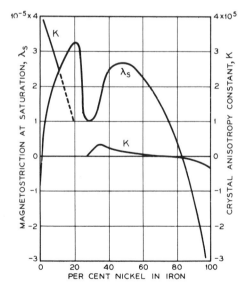

Fig. 2-19. Variation of saturation magnetostriction, λ_s, and crystal anisotropy, K, with the composition of iron-nickel alloys.

sible for the breaks at this composition. The permeabilities μ_0 and μ_m (Fig. 20) show characteristically the effect of heat treatment. The maxima are closely related to the points at which the saturation magnetostriction λ_s and crystal anisotropy K pass through zero (Fig. 2-19).

Additions of molybdenum, chromium, copper, and other elements are made to enhance the desired properties of the iron-nickel alloys.

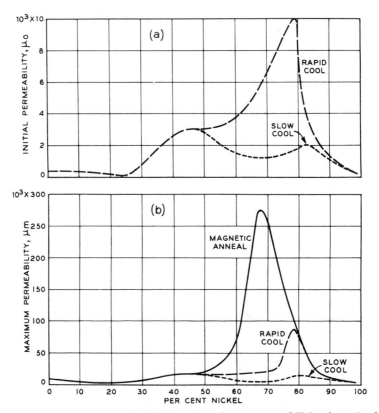

Fig. 2-20. Dependence of the initial and maximum permeabilities (μ_0, μ_m) of iron-nickel alloys on the heat treatment.

The iron-cobalt alloys, some properties of which are shown in Figs. 2-21 and 22, are usually used when high inductions are advantageous. The unusual course of the saturation induction curve, with a maximum greater than that for any other material, is of obvious theoretical and practical importance. The sudden changes in the Curie point curve are associated with α, γ phase boundaries, as mentioned earlier in this chapter. The peak of the permeability curve (Fig. 2-22) occurs at the composition for which atomic ordering is stable at the highest temperature (see also Fig. 2-2).

The sharp decline near 95% cobalt coincides with the phase change α,ϵ at this composition. Additions of vanadium, chromium, and other elements are used in making commercial ternary alloys.

Some useful alloys based on the binary iron-silicon, iron-nickel, and iron-cobalt alloys are described in Table 3.

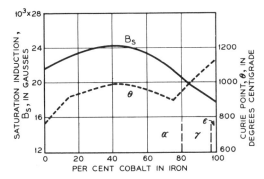

FIG. 2–21. Variation of B_s and θ on the composition of iron-cobalt alloys.

The *hardening* of material resulting from the precipitation of one phase in another is often used to advantage when magnetic hardness (as in permanent magnets) or mechanical hardness is desired. To illustrate this process, consider the binary iron-molybdenum alloys, a part of the phase diagram of which is given in Fig. 2–23. The effect of the boundary between the α and $\alpha + \epsilon$ fields is shown by the variation of the properties with composition (Fig. 2–24a). Saturation magnetization and Curie point [29T1] are affected but little, the principal change in the former being a slight change in the slope of the curve at the composition at which the phase boundary crosses 500°C, the temperature below which diffusion is very slow. The

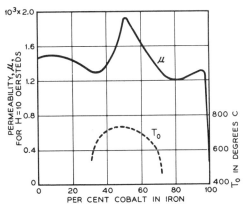

FIG. 2–22. Variation of permeability at $H = 10$ oersteds, and of the critical temperature of ordering, with the composition of iron-cobalt alloys.

Curie point curve has an almost imperceptible break at the composition at which the phase boundary lies at the Curie temperature. The changes of maximum permeability and coercive force are more drastic;

TABLE 3. SOME PROPERTIES OF SOME USEFUL ALLOYS BASED ON THE FE-SI, FE-CO AND FE-NI BINARY SYSTEMS

Name	Composition (Per cent)	Heat Treatment (°C)	Initial Permeability μ_0	Maximum Permeability μ_m	Coercive Force H_c (oersteds)	Saturation Induction B_s (gausses)	Curie Point θ (°C)
Hot Rolled Silicon Iron	4 Si, 96 Fe	800	500*	7 000	0.5	19 700	690
Grain Oriented Silicon Iron	3 Si, 97 Fe	1200	1 500*	40 000	0.15	20 000	700
Sendust	9 Si, 85 Fe, 5 Al	Cast	30 000	120 000	0.05	10 000	500
45 Permalloy†	45 Ni, 55 Fe	1200, H_2	3 500	50 000	0.07	16 000	440
4-79 Permalloy	79 Ni, 17 Fe, 4 Mo	1100	20 000	100 000	0.05	8 700	420
Mumetal	75 Ni, 18 Fe, 2 Cr, 5 Cu	1175, H_2	20 000	100 000	0.05	6 500	430
Supermalloy	79 Ni, 16 Fe, 5 Mo	1300, H_2	100 000	1 000 000	0.002	8 000	400
Permendur	50 Co, 50 Fe	800	800	5 000	2.0	24 500	980
2V-Permendur	49 Co, 49 Fe, 2 V	800	800	4 500	2.0	24 000	980
Hiperco	34 Co, 64 Fe, 1 Cr	850	650	10 000	1.0	24 200	...

* Measured at $B = 20$ instead of $B = 0$.
† Similar alloys: Hipenik, 4750, and others.

μ_m drops rapidly as the amount of the second phase ϵ increases and produces more and more internal strain (Fig. 2–24b), and H_c increases at the same time [32S6]. The experimental points correspond to a moderate rate of cooling of the alloy after annealing.

When the amount of the second phase is considerable (as in the 15% Mo alloy) it is common practice to quench the alloy from a temperature at which it is a single phase (e.g., 1100 or 1200°C) and so maintain it temporarily as such, and then to heat it to a temperature (e.g., 600°C) at which diffusion proceeds at a more controllable rate. During the latter step the second phase separates slowly enough so that it can easily be stopped at the optimum point, after a sufficient amount has been precipitated but before diffusion has been permitted to relieve the strains caused by the precipitation. A conventional heat treatment for precipitation hardening of this kind, used on many permanent magnet materials, has already been given in Fig. 2–16.

In some respects the development of atomic order in a structure is like the precipitation of a second phase. When small portions of the material become ordered and neighboring regions are still disordered, severe local strains may be set up in the same way that they are during the precipitation

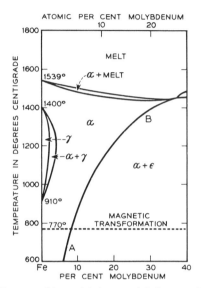

FIG. 2–23. Phase diagram of iron-rich iron-molybdenum alloys, showing solid solubility curve AB important in the precipitation-hardening process.

hardening described above. The treatment used to establish high strains is the same as in the more conventional precipitation hardening. The decomposition of an ordered structure in the iron-nickel-aluminum system has been held responsible, by Bradley and Taylor [38B1], for the good permanent magnet qualities of these alloys.

Some of the common permanent magnets, heat-treated to develop internal strains by precipitation of a second phase or by the development of atomic ordering, are described in Table 4.

The changes in properties to be expected when the composition varies across a phase boundary of a binary system are shown schematically by the curves of Fig. 2–25.

Impurities.—The principle of precipitation hardening, as just described, applies also to the lowering of permeability by the presence of accidental impurities. For example, the solubility curves of carbon, oxygen, and nitrogen in iron, described in Fig. 2–26, are quite similar in form to the curve separating the α and $\alpha + \epsilon$ areas of the iron-molybdenum diagram of Fig. 2–23; the chief difference is that the scale of composition now corresponds to concentrations usually described as impurities. One expects, then, that the presence of more than 0.04% of carbon in iron will cause the permeability of an annealed specimen to be considerably below that of pure iron. The amount of carbon present in solid solution will also affect the magnetic properties.

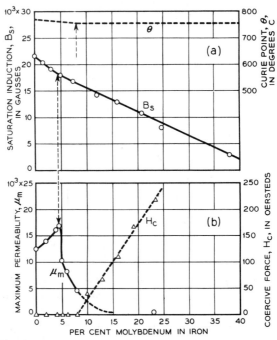

Fig. 2–24. Change of structure-insensitive properties (θ and B_s) and structure-sensitive properties (μ_m and H_c) with the composition when precipitation-hardening occurs.

Because the amounts of material involved are small, it is difficult to carry out well-defined experiments on the effects of each impurity, especially in the absence of disturbing amounts of other impurities. Two examples of the effect of impurities will be given, in addition to Fig. 2–8. In Fig. 2–27 Yensen and Ziegler [36Y3] have plotted the hysteresis loss as dependent on carbon content, the curve giving the mean values of many determinations. The hysteresis decreases rapidly at small carbon contents, when these are of the order of magnitude of the solid solubility at room temperature.

Cioffi [32C2] has purified iron from carbon, oxygen, nitrogen, and sulfur by heating in pure hydrogen at 1475°C, and has measured the permeability

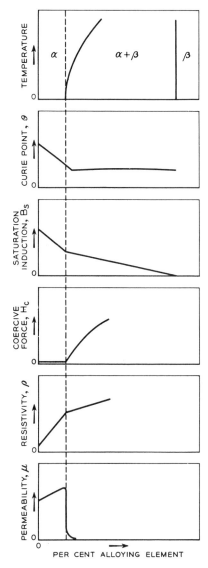

FIG. 2–25. Diagrams illustrating the changes in various properties that occur when a second phase precipitates.

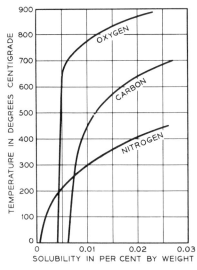

FIG. 2–26. Approximate solubility curves of carbon, oxygen, and nitrogen in iron.

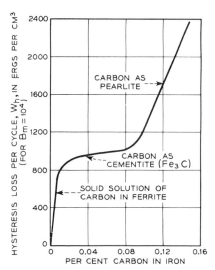

FIG. 2–27. Effect of carbon content on hysteresis in iron.

at different stages of purification. Table 5 shows that impurities of a few thousandths of a per cent are quite effective in depressing the maximum permeability of iron.

Carbon and nitrogen, present as impurities, are known to cause "aging" in iron, i.e., the permeability and coercive force of iron containing these elements as impurities will change gradually with time when maintained somewhat above room temperature. As an example, a specimen of iron

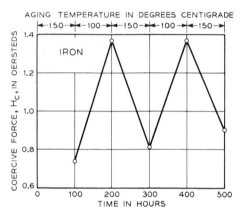

FIG. 2–28. Effect of nitrogen impurity on the coercive force of iron annealed successively at 100 and 150°C.

was maintained for 100 hours first at 100°C, then 150°C, then 100°C, and so on. The corresponding changes in coercive force are given in the diagram of Fig. 2–28. A change of about twofold is observed.

SOME IMPORTANT PHYSICAL PROPERTIES

There are many physical characteristics that are important in the study of ferromagnetism from both the practical and theoretical points of view. These include the resistivity, density, atomic diameter, specific heat, expansion, hardness, elastic limit, plasticity, toughness, mechanical damping, specimen dimensions, and numerous others. In a different category may be mentioned the effects of corrosion, homogeneity, and porosity. Most of these properties will be discussed in connection with specific materials or in special chapters when there is a direct relation between the property and ferromagnetism; the most important characteristics will be mentioned here. A table of the atomic weights and numbers, densities, melting points, electrical resistivities, and coefficients of thermal expansion of the metallic elements is given in Appendix 2.

Dissolving a small amount of one element in another increases the *resistivity* of the latter. To show the relative effects of various elements,

SOME IMPORTANT PHYSICAL PROPERTIES

TABLE 4. SOME USEFUL PERMANENT MAGNETS AND THEIR PROPERTIES

Name	Composition (per cent)*	Fabrication	Heat Treatment	H_c	B_r	Mechanical Properties
Carbon Steel	1 Mn, 0.9 C	HR, PM	Q800	50	10 000	H, S
Tungsten Steel	5 W, 0.3 Mn, 0.7 C	HR, PM	Q850	70	10 300	H, S
Chromium Steel	3.5 Cr, 0.3 Mn, 0.9 C	HR, PM	Q830	65	9 700	H, S
Cobalt Steel	36 Co, 4 Cr, 5 W, 0.7 C	HR, PM	Q950	240	9 500	H, S
Remalloy (Comol)	17 Mo, 12 Co	HR, PM	Q1200, B700	250	10 500	H
Alnico 2	12 Co, 17 Ni, 10 Al, 6 Cu	C, G	A1200, B600	550	7 200	H, B
Alnico 5	24 Co, 14 Ni, 8 Al, 3 Cu	C, G	A1300,† B600	550	12 500	H, B
Alnico 12	35 Co, 18 Ni, 6 Al, 8 Ti	C, G	Cast, B650	950	5 800	H, B
Alcomax	21 Co, 11 Ni, 5 Al, 4 Cu	C, G	A1300,† B600	550	12 500	H, B
Vicalloy	52 Co, 10 V	C, CR, PM	B600	300	8 800	D
Cunife	20 Ni, 60 Cu	C, CR, PM	B600	550	5 400	D
Platinum-Cobalt	77 Pt, 23 Co	C, CR, PM	Q1200, B650	2600‡	4 500	D
Silmanal	87 Ag, 9 Mn, 4 Al	C, CR, PM		6000‡	550	D

* Remainder iron.
† Cooled in magnetic field.
‡ Coercive force for $I = 0$.
B —baked at indicated temperature.
HR—hot rolled.
CR—cold rolled.
PM—punched or machined
Q —quenched from indicated centigrade temperature in oil.
A —cooled in air from indicated temperature.

C —cast.
G —ground.
H —hard.
B —brittle.
D —ductile or malleable.
S —strong.
W—weak.

TABLE 5

Maximum permeability of Armco iron with different degrees of purification, effected by heat treatment in pure hydrogen at 1475°C for the times indicated (P. P. Cioffi). Analyses from R. F. Mehl (private communication).

Time of Treatment (hours)	μ_m	Composition (%)					
		C	S	O	N	Mn	P
0	7 000	0.012	0.018	0.030	0.0018	0.030	0.004
1	16 000	.005	.010	.003	.0004	...	...
3	30 000	.005	.006	.003	.0003	...	...
7	70 000	.003	...	.003	.0001	...	...
18	227 000	.005	<.003	.003	.0001	.028	.004
Precision of analysis:		.001	.002	.002	.0001	...	...

the common binary alloys of iron and of nickel are shown in Figs. 2-29 and 30. From a theoretical standpoint it is desirable to understand (1) the relatively high resistivity of the ferromagnetic elements compared to their

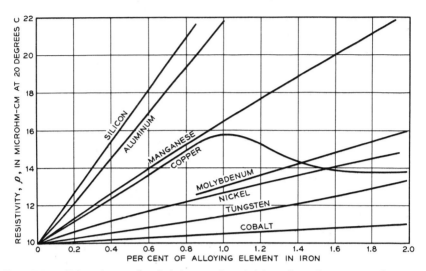

FIG. 2-29. Dependence of resistivity on the addition of small amounts of various elements to iron.

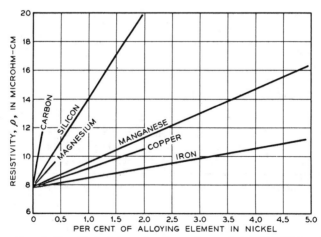

FIG. 2-30. Resistivity of various alloys of nickel [42W2].

neighbors in the periodic table and (2) the relative amounts by which the resistivity of iron (or cobalt or nickel) is raised by a given atomic percentage of various other elements. From a practical standpoint, a high resistivity is usually desirable in order to decrease the eddy-current losses in the material and thereby decrease the power wasted and the lag in time between the cause and effect, for example, the time lag of operation of a relay.

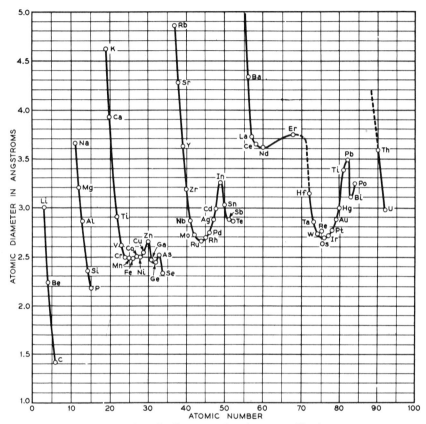

FIG. 2–31. Atomic diameter of various metallic elements.

Knowledge of the *atomic diameter* is important in considering the effects of alloying elements, and values for the metallic and borderline elements are shown in Fig. 2–31. Most of the values are simply the distances of nearest approach of atoms in the element as it exists in the structure stable at room temperature. Atomic diameter is especially important in theory because the very existence of ferromagnetism is dependent in a critical way on the distance between adjacent atoms. This is discussed more fully in Chap. 10.

Even when no phase change occurs in a metal, important *changes in structure* occur during fabrication and heat treatment, and these are complicated and imperfectly understood. When a single crystal is elongated by tension, slip occurs on a limited number of crystal planes that, in general, are inclined to the axis of tension. As elongation proceeds, the planes on which slip is taking place tend to turn so that they are less inclined to the axis. In this way a definite crystallographic direction approaches parallelism with the length of the specimen. In a similar but more complicated way, any of the usual methods of fabrication cause the many crystals of which it is composed to assume a non-random distribution of orientations,

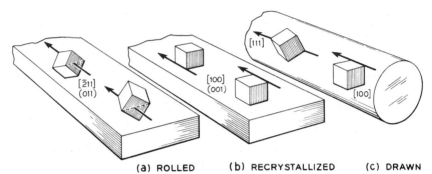

(a) ROLLED (b) RECRYSTALLIZED (c) DRAWN

FIG. 2–32. The preferred orientations of crystals in nickel sheet and wire after fabrication and after recrystallization.

often referred to as preferred or *special orientations*, or *textures*. Some of the textures reported for cold-rolled and cold-drawn magnetic materials are given in Table 6, taken from the compilation by Barrett [43B6]. The orientations of the cubes, which are the crystallographic units, are shown in Fig. 2–32(a) and (c) for cold-rolled sheets and cold-drawn wires of nickel.

TABLE 6

Preferred orientations in drawn wires and rolled sheets, before and after recrystallization, and in castings. The rolling plane and rolling direction, or wire axis, or direction of growth, are designated.

Metal	Crystal Structure	Drawn Wires		Rolled Sheets		As Cast
		As Drawn	Recrystallized	As Rolled	Recrystallized	
Iron	Body-centered cubic	[110]	[110]	(001), [110] and others	(001), 15° to [110]	[100]
Cobalt	Hexagonal close-packed	...	...	(001)	...	...
Nickel	Face-centered cubic	[111] and [100]	...	(110), [112], and others	(100), [001]	...

Since the magnetic properties of single crystals depend on crystallographic direction (anisotropy), the properties of polycrystalline materials in which there is special orientation will also be direction-dependent. In fact it is difficult to achieve isotropy in any fabricated material, even if fabrication involves no more than solidifying from the melt. The relief of the internal strains in a fabricated metal by annealing proceeds only slowly

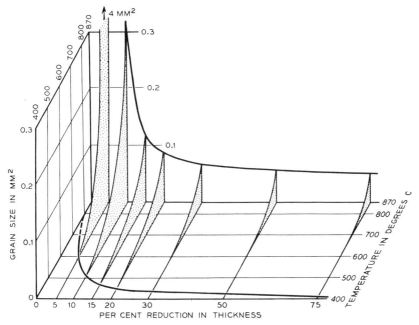

FIG. 2–33. Dependence of the grain size of iron on the amount of deformation and on the temperature of anneal.

at low temperatures (up to 600°C for most ferrous metals) without noticeable grain growth or change in grain orientation and is designated *recovery*. The principal change is a reduction in the amplitude of internal strains, and this can be followed quantitatively by X-ray measurements. Near the point of complete relief, distinct changes occur in both grain size and grain orientation, and the material is said to *recrystallize*. At higher temperatures grain growth increases more rapidly. The specific temperatures necessary for both recovery and recrystallization depend on the amount of previous deformation [48K4], as shown in Fig. 2–33. Special orientations are also present in fabricated materials after recrystallization, and some of these are listed in Table 6 and illustrated for nickel in Fig. 2–32(b).

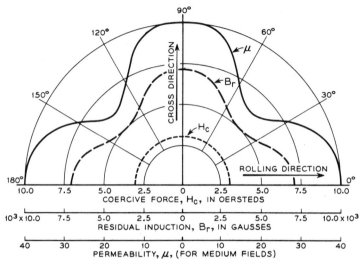

Fig. 2–34. Variation of magnetic properties with the direction of measurement in a sheet of iron-nickel alloy (40% Ni), severely rolled (98.5%) and annealed at 1100°C.

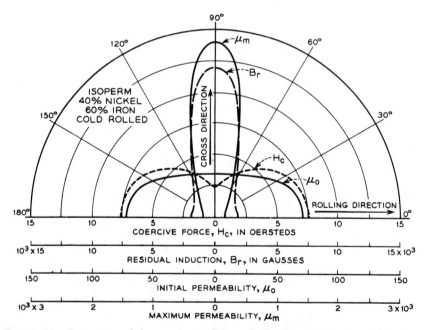

Fig. 2–35. Properties of the same material as that of Fig. 2–34, after it has been rolled, annealed, and again rolled.

As an example of the dependence of various magnetic properties on direction, Fig. 2-34 gives data of Dahl and Pawlek [36D2] for a 40% nickel-iron alloy reduced 98.5% in area by cold-rolling and then annealed at 1100°C. After further cold-rolling (50% reduction) the properties are as described in Fig. 2-35. Fuller discussion is given in Chap. 12.

The mechanical properties ordinarily desirable in practical materials are those which facilitate fabrication. Mild steel is often considered as the nearest approach to an ideal material in this respect. Silicon-iron is limited by its brittleness which becomes of major importance at about 5% silicon; this is shown by the curve of Fig. 2-36. Permalloy is "tougher" than iron or mild steel and requires more power in rolling and more frequent annealing between passes when cold rolled, but it can be worked to smaller

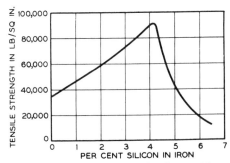

FIG. 2-36. Variation of the breaking strength of iron-silicon alloys, showing the onset of brittleness near 4% silicon.

dimensions. If materials have insufficient stiffness or hardness, parts of apparatus made from them must be handled with care to avoid bending and consequent lowering of the permeability. If the hardness is too great, the material must be ground to size. This is the case with some permanent magnets.

The effect of the size of a magnetic specimen is often of importance, as is well known in the study of *thin films* and of *fine powders* in which the smallest dimension is about 10^{-4} cm or less. Many studies have been made of thin electrodeposited and evaporated films, and generally it is found that the permeability is low and the coercive force high. The interpretation is uncertain because it is difficult to separate the effects of strains and air gaps from the intrinsic effect of thickness, though it is known that each one of these variables has a definite effect. As one example of the many experiments, we will show here the effect of the thickness of electrodeposited films of cobalt. Magnetization curves are shown in Fig. 2-37 according to previously unpublished work.

The high coercive force obtained in fine powders by Guillaud [43G2] is one of the most clear-cut examples of the intrinsic effect of particle size. The coercive force increases by a factor of 15 as the size decreases to

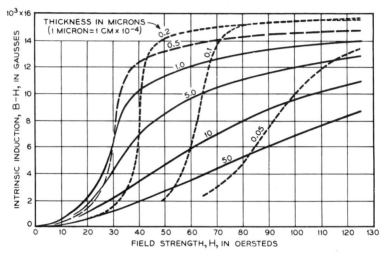

Fig. 2–37. Dependence of the magnetization curves of pure electrodeposited cobalt films on the thickness.

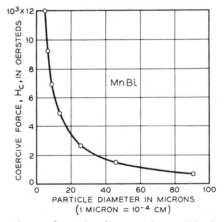

Fig. 2–38. Dependence of coercive force on the particle size of MnBi powder.

5×10^{-4} cm (Fig. 2–38). The theory of fine particles is discussed in Chap. 18.

Properties Affected by Magnetization.—In addition to the magnetization, other properties are changed by the direct application of a magnetic field. Some of these, and the amounts by which they may be changed, are as follows:

Length and volume (magnetostriction) (0.01%).
Electrical resistivity (5%).
Temperature (magnetocaloric effect; heat of hysteresis) (1°C).
Elastic constants (20%).
Rotation of plane-of polarization of light (Kerr and Faraday effects) (one degree of arc).

In addition to these properties there are others that change with temperature because the magnetization itself changes. Thus there is "anomalous" temperature-dependence of:

Specific heat
Thermal expansion
Electrical resistivity
Elastic constants
Thermoelectric force

and of other properties below the Curie point of a ferromagnetic material, even when no magnetic field is applied.

Also associated with ferromagnetism are galvanomagnetic, chemical, and other effects that will not be discussed in this book.

CHAPTER 3

TECHNICAL AND PURE IRON

In 1946 the world production of iron and steel for all purposes was estimated to be about 100 million tons, of which perhaps half was made in the United States. Although only a small fraction was used in apparatus in which the magnetic properties of the material were essential to the performance, nevertheless the tonnage of material so used was large. Specific figures are not available but the amount has been estimated to be almost a million tons.

Although many elements besides iron are purposely present in magnetic materials, nevertheless iron is present as the major constituent in by far the greatest number of useful ferromagnetic materials. These may be divided into two classifications: magnetically "soft" materials used in transformers, motors, relays, and other electromagnetic apparatus; and magnetically "hard" materials—permanent magnets—used in loud-speakers, relays, telephone receivers, and a variety of other instruments. In the magnetically soft materials a substantial portion of the iron is used unalloyed as "magnetic iron." Larger quantities are consumed in the iron-silicon alloys containing up to 6% silicon, and smaller amounts in the iron-nickel alloys (Permalloys) and some iron-cobalt and iron-cobalt-nickel alloys. The permanent magnet materials include the steels containing some essential carbon and the alloying elements cobalt, chromium, tungsten, and manganese in various proportions; also the Alnicos containing iron, cobalt, nickel, and aluminum in various proportions and sometimes also copper and titanium; and finally iron-cobalt alloys with additions of molybdenum, chromium, vanadium, tungsten, and others. The last two classifications, although often referred to as steels, do not contain essential carbon and, indeed, carbon is usually harmful; the designation "steel" is to be avoided here.

Manufacture of Iron and Steel.—By far the major portion of the iron used in magnetic alloys is first reduced from the ore in the blast furnace and refined in the basic open hearth and, except for the final steps, is processed in the same way as the common steels of commerce. A brief outline of the chemistry of these processes will be given here as a basis for understanding the presence and effects of the various impurities in the commercial product.

MANUFACTURE OF IRON AND STEEL

The *blast furnace* is a cylindrical structure of fire brick 75–100 ft high and 25–30 ft in diameter into the top of which are fed the raw materials—limestone, coke, and ore. Air is blown continually into the mass near the bottom of the furnace, and the pig iron and slag are tapped periodically from the base, the iron every 5 or 6 hours. In making 1 ton of pig iron, approximately 2 tons of ore (predominately Fe_2O_3), 1 ton of coke, $\frac{1}{2}$ ton of limestone, and 4 tons of air are required. In addition to the pig iron produced, about 6 tons of gases are evolved and $\frac{1}{2}$ ton of slag is formed. The temperature of the material in the furnace increases as it moves slowly downward and is about 2000°C in the hottest zone near the base.

Practically all of the heat necessary in bringing about the chemical changes is furnished by the burning of the coke:

$$2\,C + O_2 = 2\,CO.$$

The ore is reduced first by the action of the resultant CO according to the equation

$$Fe_2O_3 + 3\,CO = 2\,Fe + 3\,CO_2,$$

but at higher temperatures the reaction

$$Fe_2O_3 + 3\,C = 2\,Fe + 3\,CO$$

also occurs. The limestone is decomposed by heat:

$$CaCO_3 = CaO + CO_2,$$

and the lime combines with the silica and alumina present as impurities in the ore to form relatively low melting compounds that are liquid near the melting point of iron and so may be readily tapped:

$$2\,CaO + Al_2O_3 + SiO_2 = 2\,CaO \cdot Al_2O_3 \cdot SiO_2.$$

Some of the silica, however, is reduced by carbon to form metallic *silicon* which dissolves in the iron

$$SiO_2 + 2\,C = Si + 2\,CO,$$

but the calcium, magnesium, and aluminum oxides present are not so reduced.

Sulfur, present originally as an impurity in the coke, is converted into iron sulfide and then largely into calcium sulfide which is absorbed by the slag. Some sulfur remains in the iron.

$$FeS + CaO + C = CaS + Fe + CO.$$

Manganese oxides often present in the ore are partially reduced according to the equation

$$MnO_2 + 2\,C = Mn + 2\,CO,$$

and the manganese dissolves in the iron. *Phosphorus* present as calcium phosphate is also reduced,

$$P_2O_5 + 5\,C = 2\,P + 5\,CO,$$

and remains in solution in the iron.

A typical analysis of the main constituents of pig iron is given in Table 1.

TABLE 1. TYPICAL ANALYSIS OF PIG IRON

Element	Per Cent	Element	Per Cent
Fe	93	Si	1.0
C	4	P	0.2
Mn	1.5	S	0.03

There is also a certain amount of slag present as inclusions.

About 90% of the total amount of pig iron is refined in the *basic open hearth*, which removes calcium, manganese, phosphorus, and silicon by oxidation and the suspended matter, including oxides, by the scrubbing action of the gases evolved during the process. Basic open hearth furnaces are shallow rectangular structures that may have capacities of 100 tons or more of product. Lined with magnesium and calcium carbonates, they are roofed with silica brick and are charged with scrap iron, pig iron, limestone, and some ore. They are heated with atomized tar and oil and with preheated gases from blast furnaces or coke ovens. During the first few hours of heating the scrap oxidizes, somewhat later the limestone calcines and emits carbon dioxide, and in the final stage the refining action takes place underneath the slag formed by the lime and silica.

During the oxidation brought about by the furnace gases, and to some extent by the carbon dioxide, phosphorus is oxidized to calcium phosphate, $3\,CaO \cdot P_2O_5$, the manganese to MnO, and the sulfur to $CaSO_4$, and these combine with the slag on the surface of the melt. The first slag to form is composed of the lime resulting from the calcining of the limestone and the iron oxide formed from the scrap. The oxide of the slag reacts with the silicon and manganese in the melt to form metallic iron and silica and manganese oxide, and the last two form silicates that are added to the slag. These reactions are exothermic and proceed rapidly. A little later phosphorus is removed by a similar process and becomes calcium phosphate, and finally, when these reactions are well on their way toward completion, the removal of carbon begins. This is an endothermic reaction,

$$FeO + C = Fe + CO,$$

and is forced toward completion by the addition of iron ore or oxygen in other form, under carefully controlled conditions regulated by chemical analysis.

For much of the iron used for magnetic purposes final deoxidation of the

100-ton molten bath is carried out by adding ferro-silicon. The complete furnace cycle takes about 25 hours.

Chemical Composition.—Analyses of two typical specimens of basic open hearth, low-carbon iron are shown in Table 2 [35C3, 40C1]. The

TABLE 2. ANALYSES OF BASIC OPEN HEARTH IRON

Element	Basic Open Hearth Per Cent	Per Cent	Swedish Per Cent
C	0.02	0.02	0.018
Mn	.04	.05	.002
Si	.003	.01	.022
P	.005	.01	.027
S	.03	.03	.002
O	.08		.13
N	.005		.004
Al	.002		

final column shows the composition of a good grade of Swedish electrical iron of unknown manufacture.

Thus the purification effective in the refining process increases the iron content from about 93% to about 99.7–99.9%, and this product finds many uses after appropriate shaping and annealing.

Preparation of Pure Iron.—Iron of high purity may be obtained upon further refining by electrolysis, by reduction of purified oxide with hydrogen, or by the formation and decomposition of iron carbonyl.

Analyses of *electrolytic* iron from three commercial sources are given in the following table. These are Bell Laboratories analyses of the Niagara product and the analyses by others of Westinghouse [17Y1] and National Radiator [35C3] materials. Apparently because of the intrinsic expense

TABLE 3. ANALYSES OF ELECTROLYTIC IRON

Element	Niagara Per Cent	Westinghouse Per Cent	National Radiator Per Cent
C	0.006	0.02	0.010
Mn	.000	.01	.001
Si	.005	.01	.004
P	.005	.05	.003
S	.004	.003	.003
Al		.01	.009
Cu	.015		.001

of this process, no continuing supply of electrolytic iron has been commercially available.

In purifying on a commercial scale by the *carbonyl* process impure sponge iron is treated with carbon monoxide under pressure, and the resulting liquid or gaseous compound, Fe(CO)$_5$, is decomposed by injecting it into a hollow chamber at about 350°C. The iron so formed [28M1] contains about 1% carbon and is mixed with the requisite amount of FeO, obtained by combustion of some of the purified iron, or is decarbonized by hydrogen and remelted in vacuum. The final product is stated to be exceedingly pure, but several analyses reported in the literature show variable and sizable impurities, as shown in Table 4 [37Y1, 35J1].

TABLE 4. ANALYSES OF CARBONYL IRON

Element	First Source Per Cent	Second Source Per Cent
C	0.02 to 0.10	0.007
Mn	.08 to .10	tr
Si	.04 to .05	.004
P	.005 to .007	...
S	.005 to .007	.004
O	.1 to .6	.004
N	.005	

Analyses at Bell Laboratories showed the nickel content of several samples to vary from 0.008–0.1%.

A small amount of iron of exceptionally high purity has been prepared in the laboratory by Thompson and Cleaves by *reduction of the oxide* [35C3, 39T1]. The starting point was carefully recrystallized iron nitrate from which the oxide was precipitated with anhydrous ammonia and dried to Fe$_2$O$_3$. The latter was reduced with hydrogen at 500°C, sintered in hydrogen at 1000°C, pressed, melted in beryllia crucibles in an atmosphere of carbon monoxide, remelted in an induction furnace in purified hydrogen, and allowed to freeze under reduced pressure. Analysis for fifty elements showed less than 0.01% total impurities, most important among which were 0.003% oxygen, 0.002% copper, and 0.001% carbon.

The impurities which have the greatest effects on the magnetic properties are usually the non-metallic elements carbon, oxygen, sulfur, and nitrogen, and these may all be removed by treatment with pure hydrogen at elevated temperatures. The removal of carbon by hydrogen takes place according to the equation

$$Fe + C + 2H_2 = Fe + CH_4$$

which is abnormal in that it becomes less effective at high temperatures. Thus for decarburization there is an optimum temperature which, on account of the change of the diffusion rate with temperature, depends on the thickness of the material and various other factors and may vary from 400–1200°C or higher. It obviously occurs best under pressure. Oxygen, sulfur, and nitrogen are removed in the form of H$_2$O, H$_2$S, and NH$_3$, respectively, the more rapidly the higher the temperature except that sulfur is removed less effectively in the melt than in the solid phase.

An example of the purification of iron by such treatment is shown by the data of Table 5 of Chap. 2.

Physical Properties.—Below the 900° transformation point, iron exists in the α phase, having the well-known body-centered cubic structure. A similar structure exists in the δ phase stable above 1400°C, and between these two transformation temperatures it is face-centered cubic. The latter structure is more dense, consequently an increase in density occurs when the metal transforms to the γ phase from either α or δ.

A summary of the more important physical properties of iron is given in Table 5; Table 6 contains some of the more important magnetic constants. Unless otherwise noted the constants are those of high purity iron (see especially [42C1]). The properties of commercial ingot iron have been summarized by Kenyon [48K4].

TABLE 5. SOME PHYSICAL PROPERTIES OF IRON

Density, g/cm³ at 20°C	7.874
Thermal expansion coefficient $\times 10^6$ at 20°C	11.7
Lattice constant, 10^{-8} cm	2.861
Melting point, °C	1539
Boiling point, °C (approx.)	3200
Temp. of α,γ transformation (A_3) on heating, °C	910
Temp. of γ,δ transformation (A_4) on cooling, °C	1400
Resistivity at 20°C, ohm-cm $\times 10^6$	9.7
Same for commercial magnetic iron	11
Temperature coefficient of resistance	0.0065
Compressibility, cm²/kg $\times 10^6$	0.60
Specific heat, cal/g/°C	0.105
Heat of fusion, cal/g	64.9
Heat of α,γ transformation, cal/g	3.86
Heat of γ,δ transformation, cal/g	1.7
Linear contraction at α,γ on heating	0.0026
Linear expansion at γ,δ on heating	0.001 to 0.003
Modulus of elasticity, lb/in.² $\times 10^{-6}$	30
Modulus of elasticity, dynes/cm² $\times 10^{-11}$	21
Modulus of rigidity, lb/in.² $\times 10^{-6}$	12
Proportional limit for annealed iron, lb/in.² $\times 10^{-3}$	19
Tensile strength, lb/in.² $\times 10^{-3}$	25–100
Brinell Hardness number (annealed)	50–90

Some data on *thermal expansion* as dependent on temperature are shown in Fig. 3–1. The corresponding densities, based on 7.87 at 20°C, are shown on the scale at the right. The earlier data obtained by Schmidt [33S5] are derived from X-ray data and show the course over the whole temperature range in which iron is solid; the accuracy is probably not as great as that of Hidnert and of Esser and Eusterbrock [42C1, 41E3]. Densities in the δ region and in the melt, and the changes at the critical points, are still quite uncertain.

TABLE 6. SOME MAGNETIC PROPERTIES OF IRON

Curie temperature, °C	θ		770
Saturation magnetic moment/g at 20°C [36W2]	σ_s		217.75
Saturation magnetic moment/g at 0°K	σ_0		221.89
Saturation intensity of magnetization at 20°C	I_s		1 714
Saturation induction at 20°C [37S4]	B_s		21 580
Saturation induction at 25°C [41S1]	B_s		21 580
Saturation intensity of magnetization at 0°K	I_0		1 735.2
Saturation induction at 0°K	$4\pi I_0$		21 805
Bohr magnetons per atom	n_0		2.218
Maximum permeability, commercial product	μ_m		5 000
Highest maximum permeability (polycrystalline)	μ_m		350 000
Coercive force, commercial product	H_c		0.9
Lowest coercive force	H_c		0.01

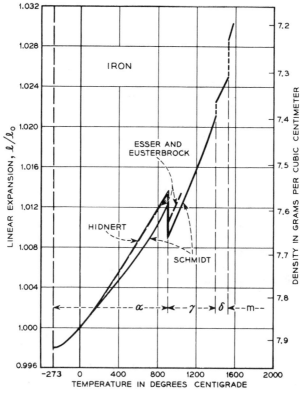

FIG. 3-1. Thermal expansion of iron according to various authors, showing breaks accompanying change of phase.

The lattice parameters in the α and γ regions are given in Fig. 3-2. Here again Schmidt's data include values lower than those of the best recent determinations [37O3, 38E2] but indicate the trend at higher temperatures.

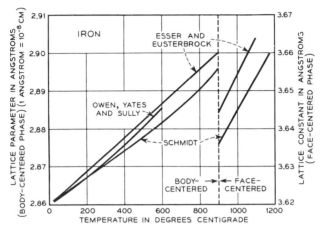

FIG. 3-2. Lattice parameter of iron as dependent on temperature.

Representative curves showing the *resistivity* as dependent on temperature up to the melting point are given in Fig. 3-3 for specimens of ingot iron [39P3] and electrolytic iron [26R1]. Values at low temperatures are averages of those of a number of authors [29O1].

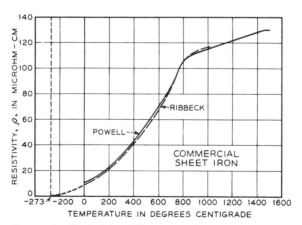

FIG. 3-3. Resistivity of iron at various temperatures.

Magnetic Properties of Commercial Material.—Figure 3-4 shows the μ vs B curves of two typical specimens of commercial sheet iron refined in the basic open hearth. Hysteresis loops at various maximum inductions are shown in Fig. 3-5 for material having a maximum permeability of

9000 and a coercive force of 0.81. This material, having a maximum permeability somewhat higher than average, was annealed for 1 hour at 925°C and cooled slowly with the furnace.

Figure 3–4 indicates that the permeability drops to rather low values at inductions of 16 000–18 000, but saturation does not occur until $B-H$ is over 21 000. The peculiar approach to saturation is best shown by the data of Steinhaus, Kussmann, and Schoen [37S4] plotted in Fig. 3–6 as

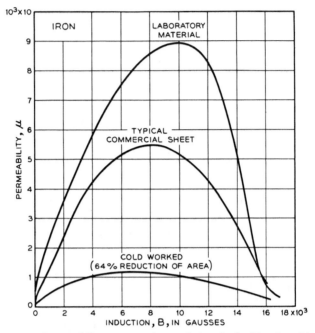

Fig. 3–4. Permeability curves of annealed and of cold-reduced iron.

$\mu-1$ vs $B-H$. This material has a saturation somewhat below that for pure iron.

The initial permeability is usually below 200, and the permeability at $B = 20$ gausses is about 250–300. For two representative specimens the relations between μ and H_m are

$$\mu = 150 + 2200 H_m,$$
$$\mu = 180 + 1400 H_m,$$

between $H_m = 0$ and 0.3 oersted. Figure 3–7 shows the reversible permeability as dependent upon the magnetizing field, for a representative specimen.

Temperature.—The effect of temperature on the induction at high field strengths ($H = 10\,000$) was accurately determined by Hegg in 1910

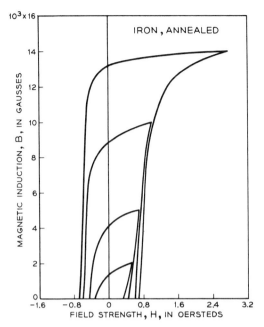

FIG. 3–5. Upper halves of hysteresis loops of ordinary annealed iron.

[10H1]. Change of B_s below room temperatures is represented [37W4] by the $T^{3/2}$ or T^2 law, e.g.,

$$B_s = B_0(1 - CT^2),$$

in which B_0 is the saturation ferric induction at 0°K, and T is the absolute

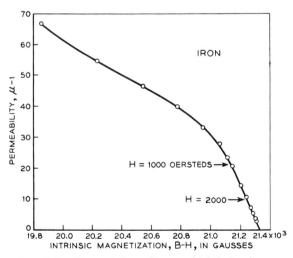

FIG. 3–6. Permeability of iron at high inductions.

temperature. In 1910, also, magnetization curves were recorded at various temperatures by Terry [10T1] for specimens of electrolytic iron of high purity (0.012% C, 0.072% H, 0.001% S, 0.003% Si, 0.004% P) annealed at 800°C. Although the maximum permeabilities of his specimens were definitely low as judged by present-day standards, probably because of the low annealing temperature, his magnetization curves, and data showing the permeability and other properties at temperatures up to

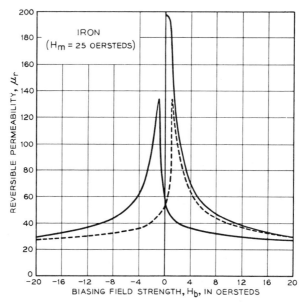

Fig. 3-7. Reversible permeability of iron at various points on the magnetization curve and hysteresis loop. "Butterfly curve."

the Curie point are the best available today (Fig. 3-8). They show the characteristic abrupt increases in μ_0 and μ_m at about 50° below the Curie point, and the continual decrease of H_c, W_h, and B_m with increasing temperature. The shrinking of the area of the hysteresis loop as the Curie point is approached and its practical disappearance when the permeability is still high are shown by the data of Kühlewein [32K2]. Curves illustrating these and other changes with temperature are given in Chap. 14.

Heat Treatment of Ordinary Iron.—The properties of ordinary iron are affected by the temperature and time of anneal and by the cooling rate. The latter is especially critical in the range 900–800°C on account of the α,γ transformation. If the α,γ transformation temperature is exceeded, the cooling rate must be rather slow, about 5°C/min. Higher maximum permeability is obtained by exceeding the transformation temperature than by not allowing the material to transform to γ at all. Hence, for high μ_m, an anneal at 925–1000°C with cooling at less than 5°C/min is recommended. If high permeability at high inductions ($B > 12\,000$) is desired, it is advisable to anneal at 800°C or below and to cool slowly [43A3]. Cooling at a much slower rate than 5°C/min sometimes causes embrittlement, but this apparently depends on the impurity content.

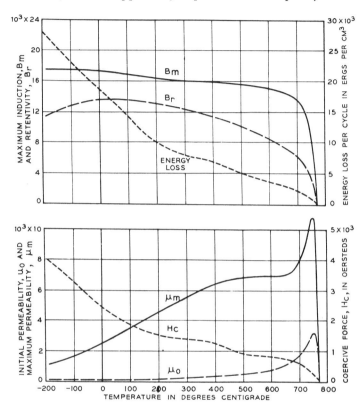

FIG. 3–8. Variation of some properties of iron with temperature. Material was annealed at 800°C.

Impurities may seriously affect the *aging* of certain magnetic properties—the coercive force and hysteresis loss of some specimens at room temperature will increase 100% or more after holding at 100°C for 200 hours and will even change appreciably at 25°C, while others are not changed measurably by this treatment. This aging is caused by the precipitation of an impurity such as carbon or nitrogen which is present in an amount exceeding its solid solubility at 100°C or at room temperature, and is the dispersion hardening well known for its effects on mechanical properties. This has already been discussed.

The solubility of any one of the common impurities in iron is affected by the presence of the others and by still other elements that may be added for special purposes. Thus, manganese is known to reduce the solubility of sulfur, and silicon that of carbon. Köster [30K4] has added aluminum to decrease the solubility of nitrogen and has found that 0.2% of this element prevented aging. Other metallic elements, e.g., titanium and vanadium, have been found to "counteract" the common impurities that cause aging.

Meager data are available on the *recovery* of the magnetic properties of iron, upon annealing at low temperatures after cold reduction. Figure 3-9, taken from Tammann and Rocha [33T2], shows the per cent drop in induction caused by drawing a well-annealed wire to 4% of its original cross-sectional area and annealing for 45 minutes at various temperatures up to about 500°C. Recovery is apparently about half completed at the relatively low temperature of 250-350°C.

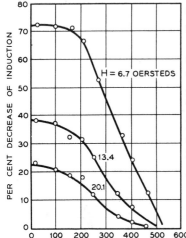

FIG. 3-9. "Recovery" of properties of cold worked iron with annealing at various temperatures.

Hydrogen Treatment.—Although the maximum permeability of commercial magnetic iron is usually 5000-10 000, measurements have been made on laboratory specimens having permeabilities of a much higher magnitude and having other properties to correspond. Yensen [29Y1] has recorded the highest values of μ_m reported in various years and pointed out that, when plotted, these data give a good picture of the progress in magnetic research made during the last half century. Figure 3-10 shows Yensen's figure to which have been added the results of more recent determinations [29Y1, 29R1, 30C1, 32C2, 34C1, 37C2, 37B1], including a final point for a pure single crystal for which μ_m is almost 1000 times that recorded by Rowland in 1873. The advances in the period 1900-1915 may be attributed chiefly to the work of Yensen who melted his materials in vacuum and at times annealed them in hydrogen at temperatures up to 1100 to 1200°C. Improvements during 1925-1940 were due primarily to the work of Cioffi [32C2, 34C1, 37C2] who treated his material in pure hydrogen at temperatures above 1300°C.

Using Cioffi's technique, maximum permeabilities of about 250 000 can be obtained repeatedly, and corresponding values of H_c and $\mu_{B=1}$ are 0.04 and 25 000, respectively. The method of treatment is to enclose the specimen in an iron box into which, by means of an iron tube, is passed a stream of hydrogen purified by diffusing the compressed gas directly through the walls of hot palladium tubes. The box is placed in a furnace through which moist hydrogen flows, the moisture being present to prevent silicon and other impurities present in the furnace from diffusing through the walls of the box into the specimen. (Pure dry hydrogen reduces silica to silicon which diffuses readily at the high temperatures employed, 1300–1500°C.)

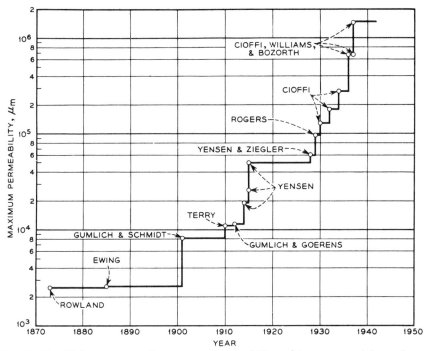

Fig. 3–10. Highest values of maximum permeability of iron reported in various years.

After cooling from the high temperature it is necessary to pass through the α,γ point showly or reanneal at about 880°C for several hours. Representative magnetic data for high and low inductions are given in Figs. 3–11 to 13, and the effect of varying the temperature of the high-temperature heat treatment is shown in Fig. 3–14. Figure 3–12 shows also a μ vs B curve for the specimen having the highest μ_m of any polycrystalline iron so far known.

Removal of Impurities.—The effectiveness of hydrogen treatment lies in the fact that magnetic properties are affected by small amounts of the common impurities only if they are non-metallic in character, and that such impurities can be removed by the treatment. These elements are probably injurious because they have only limited solid solubility in iron

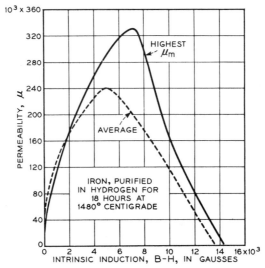

Fig. 3–11. Permeability curves of iron heat-treated at a high temperature in hydrogen.

and cause dispersion hardening; and they can be removed in a reasonable time because they diffuse through the metal and combine with hydrogen at high temperatures. Oxygen, carbon, and nitrogen are removed in this way more readily than sulfur [34C1], and carbon is removed probably more

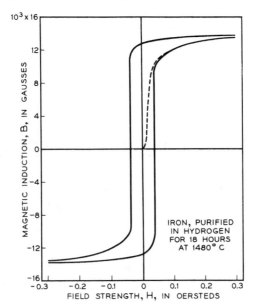

Fig. 3–12. Hysteresis loop of hydrogen-treated iron.

readily in moist than in dry hydrogen. Phosphorus is not appreciably affected by hydrogen nor does it affect the magnetic properties of otherwise

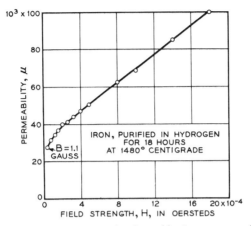

FIG. 3-13. Initial magnetization of hydrogen-treated iron.

pure iron if it is present in small amounts; in iron it may be regarded as metallic in character as is indicated by the fact that it forms a solid solution

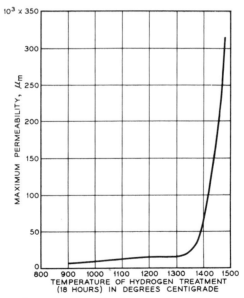

FIG. 3-14. Effect of temperature of hydrogen treatment on maximum permeability.

of the substitution type when present in amounts less than 1%. The negligible action of hydrogen on phosphorus in iron has been shown by an experiment in which iron containing 0.8% phosphorus showed no change in resistivity after treatment for 18 hours at 1400°C in pure hydrogen.

Before a given impurity can cease to be harmful it must *diffuse* through the iron to the surface of the specimen, then it must *leave the surface* either by evaporation or combination in a reasonable time when a practically attainable flow of hydrogen or other suitable gas is passed over the surface. The residual impurity must be less than its solid solubility at or somewhat above room temperature. The solubilities of the more important non-metallic elements are not known with much precision but estimated values are given in Table 7.

TABLE 7. ESTIMATED SOLID SOLUBILITY LIMIT OF SOME COMMON NON-METALLIC IMPURITIES IN IRON AT ROOM TEMPERATURE

Element	*Per Cent*
Carbon	0.007
Oxygen	.01
Sulfur	.02
Nitrogen	.001
Phosphorus	1.0

It will be noted that these are all, with the exception of phosphorus, less than the amount usually present in a good grade of open hearth ingot iron.

The rate of *diffusion* of the impurity in the iron is measured by the diffusion constant D expressed in the units $cm^2\ sec^{-1}$. Values of the constants for several temperatures are given in Table 8 [23F1, 35B8, 40W1, 41B1].

TABLE 8. DIFFUSION COEFFICIENT, $D \times 10^{10}$, IN $CM^2\ SEC^{-1}$, FOR SOME ELEMENTS IN IRON

Temperature (°C)	C	O	N	S	P	Ni	Si
900	1 000	..	600	(3)	(2)	..	(60)
1000	2 700	8	1350	6	13	..	85
1200	13 500	..	(9000)	(17)	..	1	(165)
1400	(47 000)	..	...	(40)	..	..	...

In the table, the numbers in parentheses denote interpolated or extrapolated values. Diffusion constants may be expressed in terms of the time necessary to reduce an impurity to one-tenth of its original value in a sheet of given thickness, assuming that all of the material diffusing to its surfaces is immediately carried away by combination or evaporation. For a sheet thickness of 0.036 cm (0.014 in.) the time required in hours is $760/(D \times 10^{10})$ and, for other thicknesses, is proportional to the square of the thickness. When the thickness d is expressed in inches, the time in hours is

$$\frac{760}{(D \times 10^{10})} \left(\frac{d}{0.014}\right)^2$$

(adapted from expressions given by Barrer [41B1]). It is thus apparent that to effect a tenfold reduction in the concentration in a 0.014 in. sheet in a reasonable time, say 24 hours, $D \times 10^{10}$ must be 30 or more. A

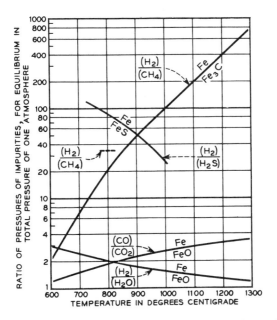

FIG. 3–15. Ratio of pressures of hydrogen and other gases for equilibrium of iron in contact with Fe_3C, FeS, and FeO.

glance at the table shows that because of its low diffusivity alone, sulfur is one of the most difficult elements to remove and that a temperature of 1300–1400°C is necessary to remove most of it in one day's time.

The diffusion constants of some elements in iron depend markedly on the presence of other elements. For example the presence of 0.5% of carbon will decrease the constant of sulfur to about one-tenth of its value for carbon-free iron [35B8]. Similarly sulfur depresses the diffusion of carbon, but oxygen increases the diffusion of nitrogen [35B8] and carbon the diffusion of carbon itself [40W1].

A further consideration in purification with hydrogen is the combination of the hydrogen with the impurity after the latter has diffused to the surface. When the impurity exceeds the solid solubility, so that some such solid phase as FeS, Fe_3C, or FeO is present, the minimum amount of hydrogen necessary to combine with a given amount of the impurity may be calculated from the equilibrium constant of the reaction. Reactions with sulfur and oxygen require less hydrogen the higher the temperature, but the reaction with carbon is unusual in that it requires an increasing amount of hydrogen. The ratio of the pressure of H_2 to that of CH_4 or H_2O in the atmosphere surrounding the impure iron, for equilibrium with a total pressure of one atmosphere, is given in Fig. 3–15 [42A3]. It is estimated that at a temperature between 1300 and 1400°C about 1000 volumes of H_2 are necessary to remove one volume of CH_4, whereas at 700°C the ratio is only 5 to 10.

When the amount of impurity is small enough so that it is held in solid solution at the temperature of treatment, the ratio of the gas pressure of the hydrogen to that of the gaseous compound is even larger than that just considered, and as the amount of impurity approaches zero the ratio becomes indefinitely large.

Nitrogen is readily removed from iron with hydrogen. At 400°C more than 35% of ammonia in hydrogen can be tolerated without formation of the nitride (Fe_4N), and 15% at 575°C [31B3]. Lower temperatures are therefore more favorable for nitride decomposition, provided diffusion proceeds at a practical rate.

The removal of sulfur is peculiar in that the vapor pressure of sulfur over FeS [30B2, 37C3], estimated to be about 0.2 mm of mercury at 1250°C, is high enough to permit the sulfur to evaporate more readily than it will combine with hydrogen, even though the evaporation will still be slow. Sulfur is thus a difficult impurity to remove from the standpoint of both diffusion through the iron and removal from the iron surface. A calculation indicates that 30 hours would be necessary to remove 90% of the sulfur from a 0.014-in. sheet at 1250°C, and that the time necessary for this purification increases rapidly with decrease in temperature. This calculation is supported qualitatively by the experiments of Cioffi [34C1].

CHAPTER 4

IRON–SILICON ALLOYS

Alloys of iron and silicon are of prime importance in the electrical industry, which consumes hundreds of thousands of tons annually—about 400 000 tons in the United States alone in 1946. The low-percentage alloys, containing 1.5–3.5% silicon, are used primarily for motors, generators, and relays. The higher-percentage alloys 3–5%, are used for high-efficiency motors and for power transformers. It was estimated [36Y1] some years ago that about 0.4% of the total electrical power used in the United States is dissipated as heat in transformer cores, and the value of this loss is of the order of 100 million dollars annually.

Outside of the electrical industry, silicon is used as a minor constituent in the manufacture of many steels. Silicon-cast iron, containing about 15% silicon and up to 1% each of carbon and manganese, is used as a corrosion-resistant material in the chemical industry.

History of Magnetic Alloys.—The usefulness of iron-silicon alloys was made known to the world as the result of researches begun by Hadfield. In 1882 he noticed [26H1] the hardness of an alloy, accidentally produced, containing over 1.5% silicon. The mechanical properties of alloys containing various amounts of silicon were investigated in an attempt to find useful applications of the material, and the results were published in 1889 [89H2]. This led to many investigations by others, and especially to the report on magnetic properties by Barrett, Brown, and Hadfield in 1900 [00B1, 02B1, 03H3].

Although the magnetic properties and resistivity of some alloys had previously been measured by J. Hopkinson [85H1], the carbon which his alloys contained obscured their intrinsic good qualities, and it was the 1900 paper which set the stage for commercial use. It stated that the addition of 2–2.5% silicon increased the magnetic softness to such an extent that the coercive force was about one-half that of the standard iron then employed in transformer cores. This report stimulated Gumlich [12G2] and the Physikalisch-Technische Reichsanstalt to foster production in Germany, and material was produced in quantity by German firms in 1903. The first commercial production in the United States was made in the same year [48M1]. The improvement over the iron previously used was fourfold: (1) the permeability was increased, (2) the hysteresis loss decreased, (3) the eddy-current loss was decreased because of the higher resistivity, and (4) there was no deterioration with time (aging); consequently, the new alloy "could not fail almost entirely to replace the usual material within a short time, in spite of the initial difficulty of production and of the much higher price" [12G2]. Silicon-iron "electrical sheet" was used commercially in the United States in 1905 and in England in 1906.

The losses in such sheet were about 1–2 watts/lb [26H1, 35C2], measured in 0.014-in. sheet at $B_m = 10\,000$ and $f = 60$ cps; this was a decrease, on the average, by about a factor of 2 as compared with the iron previously used, even before it aged. In 1925 the losses in hot rolled commercial sheet had decreased to about 0.7 watt/lb, and at present are about 0.5 watt/lb [47H2]. Cold rolled sheet can now be purchased with a loss of about 0.3 watt/lb, measured as stated above with the magnetic field applied parallel to the direction in which the sheet was rolled.

After the original work of Barrett, Brown, and Hadfield, improvements in laboratory specimens were due mainly to the work of Gumlich [18G2] in Germany, and Yensen [15Y1, 24Y1] in the United States, on the effect of impurities. Pioneer work on the effect of cold rolling was reported by Goss [34G4] in 1934. All of these researches were followed by improvements in the commercial product. Some properties of laboratory specimens prepared at various times since the work of Hadfield are given in Table 1.

Methods of Production.—Commercial iron-silicon alloys are usually melted in the basic open hearth, although occasionally electric arc or induction furnaces are used. In all cases a careful selection of scrap and rigid control of furnace conditions, particularly during the refining period, are important. Recently the introduction of oxygen during the refining operation has been found to have advantages, as pointed out by Slottman and Lounsberry [47S4]. In the basic open hearth furnace the alloying is done by adding ferro-silicon (50–90% silicon content) in the ladle. When the higher ferro-silicons, such as 90%, are used, the increased heat caused by the exothermic reaction makes it necessary to hold the material in the ladle for a considerable length of time before pouring. The pouring temperature must not be too high, for iron-silicon alloys have a tendency to develop large shrinkage cavities or pipe in the ingot.

After casting and solidification of the ingots, which are often over 5 tons in weight, they go to soaking pits and then to the blooming or breakdown mills at 1200–1250°C. The rolling speed and the per cent reduction per pass differ only slightly from the processes normally used for low-carbon steel products. Somewhat higher temperatures are permitted, and slow cooling of the thick sections is desirable when the silicon content is over $2\frac{1}{2}\%$ in order to avoid internal rupturing from cooling stresses. When slabs are rolled into hot strip 0.125–0.05 in. thick, special care must be exercised in both the handling and the heating of the slabs since silicon irons are more susceptible to breaking and cracking than are low-carbon steels.

Most of the electrical sheet is supplied at thicknesses between 0.014–0.025 in. The final thickness is attained by either hot rolling or cold rolling, as described later. In the later stages of *hot rolling* two, three, or

TABLE 1. SOME MAGNETIC PROPERTIES OF LABORATORY SPECIMENS OF IRON-SILICON ALLOYS

Observer	Year	Silicon Content (%)	Maximum Permeability	Hysteresis Loss, $B_m = 10\,000$ (ergs/cm^3/cycle)	Remarks
Barrett, et al.	1902	2.9	4 500	1800	
Gumlich [09G1]	1909	3.9	6 100	...	
Yensen [15Y1]	1915	3.4	63 000	280	
Cioffi [32C5]	1932	4.0	16 000	...	$\mu_0 = 4000$
Ruder [34R3, 36Y1]	1934	...	...	70	Single crystal
Goss [34G4]	1934	3	18 000		$\mu = 8000$ at $B = 16\,000$
Yensen [36Y1]	1936	3	83 000		
Williams [37W5]	1937	3.8	1 400 000	<30	Single crystal
Boothby [BTL]	1942	6.3	500 000	45	Heat-treated in field
(Commercial)	1950	3	50 000	250	Grain-oriented

four sections of the hot strip material (0.050–0.125 in. thick) are matched together, rolled, doubled, and then heated in a continuous pack furnace at a temperature of 800–1000°C; then they are rolled to final thickness in three to five passes and finished at about 750°C, or, alternatively, heated to 800–850°C again before the final pass. After rolling, the packs are sheared and separated into individual sheets. The sheets are given a cold pass to flatten, and the low-silicon alloys are box-annealed at 700–800°C. Surface oxide may be removed by pickling either before or after box annealing. For the higher silicon contents the annealing temperature is usually in the range 800–1150°C. When temperatures are above about 900°C, it is common practice to employ a controlled atmosphere to prevent excessive oxidation.

A recent development in the United States is the use of a final *cold rolling* or a series of cold reductions with intermediate anneals to final thickness. Sheets prepared from open hearth melts are usually processed finally to 0.014-in. thickness, but some material, usually coiled strip, is reduced commercially [46C1] to as thin as 0.002 in. with the Sendzimir mill [46S3]. Cold reduced materials fall into two groups: (1) *isotropic* materials which have almost the same magnetic characteristics as the hot rolled sheets, and (2) *anisotropic* or *grain oriented* materials which have greatly improved characteristics when measured in the direction of rolling. The cold rolling and annealing given to group (2) cause a special orientation of the crystals in the sheet, and this is the primary reason for the unusually good magnetic properties.

For the randomly oriented materials of group (1) the hot strip product may be treated in various ways, depending on the composition and available equipment. The mode used for 1.5% silicon is to cold reduce from about 0.1 in. to within a few per cent of final gauge, continuous anneal in the range 800–900°C in a controlled atmosphere, roll to final gauge, and continuous anneal a second time in the same way.

Materials having the best magnetic properties are produced today by cold rolling and annealing in such a way that the material is recrystallized with favorable crystal orientation and purified at the same time. Following is an outline of one such process that yields a high-quality, *grain-oriented* product of 0.014-in. thickness: A melt containing 3% silicon and less than 0.1% manganese is refined until the carbon content is less than about 0.03%. The hot rolling to about 0.100 in. is similar to that described previously. At this stage the strip is pickled and oiled, and cold rolled in two stages with an intermediate anneal at about 900–950°C [34G4] for recrystallization. It is then decarburized by annealing in a moist reducing atmosphere [42C2] at about 800–850°C. The crystal orientation is developed by heating slowly to 1100–1200°C in pure dry hydrogen, and this affects some additional purification. Analyses are given in Table 2. The process and resulting material are discussed further in a later section.

TABLE 2. ANALYSES OF TYPICAL AMERICAN COMMERCIAL SHEET (1947)
(Figures are for the ingot unless otherwise specified.)

Material	Composition (%)				
	Si	Mn	C	S	P
Hot rolled, electrical grade	1.1–1.3	0.2	0.05	0.02	0.02
Hot rolled, transformer grade	4–5	.2	.03	.01	.01
Cold rolled* (isotropic)	3.0	.2	.03	.01	.01
Cold rolled* (oriented)	3.0	<.1	.005	.003	.01

* Analysis of final product.

Often a coating of magnesia is applied [45C2, 49C4] before the moist hydrogen treatment, during which some silica is formed by surface oxidation of the sheet, and the two oxides combine during the high temperature anneal to form a glassy surface which acts as an insulator when the sheet is used in transformers.

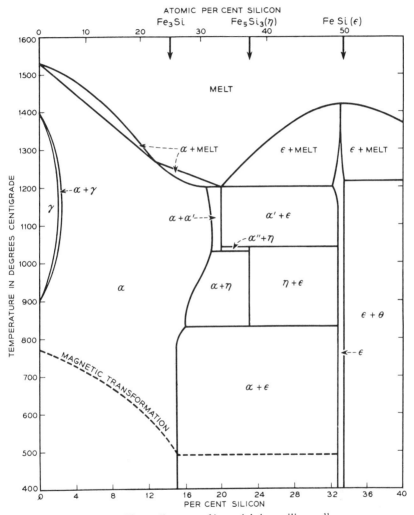

Fig. 4–1. Phase diagram of iron-rich iron-silicon alloys.

Phase Diagram.—Iron-silicon alloys are magnetic for silicon contents up to about 33% silicon, corresponding to the composition FeSi. Below 800°C the solid solubility of silicon in α iron extends to about 15%, beyond which FeSi is precipitated (Fig. 4–1). Above 830°C the phases α', α'', and η have been detected; η has been identified [43W1] as Fe_5Si_3 (23% Si), stable only between 825–1030°C, and the formula $Fe_{11}Si_5$ has been proposed for α'' [40O1]. The phase diagram of iron-rich alloys, given by Farquhar, Lipson, and Weill [45F1] and based on previous work by Greiner and Jette [37G4] and others, is shown in Fig. 4–1; it is still to be regarded as provisional in detail.

In the range of composition 15–33% silicon, a Curie point has been observed at about 90–120° as well as at 450–540° [21M1, 36F1, 45S2].

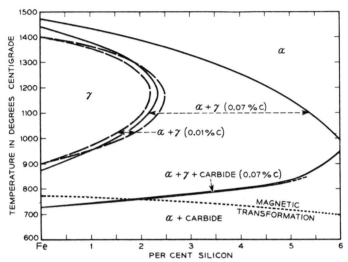

Fig. 4–2. Effect of small carbon contents on the α,γ boundaries of the iron-silicon diagram.

The lower point is ascribed [45S2] to small amounts of the η phase present under non-equilibrium conditions. The ϵ phase is non-magnetic.

In the solid solutions area, of greatest practical importance for magnetic phenomena, the γ phase characteristic of pure iron does not exist in *pure alloys* containing more than 2.5% silicon [29O2]. However, the presence of a few hundredths of a per cent of carbon may widen the $\alpha + \gamma$ region and extend the boundary between $\alpha + \gamma$ and α beyond 5% [46R1, 32K10, 31S7], as indicated in Fig. 4–2. Therefore, in most commercial alloys, which contain 5% or less, a certain small fraction of material transforms to the γ phase above about 800°C. For this reason the conventional anneal is carried out just below this temperature. When 0.1% carbon is present, the limit of the $\alpha + \gamma$ region extends to 7 or 8% silicon, according to Sato [31S7] and Kříž and Poboŕil [32K10].

The effect of carbon on the 4% silicon alloy is shown in Fig. 4–3, a compromise between the diagrams of Sato, Kříž and Poboŕil, and Rickett and Fick [46R1], which appear to differ from each other because of the differences in content of manganese and other elements. (The eutectoid point E is placed by Sato at 0.45% C and 850°C instead of at 0.65% C as shown.) The solubility of carbon in γ iron is increased by the silicon addition, and the effect of temperature on the solubility is diminished. No direct data are available for the solubility of carbon in α solid solutions. Yensen [29Y2] has inferred from magnetic behavior that the solubility

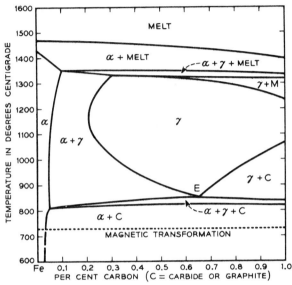

FIG. 4-3. Effect of carbon on the constitution of iron-silicon alloys containing 4% Si.

at room temperature is not greatly influenced by silicon, but in agreement with earlier work [12G1] he concludes that the presence of this element causes the carbon to be precipitated largely as graphite rather than as Fe_3C and thus to have a relatively small effect on the magnetic properties.

X-ray examination of iron-silicon alloys of high purity shows that the lattice parameter is a linear function of the atomic percentage of silicon from 0–5% silicon (about 10 atomic per cent), and another linear function from 5–18% silicon, as shown in Fig. 4-4 according to data of Jette and Greiner [33J2] and Farquhar, Lipson, and Weil [45F1]. Superstructure lines, observed originally by Phragmén [26P1], first appear at 6.7% silicon and correspond to the kind of ordering found in Fe_3Al; therefore one expects the ideal composition of the ordered structure to be Fe_3Si (14.35% Si). This kind of ordering apparently replaces a random arrangement of atoms at the break in the lattice parameter curve (5% Si), and continues somewhat beyond the ideal composition.

74 IRON–SILICON ALLOYS

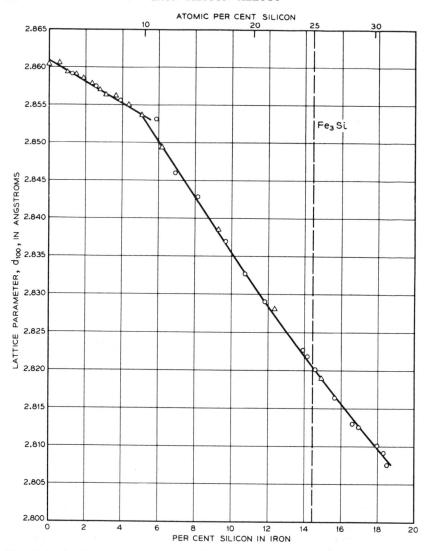

Fig. 4–4. Lattice parameter of iron-rich iron-silicon alloys, showing break at 5% Si.

Physical Properties.—The *densities* of low-silicon alloys are shown by the curve of Fig. 4–5, calculated from X-ray measurements [33J2, 45F1] and consistent with direct determinations. The change in slope at 5% reflects the onset of ordering discussed above. The densities for the whole range of alloys from 0–100% silicon may be calculated from the X-ray data and compared with the direct measurements [14S1, 25S4, 26P1].

In commercial electrical sheet it is customary to assume standard densities for alloys lying within certain limits of composition, as follows:

PHYSICAL PROPERTIES

Silicon Content (%)	Assumed Density (g/cm^3)
0 to 0.5	7.85
>0.5 to 2.0	7.75
>2.0 to 3.5	7.65
>3.5 to 5.0	7.55

These are used for calculating cross-sectional areas from weights, in determinations of B.

Sheet from which mill scale has not been removed has an average density of about 0.05–0.08 g/cm^3 less than that of the alloy. Equations commonly used for densities of sheet at 25°C, in g/cm^3, are:

(a) $7.80 - 0.069 \times \%\ Si$ (scale not removed),
(b) $7.865 - 0.065 \times \%\ Si$ (scale removed).

Thermal expansions have been determined by Goerens [24G1], Schulze [28S8], and others for alloys containing a few per cent of silicon. The effect of silicon on the expansion of iron is very small and is less than that of any other common alloying agent. This statement holds even for the chemical resistant alloys [31O1] and for the alloy Fe_3Si [45F1]. The temperature coefficients of expansion from 0–100°C all lie in the range of $11–13 \times 10^{-6}/°C$.

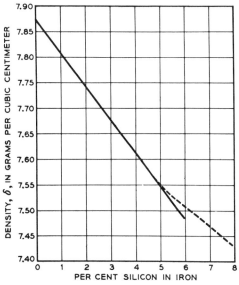

FIG. 4–5. Densities of iron-rich iron-silicon alloys, showing break corresponding to that of Fig. 4–4.

The *resistivity* of iron is increased unusually rapidly by the addition of silicon. Yensen's data [15Y1, 24Y1], given in Fig. 4–6, extend over the range of composition of useful magnetic alloys, and Corson's results [28C5] are for higher silicon contents. A minimum, not shown in the figure, occurs at about 15% silicon at the limit of solid solubility. The increase in resistivity is less rapid beyond 5% silicon, when atomic ordering begins to occur, and as ordering increases the resistivity goes through a maximum at 11–12% silicon and then falls rapidly to its minimum value.

It is to be expected that the rapid initial increase in resistivity with silicon content is to be accompanied by a corresponding decrease in the temperature coefficient. According to Gumlich [18G2] the coefficient falls from its value for pure iron (about 0.0060/°C) to 0.0015/°C at 2% silicon and to 0.008/°C at 4% silicon.

The *tensile and yield strengths* of iron are increased by the addition of silicon up to about 4.5%, and thereafter they decrease sharply. The *ductility* at room temperature is little affected up to 2.5% silicon, after which it lessens rapidly and becomes nearly zero at 5%. A rise in tem-

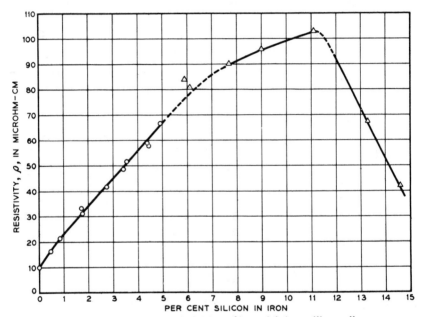

Fig. 4-6. Electrical resistivities of iron-rich iron-silicon alloys.

perature causes an increase in ductility, so that the alloy containing 5% can be bent without breaking at about 200°C, according to Pilling [23P1].

Measurements on the mechanical properties of relatively pure alloys containing less than 7% silicon have been reported by Yensen [15Y1] and are reproduced in Fig. 4-7. Results are for materials annealed at 970°C. The breaks found at 2.6% silicon in these curves have not been confirmed by others [28C5] and are believed to be due to a small amount of some unidentified impurity in Yensen's alloys.

The corrosive-resistant alloys containing 15% silicon are quite brittle and cannot be machined but must be cast, or cast and ground, to final size.

Some mechanical properties of commercial electric sheet of various grades are listed in Table 3 on p. 78.

Magnetic Saturation and Curie Point.—Magnetization at saturation at room temperature is well established as the result of many determinations,

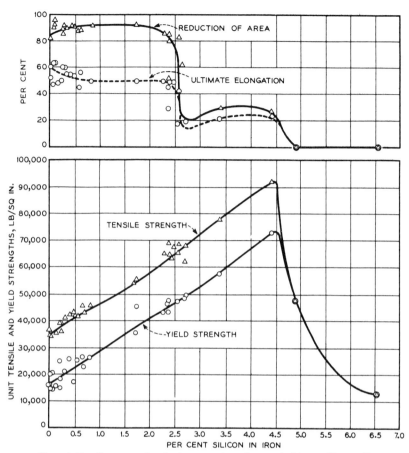

FIG. 4-7. Some mechanical properties of annealed iron-silicon alloys.

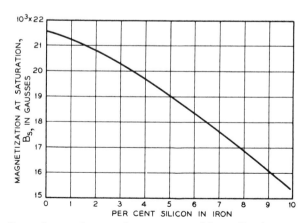

FIG. 4-8. Dependence of saturation on composition. Specimens contained 0.1–0.3% C, about 0.1% Mn.

IRON–SILICON ALLOYS

TABLE 3. SOME TYPICAL MAGNETIC PROPERTIES OF COMMERCIAL GRADES OF ELECTRICAL SHEET STEEL, ANNEALED

Silicon Content (%)	Commercial Grade	Maximum Permeability	Coercive Force* (oersteds)	Hysteresis Loss, $B_m = 10\,000$ (ergs/cm³/cycle)	Total Loss Watts/lb for 0.014″ 60 cps		Tensile Strength (lb/in.²)	Elongation (%)	Rockwell Hardness (B2 Scale)
					$B_m = 10\,000$	$B_m = 15\,000$			
0.5	Armature	5 500	0.9	2300	1.30	4.30	44 000	25	38
1.25	Electrical	6 100	.85	2200	1.17	3.60	50 000	22	50
2.75	Motor	5 800	.75	1900	1.01	2.65	68 000	14	73
3.25	Dynamo	5 800	.65	1600	.82	2.15	70 000	12	75
3.75	Transformer (72)	7 000	.5	1200	.72	1.00	80 000	8	84
4.25	Transformer (52)	9 000	.3	700	.52	1.40	67 000	4	86
3	Grain Oriented (1)†	20 000	.2	400	.45	1.00	59 000		
3	Grain Oriented (2)†	40 000	.15	300	.3	.70			

* For $B_m = 10\,000$.
† Properties measured parallel to grain.

among which only those of Gumlich [18G2] and Fallot [36F1] will be mentioned. Results are given in Fig. 4–8. Saturation has been determined by Fallot at 0°K by extrapolation from 110°K and $H = 16\,000$ oersteds, and the curve of Fig. 4–9 shows the results expressed as Bohr magnetons per atom, n_0. The broken line indicates the effect expected if the silicon behaves simply as a diluent, without effect on the moment of the iron atoms; the initial tangency of this line shows that silicon at

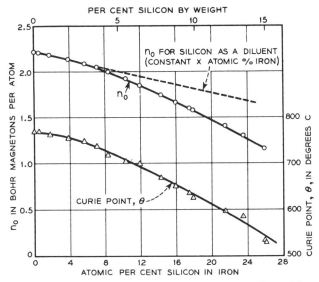

Fig. 4–9. Bohr magneton numbers and Curie points of iron-silicon alloys.

first acts in this manner, and its continuing departure means that with increasing silicon content the average atomic moment of the iron atoms is reduced.

Fallot's determinations of the Curie points are indicated in the same figure.

Permeability and Losses of Hot-Rolled Sheet.—Some magnetic properties of typical, commercial, hot-rolled sheet as annealed are shown graphically in Fig. 4–10. The curves show that addition of silicon increases the magnetic permeability and decreases the hysteresis and total losses (partly because the annealing temperature is higher for higher silicon contents). The magnetization curves of Fig. 4–11 [43B2] show that the lowering of the saturation induction with increasing silicon causes the curves to cross in the neighborhood of $B = 12\,000$.

Ruder [42R1] measured the total loss in 0.014-in. sheet of alloys containing up to 8% silicon (Fig. 4–12) and found a minimum near 6.5% silicon. Alloys of this composition have not found large commercial application, probably on account of their poor mechanical properties; but some high-silicons have been used for operation at low inductions and can be punched if heated above 100°C.

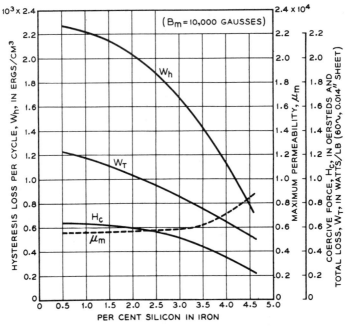

Fig. 4–10. Some magnetic properties of hot-rolled commercial silicon-iron sheet. Some of the change with composition is due to the higher annealing temperature of the material having higher silicon contents.

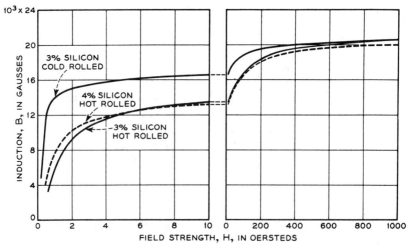

Fig. 4–11. Comparison of hot rolled and cold rolled (grain-oriented) material.

Permeabilities in low fields are shown in Fig. 4–13 for an alloy containing 3% silicon; they are noticeably higher than for unalloyed iron. Various magnetic and mechanical properties are summarized in Table 3 [47B4]. Measurements just cited are for specimens in the annealed condition. The annealing temperature depends on the silicon and carbon contents,

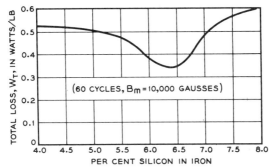

FIG. 4–12. Minimum loss occurs near 6.5% Si.

for reasons that are apparent from a study of Fig. 4–2 and may vary from less than 800°C to more than 1100°C. When the carbon content is sufficiently low (about 0.03%), no phase transformation occurs in the 3% silicon alloy, and this may be annealed at as high a temperature as is practicable—in practice it may be 1100–1200°C. After the material has been annealed once and then has been subjected to strain as in cutting, it is common practice to reanneal at 750–800°C.

A substantial amount of *aging* may occur in low-silicon sheet, and perceptible aging may occur in any commercial silicon alloy if impurities occur in certain proportions. "Aging" means here a change of permeability or core loss with time, and almost invariably the change is a deterioration. The standard test for aging is to maintain at 100°C for 600 hours and note the resulting increase in core loss measured at room temperature. The behavior of alloys with a given silicon content may be quite erratic because of the variable amounts of impurities they contain, but losses in alloys containing 0.5–1.0% silicon may increase 50 and 20%, respectively.

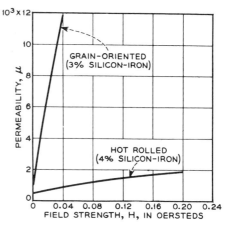

FIG. 4–13. Permeability in low fields, showing non-linear relation.

Alloys containing 1.5% silicon are generally considered to be non-aging, and are so if they are low in carbon and are slowly cooled, but an increase of loss of 5 or 10% with time is not uncommon in the usual product.

In the iron-silicon alloys the variation of hysteresis loss with induction is unlike that of most magnetic materials in that the loss does not plot linearly when log-log scales are used. The hysteresis losses at $B = $ 10,000

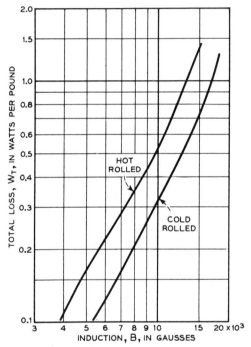

Fig. 4–14. Total losses as dependent on maximum induction of 0.014-in. sheet, 4% Si-Fe, at 60 cps.

and above are greater than would be expected from extrapolation of the loss at lower inductions, using the Steinmetz 1.6 power law (see Fig. 11–26). The upward bend of the loss curves is shown in Fig. 4–14 for both hot rolled and cold rolled material; although these are typical, a considerable amount of variation from lot to lot is to be expected.

In considering materials for use in transformer cores it is desirable to know the relation between the alternating current in, and the alternating voltage across, a given winding on the core. The product of voltage and current, measured without reference to phase, is approximately proportional to the weight of material in the core. Consequently the quantity:

$$\frac{\text{Volts} \times \text{Amperes}}{\text{Weight in pounds}}$$

is plotted against the maximum value of induction occurring during the cycle. Such a *volt-ampere* or *excitation* curve, showing this quantity as dependent on induction, is characteristic of material having certain magnetic properties and having a given thickness of lamination. The curve of Fig. 4–15 [49C1] is for 4.5% silicon-iron, 0.014 in. thick, having a core loss of 0.52 watt/lb at $B = 10\,000$, $f = 60$ cps. As might be expected, the curve bends upwards for inductions beyond the knee of the magnetization curve.

The incremental permeability μ_Δ for a small alternating flux (60 cps, $B_\Delta = 50$), superposed on a constant induction B_d, is given in Fig. 4–16 for commercial, hot rolled, 0.014-in. sheet (grade "58").

Effect of Non-metallic Impurities.—Improvements in the quality of commercial electrical sheet have paralleled closely the control of impurities. Hopkinson's [85H1] alloys were inferior to the transformer iron then in use because they contained over 0.6% carbon, and the improved properties of the alloy patented by Hadfield [03H3] were in material containing about 0.1% carbon. In 1918 Gumlich [18G2] reported analyses of 0.2–0.3% carbon, and also some as low as 0.06.

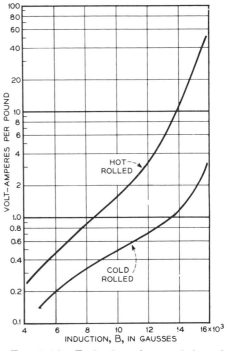

FIG. 4–15. Excitation characteristics of hot rolled (4%, Si) and cold rolled (3% Si, grain-oriented) sheets, 0.014 in. thick, at 60 cps.

At present, typical, annealed, commercial sheet which is hot rolled contains about 0.03%; cold rolled, grain-oriented sheet, less than 0.01%. Some analyses are given in Table 2 of this chapter.

Following the earlier work of Gumlich, an extensive investigation of the effect of various impurities was begun by Yensen about 1910 and continued over a period of years [14Y1, 24Y1, 36Y1]. Later German work has been reported by Eilender and Oertel [34E3]. In the laboratory, Yensen reduced impurities by vacuum melting of relatively pure raw materials and by heat treating in appropriate atmospheres, sometimes oxidizing and sometimes reducing. It appeared from his work that the effect of any impurity became pronounced when its solid solubility was exceeded; as we know now, precipitation hardening then occurs and the

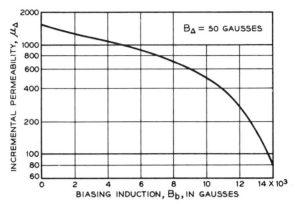

Fig. 4-16. Dependence of incremental permeability ($B_\Delta = 50$) on induction for 4% Si-Fe (58 grade).

permeability decreases. For convenience, Yensen's results are summarized in Fig. 4–17, and the effects of the important impurity elements will now be considered.

In studying the effect of *carbon* Yensen found [24Y1] that by heating specimens in a current of air the permeability was increased at the same time that the carbon was diminished by oxidation to CO (Fig. 4–18). When the carbon was so reduced to less than 0.06%, however, the im-

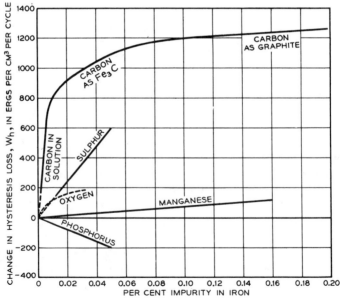

Fig. 4-17. Effect of various impurities on the hysteresis losses of 4% Si-Fe at $B = 10\,000$.

purities absorbed from the air neutralized any beneficial action of the further removal of carbon. In his experiments on vacuum melting Yensen found an occasional alloy having very high permeability, and he ascribed this to the chance occurrence in the melting charge of equivalent

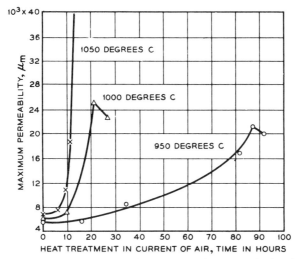

FIG. 4–18. Increase in permeability caused by removal of carbon by air.

amounts of carbon and oxygen so that both were eliminated as CO during melting and annealing. This view was substantiated by Eilender and Oertel [27E3] who measured the carbon and oxygen contents of many commercial ingots and found, as shown in Fig. 4–19, that minimum loss was

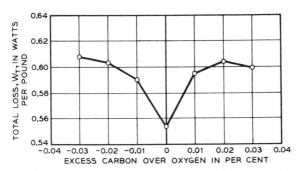

FIG. 4–19. Low losses caused by balancing of impurities of carbon and oxygen.

obtained in the finished sheet when the difference between the oxygen and carbon was zero (to the nearest 0.01%). The effect of either in excess is to increase the hysteresis loss.

As already mentioned, the condition of the carbon affects its action, and it is much less harmful when present as graphite rather than as cementite (or pearlite). According to Fig. 4–17, carbon in solid solution increases the hysteresis loss at the highest rate.

A number of methods have been used, on laboratory and commercial scales, to remove carbon from rolled sheet. Yensen [36Y2] reduced it to less than 0.005% by first oxidizing the sheet a controlled amount and then heating to about 1000°C in hydrogen. Mill scale is sometimes left on hot rolled sheet during process-annealing for the same purpose [39C2, 41C1]. Carpenter and Jackson [42C2] decarburized by heating for a few minutes at 800°C in moist hydrogen.

The effects of excess *oxygen* and of *sulfur* are also given in Fig. 4–17. Oxygen is believed to be responsible for the mechanical aging of iron that follows mechanical deformation [35D7]. *Nitrogen* is believed to be responsible for magnetic aging (see Fig. 2–28). Morrill [48M1] has shown that it has an inhibiting effect on grain growth when present to the extent of 0.0015%. The effect of *phosphorus* is peculiar in that its presence reduces the losses; this occurs only when other impurities are present in the usual proportions and may be due to the strong affinity of this element for oxygen.

Following and extending the earlier work of Ruder [14R2] and Yensen [31Y2], Cioffi [38C1] heat-treated at a high temperature—over 1300°C—in hydrogen and so removed carbon, oxygen, sulfur, and nitrogen. Maximum permeabilities of over 200 000 were obtained. Of these elements sulfur [34C1] requires the highest temperature for removal, probably because sulfur diffuses slowly and has a mobility in iron of practical magnitude only above 1250°C. In attempting to eliminate sulfur in another way J. H. Scaff of the Bell Laboratories treated the melt with a desulfurizing slag, such as sodium carbonate, and then annealed the rolled metal in hydrogen at 1000°C. Cioffi melted in an atmosphere of hydrogen. Using these methods Scaff and Cioffi were able to prepare laboratory specimens having maximum permeabilities as high as 30 000–40 000.

Starting with material already carefully refined in the open hearth, Morrill [48M1] has been able to heat treat cold reduced material at a temperature as low as 1150°C for 6 hours and obtain maximum permeabilities as high as 100 000. The development of properly oriented crystals in this material is particularly sensitive to impurities, and carbon should be reduced to 0.006% or less. Morrill also reports that fine-grained aluminum oxide, formed during deoxidation with aluminum, may inhibit proper grain growth of favorably oriented crystals. The effect of nitrogen, which is usually not present in amounts sufficient to cause trouble, has already been mentioned.

Grain Size.—The effect of the grain size of silicon-iron on the magnetic properties, especially on the hysteresis and eddy-current losses, has been, and still is, the subject of controversy. There is little doubt that variations in treatment that cause an increase in grain size often cause also a decrease in losses. It is supposed, however, that the grain size may be affected by the amount of impurities, which themselves have a direct affect on the magnetic properties. The experiments of von Auwers [28A1] showed also that materials having the same grain size may have widely different properties depending on the amounts of impurities absorbed during annealing. In addition to differences in purity, there are often variations in crystal orientation, thickness of specimen in relation to crystal size, and perhaps other factors resulting from differences in method of preparation. The complexity of the situation is increased by the fact that cold reduced sheet, in which the losses are low, frequently has a relatively small grain size. However, no appreciable effect of grain size *per se* has been established in this material. Further discussion of the subject may be found in articles by von Auwers [28A1], Ruder [34R3], Yensen [24Y1], and Daeves [29D3].

Metallic Additions.—Since the manufacture of electric sheet began, *manganese* has been an important if not essential addition to the melt. Its function has been to increase the ductility and hence aid in fabrication; it is relatively inert in its effect on magnetic properties. With the percentages usually used, 0.1–0.5, the change in the Fe-Si equilibria is shown principally by a slight enlargement of the γ region. However, manganese is known to form stable carbides, and its effect on the Fe-Si-C system is believed to be important. Beyond this, little is known of the structure of the system containing all four elements.

Wever and Hindrichs [31W2] have made an extensive investigation of the effect of additions of *aluminum* to iron-silicon alloys. Melts containing 0–4.3% silicon and 0–3.4% aluminum were made in a high-frequency furnace, cast in 100-lb ingots, reduced to sheet by hot rolling, and annealed at 880°C. Loss measurements were then made at $B = 10\,000, f = 50$, in 0.014-in. sheet, and permeabilities were recorded for $H_m = 25$ ($B = 14\,300$–15 750) and for higher fields. An 8-ton melt of one alloy (1.34% Si, 0.59% Al) was made in a basic arc furnace, and this showed that the laboratory tests could be duplicated on a commercial scale. Average losses of this lot were 1.67 watts/kg at $B = 10\,000$ for $f = 50$ and 0.014-in. sheet. These were about equal to the losses in iron-silicon sheet containing slightly less total alloying metal. The greatest benefits were noted in low-alloy sheet such as that just mentioned. However, too broad conclusions cannot be drawn from the limited data available. Wever and Hindrichs expressed the opinion that the beneficial action of aluminum is due to the superior deoxidizing power of this element.

Some commercial sheet now contains 0.6% aluminum as well as 5.0% silicon. This is more ductile than material containing 5.6% silicon alone.

Kussmann, Scharnow, and Messkin [30K5] studied the effect of *copper* on the hysteresis loss of forged bars containing 1.5 and 4% silicon. They found little effect for copper contents less that 0.7%; 1% caused a definite increase in loss.

The effect on the losses of small additions of other metallic elements has been reported by Messkin and Margolin [39M4]. Amounts were up to about 0.1% of Al, Be, Ti, and V. The lowest coercive force, observed for 0.05% vanadium, showed that no very important change in properties was effected. The changes observed may be attributable to the presence of other impurities that combine with the added elements. An influence on recrystallization was noted.

The effects of impurities on the properties at low inductions, and on the characteristics of cold rolled sheet at high and low inductions, are discussed in the appropriate sections that follow.

Preferred Crystal Orientation.—The highest quality transformer sheet now commercially available is the so-called "grain-oriented" sheet, in which the majority of the crystals are favorably oriented. Trade names of such material are Trancor 3X, Silectron, Hipersil, and others. As studies of single crystals have shown, iron-silicon is most easily magnetized in the direction of any one of the crystal axes [100], and hysteresis losses are lowest when the field is so directed. In grain-oriented sheet, as a result of the cold rolling and annealing to which it has been subjected, the crystals of which the sheet is composed are oriented so that one of the axes of each crystal is approximately parallel to the direction in which the sheet has been cold rolled, and a plane perpendicular to a face diagonal [a (011) plane] is parallel to the plane of the sheet [35B9]. Sheet so made has a maximum permeability two or more times that of the same sheet hot rolled and annealed, provided the measurements are made in the direction of rolling. Hysteresis losses are lower than in hot rolled sheets, especially at high inductions, e.g., at $B = 15\,000$. Permeability vs induction curves for hot rolled and cold rolled sheet are given in Fig. 4–20, and other data are included in Fig. 1–12 and Fig. 4–14, and Table 3 of this chapter. The outstanding feature of the new material is the increased induction that can be attained without increase in magnetic field (H_m) or core loss.

Interest in the commercial production of cold reduced silicon sheet was aroused by the work of N. P. Goss, reported in the years 1934–1937 34G4, 35G5, 37G2]. His first reported procedure was to anneal the hot rolled sheet at 900°C, cold roll to reduce thickness by 60% anneal at 900°C, cold reduce again by 60% to final size, anneal at 1100°C, and then at 700°C to reduce aging. Following this kind of procedure, material was produced having a maximum permeability of over 20 000, and a permeability of 2200 at $B = 16\,000$. Watt losses were correspondingly low. Variations were made in annealing temperature and amount of cold reduction. Previous to Goss's work, a patent relating to the improvement of magnetic properties by cold rolling was issued to Smith, Garnett, and Randall [33S8]. This refers to iron-nickel alloys, and silicon-iron is not mentioned specifically, although the method is claimed for magnetic alloys generally.

Since Goss's first report many methods have been described for producing favorable crystal orientations by cold rolling and annealing. Goss [37G2] himself produced good material by rolling at about 800°C, annealing, cold rolling, and annealing finally at about 1100°C. Frey and Bitter [38F2] used a number of cold rolling operations separated by anneals. Hiemenz

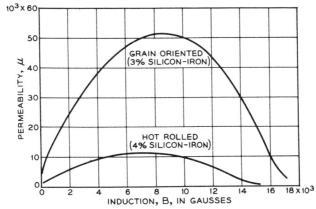

Fig. 4–20. Permeability curves of hot rolled and grain-oriented material measured parallel to the direction of rolling.

[38H2] rolled at successively lower temperatures and finished by rolling cold, and then annealing at 1250°C. Cole and Davidson [39C2] combined two cold rolling operations, each 60–70%, separated by anneals and followed by a decarburizing treatment with oxide at about 1200°C. Carpenter [41C1] decarburized before cold rolling by annealing the hot rolled sheet with its scale unremoved, and finished by annealing at 800°C in a controlled atmosphere to prevent contamination or to effect further purification. In a more recent patent Carpenter [42C3] specified a single cold reduction of 70–80%, and an anneal with time and temperature depending on silicon content.

The best materials are produced today by combining preferred orientation with purification at various stages from melt to final anneal. The manufacture of material of this type already has been described.

Some directional properties of grain-oriented sheet are illustrated in Figs. 4–20 and 21. The permeability is a maximum when the measuring field is applied in the direction in which the material was cold rolled, and a minimum when it is inclined at 55° to this direction. The ratio of per-

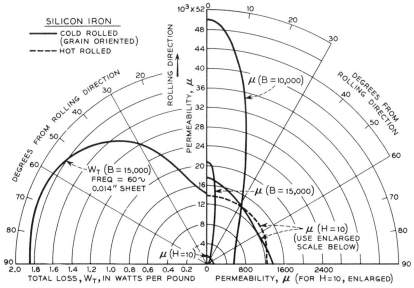

Fig. 4–21. Dependence of permeability and losses of hot rolled and grain-oriented sheet on direction of measurement in sheet.

meabilities for $H = 10$, measured parallel and at right angles to the rolling direction, is only 1.2 to 1, but the ratio of permeabilities at $B = 15\,000\text{–}16\,000$ is commonly 15 or 20 to 1. The ratio of losses at $B = 15\,000$ is about 2 to 1.

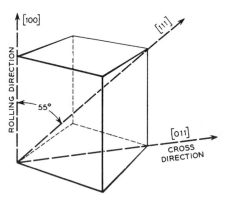

Fig. 4–22. Orientation of crystal axes with respect to rolling direction in grain-oriented sheet.

The directional properties can be understood easily in terms of the properties of single crystals of silicon-iron. The orientation of the crystals of the sheet is shown schematically in Fig. 4–22 [35B9], and the magnetic properties of a single crystal measured in the three indicated principal directions, [100], [011], and [111], are indicated in Fig. 4–23, taken from the work of Williams [37W5]. The permeabilities in these three directions are seen to be in the same order in the single crystal and in the commercial sheet: [100] highest, [111] lowest, and [110] intermediate. The data on single crystals are especially instructive because they provide a goal of permeability and core loss to which makers of commercial sheet may aspire.

Williams observed a maximum permeability of 1 400 000 in the [100] direction, and for commercial sheet the permeability is rarely over 50 000; and the hysteresis losses at $B = 15\,000$ are respectively about 50 and

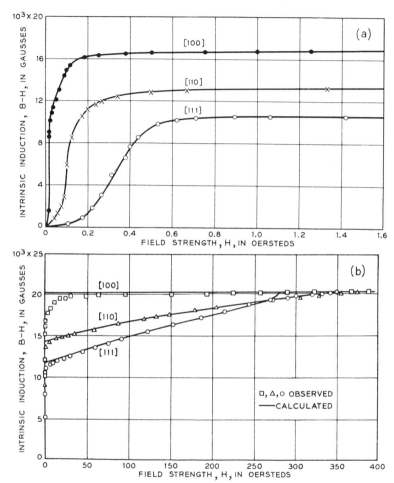

FIG. 4–23. Magnetization curves measured in various directions in single crystals of 3.8% Si-Fe.

600 ergs/cm^3 per cycle. Consequently by more perfect crystal alignment and reduction of impurities the electrical sheet industry can look forward to a gain of the order of 10 to 1 in these properties. Of course, eddy-current losses cannot be decreased in this manner, but they can be decreased by reducing the thickness of the sheet.

Young's modulus of cold rolled material varies with direction in the sheet, and the values in the three directions making angles of 0°, 90°, and 55° with the rolling direction correspond closely to the moduli for single crystals in the [100], [110], and [111] directions. Variation with angle is by more than a factor of 2, as shown in Table 4 due to Benford [46B3].

TABLE 4. VARIATION OF SOME MECHANICAL PROPERTIES WITH DIRECTION IN GRAIN-ORIENTED SHEET (3.2% Si)

	Angle with Direction of Rolling		
	0°[100]	90°[110]	55 [111]
Elastic limit (lb/in.^{2})	41 000	42 000	54 000
Yield point (lb/in.^{2})	53 000	54 000	62 000
Ultimate strength (lb/in.^{2})	59 000	64 000	68 000
Elongation in 2 in. (%)	7	33	2
Young's modulus (lb/in.^{2})	18.0×10^6	30.0×10^6	38.6×10^6
Modulus for single crystal*	19.0×10^6	30.0×10^6	41.0×10^6

* Values obtained by Astbury [48A1] are somewhat lower; 17.0, 28.6 and 37.0×10^6, respectively.

The way in which the rolling and annealing bring about the special crystal orientation is not understood at all well. It is known that after the cold rolling the crystals are oriented to some extent in a preferred manner, but this texture is quite different from that existing after the material has recrystallized during the annealing. Studies of the mechanism of orientation have been made by Dunn [45D1] and Harker [47H1]. The rate of crystal nucleation and growth, occurring during the recrystallization of a 1% silicon alloy, has been investigated by Stanley [45S1].

Properties at Low Inductions.—For some purposes, especially in communication equipment, a high permeability at low inductions is of prime importance. Although iron-nickel alloys are commonly used when this quality is desired, there has always been the possibility of using iron-silicon alloys, which are more economical because the cost of raw materials is less. Development in this direction was stimulated during World War II by the scarcity of nickel. The feasibility of increasing the initial permeability by treatment at a relatively high temperature in hydrogen was pointed out by Cioffi [38C1], who reported a μ_0 of 4000 in a 4% silicon alloy and 10 000 to 20 000 in unalloyed iron. Ashworth and Chegwidden, in unpublished work carried out at the Bell Laboratories, were able to produce a material having a permeability (at $B = 20$) of 1500–3000 by annealing commercial hot rolled sheet in hydrogen for 6 hours at 1250°C.

A more systematic study of the effect of removal or addition of impurities of carbon, nitrogen, and oxygen was made by Pawlek [43P1]. He found that 1300°C was much better than 1200°C as an annealing temperature for 3% silicon sheet 0.014 in. thick, held for 4 hours in a stream of pure electrolytic hydrogen. Only slightly higher permeability was attained by using higher temperatures, as noted in Fig. 4–24. A good vacuum was just as effective as pure hydrogen, but unpurified hydrogen and hydrogen mixed with nitrogen (decomposed NH_3) gave very poor results. After the higher permeability was attained, small amounts of water, nitrogen, or naphthalene were added to the hydrogen atmosphere during treatment at

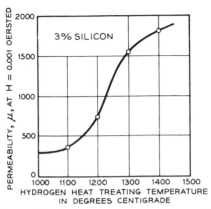

FIG. 4–24. Dependence of permeability on temperature of anneal in electrolytic hydrogen. Annealing time, 4 hours.

1350°C, and the change in permeability was noted. His data, plotted in Fig. 4–25, show that 1% by volume of water vapor is harmless, but that even $10^{-5}\%$ of naphthalene is sufficient to lower the permeability by 20%. Nitrogen is intermediate in its action. Unfortunately, results on sulfur

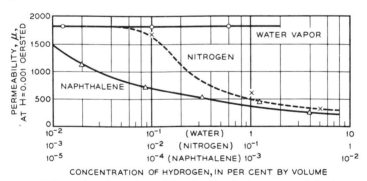

FIG. 4–25. Effect of various impurities in the hydrogen used in heat-treating 3% Si-Fe at 1350°C.

were not reported. The effect of different amounts of naphthalene on the shape of the μ,H curve, plotted in Fig. 4–26, indicates that the non-linear course often found in silicon-iron alloys is enhanced by the presence of impurities and suggests that the curve for the pure material is linear all the way from $H = 0.1$ to $H = 0$. This is substantiated by the data of Williams

[37W1] for single crystals of high purity. His results, given in Fig. 4–27, show also that the initial permeability depends on crystal orientation, and that values higher than about 4000 are not to be expected in polycrystalline material containing about 4% silicon.

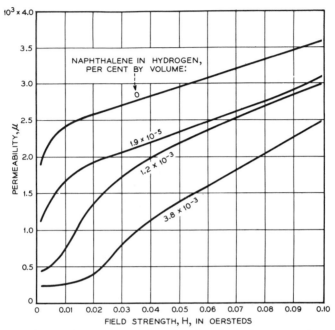

Fig. 4–26. Effect of small impurities of carbon in hydrogen on the permeability of 3% Si-Fe at low fields.

Silicon-Cast Iron.—The frames of generators and motors are occasionally made of machined cast iron containing a certain amount of silicon. The material is inexpensive, and it is required only that the strength be reasonably high, that the parts be readily machined, and that the permeability exceed some rather low value. The section is usually large so that its flux-carrying capacity is sufficient even with low permeability. Carbon contents vary from 2–4%; silicon, from 0.5–3.5%. Magnetic properties are better if the carbon is in the form of graphite rather than in combination as Fe_3C. The addition of silicon aids graphitization and improves the homogeneity of the casting. Properties are improved by tempering at 800–900°C.

A gray cast iron of typical composition (1.7% Si, 3.3% C, 0.5% Mn) has the following magnetic characteristics after tempering: $\mu_0 = 200$, $\mu_m = 700$ (at $B = 2000$), $H_c = 2.7$, $B_r = 9000$ for $H_m = 100$. Materials of other compositions have been listed by Messkin and Kussman [32M3], and an extensive study of the effects of variation in composition has been reported by Partridge [28P2].

Structural Steels.—In some grades of steel, silicon is added in excess of the amount required for deoxidation; structural steels may contain from 0.2% to more than 2% of silicon, average about 1%, and have in addition

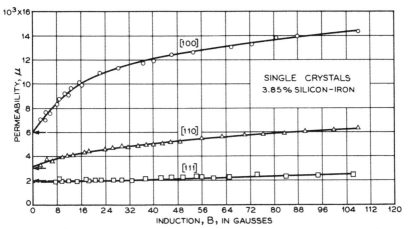

FIG. 4-27. Magnetization curves in low fields, measured in various directions in single crystals of 3.8% Si-Fe.

0.5% or more of manganese and a few tenths of a per cent of carbon. As an alloying agent, silicon is useful primarily in increasing the tensile strength.

It is impossible to give here any comprehensive survey of the magnetic properties of steels of this kind. Gerold [31G1] has estimated the effects of various alloying elements on the induction in high fields and has shown that they are roughly proportional to the amount of any element present up to about 2%. The effect of various elements is given in Table 5. In low fields the permeability generally runs parallel to the mechanical properties, and in a rough way one can say that the initial permeability of soft iron—150-250—is reduced to about 75 in a hard, tough material such as that used for ship armor. Maximum permeability is affected in a similar manner.

Iron-Silicon-Aluminum Alloys.—*Sendust.*—Some remarkable properties of alloys in this system have been uncovered by Masumoto [36M8], who investigated specimens containing up to 14% aluminum and 14% silicon. The electrical measurements show that low resistivity exists near the compositions Fe_3Si and Fe_3Al, and in ternary alloys lying between them.

TABLE 5. EFFECT OF 1% OF CERTAIN ELEMENTS ON THE INDUCTION
OF STRUCTURAL STEEL IN HIGH FIELDS

(Values are to be subtracted from the "normal value" at bottom of table.)

Element	Decrease in Induction for Following Values of H			
	31	63	126	377
C	5 000	3 800	3 000	2 800
Si	550	440	310	250
Al	750	700	690	650
Cu	350	310	150	100
Mo	1 500	950	780	600
Mn	800	400	400	300
Cr	900	580	300	200
Normal value	16 500	17 500	19 000	20 800

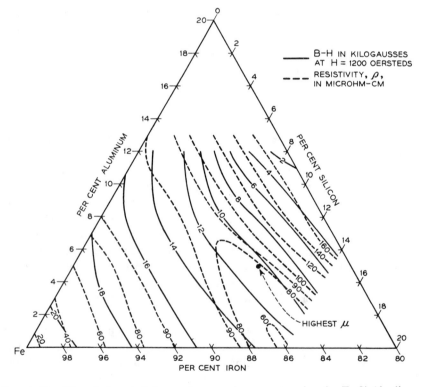

FIG. 4-28. Electrical resistivity and approximate saturation in Fe-Si-Al alloys.

This "trough" of low resistivity must be connected with the ordered atomic arrangement well known to exist in the binary systems and shown by Selissky [41S6] to occur in the ternary alloys. He has also measured [46S2] the lattice spacings of the ternary alloys and found them to change in a way consistent with the existence of superstructure; that is, the spacings undergo relatively large changes when the composition approaches Fe_3Si or Fe_3Al or any combination of them. He concludes that the compositions exhibiting superstructure extend to and just include those for which unusually high permeabilities are observed.

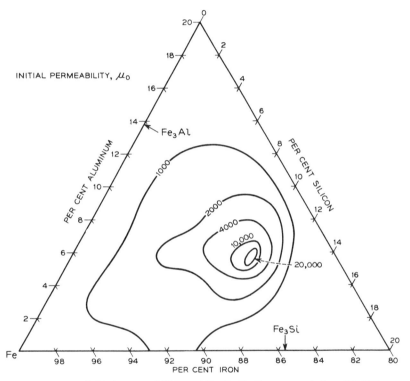

FIG. 4-29. Initial permeabilities of Fe-Si-Al alloys annealed at 1000°C.

Masumoto's alloys were vacuum-melted from electrolytic iron and pure aluminum, using silicon containing about 0.5% each of magnesium and antimony. Specimens were cast in ring form, annealed for one hour at 1000°C, and slowly cooled. Saturation magnetization (actually, $B-H$ for $H = 1200$) drops rather smoothly toward zero at about 75–80% iron, as shown by the contour lines of Fig. 4–28. Broken contour lines on the same figure relate to the resistivity already discussed.

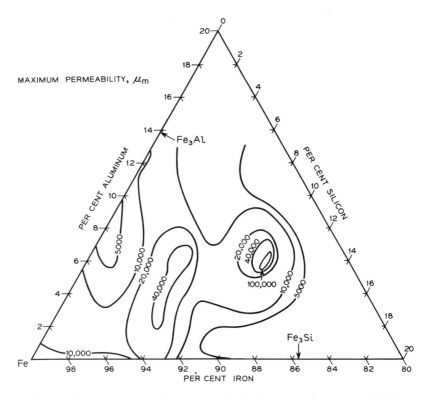

FIG. 4–30. Maximum permeabilities Fe-Si-Al alloys annealed at 1000°C.

IRON–SILICON–ALUMINUM ALLOYS 99

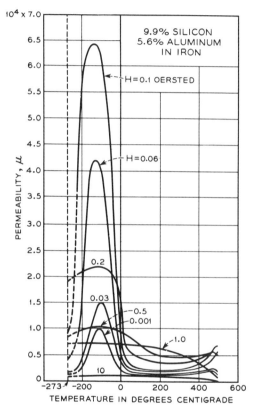

Fig. 4–31. Effect of temperature on the permeabilities of Sendust containing 9.9% Si and 5.6% Al.

The diagrams of Figs. 4–29 and 30 show initial and maximum permeabilities, which are found to be extraordinarily high near the composition 9.5–10% silicon, 5–5.5% aluminum. In one specimen Masumoto found $\mu_0 = 35\,000$, $\mu_m = 118\,000$, $H_c = 0.02$, and $W_h = 28$ (for $B_m = 5000$); in another, $\mu_m = 160\,000$.

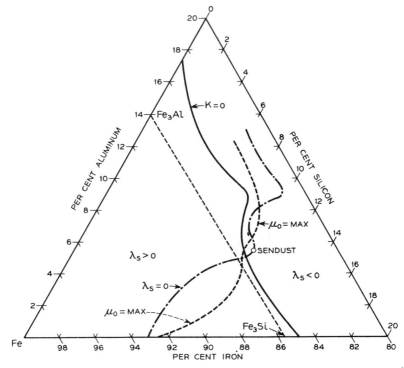

Fig. 4–32. Lines of highest initial permeability μ_0, and lowest magnetostriction λ, and crystal anisotropy K, in Fe-Si-Al alloys.

The material of high permeability is quite brittle, is easily reduced to powder, and is appropriately named *Sendust*, as well as *Feralsi*. The pressed dust has found some use in loading coil cores, where it has a permeability of about 60. The magnetic properties are affected but slightly by any of a number of impurities, which, however, increase the electrical resistivity substantially. Sendust is unusual in that the permeability in low fields *decreases* with increasing temperature up to about 200°C, as pointed out by Zaimovsky [41Z1] and shown graphically in Fig. 4–31. The highest values of permeability were found, in fact, at about −100°C. It is surmised that this is connected in some way with superstructure, but no explanation has been given.

Magnetic anisotropy and magnetostriction have been measured by Zaimovsky and Selissky [41Z2] and by Snoek and Went [47S2]; the anisotropy constant K was determined not on single crystals but by measuring the rate of approach to saturation of polycrystalline specimens. Sendust lies near the composition for which both K and λ_s are zero, as indicated in Fig. 4–32 [41Z2]. The theory of the high permeability of Permalloy thus applies also to alloys of the Sendust type.

CHAPTER 5

IRON-NICKEL ALLOYS

BINARY IRON-NICKEL ALLOYS

Before the iron-nickel alloys were used as magnetic materials, they were already known to possess certain other peculiar and valuable properties. The low thermal expansion of Invar (36% nickel) has been known for more than 40 years. Other compositions have been used to match the coefficient of expansion of glass (46% nickel) or ordinary steel (56% alloy). Still other alloys have a zero or even positive temperature co-

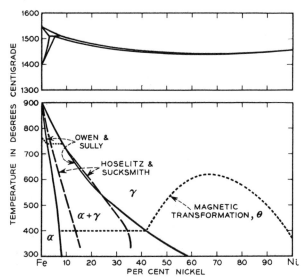

FIG. 5-1. Phase diagram of the iron-nickel system. See also Fig. 5-109.

efficient of the elastic constants. The addition of nickel to iron increases greatly the strength and corrosion resistance of iron, especially if carbon is present, and the largest use of nickel is for this purpose—in structural steel.

Structure.—Results of recent investigations of the structure of iron-nickel alloys have been incorporated in the phase diagram of Fig. 5-1. The alloys important for ferromagnetism lie in the composition range from 40–90% nickel and form a continuous series of solid solutions having a face-centered structure. When the temperature is maintained at 450°C for about a week, a superstructure is developed in the alloys near 75% nickel; this has been detected by X-rays [39L1, 39H2] and is supported less directly by measurements of resistivity [36D1, 38K2, 39B19], of magnetic saturation [39M1, 40G1, 39F4, 39P1], and of specific heat [38K2]. Källbäck [47K1] places the order-disorder transformation point of FeNi$_3$ at 506°C. Josso [51J1] has placed the Curie point of FeNi$_3$ at 611°C when ordered, 600°C when disordered; at 70% nickel the Curie point for both ordered and disordered states is 615°C.

Recent studies [39O1] of the phases containing 10–30% nickel, by Owen, Sucksmith, Hoselitz and their collaborators [37O3, 39O1, 41O1, 40P1, 43H1, 44H1, 49O1], are of importance in the study of meteorites as well as of magnetic materials. The phase boundaries have been detected only when cooled more slowly than about 10°C/day; with higher cooling rates, the diagram of Fig. 5-2 is a more practical guide to the structures present. The line indicating the magnetic transformations of the α (body-centered cubic) alloys has been drawn in Fig. 5-1 according to Peschard [25P1]. The boundaries of the δ (body-centered) region have been taken from the paper by Bristow [39B2]. The minima of the liquidus and solidus curves and the maximum of the magnetic transformation curve lie near 68% nickel, therefore just slightly below the composition corresponding to FeNi$_3$.

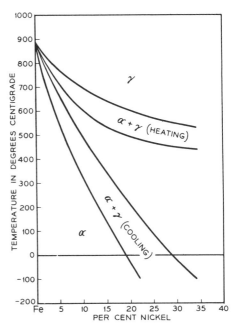

Fig. 5-2. Practical diagram of the irreversible alloys.

More extended descriptions of the structure of this binary system and further references to the original literature are given by Hansen [36H1], Marsh [38M2], and Sachs [48S4].

Density, Lattice Parameter and Thermal Expansion.—The lattice constants of both the α- and γ-phase alloys are given for various temperatures in Fig. 5-3 [36J1, 37O3, 37B2]. Densities calculated from these X-ray data are shown in Fig. 5-4 and agree well with the earlier direct determinations. Values have been determined for 0°K by extrapolation. Thermal expansion coefficients are given in Fig. 5-5, according to Chevenard [28C3], for all compositions and for various temperatures between −100 and 900°C.

Resistivity.—The resistivities at various temperatures are shown by the curves of Fig. 5–6, determined by Shirakawa [39S2]. There are large

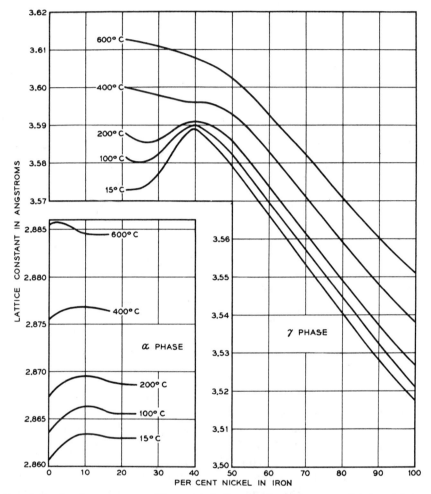

Fig. 5–3. Lattice constants of iron-nickel alloys in the body-centered (α) and face-centered (γ) condition.

variations between resistivities measured in various experiments, and these are probably caused by the different impurity contents and different degrees of atomic ordering. Differences of 20% from the drawn curve are not uncommon for the γ-phase alloys. The steep rise near 30% nickel accompanies the transition from the α to the γ phase; the break may be displaced to either lower or higher nickel contents depending on the treatment the specimen has received. The alloys used in obtaining these data were

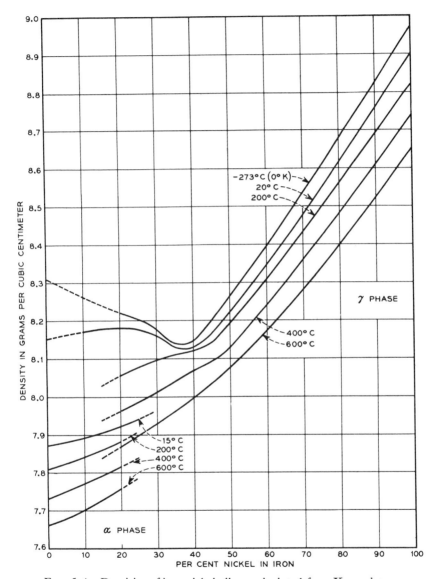

Fig. 5–4. Densities of iron-nickel alloys calculated from X-ray data.

relatively pure and were made by melting in hydrogen the constituent metals which contained as principal impurities 0.05% each of carbon, silicon, and phosphorus, and 0.02% manganese. Alloys used for magnetic purposes usually contain several tenths of a per cent of manganese and thus have slightly higher resistivities.

The effect of atomic ordering on the resistivity is noticeable in the alloys containing about 50–90% nickel. Data for different rates of cooling of the alloys are shown in Fig. 5–7, according to Kaya [38K2].

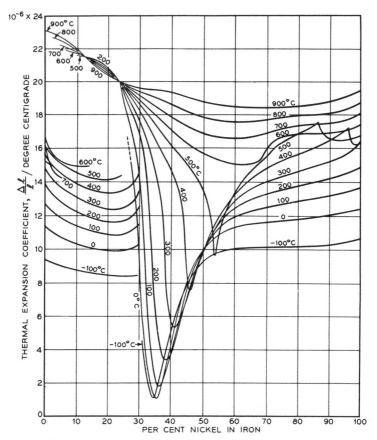

FIG. 5–5. Thermal expansion coefficients at various temperatures. See also Fig. 10–15.

Elastic and Mechanical Properties.—Large uncertainties characterize the mechanical properties; the values of Table 1 are taken from various sources [38M2, 39A2, 48M2], some unpublished. All data refer to annealed alloys. Poisson's ratio is usually 0.28 to 0.30, the bulk modulus about 17×10^6 lb/in^2.

History of Magnetic Investigations.—Hopkinson was the first to investigate systematically the magnetic properties of iron-nickel alloys. In two papers [89H1, 90H1] in 1889 and 1890, he reported on alloys containing 1, 5, 22, 33, and 73% nickel. His principal observations were that the 25% alloy was non-magnetic when received, became magnetic after cooling to −50°C, and that the induction of the 73% alloy was greater than that of the 33% alloy and also greater than that calculated for a mechanical mixture of iron and nickel in the proportions of the

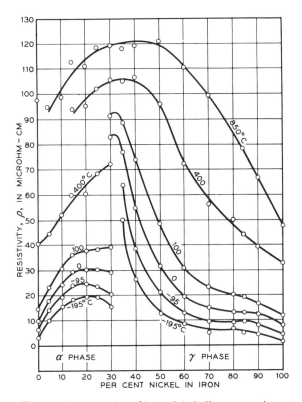

FIG. 5-6. Electrical resistivities of iron-nickel alloys at various temperatures.

analysis. A second investigation, reported by Barrett, Brown, and Hadfield [02B1] in 1902, was confined to alloys with 31% nickel or less. The 31% alloy had a coercive force of 0.5 oersted, which was less than that of iron. Their work did not include a determination of permeability in low fields as we know them today, for the lowest point on any of the normal magnetization curves was for a field strength of about 0.5 oersted.

Many of the important magnetic characteristics of the iron-nickel alloys were established in 1910 by three independent investigations. In the one by Burgess and Aston [10B1], the alloys contained 0, 25, 26, 28, 35, 47, 75, and 100% nickel and were tested at room temperature in fields of 10–400

TABLE 1. ELASTIC AND MECHANICAL PROPERTIES OF ANNEALED
IRON-NICKEL ALLOY OF HIGH COMMERCIAL PURITY

E is Young's modulus; G, the shear modulus; TS, the tensile strength; YS, the yield stress (all in lb/in.^{2}); RA, reduction in area; El, elongation at rupture; BHN, Brinell hardness number. 1000 lb/in.^{2} is 0.70 kg/mm^2.

Ni (%)	$E \times 10^{-6}$	$G \times 10^{-6}$	$TS \times 10^{-3}$	$YS \times 10^{-3}$	RA (%)	El (%)	BHN
0	29–31	11.5–12.0	50–70	30–40	60–70	35–45	100
20	23–25	9.0–9.5	150–200	100–140	15–40	5–10	300
45	21–23	8.5–9.0	65–80	20–30	70	40	80–100
78	26–30	10.5–11.5	75–90	15–30	80	50	125
100	28–31	10.5–11.5	50–80	20–30	70	50	90–100

oersteds. The inductions for all field strengths used decreased as nickel was added to iron up to 20%. At 30% the induction began to increase, then it passed through a maximum somewhere near the 47% alloy tested, and declined again toward pure nickel.

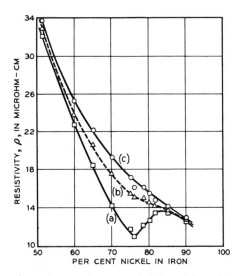

FIG. 5-7. Effect of atomic ordering on resistivity; (a) cooled slowly, (b) cooled at 200°C per hour, (c) quenched from 600°C.

In the second of these investigations, by Hegg [10H1], the nickel contents were in steps of 10%, the field strengths were high—up to 10 000—and the temperatures of the specimens during tests varied from −190 to 760°C. He determined the saturation for various temperatures and also extrapolated to absolute zero. At room temperature, the maximum in the saturation magnetization vs composition curve at 47% nickel was well defined for the first time (see Fig. 5–8). Hegg also established with some precision the Curie point curve for the whole series of iron-nickel alloys.

In neither of these investigations was any attention paid to inductions in fields less than 10 oersteds. Although Hopkinson had made measurements to $H = 0.06$ in alloys having less than 5% nickel, it was left to Panebianco [10P1] to test the γ-phase alloys of 36% and of 49% nickel in fields as low as 0.016. Initial permeabilities of 700 and 550, respectively, were obtained; these values are considerably greater than those previously

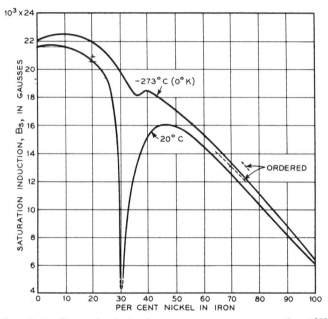

FIG. 5–8. Saturation induction at room temperature and at 0°K.

attained in iron. However, the article was published in a journal unfamiliar to later investigators and was obviously unknown to them for many years.

The first comprehensive study of the iron-nickel alloys in weak and medium fields was begun in 1913 by Elmen, who was trying to find a material superior to silicon steel for use in telephone apparatus operating at inductions of less than a few hundred gausses. As a result of his work, in which Arnold and others collaborated, it became generally recognized for the first time that the alloys containing 30–90% nickel had initial permeabilities greater than those of iron or any other material then known. Elmen also discovered a special heat treatment ("Permalloy treatment") that increased still further the initial permeabilities of the alloys from

50–90% nickel. The value for the 78% nickel alloy (78 Permalloy) was increased the greatest amount—from about 2000 to 7000. A similar increase was found in the maximum permeabilities, which attained values of over 50 000. Practical use of 78 Permalloy was made in telephone relays as early as 1921. Elmen's first Permalloy patent was filed in 1916 and issued the following year [17E1] (a previous application was withdrawn). The first patent covering the Permalloy heat treatment was filed in 1921 and granted in 1926 [26E1]. The first scientific report on the properties of the Permalloys was made by Arnold and Elmen in 1923 [23A2].

In addition to 78 Permalloy, another alloy, 45 Permalloy, found early application in telephone apparatus as a result of the experimental work of Elmen and his collaborators. This material was useful because it had a higher saturation than any of the other Permalloys and thus could be operated at higher inductions. It also possessed a high resistivity and high maximum permeability. Although high saturation and high resistivity had been reported in the literature as early as 1910, it was not until 1924 that application was made as a result of the new research.

Another investigation was begun independently by Yensen in 1915. This resulted in the new alloy "Hipernik" on which a patent was filed in 1924 and issued in 1931 [31Y1]. By heat-treating the 50% alloy in hydrogen he attained about the same maximum permeability as Elmen found in 78 Permalloy. For the transmission of power as distinguished from telephone currents, Hipernik has the advantages already mentioned in connection with 45 Permalloy. Although improvements in magnetic quality by vacuum melting were reported by Yensen in 1920 [20Y1], his important contribution as judged by present-day practice is the use of hydrogen annealing [25Y1], which raises the maximum permeability of the 45 or 50% alloy by a factor of 2 to 5, depending on the temperature and on the care taken to avoid contamination of the material during the fabrication and heat treatment. Hydrogen annealing also raises the initial permeability by a considerable factor [35E1, 39Y1]. In 1927 Cioffi annealed Permalloy in hydrogen at higher temperatures and increased the permeability considerably.

In 1921 Elmen first added a third element, chromium, to the iron-nickel alloys to increase the resistivity of materials having high initial permeability [22E1, 26E1]. In the same year, he found [28E1], that the addition of a considerable amount of cobalt gave the alloy "Perminvar" properties: a permeability practically independent of field strength up to $H = 1$ or 2 ($B \approx 1000$), and a "wasp-waisted" hysteresis loop characterized by low remanence and coercive force. Smith and Garnett [24S1] invented "Mumetal" when they added several per cent of copper. Now molybdenum, and also chromium and copper, are extensively used [30E2].

In addition to increasing the resistivity, they increase the permeability or simplify the heat treatment required for the best results. The more recent improvements in the properties of the magnetically soft iron-nickel alloys have been brought about by the control of impurities and by the use of certain heat treatments or mechanical treatments, as described on later pages. The latter involve heat treatment in a magnetic field, or controlled rolling and recrystallization to effect special crystal orientation.

Magnetic Saturation.—Saturation values of the ferric induction B_s, obtained by extrapolating $B-H$ to its value in infinite field, are given by the curves of Fig. 5–8 for room temperature and for absolute zero. The curve for the latter involves a double extrapolation, one of temperature as well as field strength; the techniques of both of these extrapolations have been developed by the Weiss school [34W2]. The broken lines near 30% nickel indicate that the curves may be displaced by varying the proportions of the α and γ phases by changing the heat treatment. There are also considerable uncertainties in the positions of the steep portions of the curve for 20°C. The dotted lines near 75% nickel show that there are increases in saturation accompanying the formation of an ordered structure brought about by very slow cooling below 500°C [39M1].

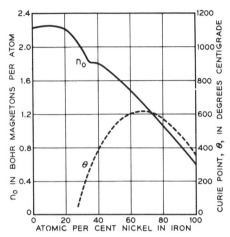

FIG. 5–9. Bohr magneton number and Curie point of iron-nickel alloys.

Saturation in Bohr magnetons per atom and Curie points are plotted in Fig. 5–9 against atomic per cent nickel.

The variation of saturation magnetization with temperature is shown for several alloys in Fig. 5–10. The data are those of Hegg [10H1], converted by Marsh [38M2] to B_s from magnetic moment per gram. More recent data for iron and nickel have been added for comparison. Results of Stäblein [34S3] for the 30% alloy are given in Fig. 5–37 for a limited temperature range. The curves of Fig. 5–10 follow the same general trend from absolute saturation at 0°K to almost zero magnetization at the Curie point, except that the curve for 40% nickel and, to a lesser degree, that for the 50% alloy show a more rapid decrease of B_s with increasing temperature when the temperature is low.

Saturation magnetization is not sensitive to chemical purity or heat treatment; consequently, the presence of slight impurities or the method of preparation of the material is not ordinarily of importance. Heat treatment is of importance only when it changes the distribution of phases, as it does near 30% nickel, or in lesser measure when it affects an ordering of the atoms, as near 75% nickel.

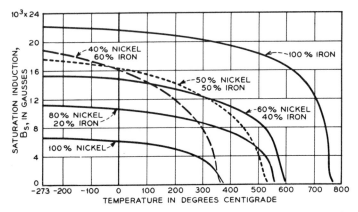

FIG. 5–10. Dependence of saturation induction of various iron-nickel alloys on the temperature.

Preparation and Annealing of Alloys.—The properties described in the next sections were determined over a period of 25 years at the Bell Telephone Laboratories by Elmen and his associates [29E1]. The alloys were prepared by melting Armco iron and high-purity, commercial electrolytic nickel in a high-frequency induction furnace in air with a covering of borax. About 0.3% of manganese (for desulfurization) and 0.1% aluminum or magnesium (for deoxidation) were added to make fabrication easier. The 7-lb ingots were swaged from 0.75 to 0.125 in., drawn to 0.040 in., rolled to 0.006-in. tape and trimmed to 0.125 in. About 30–40 turns of tape were wound circumferentially about a disk 3 in. in diameter, and the disk removed. This flat coil was heat-treated and wound with suitable windings for measurement with a ballistic galvanometer or an inductance bridge, often with the help of a Kelsall permeameter.

As shown in several of the following figures, many of the properties of the alloys containing 45–90% nickel are sensitive to the heat treatment to which the alloys have been subjected. Recognizing this, Elmen selected three heat treatments for his investigations and designated them:

1. Pot anneal, furnace cool.
2. Double treatment.
3. Bake.

The first consisted of heating to 900–950°C, maintaining at that temperature for about an hour, and cooling at the maximum rate of 100°C per hour. Before this treatment, the specimen was dusted with finely powdered silica and placed in an iron pot with the cover sealed with a mixture of powdered, low-melting glass and powdered iron. The second treatment consists of heating for 1 hour at 900–950°C and cooling as mentioned above, then heating to 600°C and cooling in the open air by placing on a copper plate at room temperature (or the first cooling to room temperature could be omitted). This procedure is also known as "Permalloy treatment," or "air quench"; the maximum rate of cooling is about 1500°C/min. For the third treatment, the specimen was first heated in a closed pot to 900–950°C and then maintained for many hours at a lower temperature, usually 20 hours at 450°C. In later experiments the maximum temperature was raised to 1050–1100°C.

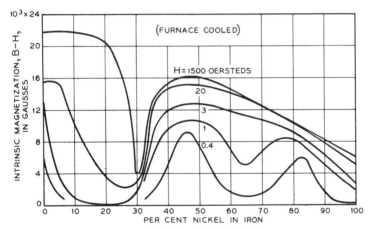

FIG. 5–11. Intrinsic induction of iron-nickel alloys at various field strengths after heat treating at 1000°C and cooling with the furnace.

Dependence of Induction on Field Strength.—In Fig. 5–11, the ferric induction, $B-H$, is plotted against the percentage of nickel for several values of the field strength. The inductions attained at the lower field strengths depend markedly on the chemical impurities present and on the fabrication and heat treatment to which the material has been subjected; the data of the figure refer only to samples furnace-cooled. The effects of other heat treatments are given special consideration later. For field strengths of 2 oersteds or less, two distinct maxima are present, one near 45% nickel and one close to 80%. Each of these compositions is near that of several commercial alloys. Similar data have been reported by Yensen [20Y1] and by Masumoto [29M1]. It should be emphasized that rather large variations in the properties occur from sample to sample and

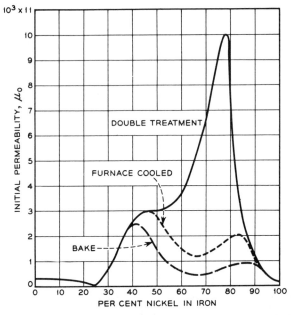

Fig. 5-12. Effect of heat treatment on initial permeability, which increases with cooling rate in the range 50–90% nickel.

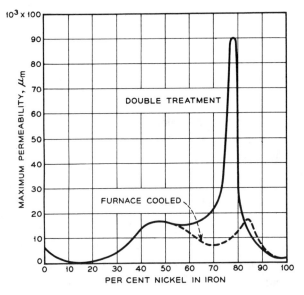

Fig. 5-13. Maximum permeability is enhanced by rapid cooling.

that a considerable amount of judgment must be exercised in drawing the curves through the large number of points.

Permeability, Coercive Force, and Hysteresis Loss.—Initial and maximum permeabilities are given in Figs. 5–12 and 13. The middle and lower curves of Fig. 5–12 show near 45% and 80% nickel the same maxima as those evident in the previous figure. In 1915, Elmen [23A2] showed that with the double treatment the maximum at the high-nickel end could be

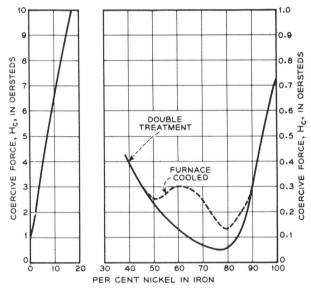

FIG. 5–14. Coercive force of alloys of ordinary purity.

raised to the then remarkable height of 10 000. The same treatment raised the maximum permeability (Fig. 5–13) and lowers the coercive force (Fig. 5–14) and hysteresis loss of the 78% nickel alloys and most of the alloys containing 40–90%. The data used in these figures were obtained some years ago and are not representative of present-day materials, but they do show the characteristic changes caused by variation in composition and heat treatment.

Annealing Temperature and Time.—After a material has been work-hardened by fabrication, the permeability is, of course, greatly reduced. The subsequent increase of μ_0 upon annealing for 1 hour at various temperatures is illustrated in Fig. 5–15 [35P3, 35B1]. The maximum permeability shows a similar increase with annealing temperature. It is to be noticed that all of the curves have inflection points near 500°C and that they are still rising at the highest temperature used.

Figure 5–16 shows [29E1] the effect of cooling rate on the initial and

IRON-NICKEL ALLOYS

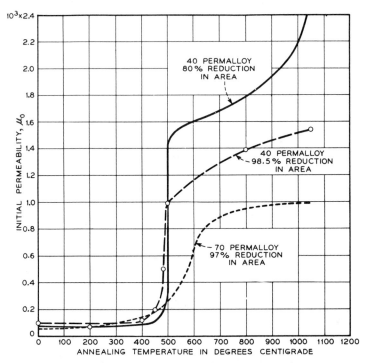

Fig. 5-15. Recovery of initial permeability after annealing of cold-worked material. Measurements at room temperature.

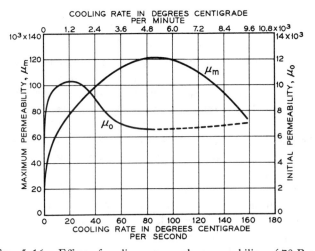

Fig. 5-16. Effect of cooling rate on the permeability of 78 Permalloy.

maximum permeabilities of 78 Permalloy, the composition most affected by it. The maxima of the curves fall at 20 and 80°C/sec, respectively. After a specimen has been cooled rapidly, the permeability and other related properties may be changed by heating and cooling slowly, i.e., the effect of rapid cooling is wiped out by a later heat treatment, as is to be expected. The manner in which this takes place is indicated in Fig. 5-17,

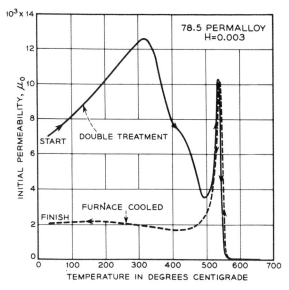

Fig. 5-17. Initial permeability measured at rising and falling temperatures.

in which the initial permeability is plotted against the temperature at which it is measured [29E1]. Coercive force is also affected by heat treatment; for the 78% alloy it is about 0.15 oersted after furnace cooling and 0.05 after double treatment.

The effect on the initial permeability of the nature of the *atmosphere* (air, hydrogen, vacuum, nitrogen) and the refractory in contact with the sheet has been investigated by Hagemann and Hiemenz [33H1] for 47 Permalloy containing 1% manganese.

Heat Treatment in a Magnetic Field.—Kelsall [34K1] first showed that the presence of a magnetic field of a few oersteds during the heat treatment of the iron-nickel alloys will cause a large increase in maximum permeability. For example, the mere presence of a field of 10 oersteds during the complete cycle of heating and cooling of 65 Permalloy [34B3] is sufficient to increase the maximum permeability from 20 000 to 200 000. The essential factor is the presence of the field during the cooling from the Curie point to about 400°C. To attain a large effect, the material must be put through a regular annealing cycle and the field applied at least during the cooling in the manner described.

Dillinger and Bozorth [35D4] determined the maximum permeabilities of a series of alloys heat-treated in a magnetic field. Single turns of 0.006-in. tape were annealed in hydrogen at 1000°C, cooled slowly, reheated to above the Curie point, and again cooled slowly in the presence of a field of 16 oersteds which was applied in the same direction in the specimen as the field was applied for measurement later at room temperature. Results are given in Fig. 5–18, curve BD. Measurements by Dahl

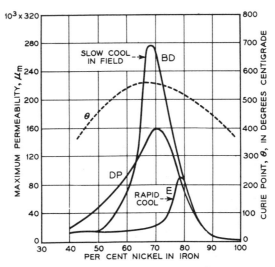

Fig. 5–18. Maximum permeabilities of iron-nickel alloys cooled in different ways after annealing at 1000°C. BD, Bozorth and Dillinger; DP, Dahl and Pawlek; E, Elmen. Curie points are shown for the same alloys, which contain some manganese.

and Pawlek [35D1] of alloys of similar purity, 0.012 in. thick, annealed in hydrogen at 1100°C, are also given as curve DP. For comparison, the results of Elmen [29E1] for rapidly cooled alloys are shown in curve E. Comparison with the Curie point curve indicates that there is a substantial effect of the field only when the Curie point lies above 400–450°C.

Experiments with 65 Permalloy show that high permeability may be attained by heat treatment in a field without giving the usual preliminary anneal at 1000°C or higher. The data [35B6] of Fig. 5–19 were taken on specimens hard rolled and then annealed only once at the temperature indicated, with the field present during cooling. The maximum permeabilities attained by heating 65 Permalloy to only 600–700°C are as high as any measured for this composition after it has been heat-treated without the help of a field.

When the field applied during heat treatment and that applied for measurement are mutually perpendicular, the permeability in intermediate fields is considerably less than it is when no field is applied during treatment. The coercive force is similarly (oppositely) affected.

The lowest temperature at which an applied field is effective during annealing is evident from Kelsall's unpublished experiments illustrated in Fig. 5-20. When the field is present at 500°C, the maximum permeability of 78 Permalloy is increased to above 100 000, about the same value as that obtained by ordinary double treatment. The effect at 400°C is negligible. The initial permeability is not increased by heat treatment in a field in this or any of the Permalloys: in fact, the rather exceptional value of 3500 found for this specimen after slow cooling is somewhat lowered by the presence of the field when the temperature is above 400°C.

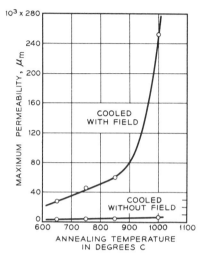

FIG. 5-19. Maximum permeabilities of samples of 65 Permalloy, originally hard worked, after annealing at various temperatures for one hour with or without an applied field.

The magnitude of the field necessary for the development of high permeability in 78 Permalloy is given in Fig. 5-21 [34K1]. Most of the effect is attained when the field is 0.5-1.0 oersted. The hysteresis loop (a) of Fig. 5-22 has the steep sides characteristic of materials whose properties have been changed by heat treating in a field. As nearly as can be determined, the sides of the loop are vertical and two of the corners perfectly sharp. It is difficult or impossible to demagnetize some of these specimens. Loop (b), heat-treated in the usual way with a field absent, has developed to some extent the wasp-waisted character of the Perminvar loops.

Two more experiments on heat treatment of the permalloys in a magnetic field will be mentioned in order to show something more about the nature of the phenomenon. The curve and inserted table of Fig. 5-23 show [35B2] how essential it is to maintain the field when cooling from the Curie point, 610, to 400°C, the temperature at which diffusion and relief of strain has practically ceased, and how irrelevant is the presence of the field at other temperatures. In an experiment [35B2] made to determine more definitely the effective range of temperature, the field was applied at or above the Curie point while the specimen was cooling, and was maintained

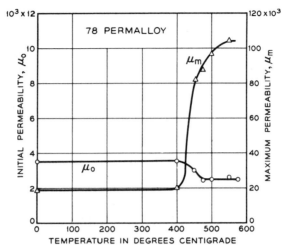

Fig. 5–20. Dependence of initial and maximum permeabilities of 78 Permalloy on temperature of anneal in a magnetic field.

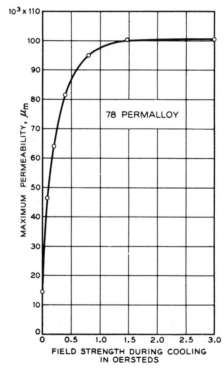

Fig. 5–21. Dependence of maximum permeability of 78 Permalloy on the magnitude of the magnetic field used during annealing.

until the temperature reached a certain point, t, after which the specimen was cooled to room temperature in zero field. As shown in Fig. 5-24, to

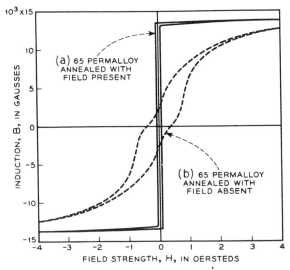

Fig. 5-22. Hysteresis loops of 65 Permalloy annealed with and without a magnetic field.

obtain the maximum effect it is important to maintain the field until the specimen has cooled to about 350°C.

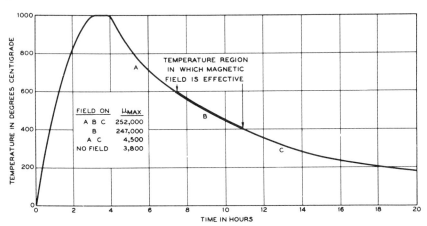

Fig. 5-23. Heating and cooling used in heat treating 65 Permalloy in a magnetic field. Maximum permeability is increased about 50 fold if field of 10 oersteds is applied during cooling.

Heat Treatment in Hydrogen.—In 1925, Yensen [25Y1] reported distinct improvements in the maximum permeability and hysteresis loss of the

IRON–NICKEL ALLOYS

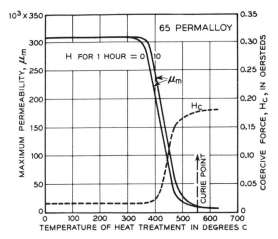

Fig. 5-24. Values of μ_{max} and H_c obtained in specimen cooled in magnetic field to indicated temperature, held for one hour, then cooled to room temperature in zero field.

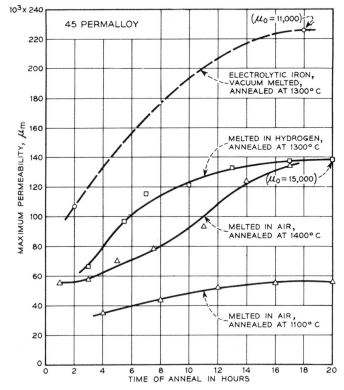

Fig. 5-25. Effect of various hydrogen treatments on the maximum permeability of 45 Permalloy.

50% nickel alloy when it was annealed in hydrogen at about 1000°C. Later [31Y2], he reported obtaining a μ_m as high as 80 000 and a hysteresis loss for $B_m = 10\,000$ (designated W_{10}) as low as 250 ergs/cm^3 by annealing for 8 hours at 1000°C in hydrogen. More recently [39Y1], he has reported for Hipernik a μ_m of 100 000, a μ_0 of 4500 and a W_{10} of 100, but no details of heat treatment are given.

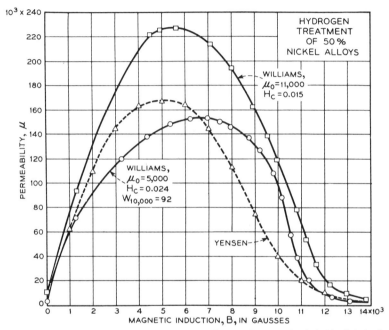

FIG. 5–26. Permeability vs induction curves for specimens of 45–50% nickel alloys, after annealing at 1300°C in hydrogen.

In 1932, Cioffi [32C2] reported large increases in the permeability of iron resulting from heat treatment in hydrogen at temperatures of 1300–1500°C. This process has since been applied to 45 Permalloy by H. J. Williams, some of whose results, not hitherto published, are given in Fig. 5–25. They show that hydrogen treatment is very effective when this material is maintained for many hours at 1300°C or higher. The μ,B curve for Williams' best result is reproduced in Fig. 5–26 where it is compared with the best specimen reported by Yensen [31Y1]. A third curve, by Williams, is for a specimen having an initial permeability of 5000.

By combining this high-temperature treatment in hydrogen with heat treatment in a magnetic field, still higher permeabilities have been attained. The highest curve B in Fig. 5–27 is for a specimen of 65 Permalloy maintained in pure hydrogen at 1400°C for 18 hours, cooled to room temperature, reheated for 1 hour at 650°C, and cooled again in the presence of a field of 15 oersteds. Before the treatment in a field, the μ,B curve was the lowest one in the figure (*A*). The properties of this alloy, unlike 45 Permalloy, are not markedly changed by the hydrogen treatment alone, but such treatment makes the material more susceptible to the subsequent treatment in a field. Curves are also given (*C* and *D*) for several other treatments of alloys having about the same composition. Notice that all of the curves have maxima pushed farther to the right than usual. Initial

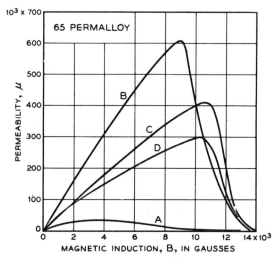

FIG. 5-27. Effect of heat treatment in a magnetic field on the permeability curves of 65 Permalloy: A, 0.125-in. sheet heat-treated at 1400°C and cooled in zero field; B, same specimen cooled in a magnetic field; C, different specimen heat-treated at 1000°C and cooled in field; D, 0.006-in. tape annealed at 1000°C and cooled in field.

permeabilities and coercive forces for these alloys, as well as the thicknesses of the specimens, are given in the figure caption. The hysteresis loops corresponding to curves *A* and *B* constitute Fig. 5-28. After treatment in a field the maximum permeability of this material is the highest, and the hysteresis loss for $B_m > 10\ 000$ is the lowest so far reported for any polycrystalline material except Supermalloy.

Mechanical Treatment.—So far little has been said about the mechanical treatment which the investigated alloys received before the heat treatment or about the effect of working the alloy after the heat treatment. Physically the working before annealing is significant because, to cite one example, it affects the orientations of the crystals—the "texture"—of the final product. Working after annealing lowers the permeability, but it also has the effect of lessening the dependence of permeability on field strength, and this property has been of industrial importance in the manufacture of loading coil cores; consequently it has been the subject of many investigations, especially in Germany. In this country, on the contrary, the method generally followed in the manufacture of loading coil cores has been to pulverize embrittled material, insulate the particles from each other, and form a compact ring of this powder under high pressure; annealing then raises the permeability and increases the mechanical strength of the core [28S1, 40L1]. In Germany, the loading coil cores have often been made of coiled strips of *Isoperm*, a name given to a heterogeneous group of

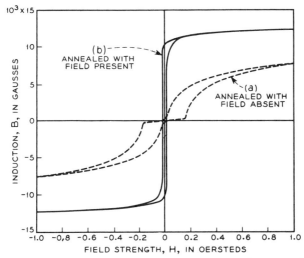

Fig. 5-28. Effect of magnetic anneal on hysteresis loops of 65 Permalloy previously annealed in hydrogen at 1400°C.

materials made of iron-nickel alloys with or without the addition of other elements (e.g., copper). A typical Isoperm [36D5] contains 40-50% nickel and the remainder iron, and one of the common treatments consists of the following steps:

(1) Reducing its thickness 98.5% by cold rolling.
(2) Annealing for an hour at 1000-1200°C.
(3) Reducing its thickness by 50%.

After the final rolling the magnetization curve and hysteresis loop, measured with the field parallel to the direction in which the sheet was rolled, are as shown in Fig. 5-29, curve B [35D6]. The magnetization curve is nearly a straight line and the material thus has fair constancy of permeability as the field strength is varied. The magnetic properties are sensitive to heat treatment and fabrication, and distinctly directional in character. Curve A in the figure shows that if part (1) of the treatment is changed from

98.5% to 80% reduction, the best Isoperm characteristics are lost. The final properties depend on the crystal orientations produced by the first rolling and the recrystallization, and on the strains produced by the final rolling. The change of magnetization by tension is of opposite sign in the Isoperm (curve B) and in the material that is given the less drastic rolling before annealing (curve A). The dependence of properties on the direction in the rolled sheet has already been shown in Figs. 2–30 and 31. In Isoperm μ_0 is a maximum and μ_m a minimum in the direction of

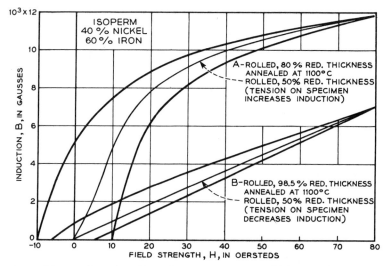

Fig. 5–29. Texture Isoperm after different treatments. High reduction is necessary to develop good Isoperm properties (curve B).

rolling, a desirable combination for the use for which the material is intended.

The magnetization curves and hysteresis loops of an Isoperm containing 50% nickel, shown in Fig. 5–30, illustrate once more the directional properties [36D5].

Dahl and his co-workers have carried out experiments to find the best treatments for obtaining high stability, low loss (especially hysteresis loss), and a permeability as high as possible. As a measure of magnetic instability they use

$$s = \frac{\mu_0 - \mu_r}{\mu_0}$$

where μ_0 is the initial permeability and μ_r the reversible permeability after the application and removal of a high field. The losses at low inductions can be expressed [36L1] with reasonable accuracy by the equation (see Chap. 17):

$$\frac{R}{\mu L f} = c + aB + ef$$

in which R is that part of the resistance of the coil (in ohms) that is due to the presence of the core, L is the inductance in henries, μ the permeability, B the maximum induction due to the alternating field of frequency f, c the residual loss coefficient, a the hysteresis loss coefficient, and e the eddy-current loss coefficient.

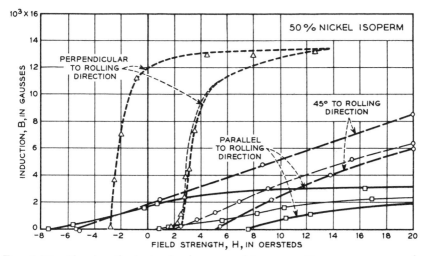

FIG. 5–30. Showing the isotropic character of Isoperm. Treatment same as for curve B of previous figure.

Since e is determined by the resistivity and the dimensions and subdivision of the core material, the important factors to study in their dependence on heat treatment and mechanical treatment are μ_0, a, and s. Minimum hysteresis loss coefficient is found to occur at 45–50% nickel and about 40% final reduction in thickness; maximum stability at 50–60% nickel. The highest μ_0 for the γ-phase alloys with 90% reduction is found at the low-nickel end. The alloy used is a compromise between the various factors.

During World War II there arose in Germany a need for an alloy with a square hysteresis loop to be used as a core in a transformer in the electrical circuit of a mechanical rectifier. Such material makes it possible to hold the current momentarily at zero and thus permit the separation of the contacts without sparking. The result was the development of *Permenorm 5000 Z*, which is basically a 50% nickel-iron alloy of high purity with carefully controlled mechanical and heat treatments. According to Both

[47B7] it is cold reduced 98% in thickness to a strip about 0.05 mm thick, then annealed at 1000–1200°C. There is an optimum temperature of anneal for each lot of material of given purity and cold reduction, and this temperature must be determined by experiment within about 25°C. The critical anneal is to obtain the best possible crystal texture, which is of the "cube" variety with (100) in the rolling plane and [001] in the direction of rolling. The magnetic properties are described later.

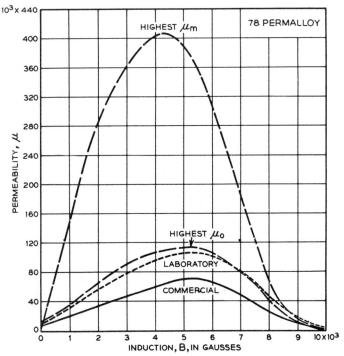

FIG. 5–31. Permeability vs induction curves of various specimens of 78 Permalloy.

COMMERCIAL BINARY ALLOYS

78 Permalloy.—This alloy was the first to find commercial application as the result of Elmen's investigation of the iron-nickel system. It is used in telephone transformers and coils and in sensitive relays on account of its high initial and maximum permeabilities and low coercive force. For most purposes it has been supplanted by the Permalloys to which a third element has been added to increase the resistivity as well as the permeability. It is prepared by melting in an arc furnace under an alkaline slag. About 0.5% of manganese is added for ductility (desulfurization) and 0.2% aluminum for deoxidation. Carbon is kept below 0.1%. The ingots are heated to 1200°C and rolled to $\frac{3}{4}$ in. without reheating. This stock is

45 PERMALLOY. HIPERNIK

rolled cold to 0.11 in. for relay stock and, after intermediate anneals in closed pots, to sheet as thin as 0.002 in. The temperature-time cycle during heat treatment has already been described (Fig. 2-16); the first treatment is carried out in closed pots, and the "air quench" is accomplished by removing the completely fabricated material from the furnace at 600°C and cooling it in still air.

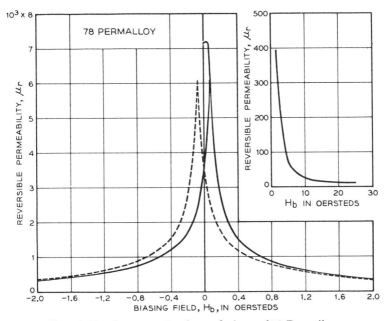

FIG. 5-32. Superposed or butterfly loop of 78 Permalloy.

The μ vs B curve for the representative commercial product is shown in Fig. 5-31. For comparison, a representative curve is given for test specimens prepared in the laboratory. The specimens having the highest μ_0 and the highest μ_m are described by the two remaining curves. The reversible permeabilities corresponding to the magnetization curve and the hysteresis loop are given in Fig. 5-32 for a representative laboratory specimen. Figure 5-33 shows hysteresis loops after "Permalloy treatment" and also one loop obtained after the simpler anneal, followed by "furnace cooling."

45 Permalloy. Hipernik.—The chief differences between these materials result from refinements of the melting, rolling, and annealing techniques. The manufacture of *45 Permalloy* is similar to that of 78 Permalloy, and all operations before the final anneal are carried out with no special protection from the atmosphere. If the final anneal is carried out at 1100°C in a sealed pot or in hydrogen, the magnetic properties are those given by curve *A* of Fig. 5-34, and the loops and curves of Fig. 5-35. Performing the final anneal at the same temperature in hydrogen of ordinary purity raises the permeability as indicated in curve *B* of Fig. 5-34. As already mentioned, a long anneal in hydrogen at higher temperatures raises the permeability still further.

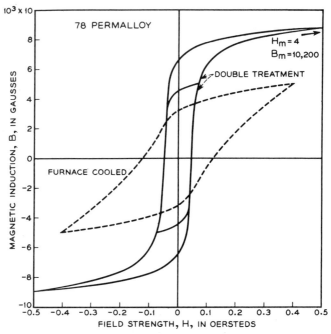

FIG. 5-33. Hysteresis loops of 78 Permalloy with different heat treatments.

In the manufacture of *Hipernik* [25Y1, 37M3] great care is taken to protect the alloy, containing 45–50% nickel, from the corrosive effects of the atmosphere. Melting is carried out in small furnaces curtained with hydrogen. Casting is in air. After fabrication, the material is annealed for 20 hours at 1200°C or higher in hydrogen carefully dried and otherwise purified. The μ vs B curves as reported by Yensen in 1931 and in 1939 [31Y1, 39Y1] are reproduced in Fig. 5-34, curves *B* and *C*. In 0.014-in. sheet, the hysteresis loss reported in 1931 was 136 ergs/cm^3/cycle for $B_m = 10\,000$; in 1939 it was 100 (Fig. 5-36). The total losses when subjected to 60-cycle fields were 0.7 watt/kg for the same B_m.

Little information is available concerning the manufacture of the alloys *Allegheny Electric Metal*, *Permenorm 4801*, *Carpenter N49*, *Nicaloi*, *Conpernik* and other alloys similar to 45 Permalloy. The composition of Allegheny Electric Metal (sometimes called Allegheny 4750) is 47–50%

nickel and the remainder iron except for impurities of less than 0.1% carbon and 0.02% sulfur. After annealing for 4 hours in hydrogen at 1100°C, the μ vs B curve is that labeled D in Fig. 5–34 [39A3]. The coercive force is about 0.1 (for $B_m = 10\,000$). The initial permeability is higher than that of 45 Permalloy and the induction for maximum permeability is lower, probably because it is annealed in hydrogen at a higher temperature. These latter properties suit the purpose for which the alloys

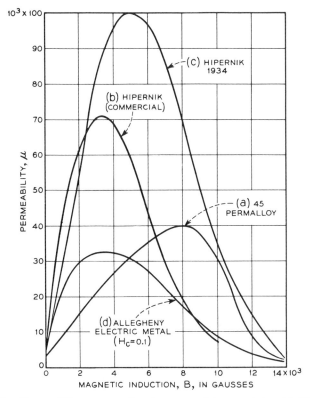

FIG. 5–34. Permeability curves of various materials containing 45–50% nickel.

are made: they are used in apparatus such as current transformers and audio frequency transformers, with or without superposed steady fields, so that permeabilities in low fields are of paramount importance. On the contrary, a considerable portion of the 45 Permalloy manufactured is used in sensitive telephone relays and receivers where the higher permeability at high inductions increases the tractive force.

Compressed Permalloy Powder.—As already mentioned the development of this material has been due to the exacting requirements of the loading coil core that have been given previously. The ideal powdered core is

132　　　　IRON–NICKEL ALLOYS

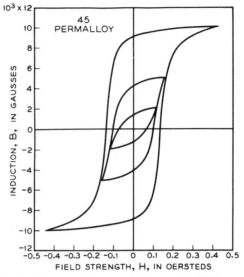

Fig. 5–35. Hysteresis loops of commercial 45 Permalloy.

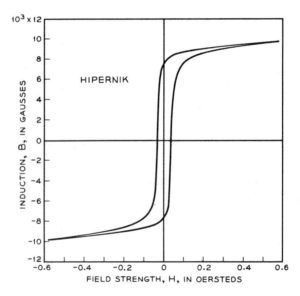

Fig. 5–36. Hysteresis loop of exceptionally good Hipernik ($\mu_0 = 4500$, $\mu_m = 100\,000$).

composed of small particles electrically insulated from one another to cut down the eddy currents induced by changes of flux. The particles are packed together so that the effective permeability will not be reduced too much by the air gaps, and the mass is heat-treated so that the permeability will be high and the hysteresis losses will have the low values characteristic of the solid, annealed material. The manufacture of cores of powdered 80 Permalloy has been described [28S1], but the account will not be repeated here because this material has been superseded in use by 2-81 Molybdenum Permalloy described later. A table of the characteristics of Isoperm, 80 Permalloy compressed powder, and 2-81 Molybdenum Permalloy compressed powder, is given on p. 146.

Isoperm.—Although there have been many reports describing the mechanical treatment necessary to give the constancy of permeability and other properties characteristic of Isoperm, no report of any specific procedure for the preparation of the commercial product has been found. Judging from the experiments of Dahl and Pfaffenberger [35D6, 36D5], the best procedure is to cold roll with 98.5–99% reduction, anneal at 1000°C, and cold roll with 50 or 60% reduction. The final sheet thickness is about 0.03 mm (0.011 in.); the constants are as given in Table 2 of this chapter (p. 146) and Table 2 of Appendix 4.

Permenorm 5000 Z.—The preparation of this alloy, as produced in Germany, has been described previously under "Mechanical Treatment." Similar treatment is given in the United States to *Deltamax* [49C3], *Orthonol*, *Orthonik*, and *Hipernik 5* [49G3] and in England to *H. C. R.* [49S10]. The squaring of the loop is enhanced by the presence of a magnetic field during cooling. The coercive force is about 0.06 oersted, and the residual induction and the corners of the hysteresis loop lie near $B = 12\,000$ to $15\,000$ gausses, $H = 0.06$ to 0.10 oersted. When magnetic annealing is used, the residual induction may rise to 17 500.

Thermoperm.—This alloy [34S3] contains about 30% nickel; consequently, its Curie point is just above room temperature and its magnetization decreases rapidly as the temperature is raised in the interval 0–50°C. This characteristic makes it suitable for compensating the change in flux occurring in other materials having the normal positive temperature coefficient of magnetization. Values of $B-H$ for various field strengths from 2 to 1000 at temperatures up to 100°C are shown in Fig. 5–37 [34S3]. Other alloys used for similar purposes are those of iron, nickel, and molybdenum, of iron, nickel, and chromium, and of nickel and copper.

There are a number of other alloys similar in composition and properties to the Permalloys, and a list of some of these and the names of the manufacturers are given in Appendix 4, Table 1. The magnetic properties of various iron-nickel alloys are summarized in Appendix 4, Table 2.

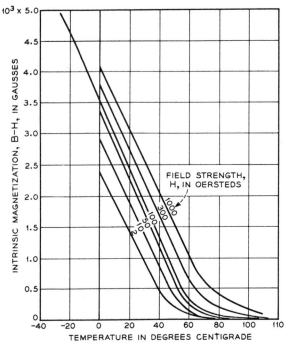

Fig. 5-37. Dependence of the magnetization of Thermoperm on the temperature of measurement.

IRON-NICKEL-MOLYBDENUM ALLOYS

Additions of molybdenum and chromium to the iron-nickel alloys were first made to increase the electrical resistivity and thus to decrease the eddy-current losses. It was soon evident that in addition to doing this they increased the initial permeability and simplified the heat treatment necessary to obtain high initial permeability. Other elements, such as Cu, Mn, Si, and Al, have also been used in alloys of commercial importance, and data are available on laboratory specimens containing Ti, V, W, Sn, Be, Ag, and others. Cobalt is used in alloys containing iron and nickel when constancy of permeability with varying field strength is desired. These added elements will be discussed in the order mentioned.

Structure.—The phase diagram of the Fe-Ni-Mo system has been partially worked out by Köster [34K5], who gave special attention to age-hardening and magnetic properties. Figure 5-38 shows the phase boundaries for the alloys cooled slowly to room temperature; in the region bounded by Fe, Ni, NiMo, and Fe_3Mo_2. In the quenched alloys the boundary line BC separating the magnetic from the non-magnetic alloys at room temperature is displaced markedly from that of the slowly cooled alloys

IRON-NICKEL-MOLYBDENUM ALLOYS

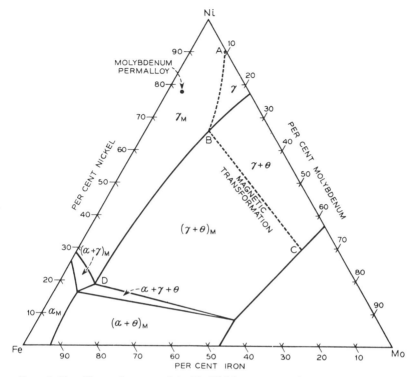

Fig. 5-38. Phase diagram of the Fe-Ni-Mo system at low temperatures.

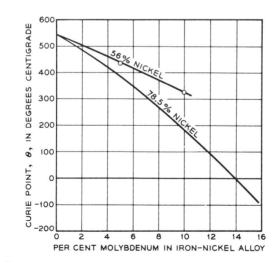

Fig. 5-39. Effect of molybdenum on the Curie points of Permalloys containing 56 or 78% Ni.

because of suppression of the θ phase, and it is a continuation of the line AB extending into the $\gamma + \theta$ region and terminating at D. Curie points for alloys containing 78% nickel and up to 15% molybdenum are given in Fig. 5–39, drawn from Elmen's data [31E2] which are consistent with the line AB of Köster's diagram.

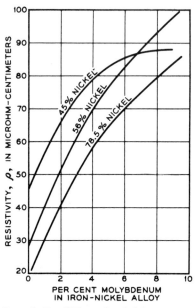

Fig. 5–40. Effect of molybdenum on the resistivity of some iron-nickel alloys.

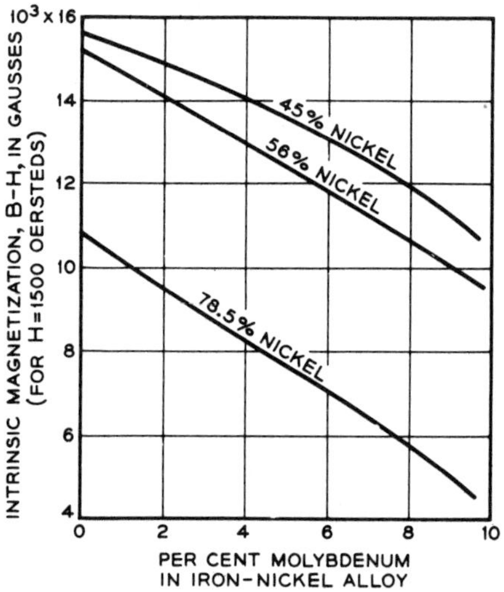

Fig. 5–41. Dependence of saturation on molybdenum content.

Resistivities of alloys containing 45 to 78% nickel, with molybdenum contents up to 10%, have been determined at the Bell Laboratories and are given in Fig. 5–40. The alloys contain 0.3–0.5% manganese.

Magnetic Properties.—These are shown graphically in Figs. 5–41 to 43, according to data obtained by Elmen, Kelsall, and others at the Bell Laboratories. The alloys were of the same quality as those used in the investigations of the other Permalloys. The molybdenum was added in the form of high-grade, commercial ferro-molybdenum, and the material was cast, swaged, drawn, and rolled in the usual way and tested in the form of 0.006-in. strip after heat treatment with furnace cooling or air quenching. The inductions in high fields, at room temperature, decrease uniformly with molybdenum content, as would be expected, and approach zero at about 15% molybdenum (Fig. 5–41). When the molybdenum content is below 2.5 or 3%, double treatment produces higher values of

μ_m and μ_0 and lower values of H_c and W_h than can be obtained by using the single furnace cool from 1000°C. Above this percentage the reverse is true.

A more extended study of the *heat treatment* was carried out in unpublished work by Chegwidden and Ashworth at the Bell Laboratories. This showed that the initial permeability of the 4% alloy is raised considerably by cooling at a rate intermediate between that used in double treatment and that used in furnace cooling. Initial permeabilities of 25 000 were obtained by withdrawing the pot from the furnace and placing it in an air stream

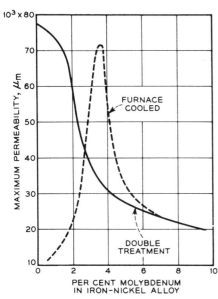
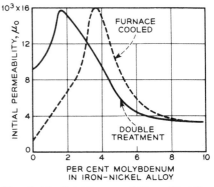

FIG. 5-42. Maximum permeability of Fe-Ni-Mo alloys with the two standard treatments. Nickel content about 78%.

FIG. 5-43. Initial permeability of Fe-Ni-Mo alloys with the two standard treatments. Nickel content about 78%.

so that the material cooled through 450°C at a rate between 20 and 70°C/min. The best cooling for a high μ_0 is about 50°C/min at 450°C, a rate intermediate between the rates for furnace cooling and for double treatment (1°C and 1000°C/min, respectively, at 450°C).

It was found that the optimum cooling rate (for high μ_0) varied with the nickel content when the molybdenum content was held at 3.8–4.0%, as shown in Fig. 5-44. Accordingly μ_0, μ_m, and H_c were determined for varying nickel contents using the optimum cooling rate for each alloy, and the results, for μ_0, given in Fig. 5-45 for two different annealing temperatures, show that μ_0 is as high as 30 000. The other properties are correspondingly good at the optimum nickel content of 79%.

A similar series of experiments was made on alloys in which the nickel was held constant at 79% and the molybdenum was varied up to 6%. It was found that the optimum cooling rate was the lower the greater

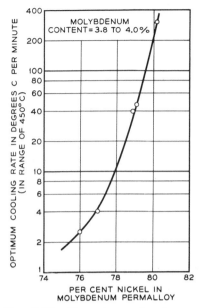

FIG. 5-44. Dependence of optimum cooling rate (for highest μ_0) on the nickel content of Permalloys containing 3.8–4.0% Mo.

the amount of molybdenum—about 30°C/min at 4% Mo, 1.5°C/min at 5% Mo.

Boothby, Bozorth, and Wenny [47B1, 47B6] have varied the cooling rate of pure alloys, prepared without the addition of any deoxidizer, and annealed in pure dry hydrogen for several hours at a temperature of 1300°C. The permeability of this material at low inductions ($B = 20$) was found to be dependent on cooling rate in much the same way as the unpurified alloys, but the permeabilities were much higher than any previously obtained. The effect of cooling rate on the purified alloys is shown in Fig. 5-46. The magnetic properties of representative specimens are given in Figs. 5-50 to 52.

Alloys prepared in the usual way, with a deoxidizer added in the melt, did not respond to any great extent to hydrogen treatment at high temperatures, even when the optimum cooling rate was used. The highest initial permeability previously reported for material made with addition of a deoxidizer was 33 000, obtained by Cioffi [38C1], and in the more recent investigation a maximum of about 50 000 was observed. The specific effect of the addition of 0.3% of calcium silicide as deoxidizer is shown in Fig. 5-46 where it is compared with similar material made without the addition of any deoxidizer. The difference is striking and shows plainly

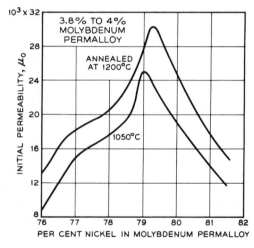

FIG. 5-45. Dependence of initial permeability on nickel content, in alloys containing 3.8–4.0% Mo. Specimens annealed at 1050 or 1200°C and cooled at the optimum rate.

that the absence of a strong deoxidizer is essential for obtaining the high permeability. This idea was confirmed by experiments in which silicon was added to the melt, and high permeability was not obtained. The action of strong deoxidizers, such as those commonly used in commercial production, is to combine strongly with the oxygen of the melt and thereby to prevent diffusion of oxygen in the solid alloy. This prevents in turn the subsequent combination of the oxygen with the hydrogen during the purification process at 1200–1300°C, and thus leaves a very fine-grained precipitate in the alloy, which depresses the permeability.

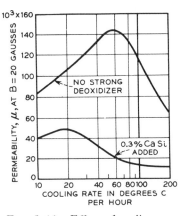

FIG. 5-46. Effect of cooling rate on the low-field permeability of an Fe-Ni-Mo alloy (16% Fe, 79% Ni, 5% Mo) prepared with or without addition of a deoxidizer and annealed in pure hydrogen at 1300°C.

Addition of manganese, commonly used as a desulfurizer, is not harmful. Apparently its combination with oxygen is so weak that purification with hydrogen is not prevented.

The dependence of the permeability on cooling rate is closely associated with atomic ordering. At any temperature below the critical temperature of ordering, about 500°C in the alloy most investigated (5% Mo, 79% Ni), both the equilibrium degree of ordering and the speed of ordering depend on the temperature. It is believed that for a certain degree of ordering both the saturation magnetostriction and the magnetic crystal anisotropy of the alloy become very small. This optimum degree of ordering may be obtained either by cooling at a certain critical rate or by cooling rapidly to a certain temperature below the critical ordering temperature, maintaining there for a definite time and then again cooling rapidly. Data obtained by following the latter procedure are shown in Fig. 5-47 and support the hypothesis described above.

Commercial Alloys. Molybdenum Permalloy.—The high initial permeability of 4-79 Molybdenum Permalloy, combined with its high resistivity, makes it one of the most useful magnetic materials. It is used in transformers transmitting weak signals at audio and carrier frequencies. A typical μ vs B curve for the commercial alloy is given in Fig. 5-48, and at low fields in Fig. 11-11. Reversible permeabilities are shown in Figs. 5-49 and 11-62, and hysteresis loops in Figs. 11-24 and 11-70. The density is 8.72. Some mechanical properties are: ultimate tensile strength, 85 000 lb/in.^{2}; Young's modulus, 30 × 10^6 lb/in.^{2} (ca. 2 × 10^{12} dynes/cm^2).

The manufacture of Molybdenum Permalloy is similar to that of 45 and

78 Permalloys. Some experience is necessary in adjusting the amount of molybdenum in the melt. Rolling technique is similar to that of 78 Permalloy and is less difficult than that for Chromium Permalloy. As discussed previously, the heat treatment is simple.

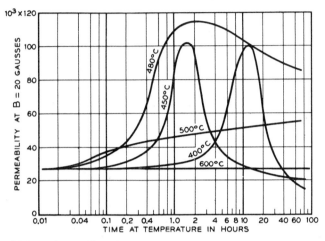

FIG. 5–47. Effect of annealing time at various temperatures on the initial permeability of Supermalloy previously annealed at 1300°C and cooled rapidly from 600°C.

Supermalloy.—The method of manufacture is based on the experiments on the hydrogen treatment and controlled cooling rate of Molybdenum Permalloy. The composition of Supermalloy is about 79% nickel, 5% molybdenum, 15% iron, and 0.5% manganese. Impurities such as silicon, carbon, sulfur, etc., are lower than in most commercial alloys. Materials are melted in vacuum in an induction furnace and poured in helium or nitrogen at atmospheric pressure. Ingots are hot and cold rolled by commercial methods to any thickness down to 0.00025 in. The tape is wound spirally to form toroidal specimens. When insulation is desired, a thin film of magnesia is applied in carbon tetrachloride suspension so that a film about 0.00005 in. in thickness is left on each side of the tape. Transformer cores are made in this manner.

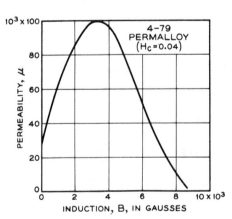

FIG. 5–48. Permeability curve of commercial 4-79 Permalloy.

SUPERMALLOY

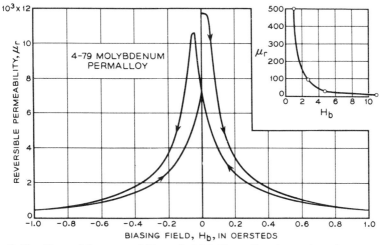

FIG. 5–49. Reversible permeability of 4-79 Permalloy in various biasing fields (butterfly loop).

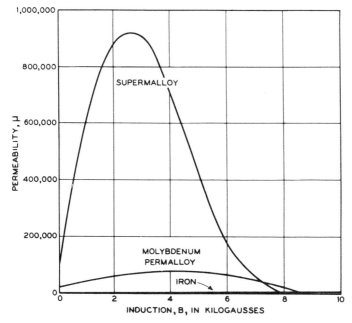

FIG. 5–50. Permeability curves of Supermalloy sheet 0.014 in. thick. Curves for other materials are for comparison.

Heat treatment consists of maintaining at 1300°C in pure dry hydrogen, and cooling through the temperature range from 600 to 300°C at a critical rate appropriate to the composition.

Magnetic properties will be given for (a) 0.014-in. uninsulated and (b) 0.001-in. insulated material. Figure 5–50 shows typical permeability vs induction curves for 0.014-in. Supermalloy, Molybdenum Permalloy, and iron. The ratio of permeabilities of Supermalloy and Molybdenum Permalloy varies with induction from about 5 at $B = 20$ to about 10 at

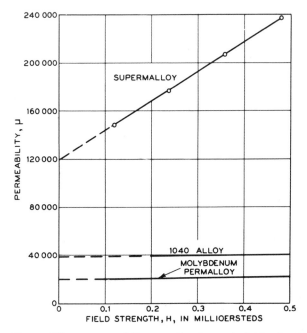

Fig. 5–51. Permeability curves of Supermalloy in weak fields, for 0.014-in. sheet, in comparison with other materials.

$B = 3000$ (maximum permeability). At inductions of 7000–8000 the permeability of Supermalloy has decreased markedly and is less than that of Molybdenum Permalloy or iron; saturation is at $B_s = 7900$. Maximum permeabilities as high as 1 500 000 have been observed.

The permeability in very low fields is shown in Fig. 5–51 for a representative specimen of Supermalloy and for other materials. Hysteresis loops for Supermalloy, and Molybdenum Permalloy for comparison, are given in the original article [47B1]. The hysteresis loss for $B_m = 5000$ is about 3–5 ergs/cm^3.

When thin tape is insulated before heat treatment, the permeability in low fields is affected but slightly. However, as the curves of Fig. 5–52 indicate, the maximum permeability of 0.001-in. insulated tape is considerably lower than that of thicker tape. Insulated 0.004-in. tape has a maximum permeability about twice as large as that of insulated 0.001-in. tape; its permeability is therefore intermediate between 0.001-in. insulated

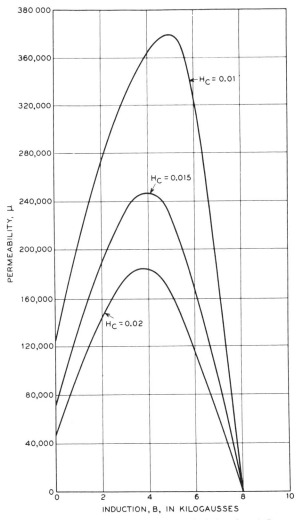

FIG. 5-52. Permeability curves of thin (0.001 in.) insulated Supermalloy tape of good, average, and marginal specimens.

and the 0.014-in. uninsulated tape. The middle curve of Fig. 5-52 is near to the average of a lot of several thousand cores of 0.001-in. tape made in the early stages of development of the material. The resistivity of Supermalloy is about 65 microhm-cm.

Other alloys are formed by addition of molybdenum to permalloy containing 45–50% nickel. These include *Monimax* (47% Ni, 3% Mo) and *4-43 Molybdenum Permalloy*, and also the alloy *Moniseal* (44% Ni, 4% Mo) made for sealing leads into glass. The first two alloys were developed during World War II as high-resistivity materials with moderately high saturation and good permeability, and were used in the form of thin sheet or ribbon ($\sim$0.004 in.) in coils operating at high radio frequencies. Monimax has a saturation induction of about 14 500, a maximum permeability of 40 000 to 50 000, an initial permeability of about 2000, a coercive force of about 0.06, and a resistivity of 70 microhm-cm. It is prepared in much the same way as the other permalloys and is preferably annealed finally at 1100°C in hydrogen.

Compressed Molybdenum Permalloy Powder.—The development of this material has been due to the exacting requirements of the loading coils which are extensively used in telephone circuits. The innate high permeability, high resistivity, and low hysteresis loss of Molybdenum Permalloy make it superior to other materials when reduced to powder, insulated,

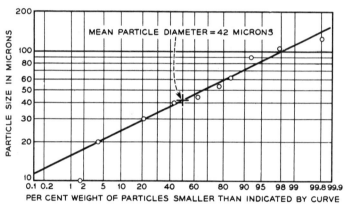

FIG. 5-53. Distribution of particle size in 120-mesh powder.

pressed, and annealed. The finished cores are, therefore, smaller in size, have lower eddy-current and hysteresis losses, and produce only a small amount of modulation.

In the preparation of *2-81 Molybdenum Permalloy* powder [34E1, 40L1], 200-lb ingots are cast which contain 2% molybdenum, 81% nickel, a few hundredths of a per cent of sulfur, and no more than a trace of manganese. These are hot rolled to 0.25 in. and then cold rolled. The embrittling action of the sulfur is adjusted to permit satisfactory hot rolling and to cause the sheet to break up into small pieces when cold rolled. During the hot rolling, a grain size is developed that is nearly equal to the final particle size desired; then, when cold, a single pass starts the process of pulverization that is continued in hammer and attrition mills until the grains, held together loosely by the sulfide coating, are broken apart. The powder is sifted, annealed at 850°C, again pulverized, insulated with a few per cent of fine ceramic powder, and then formed into rings under a pressure of 100–150 tons/in.² The rings are annealed in an atmosphere of hydrogen at 600–650°C to remove strains and develop suitable magnetic properties. After the application of a baked insulating finish the cores are wound with copper wire to obtain the proper inductance.

The distribution by weight of the particle sizes of the powder is shown in Fig. 5-53. The relations between permeability, density, dilution, and forming pressure are shown in Fig. 5-54. At the normal forming pressure and the usual dilution of 2.5%, the density, 7.7, is about 90% of that of the bulk material, 8.65. Under these conditions, the tensile strength of the finished core is 300–400 lb/in.²

The magnetic stability of the final product is remarkable—the instability, $s = (\mu_0 - \mu_r)/\mu_0$, is less than 0.3%. The extreme change in reversible permeability is also small for biasing field strengths as high as 120 oersteds, as shown in Fig. 5-55.

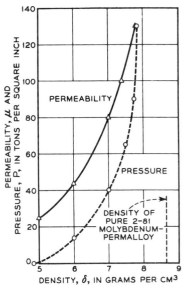

FIG. 5-54. Dependence of the permeability on the density of cores pressed at 100 tons/in.² from 120-mesh powder containing various amounts of diluent. Also, effect of density on the pressure with 2.5% of diluent.

The permeability and energy loss constants of these cores, expressed in terms of the constants of the equation given on p. 127, are compared with those of Isoperm and 78 Permalloy compressed powder in Table 2. In spite of the higher quality of the 2-81 Molybdenum Permalloy powder, the cost of manufacture is believed to be less than that of Isoperm [40L1].

The alloy *12-80 Molybdenum Permalloy* [40L1] has a Curie point near 70°C and, therefore, at room temperature has rapidly changing magnetic properties that may be used to compensate undesirable changes in other materials. In principle, its use is similar to that of Thermoperm and the iron-nickel-chromium alloy mentioned below. It has been used especially in powdered form mixed to the extent of a few tenths of a per cent with 2-81 Molybdenum Permalloy in compressed powder cores. By this means, the temperature coefficient at a given temperature has been adjusted to

TABLE 2. PERMEABILITY INSTABILITY AND ENERGY LOSSES IN STANDARD MATERIALS

Material	μ_0	$s(\%)$	$c \times 10^6$	$a \times 10^6$	$e \times 10^9$
80 Permalloy powder (compressed)	75	0.3	37	5.5	51
	26	.15	108	11.5	27
2-81 Molybdenum Permalloy powder (compressed)	125	.25	30	1.6	19
	60	.20	50	2.5	10
	26	.15	96	6.9	8
	14	.1	143	11.4	7
Isoperm (47% nickel), 0.002 in.	87	.6	15	2.1	17
Isoperm (11% copper, 30% nickel)	54	.9	32	4.6	240
Carbonyl iron powder (compressed)	58	2.0	79	9.3	7
	13	.3	60	5	1

desired positive or negative values. The initial permeability of this alloy is shown in Fig. 5-56 as a function of temperature for increasing and decreasing temperatures. (See also Table 6 of Chap. 7 for data on ferrites.)

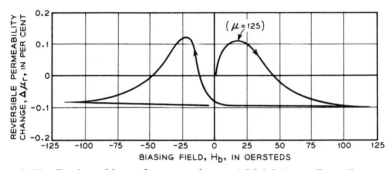

FIG. 5-55. Portion of butterfly curve of pressed Molybdenum Permalloy powder.

IRON-NICKEL-CHROMIUM ALLOYS

Structure and Magnetic Properties.—In addition to 3.8-78 Chrome Permalloy (3.8% Cr, 78% Ni), important for its magnetic properties, the well-known alloys containing these three elements as major constituents are the structural steels (e.g., 1.5% Cr, 3.5% Ni, and 0.3% C), the stainless steels (e.g., 18% Cr, 8% Ni), the heat-resistant alloys used as furnace elements, and the Elinvars having low temperature coefficients of elastic moduli.

The phase diagram of Fig. 5-57, due to Rees, Burns, and Cook [49R1], is revised from the earlier diagram of Schafmeister and Ergang [39S4]. The solid lines represent equilibrium phase boundaries at 650°C; the dashed lines, at 1200°C. The phase σ is a brittle constituent, supposedly a superstructure of ideal composition FeCr that transforms into the disordered α phase at about 1000°C on heating [40H1]. Sections at 18% chromium and at 8% nickel, given in Fig. 5-58 and due to Bain and Aborn [39A2], show the important phase relations of the stainless steels.

The *structural steels* containing chromium are magnetic, but their magnetic properties are not often of importance. Saturation values are around 20 000, independent of mechanical or heat treatment [27D1, 33H2].

The cheaper grades of *heat-resistant alloys* contain about 35% nickel and 15% chromium, and to withstand higher temperatures 80% nickel and

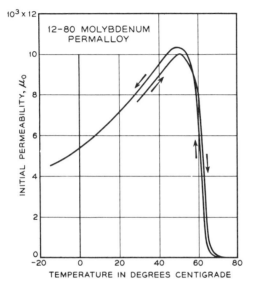

FIG. 5-56. Variation of permeability of 12-80 Molybdenum Permalloy with temperature.

15-20% chromium are used. Of these two types, the first has a Curie point near room temperature and so may or may not be magnetic, depending on the exact composition; the second is non-magnetic except perhaps at temperatures near absolute zero.

The *Elinvars* contained originally 36% nickel and 12% chromium [20G1], but improved elastic properties and increased ease of manufacture have been obtained with an alloy containing 31-33% nickel, 4-5% chromium, 1-3% tungsten, and 0.5-2% each of manganese, silicon, and carbon [32S5]. The alloy with 12% chromium is magnetic at room temperature, the Curie point lying near 100°C; the improved alloy is nonmagnetic at room temperature.

The *Stainless Steels* are ordinarily non-magnetic. When the 18-8 alloy (18% chromium, 8% nickel) is in the "soft" condition after cooling rapidly from 1200°C, it is non-magnetic at room temperature and remains non-

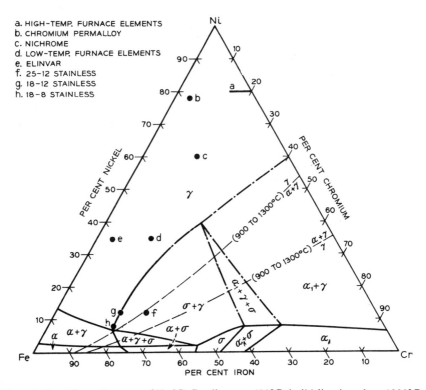

Fig. 5-57. Phase diagram of Fe-Ni-Cr alloys at 600°C (solid lines) and at 1200°C (dotted lines).

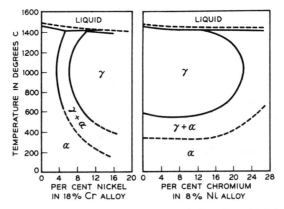

Fig. 5-58. Effect of addition of nickel (left) or of chromium (right) on the constitution of the 18% Cr, 8% Ni alloy.

magnetic when cooled to liquid nitrogen temperatures. On the contrary, when it is hardened by cold work, some of the α phase precipitates, as may be inferred from the phase diagram of Fig. 5-57; the material is then magnetic at room temperature and remains so until it is heated again to a high temperature and cooled rapidly. The term "non-magnetic" means here that the material is below the Curie point. At low temperatures 18-8 is strongly paramagnetic and the tractive force can be detected by hand in

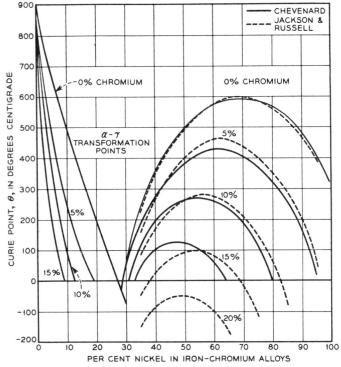

FIG. 5-59. Effect of chromium on the α,γ and the Curie points of Fe-Ni alloys.

the strong field-gradients of a large electromagnet. Maximum permeabilities measured after 90% cold reduction were observed by Horwedel [40K2] as follows: 16.7% Cr, 9.3% Ni, 0.07% C, $\mu_m = 120$; 16.6% Cr, 8.3% Ni, 0.15% C, $\mu_m = 15$; 17.7% Cr, 10.6% Ni, 0.09% C, $\mu_m = 1.4$.

The alloys that are useful for magnetic purposes lie in the large region of homogeneous γ solid solutions of face-centered cubic structure (austenite).

Figure 5-59, showing the *Curie points* of the low-chromium alloys, has been reconstructed from the data of Chevenard [28C1]. The broken lines of the figure refer to data of Jackson and Russell [38J1] for alloys containing 0.3% silicon. Chromium is particularly effective in lowering the Curie points or iron-nickel alloys, especially when the nickel content is high.

Use is often made of this fact in making non-magnetic alloys for special purposes.

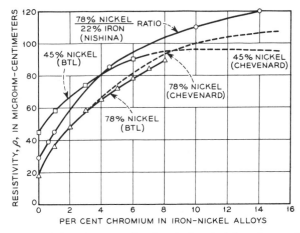

FIG. 5-60. Effect of chromium on the electrical resistivity of some Fe-Ni alloys.

The *densities* of the reversible alloys containing up to 15% chromium have been determined by Chevenard [28C2]. The reported densities of alloys having less than 30% nickel and 30% chromium, including the stainless steels, vary from 7.5 to 8.1, depending on the gross composition, treatment, and impurities. In the soft condition, commercial 18-8 stainless has a density of 7.93; for specific data on other commercial alloys the reader is referred to the *Metals Handbook*. The densities of the heat-resistant alloys mentioned above are approximately 8.0, 8.2, and 8.4, respectively. The density of structural steel is about 7.85; that of Elinvar, about 8.0. Some other physical properties, including *hardness* and *thermal expansion*, have been reported by Chevenard [28C2] and by Dean [30D1].

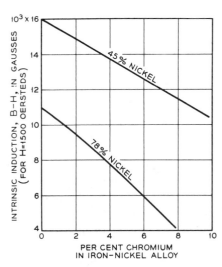

FIG. 5-61. Intrinsic induction (approximate saturation) of some Fe-Ni-Cr alloys.

The *resistivities* of alloys containing 45 or 78% nickel are shown in Fig. 5-60, drawn from data of Chevenard [28C2] and unpublished data for alloys containing some manganese. Data by Nishina [36N1] are also shown for alloys containing iron and nickel in the ratio 22:78, and 0.5% manganese, 0.5% cobalt, and 1% silicon. Other ternary alloys covering a wide range in composition have been studied by Dean [30D1], whose values appear to be somewhat higher than those of Fig. 5-60, where comparison is possible. His alloys were made from electrolytic iron, cobalt, and chromium; the only statement about impurities is that they contain less than 0.1% carbon.

Magnetization in high fields ($H = 1500$) for alloys containing 45 and 78% nickel and varying amounts of chromium is shown in Fig. 5-61. Measurements by Jackson and Russell [38J1] on the alloys with 0.3% silicon are shown in Fig. 5-62. The highest field used by them was 30 oersteds. The saturation appears to decrease linearly with chromium content.

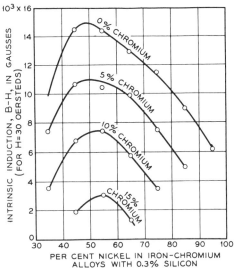

FIG. 5-62. Intrinsic induction of some Fe-Ni-Cr alloys at $H = 30$.

The *initial permeabilities* are shown in Fig. 5-63. The two standard heat treatments for Permalloy—furnace cool and double treatment—were used. The first of these figures shows definitely that the curves for the two treatments cross at about 3% chromium; thus the better heat treatment for the 3.8 alloy is simpler than that of 78 Permalloy containing no added element. The data for maximum permeability and for coercive force indicate also a crossing at 3%.

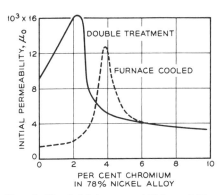

FIG. 5-63. Initial permeabilities of Fe-Ni-Cr alloys after furnace cooling or double treatment (air quench from 600°C).

The effects on the initial permeability of small additions of titanium and tin to 3-80 Chrome Permalloy have been studied by Nishina [36N1]. Titanium especially appears to have a beneficial action in increasing μ_0.

IRON–NICKEL ALLOYS

Commercial Alloys.—The properties of one of the best laboratory specimens of 3.8 *Chrome Permalloy* and of the representative commercial product

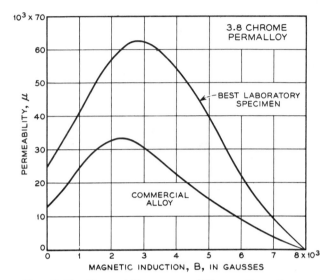

Fig. 5-64. Permeability curves of Chrome Permalloy.

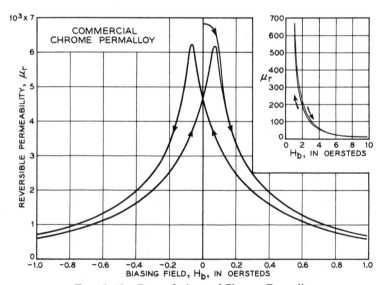

Fig. 5-65. Butterfly loop of Chrome Permalloy.

are compared in Fig. 5-64, which shows the permeability as dependent upon induction. Reversible permeability curves for the commercial product are reproduced in Fig. 5-65.

The manufacture of this alloy is similar to that of 4-79 Molybdenum Permalloy. Although Chrome Permalloy was used before Molybdenum Permalloy and Mumetal, the latter two have supplanted it commercially. The principal use has been in transformers where high initial or reversible permeability is required. The high resistivity is also important, and in high-frequency coils this has been increased by adding 8% of chromium.

Iron-nickel-chromium alloys have also been used where it is desired to have magnetic properties that change rapidly with change in temperature. A specific alloy containing 35% nickel, 5% chromium, 0.3% silicon, and the remainder iron has been suggested by Jackson and Russell [38J1, 40B1] for use in relays that open or close at required temperatures and in transformers that control the operation of small motors or other equipment. The Curie point of this alloy is about 160°C, as indicated in Fig. 5-59, and its saturation is above 7000 at 25°C, so that the induction decreases rapidly as the temperature is raised. Binary iron-nickel and nickel-copper alloys with similar properties are used for compensating undesirable magnetic changes in apparatus with which they are associated.

The commercial alloy Mumetal, containing copper as well as chromium, is discussed in the following section.

IRON-NICKEL-COPPER ALLOYS

The magnetically soft alloys of iron, nickel, and copper became of importance with the invention by Smith and Garnett [24S1] of Mumetal, which originally contained about 5% copper and 75% nickel. Isoperm also often contains several per cent of copper as well as 40–50% of nickel. Both of these may be regarded as modifications of Permalloy. Alloys having compositions near 30% nickel and 60% copper are precipitation hardenable and have the properties of permanent magnets. Thermalloy [28A4], the magnetic properties of which change rapidly with temperature because its Curie point is near room temperature, contains 58–68% nickel, 2% iron, and the rest copper. Among the alloys important for properties other than magnetic ones may be mentioned the nickel-copper alloys Monel, Advance, and Constantan.

Structure and Physical Properties.—The phase boundaries at room temperature, after slow cooling, are shown in Fig. 5-66, according to Bradley, Cox, and Goldschmidt [41B4]. The face-centered γ region in which there is a single phase extends from 35 to 100% nickel on the iron-nickel side and along the whole of the nickel-copper side, but in the large central portion of the diagram the alloys break up into two γ phases, one rich in copper and the other in iron and nickel, as shown by the connodes. This decomposition may be suppressed to a considerable extent by quenching. At 1000°C and above, there is no α (body-centered) phase, and the γ solubility curve extends only to 35% nickel. In the course of the study of the phase diagram many lattice spacings of single-phase alloys were determined.

A region of instability is noted on the diagram of Fig. 5-66, near the composition $FeNi_3Cu_4$. Daniel and Lipson [43D1, 44D1] have made an X-ray study of the alloys in this concentration area and found that just before decomposing into the γ_1 and γ_2 phases the original lattice shows

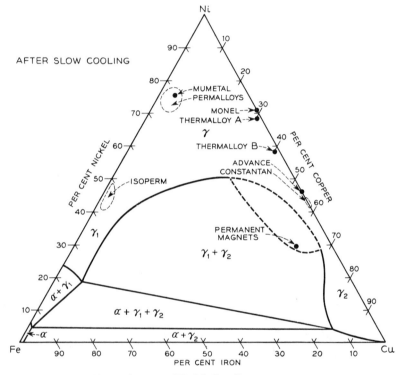

FIG. 5-66. Phase diagram of Fe-Ni-Cu alloys near room temperature.

periodic changes in its lattice dimensions, corresponding to local concentration gradients of wave length about 10^{-6} cm. See also p. 396.

The *Curie points* for alloys in equilibrium near room temperature, as determined by Köster and Dannöhl [35K3], are shown in Fig. 5-67 as solid lines; those for quenched alloys, by the dotted lines. The phase boundaries, shown by the dashed lines, are somewhat different from those given in Fig. 5-66.

Resistivities of slowly cooled alloys at 20°C have been reported by Köster and Dannöhl and by Kosting [30K1]. In the regions that are important magnetically, there is an almost linear, but only a small, increase in resistivity as copper is added to the iron-nickel alloys.

Magnetic Properties.—An extensive series of measurements on a wide variety of compositions has been carried out *in high fields* by von Auwers and Neumann [35A1]. Figure 5-68 shows the values of B for $H = 10$; these measurements give the best indication of the way that saturation magnetization changes with composition, since specific data are lacking. Materials used were in the form of rings punched from sheet rolled from

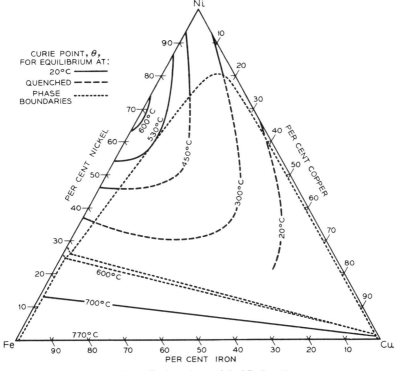

FIG. 5-67. Curie points of Fe Ni-Cu alloys.

induction melts made from Mond nickel, charcoal iron, and electrolytic copper. The heat treatments consisted in (1) Permalloy treatment: 1 hour in dry hydrogen at 900°C, quenched from 625°C in air; and (2) 2 hours in dry hydrogen at 1100°C and furnace-cooled. Although these treatments caused but slight differences in the values of B for $H = 10$, the *initial permeabilities* were markedly affected. Values for treatment (1) are higher and are given in Fig. 5-69. The highest μ_0 attained was in each case about 12 000. The different effects of the heat treatments are shown also in Fig. 5-70 in which the two curves give the values of μ_0 on the ridges of the three-dimensional μ_0 vs composition diagram.

The *coercive forces* for these alloys were also determined. Values for the quenched alloys are somewhat lower, and a minimum value of 0.025 was found for the alloy containing 27% copper and 60% nickel. Similar

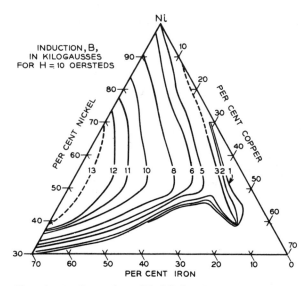

FIG. 5-68. Induction of Fe-Ni-Cu alloys at $H = 10$.

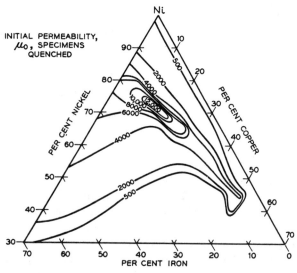

FIG. 5-69. Initial permeability of Fe-Ni-Cu alloys quenched from 625°C in air.

data have also been reported by Dahl, Pfaffenberger, and Schwartz [35D3]. An interesting illustration of the sudden increase in H_c that occurs when the solubility limit is exceeded is afforded by the data of Kussmann and Scharnow [29K1] plotted in Fig. 5–71. The change in the slope of the curve at about 16% copper is consistent with the solubility line indicated in Fig. 5–66, provided the high temperature distribution of phases is maintained by a sufficiently rapid cooling rate.

Mechanical treatment of the iron-nickel-copper alloys plays an important part in the preparation of the *Isoperms*, some of which contain up to 15% copper. As reported by Kersten [34K9] the effect of cold rolling (97% reduction) on material previously annealed for 2 hours at 900°C and slowly cooled is shown in Fig. 5–72 for the alloy made by adding 9% of copper to 45 Permalloy. Compared with the same alloy with no added copper, the permeability is lower but more independent of superposed

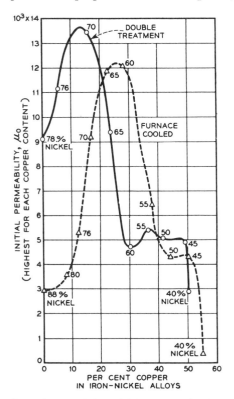
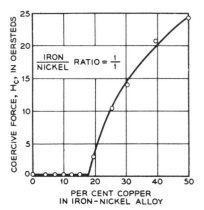

FIG. 5–70. Highest initial permeabilities of Fe-Ni-Cu alloys at given copper contents.

FIG. 5–71. Increase of coercive force accompanying precipitation of second phase. Material annealed at 900°C.

or previously applied field strength, and the hysteresis is lower. The permeability again increases on annealing and the Isoperm character is lost [33K4].

The heat treatment, to which the material is subjected before the cold rolling, affects the final properties, especially B_r, in a surprising way when the copper content is 10% or above [34D1]. This is brought out in Fig. 5–73, which shows the specific effect of reduction by rolling, for the alloy with 13% copper.

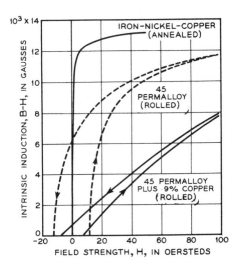

Fig. 5-72. Hysteresis loop of hard rolled alloys with and without addition of 9% copper. Ratio Ni/Fe = 45/55.

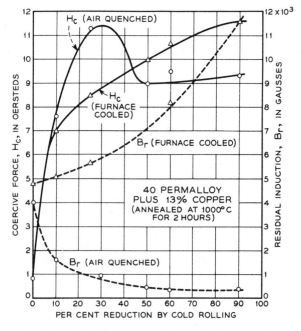

Fig. 5-73. Effect of cold reduction on some Fe-Ni-Cu alloys previously heat-treated at 1000°C and cooled as indicated.

Data by Dahl and Pawlek [34D1], not shown here, prove that the directional properties of the final product are marked and depend, as already indicated, on the rate of cooling used in the heat treatment immediately preceding the final rolling as well as on the extent of the final reduction. The initial permeability of useful Isoperms containing copper usually lies between 40 and 100; loss characteristics for various compositions and treatments are given in some detail by Dahl, Pfaffenberger, and Sprung [33D2]. Characteristics of a representative specimen of Isoperm containing 11% copper are given in Table 2 where they may be compared with other materials used for the same purpose. Isoperm containing copper has generally been superseded commercially by that containing only iron and nickel.

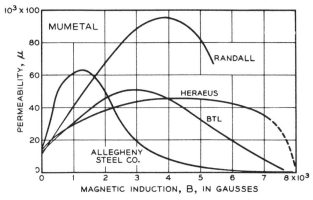

FIG. 5–74. Permeability curves of various specimens of Mumetal, as reported by Randall, by Allegheny-Ludlum Steel Corporation (annealed 4 hours at 1100°C in hydrogen), by Heraeus Vacuumschmelze, and as measured by Chegwidden (BTL) (pot-annealed at 1050°C). Allegheny specimens sometimes have $\mu_0 = 30\,000$, $\mu_m = 100\,000$.

Commercial Alloys. Mumetal.—As now produced by some manufacturers, this alloy contains 5% copper, 2% chromium, and a fraction of a per cent of manganese, and thus has a composition that puts it outside of the limits of the original patent [24S1] which specifies "not over one per cent of the elements W, Cr, Si, V, Ti, Mo, or Al." Other manufacturers give the name Mumetal to the alloy described above as 4-79 Molybdenum Permalloy. The nickel content of the copper-chromium alloy is 72–76%. The important properties of Mumetal are its high initial permeability and resistivity (40 to 45 $\times 10^{-6}$ ohm-cm) and attendant low loss in low fields, and its good mechanical properties that permit easy fabrication. In some specimens initial permeabilities as high as 30 000 have been measured but the average value is 15 000–20 000 [37R1]. Figure 5–74 shows μ,B curves for typical specimens made in the United States and in England [39A3,

37R1, 31H1]. The saturation is B_s = 8000 (BTL data). Contamination by impurities during the hot rolling and heat treating is particularly to be avoided. The American manufacturers particularly recommend that dry hydrogen (dew point, $-70°C$) be used during the final anneal, which should be carried out at 1100–1200°C.

Some of its electrical and mechanical properties are as follows: resistivity 42 microhm-cm, elongation 25%, yield point 45 000 lb/in.^{2}, tensile strength 70 000 lb/in.^{2}, Rockwell B hardness 30–40, density 8.6 g/cm^3.

Mumetal has found widespread use in the cores of current transformers and inductances used in radio apparatus. The high initial permeability and resistivity and its ease of fabrication to thin sheet permit its extensive use at audio frequencies and at frequencies of 100 000 or higher. It is effective material for screening from magnetic disturbances and is suitable for use also in the loading of telegraph cable.

1040 Alloy.—This alloy contains 14% copper and 3% molybdenum together with 72% nickel and the rest iron [34N1]. Manganese may also be present to the extent of 1%. Its most remarkable property is the initial permeability of about 40 000 (highest value reported, 51 000). The μ,B and μ_r,B curves are shown in Fig. 5–75 and the hysteresis loop in Fig. 5–76. Other characteristics are: H_c = 0.015, B_r = 2400, B_s = 6000, W_h = 0.013 watt/kg, ρ = 56 × 10^{-6} ohm-cm, θ = 290°C, density = 8.76 g/cm^3. It is prepared by careful melting in vacuum or hydrogen and can be readily worked.

Other commercial alloys are listed in Tables 1 and 2 of Appendix 4.

IRON-COBALT-NICKEL ALLOYS

Interest in the magnetic properties of these alloys was first aroused by Elmen in 1921 [28E1] when he discovered that many of them possessed a remarkable combination of low hysteresis and high permeability at low flux densities, and peculiarly shaped hysteresis loops at higher flux densities. These alloys were called "Perminvars" because of the constancy of permeability at low flux densities, a property always associated with low hysteresis loss. The usual composition is 25% cobalt, 45% nickel, and the rest iron (45-25 Perminvar), but this may vary over wide limits and sometimes molybdenum or chromium is added to increase the resistivity [29E3].

Other alloys of iron, cobalt, and nickel, useful because of properties other than magnetic properties, include Kovar [30S1] and Fernico [41H1], which have expansion characteristics similar to those of certain glasses into which they may be sealed. The composition is approximately 15% cobalt, 31% nickel, and the rest iron. For sealing into glasses of special types, Ferrichrome, containing 25% cobalt, 30% nickel, 8% chromium, and the rest iron, is sometimes used [34H2]. An alloy of low expansivity, having 31% nickel, 4–6% cobalt, and the remainder iron, has been named Super

IRON–COBALT–NICKEL ALLOYS 161

Nilvar [31M2]. Konel [31T1] is an alloy used for filaments of vacuum tubes, contains about 20% cobalt, 70% nickel, 2.8% titanium, and the rest iron.

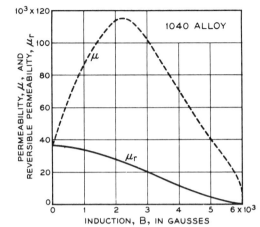

FIG. 5-75. Normal and reversible permeabilities of 1040 Alloy.

Structure and Physical Properties.—Comprehensive work on the structure of this three-component system has been done by Kasé [27K1], who made thermal, magnetic, dilatometric, microscopic, and hardness measure-

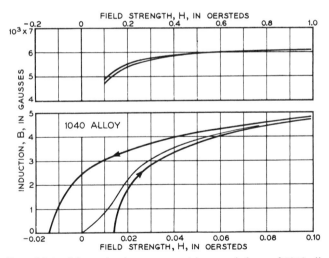

FIG. 5-76. Magnetization curve and hysteresis loop of 1040 alloy.

ments. Alloys were made from Armco iron, Mond shot nickel (containing 0.4% carbon), and granular cobalt. No impurity except carbon was contained in the raw materials to the extent of 0.1%. The liquidus and solidus surfaces were determined with alloys purified by electrodeposition and hydrogen annealing. The solid lines of Fig. 5–77 indicate the beginning of a phase transformation when the alloy is cooled, and the broken lines indicate Curie temperatures.

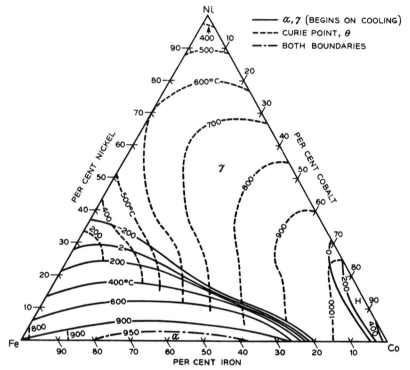

FIG. 5–77. The α,γ transitions and the Curie points observed during the cooling of Fe-Co-Ni alloys.

As in the binary iron-nickel alloys, the α phase has a body-centered cubic structure, the γ phase a face-centered cubic structure. The ϵ-phase alloys, including pure cobalt, are hexagonal close-packed. It is well to bear in mind that the phase boundaries of Fig. 5–77 have been determined by only the one investigator, and at a time when there was no evidence of superstructure in the iron-nickel [39L1] and iron-cobalt [39R1, 41E1] systems. It is therefore to be expected that the diagram will be modified in these respects and possibly in others. Kaya and Nakayama [39K3] have detected superstructure in alloys containing 20–25% iron and 0–75% cobalt. The amount of ordering, as measured by the change in specific heat with temperature, gradually decreases as cobalt is added.

Thermal expansion coefficients [31M2] and sceroscope hardness have been determined by Kühlewein. Densities of a few alloys are given in Table 3 [37M1]. The density of Perminvar containing 7.5% molybdenum (25% Co, 45% Ni) is 8.66; that of Kovar is 8.36.

In plotting the lines of equal *resistivity*, shown in Fig. 5-78, data from several sources have been combined [31M2, 27E2], including unpublished data of the Bell Laboratories for alloys containing 0.3-0.7% manganese.

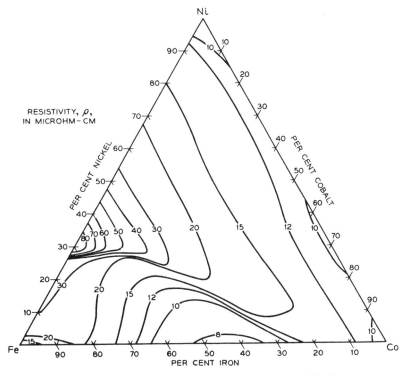

FIG. 5-78. Electrical resistivities of annealed Fe-Co-Ni alloys.

A few data are available on the effect of heat treatment on resistivity and hardness [32D1] and on specific heat [39K3].

Magnetic Properties.—The most complete investigation of the magnetic properties has been reported by Elmen [28E1, 29E1]. A second extensive investigation was reported almost simultaneously by Masumoto [29M1]. The compositions used by both investigators are shown, by dots and crosses, in Fig. 5-79. Elmen studied especially the ternary alloys near 45-25 Perminvar; Masumoto studied those near the phase boundaries between the α and γ phases and between the γ and ϵ phases. The chief practical result has been the discovery by Elmen of the typical Perminvar characteristic already mentioned, characteristics that can be developed to a greater

IRON-NICKEL ALLOYS

TABLE 3. DENSITIES OF SOME FE-CO-NI ALLOYS

Composition (%)				Density (g/cm³)
Fe	Co	Ni	Mn	
10.3	10.2	79.3	0.5	8.80
9.9	20.2	69.4	.4	8.79
20.5	14.8	64.6	.3	8.63
10.4	28.4	61.2	.2	8.75
15.1	24.8	60.1	.3	8.70
10.4	38.7	50.8	.3	8.74
24.6	25.0	50.3	.3	8.57
49.6	10.2	40.2	.3	8.31

or lesser degree in all alloys having compositions between the curved lines of the figure.

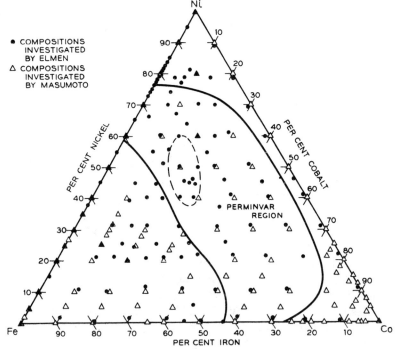

FIG. 5-79. Alloys used in the two principal investigations of the Fe-Co-Ni alloys.

Elmen's alloys were made from Armco iron, electrolytic nickel, and high-grade commercial cobalt, and the melting and reduction were carried out as has been already described. A typical analysis follows:

IRON–COBALT–NICKEL ALLOYS 165

	(%)		(%)
Fe	24.3	Si	0.06
Co	26.0	C	0.02
Ni	49.4	S	0.02
Mn	0.5	P	0.01

Al, Mg, As, Cr, Cu <0.01

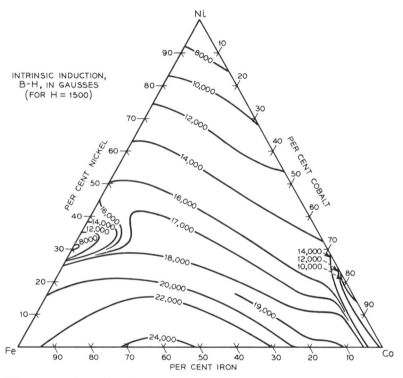

FIG. 5–80. Saturation induction (approximate) of annealed Fe-Co-Ni alloys.

Values of *saturation magnetization* as such have not been determined but $(B-H)$ at field strengths of 1500 have been measured by both Elmen and Masumoto and their data have been combined in constructing Fig. 5–80. Specimens were annealed for 1 hour at 1000°C in a closed pot (Elmen) or hydrogen (Masumoto), cooled with the furnace (about 100°C/min at 500°C) and measured at room temperature. Saturation is practically complete for all alloys except those having high cobalt content.

Initial permeabilities were reported by Elmen. Data obtained by various members of the staff of the Bell Laboratories have been used in constructing Figs. 5–81 and 82 for alloys in the annealed (furnace-cooled from 1000°C) and rapidly cooled (air quenched from 600°C) conditions, respectively. Annealed alloys show two maxima on the iron-nickel side and one on the iron-cobalt side, and minima at the three corners and near the alloy having 25% nickel and 75% iron. Rapidly cooled alloys have similar maxima and minima except that the peak near 79 Permalloy is much more prominent. The effect of rapid cooling is noticeably large in the region coinciding roughly with the Perminvar region outlined in Fig. 5-79.

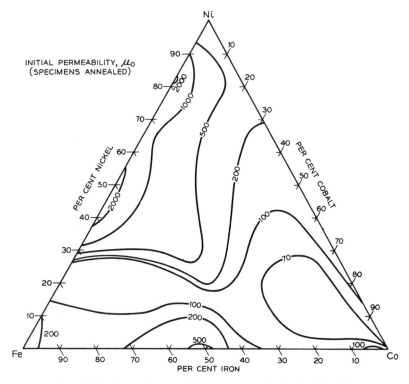

FIG. 5-81. Initial permeabilities of annealed Fe-Co-Ni alloys.

The *maximum permeabilities* shown in Fig. 5-83 are for annealed alloys. The general character of the diagram is similar to that for initial permeabilities except that the values are, of course, much higher. A few data for rapidly cooled alloys (Table 4) show that μ_m is increased by rapid cooling in even greater ratio than μ_0 is increased by the same treatment. The limited data indicate that the ratio of the values of μ_m obtained by the two treatments is greatest for 45-25 Perminvar.

The coercive force and the hysteresis loss for $B_m = 5000$ for the same alloys are lower in the rapidly cooled than in the annealed condition (Fig. 5-84).

Perminvar characteristics are constancy of permeability in low fields and a peculiar form of the hysteresis loop in intermediate fields (e.g., $H_m = 10$). These properties are developed in highest degree when the alloy is heated

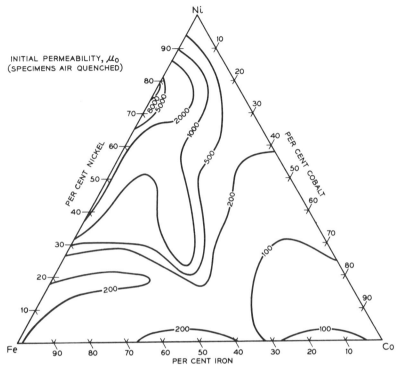

FIG. 5-82. Initial permeabilities of air-quenched Fe-Co-Ni alloys.

TABLE 4. MAXIMUM PERMEABILITIES OF SOME PERMINVARS AFTER SLOW COOLING OR QUENCHING

Composition (%)			μ_m after Slow Cooling	μ_m after Quenching	Quenching Temperature (°C)
Fe	Co	Ni			
30	15	55	2800	12 400	700
30	20	50	2400	19 500	725
25	20	55	2500	18 200	725
25	40	35	1200	5 400	810
30	25	45	2000	25 000	725
30	35	35	1300	8 400	770
35	20	45	4200	22 300	700
35	25	40	2400	25 800	715
35	30	35	2000	13 800	725
40	25	35	3800	7 700	670
40	20	40	4700	10 100	650

or "baked" for a long time at a temperature that is as low as possible and still will permit any change in properties to take place. This temperature is between 400–450°C; the treatment usually used in 425°C for 24 hours. Curves showing permeability vs field strength for the three heat treatments —annealing, air-quenching, and baking—are reproduced in Fig. 5–85 for 45-25 Perminvar. The remarkable constancy of the permeability for the

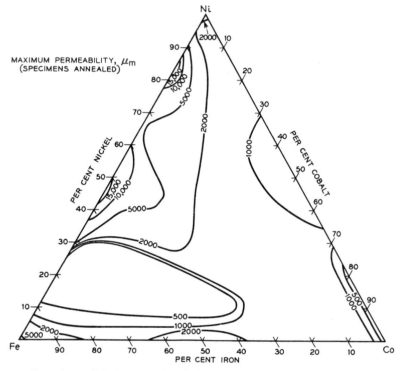

FIG. 5–83. Maximum permeabilities of annealed Fe-Co-Ni alloys.

baked specimen is shown on an extended scale in Fig. 5–86. In some specimens no change in permeability can be detected using an alternating-current permeameter [24K1] when the field strength is changed by 1 oersted or more. A change in permeability of 1% usually takes place when the maximum field strength is increased from zero to 1.5 or 2 oersteds.

Such small changes in permeability mean that hysteresis losses must also be small. On typical, baked specimens these losses are of the order of 0.01 erg/cm^{-3} per cycle for $B_m = 500$ and sometimes are considerably less than this. These losses may be compared with losses of over 2500 ergs/cm^{-3} per cycle at $B_m = 10\,000$.

As H_m increases from 2 to 3 the hysteresis loop begins to have a noticeable area. Development of the constricted loop is complete when H_m is 4 or 5 oersteds as illustrated in Fig. 5-87. The shapes of the loops are almost normal when the maximum field is high. The way in which the hysteresis losses vary with B_m is shown in Fig. 5-88 for baked and for annealed alloys. It is apparent that the usual 1.6 power law $(W_h = \eta B^{1.6})$ is not obeyed here.

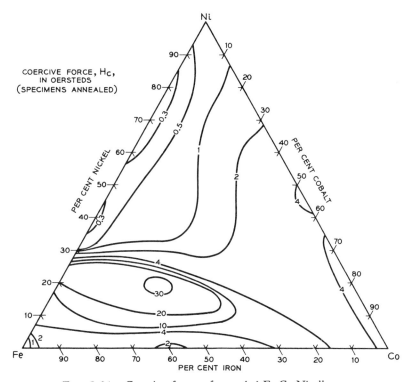

Fig. 5-84. Coercive forces of annealed Fe-Co-Ni alloys.

A variety of shapes of loops may be obtained by varying the composition and heat treatment of the alloys in the Perminvar region, as has been shown by Kühlewein [31K4, 33A3]. In Fig. 5-89(a) are shown eight of his loops for a series of different compositions extending from the edge to the middle of this region. In Fig. 5-89(b) are three loops of unusual shape obtained by heat treatment of three different compositions in a magnetic field (see below). It may be noticed in Fig. 5-89(a) that the normal magnetization curve often lies outside of the hysteresis loop.

The Perminvar character of the hysteresis loop—the constricted form— is lost at *elevated temperatures*. The loops of Fig. 5-90, recorded by Kühlewein [32K2], indicate that this change takes place in a rather narrow temperature region between 500 and 600°C. This is not surprising in view of the fact that baking below 500°C is necessary in order to develop

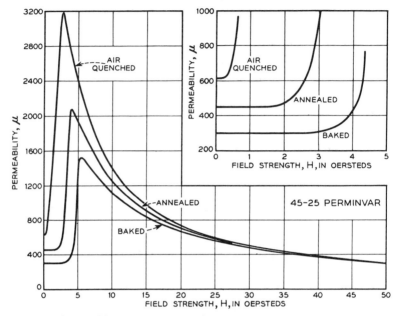

Fig. 5-85. Effect of heat treatment on the permeability of 45-25 Perminvar (30% Fe, 25% Co, 45% Ni).

the most extreme Perminvar properties and it is probably connected with the order-disorder transformation. Measurements of coercive force and

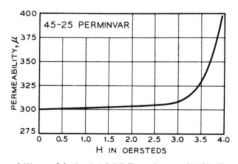

Fig. 5-86. Permeability of baked 45-25 Perminvar (30% Fe, 25% Co, 45% Ni) in low fields.

remanence measured at various temperatures [32K2] show that above 500°C the remanence increases rapidly to what would be expected of a more normal material, before it drops again toward the small values it

IRON-COBALT-NICKEL ALLOYS   171

must have as the temperature approaches the Curie point. Finally, Fig. 5-91, also due to Kühlewein, shows that the initial linear parts of the magnetization curves almost disappear in the same region of temperature. The effect of *tension* [31K4] is similar to that of temperature in that it causes the disappearance of the constricted nature of the loop (Fig. 5-92). This happens when the tension has the order of magnitude of 5 kg/mm² (7000 lb/in.²).

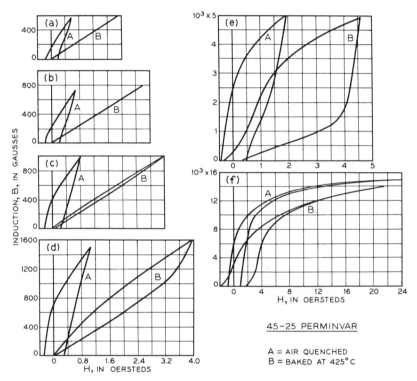

FIG. 5-87. Hysteresis loops of 45-25 Perminvar at various maximum field strengths, showing the characteristic low hysteresis at low inductions and wasp-waisted loops at high inductions.

A "Super-Perminvar" has been reported by Nishina [36N1] who has increased the range of constant permeability by sacrificing the high permeability by cold drawing. After 87% reduction in diameter the permeability remains practically constant at 93 until $H = 7$ ($B = 650$). The material contains 23% cobalt and 9% nickel.

Heat Treatment in a Magnetic Field.—The Perminvars are susceptible to heat treatment in a magnetic field in much the same way as are the Permalloys containing 55-80% nickel [34K1, 35D4]; in fact the latter alloys lie in the region designated as the Perminvar region. The hysteresis loops [35D4] for 45-25 Perminvar heat treated in longitudinal and transverse fields are shown in (1) and (2) respectively of Fig. 5–93(a); in (b) the μ vs H curve for the transverse treatment is shown with greater clearness.

Most of the experiments [35B2] on the heat treatment of Perminvar in a field were performed on 20-60 Perminvar (20% Ni, 60% Co, and 20% Fe), chosen because its Curie point (ca. 880°C) is considerably higher than the

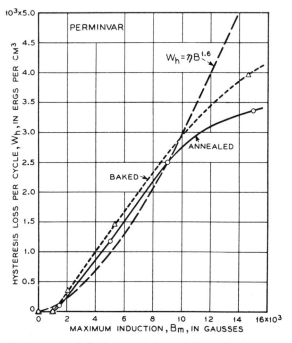

Fig. 5–88. Dependence of the hysteresis loss of 45-25 Perminvar on the maximum induction, after annealing or baking. Note that the Steinmetz low is not followed.

temperature (450°C) at which such heat treatment is effective. The maximum permeability was increased about tenfold, the coercive force somewhat decreased, the shape of the hysteresis loop changed so that its sides were vertical for an interval of 20 000 to 25 000 in induction, as illustrated in Fig. 5–94(a). Part (b) of the figure shows that the field applied during heat treatment is of some effect if it is present only while the material is at 500–800°C, but that for the full effect it must be maintained until the temperature is as low as 400°C. The length of time that the field must be applied at 480°C for full effect can be inferred from Fig. 5–95, which shows the descending branches of the hysteresis loops measured

IRON-COBALT-NICKEL ALLOYS

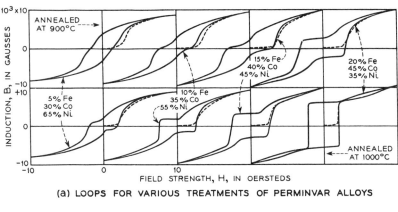

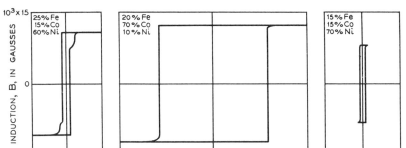

FIG. 5-89. (a) Change of loop-forms of Fe-Co-Ni alloys with composition and heat-treatment (in zero field). (b) Loops of some materials heat-treated in a magnetic field.

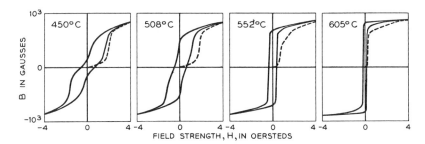

FIG. 5-90. Hysteresis loops measured at temperatures of 450–605°C, showing disappearance of wasp-waisted form in this range of temperature. Composition: 30% Fe, 25% Co, 45% Ni

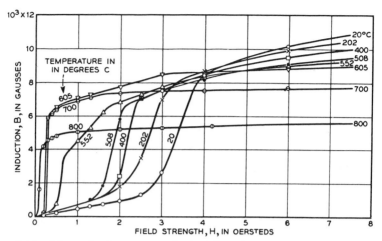

Fig. 5–91. Magnetization curves of 45-25 Perminvar measured at temperatures up to 800°C. Characteristic straight initial portion disappears at about 550°C.

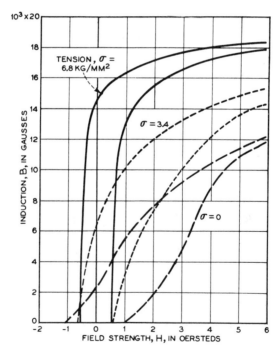

Fig. 5–92. Hysteresis loops of Perminvar (35% Fe, 30% Co, 35% Ni) under various tensions, showing disappearance of loop-constriction under moderate tension.

at this temperature at various times after the application of the field. Almost all of the effect is attained in 15 minutes at this temperature. Some of the magnetic characteristics of other compositions of Perminvar heat-treated in a field are given in Table 5 [35D4] and Fig. 5–96 according

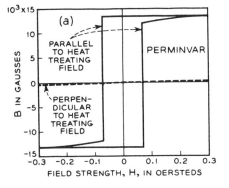
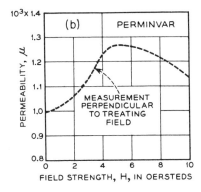

FIG. 5–93. (a) Hysteresis loops of Perminvar (30% Fe, 25% Co, 45% Ni) heat-treated in a magnetic field and measured parallel or perpendicular to the direction in which the field had been applied during heat treatment. In the permeability curve (b) the two fields were perpendicular.

TABLE 5. MAGNETIC CHARACTERISTICS OF IRON-COBALT-NICKEL ALLOYS

Composition (%)	Annealed without a Field		Annealed with a Field	
Fe-Co-Ni	H_c	μ_{max}	H_c	μ_{max}
10-20-70.......	1.16	1 490	0.28	11 600
10-39-51.......	2.54	2 440	.69	7 100
11-50-30.......	3.86	1 110	2.28	5 300
11-78-11.......	3.56	1 730	2.54	4 100
25-15-60.......		2 680	.052	166 000
25-20-55.......		2 550	.055	165 000
30-15-55.......		2 770	.042	257 000
30-25-45.......	0.90	2 000	.090	115 000
30-44-26.......	2.90	1 150	.74	12 500
34-23-43.......		2 380	.033	427 000
35-25-40.......		2 390	.039	311 000
35-30-35.......		2 030	.077	176 000
40-60-0........		3 300	.53	31 000
45-10-45.......	0.69	7 550	.44	18 300
48-52-0........	0.70	11 000	.55	22 400

to unpublished measurements by Miss M. Goertz. Figure 5–97 brings out the difference in the properties resulting from the different heat treatments of one alloy. Important characteristics to be noted, especially in

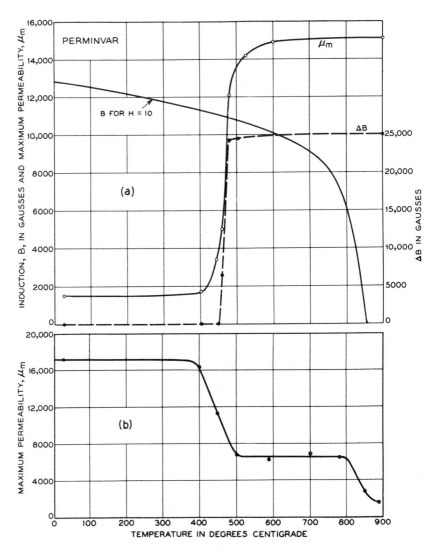

Fig. 5-94. (a) Some properties of a specimen (20% Fe, 60% Co, 20% Ni) heat-treated in a magnetic field which was *applied* at the indicated temperature on cooling. ΔB is the height of the vertical part of the hysteresis loop. Measurements of μ_m and ΔB are at room temperature; B for $H = 10$ is measured at temperature. In (b) the magnetic field was applied during cooling from 950°C and *removed* at the temperature indicated.

Fig. 5-96, are the high magnitudes of the permeability and of the induction at which the permeability is a maximum. This is equivalent to a rapid and complete saturation of the material as the field is applied.

The alloys which respond to heat treatment in a field are, roughly speaking, those that have Perminvar characteristics when treated in the usual way (Fig. 5-79). A high ratio (50 or more) of the maximum permeabilities

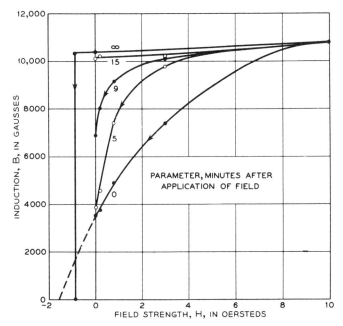

FIG. 5-95. Portions of hysteresis loops measured at 480°C at various times after the application of a field of 10 oersteds.

heat-treated in a field and in zero field, respectively, is found in alloys represented on the triangular diagram by a narrow, elliptical region lying between the points 25% iron, 15% cobalt, 60% nickel, and 35% iron, 30% cobalt, and 35% nickel.

Reversible Permeability.—After observing the unusual μ,H curves and hysteresis loops of the Perminvars, it is not surprising to find peculiar changes in the reversible permeability as the superposed field is increased. A typical curve of this sort is given in Fig. 5-98 [28E1]. A most unfortunate property of this material, from the point of view of engineering applications, is the difference between the initial permeability (at a) and the reversible permeability at remanence (b), i.e., the value of the instability, $s = (\mu_0 - \mu_r)/\mu_0$ is high. This means that the characteristics of a transformer used to transmit weak signals and working at a will be radically changed if it has been momentarily subjected to a large, magnetizing force. The permeability is increased by 75%, and the hysteresis losses are much higher.

A number of methods have been tried to reduce the instability, and some success has been achieved. One of the first experiments was heat-treating the material after it had been magnetized and the magnetizing

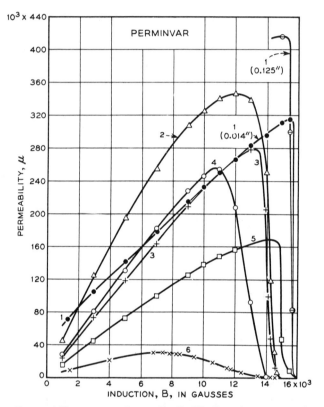

FIG. 5–96. Permeability curves of some Fe-Co-Ni alloys heat-treated in a magnetic field. Compositions are: (1), 30% Fe, 20% Co, 50% Ni; (2), 34% Fe, 23% Co, 43% Ni; (3), 30% Fe, 20% Co, 50% Ni; (4), 30% Fe, 15% Co, 55% Ni; (5), 35% Fe, 25% Co, 40% Ni; (6), 25% Fe, 30% Co, 45% Ni.

field removed, i.e., with residual induction. This had the effect of stabilizing the permeability except when the magnitude of the magnetic "shock" lay within rather narrow limits. Further substantial increase in this kind of stability has been obtained by first heat-treating in a moderate magnetic field and then removing the field during further heating, and finally heating at a lower temperature in the absence of a field. By this rather complicated treatment, carried out by H. L. B. Gould, the whole range of variation of permeability before and after "shock" has been reduced from 75% to 0.5%.

Commercial Alloys.—Although most attention has been paid to 45-25 Perminvar (see Figs. 5-85 to 98), it is doubtful whether it may be classed

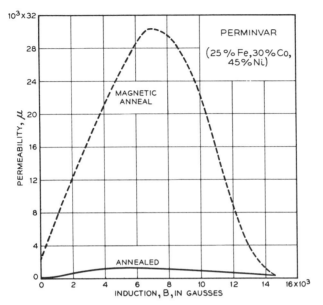

FIG. 5-97. Permeability of Perminvar (25% Fe, 30% Co, 45% Ni) as annealed and as subsequently heated and cooled in a magnetic field.

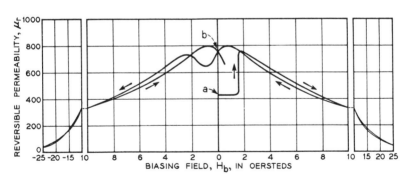

FIG. 5-98. Reversible permeability of 45-25 Perminvar (25% Co, 45% Ni), after baking.

as a commercial alloy since it has found so few applications thus far. On account of the higher initial permeability and lower cost, more practical applications have been made of *7-70 Perminvar* containing 7% instead of 25% of the most costly element. The μ vs B curve for this material is given in Fig. 5-99. It is annealed at 1000°C and cooled to room temperatures at a rate of less than 5°C/min. The initial permeability of 850 is maintained within 1% until $B = 600$ ($H = 0.5$ to 1.0). The coercive force after the application of a high field is 0.5, and the resistivity is 17 microhm-cm. After slow cooling this alloy has been used in transformers when very low modulation is required. It is subject to magnetic shock and after application of a strong field must be demagnetized in order to restore its original properties in weak fields.

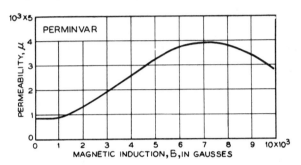

FIG. 5-99. Typical permeability curve of 7-70 Perminvar (23% Fe, 7% Co, 70% Ni).

Of the ten *elements added to the Perminvars* by Elmen and his colleagues at the Bell Laboratories the most important is molybdenum, and the composition chosen [35E1] for applications is 7.5% molybdenum, 45% nickel, 25% cobalt and the rest iron (designated 7.5-45-25 Mo-Perminvar). This material has a resistivity of 80 microhm-cm and has been used in an experimental deep-sea telephone cable. After baking at 425°C for 24 hours it has an initial permeability of 420, and this increases only 1% when the field strength is 1 oersted. When furnace-cooled from 1100°C, $\mu_0 = 100$, $\mu_m = 4500$, $H_c = 0.5$, and $B_r = 7500$. When quenched from 600°C, $\mu_0 = 3500$.

Gumlich, Steinhaus, Kussmann, and Scharnow [28G1] measured μ_0 and H_c of two alloys containing 50% nickel, 8% cobalt, and 2 and 4% manganese. The highest initial permeability was 2060 and the lowest coercive force 0.12 oersted, obtained after rapid cooling.

To summarize the names of various magnetic alloys containing iron, nickel, and other elements, Table 1 of Appendix 4 has been prepared. This lists the trade names and the names of the manufacturers and gives approximate compositions. The following Table 2 of Appendix 4 summarizes the properties of many of the useful iron-nickel and other alloys.

OTHER IRON-NICKEL ALLOYS

Iron-Nickel-Manganese Alloys.—Little is known of the structure of these ternary alloys. Binary alloys of iron and manganese [36H1] are complicated by the two-phase transformations in each of the elements in the solid state. In the iron-rich alloys, manganese behaves somewhat as nickel does in depressing rapidly the α,γ transformation. Nickel-manganese alloys form one superstructure, Ni_3Mn [40T1], and perhaps another, NiMn, about which hardly anything is known. Both nickel and manganese increase tremendously the sluggishness of the transformations of

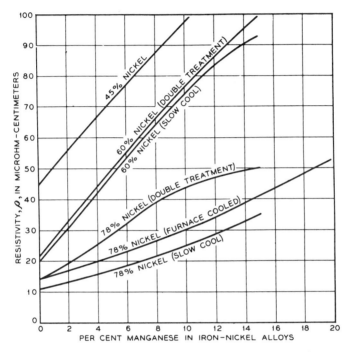

Fig. 5-100. Electrical resistivity of some Fe-Ni-Mn alloys.

their iron-rich alloys in the solid state and make the determination of the phase diagram difficult.

Two magnetic alloys have been made commercially (see Fig. 5-103), but it is doubtful if these have been used to any considerable extent. There are no important applications of other alloys of this system, apart from the non-magnetic austenitic manganese steels containing 5-15% manganese and about 1% carbon and sometimes 3-5% nickel.

The alloys lying between Ni_3Mn and Ni_3Fe in composition have been carefully studied by Kaya, Nakayama, and Sato [43K13], who measured resistivity, saturation, and specific heat after quenching from various temperatures and after aging to produce the maximum amount of ordering. In the aged alloys the saturation magnetization is higher ($B_s \approx 13\,000$) in the ternary alloy containing about 10% manganese than it is in either

182 IRON-NICKEL ALLOYS

Ni$_3$Fe ($B_s \approx 12\,100$) or Ni$_3$Mn ($B_s \approx 11\,300$). The energy of ordering, as measured by the area under the specific heat curve, also shows a maximum at 10–15% manganese. The Curie point does not change much with composition along the line Ni$_3$Fe–Ni$_3$Mn; Curie points of the other ternary alloys have not been investigated.

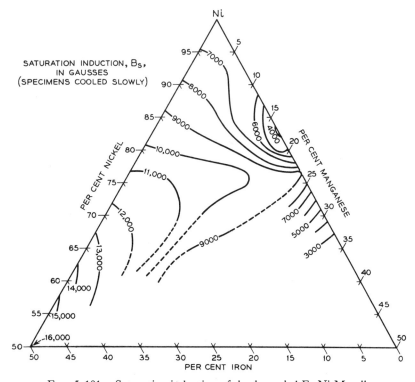

FIG. 5–101. Saturation induction of slowly cooled Fe-Ni-Mn alloys.

The *resistivities* of iron-nickel alloys are increased considerably by addition of manganese. As shown in Fig. 5–100, the resistivities of those containing 78% nickel are especially susceptible to heat treatment. Triangular diagrams of resistivities of alloys containing 50–100% nickel and 0–25% manganese have been constructed by von Auwers [36A1] from the data of others [33K4]. Changes of resistivity and Brinell hardness, upon annealing at various temperatures after quenching, have been reported by Dahl [32D1] for alloys containing 74% nickel and 0.5–12% manganese; the greatest effect is observed at about 500°C, which is near the order-disorder transformation temperature.

The *saturation intensity of magnetization*, expressed as $B - H$, is shown for slowly cooled alloys in Fig. 5–101, drawn from data of Kussmann,

Scharnow, and Steinhaus [33K4]. The effect of quenching on the saturation of alloys having 78% nickel is shown in the next figure; the effect of heat treatment is marked in the range of composition shown, on account of the variable amount of atomic ordering.

There is not good agreement as to the *initial permeabilities* of rapidly cooled alloys containing 78% nickel. The curve for slowly cooled alloys has the same general shape as that for chromium and molybdenum alloys. The *coercive force* decreases continually with addition of manganese.

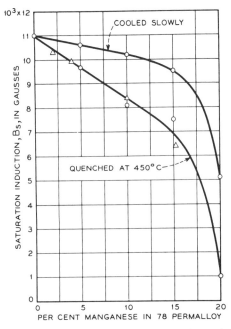

FIG. 5–102. Effect of cooling rate on the saturation induction of some Fe-Ni-Mn alloys. (Circles, Kussmann, Scharnow, and Steinhaus; triangles, G. A. Kelsall, unpublished).

Permeability curves [30G2] for the commercial alloys *Megaperm* 4510 (45% nickel) and *Megaperm* 6510 (65% nickel), each having 10% manganese, are given in Fig. 5–103. For these alloys, the resistivities are respectively 97 and 58 $\times$ 10^{-6} ohm-cm. Loss data have been published by Keinath [31K5].

Iron-Nickel-Aluminum Alloys.—Aluminum as an addition to iron-nickel alloys is valuable chiefly in the permanent magnet alloys discovered by Mishima [32M1] and containing approximately 25% nickel and 13% aluminum. But additions of 5% or more of aluminum have been made to iron-nickel alloys of the Isoperm type [33D2] in attempting to improve loading coil cores, and some silicon has also been added for the same purpose. Both elements have been added to 78 Permalloy to study their effects on the magnetic properties and resistivity. There are no non-magnetic alloys of these systems that have found important applications.

Because of its close relation to the permanent magnet alloys, the *phase diagram* of the ternary Fe-Ni-Al alloys is shown in Chap. 9. The structure is more complex than was commonly supposed before the X-ray work of Bradley and Taylor [38B1]; the complications arise from the presence

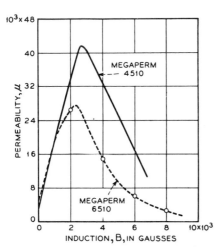 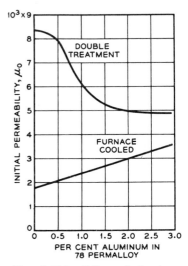

Fig. 5–103. Permeability curves of Megaperm 4510 (45% Ni, 10% Mn) and Megaperm 6510 (45% Ni, 10% Mn).

Fig. 5–104. Effect of aluminum on the initial permeability of alloys containing 78% Ni.

of superstructures in each of the binary systems. Regions in which superstructures are stable are indicated by boundary lines in Fig. 9–33. The positions of the boundaries proposed by Bradley and Taylor may have to be changed to take account of the superstructure in the binary iron-nickel alloys [39L1] which now may be regarded as definitely established in slowly cooled alloys.

The *initial permeabilities* of these alloys, obtained after the usual heat treatments, are shown in Fig. 5–104 and 105.

The effect of aluminum on various properties important in alloys of the Isoperm type is summarized in Fig. 5–106 [33D2]. The effect on these properties of various amounts of cold-rolling for the alloy containing 3% aluminum has also been studied.

Iron-Nickel-Silicon Alloys.—The phase structure of this system has been studied by Greiner and Jette [43G1], who have drawn a phase diagram for

600°C for alloys containing up to 40% of nickel and silicon. X-ray methods were used, and four phases were detected in addition to the α phase of iron.

Dahl [32D1] added 0.5–2% silicon to 74 Permalloy and measured the change in resistivity with annealing temperature after first quenching

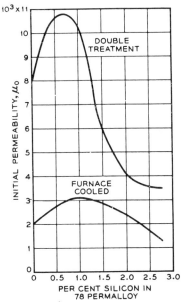
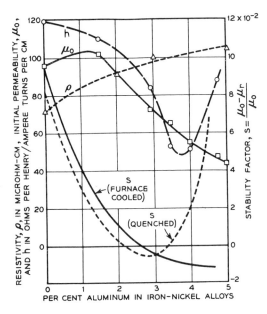

FIG. 5-105. Effect of silicon on the initial permeability of alloys containing 78% Ni.

FIG. 5-106. Some magnetic properties of Fe-Ni-Al alloys at low fields: ρ, resistivity; μ_0, initial permeability; s, instability (furnace-cooled or quenched); h, hysteresis loss coefficient. Ratio Ni/Fe = 40/60. Final treatment, cold-reduction of 90%.

from 950°C. A dip in the resistivity amounting to about 5% was observed after annealing the 1% alloy at 500°C, and a smaller effect in the 2% alloy.

An alloy called "Superpermalloy #1," containing 77% nickel, 1% silicon, 0.5% manganese, and the rest iron, is reported [36N1] to have a permeability of 12 400 at $H = 0.001$, of 14 000 at $H = 0.01$, and a resistivity of 26×10^{-6} ohm-cm. The heat treatment is annealing at 1000°C for 1 hour, cooling slowly, then quenching in air from 580°C.

In *Sinimax* 3–3.5% silicon is combined with 43% nickel (remainder iron) to increase the resistivity of the binary alloy. Otherwise its properties are much like those of 45 Permalloy.

Dahl, Pfaffenberger, and Sprung [33D2, 33P2] carried out exploratory experiments on the addition of silicon to alloys containing 40 and 71% nickel.

In Japanese publications, T. Yamamoto [47Y2] reports the study of iron-rich alloys containing 0–25% nickel and 0–20% silicon, with special reference to the alloy containing 73% Fe, 15% Ni, and 12% Si, which has been named Senperm. These alloys have high permeability, respond to

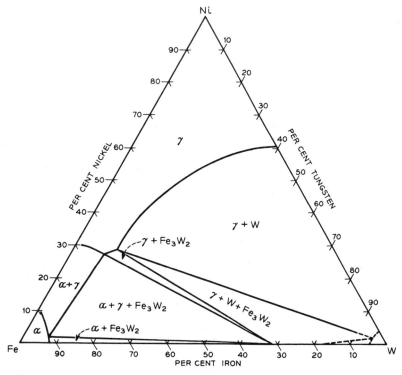

Fig. 5–107. Phase diagram of Fe-Ni-W system near room temperature.

heat treatment in a magnetic field, and on prolonged heat treatment at a relatively low temperature (about 500°C) show the Perminvar characteristics of constant permeability and constricted hysteresis loop. These properties are believed to be associated with the formation of superstructure.

Iron-Nickel Alloys Containing Titanium, Vanadium, Tantalum, or Tungsten.—The *structures* and magnetic properties of these systems have not been extensively studied. The structure of the iron-nickel-tungsten system is best known and the diagram, due chiefly to Winkler and Vogel [32W1], is reproduced in Fig. 5–107. However, there are but few available data on magnetic and physical properties—the most important of these are included in Table 6.

A schematic diagram of the *iron-nickel-vanadium* system has been proposed by Marsh [38M2]; this is based on data taken by Störmer and reported by Kühlewein [34K7], and on the binary systems iron-vanadium

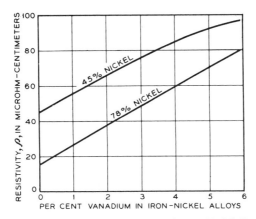

FIG. 5–108. Electrical resistivity of some Fe-Ni-V alloys.

[30W1] and nickel-vanadium [15G2]. Data relating to the structure of *iron-nickel-titanium* alloys are practically non-existent. Meager data on

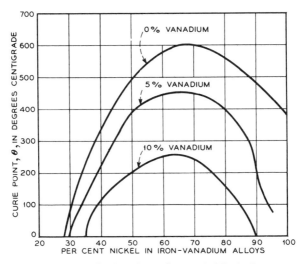

FIG. 5–109. Curie points of some Fe-Ni-V alloys.

the iron-rich and nickel-rich binary alloys have been reported [36H1]. The γ loop of iron is closed by about 1% titanium, and the Curie point is lowered to about 690°C by 2–20% titanium. Dahl and Pfaffenberger [34D1] added 1–4% to alloys of the Isoperm type.

TABLE 6. MAGNETIC PROPERTIES OF SOME IRON-NICKEL ALLOYS CONTAINING
VANADIUM, TUNGSTEN, TITANIUM OR TANTALUM

Composition (%)	ρ (microhm-cm)	$B_H = 50$	μ_0	μ_m	Source
65 Ni, 6 V	90		1 140		[30E3]
45 Ni, 6 V	91		630		[30E3]
80 Ni, 7 V	92		12 000	38 000	[34K7]
65 Ni, 6 W	51		1 855		[30E3]
78 Ni, 2 W	25	10 100	6 100	37 000	*
78 Ni, 0.4 Ti	21	10 700	9 200		*
78 Ni, 1.5 Ta	22	10 400	1 700	35 000	*

* Unpublished work carried out by Elmen, Kelsall and others at the Bell Laboratories.

A rather extensive investigation of the magnetic properties of alloys containing up to 20% vanadium has been reported by Kühlewein [34K7]. These data together with some from the Bell Laboratories make possible

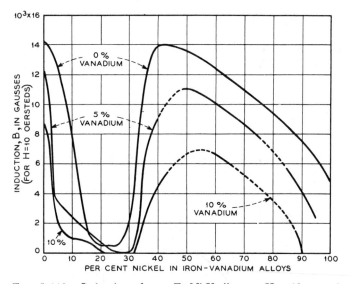

FIG. 5–110.   Induction of some Fe-Ni-V alloys at $H = 10$ oersteds.

Figs. 5–108 to 110, showing the *resistivities*, *Curie points*, and *saturation magnetizations*. Kühlewein's alloys were made from Swedish charcoal iron, Mond nickel, and commercial ferro-vanadium, were rolled to 0.35–0.50 mm sheets, and specimens were annealed usually for 1 hour at 900°C in hydrogen. In addition to the data here reproduced, the original paper records densities determined from X-ray data and some other measurements on alloys containing up to 40% vanadium.

Initial permeabilities are affected by heat treatment [34K7] in the same general manner as those of other ternary alloys. The variation of *coercive force* with nickel content [34K7] is given in Fig. 5–111 for alloys containing 5 and 10% vanadium.

The magnetic properties of some alloys containing vanadium, tungsten, titanium, or tantalum are recorded in Table 6. None of these alloys has been found to be especially interesting or useful. Yensen [20Y1] added up to 1% of titanium to various iron-nickel melts but did not determine how much remained unoxidized in the final alloy.

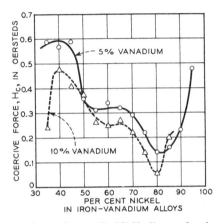

FIG. 5–111. Coercive force of some Fe-Ni-V alloys after heat treatment at 900°C.

Alloys Containing Silver or Beryllium.—Silver and beryllium are each but slightly soluble in iron-nickel alloys; consequently, small additions of each make the alloy precipitation hardenable. Kroll [29K2] experimented with a low nickel alloy containing 1.3% beryllium. Preisach [35P4] applied various heat treatments to 55 Permalloy containing 0.5% beryllium. Dahl and Pfaffenberger [34D1] added beryllium as well as copper and aluminum in their experiments on the Isoperms. Dahl [36D1] later studied age-hardening in an alloy (39% Fe, 1.5% Be) by measuring resistivity and Brinell hardness after various heat treatments.

Gumlich, Steinhaus, Kussmann, and Scharnow [30G2] added up to 5% silver to alloys with 67–79% nickel and measured the initial permeability and coercive force after different rates of cooling. Little of importance has been found in any of these investigations.

CHAPTER 6

IRON–COBALT ALLOYS

Perhaps the most interesting property of the iron-cobalt alloys is the high saturation induction of 24 300, a value 13% above that of pure iron at room temperature. This was first reported by Preuss [12P1] and Weiss [12W1] in 1912 and refers to the alloy of approximate composition Fe_2Co. The alloy containing 50% of each element was found later to have higher permeabilities at high flux densities and a saturation only slightly lower; this more useful alloy was first reported by Ellis [27E2] in 1927 but had been invented in the previous year by Elmen [29E4] and named by him "Permendur." The addition of vanadium to make an alloy that could be cold worked was patented in 1932 [32W2] by White and Wahl. Superstructure in the iron-cobalt and iron-cobalt-vanadium alloys was first definitely observed in 1939 [39R1, 41E1, 41E2]. Except for magnetic purposes there are no important uses of these alloys.

Structure and Physical Properties.—Since the publication of Hanson's summary [36H1] of binary alloy systems, the *phase diagram* has been studied by Ellis and Greiner [41E1]. Figure 2–2 shows their results combined with previous data. The order-disorder transformation temperature has a maximum, 730°C, near 50% cobalt, the composition most useful from a magnetic standpoint. No order-disorder transformations were observed in alloys containing less than 30% or more than 70% cobalt, or in any of the alloys when cooled rapidly from 800°C or above.

The *Curie point* in the α phase rises with increasing cobalt content and intersects the $\alpha + \gamma$ region near 17% cobalt. The form of the magnetization vs temperature curves of alloys containing 20–75% cobalt (see Fig. 6–6 below) indicates that "virtual" Curie points of the α alloys lie above the α,γ transition temperatures. Forrer [30F2] extrapolated these B_s vs T curves and determined the virtual Curie points (called by him the ferromagnetic Curie points), which we may define as the temperatures at which the alloys would lose their ferromagnetism if they had not transformed from the body-centered (α) to the face-centered (γ) structure.

The *densities* of the alloys, plotted in Fig. 6–1, are based on direct determinations by Weiss and Forrer [29W1] and on calculations from measurements of lattice spacings by X-rays carried out by Ellis and Greiner [41E1]. These lattice spacings of alloys containing up to 85% cobalt are given in Fig. 6-2, and show a definite expansion of the lattice on ordering, whereas most alloys show a change in the opposite direction. Coefficients of expansion are shown also in Fig. 6-1, according to Fine and Ellis [48F1].

The *resistivities* at various temperatures and the temperature coefficients at 20°C are given in Fig. 6-3. Data are by Kussmann, Scharnow, and

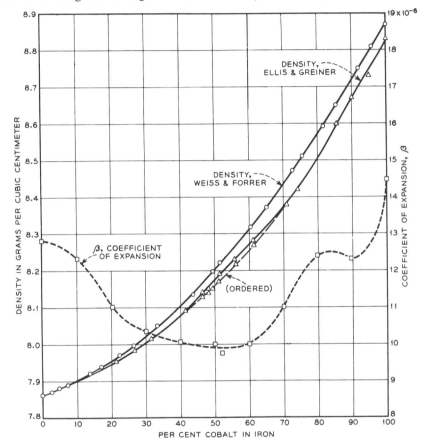

Fig. 6-1. Densities and coefficients of expansion of iron-cobalt alloys.

Schulze [32K3] and by various experimenters at the Bell Laboratories. The minimum near 50% cobalt had been observed previously at room temperature [19H1], and Ellis [27E2] showed that it persisted to temperatures as high as 900°C. Since the detection of superstructure in this alloy, it is reasonable to ascribe this minimum to ordering, which is known to lower the resistivity of other alloys; however, it is difficult to explain on this basis the continued existence of the minimum at temperatures over 100°C above the order-disorder transformation. The persistence of short-range order above the transformation point hardly seems adequate to account for the low values of the resistivity at temperatures as high as 900°C. There are no data available on the variation of resistivity with rate of cooling in the temperature range in which the α phase is stable.

Shirakawa [39S3] has determined the effect of magnetic fields up to $H = 1600$ on the resistance of various iron-cobalt alloys at temperatures ranging from $-195-1150°C$.

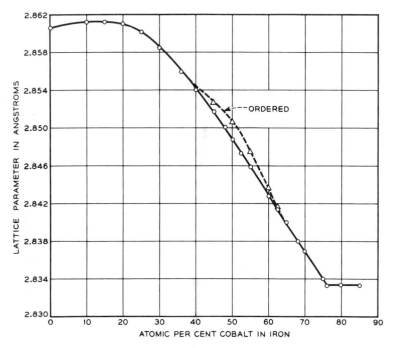

FIG. 6–2. Lattice constants of iron-cobalt alloys.

Some data on the moduli of *elasticity* and *rigidity* and on *thermal conductivity* have been reported by Honda [19H1].

Care must be exercised in the *fabrication* of the iron-cobalt alloys that have useful magnetic properties. The early experiments were made on ellipsoids turned from the castings. Ellis [27E2] was successful in hot forging and swaging the whole series of alloys and found that after such treatment the alloys containing 0–30% and 80–100% cobalt could be cold drawn to wire. Elmen in his investigation used $\frac{1}{8}$-in. rods hot swaged from ingots made of ingot iron and cobalt reduced from the oxide with carbon (rondelles). For proper reduction the sulfur content of the alloy must be kept below about 0.05% to prevent hot shortness, and in

Elmen's alloys it was about 0.02%. The carbon and silicon were about 0.04% and 0.08%, respectively, and manganese 0.3–0.4%. Alloys so prepared may be hot rolled, and good results are obtained by rolling first at 1000°C or above (in the γ phase) and finishing by a reduction in thickness of 50% at a temperature below 950°C (in the α phase).

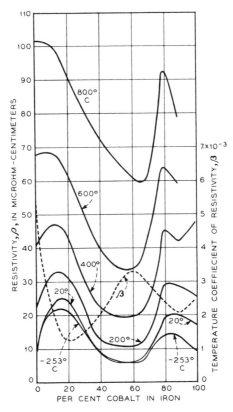

FIG. 6–3. Electrical resistivity and its temperature coefficient for iron-cobalt alloys.

Magnetic Properties.—These have been measured in high fields by Preuss [12P1] and by Weiss and Forrer [29W1] at temperatures ranging from 90°K to the Curie points, and in fields as high as $H = 17\,000$. Saturation at absolute zero was obtained by extrapolation, and the values [29W1], when plotted, lie on the upper curve of Fig. 6–4. Here the original results have been presented after conversion of magnetic moment per gram to ferric induction, using the density obtained by extrapolation to 0°K from room temperature (this amounts to an increase of about 0.05 g/cm^3). The maximum of 24 600, lying at about 35% cobalt, is the highest for any magnetic material. The saturation in Bohr magnetons is given in Fig. 6–5.

The second highest curve [29W1] in Fig. 6–4 shows $B-H$ at room temperature when $H = 17\,000$, and is substantially the B_s curve. Here the Fe$_2$Co alloy shows itself to have the highest saturation of any material at room temperature. In addition to the measurements recorded here, there have been several other reports of the saturation of this alloy at room temperature [15Y1, 19H1, 29M1, 32K3], and notable among these is that

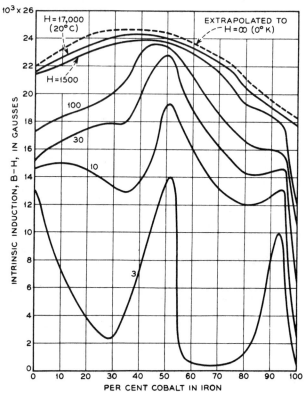

FIG. 6–4. Intrinsic induction of annealed iron-cobalt alloys at various field-strengths.

of E. H. Williams [15W1], who measured 25 800 in a specimen that had been vacuum-melted and hot-forged. A specimen sent to him by P. Weiss was remelted and forged and the saturation increased from 22 000 to 25 600. For these two specimens the same method of measurement was used so that the comparison would be more definite. The values so determined may be compared with the value of 24 300 shown in Fig. 6–4 and reported by Weiss and Forrer [29W1], who cast their specimens at atmospheric pressure and measured them in more intense fields. However, no confirmation of the high result has been given.

Other curves in Fig. 6–4 give $B-H$ for various values of H according

MAGNETIC PROPERTIES 195

to Elmen and others at the Bell Laboratories; curves of similar shape may also be drawn from data given by Masumoto [29M1]. The effect of the γ,ϵ transformation, occurring at about 95% cobalt and 200°C, is easily

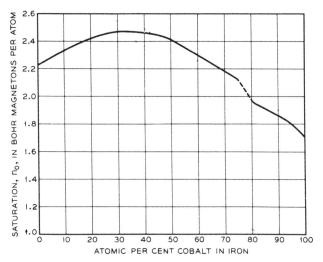

FIG. 6–5. Bohr magneton numbers of iron-cobalt alloys.

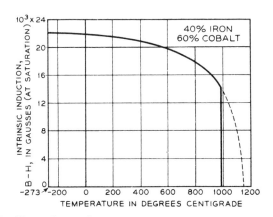

FIG. 6–6. Dependence of saturation induction B_s on temperature.

discernible in the curves for $H = 1500$ and less. The prominence of the maximum near 50% cobalt is apparent; its shift from the 35% position occurs at about $H = 100$ and persists to the lowest field strengths.

The effect of *temperature* on the magnetization in high fields was investigated in the original work of Preuss [12P1, 29W2] and is illustrated in Fig. 6–6, which shows also Forrer's [30F2] extrapolation to the virtual Curie point. The sharp drop in saturation at the α,γ transformation point, 980°C for the 60% cobalt alloy, is plainly visible in the curves for alloys containing 30–70% cobalt.

In the 50% alloy, Forrer [31F1] has observed a peculiar effect of temperature that is probably connected with the order-disorder transformation. He measured the coercive force as dependent on temperature while heating and cooling rapidly, and again while going through the same temperature

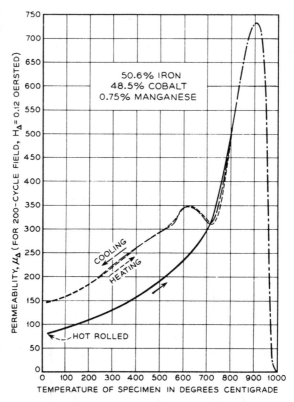

FIG. 6–7. Incremental permeability of an iron-cobalt alloy (50.6% Fe, 48.5% Co, 0.8% Mn) as dependent on temperature, showing order-disorder transformation at about 720°C.

cycle slowly, and found a marked difference in the behavior at temperatures between 500 and 800°C. The time necessary to establish equilibrium in the order-disorder transformation will possibly account also for the magnetic viscosity of this alloy observed by Weiss and Freudenreich [16W1].

More definite location of the order-disorder transformation temperature by magnetic means has been made by J. A. Ashworth (previously unpublished) in an alloy 0.10 in. thick containing 48.5% cobalt and 0.7% manganese. Figure 6–7 shows the permeability he measured in a Kelsall permeameter [24K1] using a field strength of 0.12 oersted alternating at 200 cps. The specimen was originally hot rolled to size and then, during measurement, was heated to 830°C, cooled to room temperature, and reheated to 830°C. The hot rolling presumably caused pronounced disordering which was maintained during the rather rapid cooling that took place after the rolling. During the slower cooling that occurred during measurement, ordering set it at 720°C, and the permeability rose as the temperature fell. The same curve-path was followed during the

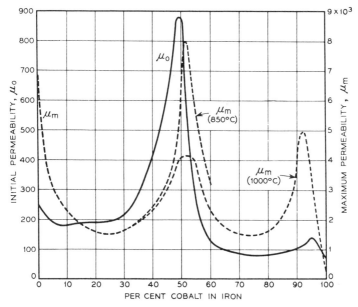

FIG. 6–8. Initial and maximum permeabilities of iron-cobalt alloys. Near 50% Co the higher μ_m is found for the lower annealing temperature.

next slow heating and cooling cycle. Variations of μ_0, H_c, and B_r with temperature have also been studied by Kühlewein [32K2].

Elmen [29E4] was the first to measure the *initial permeabilities* of the iron-cobalt alloys and to find them greater than that of iron. More recent measurements at the Bell Laboratories show the pronounced high values of both initial and *maximum permeabilities* of the 50% alloy (Permendur) after it has been annealed at either 850 or 1000°C (Fig. 6–8). Permeability as dependent on induction is shown in Fig. 6–9 for representative specimens of Permendur annealed at 850 and at 1000°C and cooled with the furnace. A curve for annealed ingot iron is shown for comparison. High permeabilities at high inductions are apparent.

No systematic investigation has been made to determine the best *annealing temperature* or rate of cooling for any of these alloys. It has been established that 850 and 950°C are both superior to 1000°C for

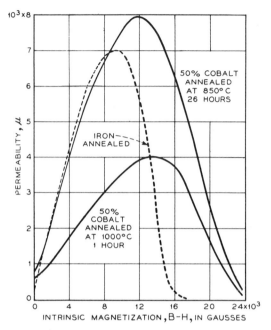

FIG. 6–9. The 50% cobalt alloy has much higher permeabilities at high fields than iron.

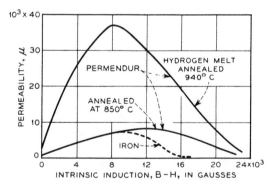

FIG. 6–10. Permeability curves of Permendur and iron.

producing high values of μ_0 and μ_m in alloys containing 30–60% cobalt. Yensen [15Y1] found 900°C a better annealing temperature than 1100°C.

Some experiments on material melted in hydrogen have been carried out by Cioffi [35E1]. Using ingots made from cobalt rondelles and electrolytic iron, his specimens were hot rolled to sheet and then annealed for 18 hours at 940° in hydrogen; he found $\mu_m = 37\,000$, $\mu_0 = 1000$ to 1400, $H_c = 0.20$, and $\mu = 5000$ at $B = 21\,000$. In Fig. 6–10 his results are compared with iron and with Permendur made and heat-treated in the usual way.

The maximum permeability is increased and the coercive force reduced by *heat treatment in a magnetic field*. This has already been discussed in Chap. 5, and some data are included in Table 5 of that chapter (p. 175). Other data have been obtained by Libsch, Both, Beckman, Warren, and Franklin [50L1], who have prepared specimens by the methods of powder metallurgy and have been interested in obtaining material with a rectangular hysteresis loop. In the 50% alloy they measured $B_s = 22\,400$, $B_r = 19\,000$, $H_c = 0.68$, and a sudden change of induction on the steep part of the loop of $\Delta B \approx 35\,000$.

Permendur is useful in the magnetic circuits of electromagnets and permanent magnets when the induction is higher than 10 000 to 15 000. Although the quantity of material so used has not been large, the reduction in the size and weight of apparatus that can be accomplished by its use has had an important effect on technical developments.

IRON-COBALT ALLOYS WITH ADDITIONS

The most extensive work on iron-cobalt alloys to which another element has been added has been carried out by Elmen and others at the Bell Laboratories. They have studied the effect of a few per cent of each of ten or more elements on the resistivity and magnetic properties of alloys containing about equal proportions of iron and cobalt. Most extensive investigations have been made of iron-cobalt-vanadium alloys. Some work on the constitution of alloys with molybdenum, tungsten, titanium, and others has been reported in connection with studies of materials for permanent magnets (Chap. 9).

The data on *resistivities* are given in Fig. 6–11. Alloys contained equal proportions of iron and cobalt and were annealed at 1000°C. In most of the specimens it was not known whether or not the α or γ phase predominated. In general, the increase of resistivity is more, the higher the valence of the element present in a given percentage by weight. The curves for molybdenum and tungsten begin to bend downward at 0.5–1%, and this indicates that in these alloys a solid solubility line on the phase diagram has been crossed.

The *saturation magnetization* of the same alloys was measured but not with high precision. All of the elements added caused a decrease in saturation that was linear within the accuracy of the measurements. Approximate values of B_s for alloys containing 2% of the added element are given in Table 1.

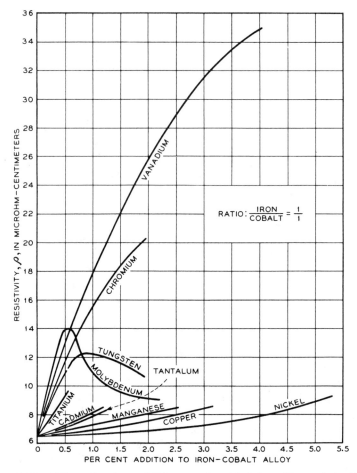

Fig. 6–11. Effect of various additions on the resistivity of iron-cobalt alloys.

Table 1. Approximate Saturation Induction, B_s, for Fe-Co Alloys Containing 2% of the Added Element. Fe and Co in Equal Amounts

Added Element	B_s	Added Element	B_s
None	24 000	Ti*	23 300
W	23 700	Nb*	23 200
Mn	23 500	V	23 000
Ni	23 500	Cr*	21 600
Mo	23 400	C*	21 600

* Extrapolated.

Iron-Cobalt-Vanadium Alloys.—From a practical point of view these are the most important ternary alloys of iron and cobalt having high permeability, and they have been investigated more thoroughly than the others. Data on the phase equilibria have been reported by Köster and Lang [38K3]. In some respects they are in disagreement with unpublished data of Greiner and Ellis, shown in the pseudo-binary diagram of Fig. 6-12,

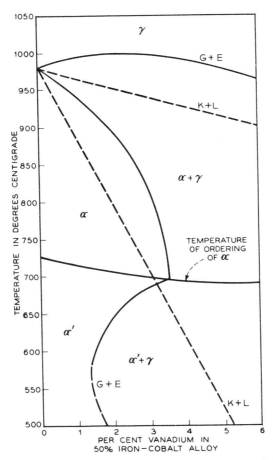

FIG. 6-12. Quasi-binary phase diagram of Fe-Co-V alloys containing Fe/Co = 1/1 G + E, Greiner and Ellis; K + L, Köster and Lang.

for alloys containing equal amounts of iron and cobalt. The latter workers have found that the temperature of ordering of the pure iron-cobalt alloy is lowered slightly from 725°C upon the addition of vanadium, and their results, from X-ray diffraction on equilibrated alloys, indicate that the solubilities of vanadium in the α (disordered) and α' (ordered) body-centered cubic phases are greatest at or near the temperature of ordering. Nesbitt, in unpublished work, found by X-rays that the cold-worked alloys having 50% cobalt and 8% vanadium, contained no γ phase at room temperature. This indicates that the α', $\alpha' + \gamma$ boundary bends toward the right at temperatures below 500°C. These investigations emphasize the need for further study of the solubility relations in this system.

The alloys containing vanadium are easier to fabricate than the binary alloys containing 30–80% cobalt. A cold-workable alloy containing 1.5–4% vanadium and about equal proportions of iron and cobalt was patented by White and Wahl [32W2] and the 2% alloy has found extensive use. Ingots

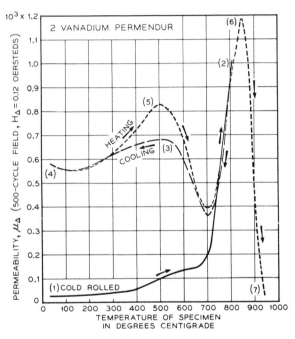

FIG. 6–13. Incremental permeability of 2V Permendur as dependent on temperature, showing order-disorder transformation at about 720°C.

are broken down at 1000–1200°C and reheated when the temperature drops below 1000°C, the α,γ point. Hot rolling is continued until the sheet is reduced to about 0.1 in., when it is annealed for a short time at about 850–1000°C and quenched rapidly in cold brine. The sheet may then be cold rolled to any reasonable thickness and, after punching, is annealed for an hour at 850°C. The rate of cooling from 850°C should be neither too fast nor too slow.

This procedure can be interpreted in terms of Greiner and Ellis' diagram of Fig. 6–12. After cold rolling and annealing, the 2% alloy is entirely in the ordered α' phase, provided cooling has been rapid enough to prevent precipitation of γ. Thus α' seems to be stable only in a narrow temperature range, the limits of which may be provisionally estimated as 700 and 650°C. Apparently the cooling below 700°C must be slow enough to effect ordering but not slow enough to permit precipitation of the γ phase.

The existence of the order-disorder temperature at about 700°C is substantiated by the curves of Fig. 6-13, recorded by Ashworth, which show the effect of temperature on the permeability (at $H = 0.12$ oersted) of the alloy containing 49% vanadium. Starting with cold rolled material (point 1), the permeability increases in a perfectly normal manner to a maximum and decreases to zero at the Curie point. Upon cooling from

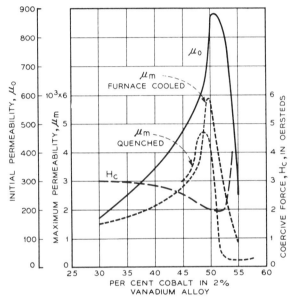

FIG. 6-14. Some magnetic properties of iron-cobalt alloys containing 2% V, heat-treated at 850-860°C and quenched or furnace-cooled.

the Curie point, or from 810°C (point 2), the curve departs from the normal course at about 700°C and then passes through a second maximum (point 3); upon reheating again from room temperature (points 4, 5, 6, and 7) this critical point is again evident. Apparently the ordering of this alloy occurs at 700-730°C and raises its permeability in weak fields.

The effect of the cobalt content on the permeability (μ_0 and μ_m) and coercive force of alloys containing 2% vanadium is given in Fig. 6-14. The maximum permeability is not very sensitive to the rate of cooling from the annealing temperature of 850°C, as the two curves show. The maximum near 50% cobalt is evident. The optimum annealing temperature of 850°C is established by the data of Fig. 6-15; apparently this temperature is optimum because precipitation of γ phase occurs at slightly higher temperatures. The properties are badly deteriorated by annealing the specimen at 1000°C, especially if the specimen is rapidly cooled.

As vanadium is added to the binary iron-cobalt alloy, the magnetic properties deteriorate continually. In the alloys containing iron and cobalt in equal proportions the addition of 2% vanadium causes, as nearly

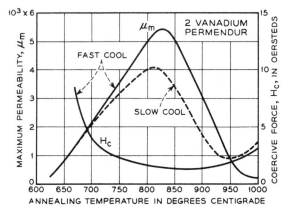

FIG. 6-15. Effect of annealing temperature on some magnetic properties of 2V Permendur.

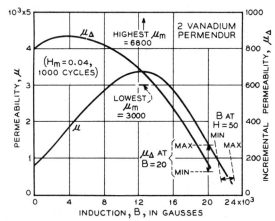

FIG. 6-16. Typical normal and incremental permeability curves of 2V Permendur (49% Fe, 49% Co, 2% V).

as can be determined, the following changes in the more important properties: in μ_0, a decrease of 25%; μ_m, decrease of 50%; H_c, increase of 10%; B_s, decrease of 4% (to 23 000). The resistivity and proportional limit increase somewhat. Selection of 2% as the vanadium content of the commercial alloy, known as *2V Permendur*, is a compromise between the gradual deterioration of the magnetic properties, and the gain in ease of fabrication and in mechanical properties and resistivity as the vanadium content is increased. As used in receiver diaphragms, one of the most important of its applications, the alloy must have a high normal permeability and a high incremental permeability at a high induction, as well as suitable mechanical properties. Representative permeability data are reproduced in Fig. 6–16, and several hysteresis loops are shown in Fig. 6–17. In electromagnets of the more conventional types only the normal permeability at high inductions is of considerable importance. In this

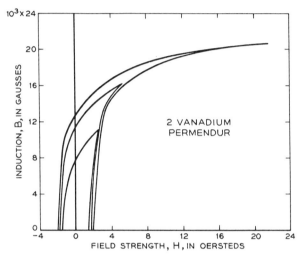

FIG. 6–17. Typical hysteresis loops of 2V Permendur.

respect 2V Permendur is not as good as the binary alloy, but its ease of fabrication, accomplished as outlined previously, often makes it the most suitable material.

Other Iron-Cobalt Alloys.—Several other ternary iron-cobalt alloys have been investigated by Köster and his co-workers, who were especially interested in precipitation-hardenable alloys for permanent magnets. Tentative phase diagrams covering all compositions have been published for systems with chromium [32K6, 35K5], molybdenum [32K4, 32K8], tungsten [32K8, 32K5], tantalum [39K4], manganese [33K6, 35K5], antimony [39G3], and part of the field has been mapped for alloys with beryllium [39K5] and aluminum [33K5]. Alloys of about 1 kg were melted in a high-frequency furnace, using raw materials containing less than 0.05% carbon, often with the addition of 0.5% manganese. Heating and cooling curves were determined, and the arrests of the former used. Microscopic and occasionally dilatometric and resistivity data were also employed.

The *iron-cobalt-molybdenum* diagram [32K4, 32K8], reproduced in Fig. 6–18, shows the boundaries as determined for 20°C (solid lines) and for 1300°C (broken lines). This single-phase diagram is reproduced as an example of the results obtained by this method for a number of the ternary alloys. The type of diagram is much the same, the phases present being α, γ, θ, and ϵ, although the last is not shown in Fig. 6–18. Areas marked α, θ, and $\alpha + \gamma + \theta$ are long and narrow, and as the temperature increases the $\alpha + \gamma + \theta$ and α regions swing to the left (see dotted lines) and α eventually disappears. After quenching from 1100°C the alloys were

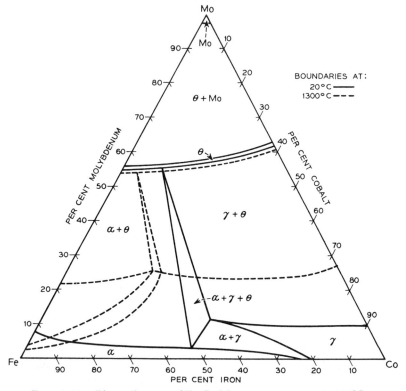

FIG. 6–18. Phase diagram of Fe-Co-Mo system at 20 and 1300°C.

aged at various temperatures up to 800°C and measurements made of Brinell hardness, magnetic saturation, residual induction, and coercive force. These alloys are properly considered in more detail in a discussion of materials for permanent magnets.

In the *iron-cobalt-chromium* alloys, the phase diagram [32K6] and the resistivities [35K5] of the α-phase alloys have been plotted, and the Brinell hardness, B_s, B_r, and H_s have been determined for several alloys as dependent on heat treatment. At 50% cobalt about 1% of chromium is held in solid solution in the α phase at room temperature.

Masumoto [34M4] measured the thermal expansion of alloys containing up to 20% chromium and found a minimum at 9% chromium and 37% iron, an alloy he has named *Stainless Invar* on account of its high corrosion resistance. He also has published curves showing the induction and magnetostriction as a function of field strength. Magnetostrictive expansion is 6×10^{-6} at $H = 1400$.

Stanley and Yensen [47S5, 48S14] have added chromium to the 35% cobalt-iron alloy having high saturation and have been able to make a ductile material which can be rolled to thin sheets and which, after annealing, has high permeability at high inductions. Applications of this material and of Vanadium Permendur are similar. The composition is 35% cobalt, 0.5% chromium, and the rest iron, and it is named *Hiperco*. It is prepared by melting iron and cobalt of high commercial purity, with the addition of 0.1% of carbon (to improve forgeability), and is deoxidized with silicon and titanium. The ingot is hot rolled, and the strip is quenched from 910°C and cold rolled to 0.025 in. or less—in the laboratory 0.001-in. strip has been produced. The sheet is annealed at 875–925°C, below the phase transformation at 950°C, and at the same time the carbon is reduced, preferably below 0.005%.

The following properties are obtained ($B_m = 15\,000$):

$B_s = 24\,000$ $H_c = 0.63$
$\mu_0 = 650$ $B_r = 11\,500$
$\mu_m = 10\,000$ $\rho = 20 \times 10^{-6}$ ohm-cm
Losses: 3 watts/lb at 60 cps, 0.017-in. strip
100 watts/lb at 600 cps, 0.017-in. strip

A magnetization curve and hysteresis loops are shown in Fig. 6–19. In the range $H = 10$ to $H = 400$ the permeability is 1800 to 61.

Data relating to the structure and to physical and magnetic properties have been reported for iron-cobalt alloys containing the following elements:

Aluminum.—Alloys containing less than 50 atomic per cent aluminum and quenched from 800°C have been investigated by Edwards [41E2] by means of X rays. The single-phase, body-centered cubic α alloys extend over a large region from pure iron to 70% cobalt and to 30% aluminum, and include the composition CoAl (31.4% Al). Fe_3Al, FeAl, CoAl, and FeCo have previously been shown to have superstructures. In the two-phase region between the α and the face-centered γ alloys (including pure cobalt) is a new face-centered cubic phase, α'. Köster [33K5] determined the Curie points and found them to decrease rapidly with addition of this element; alloys become non-magnetic at room temperature at about 20% aluminum on the iron side and 27% on the cobalt side.

Antimony.—Geller [39G3] has drawn triangular diagrams, based on thermal, dilatometric, and microscopic data, for liquidus and solidus and for phases present at 20°C. No Curie points or magnetic data are reported.

Beryllium.—The phase diagram [39K5] extends to CoBe and FeBe$_2$. The saturation and the Curie points drop rapidly with addition of beryllium, e.g., B_s drops from 20 000 to 13 000 when the beryllium content of the 25% cobalt alloy is increased from 0 to 4%. Somewhat different values of B_s are obtained in alloys quenched from 1100°C and in those aged at 700–800°C.

Manganese.—Köster [33K6] has investigated these alloys also, and has traced the α,γ and γ,ϵ (hexagonal) boundaries and plotted resistivities

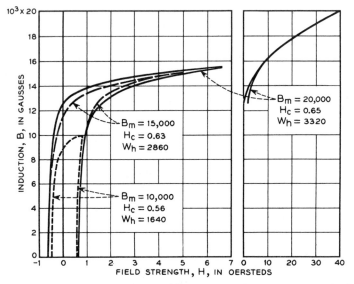

FIG. 6–19. Hysteresis loops of Hiperco (64% Fe, 35% Co, 0.5% Cr).

[35K5] of alloys containing up to 50% manganese. X-ray analysis was used to determine lattice parameters and to aid in locating boundaries. A few magnetic data, including Curie points, are given also. White and Wahl [32W2] have found that manganese, like vanadium, makes a cold-workable alloy when a few per cent are added to the 50% binary alloy.

Tantalum.—The ternary diagram is similar to that of the iron-cobalt-molybdenum system; according to Köster and Becker [39K4] the corner of the $\alpha + \gamma + \theta$ region corresponding to the least amount of tantalum lies at about 22% iron. No specific magnetic data are given.

Tungsten.—This system [32K5, 32K8] is also similar to that of iron-cobalt-molybdenum, and phase boundaries and connodes have been drawn for 20 and 1300°C. The low-tungsten corner of the $\alpha + \gamma + \theta$ region lies at about 38% cobalt. Some data are given for hardness, resistivity, density, B_s, B_r, and H_c. Measurements of B_s, B_r, H_c, density, conductivity, hardness, and Young's modulus have been reported for 20 alloys by Rogers [33R3], but, like Köster's data, his results are of special significance for studies of permanent magnetism and age-hardening.

The best-known magnetic iron-cobalt alloys are listed in Tables 1 and 2 of Appendix 4.

CHAPTER 7

OTHER IRON ALLOYS OF HIGH PERMEABILITY

Except for the alloys of iron with silicon, nickel, or cobalt, no binary alloys of iron have high permeabilities that are of great technical use. The iron-aluminum alloys have been considered for commercial use and have been rejected. The iron-carbon alloys have been of importance in permanent magnets, and their complex alloys are still used (see Chap. 9).

In this chapter a brief description is given of the best-known binary systems; the phase structure, physical properties, and magnetic properties are considered and references are given for further investigation. Generally the Curie point and saturation magnetization of iron are both lowered by addition of another element, but vanadium raises the Curie point by about 45°C. Precipitation of a second phase occurs almost invariably when a sufficient amount of the second element has been added, and it is often important to know at what composition this first occurs and how the critical composition changes with temperature. The precipitating phase is generally non-magnetic, but some compounds and ordered structures are known to be ferromagnetic. Those that have been reported as ferromagnetic are:

Fe_3Al Fe_2Ce
Fe_2B Fe_3P
$FeBe_2$ FeS
$FeBe_5$ $FeSn_2$

A systematic study of the change of Curie point and saturation magnetization with composition has been made for a number of alloys by Fallot [36F1]. A comprehensive résumé of the magnetic properties of iron alloys has been published by von Auwers [36A1]. Haughton has compiled a bibliography of papers relating to phase diagrams [42H2, 44H4].

Iron-Aluminum Alloys.—The magnetic properties of these alloys were investigated as long ago as 1900, when Barrett, Brown, and Hadfield [00B1] studied three specimens containing as much as 5.5% aluminum. They were considered in the construction of transformer cores at the same time as iron-silicon alloys, but were abandoned in favor of the latter, apparently because of their higher cost and greater difficulty of manufacture.

Later investigations showed that in spite of certain advantages—greater ductility and resistivity—the disadvantages of these alloys are controlling for the commercial production of transformer sheet. When the hot metal is exposed to air the aluminum oxidizes with great rapidity, and the hardness of the oxide causes undesirable wear on machinery used for fabrication. Limited use may yet be found for material melted in a controlled atmosphere and annealed after being reduced substantially by cold rolling or drawing.

The film of oxide formed on the hot metal offers moderate protection from further oxidation, and the alloys are used for electrical heating elements and as heat-resistant alloys [32Z1], usually with the addition of chromium.

The *phase diagram* of Fig. 7-1 shows that the γ loop is closed at about 1.0% aluminum. The *Curie point* drops with addition of aluminum, and alloys containing more than 16–18% aluminum are no longer ferromagnetic at room temperature. It is estimated that the Curie point is reduced to 0°K at about 19–20% aluminum.

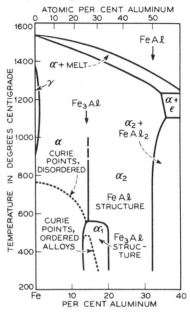

FIG. 7-1. Phase diagram of iron-aluminum alloys.

In the range of solid solubility, 0–32% aluminum, two types of *superstructure* are formed; they are designated Fe_3Al and $FeAl$ corresponding to 13.9 and 32.6% aluminum by weight. As indicated in Fig. 7-2, in the Fe_3Al structure the atoms of aluminum avoid adjacent positions and so occupy the centers of the alternate small cubes (each $\frac{1}{8}$ of the true unit of structure) that show the structure of pure iron. In the FeAl structure the center of each such cube is occupied by Al atoms so that the crystal is composed of alternate layers of Fe and Al, in (100) planes. Then each Al is surrounded first by 8 Fe atoms and then by 6 Al atoms at a somewhat greater distance. The change of atomic distribution with composition, for alloys slowly cooled or quenched, has been determined by Bradley and Jay [32B7] and is shown at the bottom of Fig. 7-2. Ordering is first detected in slowly cooled alloys at 10% (about 18 atomic per cent), and is accompanied by a slight dependence of lattice parameter on cooling rate, as shown in Fig. 7-3. The FeAl structure occurs in alloys containing 50 atomic per cent of aluminum even when they are quenched from 1000°C, and no random arrangement of atoms has been observed in alloys near this composition.

The phase diagram of Fig. 7-1 is taken from the paper by Bradley and Taylor [38B1] and is based on the earlier diagram of Ageew and Vher [30A4], the X-ray work of Bradley and Jay [32B7], and the study of transformations by Sykes and Evans [34S9, 35S8].

The effect of aluminum (like silicon) on an impurity of carbon is to force it into the graphitic form. A tentative Fe-Al-C diagram has been

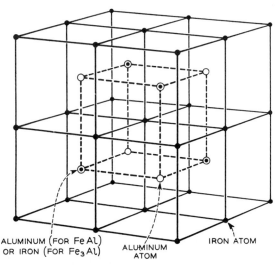

THERMAL PREPARATION	ATOMIC PER CENT ALUMINUM	POSITIONS OF ALUMINUM ATOMS
COOLED SLOWLY	0 TO 18	AT RANDOM, ●,○,◉
	18 TO 25	○ PREFERRED
	25	ALMOST ALL AT ○
	25 TO 38	MOST AT ○, SOME AT ◉
	38 TO 50	○ AND ◉ EQUALLY
	50	○ AND ◉ FILLED
QUENCHED FROM 600 – 700° C	0 TO 25	AT RANDOM, ●,○,◉
	25 TO 50	○ AND ◉ EQUALLY

FIG. 7-2. Structure of ordered Fe₃Al and Fe-Al alloys: solid circles, Fe atoms; open circles, Al atoms; circles with solid centers, Fe atoms in Fe₃Al and Al atoms in FeAl.

suggested by Morral [34M8], and some magnetic data on ternary alloys have been reported by Lönberg and Schmidt [38L3].

Densities calculated from the X-ray data of Bradley and Jay agree well with the direct determinations of Sykes and Bampfylde [34S10] and are given in Fig. 7-4. Thermal expansivity, shown in Fig. 7-5, increases only slightly with increase in aluminum content, according to Schulze [28S8]. Sykes and Evans [34S9, 35S8] found slight irregularities in the thermal expansion of alloys near the composition of Fe₃Al, at temperatures

IRON–ALUMINUM ALLOYS 213

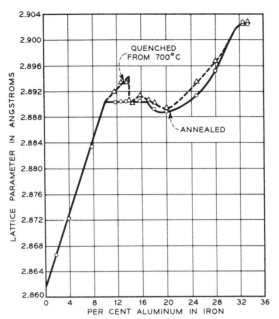

FIG. 7–3. Lattice spacing of iron-aluminum alloys in the quenched and annealed states (32B7).

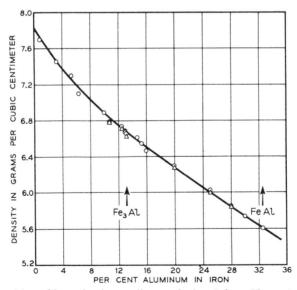

FIG. 7–4. Densities of iron-aluminum alloys, calculated from X-ray data; circles, annealed; crosses, quenched.

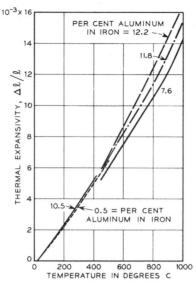

FIG. 7-5. Expansivity of three iron-aluminum alloys.

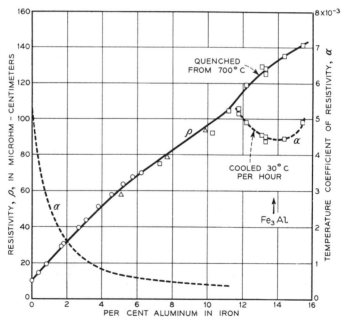

FIG. 7-6. Resistivity of iron-aluminum alloys as quenched and as cooled at 30°C/hr from 700°C. Also temperature coefficient.

between 500 and 600°C, and identified them with the order-disorder transition which lies at about 550°C.

Electrical resistance increases rapidly with aluminum content (Fig. 7-6), and at 12% aluminum and above it is sensitive to cooling rate [34S9, 35S8, 45M1], on account of atomic ordering. The temperature coefficient [18G2] drops rapidly with increase of aluminum.

Mechanical properties are important in relation to the possible commercial uses of the alloys. Yensen and Gatward [17Y1] measured tensile strength and other properties and showed that aluminum increases the

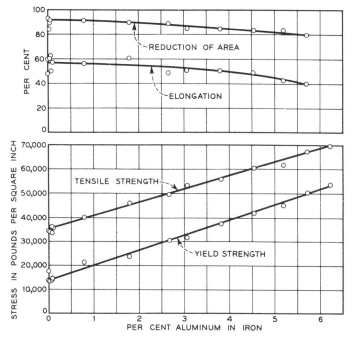

Fig. 7-7. Some mechanical properties of iron-aluminum alloys.

strength of iron in proportion to the amount added and does not increase the brittleness, at least up to 6%, when carbon and other impurities are kept low (Fig. 7-7). This lack of embrittlement is in contrast to the effect of silicon, which has a decided embrittling effect beyond 4.5%. Sykes and Bampfylde [34S10] state that iron-aluminum alloys can be cold worked up to 5% aluminum, and hot worked up to 16%. After some working of the ingot, it may be drawn to wire at 12% aluminum.

Saturation magnetization at room temperature, shown in Fig. 7-8 according to data by Fallot [36F1], decreases at first linearly with aluminum content, then changes irregularly near the composition Fe_3Al and finally drops rapidly to zero near 18%. A similar course is followed by the Bohr magneton number n_0, derived from the extrapolated value of the saturation magnetization at 0°K. The broken line in the figure indicates the change in n_0 to be expected if aluminum behaves simply as a diluent, without changing the moment of the iron atom. The data show that the moment of the iron atom begins to change markedly only when Fe_3Al ordering becomes possible. Sucksmith [39S13] found in some alloys a difference in n_0 for the annealed and quenched states. This amounts to about 3% at 12% aluminum, the quenched (unordered) alloys having the higher moments per atom.

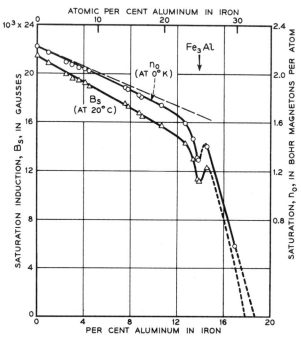

FIG. 7–8. Saturation induction at 20°C, and Bohr magneton number, of iron-aluminum alloys.

In 1902, Barrett, Brown, and Hadfield reported measurements of the permeability and hysteresis of specimens containing 2.25 and 5.50% aluminum, and found the 2.25% alloy to compare favorably with the best iron-silicon alloys then known. Their best results, and those of later investigators, are given in Table 1. The commercial use of the silicon instead of the aluminum alloys was determined partly by the current prices of these two elements, 12 cents and 2 dollars per pound, respectively.

In the investigation of Yensen and Gatward [17Y1] higher maximum permeability and lower hysteresis were obtained. An important result of this work was the general view that in alloys containing 1–6% aluminum

TABLE 1. SOME PROPERTIES ATTAINED IN SEVERAL INVESTIGATIONS
OF FE-AL ALLOYS

Investigation	μ_m	H_c	μ for $B = 10\,000$	μ for $B = 14\,000$
Barrett, Brown, and Hadfield, 1902 [02B1]	6 000	1.0	1 700	1 000
Gumlich, 1918 [18G2]	6 000	0.8	3 000	1 200
Yensen and Gatwood, 1917 [17Y1], annealed 1100°C	20 000	0.4	1 000	(2 000)
Bozorth, Williams, and Morris, 1940 [40B4, 42B2]				
annealed 1000°C*	25 000	0.3	21 000	12 000
annealed 1330°C	20 000	0.08	300	600

* Rod measured in direction of drawing.

the magnetic properties deteriorate slowly with increasing aluminum. However, the resistivity increases markedly, and the aging, common in iron, is practically eliminated. Gumlich's results [18G2] showed that μ_0 as well as μ_m gradually decreases with increasing aluminum. This is probably associated with the high magnetostriction of these alloys.

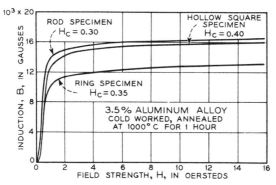

FIG. 7–9. Magnetization curves of 3.5% aluminum-iron specimens cut from sheet and rod having preferred crystal orientation produced by cold working and annealing.

In an investigation at the Bell Laboratories [40B4] it was found that directional properties could be imparted by cold rolling or cold drawing so that permeabilities at high inductions could be raised considerably over the values previously reported. This characteristic, as well as the ductility and non-aging qualities, makes it appropriate for use in electromagnetic apparatus such as relays. The curves of Fig. 7–9 show the directional properties as measured in rod and in sheet that have been cold reduced 70% in area and then annealed at 1000°C. The permeability in the sheet is high in directions parallel and at 90° to the direction of rolling; hence the high permeability can be measured in a hollow square specimen having sides parallel to these directions. In the various portions of a ring specimen the induction lies in all directions in the plane of the sheet, and the average

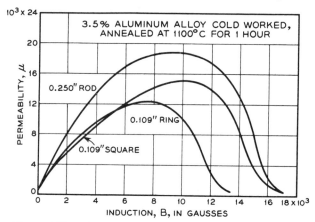

Fig. 7–10. Permeability curves of 3.5% aluminum-iron having preferred crystal orientation.

permeability is much lower, as is evident in the figure. This lower curve is about the same as that found for hot rolled sheet. Figure 7–10 gives μ vs B curves for specimens cut in the same ways. Some representative properties of material fabricated from 250-lb melts are shown in Table 2.

Table 2. Representative Properties of Some Hot Rolled and Cold Rolled Fe-Al Alloys
(Hysteresis loss in ergs/cm³ for $B_m = 10\,000$)

Treatment	μ_m	μ at $B = 14\,000$	H_c	W_h
Hot rolled sheet, annealed 1000°C.......	13 000	1600	0.4	...
Cold rolled sheet,* annealed 1000°C.....	14 000	5000	.4	1200
Cold drawn rod,* annealed 1000°C......	16 000	7000	.4	800
Cold drawn rod,* annealed 1330°C......	20 000	600	.1	350

* Measured parallel to the direction of rolling or drawing.

The effect of crystal orientation is greatest after annealing the cold rolled material at 1200°C, and it then declines with increasing annealing temperature. Measurements of this effect were made by plotting the torque acting on a disk in a magnetic field of 2000 oersteds. After cold rolling, before annealing, the directions of easy magnetization are at 45° to the rolling direction, as shown by the positions of greatest negative slope of the torque curve of Fig. 7–11. Annealing at 600–1400°C causes

IRON–ALUMINUM ALLOYS

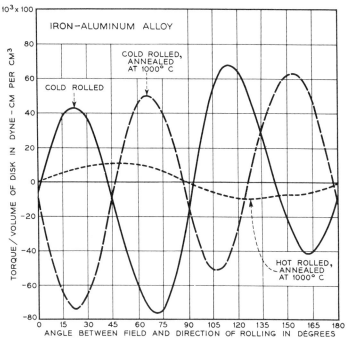

Fig. 7–11. Torque curves of aluminum-iron with various mechanical and heat treatments.

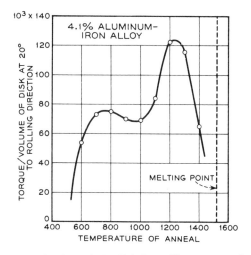

Fig. 7–12. Torque on aluminum-iron disk in uniform magnetic field. Magnetic anisotropy increases with annealing temperature to 1200°C, then decreases.

the directions of easy magnetization to lie at 0° and 90° to the direction of rolling, and these two directions are then about equally "easy" as measured by the depth of the minima of the torque curve that follow these angles. The effect of the temperature of anneal was determined by measuring the torque at 20° to the rolling direction after annealing for 1 hour at various temperatures between 600 and 1400°C. According to the results, plotted in Fig. 7–12, the special orientation rises to a maximum and then declines toward zero for annealing temperatures approaching the melting point. Special orientation may also be made small by annealing for many hours at 1300°C.

Masumoto and Saito [45M1] studied the effect of heat treatment on alloys containing up to 17% aluminum and found the maximum and initial permeabilities, and coercive force and residual induction, to be sensitive to cooling rate, especially in the neighborhood of 14–16% aluminum. The highest permeabilities were obtained with the alloy containing about 16%, and this was named *Alperm*. After the rolled material is quenched from 600°C the following properties were measured: $\mu_m = 55\,000$, $\mu_0 = 3100$, $H_c = 0.04$, $B_r = 2100$, $W_h = 41$ (for $B_m = 3000$). The related *Alfer*, containing 13% aluminum, has high magnetostriction (Chap. 13).

Iron-Antimony.—Like Fe-As, this system is reported to have a closed γ loop, extending to about 2% antimony. A eutectic with FeSb (ϵ) forms at 1000°C and contains 50% antimony. At this temperature 8–10% of antimony is soluble, and at room temperature this decreases to 6–7%. FeSb has the FeAs structure [36H1].

The Curie point appears to remain constant with antimony content in the solid solution range and in the $\alpha + \epsilon$ region.

According to an early report of Weiss [96W1], the ferric induction in a field of 500 oersteds falls from over 1000 at 50% antimony to about 10 at 60% antimony; therefore FeSb (68% Sb) may be considered non-magnetic.

Burgess and Aston [09B2], as well as Dahl, Pawlek, and Pfaffenberger [35D2], have measured losses in sheets containing about 1% antimony. The hysteresis loss is considerably greater than in iron, and the total loss is also greater in spite of the increase of resistivity with antimony content. The resistivity is about 20 microhm-cm for the 1% alloy. After annealing at 1000°C, $H_c = 4$ [09B2].

Iron-Arsenic.—This system has a closed γ loop extending to 1.5% arsenic (see Fig. 7–13). The eutectic with Fe_2As melts at 827°C and contains 30–31% arsenic. At this temperature the solid solubility of arsenic is 7 or 8%; no quantitative data are available regarding its decrease with decreasing temperature. The Curie point of iron is not lowered substantially by addition of arsenic and presumably is constant to the composition corresponding to Fe_2As (40% arsenic). This compound is tetragonal

[29H1]; it is not known if it is ferromagnetic. The compound FeAs contains about 57% arsenic and is rhombic.

The alloys can be hot worked to about 3% arsenic, but are brittle at room temperature when more than about 1.5% is present [12L1]. The tensile strength increases from 30 to 43 kg/mm² at 2.5% and then begins to decrease. Sheets containing up to 3.5% have been made by Dahl, Pawlek, and Pfaffenberger [35D2] by sprinkling arsenic powder on sheet

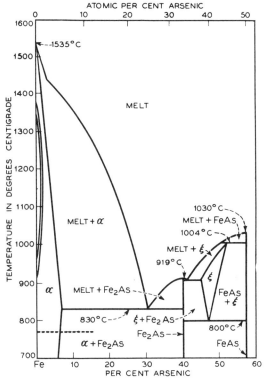

Fig. 7-13. Phase diagram of iron-arsenic alloys.

iron (0.014 in.) and baking, first at 500°C to fix the arsenic, and then for several hours at a higher temperature (e.g., 4 hours at 1100°C) to allow diffusion to effect homogeneity.

The resistivity increases moderately with arsenic content [35D2] to almost 40 microhm-cm at 4% (Fig. 7-14).

No data are available for saturation induction. Figure 7-15 shows variation of induction with arsenic content for given values of field strength (300, 100, ... 5), as determined by Dahl, Pawlek, and Pfaffenberger [35D2]. The initial upward slopes of the curves may be associated with the higher temperature of α,γ transformation accompanying increase in arsenic content, or it may be due to the deoxidizing power of the arsenic.

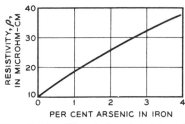

FIG. 7-14. Electrical resistivities of iron-arsenic alloys.

Arsenic was found to have a beneficial effect on hysteresis losses by Burgess and Aston [10B1] as early as 1910. Liedgens [12L1] found a slight increase in maximum permeability and decrease in coercive force for arsenic contents up to 3.5% and confirmed the decrease in hysteresis loss, which he found, however, to be only about 25% below that of iron (at $H_m = 300$). The total losses (in 0.014-in. sheet for $B_m = 10\,000, f = 50$) are decreased by about a factor of 2, when 1–3.5% of arsenic is added to iron, according to the results of Dahl, Pawlek, and Pfaffenberger shown in Fig. 7-16. The losses are then not far from those in 3–4% silicon-iron.

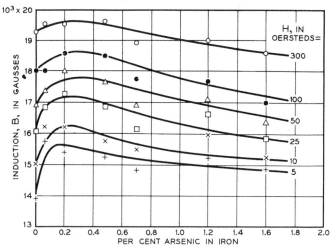

FIG. 7-15. Induction of iron-arsenic alloys at various field strengths.

Iron-Beryllium.—The structure of this system, to about 25% beryllium, has been investigated by Wever and Muller [30W4]. The compound $FeBe_2$ (24.4% Be) forms a eutectic in equilibrium with 9% beryllium in α iron at 1155°C (Fig. 7-17). At this temperature the solid solubility is 6.5%, decreasing with temperature until at room temperature it is 1 or 2%. The γ loop closes at about 0.5%. The Curie point drops rapidly to 650°C (at about 2% beryllium) and then remains constant in the α + $FeBe_2$ area. This compound has the hexagonal $MgZn_2$ structure and is ferromagnetic below 520°C [35M2]; it has strong magnetic anisotropy. Another compound, $FeBe_5$ (44.7% Be), is cubic and is ferromagnetic at liquid

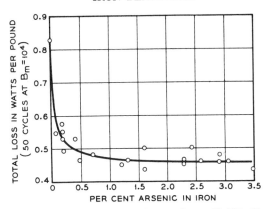

FIG. 7-16. Power losses in iron-arsenic alloys at $B_m = 10\,000$, 50 cps, 0.014-in. sheet.

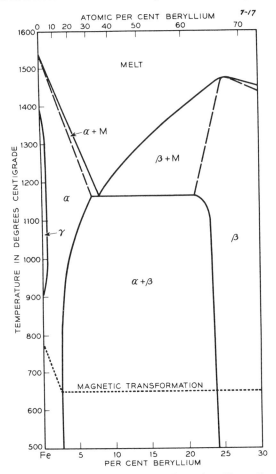

FIG. 7-17. Phase diagram of iron-beryllium alloys.

air temperatures. X-ray measurements of both compounds have been made by Misch [35M2].

Alloys containing up to about 4% can be forged at 900 or 1000°C and have very large grains which, however, can be reduced in size by the addition of nickel [36K6]. Masing [28M5] and Kroll [29K2] found mechanical hardening in alloys containing over 2% beryllium. Beryllium is a powerful desulfurizer of iron.

Seljesater and Rogers [32S6] observed some magnetic aging in the 1.4% alloy and decided aging in the 2.8 and 5.2% alloys (H_c = 42 and 63, respectively, after quenching and drawing).

Von Auwers [29A2] measured the magnetic properties of several alloys containing up to 4% beryllium, after annealing for 1 hour in hydrogen and

TABLE 3. SOME MAGNETIC PROPERTIES OF FE-BE ALLOYS CONTAINING 0.5–4% BE (VON AUWERS [29A2])

Per Cent Be	B for high H	μ_m	H_c	B_r	$\rho \times 10^6$
0.5	16 400	2600	1.6	8500	19
1.0	12 000	1200	2.5	5700	49
1.5	14 200	1000	2.5	5300	46
2.0	17 800	900	4.8	7800	48
3.0	10 800	540	4.3	4700	65
4.0	13 300	230	23.3	7600	54

cooling slowly with the furnace. The results, given in Table 3, show that the permeability is definitely reduced by addition of beryllium. The coercive force rises rapidly when about 4% has been added; this composition corresponds approximately to the solid solubility limit at the temperature of anneal. Kroll also found a decrease of permeability and noted a linear increase of resistivity up to 57 microhm-cm at 2.4% beryllium.

Iron-Boron.—According to the phase diagram of Fig. 7–18, taken from Wever and Muller [30W5], the compound Fe_2B forms a eutectic containing 3.8% boron and melting at 1174°C. In both the α and the γ phases of iron, boron has only slight solid solubility, estimated as 0.15% at the eutectic temperature, 0.1% just above the α,γ point, and 0.15% just below it. The α,γ point is raised slightly by the addition of this small amount of boron. The Curie point is affected but little up to 8.9% (Fe_2B). Alloys containing up to 15% have been reported magnetic [12B2]. The magnetic character of FeB (16.2% B) has not been established, but it is no more than weakly magnetic at room temperature.

Binet du Jassoneix [12B2] found Fe_2B to be magnetic, and Weiss and Forrer [29W1] measured the saturation magnetization at room temperature and below and by extrapolation found the Bohr magneton number to be

1.91 per atom of iron. The magnetization per gram was found to be 165 at 0°K and 160 at room temperature, in a field of about 17 000 oersteds. Yensen [15Y2] found that additions of boron first increased somewhat the magnetic softness of iron, and that the permeability was greatest when only a trace of boron was present. This effect is attributed to its deoxidizing power. Boron contents of more than 0.05% resulted in a definite increase in coercive force and hysteresis loss, and at 0.5% boron the coercive force had risen to 2 from its minimum of 0.3. At the highest boron content

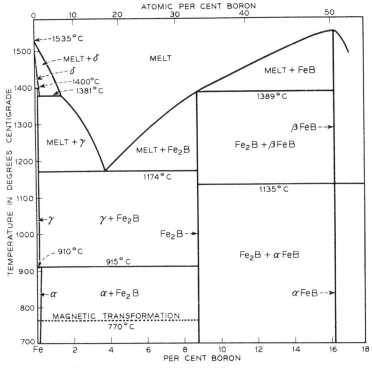

FIG. 7-18. Phase diagram of iron-boron alloys.

investigated, 0.45%, the resistivity had increased to 12.2 microhm-cm; the tensile stress and elongation at rupture of material annealed at 1100°C were about the same as for unalloyed iron.

Iron-Cerium.—Both the α,γ point and the Curie point are raised about 20° by the addition of 10–12% cerium—the limit of solid solubility of cerium in iron at these temperatures—and then remain constant to 50%. At higher cerium contents Fe_2Ce (56% Ce) is formed, and this has a Curie point of 116°C, according to Vogel [17V1], as shown in Fig. 7-19. Clark, Pan, and Kaufmann [43C1], however, measured 2°C as the Curie point in alloys containing over 85% cerium, and 202°C as the Curie point of the 80% alloy. Vogel found the alloys ferromagnetic up to 100% cerium; La Blanchetais [45L1] has found that cerium of high purity is not ferromagnetic, although its magnetic properties are strongly modified by even 0.01% or iron.

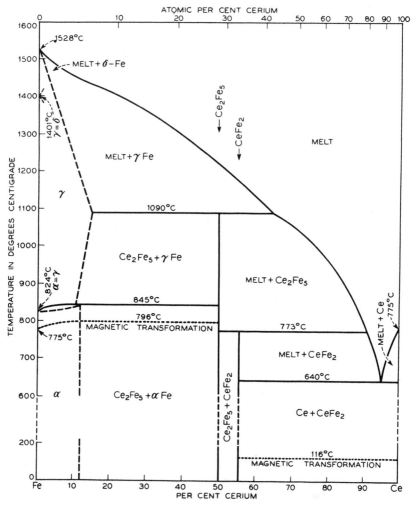

FIG. 7-19. Phase diagram of iron-cerium alloys.

Iron-Chromium.—Chromium is used in various ferrous alloys, with or without carbon, to increase mechanical strength and toughness and to increase oxidation and corrosion resistance at atmospheric and higher temperatures. These properties are used in tool steels and cutlery and in many of the applications where corrosion is important, as in stainless steel of the 18-8 variety. In the presence of carbon, chromium increases the effectiveness of permanent magnets, and in the absence of effective carbon it increases the permeability of the high-nickel alloys.

No specific use has been found for the magnetic properties of binary iron-chromium alloys, containing no essential carbon. However, this system has some unusual and interesting features, and the alloys are magnetic at room temperatures up to 70% chromium.

The phase diagram, modified according to Cook and Jones [43C2], shows a γ loop of peculiar shape (see Fig. 7–20) closed at about 12% chromium in carbon-free alloys, and 23% in alloys containing 0.25% carbon. Alloys containing more chromium form a continuous series of solid solutions,

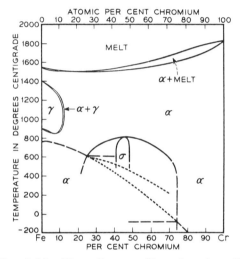

FIG. 7–20. Phase diagram of iron-chromium alloys.

provided they are cooled below 900°C at a moderately rapid rate. Very slow cooling produces a brittle phase, σ, of unknown structure, stable near 50% in the area indicated. The Curie point first rises slightly with addition of chromium to iron, then falls gradually to 0°C at about 70% chromium and to less than −200°C at 80%, according to Adcock [31A5]. The σ phase itself is non-magnetic.

The plasticity decreases with increasing chromium, but the alloys are hot forgeable up to 60 or 70% and, to a certain extent, to higher chromium contents. Hardening is effected by the production of martensitic structure in alloys containing carbon; a tensile strength of 250 000 lb/in.^{2} is attained in the alloy containing 13% chromium and 0.4% carbon.

Chromium expands the iron lattice in a non-linear fashion as shown by the *lattice spacings* determined by X rays, and this is reflected in the curve showing the *density* as a function of composition [31A5, 29S4, 31P5]

(Fig. 7-21). The thermal expansion and thermal conductivity of various commercial alloys have been tabulated [43A1].

The resistivity increases moderately with addition of chromium (Fig. 7-22) and passes through a maximum of 53 microhm-cm near 60% chromium [31A5].

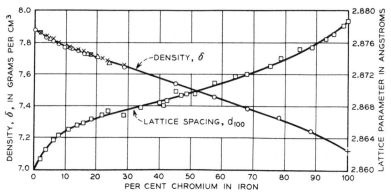

Fig. 7-21. Densities and lattice parameters of iron-chromium alloys.

Saturation magnetization at room temperature has been determined by Stäblein [29S4] and by Fallot [36F1], and their results are given in Fig. 7-23. Saturation at 0°K was estimated by Fallot, whose data, expressed as Bohr

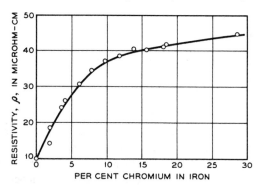

Fig. 7-22. Electrical resistivities of iron-chromium alloys.

magnetons per atom, are plotted against atomic per cent chromium in the same figure. Chromium appears to reduce the atomic moment slightly more than it would if acting as a simple diluent.

Webb's measurements [31W3] show that the permeability of iron is lowered rather rapidly by addition of chromium up to about 10%. Further addition causes a slight rise in permeability to a weak maximum at around 20% [31W3, 30F1], which Fischer has associated with the low magnetostriction known to exist at this composition [30F1]. Inductions for various field strengths are shown in Fig. 7–24, due to Webb [31W3].

Iron-Copper.—These alloys are characterized by limited solubility of each solid element in the other, even in the temperature range 830–

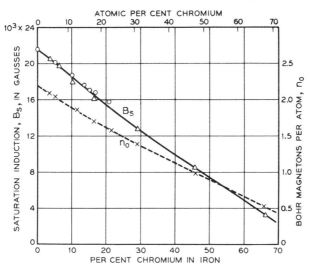

Fig. 7–23. Saturation induction (20°C) and Bohr magneton numbers of iron-chromium alloys.

1095°C, when each has the face-centered cubic structure. Whether or not there is complete miscibility in the liquid state is still a matter of con-

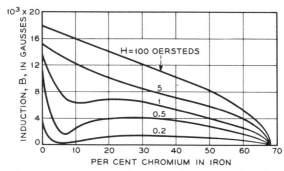

Fig. 7–24. Induction of iron-chromium alloys at various field strengths.

troversy; for many years the diagram (Fig. 7–25) has been drawn with an area in which two immiscible liquids are stable, but two of the more recent reports [36M9, 38I3] state that this occurs only when carbon is present, and that, when carbon is below 0.02%, miscibility is complete.

Solubility of copper in iron is about 10% at maximum, at 1475°C.

The alloy of this composition is peculiar in that it becomes partially liquid when heated to 1095°C, then solidifies completely above 1400°C before becoming entirely liquid near 1500°C. Solid solubility decreases with decreasing temperature, abruptly at the γ,α transformation point and then gradually to about 0.3% at 600°C, according to Norton [35N2]. The solubility of iron in copper decreases from about 4% at the melting point to about 0.1% at 600°C, and to a very small amount at room temperature.

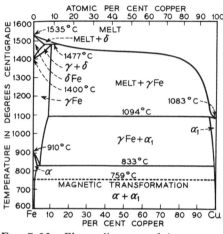

FIG. 7–25. Phase diagram of iron-copper alloys.

The Curie point of iron decreases about 10° at the limit of solid solubility, and then remains constant to beyond 99.5% copper. The magnetism of copper containing traces of iron is discussed below.

Lattice parameters of the solid solutions of copper in iron have been determined by Norton, and those of iron in copper by Anderson and Kingsbury [43A2], and are given in Fig. 7–26. Each element expands the lattice of the other. The density of iron is affected only slightly by the

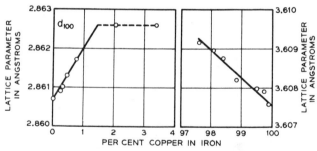

FIG. 7–26. Lattice parameters and densities of iron-copper alloys.

addition of heavier atoms of copper in the expanding lattice, but the density of copper is substantially decreased by addition of iron. Quenched copper-rich alloys may then be expected to be less dense than those slowly cooled. In the two-phase region the calculated density is represented by a straight line; no data have been found in the literature. The thermal expansion is intermediate between that of iron and that of copper; the

IRON–COPPER

expansivity is linear between 20° and 800°C, and is 0.013 per degree C at the latter temperature, according to Simpson and Bannister [36S7].

The resistivity is observed to depend markedly on heat treatment. The 50% alloy has been most studied, and its resistivity has been observed

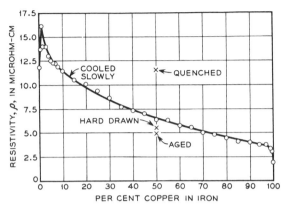

FIG. 7–27. Electrical resistivities of iron-copper alloys and their change with heat treatment.

by Schumacher and Souden [36S8] to vary by a factor greater than 2. As expected, the quenched specimen has a higher resistivity than the hard drawn or aged material. The data of Fig. 7–27 show also the resistivities of the whole series of iron-copper alloys, after being slowly cooled, as determined by Ruer and Fick [13R1].

All of the alloys can be hot or cold worked, if care is exercised. The tensile strength of the 50% alloy can be raised to over 200 000 lb/in.^{2} by cold drawing or by quenching and aging [36S7, 36S8]. Age hardening has also been studied in alloys containing 5% [30K6] and 2.4% [40G4] copper.

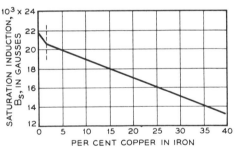

FIG. 7–28. Saturation induction of iron-copper alloys at room temperature.

The saturation induction of slowly cooled alloys has been determined by Kussmann and Scharnow [29K1] and, as expected, decreases almost linearly with copper content to zero at pure copper. A slight departure from such a relation for low copper contents, when solid solution is present, may be noted in Fig. 7–28.

The magnetization of the alloys containing only small amounts of iron is important in the study of its solid solubility and has been measured by Tammann and Oelsen [30T1] with good effect. Alloys containing up to 5% iron were annealed for several hours at various temperatures and the magnetization per gram, σ, measured in a high field (8300 oersteds). A plot of σ vs per cent iron, for a given temperature of anneal, is a straight line (see Fig. 7–29) whose intercept on the axis of abscissae is the solid solubility of iron at that temperature, provided the temperature is high enough for equilibrium to occur by diffusion. Magnetization of cold

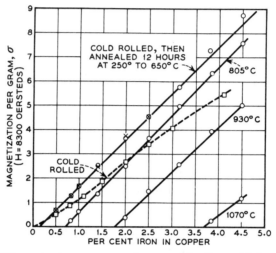

FIG. 7–29. Magnetization per gram of copper-rich alloys with iron, after annealing at various temperatures. Intercepts show solid solubilities at the specified temperatures.

worked alloys, similarly plotted, showed an intercept indistinguishable from zero, as indicated by the broken line. Thus the slightest amount of iron is thrown out of solution by cold working, and its presence is indicated by a slight ferromagnetism. This has been studied by Constant and others [43C3] and is important in the determination of the diamagnetic susceptibility of copper. It is taken into account in the way described, for example, by Bitter and Kaufmann [39B10]. Drigo and Pizzo [49D4] suggest that the impurities behave as single domain particles.

In 1909, Burgess and Aston published magnetization curves of alloys containing 0–7% copper. They found the permeability to be lower the greater the amount of copper present, and to fall rapidly when more than 4% was present. The permeability was halved and the coercive force doubled for 1–2% copper. This magnetic hardening was studied in more detail by Kussmann and Scharnow [29K1], who correlated it with the mechanical hardness and the limit of solid solubility. The coercive force and the Rockwell B hardness are shown in Fig. 7-30, for alloys cooled slowly from 800°C. Both curves rise at the solid solution limit; the mechanical hardness, however, attains its highest value rather quickly while H_c has a broad maximum near 50% copper. The authors came to the conclusion that H_c is affected not only by the simultaneous precipitation of two phases, but also by the difference in the coefficients of expansion of the two constituents. Later Köster [30K7] expressed the opinion that both expansion coefficients and particle size contribute to the change of H_c with composition and heat treatment. Recent work suggests that particle shape may also affect the result.

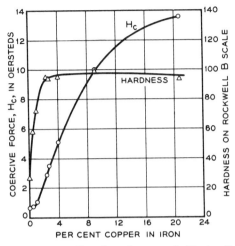

FIG. 7-30. Coercive force and Rockwell B hardness of iron-rich iron-copper alloys.

Iron-Gadolinium.—No investigation of this system has been reported. Since gadolinium is ferromagnetic at temperatures below 16°C, any of its alloys with iron might be ferromagnetic. The compound Fe_2Gd has been prepared and found to be cubic with a structure of the $MgCu_2$ type [44E2].

Iron-Germanium.—A provisional phase diagram [40R2] indicates a eutectic of Fe and Fe_2Ge at 1125°C and 35% germanium. The γ loop is reported to be completely closed [29W4], probably [40R2] at a few per cent of germanium. The solid solubility in iron has not been determined; it is presumably considerably less than 35%. No data on Curie temperature have been found.

Jaffee, McMullen, and Gonser [46J1] prepared a 0.014-in. transformer sheet containing 2.1% silicon and 0.5% germanium and compared it with sheets containing silicon alone (2.1%). The addition of germanium caused some improvement in material annealed at 900°C; coercive force was then 0.65 (for $B_m = 10\,000$) as compared with 0.82 for the pure silicon addition. After annealing at 1200°C the difference became negligible ($H_c = 0.26$).

Iron-Hydrogen.—The solubility of hydrogen in iron, as dependent on temperature, has already been plotted in Fig. 2-12. The amount of hydrogen absorbed, measured in cm^3 at 0°C and 1 atm of pressure, generally increases with temperature. At the α,γ point and the melting point it increases by an especially large amount (almost by a factor of 2 at the melting point), and at the γ,δ point it decreases slightly. The solubility is proportional to the square root of the pressure. The presence of a few per cent of nickel increases the solubility substantially. A summary of data has been given by Dushman [49D1].

Iron-Manganese.—These alloys have been used in steelmaking for more than 100 years and have been repeatedly investigated, but still their structure is imperfectly known. In 1888 Hadfield reported the extraordinary properties of the carbon-containing alloys (about 12% Mn, 1.2% C) which have such great resistance to abrasion. The alloys containing no essential carbon have not found commercial use.

The binary system resembles somewhat the iron-nickel system in that the α + γ area of the phase diagram is broadened and pushed to lower temperatures by the alloying, and the transformations become sluggish. The iron-rich end of the diagram, as modified by Troiano and McGuire [43T1], is shown in Fig. 7–31. It is still a matter of discussion [36H1, 43T1] whether or not the ε phase, a hexagonal, close-packed structure with $c/a = 1.61$, is stable at any temperature, but it is formed under a variety of conditions and contributes to the hardening of alloys in the range 10–25% or more of manganese.

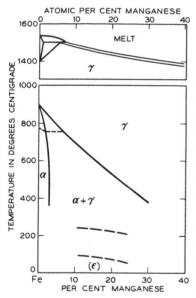

FIG. 7–31. Phase diagram of iron-manganese alloys.

Both the γ and ε phases are non-magnetic, and the temperature of loss of ferromagnetism of any alloy is usually governed by the temperature at which the α phase transforms to γ or ε. In pure α alloys the Curie point is lowered 10–15°C for each per cent of manganese present. The Curie points of alloys, cooled moderately slowly and thus containing unknown proportions of α and γ, are given in Fig. 7–32 according to Gumlich [18G2]. Alloys containing more than about 16% manganese are not obtained in the magnetic condition.

Iron-rich alloys can be forged until the manganese content is 40% or more, and those containing up to about 30% can be cold rolled after being forged and hot rolled [33W2, 42W6, 46D1]. The alloys having more than 70% are decidedly brittle.

Density, thermal expansion, and resistivity have been given in a number of reports. These are so dependent on heat treatment that consistent values are difficult to report and are not of great usefulness. Lattice parameters, however, are precisely known and can be used for calculation of densities, provided the proportions of the phases are known. Schmidt [29S5] measured the parameters of alloys containing 0–100% manganese, and in the investigation discovered the existence of the ε phase, which was found also, independently, by Ishiwara [30I1]. Walters and his colleagues have also measured lattice parameters [35W4], and resistivity and thermal expansion [33W2] as dependent on temperature, and in a series of papers have reported studies of constitution and of mechanical and other physical properties [42W6]. The lattices of α, γ, and ε phases are all expanded by an increase in manganese content. The resistivities of (slowly cooled)

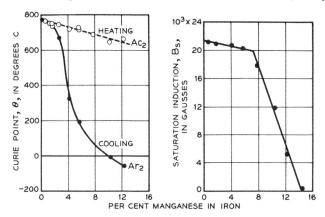

FIG. 7–32. The α,γ transitions on heating and cooling, and the saturation induction, of iron-manganese alloys.

alloys rise with manganese rather rapidly to about 50 microhm-cm at 10% manganese [18G2], and then increase less rapidly. Resistivity increases and the alloy contracts upon conversion of α to γ [33W2].

Saturation magnetization at room temperature has been measured by Gumlich [18G2] and is recorded in Fig. 7–32. Alloys contained about 0.1% carbon and were slowly cooled. The sudden change in slope at 7% manganese is presumably the result of α,γ transformation. The practical limit of ferromagnetism is at about 14% manganese.

The saturation at 0°K has been estimated by Sadron [32S3] for three alloys containing less than 4% manganese. In this range, when the alloys are presumably in the α phase, the moment per atom drops only slowly and manganese appears to behave merely as a diluent. At higher manganese contents, the moment drops more rapidly, but this is probably associated with the presence of a non-magnetic γ or ε phase.

In alloys containing about 0.1% carbon, both Gumlich and Hadfield

[27H1] observed that the coercive force first increased slowly with addition of manganese, and then more rapidly as hardening occurred as a result of the precipitation of new phases. The permeability is similarly lowered by addition of manganese; Gumlich found μ_m to be reduced from 3400 to 1000 by about 2% manganese. The high coercive forces obtained in alloys containing substantial amounts of carbon are discussed in Chap. 9.

Iron-Mercury.—Iron has only a small solubility in mercury, and it is so small that no agreement has been reached as to its amount. However, reports from early investigations indicated that amalgams possessed considerable magnetizability and rather high coercive forces—as high as 370 oersteds [96N1] in an amalgam containing 2.3% iron. It is believed now that these so-called amalgams are colloidal solutions of iron in mercury. Sometimes the particles are fine enough to filter through leather; in one case the iron content of such a colloidal solution was found to be 0.0014%. Solid iron separates from some of the more concentrated mixtures when they are subjected to a strong inhomogeneous field.

Bates and Illsley [37B10] have measured the magnetic susceptibilities of "amalgams" of this kind, prepared by electrodeposition of iron onto mercury and containing 0.001–0.02% iron. The apparent susceptibility was found to be proportional to the iron concentration, except when this was very small, and to be nearly that expected on the assumption that the saturation magnetization of iron has its usual value. The most intense field used was 7100 oersteds, and the highest volume susceptibilities observed were about 30×10^{-6}.

The high coercive force of amalgams reported years ago may be attributed to the small size of the iron particles and apparently has the same origin as the coercive force of the commercial permanent magnets composed of pressed aggregates of fine iron particles. In fact the first such magnets were produced by Dean and Davis [41D1] by electrolysis of iron onto mercury and subsequent separation of the iron from the mercury. Coercive forces of 500 were attained; this work is referred to in Chap. 9. Antik and Kubyschkina [34A3] had previously studied magnetization and hysteresis in liquid iron "amalgams." They found that the small iron particles behave as separate domains and have the high coercive force associated with material having no moving boundaries between domains (see Chap. 18). It may be inferred from their work that the coercive force was equal to the field strength necessary to saturate in the direction of difficult magnetization, namely, about 500 oersteds.

Iron-Molybdenum.—A portion of the phase diagram is shown in Fig. 7–33, taken from Gregg [32G2]. The γ loop is closed at 2.7% molybdenum. The solid solubility of molybdenum in α iron increases from about 5% at room temperature to 35–40% at the eutectic temperature, 1440°C. The Curie point drops from 770 to 760°C in the solid solution area, which extends to 8% at this temperature and then remains constant as Fe_3Mo_2 precipitates. This compound (53% Mo) is trigonal in crystal form [28A2] and is non-magnetic.

The lattice spacing of iron increases slowly with the addition of molybdenum, according to Bowman, Parke, and Herzig [43B3]. The resistivity

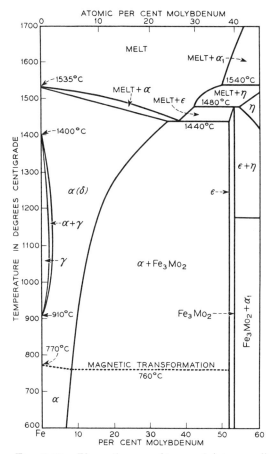

FIG. 7–33. Phase diagram of iron-molybdenum alloys.

also increases but slowly with addition of molybdenum, and for the alloy containing 10% it is 40 microhm-cm.

Measurements of Takei and Murakami [29T1] show that the saturation drops slowly with molybdenum to about 20 000 at 8%. At 18% the saturation depends on the cooling rate, as one would expect from the phase diagram, and varies from 11 500 to 17 000. Similar dependence on cooling rate occurs also at higher molybdenum contents, as the saturation falls to zero at about 50%.

The permeabilities of the α-phase alloys, annealed for 1 hour at 1000°C, increase with molybdenum to 4.5% and then decrease rapidly, as shown in Fig. 7-34, taken from unpublished data obtained by Williams and Bozorth. At about the same critical composition, the limit of solid solution at low temperature, the coercive force begins to increase rapidly, and at 23% Seljesater and Rogers [32S6] observed a coercive force of over 200 (see Fig. 2-24). An even higher value, 300, was reported by Messkin and Somin [36M10].

Directional magnetic properties can be developed in alloys containing less than 5% molybdenum by cold working and annealing. The direction of easy magnetization, fixed by the special orientation of the crystals in the material, is then parallel to the direction of rolling or drawing, and the permeabilities at high inductions are considerably higher than they are in the hot-rolled material.

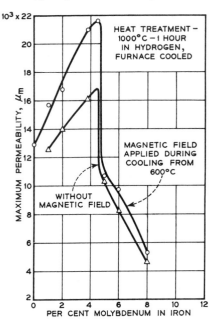

Fig. 7-34. Maximum permeabilities after annealing at 1000°C, and after subsequent annealing in a magnetic field, of iron-molybdenum alloys.

Iron-Neodymium.—Drozzina and Janus [35D5] observed a coercive force of 4600 oersteds (for $I = 0$) in one of these alloys. It was not determined whether it contained one or more phases, but possibly one phase was iron or a solid solution of neodymium in iron. The composition and saturation intensity of magnetization are uncertain, for the statements in the text are inconsistent, presumably on account of a typographical error. The high coercive force suggests that one phase was present in finely distributed form.

Iron-Niobium.—By addition of about 4% of niobium the α,γ point of iron is raised from 900 to 965°C. The γ loop is restricted, and the pure γ phase extends to a maximum of about 2%, at 1220°C. At higher niobium contents the ε phase, Fe_3Nb_2 (53% Nb) is precipitated. The area of pure δ is enlarged so that 12% niobium is soluble at the eutectic temperature of 1350°C. The Curie point of the alloys is not noticeably different from that of iron and remains constant in the α + ε region, which at this temperature extends from about 2 to 47% niobium. The diagram of Eggers and

Peter [38E6], modified by the additional details reported in iron-rich alloys by Genders and Harrison [39G6], is given in Fig. 7-35.

Precipitation hardening has been observed [42W5] when more than 0.5% niobium is present. In the 1.9% alloy, quenched from 1100°C and tempered at 600 to 650°C, Rockwell hardness of 25 is exceeded, and the creep limit at 500°C is almost 50 000 lb/in.^{2}. Unusually high creep strength has also been observed by Parker [40P4] in the 3% alloy. The hardening, as may be inferred from the phase diagram, is twofold in nature—the γ,α

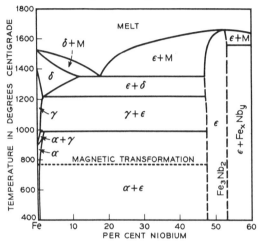

FIG. 7-35. Phase diagram of iron-niobium alloys.

transformation and the change in solid solubility in the α phase both contribute. Still higher values of hardness are obtained by additions of carbon.

Apart from the Curie point determinations, no data relating to magnetic properties have been found.

Iron-Nitrogen.—Among the iron-rich alloys the compounds Fe_4N (5.9% N), Fe_3N (7.7% N) and Fe_2N (11.1% N), have been identified with X-rays. Of these Fe_4N is known to be ferromagnetic.

The nitrides are formed by passing mixtures of purified hydrogen and ammonia over powdered iron at various temperatures. As nitrogen "opens up" the iron structure the latter is converted first to a face-centered cubic structure with nitrogen in the interstices (Fe_4N), then to a more expanded structure that is nearly hexagonal close-packed (Fe_3N), the nitrogen again lying in the interstices in a regular array. In a recent study Jack [48J2] has confirmed the positions of the iron atoms reported by others and definitely placed the nitrogen atoms in Fe_4N and Fe_2N; the latter is orthorhombic.

The general character of the phase diagram of this system is shown in Fig. 7-36, which results from the work of Paranjpe, Cohen, Bever and Floe [50P1]. The relation of Fe_3N and Fe_4N to the other phases is not yet clear. The solid solubility of nitrogen in iron is variously reported [36H1] as 0.3 to 0.02% at about 400°C, the difference perhaps being due to differences in purity. In technically pure iron Köster [30K8] estimated the solubility to be 0.001% at 100°C; his data are represented by the curve of Fig. 2-26 and show a rapid increase of solubility with temperature. The data of Paranjpe on pure iron show a somewhat smaller solubility below 300° and a higher solubility at higher temperatures —0.10% nitrogen at the eutectoid temperature, 590°C. The phase structure existing above the α,γ point of iron is practically unknown. The diagram shows that nitrogen greatly enlarges the γ field.

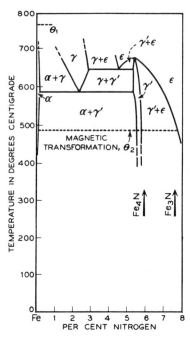

FIG. 7-36. Phase diagram of iron-nitrogen system.

No change in the Curie point of iron, caused by solution of nitrogen, has been established. The compound Fe_4N has a Curie point of 388°C (Creveaux and Guillaud [46G6]). They have measured the atomic moment and found it to be 2.22 Bohr magnetons per atom of iron, the same as in α iron. The saturation induction at room temperature is about $B_s = 17\,500$, and it has been determined at various temperature from 77°K to 85°C [46G2]. Especially high fields are necessary to saturate. Since either α iron or Fe_4N, or both, exist in alloys containing 0–8% nitrogen, alloys are ferromagnetic up to the latter composition.

The thermodynamic properties, and the reactions of iron with NH_3, have been summarized by Kelley [37K9]. Practically, Fe_4N is formed in an atmosphere of NH_3 and H_2 at 550–800°C [31B3, 46G6]. The dissociation pressure of Fe_4N is over 4000 atm at 420°.

The structures present in the iron-nitrogen-carbon system have also been studied by Jack [48J3].

Iron-Oxygen.—The phase diagram of this system, to 35% oxygen, is shown in Fig. 7-37 [48D1]. Under equilibrium conditions α iron is present when the oxygen content is less than 23%, and magnetite when it is 23–30%. In addition to these magnetic phases there is an unstable cubic oxide, γ Fe_2O_3, that is magnetic; FeO (22.3% O) is cubic (NaCl structure) and is paramagnetic. The rhombohedral form of Fe_2O_3 (α) is variously reported as paramagnetic and ferromagnetic; Chevalier [51C1] believes it to be ferromagnetic with a very small saturation magnetization. This small saturation is attributed by Néel [49N1] to small traces of Fe_3O_4, and by Snoek [50S4] to structural irregularities.

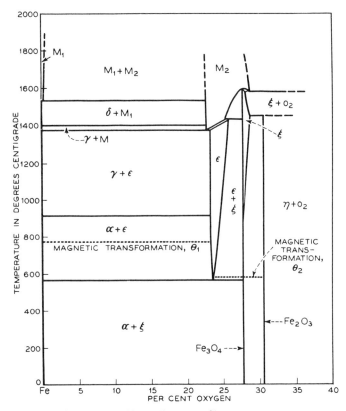

FIG. 7-37. Phase diagram of iron-oxygen system.

The solubility of oxygen in α iron is not well known, but has been estimated to be less than 0.01% at 800°C. A tentative solubility curve, estimated by Ziegler [32Z2] is given in Fig. 2–26. The solubility is reported to be greater in γ iron, and still larger (0.2% or more) in the melt, for which the data are more reliable. Neither iron nor magnetite forms a solid solution extensive enough to lower the Curie point noticeably.

The unstable γ form of Fe_2O_3 (30.1% O) is prepared by heating Fe_3O_4 (27.6% O) or its hydrate in an oxidizing atmosphere, or by dehydrating lepidokrolite or artificial $Fe_2O_3 \cdot H_2O$, or by other means, and occurs in nature as maghenite. It transforms irreversibly to the paramagnetic rhombohedral variety, haematite, on heating, the temperature of transformation (e.g., 400–800°C) depending on the method by which it has been prepared. Verwey [35V3] has shown that the cubic ferromagnetic form has the same arrangement of iron and oxygen atoms as magnetite except that some of the positions that are occupied by iron in Fe_3O_4 are unoccupied in Fe_2O_3 (see Fig. 7–39). The Curie point is generally considered to be 675°C but appears to vary with treatment.

The vapor pressures of oxygen over the three oxides are all quite small: at 1100°C it is $<10^{-12}$ mm in FeO, $<10^{-8}$ mm in Fe_3O_4, $<10^{-7}$ mm in Fe_2O_3 (haematite). Data on decomposition in various atmospheres are summarized by Dushman [49D1].

The saturation magnetization of artificial magnetite has been measured [29W2] from 20°K to the Curie point, 578°C. The saturation moment per gram, σ_s, is 92 to 93 [29W1, 29W2] at room temperature when the density is 5.25 ($B_s = 6150$). Extrapolation to 0°K gives $\sigma_0 = 98.2$ [29W1] and a Bohr magneton number per atom of iron of $n_0 = 1.36$. The change of magnetization with temperature follows a curve that lies definitely below the theoretical hyperbolic tangent curve for $j = \frac{1}{2}$ (see Fig. 10–4). About the same value of the intensity of magnetization of natural crystals at room temperature was found previously by Quittner [09Q1].

The saturation moments of Fe_3O_4 and $\gamma\ Fe_2O_3$ have been interpreted by Néel [48N1] in terms of antiferromagnetism. As shown by Verwey and de Böer [36V3] the Fe^{2+} ions in Fe_3O_4 are all in positions of one kind (octahedral positions, see Fig. 7–39) and the Fe^{3+} ions are equally divided between positions of two kinds (octahedral and tetrahedral positions). The Fe^{3+} ions in the one kind of position are assumed by Néel to lie antiparallel to those in the other so that their net moment is zero. The one Fe^{2+} ion per molecule then contributes 4 Bohr magnetons so that the whole structure should have 1.33 Bohr magnetons per atom of iron, a value near to the observed moment of 1.36.

If one forms $\gamma\ Fe_2O_3$ from Fe_3O_4 by removing one-ninth of the iron atoms, the original unit cell containing 8 Fe_3O_4, or $Fe_{24}O_{32}$, is reduced to $Fe_{64/3}O_{32}$, or $\frac{32}{3}\ Fe_2O_3$. There will still be 8 Fe^{3+} ions in the tetrahedral position and therefore $\frac{40}{3}\ Fe^{3+}$ ions in the antiparallel octahedral positions. Allowing 5 Bohr magnetons per Fe^{3+} ion, the moment should then be $5(\frac{40}{3} - 8)$, or $\frac{90}{3}$, Bohr magnetons per unit cell, or 1.25 per atom of iron. The observed moment [29W1] is 1.20 per atom.

Maxwell, Smart, and Brunauer [49M1] have measured the saturation magnetization of the cubic oxide intermediate in oxygen content between Fe_3O_4 and Fe_2O_3, and having the ratio $Fe^{2+}/Fe^{3+} = 0.352$. This corresponds to the removal of 0.027 of the iron atoms from the original Fe_3O_4 structure, and the unit cell then contains, according to the Néel theory, 8 tetrahedral Fe^{3+}, 9.27 octahedral Fe^{3+} (antiparallel to the 8), and 6.08

octahedral Fe^{2+} (also antiparallel). The magnetic moment should then be 30.7 Bohr magnetons per cell or 1.31 per atom of iron, while the observed number is 1.31, in excellent agreement.

Kopp [29W1] and Snoek [36S6] have found the magnetic saturation to be a maximum at the composition Fe_3O_4. Apparently FeO precipitates when the oxygen content is below that for Fe_3O_4.

At about $-160°C$, magnetite undergoes a transition, and sudden changes occur in the resistivity, specific heat, and magnetic properties. Domenicali [49D2] has shown that the magnetic properties below the transition point are affected by the presence of a field during the transition. Verwey and Haayman [39V1, 41V3] have suggested that at low temperatures there is an ordering of electrons among the lattice sites, so that Fe^{++} and Fe^{+++} are regularly arranged. It is possible that tetragonality in the structure, of some such kind as that suggested by Verwey and Haayman, is effected and directed by the presence of the field.

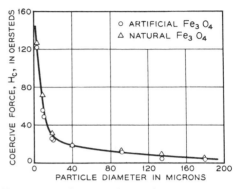

FIG. 7–38. Coercive force of magnetite as dependent on particle size.

Weiss and Forrer [29W1] observed a sudden decrease in the magnetic moment as magnetite cooled through the transition point, but noted that the discontinuity approached zero as the field became very large. At the highest field used, about 19 000, the decrease was less than 0.5%.

The permeability and coercive force of magnetite vary considerably with its source and particle size. Some values for solid specimens, examined by Herroun [43H2] lie in the ranges indicated in the following table:

Property	Range of Values
μ_m, maximum permeability	5–35
H, field strength for μ_m	35–300
H_c, coercive force ($B = 0$)	8–100
B_r, residual induction ($H_m = 1200$)	35–800

The dependence of coercive force on particle size was determined in an important investigation by Gottschalk [35G6, 41G2]. He ground natural and artificial magnetites and observed that the coercive force increased as the particle size diminished (see Chap. 18). Results are plotted in Fig. 7–38. When H_c is replotted against the reciprocal of the diameter, a straight line is obtained. Extrapolation of these data indicates that large particles should have a coercive force of about 3, a value somewhat higher than that obtained by direct experiment. Other properties of magnetite in powdered form are also reported by Gottschalk.

Ferrites.—Some of the cubic ferrites, of composition corresponding to $MO \cdot Fe_2O_3$, have been found to have useful magnetic properties. They are particularly applicable at high frequencies because their resistivities are a million or more times those of alloys commonly used, and this makes eddy-current losses low and lamination unnecessary. The saturation

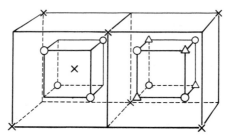

SYMBOL	NORMAL STRUCTURE		INVERSE STRUCTURE	COORDINATION NUMBER
	$MgAl_2O_4$	MFe_2O_4	MFe_2O_4	
×	$1 Mg^{++}$	$1 M^{++}$	$1 Fe^{+++}$	4 (TETRAHEDRAL)
△	$2 Al^{+++}$	$2 Fe^{+++}$	$1 M^{++}$, $1 Fe^{+++}$	6 (OCTAHEDRAL)
○	$4 O^{--}$	$4 O^{--}$	$4 O^{--}$	

Fig. 7-39. Crystal structure of spinel, $MgAl_2O_4$, and of the normal and inverse cubic ferrites, $MO \cdot Fe_2O_3$.

magnetization ($\sim$5000) is much less than that of iron, and they therefore find use at low inductions. The initial permeability is often 1000, and 4000 has been measured.

The useful ferrites have the crystal structure of spinel, $MgAl_2O_4$, illustrated in Fig. 7-39. Oxygen ions occupy the positions marked by circles and form a close-packed, face-centered structure. In "normal" spinel the crosses indicate Mg^{2+} ions, which are each surrounded symmetrically by 4 oxygen ions at the corners of a regular tetrahedron. The Al^{3+} ions, designated by triangles, are each surrounded by 6 oxygen ions located at the centers of faces of a cube (octahedral positions). In zinc ferrite, $ZnFe_2O_4$, the Zn^{2+} ions occupy the Mg^{2+} positions (crosses), the Fe^{3+} the Al^{3+} positions (triangles), of spinel.

Barth and Posnjak [32B6] showed that compounds which crystallize in a spinel structure can be divided into two groups: the "normal" spinel as described above for $MgAl_2O_4$, and $ZnFe_2O_4$ and the "inverse" spinel of which nickel ferrite is an example. In nickel ferrite half of the Fe^{3+} ions are in the Mg^{2+} positions (tetrahedral), and the other half and all of the Ni^{2+} ions are in the Al^{3+} positions (octahedral). A normal ferrite may be written $M^{2+}Fe_2^{3+}O_4$, and an inverse ferrite $Fe^{3+}(M^{2+}Fe^{3+})O_4$. Magnetite, Fe_3O_4, has the inverse spinel structure and is then $Fe^{3+}(Fe^{2+}Fe^{3+})O_4$. Verwey, Heilmann, Haayman, and Romeijn [47V4, 47V5] have applied the Barth and Posnjak principle to the ferrites. X-ray intensity data show that Zn and Cd ferrites have the normal spinel structure. By studying the changes in lattice spacing that occur when the trivalent ion is changed from Al to Cr to Fe, they found that Mn, Fe, Co, and Ni ferrites have the inverse structure. The latter are ferromagnetic; the others, $ZnFe_2O_4$ and $CdFe_2O_4$, are believed not to be magnetic (except after rapid quenching, when they may have the inverse structure [50B2]). Ferrites are also made by substituting trivalent ions (e.g., Al^{+++} for Fe^{+++}) [48P2, 50S1].

In both normal and inverse structures, the whole may be regarded as a close-packed assemblage of large oxygen atoms, the metallic atoms lying in the interstices. Of the 12 interstices for each molecule of MFe_2O_4, 8 have a coordination number of 4 (tetrahedral holes) and only 1 of the 8 is occupied by a metal atom, and 4 have a coordination number of 6 (octahedral holes) and only 2 of the 4 are occupied. In the normal spinels the atoms in any one kind of position are of the same kind, whereas in the inverse spinels they are different.

Ferrites of the type MFe_2O_4 form solid solutions with either of the components, Fe_2O_3 and MO. Van Arkel, Verwey, and van Bruggen [36V1, 36V2] have shown that at high temperatures (1300°C) the range of composition of each solution may be up to 65–75% Fe_2O_3, 35–25% MO on the one hand, and to 15% Fe_2O_3, 85% MO on the other. At lower temperatures the composition range is not so large and Fe_2O_3 or MO may precipitate in the form of haematite (rhombohedral) or the bivalent metal oxide (NaCl structure in many cases). Solid solution may be preserved by quenching from 1200–1400°C, and a second phase may be precipitated if desired by heating for some time at a lower temperature but not much below 750°C. The decomposition pressure of oxygen at these temperatures varies widely with the nature of the bivalent metal.

Solid solutions between various ferrites are usually quite stable: e.g., $MnZnFe_4O_8$ and $MnNiFe_4O_8$. Some phase equilibrium data for several of the ferrites are given in the articles by Van Arkel, Verwey, and van Bruggen and elsewhere [25F1].

The *magnetic properties* of the ferrites have been subject to investigation for many years. Of the older works those of Hilpert [09H2], Wedekind [12W2], and Forestier and Chaudron [25F1, 31F1, 39F1] should be mentioned specially. More recent results have been obtained mainly by Snoek [47S2, 48S17, 48S18, 48S19] in a systematic investigation in which

246   OTHER IRON ALLOYS OF HIGH PERMEABILITY

the bivalent metals Mg, Mn, Fe, Co, Ni, Cu, and Zn, and to a limited extent Ca and Cd, were used. His specimens were prepared, for example, by mixing together the constituent high-purity oxides in finely divided form and pressing. They were then sintered at 900°C for 2 hours, ground, pressed into test rings at about 4 tons/cm^2 and heated at 1200–1250°C in commercial nitrogen or oxygen. Many variations in preparation and heat treatment were tried, and appropriate treatments were found for each material. Considerable deviation in composition from simple stoichiometric proportions usually resulted in lower initial permeability. Magnetic characteristics of some of these materials are given in Table 4. The Curie points are taken from various sources.

TABLE 4. MAGNETIC DATA ON SOME SIMPLE FERRITES AT ROOM TEMPERATURE

(Initial permeabilities are approximate maximum values.)

Ferrite	B_s	$\theta(°C)$	μ_0
$MnFe_2O_4$	4500	295	250
$FeFe_2O_4$	6100	575	70
$NiFe_2O_4$	3000	590	10
$CuFe_2O_4$	3650	455	70
$MgFe_2O_4$	1800	310	10
$AlFe_2O_4$		339	...
$CoFe_2O_4$		520	>1

The relation of the saturation magnetization to composition has been considered by Néel [48N1, 50N3], who has interpreted the magnitude of the molecular magnetic moments and their variation with composition in terms of antiferromagnetism. The theory was designed primarily to explain the fact that the average atomic moment of the iron atoms in Fe_3O_4, and of the iron and nickel atoms in $NiFe_2O_4$, etc., is much smaller than one would expect from the known structure of the magnetic atoms. Néel points out that the two different kinds of positions for the iron atoms (marked by crosses and triangles in Fig. 7–39) are differently surrounded by other atoms, and that the atomic magnetic moments of the ions of one kind may be antiparallel to those of the other kind, the observed moment being then the difference between the two.

This theory of antiferromagnetism has already been discussed in relation to magnetite and has been successfully applied to the mixed ferrites by E. W. Gorter [50G3]. As discussed above, the theory says that the magnetic moments of the 2 Fe^{+++} ions in Fe_3O_4 (an inverse spinel) will neutralize each other and the moment will be due to the Fe^{++} ions and be 4 Bohr magnetons per molecule. In $CoFe_2O_4$ and $NiFe_2O_4$ the Co^{++} and Ni^{++} ions carry the magnetic moments and these should then be respectively 3 and 2 Bohr magnetons per molecule. The following Table 5 shows the calculated and observed moments for the homologous series of ferrites containing Mn, Fe, Co, Ni, and Cu, and of the ferrites of Mg and Li.

TABLE 5. MAGNETIC MOMENTS OF SOME FERRITES,
IN BOHR MAGNETONS PER MOLECULE
(As predicted according to the Néel theory and as observed by Gorter [50G3])

Ferrite	Calculated Moment	Observed Moment
$MnFe_2O_4$	5	5.0
$FeFe_2O_4$	4	4.2
$CoFe_2O_4$	3	3.3
$NiFe_2O_4$	2	2.3
$CuFe_2O_4$	1	1.3
$MgFe_2O_4$	0	1.1
$Li_{0.5}Fe_{2.5}O_4$	2.5	2.6

When $ZnFe_2O_4$ is added to these ferrites in solid solution, the Zn atoms apparently go into the tetrahedral positions and replace some Fe^{+++} ions which move into the octahedral positions formerly occupied by Ni^{++} or other metal ions. The calculated effect of this is to increase the moment per molecule, and the initial rate of increase is such as to extrapolate to 10 Bohr magnetons per molecule for pure $ZnFe_2O_4$. In Gorter's experiments the moment vs composition curve first increases with the theoretical slope and then bends downward toward a moment of zero, in accordance with the fact that $ZnFe_2O_4$ is non-magnetic. The maximum moment observed experimentally was 6.25, at 60 mole per cent $MnFe_2O_4$.

TABLE 6. MAGNETIC DATA ON SOME DOUBLE FERRITES

Designation	Mole Per Cent of Ferrites	Density	$\theta(°C)$	μ_0	ρ (ohm-cm)	$a \cdot 10^6$	$c \cdot 10^6/\mu$	tan δ*
Ferroxcube 1	40 Cu, 60 Zn	5.2	90	1100	10^5	10		100
Ferroxcube 2	$(MgZnFe_4O_8)$†	4.4	100	400	10^6	10	44?	130
Ferroxcube 3	45.5 Mn, 49.5 Zn, 5 Fe	5.1	110	1000	10^2	1	40	170
Ferroxcube 4	40 Ni, 60 Zn	4.8	80	82	10^6		5.4	55

* At 1 mc/sec.   † Approximate.

The mixed ferrites of manganese, cobalt and nickel with zinc have also been studied by Guillaud, Roux and Creveaux [49G8, 50G5], who observed the maximum magnetization per gram at 20 to 25 mole per cent MnO, CoO or NiO, 30 to 25 mole per cent zinc. They found the Curie point of $NiFe_2O_4$ (640°K) to decrease continually with substitution of ZnO for NiO. Néel [50N6] has pointed out that the atomic moments change in accord with theory.

Some of the double ferrites have been designated Ferroxcube and are described in Table 6; of these Ferroxcube 3 especially has found commercial uses. An important feature in their application is the low energy loss in a high-frequency alternating field. Since the resistivity is so high, the eddy-current loss is quite small, and the hysteresis loss coefficient a, the residual loss coefficient c, and the permeability μ are the important criteria. The residual loss coefficient of the Ferroxcubes is not independent of frequency and increases rapidly when the frequency is sufficiently high. Figure 7–40 shows the loss up to 10^6 cps for Ferroxcube 3, which is limited by losses to use at frequencies below about 5×10^5 cps. Other Ferroxcubes can be used at higher frequencies, but they generally have lower permeability; e.g., $NiZnFe_4O_8$ can be used to 40×10^6 cps but has a permeability of only about 50. Other compositions (15 at. % NiO, 35 at. % ZnO, 50 at. % Fe_2O_3) have initial permeabilities of 2000–4000.

Measurements of the magnetic characteristics at frequencies of a few megacycles per second have shown peculiar resonance phenomena which have been clarified in a paper by Brockman, Dowling, and Steneck [50B1]. Experiments were made on specimens of Ferroxcube 3 having cross-sec-

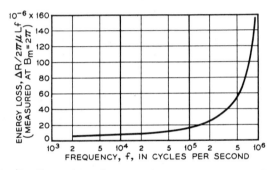

FIG. 7–40. Power loss of Ferroxcube 3 as dependent on frequency.

tional areas of 1.25×2.5 cm^2. This material, as they used it, had a permeability, μ, of the order of 10^3 and a dielectric constant, ϵ, of about 10^5, resulting perhaps from the dielectric polarization of the material around the microscopic holes that exist after the material has been pressed from powder and sintered. The velocity of propagation of an electromagnetic wave in such a medium is given by

$$v = c/(\epsilon\mu)^{1/2} = f\lambda,$$

c being the velocity of light, f the frequency, and λ the wave length of the wave in the material. If the thickness of the material is 1 wave length, a standing wave will exist and resonance phenomena will occur. For the constants given above, λ will be equal to the thickness of the material when f is 1 to 2×10^6 cps, at which frequency a pronounced maximum in the effective loss resistance is actually observed. This dimensional resonance is naturally of considerable importance in the practical application of the material.

An important property of core material for loading coils is its stability—

the inductance should vary as little as possible with superposed field and temperature. The effect of a superposed field on the permeability of Ferroxcube 3 is given in Fig. 7–41, as measured by Miss M. Goertz at the Bell Laboratories. The observed variation of 10–15% is much larger than the 0.2% found on compressed powdered cores [40L1]. A similar large variation of permeability with temperature has been noted. However, because of the large permeabilities of certain Ferroxcubes, it is possible to make useful cores by the introduction of an air gap, and the dependence of the effective permeability on temperature and field strength is then greatly reduced. The material so made has many commercial applications.

It has already been mentioned that addition of $ZnFe_2O_4$ to $MnFe_2O_4$ increases μ_0. This was explained [47S2] as follows: the addition of non-

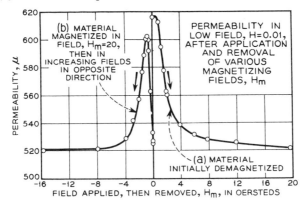

FIG. 7–41. Reversible permeability of Ferroxcube 3 as dependent on biasing field.

magnetic $ZnFe_2O_4$ in solid solution lowers the Curie point so that it is just above room temperature, and the maximum in the μ_0 vs T curve, known to lie just below the Curie point, then comes at room temperature. This may be ascribed in turn to the known variation of crystal anisotropy with temperature.

Snoek has pointed out that the saturation magnetostriction of $FeFe_2O_4$ (Fe_3O_4) is positive while that of all the other single ferrites is negative. It is to be expected, then, that addition of Fe_3O_4 to another ferrite might result in a material having a low magnetostriction and a high initial permeability. No relative data on the variation of the magnetostriction of ferrites with composition have been found in the literature. As mentioned in Chap. 13, the saturation magnetostriction of $ZnMgFe_4O_8$ is very small.

Iron Alloyed with Palladium and Platinum Metals.—The alloys of iron with palladium and platinum are discussed in some detail in Chap. 9. Fallot [38F1] has measured the saturation moments of alloys of iron with each of the six metals Ru, Rh, Pd, Os, Ir, and Pt, and a short summary of his results are given in Table 7. It will be noticed that the Bohr magneton numbers (per atom of alloy) *increase* when the added element is rhodium, iridium, or platinum. With rhodium and platinum the atomic moments

TABLE 7

SATURATION MAGNETIZATION OF ALLOYS OF IRON WITH PALLADIUM AND PLATINUM GROUP METALS

Here σ_0 and σ_s refer to 0°K and 17°C, respectively. The Curie point θ(°C) and the atomic moment n_0 in Bohr magnetons per atom of alloy, are shown. Data are by Fallot [38F1].

Element	Weight Per Cent	Atomic Per Cent	σ_0	σ_s	n_0	θ(°C)
Ruthenium........	4.0	2.2	217.7	212.9	2.22	733
	12.0	7.0	206.5	199.7	2.18	660
	16.8	10.0	186.5	178.7	2.02	606
	20.6	12.5	105.7	104.7	1.17	...
Rhodium.........	6.1	3.4	218.3	213.2	2.25	757
	17.2	10.1	214.2	208.7	2.32	...
	24.5	15.0	210.3	203.9	2.37	...
	38.1	25.0	197.9	191.7	2.39	714
	45.3	31.0	181.9	174.0	2.29	702
	55.1	40.0	168.8	161.3	2.26	624
	69.3	55.0	...	...	...	340
Palladium.........	4.0	2.2	216.9	216.2	2.21	763
	10.0	5.5	207.3	202.3	2.19	754
	56.0	40.0	138.8	129.0	1.89	756
	85.0	74.8	57.6	45.4	0.97	432
Osmium..........	12.5	4.0	195.7	190.2	2.15	725
	23.3	8.1	164.5	158.1	1.97	...
	32.8	12.5	51.7	50.5	0.69	...
Iridium...........	12.6	4.0	205.3	200.4	2.25	753
	21.9	7.5	191.6	187.0	2.27	...
	32.0	12.0	168.3	162.1	2.18	...
	37.9	15.0	122.3	119.9	1.67	...
Platinum..........	13.0	4.1	208.1	203.4	2.29	770
	23.5	8.1	196.7	191.4	2.36	...
	33.0	12.4	185.3	177.2	2.43	...
	43.0	17.8	163.7	159.1	2.36	...
	59.9	30.0	113.4	99.3	1.98	...
	73.4	44.1	40.6	39.2	0.85	...

are 7 and 10% higher, respectively, than in iron, but the weights of these elements are so great that the saturation moment per gram of alloy, σ_0, decreases continuously as they are added. However, the saturation induction at room temperature, B_s, is higher in the iron-platinum and iron-rhodium alloys than in iron.

Iron-Phosphorus.—The phase structure of this alloy system is given in Fig. 7-42. It shows a closed γ loop extending to about 0.6% phosphorus (the $\alpha + \gamma$ region extends from 0.3–0.6% phosphorus at 1100°C). A eutectic with Fe_3P solidifies at 1050°C, and at this temperature 2.8% phosphorus is held in solid solution. The solid solubility decreases to 1.1% at 750°, and probably to 0.4–0.5% at room temperature. Precipitation hardening has been observed in alloys containing more than 1.2%, but not in the alloy with 1.1% phosphorus.

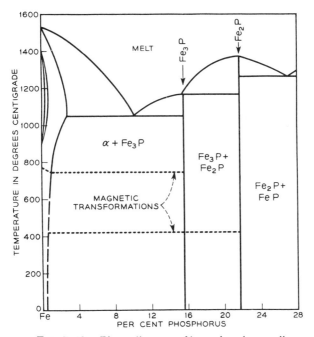

FIG. 7-42. Phase diagram of iron-phosphorus alloys.

The Curie point drops to 745° at the limit of solid solubility and then remains constant to 15.6% (Fe_3P). This compound has a Curie point at 420°C, and alloys containing up to 21.7% (Fe_2P) are magnetic below this temperature. Alloys containing more than the latter amount of phosphorus are non-magnetic. The diagram used is due to Haughton [27H2], as modified by more recent data of Lorig and Krause [36L3].

Alloys can be cold rolled readily up to about 0.5% phosphorus, and with difficulty to about 0.65%, according to Stanley [43S1]. Lorig and Krause, and Kamura [35K10], have shown that the tensile strength of annealed alloys increases with phosphorus to about 75 000 lb/in.^{2} at 0.5–0.7% [36L3]. Elongation is not much affected by phosphorus below 0.5%, but then it begins to drop rapidly and at 0.7% it is practically zero.

The resistivity increases to about 19 microhm-cm at 0.5% phosphorus and then less rapidly to 22 at 1.0% [36L3].

For small phosphorus contents the saruration induction of iron decreases approximately 300 gausses for each per cent of phosphorus [37S4].

With addition of phosphorus the maximum permeability increases, and hysteresis loss decreases by an amount depending on purity and heat treatment. Kamura [34K11, 35K11] found the highest maximum permeability in alloys containing around 0.7% phosphorus and noted that the permeability was higher the higher the temperature of anneal, when this was in hydrogen. Values greater than 20 000 were observed after annealing at 1100°C, whereas the highest value obtained for iron was about 7000.

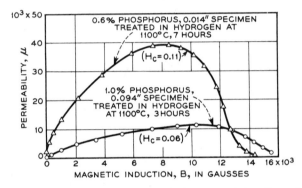

FIG. 7–43. Permeabilities of iron-phosphorus alloys containing 0.6 and 1.0 per cent phosphorus.

Hysteresis loss was as low as 720 ergs/cm^3 for $B_m = 10\ 000$. For phosphorus contents below 0.6% the optimum annealing temperature is below the α,γ point. An optimum phosphorus content of 0.7–0.8% was also reported by Lorig and Krause. Magnetic aging was observed in the low phosphorus alloys but was negligible beyond 0.5% phosphorus.

In another investigation, at the Bell Laboratories, permeabilities of 40 000 were observed in the 0.6% alloy after annealing 0.014-in. sheet at 1100°C for 7 hours. High permeability at high induction—6000 at $B = 14\ 000$—was obtained in thicker sheet containing 1.0% phosphorus. Representative μ vs B curves of these materials are given in Fig. 7–43. Similar characteristics at high inductions occur in rolled sheet containing 0.75% phosphorus and 0.5% manganese, annealed at 800°C. These properties compare favorably with those of iron for use in relays and other electromagnetic apparatus.

Initial permeabilities greater than those ordinarily found in iron are obtained only in thin sheet annealed in hydrogen for 5–10 hours at 1100°C or above. This suggests that purification is necessary. Initial permeabilities of over 2000 were so obtained in 0.014-in. sheet containing either 0.7% or 1.0% phosphorus.

Köster [31K8] investigated precipitation hardening in alloys containing more than 1.0% phosphorus. The highest coercive force observed, $H_c = 6$, is so low that a low saturation magnetostriction is indicated.

Iron-Selenium.—Selenium is soluble in iron to a small, undetermined extent. The area in which Fe and FeSe are in equilibrium extends to about 50 atomic per cent selenium (58.6 weight per cent) [36H1].

Iron-Sulfur.—Iron and FeS form a eutectic melting at 985°C, containing 30.8% sulfur. The solubility of sulfur at this temperature has been estimated as 0.3%. At higher temperatures it is less, and at lower temperatures it is much less, perhaps 0.01–0.02% at room temperature.

The compound FeS (36.5% S) dissolves sulfur and at first forms an ordered structure (50 to 51 atomic per cent S) that is paramagnetic. Above 138°C, or with the addition of more sulfur, this forms pyrrhotite which has the familiar "NiAs" structure and a Curie point of 300–325°C, depending on composition [41H6]. In the range 40–53% sulfur, FeS and non-magnetic FeS_2 are both present, and the saturation induction decreases toward zero. Beyond 53% the stable phases are FeS_2 and S. The vapor pressure of sulfur over FeS has been measured by Britzke and Kapustinsky [30B2]. The equilibrium with H_2S has already been discussed in Chap. 3.

The saturation intensity of magnetization of pyrrhotite at room temperature was determined by Ziegler [29W2] as $I_s = 62$, but lower values have been observed. The coercive force varies from 15 to 30.

The coercive force of iron increases with sulfur content to about 4 oersteds at 5% sulfur [32M3]. The effect of smaller amounts of sulfur is known to prevent attainment of high permeability and is discussed in Chap. 3. The resistivity increases linearly to about 50 microhm-cm at 10% sulfur.

Hilpert [38H5] observed ferromagnetism in the compound $FeS \cdot Fe_2O_3$, which has a Curie point of 580°C.

Iron-Tantalum.—Reports on the phase structure of this system are conflicting. Genders and Harrison's diagram [36G3] of Fig. 7–44 differs considerably from those previously published [35J2] and is inconsistent in some respects with the later work of Nemilov and Voronov [38N4]. The statement that the precipitating phase is Fe_2Ta is in agreement with the report of Wallbaum [41W2]. No measurements of Curie point have been noted.

Age hardening is observed in alloys containing up to 35% tantalum. The hardness rises rapidly with tantalum content, even in the quenched alloys. Dean [33D3] hardened an alloy containing 22% tantalum and measured $H_c = 19$, $B_r = 7600$.

254   OTHER IRON ALLOYS OF HIGH PERMEABILITY

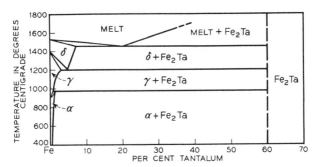

Fig. 7-44. Phase diagram of iron-tantalum alloys.

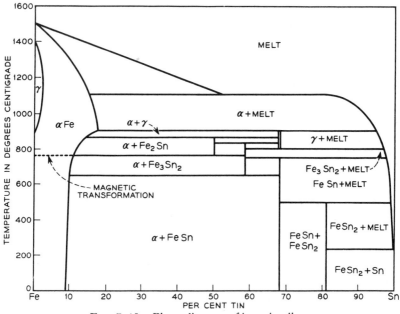

Fig. 7-45. Phase diagram of iron-tin alloys.

Jellinghaus [35J2] determined saturation induction as follows:

%Ta	B_s	%Ta	B_s
15	15 050	50	5170
25	12 950	75	80

From these data it appears that Fe_2Ta (62% Ti) is non-magnetic.

Iron-Tin.—This system has a rather complicated phase diagram in which appear five compounds, all of which have hexagonal crystal structures (Fig. 7-45). The γ loop closes at about 2% tin. Maximum solubility in the α phase is about 17% at 760°C; this is very near the Curie point of iron, which is changed only slightly by the addition of this much tin. At room temperatures the solubility has decreased to 8 or 9%.

Apparently $FeSn_2$ is magnetic, for the alloy containing 97% tin is reported [25W3] to be weakly ferromagnetic. The phase diagram is that of Ehret and Gurinsky [43E2].

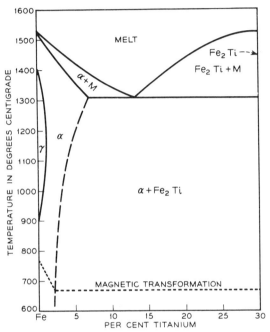

FIG. 7-46. Phase diagram of iron-titanium alloys.

Fallot has measured Curie points and atomic moments of alloys containing 12% and less of tin. In this range the Curie point drops only 2 to 5°C, and the saturation at room temperature decreases about 8%.

In some early work, Burgess and Aston [09B3] found a reduction in hysteresis by adding 2% of tin to iron. Dahl, Pawlek, and Pfaffenberger [35D2] observed a reduction in total losses in sheet which contained less than this amount of tin. In experiments at Bell Laboratories a permeability of 34 000 and a coercive force of 0.15 oersted have been attained by heat treating the 3% alloy in pure hydrogen at 1100°C.

Alloys containing carbon become brittle when the tin content exceeds a few tenths of a per cent [23W2].

Iron-Titanium.—Iron forms a eutectic with the compound Fe_2Ti, which has a hexagonal $MgZn_2$ structure and melts at about 1520°C [41W2]. The eutectic solidifies at 1310°C, and at this temperature 6-7% of α iron is soluble. Solubility is lower at room temperature and is estimated to be less than 2% [38V3]. The γ loop closes at about 0.8% titanium. (See Fig. 7-46.)

According to early data [14L1] the Curie point of iron is lowered to 690°C by addition of titanium and then remains constant to over 20% titanium. It is probable that alloys in the Fe + Fe_2Ti field all show ferromagnetism and that Fe_2Ti (30.0% Ti) itself is non-magnetic.

According to Messkin and Somin [36M10] the ferric induction at high field strength decreases approximately linearly to about 9500 at 10% titanium. Seljesater and Rogers [32S6], however, measured higher inductions, and these are consistent with a decrease of B_s to zero at about 30% titanium (Fe_2Ti).

Applegate [15A1] observed that addition of 0.5-1% of pure titanium reduced the hysteresis loss and raised the permeability slightly, and that larger additions had the opposite effect. Magnetic hardening was observed by Seljesater and Rogers in alloys containing 3.3-7.1% titanium. A coercive force of 55 and a remanence of 7300 were attained in the 7.1% alloy.

Iron-Tungsten.—In this system the γ loop closes at about 6% tungsten; the α + γ region is unusually broad, and at 1100-1200°C extends from 3 to over 6%. At 400°C the solubility limit of α solid solutions is about 6% tungsten, and at higher temperatures the solubility increases rather rapidly as shown by the diagram of Fig. 7-47, taken from Sykes [39S14]. The areas of equilibrium of Fe_2W (hexagonal) and Fe_3W_2 are also shown there. Both of these compounds are believed to be non-magnetic.

The Curie point is not noticeably lowered by addition of tungsten. Alloys are magnetic at room temperature as long as the α phase persists, that is, to 60-70% tungsten.

Saturation induction decreases linearly and slowly to 6% tungsten, when it is 21 000, according to Stäblein [29S4]. At higher tungsten contents it decreases somewhat more rapidly to 14 000 at 30% and then toward zero at 60-70% tungsten.

Magnetic softness is not much affected by tungsten additions in the α-phase area. In annealed alloys the coercive force was found to be constant at 0.5 oersted up to 6%. At higher percentages, when precipitation hardening occurs, H_c increases rapidly and is as high as 150 in the 28% alloy after favorable heat treatment [32S6]. The lattice parameters and the mechanism of precipitation have been studied by C. S. Smith [41S7].

The Brinell hardness increases continuously with tungsten addition to 30-50%. Alloys may be hot forged to about 25% and rolled to about 20% [39S14]. Densities have been reported by Nishiyama [29N2] and Stäblein [29S4]; they increase to 9.5 at 30% tungsten. Resistivity increases to about 17 microhm-cm at 5%, then to 20 microhm-cm at 10% [28S8], and then remains almost constant to 25% [29S4].

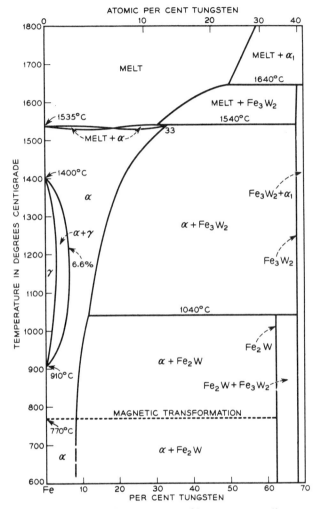

Fig. 7-47. Phase diagram of iron-tungsten alloys.

The magnetic properties of alloys containing carbon are discussed in Chap. 9.

Iron-Uranium.—Gordon and Kaufmann [50G2] have found the compounds Fe_2U (68.1% U) and FeU_6 (97.1% U) and although they are not ferromagnetic at room temperature, Fe_2U is ferromagnetic at 81°K and there has a saturation of σ_s =19.9. Co_2U and Ni_2U are not ferromagnetic at the low temperatures used. The eutectic of Fe_2U and Fe freezes at 1080°C. The solubility of uranium in iron is slight and has not been determined. The two-phase area of Fe + Fe$_2$U extends to 68% uranium.

Iron-Vanadium.—These alloys, like the iron-chromium alloys, are characterized by a closed γ loop and a series of α solid solutions continuous except for a compound of composition FeV (see Fig. 7–48). The FeV (47.7% V) phase was discovered by Wever and Jellinghaus [30W1] who obtained relative thermal, X-ray, and magnetic data. The structure, as indicated by X-rays, is complex and has not been determined. The pure compound and its solid solutions can exist between about 30 and 60% vanadium. Lattice spacings of the iron-vanadium alloys have been measured by Wever and Jellinghaus, by Nishiyama [29N2], and by Osawa and Oya [29O3].

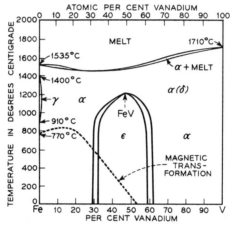

Fig. 7–48. Phase diagram of iron-vanadium alloys.

The Curie point of iron is raised by addition of vanadium to a maximum of 815°C at 5% vanadium, according to Oya [30O2]. From this point the Curie point falls slowly and then rapidly with increasing vanadium. Beyond 30% the Curie point depends on cooling rate and is higher for quenched than for annealed alloys. When quenched, the 56% alloy is strongly magnetic, whereas the 64 and 67% alloys are only weakly magnetic at liquid air temperature [30W1]. The 33.5% alloy is non-magnetic when slowly cooled. Fallot [36F1] observed a second Curie point of 98°C in alloys containing FeV; however, the alloys contained over 1% of carbon, and this transformation might be attributed to the carbide rather than to FeV.

The saturation magnetization drops to about half that of iron at 25% vanadium [36F1], and then drops even more rapidly.

Additions of a few per cent of vanadium to iron have only a small effect, if any, on the maximum permeability. Addition of 1% is enough to reduce the aging of iron to an inappreciable amount.

The resistivity of iron increases approximately linearly to 35 microhm-cm at 6% [31H4], and then increases less rapidly [25M1].

Iron-Zinc.—In this system there are six or more solid phases of which only one, the α phase, is magnetic. There is still some uncertainty in the phase structure of the iron-rich alloys, but it appears that zinc lowers the γ,α transformation temperature of iron to 623°C (at 20–73% Zn). It is soluble to the extent of 20% in α iron and 35% in γ iron (at 780°C) as shown in the diagram of Fig. 7–49, drawn according to Schramm [36S9].

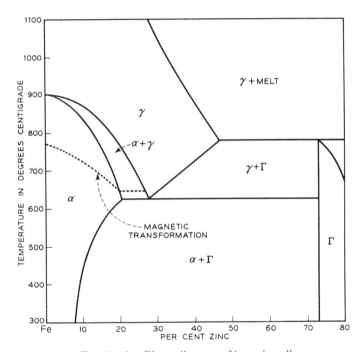

Fig. 7–49. Phase diagram of iron-zinc alloys.

In order to avoid the difficulties associated with the high volatility of zinc, his alloys were prepared by heating pressed, powdered mixtures of particles composed of iron and the 85% zinc iron alloy.

The α phase exists in equilibrium with the Γ phase (cubic, γ brass structure) to over 70% zinc, when the alloys cease to be magnetic. The Curie point of iron is lowered to 670°C by 19% of zinc [37F1, 36S9] and does not fall below 620°C, the α,γ point for higher zinc alloys. Estimates by different workers of the solubility of zinc at low temperatures lie between 7% [36S9] and 16% [37S7]. The precipitation hardening of iron-rich alloys has been studied by Schramm [48S6].

Forrer [37F1] has determined the atomic moments of alloys containing less than 20% zinc. In this range of composition zinc appears to act mainly as a diluent, the atomic moment of the iron atoms remaining approximately constant, but there is some evidence that zinc adds about 1% to the atomic moment.

In experiments on two alloys, prepared by electrodeposition and annealing and containing 0.5 and 0.6% zinc, Dahl and co-workers [35D2] found that the magnetic losses in sheets were greater than in iron.

Iron-Zirconium.—Zirconium is but slightly soluble in α or γ iron—probably less than 0.3% in α iron and less than 1% in γ iron. The α,γ point is lowered to about 835°C, and the Curie point is not noticeably changed. A diagram has been drawn by Vogel and Tonn [32V2]. Alloys are magnetic when the α phase is present, and the area containing α probably extends to the composition Fe_2Zr (45.0% Zr), according to Wallbaum [41W2]. This compound has the same structure as Fe_2Ti.

CHAPTER 8

OTHER HIGH PERMEABILITY MATERIALS

COBALT

In 1946 the world production of cobalt was about 6000 tons, most of it mined in the Belgian Congo (province of Katanga), some in Northern Rhodesia, French Morocco, the United States, and Canada. Over one-third of this was used in the manufacture of permanent magnets. On account of new uses for cobalt, developed by research, the production and consumption have increased tenfold in about twenty years.

Before World War II the cobalt residues from the Belgian Congo were shipped to Belgium for refining. When that became impossible, refineries were set up in the United States, and total shipments of ore to this country increased from 300 tons in 1939 to 5000 tons in 1940 [41K1].

In addition to its use in permanent magnets, considerable quantities of the metal are used in the manufacture of Stellite (45 to 52% Co, 28 to 32% Cr, 10 to 15% W) and tool steels containing up to 12% cobalt and several per cent of tungsten, molybdenum, chromium, or vanadium. In much smaller quantities cobalt is used in high-temperature, creep-resistant alloys, in alloys for dies, valves, cutting tools, vacuum-tube filaments, and glass-to-metal seals, and in electroplating. In the combined form it is important in the ceramic industry.

In permanent magnets cobalt is used in most of the Alnicos (in proportions up to 24%), in KS steel (36% Co), in New Honda steel (27% Co), in cobalt-tungsten steel (e.g., 17% Co), in cobalt-chrome steel (e.g., 15% Co), and in Remalloy (12% Co). The magnetically soft alloys include Permendur (50% Co) and Perminvar (7.5–25% Co).

Metallurgy of Cobalt.—In the large mines of Africa cobalt is found in ores associated with copper and iron. The higher-quality ores are smelted in electric furnaces with lime and coke to form a "white alloy" containing about 40% cobalt and 15% copper. In the refining operation this is dissolved in acid, precipitated as the hydroxide, dried, mixed with dextrin, and formed into pellets which are reduced with charcoal (carbon monoxide) in a reverberatory furnace at 1050°C. Lower-quality ores must be subjected to more complicated processes, which utilize electrolysis to an increasing extent. In any case electrolysis may be used for final purification. Typical analyses of the rondelles and of the purified product are given in Table 1.

TABLE 1. TYPICAL ANALYSES OF COBALT

Element	Rondelles (%)	Electrolytic (%)
Co	97–99	99.9
Fe	0.1–0.6	0.05
Ni	0.2–0.4	tr
Mn	0.05–0.1	tr
Si	0.03–0.1	0.02
S	0.01–0.06	0.05
C	0.02–0.3	0.00

Physical Properties.—Many of these are given in Table 2. Expansion [27S2, 37M5] and resistivity [21H1, 27S2, 36M5] are shown as a function of temperature in Figs. 8–1 and 2, and these indicate graphically the phase-transformation [26M1] that occurs on heating at 400–500°C. In two specimens of cobalt of high purity (Kahlbaum, Heraeus) the transitions on

TABLE 2. SOME PHYSICAL PROPERTIES OF COBALT

Density, g/cm³, direct	8.78–8.85
Density, g/cm³, X-ray (hexagonal)	8.84
Density, g/cm³, X-ray (cubic)	8.79
Lattice spacing, kx (hexagonal)	$c = 4.0611$ $a = 2.5020$
Lattice spacing, kx (cubic)	3.5368
Thermal expansion, $\alpha \times 10^6$ (0–400°)[36M5]	14×10^{-6}
Resistivity, microhm-cm at 20°C	6.24
γ, ϵ transformation temperature (approx), °C	425
Melting point, °C	1495
Boiling point, °C (approx)	2900
Thermal conductivity, cal g^{-1} deg^{-1}	0.16
Compressibility, cm²/kg	0.50×10^{-6}
Specific heat, cal g^{-1} deg^{-1} (20°C)	0.100
Specific heat, cal g^{-1} deg^{-1} (18°K)	0.0001
Heat of fusion, cal/g	60
Heat of vaporization, cal/g	1500
Heat of transformation (γ, ϵ) cal/g [35S5]	0.1
Increase in volume at 450°C on heating, (%)	0.54
Hardness, Brinell	125
Tensile strength, annealed, lb/in.²	30 000
Tensile strength, drawn, lb/in.²	100 000
Young's modulus, dynes/cm²	$1.8–2.0 \times 10^{12}$
Young's modulus, lb/in.²	$26–30 \times 10^6$

heating and cooling were observed [27S2] to be completed at 460–470 and at 390°C, respectively. In one specimen containing about 0.5% iron and 0.3% manganese the transition on cooling was completed only at the rather low temperature of 130°C.

There has been some discussion [35S4] of a transformation at temperatures over 1000°C, and Marick's data of Fig. 8–1 indicate a sudden increase in expansion at 1100°C; but he associated this with the Curie point, and using X-rays observed only the cubic phase between 500 and 1200°C.

COBALT PROPERTIES

In the low-temperature, hexagonal phase the atoms occupy positions very nearly those of ideal close-packing in which the inter-atomic distances in any one (00·1) plane (perpendicular to the hexagonal axis) are equal to the shortest distances between atoms in neighboring planes. Adjacent

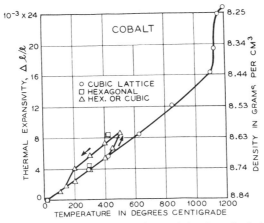

FIG. 8–1. Thermal expansion and density of cobalt.

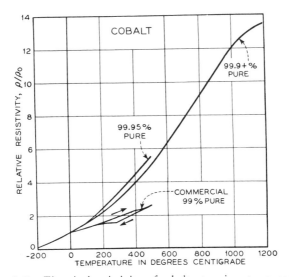

FIG. 8–2. Electrical resistivity of cobalt at various temperatures.

layers of atoms in (00·1) planes are displaced with respect to each other but the pattern repeats every second layer. In cubic cobalt the (111) planes are similar to those of the (00·1) planes of hexagonal cobalt, but the succession of planes is different so that in the cubic structure the pattern repeats after every third layer. The similarity in the structures goes hand in hand with the small heat of transition [35S5] and, according to Edwards, Lipson, and Wilson [42E1, 42W3, 43E3], is a necessary condition for the formation of faults or breaks in the perfect, two-layer, cobalt structure, faults which he postulates to explain the irregularities in widths and intensities of X-ray reflections from various types of crystal planes. Their analysis of the data indicates that at every eighth to tenth layer, on the average, the perfect regularity fails. This peculiar behavior of cobalt is probably associated with the fact that the hexagonal phase is stable only at temperatures so low that the diffusion of atoms is extremely slow and atoms cannot move easily enough to form a perfect lattice after they have once been disrupted by the transformation. A discussion of the nature of the transformation has also been given by Troiano and Lokich [48T1].

The cubic phase has a lower density and lower resistivity than the hexagonal phase has at the same temperature. The best value of the density is assumed to be 8.84 g/cm^3 at 0°C. The density at 0°K, obtained by extrapolation using Grueneisen's equation, is then 8.89.

Magnetic Properties.—A summary of data relating to the saturation magnetization of cobalt at room temperature and at absolute zero is given in Table 3. Values of magnetization per gram (σ) and per cm^3 (I) have been converted one to the other using a density of 8.84 g/cm^3 at room temperature and 8.89 at 0°K.

TABLE 3. SATURATION MAGNETIZATION AND CURIE POINT OF COBALT

Author	0°K			20°C			
	σ_0	I_0	$4\pi I_0$	σ_s	I_s	$4\pi I_s$	θ (°C)
Bloch [12B1].........	170.2	1512	19 000	169	1492	18 750	1121
Masumoto [26M1].....							1115
Weiss, Forrer, and Birch [29W3].......	162.3	1442	18 120	161	1422	17 870	1115
Honda and Masumoto [31H2]...	162.8	1446	18 170	161	1422	17 870	...
Hashimoto [32H1]......							1121
Hansen [36H1]........							1112 to 1145
Myers and Sucksmith [51S6]	162.5	1445	18 150	...	...	...	1131

Bloch's value for σ_0 was obtained by extrapolation from -181°C.

The number of Bohr magnetons per atom is

$$n_0 = 162.5 \times 58.94/5587 = 1.714.$$

Magnetization as dependent on *temperature* is shown in Fig. 8-3, derived from the original data of Bloch but corrected to bring them into agreement

COBALT PROPERTIES 265

with the low-temperature measurements of Honda and Masumoto [31H2], whose data also are shown for temperatures below 400°C. Both sets of measurements were made with field strengths up to 10 000. No change

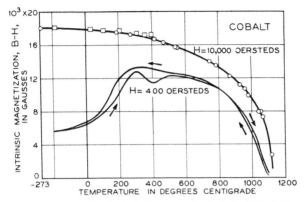

FIG. 8-3. Intrinsic induction of cobalt in high fields.

in saturation is apparent here as the polycrystalline specimen goes through the ϵ,γ transformation point near 400°C, but Myers and Sucksmith [51S6] have observed a 1.5% rise in saturation as the material is heated

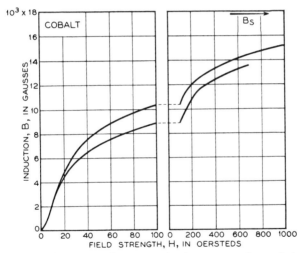

FIG. 8-4. Representative magnetization curves of annealed cobalt.

through this range of temperature. An irregular change in B at this temperature was noticed by Honda and Shimizu [05H1] when fields of 400 or lower were used. It should be stated that the permeability of the specimen of Honda and Shimizu is considerably less than that obtained by later workers.

Two magnetization curves of cobalt are given in Fig. 8-4. Miss Samuel [28S2] used a rod containing 1.5% iron and less than 0.1% total of the elements Ni, Mn, Si, and C, and measured it with a magnetometer after it had been annealed and slowly cooled. The Bell Laboratories specimen was a ring formed of many layers of 0.006-in. rolled tape prepared from a melt of rondelles to which manganese had been added. It contained 98.6% cobalt (chief impurities, 0.6 Fe, 0.9 Mn, 0.1 Si, 0.02 C) and was annealed in a closed pot at 1000°C and cooled with the furnace.

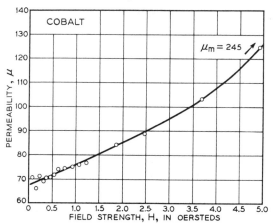

Fig. 8-5. Permeability of cobalt in low fields.

The maximum permeabilities of these specimens are about 250, a value approximately one-tenth of that of nickel and one-twentieth of that of ordinary iron. The initial permeability is correspondingly low, as indicated in Fig. 8-5, and the permeability in low fields is given by:

$$\mu = 68 + 8H.$$

Magnetic constants for this same specimen are as follows: $\mu_m = 245$, $H_c = 8.9$, and $B_r = 4900$; hysteresis losses for this and others are given in Table 4.

TABLE 4. HYSTERESIS LOSSES (ERGS/CM³ PER CYCLE) AND HYSTERESIS CONSTANT η OF EQUATION $W_h = \eta B^{1.6}$ FOR VARIOUS SPECIMENS OF COBALT

Observer	H_m	B_m	W_h	η
Ewing [00E1]	140	10 000	30 000	0.01
Samuel [28S2]	700	13 000	23 000	.006
Bell Laboratories	21	5 000	6 900	.008
Kühlewein [32K2]	8	1 620	1 000	.007
Kühlewein (380°C)[32K2]	8	11 600	7 800	.002
Bell Laboratories (calc)	1.3	100	2	...

Figures 8-6 and 7, showing the reversible permeability as dependent on induction and the coercive force at various temperatures, refer to the same specimen of cobalt as that used by Miss Samuel [28S2] for the data of Fig. 8-4. She also recorded an ideal magnetization curve.

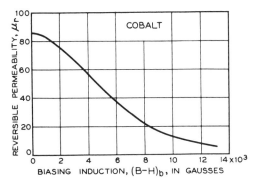

Fig. 8-6. Reversible permeability of cobalt at various biasing inductions.

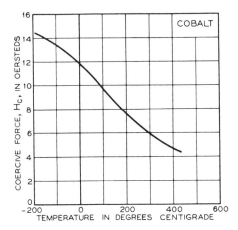

Fig. 8-7. Coercive force of cobalt at various temperatures.

NICKEL

In recent years the annual world production of nickel has been about 150,000 tons. Of this the largest proportion has come from Canada.

Over half of the nickel is consumed in the manufacture of low-nickel alloys and cast irons. These include structural steels containing a few per cent of nickel and one per cent or less of carbon, and corrosion- and heat-resistant alloys such as stainless steel (e.g. 18% Cr, 8% Ni, 74% Fe). About one-fourth is used in substantially pure form for electroplating and for corrosion-resistance in the chemical industries.

High-nickel alloys, containing, for example, 80% Ni and 20% Cr, or 60–65% Ni, 15% Cr and some iron, are employed because of their resistance to high-temperature oxidation. Monel metal, containing about 67% nickel and 30% copper, is made directly from the ore which contains these elements in these proportions, or it may be made by alloying nickel and copper in the correct proportions. It is used because of its corrosion-resistance. Invar (36% Ni) and Elinvar (36% Ni, 12% Cr) have low temperature coefficients of expansion and of elastic modulus, respectively; and Kovar and Fernico (28% Ni, 18% Co) are used for sealing into glass. All of these have rather variable compositions in the various commercial products.

Only 4–5% of the world's nickel finds its way into magnetic alloys. These are the high-permeability materials such as the Permalloys, Hipernik, and Mumetal, which contain 40–80% nickel (the remainder mainly iron), and the permanent magnets of the Alnico type containing 14–28% nickel, 0–25% cobalt, 8–12% aluminum and sometimes a few per cent of copper. Pure nickel is used in magnetostriction oscillators and microphones. Magnetic alloys, containing 10–30% copper, are used for temperature-compensation, and higher copper contents are present in non-magnetic alloys.

Metallurgy of Nickel.—The chief source of nickel is a sulfide ore containing only a few per cent of nickel and copper and a much larger amount of iron. Minute amounts of the platinum metals are also present and constitute the principal source of these materials.

In producing nickel selected high-grade sulfide ore may be charged directly into a blast furnace, but it is more general practice to beneficiate the ore by crushing and fine grinding, and to follow this by bulk flotation, or by selective flotation which removes much of the copper for separate refining. The flotation concentrate may be roasted to lower the sulfur content, sintered and charged into a blast furnace or roasted in a multiple hearth furnace and melted in a reverbatory furnace. The resulting matte is further treated in a Bessemer converter, which lowers the iron and sulfur contents. The product may be subjected to additional purifying treatments and finished as "nickel oxide sinter" for the steel industry, or converted into anodes for electrolytic refining for use in high-quality alloys. Or it may be treated by the Mond process in which the oxide is gas-reduced and then treated with carbon monoxide below 100°C. The nickel carbonyl so produced, $Ni(CO)_4$, is decomposed at 180–200°C and nickel pellets (grade A nickel) are formed.

NICKEL 269

TABLE 5. TYPICAL ANALYSES OF COMMERCIAL NICKELS [39A2]

	Electrolytic Ni (Com'l)	Wrought "A" Ni	Cast Ni
Ni + Co	99.95%	99.4%	96.7%
Fe	0.03	0.15	0.5
Cu	.02	.1	.3
Mn	...	.2	.5
C	tr	.1	.5
Si	tr	.05	1.5
S	tr	.01	

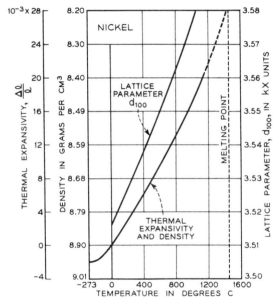

FIG. 8-8. Thermal expansion, lattice parameter, and density of annealed nickel.

Physical Properties.—Some of the physical properties of pure nickel are given in Table 6, for which data have been collected from various sources [30J1, 37W2, 39A2, 48W3]. The structure is face-centered cubic at all temperatures. The thermal expansion and lattice parameter are shown for temperatures up to 1100°C in Fig. 8-8 [38E2, 36O1, 41N1]. The dependence of the resistivity on temperature is given in Fig. 8-9 [23H1, 37P1], which includes unpublished data of S. Umbreit.

Magnetic Properties.—The best estimates of the saturation magnetization at room temperature and at temperatures down to absolute zero are due to Weiss and his students whose values [37W4] are given in Table 7. In their experiments the σ_s's were determined directly, and in the table the

270 OTHER HIGH PERMEABILITY MATERIALS

TABLE 6. SOME PHYSICAL PROPERTIES OF PURE NICKEL

Density at 20°C, g/cm^3	8.90
Thermal expansion coefficient (0–100°C)	13 × 10^{-6}
Lattice constant, kx	3.5168
Melting point, °C	1455
Boiling point, °C (approx)	2730
Curie point, °C	358
Resistivity at 20°C, microhm-cm	6.8
Temperature coefficient of resistance (0–100°C)	0.007
Thermal conductivity, cal cm^{-1} deg^{-1} sec^{-1} (20°C)	0.22
Compressibility, cm^2/kg	0.531 × 10^{-6}
Specific heat, cal g^{-1} deg^{-1} (100°C)	0.11
Heat of fusion, cal/g	73.8
Modulus of elasticity, unmagnetized, lb/in.^{2}	30 × 10^6
Modulus of elasticity, kg/mm^2	21 000
Modulus of rigidity, lb/in.^{2}	11 × 10^6
Elastic constants, c_{11}, c_{12}, c_{44} (resp.), in dyne-cm^{-2} × 10^{-12}	2.50, 1.60, 1.185
Tensile strength, lb/in.^{2} (annealed)	46 000
Yield strength (annealed), lb/in.^{2}	8500
Brinell hardness number (annealed)	75
Elongation (annealed)	30%

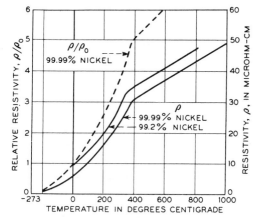

FIG. 8–9. Electrical resistivity of nickel at various temperatures.

TABLE 7. SATURATION MAGNETIZATION

σ_s (15°C)	54.39	σ_0	57.50
I_s (15°C)	484.1	I_0	508.8
$4\pi I_s$ (15°C)	6084	$4\pi I_0$	6394

other quantities have been derived from them using the densities indicated in Fig. 8–8. Older values of $4\pi I_s$ at room temperature are near 6150 [30J1]. From σ_0 may be calculated the number of Bohr magnetons per atom,

$$n_0 = \frac{57.50 \times 58.69}{5587} = 0.604.$$

NICKEL PROPERTIES 271

Determinations of the Curie point vary from 358°C [16A1] to 382°C [12B1]. The true value probably lies between 358 and 363°C [29W2, 37M2].

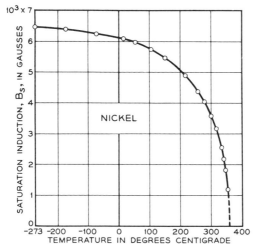

FIG. 8–10. Saturation induction of nickel at various temperatures.

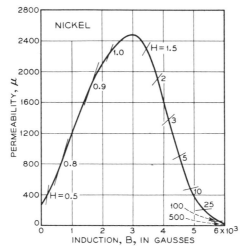

FIG. 8–11. Representative permeability curve of annealed nickel.

The variation of B_s with temperature is shown graphically in Fig. 8–10. The data are those of Weiss [29W2, 37W4]; values for temperatures between 0 and $-200°C$ have been determined from the original data by interpolation, using the $T^{3/2}$ law.

A typical μ, B curve for commercial electrolytic nickel annealed for 1 hour at 1000°C in hydrogen is shown in Fig. 8–11. B, H curves are given

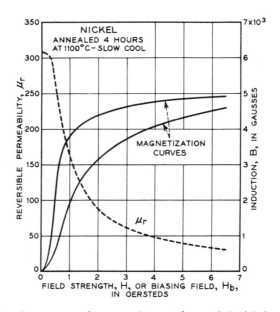

FIG. 8-12. Induction curves of two specimens of annealed nickel, and reversible permeability. Lower B,H curve is annealed at a lower temperature.

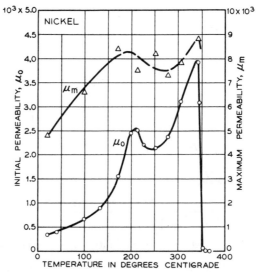

FIG. 8-13. Dependence of initial and maximum permeabilities of nickel on temperature.

in Fig. 8–12, in which is shown also the reversible permeability, μ_r, as measured by Kirkham [37K1] using a specimen containing 0.1% Fe,

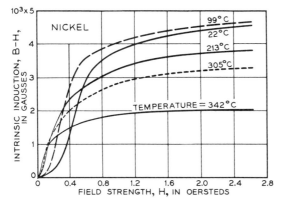

FIG. 8–14. Magnetization curves of nickel at various temperatures.

0.1% Mg, 0.05% C, and 99.7% Ni, annealed for 4 hours at 1100°C in hydrogen. He determined μ_0 and also μ_m as a function of temperature, and Fig. 8–13 shows how these pass through one maximum at about

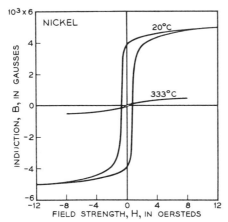

FIG. 8–15. Hysteresis loops of nickel at room temperature and at a temperature just below the Curie point.

200°C and a second maximum at about 25°C below the Curie point. The first maximum may be considered abnormal and is attributed to the change, at this temperature, of the direction of easy magnetization in the crystals from [111] to [100]. From the same set of data have been plotted B,H curves for five temperatures between 0°C and the Curie point (Fig. 8–14).

The effect of temperature on hysteresis is shown in Fig. 8–15 as determined by Kühlewein [32K2], whose specimen contained 1% manganese. He showed also that the coercive force decreases approximately linearly with temperature from 20° to the Curie point where it still appears to have

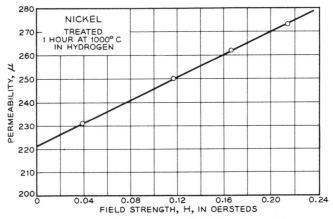

FIG. 8-16. Permeability of nickel at low fields.

a value of about 0.1 oersted. The residual induction increases less smoothly with temperature and apparently approaches zero at the Curie point.

A typical μ,H curve for low inductions is shown in Fig. 8-16. In this region the hysteresis loss per cycle is given approximately by $W_h = 26 H_m^3$

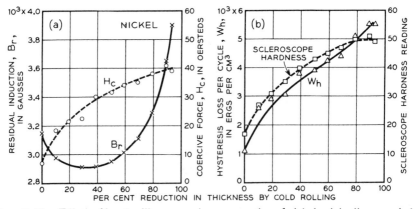

FIG. 8-17. Effect of hard rolling on various properties of nickel originally annealed.

erg/cm^3. At $B_m = 5000$ the hysteresis loss per cycle was found to be 1270 erg/cm^3. The specimen was one of high commercial purity, containing 0.1% iron and 0.4% cobalt.

An interesting set of hysteresis loops has been published by Forrer [26F1]. He took a wire of 0.0087-in. diameter, stretched it to breaking, wound it on a second wire of 0.071-in. diameter, unwound it, and kept it straight during measurement by inserting it in a capillary glass tube. The loops have exceedingly steep sides and in some cases it was established that they were absolutely vertical (see Figs. 13-20 and 21). This was the first time that such characteristics had been observed for any ferromagnetic material. More recently they have been observed under a variety of conditions, as will be discussed further later.

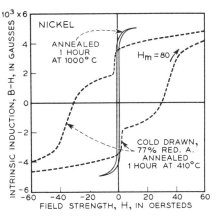

FIG. 8-18. Hysteresis loops of nickel: (a) partially annealed (410°C) after cold reduction of area by 77%; (b) fully annealed (1000°C).

The effect of different amounts of cold rolling on some of the magnetic properties and on the scleroscope hardness of commercial A nickel has been investigated by Williams [27W2], whose data are reproduced in Fig. 8-17. Specimens were 0.25 in. thick before rolling. The maximum field strength to which they were subjected was about 150 oersteds. The coercive force of 7 oersteds measured before cold rolling indicates that the material was originally not fully annealed.

The effect of the temperature of anneal of previously hard worked nickel has been most carefully investigated by Bittel [38B2]. The hysteresis loop obtained in a nickel wire after a reduction in area of 77% by drawing and a subsequent anneal for 1 hour at 410°C is given in Fig. 8-18 where it is compared with a fully annealed specimen. The peculiar shape of the loop of the partially annealed nickel indicates that at this temperature only a portion of the material has recovered. In Fig. 8-19 the coercive force is shown for various annealing temperatures for the same wire and for another wire stretched to breaking after being annealed initially at 1000°C.

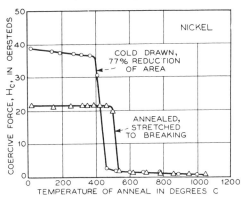

FIG. 8-19. Dependence of the coercive force of nickel on the annealing temperature: (a) annealed and drawn with reduction in area of 77%; (b) annealed and stretched to breaking.

Recovery here is seen to be very sharp and, as usual, to take place at a somewhat lower temperature when more severely worked.

COBALT-NICKEL ALLOYS

None of these alloys has found any important industrial use. Even the magnetic properties have no interesting peculiarities; they are rather the result of mixing two elements that are similar chemically and different magnetically in that one has almost three times the saturation magnetization of the other.

Structure and Physical Properties.—Early work on the *phase diagram* [12R1] indicated that these alloys constitute a continuous series of solid solutions at all temperatures below the solidus. In 1926 Masumoto [26M1] discovered the γ,ϵ transformation at 400–480°C in pure cobalt and determined the transformation temperatures in alloys containing up to 30% nickel. The ϵ phase has a close-packed hexagonal structure with an axial ratio (1.62) not far from that for ideal packing (1.63); the γ phase has the face-centered cubic structure of nickel. Figure 8–20 shows the γ,ϵ transformation temperatures on heating and on cooling, and the separation of the lines is here 70–100°C. The width of the two-phase region at equilibrium has not been determined. The figure shows also the small separation between the liquidus and the solidus [27K1], and the *Curie points* as determined by a number of workers [12B1, 26M1, 32H1, 35B7].

The structures of the γ and ϵ phases have been confirmed by X-rays by Osawa [30O1]. The alloy containing 25% nickel was found to have the hexagonal structure; that with 30% nickel, the face-centered cubic structure. It is interesting to note that the "irreversible" character of the alloys, now known to undergo a γ,ϵ transformation, was noted by Bloch [12B1] in his magnetic investigation in 1912. The possibility of a phase transformation in pure cobalt at 1000–1200°C and its effect on the cobalt-nickel diagram are still under discussion, but its existence is doubtful.

Bloch's earlier direct determinations of the densities of these alloys were of admitted inaccuracy and were not used by him in his own magnetic study. X-ray spacings determined by Osawa [30O1] do not agree with the more recent and precise determinations of Owen and Yates [36O1] for nickel, and Meyer [37M4], Marick [36M5], and Ellis and Greiner [48E1] for cobalt; consequently they cannot be used for a proper determination of the densities of the alloy series. Yamamoto, in a paper published in Japanese in 1942, found a linear relation between density and composition except in the range 20–40% nickel, when two phases are present and the density is lowered to about 8.6. The best values for the densities of pure cobalt and of pure nickel are, respectively, 8.84 and 8.90.

COBALT-NICKEL ALLOYS

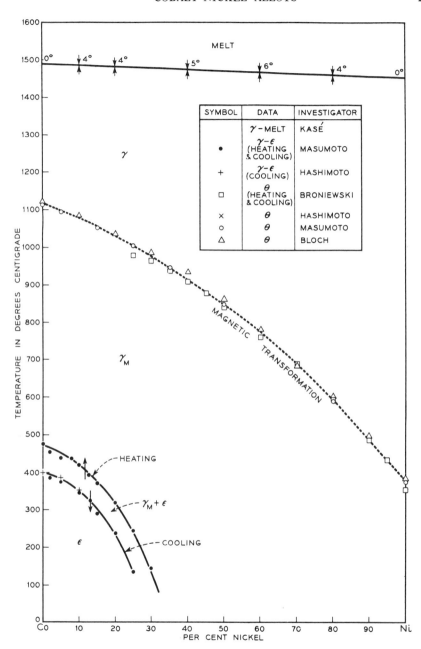

Fig. 8-20. Phase diagram of cobalt-nickel alloys.

278 OTHER HIGH PERMEABILITY MATERIALS

The coefficient of *thermal expansion* has not been determined well enough to show a trend with composition. A value of 11 to 14 × 10^{-6} applies equally well to all alloys [27S1, 31M2, 35B7].

Likewise, the *resistivities* reported by various authors do not agree well with each other. Three sets of determinations have been plotted in Fig. 8-21. The early data by Masumoto [27M2] were taken on specimens made from Mond nickel containing 0.1% iron and 0.04% carbon, and cobalt containing 0.2% iron and 0.22% carbon. Later data by Broniewski and Pietrek [35B7] for 0°C have much lower values than those of Masumoto, lower than can be accounted for by the temperature coefficient [35B7] or the impurities supposedly present in the latter's specimens. Intermediate values of the resistivities at 20°C have been obtained at the Bell Laboratories, and these may be considered representative of laboratory specimens of average purity; a typical analysis contains 0.7% iron and 0.4% manganese.

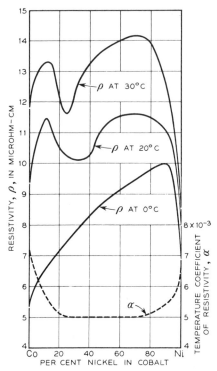

Fig. 8-21. Electrical resistivity of cobalt-nickel alloys at 0 to 30°C, according to several reports.

The data on *hardness* [28C2, 30K3, 35B7] all show, as may be expected on account of the transformation, that the alloys containing up to about 30% nickel are much harder than the others; the Brinell number [35B7] drops from 180 to 100 as the nickel increases from 20 to 40% and then remains almost constant to 100% nickel. Scant data are available for thermal conductivity [27M2] and thermoelectric power [35B7].

Many of the investigations of magnetic and other physical properties have been made on specimens that were cast, machined, and annealed. However, the nickel-rich specimens may be forged or swaged cold and drawn to wire. The fabrication of alloys containing less than 30% nickel is similar to that of cobalt, that is, they may be hot worked or perhaps cold worked with difficulty.

Magnetic Properties.—Properties in high fields were investigated first by Bloch [12B1] and more recently with but slightly different results by

Weiss, Forrer, and Birch [29W3]. The latter measured the magnetization at low temperatures and high fields—more explicit data are not available—and extrapolated to 0°K and infinite field strengths. Results, converted

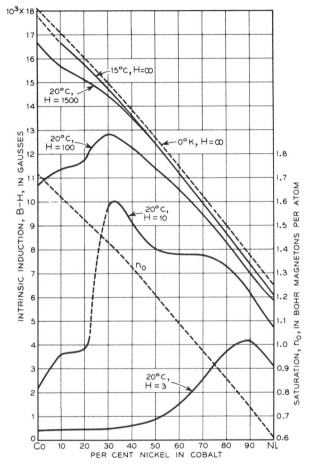

Fig. 8–22. Intrinsic induction of cobalt-nickel alloys at various field strengths, and Bohr magneton numbers.

from Weiss magnetons to ferric induction, $B-H$, are shown in the upper curve in Fig. 8–22.

Saturation at 15°C, similarly calculated, is shown in the next lower curve. Data for room temperature in fields of 1500, 100, 10, and 3 are also recorded from measurements made at the Bell Laboratories. Masumoto [29M1] has obtained similar results, except that his cobalt-rich alloys have considerably lower values of $B-H$ at the same values of H. The effect of *temperature* on the magnetization in high fields is shown in the original investigation of Bloch [12B1] and his curves have been reproduced elsewhere [29W2]. Bohr magneton numbers, n_0, are also plotted in the same figure.

The hardness factor, represented by a in the relation $I = I_s(1 - a/H)$, which shows the rate of approach to saturation at a given temperature, has been reported [29W3] for nickel-rich alloys and varies from 3 to 40, the highest value being for the lowest nickel content, 34%.

Magnetic properties in low fields have been measured by members of the staff of the Bell Laboratories, and data for μ_0, μ_m, and H_c are plotted

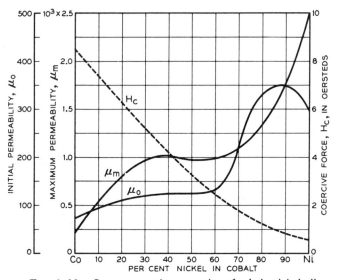

Fig. 8-23. Some magnetic properties of cobalt-nickel alloys.

in Fig. 8-23. Values of μ_m may also be obtained from Masumoto's data [29M1]. The hysteresis loss at $B_m = 5000$ decreases uniformly from 8000 ergs/cm³/cycle for cobalt to 1200 for nickel.

Ternary Alloys.—An extensive series of experiments on the ternary alloys of *cobalt, nickel, and copper* has been carried out by Dannöhl and Neumann [38D1]. In this system a single complete series of solid solutions exists near the cobalt-nickel and nickel-copper binaries, except in the cobalt corner where the ε phase occurs at low temperatures. There is limited solubility at both ends of the cobalt-copper series, and the two-phase region extends over the central part of the ternary diagram. As shown in Fig. 9-46 this region decreases in area with increasing temperature and has its boundary at M when the alloys melt. The two phases in the area are both face-centered cubic, but one is copper-rich and the other cobalt-rich; connodes are parallel to the cobalt-copper side.

Diagrams [38D1] showing Curie points and saturation ferric induction are reproduced in Fig. 8–24 and 25. In Dannöhl and Neumann's study of this sytem they have been interested primarily in the permanent magnet properties developed by the precipitation hardening. These are discussed more fully in Chap. 9.

Forsyth and Dowdell [40F1] have investigated the *cobalt-nickel-silicon* system up to 20% silicon. About 6% silicon goes into solid solution at

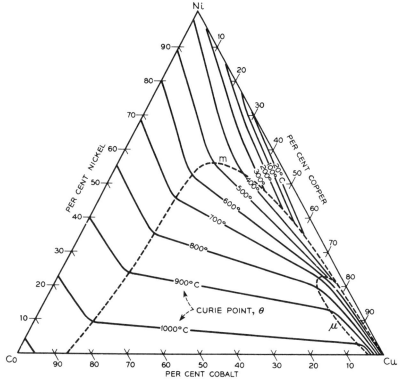

FIG. 8–24. Curie points of Co-Ni-Cu alloys. The dotted line shows the phase boundary at the Curie point.

1000°C, and the γ phase (face-centered cubic) exists up to about 15% silicon. The α phase is always ferromagnetic at room temperature. Alloys containing 10% or more of silicon are brittle.

The *cobalt-nickel-aluminum* system has been studied by Schramm [41S2], who has determined the Curie points as well as the structure, and the *cobalt-nickel-arsenic* system by Friedrich [12F1].

OTHER COBALT ALLOYS

The Curie points of a series of cobalt alloys have been reported in three papers by Köster and his colleagues [37K7, 37K8, 38K14]. They and

Hashimoto [37H4, 37H5, 37H3, 38H3, 37H4] have studied also the effect of many elements on the low-temperature (ϵ,γ) transformation of cobalt that occurs at about 400°C. This point is generally lowered by the face-centered cubic metals, which tend to stabilize the face-centered (γ) phase of cobalt. Some elements raise the ϵ,γ points of alloys when they are heated and lower them when they are cooled. Equilibrium is particularly

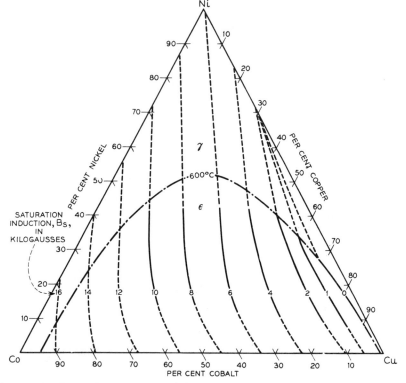

FIG. 8-25. Saturation induction of Co-Ni-Cu alloys. The dotted line shows the phase boundary at 600°C.

difficult to obtain at low temperatures in cobalt alloys. Hashimoto has constructed diagrams, not shown below, of cobalt with Ca, Cd, Si and Mg.

The saturation intensities of magnetization of four series of alloys of cobalt (with Al, Cr, Mo, W) have been determined by Farcas [37F5] and are mentioned in the appropriate places. Apart from these results and the Curie points, few data on cobalt alloys have been recorded. The following compounds (or ordered structures) have been reported to be ferromagnetic:

Co_5As_2 Co_2P CoS_2
Co_2B $CoPt$ $CoZn$
Co_4Zr

Cobalt-Aluminum.—The partial system Co-CoAl is shown in Fig. 8-26, according to Schramm [41S11]. The solid solubility of aluminum in cobalt (about 97.5% pure) decreases from 8% at the eutectic point (1400°C) to 4.5% at 840°C (the Curie point) and to 1% at the ϵ,γ transformation point (300° on cooling). Köster and Wagner [37K7] had pre-

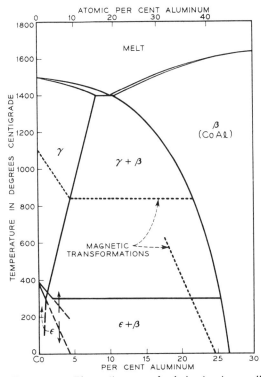

FIG. 8-26. Phase diagram of cobalt-aluminum alloys.

viously placed the solid solubility at 0-500°C at about 5% and the Curie point of saturated γ as 640°. Hashimoto [37H5] placed the Curie point of saturated γ as 725°.

CoAl (31.4% Al) has the same structure as NiAl (CsCl type). Co_2Al_5 (53.4% Al) is hexagonal [39B15]. Several other compounds, richer in cobalt than CoAl, have also been reported.

The β phase, CoAl, and solid solutions of cobalt in it have sluggish transitions and it is difficult to make them homogeneous. By measurement of magnetization during the heating and cooling of alloys that had had various heat treatments, Schramm was able to measure the Curie points of β-phase alloys (lower curve in the diagram). The pure compound CoAl is not ferromagnetic at room temperature.

Farcas [37F5] measured the magnetization per gram of alloys containing 3.05 and 3.90% aluminum and found the saturation per gram σ_s to be 151.4 and 148, respectively, at 23°C. The estimated values at 0°K are 153.5 and 150.2, corresponding to $n_0 = 1.56$ and 1.51 Bohr magnetons per atom of alloy. This means that the magnetic moment per atom of cobalt in the alloy is lower than that of pure cobalt. The phase structures of these alloys are not reported.

Cobalt-Antimony.—The phase diagram of Fig. 8–27 is that of Hansen [36H1], revised according to more recent data of Köster and Wagner

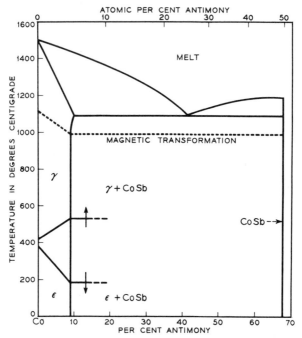

FIG. 8–27. Phase diagram of cobalt-antimony alloys.

[37K7]. These alloys are magnetic to 67% antimony, the content of the non-magnetic compound CoSb. This compound precipitates when more than 8% of antimony is present at room temperature, as shown (5% according to Hashimoto [37H5]). It has the hexagonal pyrrhotite structure [27J1].

The Curie point drops in the solid solution area to about 990°C at 8% antimony and then remains constant.

When antimony is added to cobalt, the ϵ, γ transformation occurs at higher temperatures on heating and at lower temperatures on cooling, according to Köster and Wagner. Hashimoto [37H5] found the transformation temperature to be raised by addition of antimony.

Cobalt-Arsenic.—The compounds Co_5As_2 (33.7% As) and Co_2As (38.9% As) are formed, and others have been reported. Co_5As_2 forms with cobalt a eutectic that solidifies at 920° (Fig. 8-28). The solid solubility of arsenic in cobalt at this temperature is 7%, and it decreases to about 5% at low temperatures, according to Köster and Mulfinger [38K14] (less than 3% according to Hashimoto [37H5]). The former have shown that the ϵ,γ transformation temperature of cobalt is raised by arsenic to

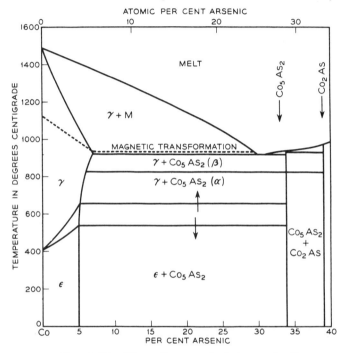

Fig. 8-28. Phase diagram of cobalt-arsenic alloys.

660°C (on heating) or 535°C (on cooling). A heat effect of unknown origin occurs in the composition range 15–45% arsenic, at temperatures 250–350°C, on cooling.

The Curie temperature decreases on addition of arsenic to 925° [38K14], then remains constant. The older work [08F2] indicates that alloys are magnetic up to 38% arsenic; therefore Co_5As_2 is presumed to be magnetic and Co_2As non-magnetic. The magnetic α phase and the eutectic are in equilibrium with the liquid from 3–50% arsenic.

Cobalt-Beryllium.—Köster and Schmid [37K8] have determined the phase diagram of the cobalt-rich alloys, as shown in Fig. 8-29. Solid solubility at room temperature is a little more than 1%. The ϵ,γ transformation of cobalt for rising temperature is increased, and that for falling temperature decreased, by addition of beryllium. The Curie point drops to 950°C at 1.5–2% beryllium and then remains constant. The magnetic properties of the compound CoBe (11.3% Be) have not been investigated. Masing [28M5] states that the alloys are similar to beryllium-nickel in structure; in that system NiBe is non-magnetic.

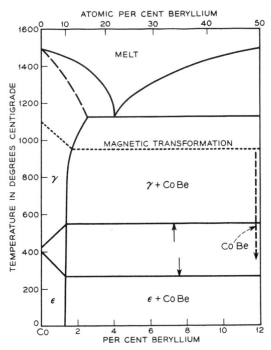

FIG. 8-29. Phase diagram of cobalt-beryllium alloys.

Cobalt-Bismuth.—Cobalt and bismuth have only limited solubility in the melt. The eutectic contains about 3% cobalt and melts at about 258°C, 10–12° below the melting point of bismuth, so that practically all alloys begin to melt at the eutectic temperature. The solid solubility of bismuth in cobalt is less than 1% [37H5], and the effects of bismuth on the Curie point and the ϵ,γ transformation are small and are not known. The solubility of cobalt in solid bismuth is estimated to be about 0.001% [30T1].

One therefore expects all alloys containing more than 0.001% cobalt to be ferromagnetic with a Curie point of about 1115°C, and to have saturation magnetization proportional to the cobalt content.

Cobalt-Boron.—Köster and Mulfinger's [38K14] diagram of the cobalt-rich alloys is shown in Fig. 8-30. The Curie point of cobalt is lowered by about 1% boron to 1090°C, 12° below the melting point of the eutectic formed with Co_2B (8.4% B). This compound is ferromagnetic with a Curie point of 510°C. The ϵ,γ transformation point of cobalt is decreased to 360° or 260° for rising or falling temperatures, respectively.

Both Co_2B and non-magnetic CoB (15.5% B) have lattice structures like the corresponding compounds of iron and nickel. X-ray data are available [33B4] to 20% boron.

Cobalt-Carbon.—The diagram is shown in Fig. 8-31 for alloys containing carbon up to 5% and is based largely on data supplied by Boecker [12B3], Hashimoto [38H4], and Köster and Schmid [37K8]. The precipitated phase in the solid alloys is mainly graphite, and in this respect they differ markedly from the iron-carbon alloys. Co_3C (6.4% C) has been reported to be stable in the range 500–800°C [37M4]. Thermodynamic properties have been summarized by Kelley [37K9].

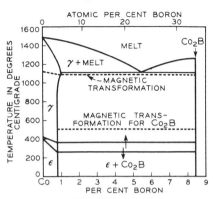

Fig. 8–30. Phase diagram of cobalt-boron alloys.

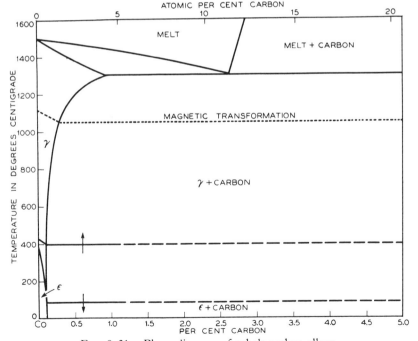

Fig. 8–31. Phase diagram of cobalt-carbon alloys.

The eutectic lies at about 2.6% carbon and 1300–1310°C, and at this temperature 0.8–1.0% carbon is soluble. At lower temperatures the solubility decreases to 0.1% according to some authorities, whereas others [47H3] find no evidence of solubility below the ϵ,γ point. The Curie points are lowered by carbon, as indicated in the diagram.

Cobalt-Cerium.—Vogel [47V2] reports the following compounds: $CoCe_3$ (12.3% Co), Co_2Ce (45.7% Co), Co_3Ce (55.8% Co), Co_4Ce (62.7% Co), and Co_5Ce_3 (67.8% Co). Co_2Ce and Co_5Ce have been confirmed with

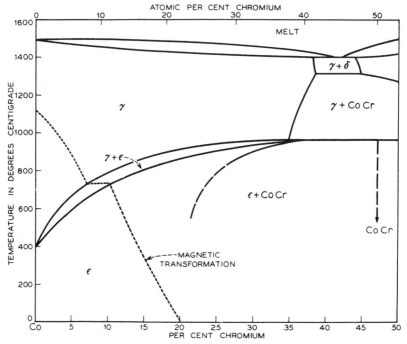

Fig. 8–32. Phase diagram of cobalt-chromium alloys.

X-rays and have respectively a cubic spinel-like structure with $a = 7.155$ and the Zn_5Ca structure, and both are isomorphous with the corresponding nickel compounds. The phase diagram shows a melting point decreasing continually from cobalt to 85% cerium. No magnetic data have been found in the literature.

Cobalt-Chromium.—Determination of the diagram (Fig. 8–32) is made difficult by the sluggishness of the diffusion and the low-temperature transformations, and the boundaries are uncertain. The recent diagram of Elsea, Westerman, and Manning [48E2] is based on the previous work of Wever and co-workers [29W7, 30W6] and of Matsunaga [31M4] and shows a rising ϵ,γ transformation with increasing chromium, in qualitative agreement with earlier work [37H5]. The Curie point falls to room temperature at about 20% chromium, and the curve shows a break at the ϵ,γ point [30W6, 31M4]. Solid solubility of ϵ in the non-magnetic phase, CoCr (46.9% Cr), is quite uncertain and estimates vary from 10 to over 30% chromium at low temperature.

Farcas [37F5] has measured the saturation magnetization per gram, σ_s, at 23°C for alloys containing up to 20% chromium, and estimated the saturation at 0°K, σ_0, and the number of Bohr magnetons per atom, n_0, as follows:

Per Cent Chromium	Atomic Per Cent Chromium	σ_s	σ_0	n_0
5.0	5.6	134	136	1.42
9.5	10.6	100	103	1.07
15.0	16.7	59.5	62	0.64
20.0	22.1	19	23	0.24

σ_0 is linearly dependent on chromium content, and extrapolation indicates that the alloys become non-magnetic at 0°K at about 25 atomic per cent.

Stellite, containing, for example, 27% Cr, 6% Mo, 2% Ni, 0.3% C and the remainder cobalt, is non-magnetic. Its hardness increases with the addition of chromium and decreases as the cobalt content is increased [48Y1]. This series of heat-resistant and corrosion-resistant alloys is used for dental alloys, turbine blades, cutlery, and other purposes.

Cobalt-Copper.—These metals are apparently miscible in the liquid phase, if pure, and may segregate into two layers if impurities are present [48S7]. In the solid state, cobalt is soluble only to the extent of 5% at 1100° and 0.2% at 400°C [30T1]. The solubility of copper in cobalt at low temperatures is about 10% [37H5]; Köster and Wagner [37K7] found 12% present after furnace cooling. Copper lowers the Curie temperature and the ϵ,γ transformation [37H4] as indicated in Fig. 8-33. All alloys are therefore ferromagnetic at room temperature if they contain more than about 0.2% cobalt.

Cobalt-Germanium.—Pfisterer and Schubert [49P1] have found that cobalt forms the compounds Co_2Ge (34.2% Ge), $CoGe$ (51.0% Ge), Co_2Ge_3 (61.0% Ge) and $CoGe_2$ (67.6% Ge), and with the first forms a eutectic melting at 1110° (30% Ge). The solid solubility of germanium is about 15% at this temperature and 10% at lower temperatures. The ϵ,γ transformation temperature is not noticeably affected by germanium addition.

Co_2Ge has the structure of pyrrhotite (NiAs type), and has a superstructure with an order-disorder point of 625°C. With excess of germanium this decreases to 400° and disappears at CoGe. Co_2Ge is not ferromagnetic even at low temperatures.

Cobalt-Gold.—This is a simple system [36H1] with a single eutectic melting at about 1000°C, 65° below the melting point of gold. At this temperature about 6% of gold is soluble in cobalt and possibly 13% of cobalt in gold. At room temperature less than 3.5% of gold is soluble. The Curie point is lowered somewhat by gold but always lies above the eutectic temperature. Hashimoto [37H5] reported that gold lowers the ϵ,γ transformation of cobalt. At room temperature the saturation mag-

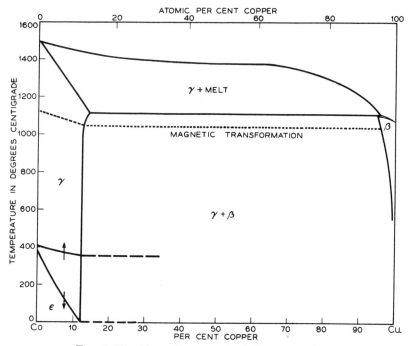

FIG. 8–33. Phase diagram of cobalt-copper alloys.

netization must be nearly proportional to the cobalt content except when this is small. Wahl [10W2] reported that the magnetization in a constant field (strength not stated) first dropped rapidly as gold was added to cobalt, then more slowly, and approached zero for 100% gold.

Cobalt-Hydrogen.—The solubility of hydrogen in cobalt has been measured by Sieverts and Hagen [34S13] from 600–1200°C. As in iron and nickel (see Fig. 2–12) the solubility in cobalt increases with temperature, but it is substantially less than in these other metals. The authors found 3.2 cm^3 (measured at 0°C, 1 atm) dissolved in 100 mg of cobalt at 1000°C.

Cobalt-Lead.—Cobalt is practically insoluble in lead in the solid and liquid states; consequently, the addition of a small amount of cobalt makes lead ferromagnetic. Tammann and Oelsen [30T1] estimated that $10^{-3}\%$ of cobalt is held in solid solution.

Cobalt-Manganese.—The data regarding the Curie point and the ϵ,γ transformation are conflicting [36H1]. It is apparent, however, that alloys become non-magnetic at room temperature when they contain between 30 and 40% manganese, and that the ϵ,γ point is depressed by manganese. For further discussion the reader is referred to Hansen [36H1] and Schneider and Wunderlich [49S5].

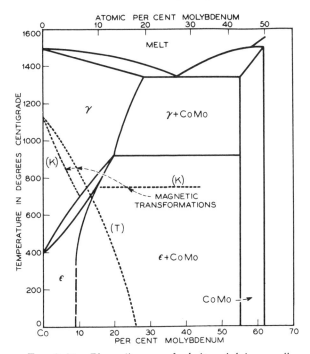

FIG. 8–34. Phase diagram of cobalt-molybdenum alloys.

Sadron [32S3] has measured the atomic moments of alloys containing up to about 20% manganese, and his results are plotted in Fig. 10–11.

A diagram of the Co-Mn-Al system and the magnetic properties of these alloys have been reported by Köster and Gebhardt [38K8, 38K12]. Data relating to saturation, Curie point, and coercive force are described in the original papers. The highest coercive force obtained was about 300 oersteds.

The Co-Mn-Cu system has been studied by Köster and Wagner [38K9], who measured saturation, coercive force and remanence.

Cobalt-Mercury.—Tammann and Kollman [27T1] have estimated the solubility of cobalt in mercury to be about 0.2% at room temperature.

Tammann and Oelsen [30T1] calculated a solubility of 0.06% from the older magnetic measurements of Nagaoka [96N1].

Cobalt-Molybdenum.—There has been some uncertainty regarding this system, and it should not be considered to be established. The ϵ,γ transformation is raised by molybdenum to 915° at 20% molybdenum, according to Köster and Tonn [32K12] (Fig. 8–34). They find also that the Curie point falls in the γ-phase region to 700° at 10% molybdenum and continues to fall to 200° in alloys quenched from 1200° to preserve their γ structure.

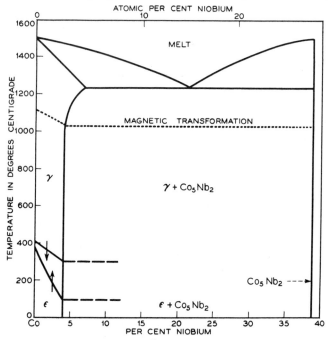

Fig. 8–35. Phase diagram of cobalt-niobium alloys.

(Curve T is due to Takei [28T1]). When the alloys are well annealed, however, the ϵ phase is stabilized, and the Curie point remains constant at 750°C from 15–50% molybdenum. These latter alloys contain precipitated CoMo (62.0% Mo) which is non-magnetic. At room temperature (below 500°C) this phase is soluble to the extent of about 10%.

Another diagram has been proposed [35S10], but no magnetic measurements were made in connection with it.

Farcas [37F5] has measured the saturation magnetization per gram at 23° for alloys containing 3 and 5% molybdenum and obtained $\sigma_s = 158.7$ and 148.5, respectively. The estimated numbers of Bohr magnetons per atom obtained by extrapolating to 0°K are 1.71 and 1.62.

Cobalt-Niobium.—Köster and Mulfinger [38K14] have studied alloys containing up to 38.8% niobium (Co_5Nb_2, 38.7% Nb) and their diagram is shown in Fig. 8–35. At the temperature of the eutectic (31% Nb, 1235°C) the solid solubility of niobium is 7%, and at lower temperatures it is less. A furnace-cooled specimen contained 4% in solid solution, while a specimen with 3% showed precipitation hardening after suitable heat treatment.

The ϵ,γ transformation point decreases as shown. The Curie point is lowered to 1030° at 4% niobium, the solid solution limit at this temperature. Apparently Co_5Nb_2 is non-magnetic.

Cobalt-Nitrogen.—No combination of cobalt with nitrogen has been detected at low temperature, though the formation of Co_3N_2 (13.7% N) at very high temperatures has been reported [19V1].

Cobalt-Oxygen.—CoO (21.4% O) has the cubic NaCl structure and melts at about 1935°C [31W4]. The vapor pressure of oxygen over CoO at 730°C is very small, about 10^{-14} mm, and increases to about 10^{-4} mm at 1430°C. Its reducibility by hydrogen is expressed by the equilibrium constant $K = (H_2)/(H_2O) = 0.03$ to 0.05 over the temperature range 700–1300°C [49D1]. The dissociation pressure of Co_3O_4 (26.6% O) is about 10 mm at 800°C and 400–500 mm at 950°C [42T1].

The solubility of oxygen in cobalt has been investigated by Seybolt and Mathewson [35S4] at temperatures between 600 and 1500°C. The solubility of oxygen was found to be 0.006% at the lowest temperature and to increase with temperature to 850°C, where a transformation was observed and the solubility decreased to a low value. With increasing temperature the solubility again increased to about 0.02% at 1500°C.

Both CoO and Co_3O_4 are believed to be paramagnetic, although there are reports to the contrary [43S2]. Bizette [46B4] found CoO to be antiferromagnetic with a Curie point of about 20°C, and Foëx [48F3] found an anomaly in the thermal expansion at this temperature.

Cobalt-Palladium and Cobalt-Platinum.—See Chapter 9.

Cobalt-Phosphorus.—The compounds Co_2P (20.8% P), CoP (34.5% P), and CoP_3 (61.2% P) are known and others have been reported. Co_2P is weakly magnetic [09Z1] and has a phase transformation at 920°C. The solid solubility of phosphorus is small, about 0.3% according to Hashimoto [38H3], and the Curie point is lowered to 1034°C. The eutectic with Co_2P melts at 1022° with 11.5% phosphorus [36H1].

Cobalt-Selenium.—The compounds Co_2Se (40.2% Si), CoSe (57.3% S), and $CoSe_2$ (72.9% S) have been produced and studied by X-rays [27J1]. Like the corresponding sulfides (see below) the two latter compounds have respectively the structures of pyrrhotite (FeS) and pyrites (FeS_2). The ϵ,γ transformation is reported to be raised on heating to 520° and to be lowered on cooling to 400°C [38H3]. Solid solubility of selenium is small, probably less than 1% [38H3].

Cobalt-Silicon.—The non-magnetic compound Co_2Si (19.2% Si) forms with cobalt a eutectic that melts at 1200°C and contains 12% silicon (Fig. 8–36). The solubility of silicon at that temperature is about 9.5%, and it decreases to less than 8% at lower temperatures.

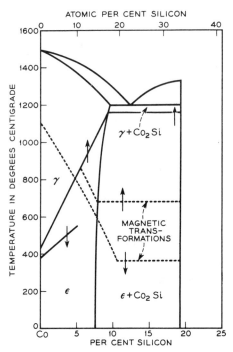

FIG. 8–36. Phase diagram of cobalt-silicon alloys.

The constitutions of the solid solution and the two-phase areas are complicated by the sluggishness of the ϵ,γ transformation, and the diagram shows the phases present and the Curie temperatures after heating or cooling at reasonable rates according to Köster and Schmid [37K8]. Hashimoto [37H3] also observed a rising ϵ,γ transformation with increasing silicon.

At higher concentrations CoSi (32.25% Si) and other compounds are formed. Co_3Si (13.7% Si) has been reported by one observer but not by others [36H1].

Both ϵ and γ phases contract with increase in silicon content [34B6].

Cobalt-Silver.—Cobalt and silver are immiscible in the liquid state. Tammann and Oelsen [30T1] found only 10^{-3}–10^{-4}% of cobalt in silver saturated with cobalt at 1000–1200°C. Thus all alloys have Curie points and saturation magnetization characteristic of the cobalt which they contain. Hashimoto [37H5] reported a solubility of about 2% of silver in cobalt, with slight change in the Curie and ϵ,γ points.

Cobalt-Sulfur.—The compounds CoS (35.2% S), Co_9S_8 (32.6% S), Co_3S_4 (42.0% S), and CoS_2 (52.1% S) have been prepared and studied with X-rays [38L6]. CoS has the hexagonal structure of pyrrhotite (FeS), Co_3S_4 the spinel structure, Co_9S_8 a cubic structure, and CoS_2 the structure of FeS_2 (pyrites). Haraldsen and Klemm [35H6] found CoS_2 to be ferromagnetic at $-180°C$, with $\sigma = 45$ to 50 when $H = 1000$ to 3700.

The solid solubility of sulfur in cobalt is quite small and its amount has not been determined. In one diagram proposed [36H4], $Co + Co_6S_5$

(or Co_9S_8) is stable below 787° to 31% sulfur; $Co + Co_3S_4$ is stable at higher temperatures. CoS appears beyond 31% sulfur. Friedrich [08F1] reports that sulfur-rich alloys are not attracted by a magnet.

Cobalt-Tantalum.—The cobalt-rich alloys have been studied by Köster and Mulfinger [38K14] whose phase diagram is shown in Fig. 8–37. The compound Co_5Ta_2 (55.2% Ta), which is non-magnetic, forms a eutectic at 1275° with 31% tantalum. The solid solubility decreases from 13% at this temperature to about 8.5% at temperatures below 500° when the

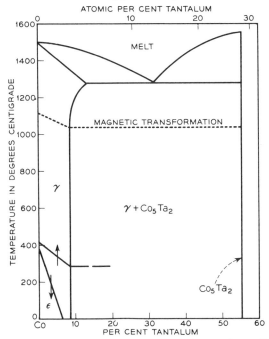

FIG. 8–37. Phase diagram of cobalt-tantalum alloys.

alloys are furnace-cooled. Alloys containing 6% tantalum age harden; consequently, the equilibrium solubility is less than this amount. The ϵ,γ point is lowered by addition of tantalum, as shown, according to Köster and Mulfinger, but raised according to Hashimoto [37H5].

The Curie point of cobalt is lowered to 1035° at 8.5% tantalum, then remains constant.

Cobalt-Tellurium.—The compound CoTe (68.4% Te) has the hexagonal structure of pyrrhotite [27O1].

Cobalt-Thallium.—This system is similar to cobalt-lead, thallium being but slightly soluble in molten cobalt [36H1].

Cobalt-Tin.—The phase diagram of Fig. 8–38 is that of Hansen [36H1], based on early work, to which have been added the more recent results of

Köster and Wagner [37K7]. The compounds Co_2Sn (50.2% Sn) and CoSn (66.8% Sn) are both non-magnetic and are respectively tetragonal and hexagonal with known crystal structures [38N6]. $CoSn_2$ has also been observed [38N6].

The solubility of tin in γ cobalt is small: 5% at 1000° according to Nial [38N6], and 8% according to Köster and Wagner [37K7]. The latter find 5% solubility at 600°C and below, for a practical rate of cooling. The

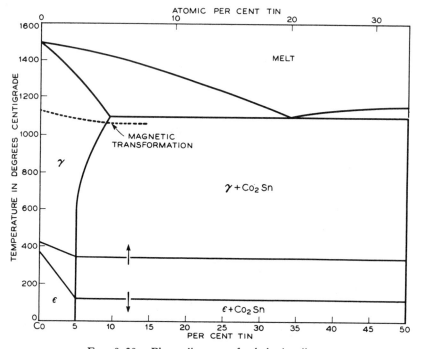

FIG. 8–38. Phase diagram of cobalt-tin alloys.

ϵ,γ transformation falls with increasing tin for both heating and cooling, according to Köster and Wagner, but rises on heating according to Hashimoto [38H3].

Alloys are magnetic to 50% tin (Co_2Sn), with a Curie point of 1033°C [38H3] to 1080°C [37K7].

Cobalt-Titanium.—Little is known of the melting points of these alloys. Köster and Wagner [37K7] found a solid solubility limit at 7.5% for furnace-cooled alloys. At this composition the Curie point has decreased to 900°C. The ϵ,γ point is lowered to about the same extent as in the cobalt-tin system.

The compounds Co_2Ti (28.9% Ti), CoTi (44.8% Ti), and $CoTi_2$ (61.9% Ti) are known [39L3]; the first has the cubic $MgCu_2$ structure;

the second, the cubic CsCl structure; and the third, a complicated face-centered cubic structure.

Cobalt-Tungsten.—Köster and Tonn [32K12] find the diagram similar to that of the cobalt-molybdenum alloys shown in Fig. 8–34 but with a solid solubility at room temperature of about 13% tungsten instead of 9% molybdenum and a Curie point in the $\epsilon + \beta$ area of 825°C. They believe the alloys to be magnetic up to CoW (75.7% W). Hashimoto [37H5] obtained a similar diagram with a solubility limit of about 30%

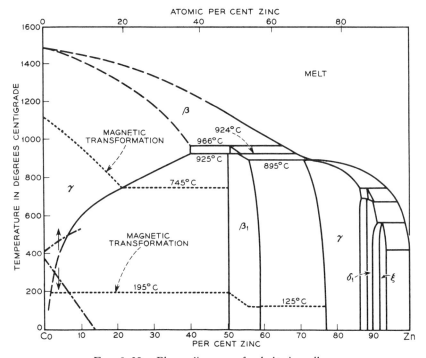

FIG. 8–39. Phase diagram of cobalt-zinc alloys.

tungsten, at which the Curie point has dropped to 500°C, and he found the ϵ,γ point raised on heating and lowered on cooling. Sykes [33S4, 48S8] proposes a quite different diagram.

Babich, Kislyakova, and Umanskii [39B17] have detected Co$_3$W (51.0% W) and CoW with X rays; the former is hexagonal and shows atomic ordering.

Farcas [37F5] has measured the magnetic moments per gram at 23°C and estimated them for 0°K. For 5 and 10% tungsten they are, respectively, $\sigma_s = 147$ and 134, and $\sigma_0 = 149$ and 136. The numbers of Bohr magnetons per atom are then 1.63 and 1.54.

Cobalt-Vanadium.—Köster and Wagner [37K7] have investigated cobalt-rich alloys containing vanadium and iron in the ratio $V/Fe = 4/1$. The ϵ,γ transformation upon heating is raised, whereas that on cooling is lowered, by additions of cobalt up to 10%. At 12% vanadium the transformation reaches room temperature. The Curie point decreases continuously, at an increasing rate, as the vanadium content increases to 20%. At this composition θ is about 400°C, and at 25% the alloys cease to be ferromagnetic at room temperature.

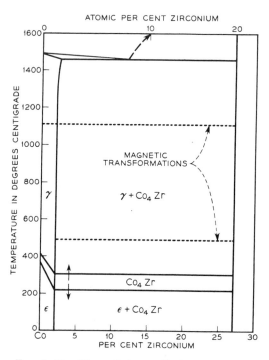

FIG. 8–40. Phase diagram of cobalt-zirconium alloys.

Cobalt-Zinc.—The diagram of Fig. 8–39 has been reported by Schramm [41S12]. The compound CoZn (52.6% Zn) is ferromagnetic and this is present up to 77% zinc. Other phases are present at higher zinc contents.

The solid solubility of zinc in cobalt decreases from 40% at 950° to 6% at 400°C and to still lower values at room temperature. The Curie point is lowered to 745° at the solution limit at this temperature (21%) and then remains constant to 50%. Alloys in this range of composition also have the Curie point of CoZn, 195°C. Addition of zinc to CoZn reduces this Curie point to 125°, and this then remains constant to 77%.

The ϵ,γ transformation on heating is raised, whereas that on cooling is lowered, by addition of zinc to cobalt, according to Köster and Wagner [37K7]. Hashimoto [37H5] observed the point to be lowered.

Schramm and Mohrnheim [48S9] observed age hardening in alloys containing 40% zinc, and followed the magnetic changes occurring after various heat treatments. They can be understood in terms of the diagram.

Cobalt-Zirconium.—According to Köster and Mulfinger [38K14], cobalt forms a eutectic containing 12% zirconium and melting at 1460°C (Fig. 8–40). The solubility of zirconium at this temperature is only about 2%, and this decreases somewhat at lower temperatures. The precipitating phase, probably Co_4Zr (27.9% Zr), is magnetic and has a Curie point of 490°C. Thus both phases present from 2 to 27% zirconium are magnetic.

Zirconium lowers the ϵ,γ point upon heating or cooling, according to Köster and Mulfinger. However, Hashimoto [38H3] reports that it is raised on heating. He also finds the Curie point to be lowered to 1029°C at the solubility limit, 0.2% zirconium.

OTHER NICKEL ALLOYS

The most general commercial use of non-ferrous nickel alloys is in nickel-copper alloys. These include *Monel*, *"nickel-silver,"* *Constantan*, and *coinage* alloys. In order to comply with certain mechanical requirements nickel is often hardened by additions of silicon or aluminum. Chromium is added to nickel to increase the corrosion resistance under oxidizing conditions.

There are no non-ferrous nickel alloys having commercially important magnetic properties. The nickel-manganese alloys are of considerable scientific interest because their properties depend markedly on atomic ordering, the saturation magnetization and Curie points varying between wide limits. Ni_2Mg, Ni_3Mn and $NiHg_3$ are reported to be ferromagnetic.

Nickel-Aluminum.—The diagram of Fig. 8–41 [48F2] is based mainly on three investigations [37B9, 37A1, 41S13]. The face-centered cubic α' phase, Ni_3Al (13.3% Al), and the body-centered cubic β phase, NiAl (31.5% Al), are both ordered, as was shown by Bradley and Seager [37B9]. The solid solubility of aluminum in nickel decreases from about 11% at 1385° to less than 5% at low temperatures. About 4.5% aluminum is used in the precipitation-hardened commercial alloy "Z-nickel."

Schramm [41S13] found the Curie point to decrease to about 80° at the solid solubility limit, 4.5%, and then to be constant in the $\alpha + \alpha'$ field. Ni_3Al is non-magnetic. In alloys containing some cobalt Gwyer [08G1] found alloys containing over 10% aluminum to be magnetic at room temperature.

Sadron [32S3] found the Curie point of the 9.7% alloy to be $-75°C$. Somewhat different points were observed by Marian [37M2].

Marian determined the specific magnetization at 17°C (σ_s) and 77°K and extrapolated to 0°K (σ_0). Values obtained are as follows:

Per Cent Aluminum	Atomic Per Cent Aluminum	σ_s	σ_0	n_0
0.90	1.96	47.06	51.80	0.54
4.50	9.31	21.18	33.81	.34

The average moment extrapolates to zero at about 20 atomic per cent aluminum (see Fig. 10–11).

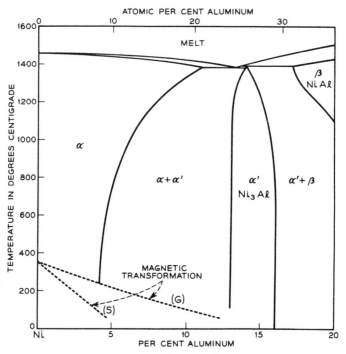

FIG. 8–41. Phase diagram of nickel-aluminum alloys.

Nickel-Antimony.—Figure 8–42 shows the nickel-rich portion of the phase diagram of Shibata [41S15]; Curie points are by Marian [37M2]. The hexagonal δ phase (Ni_3Sb, 40.9% Sb) transforms at 698° to the tetragonal β phase. The atomic moments drop almost linearly to zero at 15 to 20% Sb (Sadron [32S3], Marian [37M2], Rado and Kaufmann [41R3]).

Alloys are apparently magnetic until the α phase disappears. The Curie points in the figure are taken from the older work of Lossew [06L2].

Nickel-Arsenic.—The older data, summarized by Hansen [36H1], are supplemented by the more recent results of Köster and Mulfinger [40K3], which include measurements of the Curie point, and are shown in Fig. 8–43.

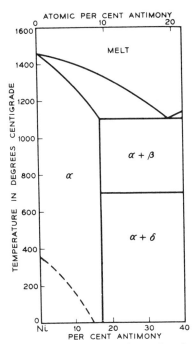

Fig. 8-42. Phase diagram of nickel-antimony alloys.

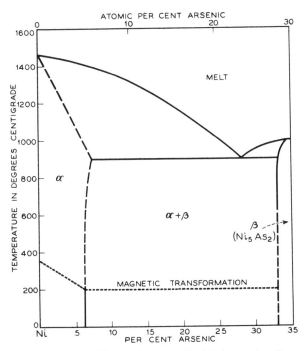

Fig. 8-43. Phase diagram of nickel-arsenic alloys.

The compound Ni_5As_2 (33.2% As) is non-magnetic. NiAs (56.1% As), also non-magnetic, has the hexagonal structure often referred to as the NiAs or pyrrhotite structure in which each atom has 6 neighbors of the opposite kind.

A ternary diagram of Ni-Cu-As alloys containing up to 33% arsenic has been given by Köster and Mulfinger [40K3]. The Curie point of nickel is

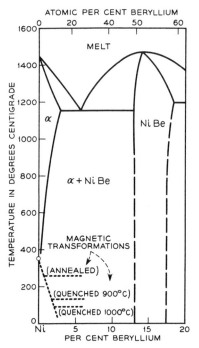

Fig. 8-44. Phase diagram of nickel-beryllium alloys.

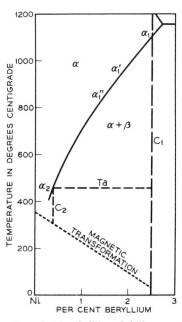

Fig. 8-45. Solid solubility and Curie points of nickel-beryllium alloys.

depressed to 0°C by addition of 34% copper, when no arsenic is present, and by 36% copper when 6–32% arsenic is present. The solubility of arsenic in nickel is not appreciably changed by addition of copper.

Nickel-Beryllium.—This system is characterized by exceptional age hardening of the alloys containing 2–3% beryllium, in which the Brinell hardness can exceed 600 [29M5, 48W4]. As shown in the diagram of Fig. 8-44, the solid solubility decreases from 3% at the eutectic, 1155°, to 0.2% at 400°C. The precipitating phase is NiBe (13.3% Be), which has a body-centered lattice of the CsCl type.

The solid solubility and the Curie temperature have been accurately determined by Gerlach [37G3], using a magnetic method, as follows. The

Curie points of quenched alloys containing 0–2.5% Be were first measured in the usual way, with the results given in the dotted curve of Fig. 8–45. Then the quenched alloys were hardened by hammering, annealed for various lengths of time at different temperatures, and the magnetization measured at various temperatures up to the Curie point. The data obtained in the 2.5% alloy after annealing ("aging") at 454°C are given in Fig. 8–46. These show that the Curie point is about 40° for the specimen before aging, and that as aging proceeds there are two constituents, one of which has a Curie point constant at 303°, the other having a Curie point that varies from 40 to 303°C depending on the time of aging.

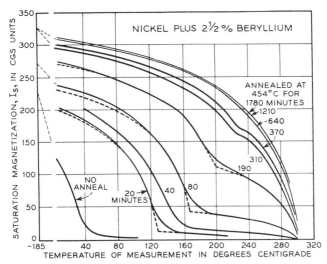

Fig. 8–46. Magnetization vs temperature curves of inhomogeneous nickel-beryllium alloys (2.5% Be) annealed at 1000°C, hard-hammered and heated for various times at 454°C.

The results are interpreted with the aid of Fig. 8–45. Let C_1 represent the concentration of beryllium in the alloy, in one case 2.5%, and let T_a be the temperature of aging, e.g., 454°. The dotted curve shows the Curie point of the quenched alloy (40°C). After aging for a long time the observed Curie point of the homogeneous alloy is 303° (Fig. 8–46), which corresponds to a concentration C_2 of 0.4% beryllium. Intersection of the lines C_2 and T_a fixes the point, 0.4% beryllium for 454°, on the solid solubility curve. Other points are determined in the same manner.

Similar experiments have been carried out by Okamoto [4002], and the Curie points he observed after annealing for a long time at 600°C, or at 900 or 100°C, are shown on the diagram of Fig. 8–44. He also measured magnetization as a function of temperature for alloys containing up to 6% beryllium.

The data of Fig. 8–46 enabled Gerlach to draw interesting conclusions on the mechanism of precipitation of this alloy. The curves corresponding to intermediate times of aging show the presence of two phases, each of which follows the same relation between saturation magnetization and absolute temperature: $I_s/I_0 = f(T/\theta)$, θ being the Curie point in °K. The fact that the θ increases with the aging time from 40 to 303°C for one phase (α_1, α_1', α_1'', $\cdots$) and remains constant at 303°C for another phase, α_2, shows that during precipitation the phase of constant composition α_2 begins to separate immediately and continues to separate in increasing amounts, while the phase of composition α_1, α_1', $\cdots$, is present at first in

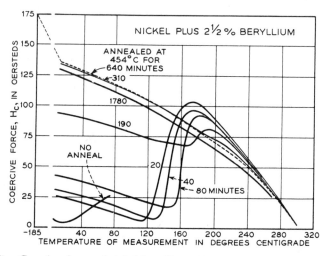

Fig. 8–47. Coercive forces of nickel-beryllium alloys (2.5% Be) annealed at 1000°C, hardened, and heated for various times at 454°C.

large amounts and then decreases in amount as its composition changes continually from α_1 to α_2. A different mechanism operates in the system nickel-gold (see below).

During the precipitation of the two magnetic phases α_1 (variable) and α_2 (constant) a non-magnetic phase NiBe is also precipitating, and its amount can be estimated by measuring the saturation magnetization at low temperatures and extrapolating to 0°K.

The mechanism of precipitation in the alloys subjected to other treatments, for example those not mechanically deformed after quenching, is not so clear-cut, and the phases present are not always homogeneous. For further discussion the reader is referred to the original papers [36G1, 37G3, 38G1] and to a more recent paper on the mechanism of precipitation by Auer [39A5].

The change in coercive force during aging has also been measured as a function of temperature, and the results are given in Fig. 8-47 for the same 2.5% beryllium alloy. The precipitating nickel-rich phase α_2 is apparently associated with the high coercive force.

Hysteresis loops were found to have the form associated with two phases of different magnetic properties.

Nickel-Bismuth.—Nickel-rich alloys have a liquid phase above 655°C (Fig. 8-48) in equilibrium with a β phase, probably NiBi (78.1% Bi), which is non-magnetic. The solid solubility of bismuth in nickel at room

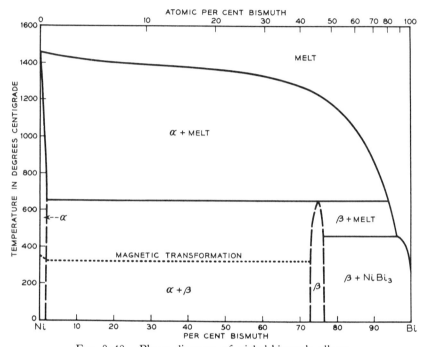

FIG. 8-48. Phase diagram of nickel-bismuth alloys.

temperature is 1 or 2%, enough to lower the Curie point of nickel about 25°C. Alloys are magnetic, with a Curie point of about 330, to about 73% bismuth.

Nickel-Boron.—The compound Ni_2B (8.4% B) has been established by X-rays [33B4], and Ni_3B_2 (10.9% B) and NiB (15.6% B) and others have been indicated by X-rays and by thermal means. Ni_2B has the same body-centered tetragonal structure as Co_2B and Fe_2B and is apparently non-magnetic. Giebelhausen's [15G2] results indicate a eutectic point at 1140°C and 4% boron, and the presence of two phases (Ni and Ni_2B) up to 8% boron.

Nickel-Carbon.—The nickel-rich alloys form a simple diagram (Fig. 8-49) which contains no compound stable at temperatures below the melting point of nickel. Ni_3C (6.4% C) has been reported to exist at higher temperatures in contact with the melt and has been produced at lower temperatures (300°C) by the decomposition of carbon monoxide. For further discussion see Hansen's summary [36H1].

The solid solubility of carbon decreases from about 0.6% at the eutectic temperature (1318°C) to 0.2% at 900–1000°C and is estimated to be 0.03%

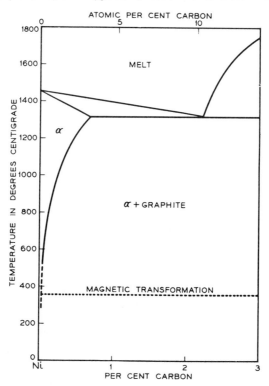

FIG. 8–49. Phase diagram of nickel-carbon alloys.

at 500°C [48K7]. This amount of carbon does not lower the Curie point of nickel by a noticeable amount; therefore all stable nickel-rich alloys have the same Curie point as nickel when they are in equilibrium. However, Gerlach and Rennenkampff [43G3] have detected indirectly a Curie point as low as 50°C in a supersaturated alloy.

Lange [38L5] and Gerlach and Rennenkampff have studied nickel supersaturated with carbon by melting in a graphite crucible and quenching and have measured the magnetization as dependent on temperature. In the range of temperature of 0°C to the Curie point the curve is abnormal in form and decreases almost linearly to zero at 350°C. Gerlach and Rennenkampff measured also the coercive force and remanence, before and after annealing at 600°C. The properties of the annealed specimen show a normal behavior with temperature, and the properties of the quenched specimen are those to be expected of an inhomogeneous alloy, having constituents of various Curie points and saturation intensities of magnetization. The nickel was originally supersaturated with carbon, and, during the precipitation of the excess carbon, alloys with various carbon contents—and therefore different magnetic properties—were formed.

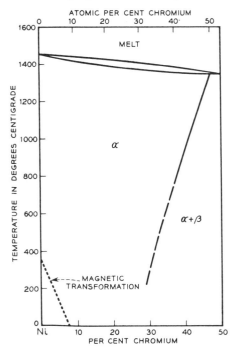

FIG. 8–50. Phase diagram of nickel-chromium alloys.

Nickel-Cerium.—Vogel [47V2] has published a diagram of this system showing a eutectic (m.p., 1210°C) of nickel with a compound to which the formula Ni_5Ce (32.3% Ce) is assigned. Other compounds are Ni_4Ce (37.4% Ce), Ni_3Ce (44.3% Ce), Ni_2Ce (54.4% Ce), $NiCe$ (70.5% Ce), and $NiCe_3$ (87.7% Ce). The solid solubility of cerium in nickel is slight, and the Curie point of nickel is not noticeably lowered. Alloys are magnetic to about the composition of Ni_5Ce.

The diagrams for nickel-lanthanum and nickel-praseodymium are quite similar to that for nickel-cerium [47V2].

Nickel-Chromium.—There is a wide range of solid solubility of chromium in nickel, over 45% at the eutectic (1343°C), as shown in Fig. 8–50, based on the work of Jette [34J2] and Jenkins [37J2] and their colleagues and on older work. The solubilities below 500°C are not well known. The second phase, β, precipitates as the chromium-rich phase having the body-centered cubic structure of chromium with contracted lattice spacing. The lattice spacings of both α and β phases have been determined by Jette and co-workers [34J2]. Chromium increases the spacing of nickel, and nickel decreases the spacing of chromium.

Sadron [30S8] and Marian [37M2] have measured the Curie points and saturation intensities of magnetization at temperatures down to 110°K and then extrapolated to 0°K. Sadron's results are shown below. Values of the numbers of Bohr magnetons per atom n_0 are appended.

308   OTHER HIGH PERMEABILITY MATERIALS

Per Cent Chromium	Atomic Per Cent Chromium	σ_s $(150°K)$	σ_0	n_0
1.51	1.70	49.75	50.9	0.53
2.92	3.28	42.40	43.7	0.46
6.02	6.74	25.40	29.0	0.30
8.00	8.76	16.3	21.9	0.23
10.0	11.2		15	0.16

At room temperature the alloys become non-magnetic at about 6% chromium, and at 0°K at about 12%. For various other determinations of Curie point, see Hansen [36H1].

The nickel-chromium alloy containing 13–20% chromium is used for corrosion-resistance and is non-magnetic.

Nickel-Copper.—This system is especially interesting from a theoretical point of view because the elements, one ferromagnetic and the other

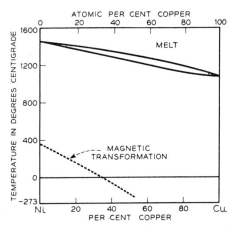

FIG. 8–51. Phase diagram of nickel-copper alloys.

diamagnetic, are neighbors in the periodic table, and because additions of copper cause a linear decrease of Curie point (in °K) [29K5] and of saturation to zero (Figs. 8–51 and 52). From the practical point of view the system contains *Monel metal*, *Constantan*, and other alloys that have various commercial uses. The properties of Monel are described later. Constantan, the 55% copper alloy, is useful because it has a high resistivity (49 microhm-cm), a low temperature coefficient of resistance (0 to 1 × 10^{-4}/°C), and a convenient thermal emf against copper. It is non-magnetic except at a few degrees above 0°K. "Nickel-silver," containing 10–20% nickel, 55–70% copper, and admixtures of zinc and other elements, is non-magnetic. The 75/25 alloy has long been used for coinage.

Some physical properties of the binary alloys are shown in Fig. 8–53. The lattice parameters, and densities calculated therefrom, are due to

NICKEL-COPPER

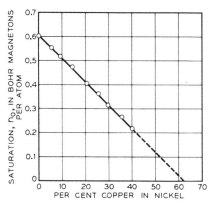

FIG. 8-52. Bohr magneton numbers of nickel-copper alloys.

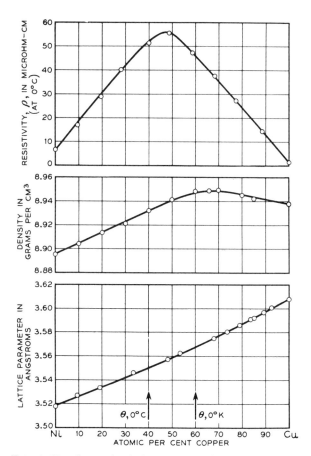

FIG. 8-53. Some physical properties of nickel-copper alloys.

Owen and Pickup [34O2], whose values are close to those of Burgers and Basart [30B7]. The resistivities were determined by Svensson [36S16] at low and high temperatures as well as at room temperature.

Alder [16A1] determined the specific magnetization in a field of 10 000 oersteds, at various temperatures, and extrapolated to 0°K. His results,

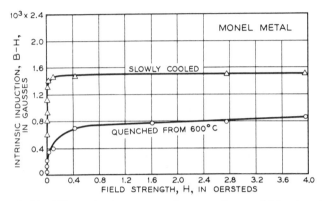

Fig. 8–54. Magnetization curves of two specimens of Monel metal.

expressed now as Bohr magnetons per atom of alloy (Fig. 8–52), show a linear decrease with copper content toward zero at about 60% copper. Kaya and Kussmann [32M3] measured B_s at room temperature and found 5000, 3450, and 1500 at 10, 20, and 30% copper, respectively.

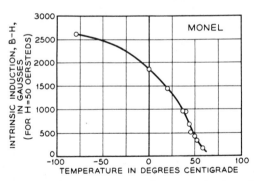

Fig. 8–55. Effect of temperature on the magnetization of Monel metal.

The alloy most studied magnetically is Monel metal, a combination of nickel and copper that has been reduced from an ore that contains them in the approximate ratio 2 to 1. A typical analysis of the alloy is as follows:

66–69% Ni	1.0% Mn
28–30% Cu	0.1% Si
1–2% Fe	0.1% C

It is ductile and corrosion-resistant. For its physical and mechanical properties the reader is referred to the *Metals Handbook* [48M2].

Burrows [21B2] found that at 200 oersteds the induction of cast Monel was about 500 gausses, and that of annealed Monel 2000. Kussmann [28K4] found a smaller but still considerable difference in the two states, as

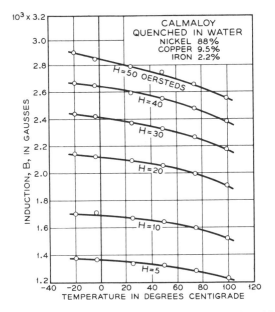

Fig. 8-56. Effect of temperature on the induction of Calmalloy No. 1 (2.2% Fe, 88% Ni, 9.5% Cu) at various field strengths.

shown in Fig. 8-54. The induction changes rapidly with temperature near room temperature, the Curie point usually lying between 25 and 100°C. Kussmann's data for Monel containing 65.8% Ni, 30% Cu, 1.9% Fe, 1.1% Mn, and 1% Zn are plotted in Fig. 8-55. Similar data have been reported by Inglis [29I2], who found also an initial permeability of 1100. Bates and Brown [28B4] found a retentivity of 670 and a coercive force of 1.8.

Alloys of two different compositions have been used with permanent magnets for compensation of change of flux with temperature (see also the iron-nickel alloy, *Thermoperm*, p. 133). These are called *Calmalloy* and are composed of 66.5% Ni, 30% Cu, 2.2% Fe and 88% Ni, 9.5% Cu, 2.2% Fe, respectively. The former is a Monel, and when used for this purpose is also called *Thermalloy* [28A4]. The relevant magnetic properties of the Calmalloys (Figs. 8-56 and 57) and the method of use have been described by Kinnard and Faus [30K9]. A similar alloy, containing 70%

nickel, 1% manganese, and 0.5% silicon, has been described by Yensen [48Y2].

Additions of aluminum, silicon, or sulfur are made to the Monel base to adapt the mechanical properties to various uses. Some of the better-known alloys of this kind are described in Table 8, compiled from publica-

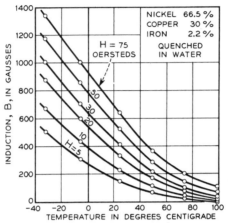

FIG. 8-57. Effect of temperature on the induction of Calmalloy No. 2 (2.2% Fe, 66.5% Ni, 30% Cu) at various field strengths.

tions of the International Nickel Company. The permeability of K-Monel at $H = 100$ has been plotted as a function of temperature by Shaw [48S10]. Some data are available on the structure of a number of *ternary systems*

TABLE 8. COMPOSITIONS AND PROPERTIES OF VARIOUS MONELS

Material	Composition (%)	Form	Curie Point (°C)
Monel	67 Ni, 30 Cu 1.4 Fe, 1 Mn	Wrought	25–100
Monel	67 Ni, 29 Cu 1.5 Fe, 1.25 Si	Cast	25–100
R-Monel	67 Ni, 30 Cu 0.035 S	Rolled Drawn	25–100
K-Monel	66 Ni, 29 Cu 2.75 Al, 0.9 Fe	Rolled Age-hardened	$\theta < -100$
H-Monel	65 Ni, 29.5 Cu 3 Si, 1.5 Fe	Cast Age-hardened	$\theta < -25$
S-Monel	63 Ni, 30 Cu 4 Si, 2 Fe	Cast Age-hardened	$\theta < -50$

containing nickel and copper. Of these the ones having most interest are Fe-Ni-Cu and Co-Ni-Cu; these are discussed on pp. 153 and 402. For modified Monels not already referred to in Table 8, the reader may consult the references of Table 9.

TABLE 9. SOME RECENT REFERENCES TO TERNARY SYSTEMS CONTAINING NICKEL AND COPPER

System	References	System	References
Ni-Cu-Al	48K5, 48S11	Ni-Cu-Mo	24D1
As	40K3	O	41B8
Au	47R1	S	40K3
Be	41I1	Sb	41S14
C	31M5	Si	39O4
Co	See p. 402	Sn	48E3, 39F3
Cr	39A1	W	38P3
Fe	See p. 153	Zn	48K6
Mn	44D2, 41A1		

Nickel-Gallium.—Kroll [32K1] reports that nickel alloys easily with gallium, and that no hardening is produced by quenching and aging the alloy containing 1.3% gallium.

Nickel-Germanium.—Ruttewit and Masing [40R2] have constructed the nickel portion of the phase diagram, using thermal and microscopic data. The β phase is presumably Ni_3Ge (29.2% Ge), and first appears at about 14% germanium. The eutectic melts at 1161°C.

Nickel-Gold.—At 900°C these alloys form a continuous series of solid solutions (Fig. 8–58). At lower temperatures they form two phases of face-centered cubic structure of different lattice parameters, which have been measured by Bain [29W8]. Westgren and Ekryan [30W7] have found NiAu (77.1% Ni) to have the ordered structure of β brass.

Sadron [32S3] and Marian [37M2] have measured the Curie points in annealed and in quenched alloys, as shown on the diagram. Their measurements of the specific saturation magnetization at 17°C (σ_s) and 0°K (σ_0) for the two treatments are as follows:

Per Cent Gold	Atomic Per Cent Gold	Annealed			Quenched		
		σ_s	σ_0	n_0	σ_s	σ_0	n_0
10.5	3.4	46.62	51.63	0.58	...	...	...
16.1	5.4	43.15	48.12	.57	43.09	48.02	0.57
29.8	11.2	35.92	39.25	.52	32.76	38.0	.50
40.0	16.6	30.24	32.94	.48	23.34	30.21	.44
52.8	25.0	28.82	25.74	.43	...	19.24	.32
59.0	30.0	20.54	22.18	.40	...	15.39	.27

The number of Bohr magnetons per atom of alloy decreases to zero at about 95% gold for the annealed alloys, the limit of solid solution, and at about 80% for the quenched alloys (when $\theta = 0°K$).

Köster and Dannöhl [36K8] measured the saturations as dependent on heat treatment and temperature and recorded the coercive force (to 220 oersteds) and remanence of a series of alloys. The mechanism of precipitation has been studied by magnetic methods by Gerlach [38G1, 49G4].

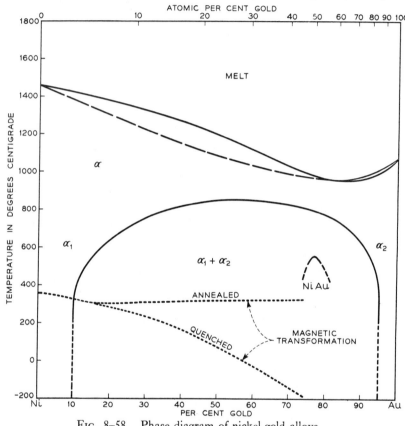

Fig. 8-58. Phase diagram of nickel-gold alloys.

Nickel-Hydrogen.—Sieverts [29S8] has measured the solubility of hydrogen in nickel at temperatures up to and above the melting point. Results are shown in Fig. 2–12, and have been confirmed by Armbruster [43A4]. The solubility is greater than in iron, and it increases rapidly at the melting point.

Armbruster has also measured the solubility of hydrogen in several nickel-iron alloys (8% Ni, 18% Cr; 28% Ni). Results of various reports have been summarized by Dushman [49D1].

Bredig and Schwartz von Bargkampf [31B1], and Büssem and Gross [33B5], observed that nickel films sputtered in hydrogen are sometimes hexagonal in structure. Colombani [44C1] confirmed this result and noted that the hexagonal form is only weakly magnetic but that it becomes strongly ferromagnetic when converted to the cubic form by annealing. Earlier work on the properties of thin films of nickel are referred to in this paper.

Nickel-Indium.—Weibke [39W5] has investigated nickel-rich alloys and found a solid solubility changing rapidly with temperature from less than 1% at low temperature to 14% at 883°, the eutectic temperature.

Nickel-Magnesium.—The solid solubility of magnesium in nickel is small and is not known. The solubility of nickel in magnesium has been estimated to be less than 0.1% [34H3] at 500°C. The β phase, Ni_2Mg (17.2% Mg), which forms the eutectic with nickel, has a hexagonal structure [35L2] and appears to be ferromagnetic with a Curie point of 235°; it is present in alloys containing less magnesium than corresponds to $NiMg_2$ (45.3% Mg) [08V1].

Kroll [44K1] found no precipitation hardening in alloys containing up to 1.8% magnesium and no carbon; definite aging was observed when 0.7% magnesium and 0.16% carbon were present.

Nickel-Manganese.—Figure 8-59 shows the phase diagram of Köster and Rauscher [48K8], who have used thermal analysis, X-rays, magnetic and elastic measurements, and microscopic data, and have had the advantage of a number of previous investigations. Near 50% manganese occurs the MnNi (δ) phase, which has a face-centered cubic structure between 650 and 1000°C, and transforms to the δ' tetragonal structure at lower temperatures. These phases are both non-magnetic.

In relation to the magnetic properties, the most interesting characteristic of this system is the ordering of the face-centered cubic solid solution of manganese in nickel, which occurs in a broad region of composition between 20 and 40% manganese, near Ni_3Mn. This was first reported by Kaya and Kussmann [31K7], who showed that the resistivity, Curie point, and saturation magnetization depend on heat treatment. The change of resistivity with composition for quenched and for annealed alloys is shown in Fig. 8-60, according to the more recent data of Kaya and Nakayama [40K5]. The change of ordering with temperature for the composition Ni_3Mn (23.8% Mn) is given in Fig. 8-61. Ordering has recently been observed by neutron diffraction by Shull and Siegel [49S2].

As indicated on the diagram, the Curie points of the disordered alloys (α) decrease with increasing manganese content to 0°C at 22% manganese [31K7]. The ordered (α') alloys, however, have higher Curie points than nickel.

The dependence of magnetic properties on heat treatment was noted by Gray [12G3] in 1912. Kaya and Kussmann showed that the saturation

316   OTHER HIGH PERMEABILITY MATERIALS

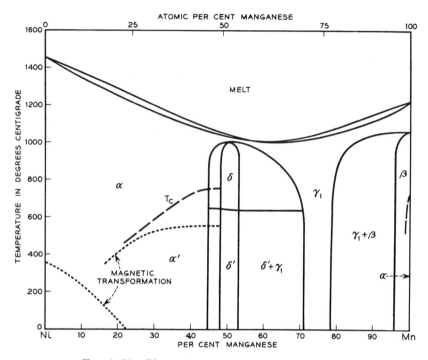

Fig. 8–59. Phase diagram of nickel-manganese alloys.

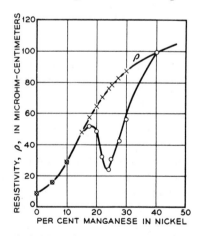

Fig. 8–60. Electrical resistivities of nickel-manganese alloys (a) quenched (upper curve, disordered structure) or (b) annealed (lower curve, ordered structure).

induction of the Ni_3Mn composition is considerably higher than that of nickel (see Fig. 8-62), and that annealing of the quenched alloy can raise

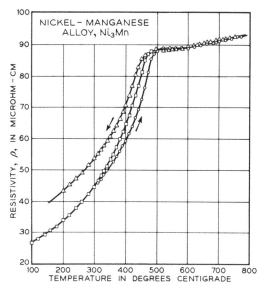

FIG. 8-61. Resistivity of Ni_3Mn (23.8% Mn) on heating and cooling, showing the effect of ordering. Equilibrium is represented by the middle curve.

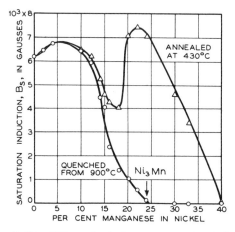

FIG. 8-62. Effect of ordering on saturation induction.

B_s from 0 to 7000. Manganese raises the saturation of the disordered alloys (0 to 10% Mn) also. Volkenstein and Komar [41V2] have reported a saturation induction of 9000 after annealing at 300–400°C.

The change of saturation induction and resistivity of Ni_3Mn with the

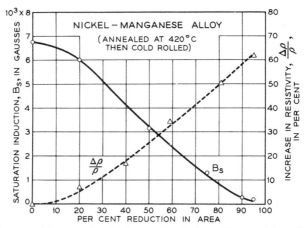

Fig. 8-63. Change of degree of order by mechanical working, and its effect on the saturation induction and resistivity of Ni_3Mn.

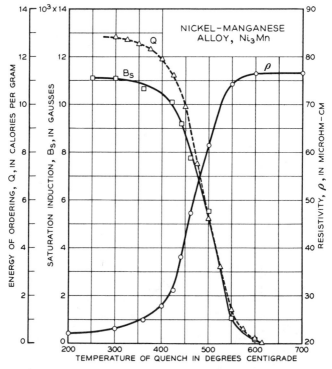

Fig. 8-64. Effect of quenching temperature (changing degree of order) on some properties of Ni_3Mn: ρ, resistivity; B_s, saturation induction; Q, energy of ordering.

disordering caused by the cold rolling of a well-annealed material is illustrated in Fig. 8-63, according to Dahl [36D1]. The similar changes brought about by quenching from various increasing temperatures are shown in Fig. 8-64 [40K5].

Guillaud [44G1] has measured the specific magnetization of pure Ni_3Mn, annealed 3 weeks at 470°C, just below the critical temperature of ordering. At 17°C and 20 000 oersteds he found $\sigma = 90$ (cf $\sigma = 54.4$ for nickel), and by extrapolating to infinite field and to the absolute zero of temperature he obtained the value $\sigma_0 = 98.16$, corresponding to an average Bohr magneton number per atom of $n_0 = 1.015$ (cf. 0.604 for nickel).

Gray [12G3] showed that the permeability of 15 and 20% manganese alloys are higher after slow cooling than after quenching and observed a permeability of 170 (at $H = 5$) in the 15% alloy. Jaffee [48J1] studied three alloys near Ni_3Mn in composition after annealing for a long time at 380°C. Some of the properties of the alloys so ordered, measured at room temperature, were as follows:

Per Cent Manganese	Maximum Permeability, μ_m	H for μ_m	B for $H = 30$	Residual Induction	Coercive Force
20.1	1800	0.75	4130	1590	0.52
21.4	5300	0.25	3910	1420	0.15
25.3	60	1.0	885	260	4.0

The highest observed permeability was 7500, at a field strength of about 0.1 oersted and a temperature of 91°C. Other data are given in the original article.

The coercive force of Ni_3Mn was measured by Volkenstein and Komar [41V2] after annealing at various temperatures from 300 to 900°C. A sharp maximum of over 90 oersteds was observed after annealing at 520°C (Fig. 8-65).

Kaya and Nakayama [40K5] have measured the specific heat of the alloy of composition Ni_3Mn, after annealing at various temperatures, and from the data estimated the energy of ordering Q. The results are shown by the curve of Fig. 8-64; Q is nearly proportional to the saturation induction. These authors have also measured the hardness, thermoelectric force, and thermal expansion through the transition temperature. Other data and a consideration of order have been given by Thompson [40T1].

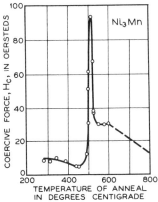

FIG. 8-65. Coercive force of Ni_3Mn as dependent on annealing temperature.

The resistivities of the alloys

containing up to 80% manganese have been reported by Valentiner and Becker [35V2].

Nickel-Mercury.—The early measurements [36H1] have indicated only a very slight solubility of nickel in mercury at room temperature, estimates varying from 2×10^{-5} to 0.1%. By decomposition of nickel carbonyl, Brill and Haag [32B11] prepared an amalgam containing 8.8% nickel and having a simple cubic structure with a lattice spacing of 3.00 kx units.

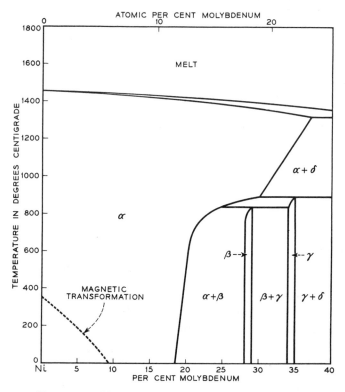

FIG. 8-66. Phase diagram of nickel-molybdenum alloys.

Bates and Baker [40B7] found that dilute amalgams (0.01–2.7%) became ferromagnetic when they had been heated to 225°C. Bates [41B7] concentrated amalgams by heating below 225°C and found ferromagnetism when the composition corresponded to the formula $NiHg_3$ (91.1% Hg).

Nickel-Molybdenum.—Figure 8-66 shows the diagram of Ellinger [42E3], based on the compounds (or ordered structures) Ni_4Mo (29.0% Mo), Ni_3Mo (35.3% Mo), and $NiMo$ (62.1% Mo). Ni_4Mo is ordered below 840°C, so that the face-centered cubic structure becomes slightly tetragonal with axial ratio 0.6231 or $0.6231\sqrt{2} = 0.981$, depending on the unit chosen [38G3, 44H3]. Ni_3Mo is hexagonal close-packed [42E3].

The solid solubility of molybdenum is 37% at the eutectic (1320°) and about 17% at room temperature. Age hardening begins at 20% molybdenum [38G3].

Curie points have been determined by Dreibholz [24D1], Grube and Schlecht [38G3], Grube and Winkler [38G4], and Marian [37M2]. Except for the older results, the data are consistent with the disappearance of ferromagnetism at room temperature at about 10% molybdenum. Marian's values of specific magnetization at saturation at 17°C (σ_s) and at 0°K (σ_0) are as follows:

Per Cent Molybdenum	Atomic Per Cent Molybdenum	σ_s	σ_0	n_0
3.00	1.88	42.28	47.58	0.51
5.20	3.25	31.84	39.44	.40
6.80	4.27	23.11	34.07	.37

The atomic moment becomes zero at about 15% molybdenum.

Nickel-Niobium.—The compounds Ni_3Nb (34.5% Nb) and $NiNb_3$ (82.6% Nb) are formed and give rise to eutectics melting at 1275° and 1175°C, respectively, as shown in Fig. 8–67 based on the work of Pogodin and Zelikman [43P2]. Somewhat lower solubilities than those shown in the diagram are reported by Kubaschewski and Schneider [49K2]. The solid solubility of niobium decreases rapidly with temperature. No data relating to Curie point or other magnetic properties have been uncovered.

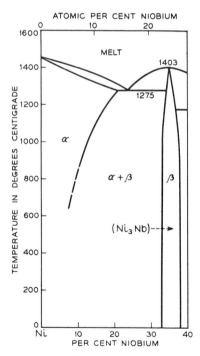

Fig. 8–67. Phase diagram of nickel-niobium alloys.

Nickel-Nitrogen.—Sieverts and Krumbaar [10S1] observed no solubility of nitrogen in nickel. Juza and Sachsze [43J2] report that 0.07% of nitrogen is dissolved in metallic nickel heated at 445°C in a current of ammonia. The compound Ni_3N (7.4% N) is formed, in which the nickel atoms are arranged in hexagonal close-packing. At high temperature Ni_3N_2 (13.7% N) appears to be stable [19V1]. Nickel sputtered in nitrogen has been observed to be hexagonal and non-magnetic [31B1].

Nickel-Oxygen.—NiO (21.4% O) has the NaCl structure and melts at about 1990°. It forms a eutectic that solidifies about 20° below the melting point of nickel and contains 0.2% oxygen. Seyboldt [36S17] reports that the solubility of oxygen in solid nickel decreases with increasing temperature: 0.020% at 600° and 0.012% at 1200°C.

The vapor pressure of oxygen over NiO has been estimated by Johnson and Marshall [40J1] as 0.6×10^{-6} mm of mercury at 1167°C, and 14×10^{-6} mm at 1293°C. The ratio of hydrogen to water vapor necessary to reduce NiO to Ni varies from $(H_2)/(H_2O) = 0.03$ to 0.04 in the temperature range 1200–1300°C. NiO is thus much more easily reduced than FeO.

Klemm and Hass [34K14] found the paramagnetic susceptibility of NiO to increase with increasing temperature in the range 100–700°C. When the oxygen falls to or below the stoichiometric ratio, metallic nickel is liberated and the material shows ferromagnetism. NiO is antiferromagnetic, with a Curie point of 250°C, deduced from measurements of the thermal expansion [48F3]. Below this temperature the susceptibility decreases on cooling, but only at a slow rate [46B4].

Nickel-Palladium and Nickel-Platinum.—See Chap. 9 and Fig. 10–11.

Nickel-Phosphorus.—The compounds Ni_3P (15.0% P), Ni_5P_2 (17.5% P), and NiP (34.6% P) have been reported repeatedly [38B9, 38N5], and others, containing more phosphorus, have been suggested. The eutectic of nickel with NiP melts at 880°C and contains 11% phosphorus [08K1]. The solid solubility of phosphorus in nickel and the Curie points of the alloys have not been determined.

Nickel-Selenium.—NiSe (57.4% Se) has the hexagonal structure of pyrrhotite; $NiSe_2$ (73.0% Se) has the cubic pyrites structure [38T3]. Other compounds reported are Ni_2Se_3 (66.9% Se) and Ni_3Se_4 (64.2% Se).

Nickel-Silicon.—This is a complex system in which eight compounds or superstructures have been identified with X-rays [39O5]: Ni_3Si (13.0% Si), Ni_5Si_2 (16.1% Si), Ni_2Si (19.3% Si, 2 modifications), Ni_3Si_2 (24.2% Si), NiSi (32.4% Si), and $NiSi_2$ (48.9% Si, 2 modifications). For the diagram the reader is referred to the *Metals Handbook* [48S12].

The solid solubility at room temperature or above is about 6% silicon, and at this composition the Curie point of nickel has been depressed below 0°C. In the alloys prepared by Iwasé and Okamoto [36I1] θ was lowered to -45°C at the solubility limit and then stayed constant with increasing silicon content. Kussmann and Scharnow [32M3] found θ constant at -75°C from 6–10% silicon. Marian [37M2] observed a θ of -33°C for 5% and 8% alloys after annealing, and -100 and -120°C after quenching. Marian's values of saturation magnetization at 17°C and 0°K are:

Per Cent Silicon	Atomic Per Cent Silicon	σ_s	σ_0	n_0	θ, °C
0.94	1.95	47.62	52.25	0.54	306
1.78	3.66	40.34	46.54	.48	234
2.77	5.67	30.29	36.69	.41	160
3.36	6.79	23.74	35.29	.36	117
4.30	8.60		27.23	.26	2
4.40	8.78		28.09	.28	19

The Bohr magneton numbers, n_0, and the Curie points in °K decrease toward zero at about 15 atomic per cent silicon (8% Si) (see Fig. 10–11). Silicon is sometimes added to nickel for mechanical hardening for commercial use.

Nickel-Silver.—Molten nickel and silver are immiscible. The solubility of silver in nickel is of the order of 3 or 4% at 900°C, and the solubility of nickel in silver is about 0.1% at the same temperature and less at lower temperatures [30T1]. Alloys of nickel and silver are therefore magnetic up to practically 100% silver.

Nickel-Sulfur.—The solubility of sulfur in nickel at room temperature is estimated to be about 0.005% [25M2], the precipitating γ phase being probably Ni_6S_5 (31.3% S). The Curie point is not noticeably lower than that for pure nickel until ferromagnetism disappears at about 30% nickel.

The nickel-rich alloys transform at 532°C. Above this temperature, especially when the sulfur content is greater than about 20%, the diagram is complex.

Naturally occurring NiS (35.3% S) is hexagonal with the FeS pyrrhotite structure, NiS_2 (52.2% S) has the cubic pyrites structure, and Ni_3S_4 (42.1% S) has the spinel structure [36H1]. Ni_3S_2 (26.7% S) is rhombohedral [38W4].

A partial Ni-Cu-S diagram has been given by Köster and Mulfinger [40K3]. Sulfur increases the range of composition in which nickel-copper alloys are magnetic.

Nickel-Tantalum.—Compounds Ni_3Ta (50.75% Ta) and NiTa (75.5% Ta) are indicated [36H1]. The solid solubility is 30% tantalum at 1000°C. The Curie point decreases to room temperature at 20–22% tantalum, according to Therkelsen [33T3] and Kubaschewski and Speidel [49K3]. The atomic moments drop to zero at about 20% Ta [41R3].

Nickel-Tellurium.—NiTe (68.5% Te) has the hexagonal structure of pyrrhotite, FeS [27O1]; $NiTe_2$ (81.3% Te) has the structure of CdI_2 [38T3]. The transition from one to the other has been studied by Klemm and Fratoni [43K12].

Nickel-Thallium.—The diagram was determined in 1908 by Voss [08V1] and shows that there is only partial miscibility of the melts. Above 302 and below 1387°C, alloys containing from about 3–100% thallium are composed of a liquid phase and a magnetic solid phase having a Curie point of 330°C. Thallium thus dissolves in nickel enough to lower the Curie point 15–20°C.

Except for the narrow range of solid solubility, the nickel-rich and thallium-rich phases are both present at room temperature; ferromagnetism thus persists to practically 100% thallium.

Nickel-Tin.—The diagram of Fig. 8–68 is that of Heumann [43H3]. The β phase, Ni_3Sn (40.3% Sn), is ordered (β') below 850–920°C in a hexagonal close-packed structure like that of Mg_3Cd.

The solid solubility of tin in nickel decreases from 20% at 1130°, the eutectic temperature, to about 2% at 500°C [37M9, 35F2], as determined by X-rays.

Determinations of the Curie point scatter considerably. Honda and Voss observed ferromagnetism to almost 40% tin; apparently β and β' are non-magnetic.

The older Curie point data [36H1] lie on the upper line in the figure; Marian's data for annealed and for quenched alloys lie on the upper and

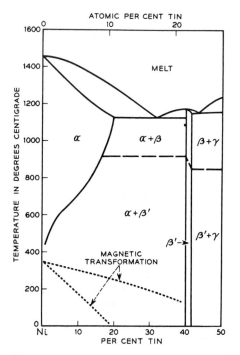

Fig. 8-68. Phase diagram of nickel-tin alloys.

lower lines, respectively. For quenched alloys the Curie points in °K and the saturation intensities of magnetization at 0°K extrapolate to zero at about 30% tin (15 at. % Sn). (See Fig. 10-10.) Values of σ_s (17°C) and σ_0 (0°K) of the quenched alloys are as follows:

Per Cent Tin	Atomic Per Cent Tin	σ_s	σ_0	n_0
5.20	2.65	40.13	45.79	0.49
14.4	7.68	20.23	31.07	.35
16.6	8.97	9.94	26.16	.30
20.2	11.1	...	18.8	.22

Nickel-Titanium.—The diagram (Fig. 8–69) has been constructed by Wallbaum [41W2]. The β phase, Ni_3Ti (21.4% Ti), is close-packed hexagonal with superstructure [39L4]. The γ phase, NiTi (44.9% Ti), is body-centered cubic with the CsCl structure. A solid solubility of only 2% titanium at 800°C is indicated by Wallbaum.

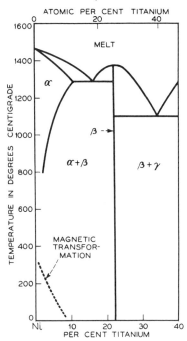

FIG. 8–69. Phase diagram of nickel-titanium alloys.

The Curie point falls to room temperature at about 9% titanium, according to Marian [37M2]. He has also determined the specific magnetization at saturation at 17°C (σ_s) and by extrapolation at 0°K, as follows:

Per Cent Titanium	Atomic Per Cent Titanium	σ_s	σ_0	n_0
3.89	4.84	34.53	41.18	0.43
5.69	7.0	27.06	31.69	.33
8.49	10.3	...	21.2	.22

The atomic moment extrapolates to zero at about 15 atomic per cent titanium.

Nickel-Tungsten.—The diagram of Ellinger and Sykes [40E1] is given in Fig. 8–70. The solid solubility of tungsten in nickel is about 30% at low temperatures and increases to about 40% at 1000–1400°C. Below 970°C the face-centered α phase and the body-centered γ phase (solid solutions of nickel in tungsten) combine to form the compound Ni$_4$W (43.9% W). Epremian and Harker [49E1] found this to be an ordered tetragonal structure. Also, they find evidence that the β field is larger than that given in the diagram.

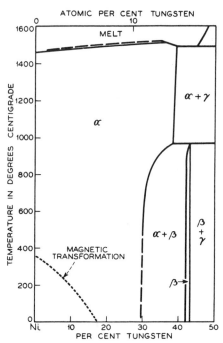

FIG. 8-70. Phase diagram of nickel-tungsten alloys.

Curie points were reported by Vogel [21V1]. The more recent determinations of Marian [37M2] are shown in the figure. His values of specific saturation magnetization at 17°C and 0°K are as follows:

Per Cent Tungsten	Atomic Per Cent Tungsten	σ_s	σ_0	n_0
6.35	2.11	39.22	44.30	0.49
7.95	2.68	33.35	40.18	.45
11.25	3.89	19.94	30.09	.34
15.05	5.35	...	21.16	.25

The Bohr magneton number, n_0, referred to one atom of alloy, extrapolates to zero at about 10 atomic per cent tungsten (26% W).

Nickel-Vanadium.—The work of Giebelhausen in 1915 [15G2] indicates that the solid solubility of vanadium in nickel is about 20% at room temperature. He found the Curie point to be lowered to about 250°C by 8% vanadium.

Sadron and Marian [37M2] observed a linear decrease of θ to 0°K at about 10% vanadium, their alloys becoming non-magnetic at room temperature at about 6%. They also measured the atomic moments which extrapolate to zero at the same composition.

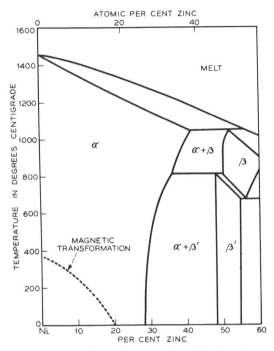

FIG. 8-71. Phase diagram of nickel-zinc alloys.

Nickel-Zinc.—The solid solubility of zinc in nickel increases from about 28% at room temperature to 41% at 1040°, the melting point of this alloy (Fig. 8-71) [48L1]. The precipitating phase at higher zinc contents is NiZn (52.7% Zn), which is cubic above 810° (β phase) and has the tetragonal AuCu structure (β') below this temperature. Other compounds exist in alloys containing greater amounts of zinc.

The Curie points of Tamaru [32T1] and Marian [37M2] are in fair agreement.

Marian's determinations of the specific saturation magnetization at 17°C (σ_s) and 0°K (σ_0) are as follows:

Per Cent Zinc	Atomic Per Cent Zinc	σ_s	σ_0	n_0
4.57	4.12	45.34	49.83	0.52
6.17	5.57	41.46	45.11	.48
11.90	10.81	25.38	35.18	.37

Additional points have been reported by Sadron [32S3]. The number of Bohr magnetons per atom of alloy, n_0, extrapolates to zero at about 30% zinc (see Fig. 10–10).

Nickel-Zirconium.—The work of Allibone and Sykes [28A5] shows the existence of a eutectic containing about 16% zirconium, formed from a nickel-rich phase and probably Ni_4Zr (28.0% Zr). Electrical resistance measurements [29S9] indicate that about 0.5% zirconium is held in solid solution.

The saturation induction of nickel is decreased from 6500 to 5000 by the addition of 6% zirconium, according to Sykes. The permeability is also reduced.

HEUSLER ALLOYS

In 1898 F. Heusler discovered that certain alloys of manganese, copper, and tin are ferromagnetic and have a saturation magnetization comparable with that of nickel. In a series of papers published in 1903 and the following years, investigations of these and other manganese alloys, particularly those of manganese, copper, and aluminum, were reported by Heusler [03H4, 04H1] and his associates, including Take [11T1], Starck and Haupt [03H5], and Asteroth [08A1], and by Ross and Gray [10R2] in Great Britain and by many others. An extensive bibliography of the work published before 1912 has been given by Heusler and Take [12H1] and by Ross [12R3], and later summaries of the magnetic properties have been published by Gumlich [27G3], Valentiner and Becker [33V1], O. Heusler [34H5], and Carapella and Hultgren [42C5].

The saturation magnetization and Curie point vary with both composition and heat treatment. The highest saturations in the two systems most investigated occur in alloys very near the compositions Cu_2MnAl and Cu_2MnSn. Ferromagnetism has also been observed in Cu_2MnIn and Cu_2MnGa. Coercive force and remanence vary over wide limits. Comparable alloys with silver in place of copper are also ferromagnetic. Unusually high coercive forces have been observed in Ag_5MnAl.

Structure.—Ferromagnetism is associated with the β phase, a solid solution with certain limits of composition estimated for the Cu-Mn-Al alloys by O. Heusler [28H4, 34H5] and shown by the solid lines of Fig. 8–72. A superstructure in this phase was found by Persson [28P3, 29P4] and Potter [29P3] who established the fact that the aluminum atoms were arranged as in Fe_3Al. O. Heusler [33H5, 34H5], and Bradley and Rogers [34B8] found that the copper and manganese atoms also occupied specific positions, which were determined by means of special X-ray technique. This highly ordered structure may be described as body-centered cubic with a face-centered superstructure, as illustrated in Fig. 8–73. The calculated density is 6.6.

HEUSLER ALLOYS 329

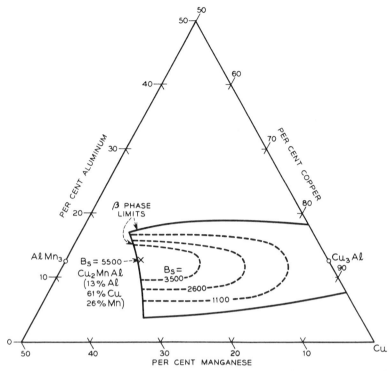

FIG. 8-72. Saturation inductions (broken lines) and phase boundaries (solid lines) of Heusler Mn-Cu-Al alloys.

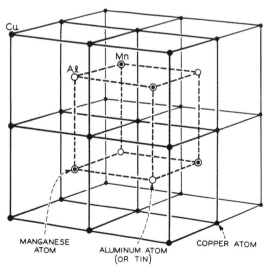

FIG. 8-73. Crystal structure of Heusler Mn-Cu-Al alloys: solid circles, Cu atoms; crosses, Al atoms, double circles, Mn atoms.

The same β-phase structure, with superstructure, is observed in the Cu-Mn-Sn alloys [29P4, 42C5], and its limits have been determined by Carapella and Hultgren [42C5]. In Fig. 8-74 the solid lines indicate the maximum extent of the β phase, when the most favorable heat treatment

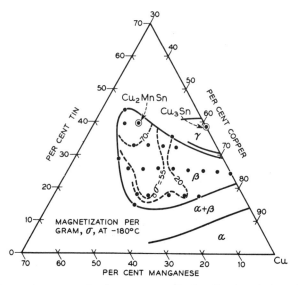

FIG. 8-74. Saturation magnetization per gram (broken lines) and phase boundaries (solid lines) of Mn-Cu-Sn alloys.

has been used. These authors found ordering to persist to a high temperature, perhaps to the melting point, and made a direct test with X-rays at 630°C.

Alloys can be mechanically worked if the manganese content is not too high; the limiting concentration has been given [27G3] as about 24%.

Saturation and Curie Point.—The effect of composition on the magnetization (for $H = 150$) is shown for the Cu-Mn-Al alloys by the points and broken lines of Fig. 8-72, according to F. Heusler and Richarz [09H1]. The highest saturation induction obtained at room temperature was about 5500 gausses.

More recent data of O. Heusler [34H5] are given in Figs. 8-75 and 76 for two cuts through the composition plane: one for alloys obtained by adding various amounts of aluminum to the composition Cu_2Mn, and the other containing 25 atomic per cent aluminum and variable copper and manganese. All of the alloys were measured first as quenched (from 50° below the melting point, after homogenizing), then after aging for 4000 hours at 110°C. The temperatures of measurement were −195°C to 110°C, and the maximum field was 1000 oersteds (assumed to give saturation). By extrapolating the B_s vs T curves, which were of normal form, the virtual Curie points were estimated and are given in the figures. The actual temperature at which ferromagnetism disappears is usually lower than the virtual Curie point; changes caused by heating the alloy in this temperature range decrease the Curie point, whereas aging at 110° increases it.

The highest ferric inductions measured by O. Heusler were 6300 gausses at room temperature and 7000 at −195°C. These maxima occur at the

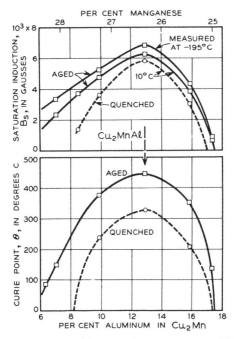

Fig. 8–75. Effect of the composition and heat treatment of Mn-Cu-Al alloys on their saturation and Curie point. Aluminum is added to the alloy for which the atomic ratio is Cu/Mn = 2/1.

composition Cu_2MnAl. The highest virtual Curie point, 450°, also occurs at the same composition. In addition, the resistivity is a minimum at 25 atomic per cent aluminum.

O. Heusler estimates that the saturation induction at 0°K, of the perfectly ordered alloy, is about 20% higher than the highest measured value. If that be true the number of Bohr magnetons per "molecule" of Cu_2MnAl is almost exactly 4.

As a result of the various magnetic and X-ray studies it seems to be well established that the magnetic properties are to be attributed to the atomic arrangement of Cu_2MnAl illustrated in Fig. 8–73. Slater [30S3] has shown that ferromagnetism is to be expected when manganese atoms are arranged on a space lattice with a certain interatomic distance, and that their separation in the Heusler alloys is equal to this distance, whereas in pure manganese the separation is too small (see Chap. 10). The aging at the low temperature, e.g., 110°C, has the effect of placing a large fraction of the manganese atoms in the ideal position.

Any increase in the proportion of the manganese atoms over that in Cu_2MnAl or any displacement of the atoms disturbs the regular arrange-

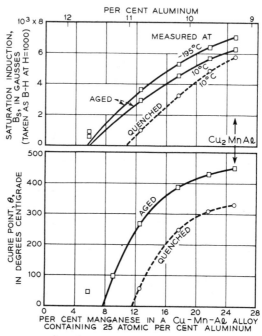

Fig. 8–76. Same as previous figure, except that composition is varied with aluminum held constant at 25 atomic per cent.

ment and decreases the atomic magnetic moment and the Curie point of the alloy. When the alloy has an excess of manganese atoms, the β phase decomposes, at temperatures above 200°C, into Cu_2MnAl and β manganese; when there is a deficiency of manganese the decomposition is to Cu_2MnAl and Cu_3Al [34B8].

The saturation magnetization of the manganese-tin-copper alloys has been measured at 25 and at −182°C by Carapella and Hultgren (see Fig. 8–74). Their results are expressed in units of magnetic moment per gram and are readily converted to Bohr magnetons per atom of manganese at −182°C. The approximate number of Bohr magnetons at 0°K may be obtained from their data by multiplying the number for −182°C by 1.02.

The highest measured value of the saturation induction is about $B_s = 7500$ at room temperature and 8300 at $-182°C$, values to be compared with 6300 and 7000, respectively, for the aluminum alloy. The estimated number of Bohr magnetons per atom of manganese is 4.1 at maximum, nearly the same as in the aluminum alloys.

F. Heusler found in his early work that arsenic, antimony, or bismuth may replace the aluminum or tin used in the original alloys. In 1947 Valentiner [47V3] reported that indium may be substituted for aluminum, and independently Coles, Hume-Rothery, and Myers [49C2], and Grinstead and Yost [49G2], also found ferromagnetism in this system in an alloy near in composition to Cu_2MnIn. Hames and Eppelsheimer [48H2] have also reported some Cu-Mn-Ga alloys to be strongly ferromagnetic.

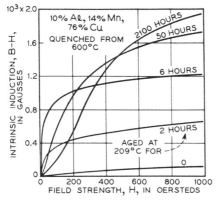

FIG. 8–77. Magnetization curves of a Heusler alloy (14% Mn, 76% Cu, 10% Al) after quenching from 600°C and aging at 209°C for various lengths of time.

Potter [31P3] observed ferromagnetism in Ag_5MnAl, which resembles the Heusler alloys in that it contains manganese and aluminum and an element similar to copper, but differs from them in the proportions contained. It has a saturation at room temperature of $B_s = 880$, and a high coercive force (see Chap. 9).

Permeability and Hysteresis.—These properties are markedly affected by composition and heat treatment. In the Cu-Mn-Al alloys permeabilities of several hundreds are not difficult to obtain, and a value of over 1000 has been reported [05G1]. The coercive force varies from about 1 oersted to 200, but lower values are found in alloys having low saturation. Magnetic softness is obtained by quenching from a red heat and aging at a low temperature, 110°C or higher. Magnetization curves of one alloy (76% Cu, 14% Mn, 10% Al), quenched from 600°C and aged for various lengths of time at 209°C, are given in Fig. 8–77, taken from the data of Take [11T1]. The permeability is increased by moderate aging but begins to decrease when the aging is long continued. Figure 8-78 shows the coercive force and residual induction of alloys of similar composition, aged for various lengths of time at 184°C. Changes in these properties with aging at other temperatures are described in Take's original article.

It is possible that the ordering first increases with time of aging, and that at a later time a separation of Cu_3Al and Cu_2MnAl begins, as suggested by Bradley and Rogers [34B8]. However, explanation of the simultaneous change of saturation, remanence, and coercive force has not been given.

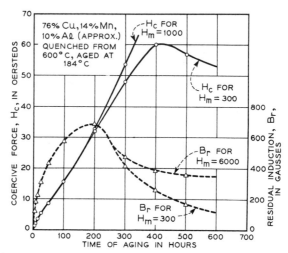

Fig. 8-78. Coercive force and residual induction of a Heusler alloy (14% Mn 76% Cu, 10% Al) quenched from 600°C and aged for various lengths of time at 184°C.

OTHER MANGANESE ALLOYS

In addition to the Heusler alloys, other alloys and compounds of manganese have been observed to be ferromagnetic. The Curie points and saturation magnetizations of some of the compounds are given in Table 10.

Summarizing articles on the magnetic properties of manganese compounds have been published by Guillaud [48G3, 43G2], Klemm [36K4], Messkin and Kussmann [32M3], Bates [29B4], Wedekind [24W7], von Auwers [20A3], Hilpert [13H1], and others. The existence of ferromagnetism in alloys containing As, Sb, Bi, and B was reported by F. Heusler [04H1] in 1904.

In the following pages no attempt is made to report the phase structure of the various phase systems, except as they relate to the magnetic data. Some information of this kind is available in Hansen [36H1] and *Metals Handbook* [48M2]. Crystal structures are given when known.

Manganese-Antimony.—After Heusler's report, these alloys were investigated by Williams [07W3] and by Wedekind [09W1]. The com-

TABLE 10. MAGNETIC DATA REPORTED FOR SOME MANGANESE COMPOUNDS

Here the saturation moment per gram at 0°K is represented by σ_0, the Bohr magnetron number per atom of manganese by n_0, and the ferric induction at room temperature by B_s. In converting σ_s to B_s, the saturation induction at room temperature, estimated densities are sometimes used.

Compound	Curie point (°C)	σ_0	n_0	B_s	H_c	Structure
MnAs	45	141.6	3.40	8400	7	Pyrrhotite
MnB	260	...	...	1850	30	Orthorhombic
MnBi	360	74.8	3.52	7800	600	Pyrrhotite
Mn-H	...	...	...	~200	650	...
Mn$_4$N	470	30	0.24	2300	200	Cubic
MnP	25	77	1.2	...	60	Orthorhombic
Mn$_2$Sb	277	45.2	0.94	2900	31	Tetragonal
MnSb	314	111.7	3.53	8900	8	Pyrrhotite
Mn$_4$Sn	150	...	...	1250	...	Hexagonal (Mn$_{11}$Sn$_3$)
Mn$_2$Sn	−11, 0	64.7	0.86	...	...	Pyrrhotite

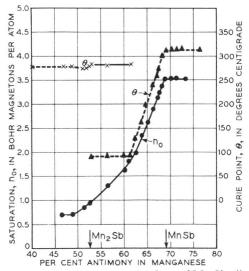

FIG. 8-79. Bohr magneton numbers of Mn-Sb alloys.

pounds MnSb (68.9% Sb) and Mn$_2$Sb (52.5% Sb) are ferromagnetic. The former has the hexagonal pyrrhotite structure with $c = 5.78$, $a = 4.12$ kx, $c/a = 1.40$; the latter has a tetragonal structure of the Fe$_2$As type, with $c = 6.56$, $a = 4.08$ kx, $c/a = 1.61$ [36H5]. Mn$_2$Sb is present in alloys containing from about 0–60% antimony; MnSb, in alloys from about 70–100% antimony [36H1, 43G2, 49G9]. All manganese-antimony alloys are ferromagnetic except those very close in composition to pure manganese and pure antimony. Guillaud's data are given in Table 10.

The variation of Curie point and Bohr magneton number with composition are shown in Fig. 8-79. An additional Curie point of 90°C coexists with that of Mn_2Sb on the antimony-rich side of Mn_2Sb, and when Mn_2Sb is no longer present as a phase this Curie point rises linearly to the value 314°C for MnSb. A solid solution of manganese in MnSb is indicated.

The crystal anisotropies of the two compounds are discussed in Chap. 12.

Wedekind [12W2] reported a coercive force of 31 oersteds in Mn_2Sb and 8 oersteds in MnSb.

Manganese-Arsenic.—After Heusler's report, the compound MnAs (57.7% As) was prepared and studied by Hilpert and Dieckmann [11H2], and their findings were confirmed by Bates [28B5, 30B8]. It has the hexagonal pyrrhotite structure, with $c = 5.691$, $a = 3.710$ kx, $c/a = 1.53$ at 20°C, and $c = 5.691$, $a = 3.659$, $c/a = 1.55$ at just above 45°C [44G4]. The sudden decrease in the dimension a at 45°C has been observed dilatometrically and coincides with the loss of ferromagnetism on heating and with an increase in resistivity by 4 times [34B7]. Temperature hysteresis occurs, so that ferromagnetism disappears at 45°C on heating and reappears at 34° on cooling [29B4, 43G2]. The virtual Curie point, obtained by extrapolating the σ_s vs T or σ_s^2 vs T curve beyond 45°C, is about 130°C.

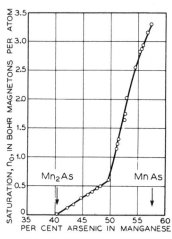

FIG. 8-80. Bohr magneton numbers of Mn-As alloys.

Guillaud [43G2, 49G9] measured the specific magnetization at low temperatures and determined σ_0 as dependent on composition from 33–49.5 atomic per cent arsenic (40–57% As). Results are shown in Fig. 8-80. In this range σ_0 increases from 0–141.6 units (3.29 Bohr magnetons) and by extrapolation to 50 atomic per cent is equivalent to 3.40 Bohr magnetons per atom of manganese. Guillaud has also plotted a curve showing σ/σ_0 vs T/θ for MnAs [46G2]. The saturation induction at 0°K is about 11 000 and at room temperature about 8400.

In the range 44–56% arsenic, Curie points of -147, 12, 28, and 40°C have been observed. The lowest point is reversible with temperature and is believed by Guillaud to be characteristic of a phase having the composition $Mn_{29}As_{21}$, stable from 48–53%. Further proposals regarding the phase structure have been made [43G2].

In a field of 760 oersteds Bates [29B5] observed a ferric induction of $B-H = 2500$ at room temperature and 2800 at liquid air in a specimen of MnAs. The coercive force of this sample was 60 oersteds, and 7 was reported later [34B7] for another sample.

Manganese-Bismuth.—Following Heusler's [04H1] report, the Curie point of MnBi (79.2% Bi) was determined by Hilpert and Dieckmann [11H2]. Its structure is like that of pyrrhotite and is hexagonal with $c = 6.12$, $a = 4.30$ kx, $c/a = 1.43$ [39H4]. According to Guillaud [43G2] it forms slowly at 445°C on cooling, probably from the bismuth-rich melt and a manganese-rich compound stable between 445 and 600°C (see Hansen [36H1]). The Curie points of the alloys containing 2–98% are all the same—360°C on heating, 340° on cooling; consequently, MnBi is stable over this wide range and forms no appreciable amount of solid solution with manganese or with bismuth. The nature of the transformation at 445° may be similar to that occurring at 45°C in MnAs.

Guillaud [43G2] measured $\sigma_0 = 74.8$, corresponding to $n_0 = 3.52$ Bohr magnetons per atom of manganese, a value somewhat higher than that (3.13) previously reported by Thielmann [40T5]. According to the latter, the ferric induction is $B-H = 7600$ at $-180°C$. At room temperature Guillaud's data led to a saturation induction of $B_s = 7800$.

The coercive force is markedly dependent on crystal size and varies from 600 oersted for particles 100×10^{-4} cm in diameter to 12 000 oersteds for 3×10^{-4} cm particles, as measured by Guillaud [43G2]. This material is discussed in more detail in Chaps. 9 and 18, and a curve showing coercive force as dependent on crystal size is given in Fig. 2–38. The crystal anisotropy is discussed in Chap. 12. Saturation and hysteresis as dependent on temperature have also been measured by Thielmann [40T5].

Manganese-Boron.—Heusler first reported the ferromagnetism of these alloys, and Binet du Jassoneix [06B1] showed that it was characteristic of MnB (16.5% B) and not MnB_2. Wedekind [07W4, 12W2] gave rough values of the magnetization and permeability of the powder in comparison with iron powder. The induction (at $H = 796$) is about 1830 at room temperature [29W2]. A coercive force of about 30 was obtained.

Manganese-Carbon.—Heusler [04H1] observed no ferromagnetism in Mn-C alloys. Wedekind [11W3] reports Mn_4C to be "weakly magnetic."

Manganese-Gallium and Manganese-Indium.—The existence of ferromagnetism in some binary alloys of these systems has been reported by Hames and Eppelsheimer [48H2, 49H3].

Manganese-Hydrogen.—When cast in hydrogen, manganese has been found to be slightly ferromagnetic by Weiss and Onnes [10W3] and by Hadfield, Chéveneau, and Géneau [17H1]. Kuh [29W2] measured a hysteresis loop of such material, for which $\sigma = 2.2$, $B_s = 200$, $H_c = 670$, after a field of 12 000 had been applied.

Manganese-Nitrogen.—Wedekind and Viet [08W2] reported some nitrides to be ferromagnetic, and Weiss and Onnes [10W3] and Ishiwara [16I1] observed manganese to be ferromagnetic under certain conditions when contaminated with nitrogen. Ochsenfeld [32O1] observed that manganese takes up nitrogen at about 1100°C and gives it off again at 1300–1320°C, and that the nitrided product is ferromagnetic with a Curie point of about 500°C, a coercive force of $H_c = 200$, a ferric induction of $B - H = 200$ when $H = 600$, and a remanence of $B_r = 110$. The work of Hägg [29H3] and Schenck and Kortergräber [33S9] showed the ϵ phase,

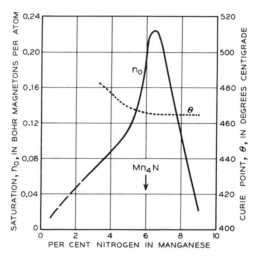

Fig. 8–81. Bohr magneton numbers and Curie points of Mn-N alloys.

Mn_4N (6.0% N), face-centered cubic in structure ($a = 3.85$ kx), to be the ferromagnetic constituent and the ζ phase (probably Mn_5N_2) to be hexagonal close-packed and nonmagnetic.

In two more recent papers, Guillaud and Wyart [44G2, 46G3] have prepared alloys by nitriding at various temperatures. When prepared above 800°C the number of magnetons per atom, n_0, and the Curie points, θ, are those shown in Fig. 8–81. The maximum of n_0 lies at a composition near Mn_4N. Compositions richer in nitrogen have a fixed θ, and n_0 decreases linearly toward zero at about the composition Mn_5N_2; therefore these alloys are probably mixtures of the ϵ and ζ phases. On the nitrogen-poor side of Mn_4N solid solution occurs. Interatomic distances in solid solutions prepared below 800°C increase linearly with atomic per cent nitrogen in the range 12–21 atomic per cent.

Manganese-Oxygen.—Of the various oxides Mn_3O_4 is the most magnetic, but no ferromagnetism has been observed. Haraldsen and Klemm

[34H4] report a volume susceptibility of 69×10^{-6} at room temperature and 73×10^{-6} at 90°K.

Ferromagnetism has been observed by Jonker and van Santen [50J1] in the binary solid solutions of $LaMnO_3$ with $CaMnO_3$, $SrMnO_3$, and $BaMnO_3$. These all have the perovskite structure, as does the ferroelectric material $BaTiO_3$. None of these manganites in the pure form is ferromagnetic, but when even a small amount of an alkaline-earth manganite is added to $LaMnO_3$ both the Curie point and saturation magnetization rise rapidly to a maximum, which occurs at 30–35% of the added manganite, and then decreases to zero. The maximum Curie points are about 0°, 100°, and 70°C, respectively, and the saturation intensities of magnetization at 90°K are about $I_s = 99$, 91, and 90, for the solid solutions containing Ca, Sr, and Ba.

The origin of the magnetic moment is associated with the incomplete $3d$ shell in the Mn^{3+} and Mn^{4+} ions, and the ferromagnetism is believed to be associated with super-exchange (see "Antiferromagnetism" in Chap. 10). If this is so, it is the first example of a positive super-exchange.

Manganese-Phosphorus.—These alloys are ferromagnetic [07W4] when they contain more than 17% phosphorus, about the composition at which MnP (36.1% P) first forms [36H1]. In alloys containing 17–34% phosphorus, Guillaud and Créveaux [47G2] found the Curie point to be constant at 25°C and the saturation to vary with the content of MnP. Extrapolation of the saturation values to that for MnP yielded $\sigma_0 = 77$, and $n_0 = 1.20$ Bohr magnetons per molecule. Variation of σ_s/σ_0 with T/θ for MnP has been plotted by Guillaud [46G2]. Berak and Heumann [50B5] find Mn_3P_2 to be more strongly ferromagnetic than either MnP or Mn_2P.

Bates [29B4] found the Curie point to be 40°C or less and to have no temperature hysteresis. The coercive force at room temperature was about 60 oersteds for the specimen measured.

Manganese Combined with S, Se, Te, Si.—Wedekind [11W3] reports natural and artificial MnS to be magnetic when heated. Haraldsen and Klemm [34H4] find it to be paramagnetic between 90°K and room temperature. MnS_2, like FeS_2 and NiS_2, is paramagnetic at room temperature whereas CoS_2 is ferromagnetic below $-180°C$ [35H6]. MnSe, MnTe, and MnSi are said to be magnetically weak [11W3], presumably paramagnetic.

Manganese-Tin.—After Heusler's [04H1] discovery of the ferromagnetism of these alloys, Williams [07W3] found ferromagnetism in all the alloys containing 8–93% tin. His work indicated the existence of MnSn (68.4% Sn), Mn_2Sn (51.9% Sn), and Mn_4Sn (35.1% Sn). Honda [10H3] found a sharp peak in the remanence vs composition curve near Mn_4Sn, and absence of remanence in alloys that contained no Mn_4Sn phase.

Later work by Potter [31P3] placed the Curie points of Mn_4Sn and Mn_2Sn at about 150 and 0°C, and the saturation ferric inductions at 1250

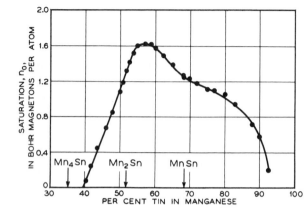

FIG. 8–82. Bohr magneton numbers of Mn-Sn alloys.

(room temperature) and 5900 gausses (liquid air), respectively. He noted that the magnetization of alloys containing less than 65% tin (including Mn_4Sn and Mn_2Sn) was increased by annealing for a long time at 500° but decreased by annealing at higher temperatures. Guillaud [43G2], on the contrary, found Mn_4Sn and all alloys containing less than 38% tin to be non-magnetic, after annealing for a long time at 500°C. He determined the magnetization per gram at 0°K, and the corresponding number n_0 of Bohr magnetons per atom of manganese, for alloys containing 38–94% tin, with the results shown in Fig. 8–82. The Curie points of all of these alloys were about −11°C, lower than previously found.

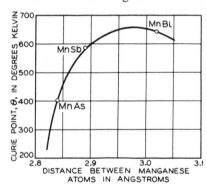

FIG. 8–83. Curie points in MnAs, MnSb and MnBi as dependent on interatomic distance. Note relation to Bethe interaction curve.

The maximum of n_0 occurs at 57% tin, somewhat on the tin-rich side of Mn_2Sn, and there $\sigma_0 = 73$, corresponding to a ferric induction of about 11 000 at 0°K.

The discrepancies between the various works seem to be connected with the effect of heat treatment on atomic arrangement. An ordered structure has been observed in the alloy of composition $Mn_{11}Sn_3$ (37.1% Sn) by Nowotny and Schubert [46N3]. They found also that Mn_2Sn has the pyrrhotite structure, and $MnSn_2$ (81.2% Sn) a centered tetragonal structure like that of FeSn$_2$ and CoSn$_2$ and similar to the face-centered cubic structure of CuAl$_2$ and FeB$_2$.

Curie Points of Arsenic Group.—In the compounds MnAs, MnSb, and MnBi the structures are all the same, only the distances between the atoms varying. The Curie points in degrees Kelvin (the virtual Curie point in the case of MnAs) are plotted against the distance between metal atoms in Fig. 8–83, according to Guillaud [43G2]. The curve is similar in form to the theoretical "interaction" curve (Fig. 10–13), though its course is not too well defined by the limited number of points. The fact that the point for MnAs lies on the steep part of the curve agrees nicely with the fact that the disappearance of ferromagnetism is associated with a contraction of the lattice. The sharp drop of the curve at lower interatomic distances is also consistent with the lack of ferromagnetism in the element manganese, for which the interatomic distance is small.

OTHER ALLOYS AND ELEMENTS

In addition to the alloys already described there are a number of chromium compounds that are weakly ferromagnetic, and these will be discussed briefly.

Among the elements, gadolinium is definitely ferromagnetic, and its properties are described later. It is then a question whether or not some of the other rare earth elements may also be ferromagnetic. Investigations have been made of metallic cerium [37T4, 45L1], lanthanum [37T4], and neodymium [37T4] of high purity, and these have been found to be non-magnetic. Trombe [45T1] has investigated *dysprosium* and reports that it is ferromagnetic below 105°K, and then shows remanence and saturation. At 88°K it then has a susceptibility per gram varying from 0.0512 at $H = 1025$ ($\sigma = 52.5$) to 0.0245 at $H = 8775$ ($\sigma = 215$), and thus has a saturation induction at this temperature of the order of 23 000. These results have not yet been confirmed.

Metallic manganese, chromium, and vanadium, when pure, show no ferromagnetism even at the temperature of solid hydrogen, 14°K [10W3].

Chromium-Antimony.—The compounds CrSb and CrSb$_2$ have been prepared. CrSb$_2$ is definitely non-magnetic, while the quality of CrSb has not been definitely established [43H4, 43N3].

Chromium-Arsenic.—Nowotny and Arstad [37N4] observed weak ferromagnetism in CrAs and Cr$_2$As, which are isomorphous with FeAs and Fe$_2$As. Haraldsen and Nygaard [39H5] investigated over the composition range 32–56% arsenic and found the highest susceptibility at the composition corresponding to Cr$_3$As$_2$, which has a rather complicated structure. The Curie point is about $-50°C$. The highest permeability observed was about 2.5 at $-183°C$, and at the high field used this corresponds to an induction of about 3200.

Chromium-Oxygen.—Cr_5O_9 was one of the first compounds of nonmagnetic elements to be reported as ferromagnetic, and several investigations have been made since its discovery in 1859 [1859W1]. Recently Guillaud, Michel, Bénard, and Fallot [44G3] have prepared CrO_2, admixed with Cr_2O_3 and oxygen. From measurements on two preparations, containing 40 and 50% CrO_2, the moment of pure CrO_2 was derived as 138 units per gram when saturated at 0°K. This corresponds to 2.07 Bohr magnetons per atom of chromium.

CrO_2 has a tetragonal structure in which the Cr atoms lie in a linear chain 2.86 kx apart.

Chromium-Platinum.—See Chap. 9. Ordered alloys are ferromagnetic with saturation induction as high as 3000.

Chromium-Sulfur and Chromium-Selenium.—Bates [39B16] prepared several specimens of various Cr/S ratios (0.5–1.4). Curie points of 30–100°C were observed. The saturation magnetization per gram σ_s was quite small, 0.20 at the maximum which occurred at CrS. CrSe is paramagnetic [35H5].

Chromium-Tellurium.—The ferromagnetism of CrTe has been noted by Goldschmidt [27G2] and Haraldsen and Kowalski [35H5]. It has the hexagonal pyrrhotite structure with $a = 3.981$, $c = 6.211$ kx [27O1]. Guillaud and Barbazat [46G4] have found the saturation moment at 0°K to be 74.57, equivalent to 2.39 Bohr magnetons per atom of chromium. The Curie point is 66°C. The variation of saturation with temperature follows closely the theoretical curve for $J = 2$ [46G5], and at room temperature B_s is about 3100. Galperin and Perekalina [49G5] found ferromagnetism to increase with tellurium from 1 to 50 atomic per cent and then to decrease rapidly.

Silver-Fluorine.—Gruner and Klemm [37G5], in a brief report, found AgF_2 to be ferromagnetic below $-110°C$, with specific magnetization of about 0.3 unit at $-150°C$. Further consideration of this apparent anomaly has not been reported.

Potassium-Sulfur.—A similar peculiar behavior has been reported by Klemm and Sodoman [35K13] for K_2S_3. Below $-50°C$ this material is drawn to the pole of an electromagnet, while at $-40°C$ it shows normal diamagnetism. The authors suggest further investigation.

Gadolinium.—In 1935 Urbain, Weiss, and Trombe [35U1] discovered that metallic gadolinium is ferromagnetic. The specimen they used contained 99.3% gadolinium, almost 0.7% silicon, and less than 0.03% iron. Using fields as high as 17 500 oersteds and temperatures down to 77°K, they determined the saturation magnetization per gram at 0°K as 253.5 units, corresponding to 7.1 Bohr magnetons per atom. The Curie point is 16°C.

A magnetization curve at 77°K is shown in Fig. 8–84. The intensity of magnetization can be estimated from the density, known from the crystal structure to be 7.97 at room temperature and estimated at lower temperatures assuming a normal thermal contraction. At 0°K the estimated B_s is 25 400, considerably above that of iron, 21 800. The number of Bohr magnetons per atom is even larger in relation to iron: 7.1 to 2.2.

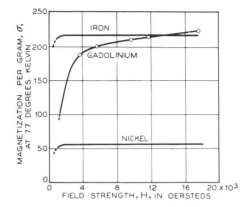

FIG. 8-84. Intensity of magnetization (per gram) of gadolinium, as compared with iron and nickel, at 77°K.

Several paramagnetic compounds of gadolinium (GdC_2, GdD_2, GdH_2) have been investigated magnetically by Trombe [44T2]. In all of these there are about 7 Bohr magnetons per atom of gadolinium; the number is unaffected by chemical combination because the inner $4f$ shell, responsible for ferromagnetism, lies too deep in the atom. The hydride is believed to have a ferromagnetic Curie point somewhat above absolute zero. The lowest temperature used in the investigation was about 80°K.

Auwärter and Kussmann [50A3] have found manganese-platinum alloys to be ferromagnetic in the range 84–90% Pt (68–82 at. % Pt). Ferromagnetism is attributed to $MnPt_3$, which has the ordered structure of $AuCu_3$ below 900°C. Superstructure was observed in all magnetic samples, the most intense superstructure lines and the highest saturation being observed at 75 at. % Pt. The value of B_s at this composition is 5000 at 20°C, 8000 at −193°C. The Curie point varies from 430°C to −100°C as the platinum content increases.

In unpublished work following the investigation of Mn-Cu-In alloys [49G2] (see p. 333), Goeddel and Yost have observed ferromagnetism in binary Mn-In alloys containing between 40 and 90% In. The existence of ferromagnetism in this binary system has also been indicated in preliminary experiments by Hames and Eppelsheimer [48H2] (see p. 337).

CHAPTER 9

MATERIALS FOR PERMANENT MAGNETS

Introduction.—Permanent magnets, like electromagnets, are used to produce external magnetic fields of considerable strength and constancy. Instruments which depend on permanent magnets for operation include loud-speakers, d-c meters, watt-hour meters, small motors and generators (magnetos), magnetons, polarized relays, and telephone receivers and ringers. In all of these the magnetic circuit containing the magnet contains also an air gap; consequently, the magnet always operates under the influence of a demagnetizing field and not at residual induction but at some

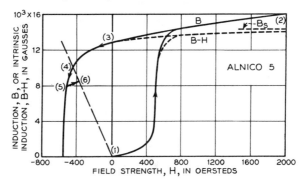

FIG. 9–1. Magnetization and demagnetization curves of a material of high coercive force (Alnico 5). In high fields $B - H$ approaches the saturation limit, B_s. In magnetizing and stabilizing a permanent magnet steps 1 to 6 are followed.

lower value of B. Thus the important curve for a permanent magnet material is that portion of the hysteresis loop that lies in the second quadrant, between residual induction and coercive force; this is called the *demagnetization curve*. Magnetization and demagnetization curves for Alnico 5 are shown in Fig. 9–1.

Single quantities most used in evaluating the quality of materials are H_c, B_r, and the products B_rH_c and $(BH)_m$, the latter being the maximum value of the product of B and H for points on the demagnetization curve (negative sign omitted). A typical (BH) vs B curve is given in Fig. 9–2. Often the reversible permeability and portions of minor loops are of importance, and these are also shown in the figure. Among the other properties that must often be taken into consideration in the use of a material are the resistivity, density, and strength. In the manufacture the workability of the alloy is an important item. The importance of physical factors such as magnetostriction, internal strain, and so on will be discussed in a later section.

After the permanent magnet has been finally assembled in the apparatus in which it is to be used, and after it has been magnetized in place, the field strength in the gap should be as insensitive as possible to shock or vibration, to variation in ambient temperature, to the presence of external magnetic fields, or to any change with time due to structural or metallurgical aging

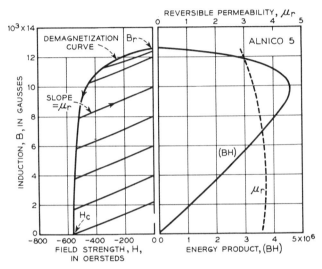

FIG. 9-2. Demagnetization curve (with portions of subsidiary loops), energy product curve and reversible permeability curve of permanent magnet material.

of the material. In order to minimize such changes, the magnetic circuit is usually "stabilized" after magnetizing, by momentarily applying a small field in the reverse direction. The magnetic changes then taking place are indicated in Fig. 9-1 by points 4, 5, and 6, the curve already having passed through points 1, 2, and 3 and arrived at point 4 determined by the demagnetizing action of the gap. Sometimes it is necessary to protect the surface of the permanent magnet from accidental contact with extraneous magnetic material, for such contact may change the state of magnetization at the point of contact and thereby the strength of the field in the gap.

History.—The common materials now used for permanent magnets are listed in Table 1. Before 1910 hardened carbon steel, containing up to 1.5% carbon, was commonly used and had been for centuries for the construction of compass needles. Manganese was often added to carbon

TABLE 1. COMPOSITIONS AND PROPERTIES OF SOME USEFUL PERMANENT MAGNET MATERIALS[†]

Name	When Used	Reference	Typical Composition[†]	H_c	B_r	$(BH)_m \times 10^{-6}$
Tungsten steel	1885	24H2	6 W, 0.7 C, 0.3 Mn	65	10 500	0.3
Low chrome steel	1916	10B2	0.9 Cr, 0.6 C, 0.4 Mn	50	10 000	0.2
High chrome steel	1916	16G1	3.5 Cr, 1 C, 0.4 Mn	65	9 500	0.3
KS magnet steel[1]	1917	20H2	36 Co, 3.5 W, 3.5 Cr, 0.9 C	230	10 000	0.9
Cobalt chrome steel[2]	1921	23G1	16 Co, 9 Cr, 1 C, 0.3 Mn	180	8 000	0.6
Remalloy[3]	1931	32S6	12 Co, 17 Mo (or W)	250	10 500} 10 000}	1.2} 1.1}
Mishima alloy[4]	1931	32M1	25 Ni, 12 Al	475	7 000	1.4
Alnico 2[5]	1934	35H2	12 Co, 17 Ni, 10 Al, 6 Cu	560	7 300	1.7
Magnetoflex[6]	1935	37N2	20 Ni, 60 Cu	600	5 800	2.0
Platinum cobalt alloy	1936	36J2	77 Pt, 23 Co	3000*	5 000	4.5
Vicalloy	1938	40N1	52 Co, 10 V	200	11 500	1.5
Alnico 5[7]	1940	40V1 41J1	24 Co, 14 Ni, 8 Al, 3 Cu	575	12 500	4.5

Values of constants taken largely from Refs. [39S5] and [42S1], and from bulletins of the Indiana Steel Products Company and experience of the Bell Telephone Laboratories.

[1] Called Koerzit in Germany.
[2] Also alloys containing 3, 6, 9 cobalt with variable chromium.
[3] Also called Comol in U.S.A. and Oerstit in Germany.
[4] Also called Alnico 3. Alnico 1 contains cobalt.
[5] Called Alnico in England and Oerstit in Germany.
[6] Also called Cunife 1. Cunife 2 and Cunico contain cobalt.
[7] English Alcomax has a somewhat different composition.
* H for $B = 0$.
† Balance iron.

steel to the extent of 0.8%. Occasionally cast iron was useful on account of its cheapness and machinability.

The origin of tungsten steel is uncertain, but it is reported [24H2] that it was made in Austria in 1855 and used as a permanent magnet in 1885. Chromium steel had been made on a practical scale in America by about 1870 [92H1], and the magnetic properties of a hardened bar had been reported by Hopkinson in 1885 [85H1], but it came into general use for the first time during the World War I [16G1] when tungsten was difficult to procure. Later cobalt was added to improve the coercive force, and various alloys became available with cobalt contents up to 35%.

A great step forward was made in 1917 when Honda and Takagi [20H2] discovered that a high-cobalt steel containing tungsten and chromium had a coercive force of 230, about three times that of any material existing at the time. This material was destined to lead the field for almost 15 years, and even now is the highest-quality material that may be classified as a true steel, containing carbon as an essential constituent.

In 1931 Seljesater and Rogers [32S6] (apparently at the suggestion of Dean [33D3]), and Köster [32K8], independently investigated the Fe-Co-Mo and Fe-Co-W systems and found useful alloys. Remalloy, one of the alloys of this group, has the advantage that it can be hot rolled and machined before the final hardening treatment. The Fe-Co-Mo alloys were the first commercial permanent magnet materials that contained no essential carbon, and since that time all of the improvements have been affected with alloys hardened by the precipitation of some other element than carbon.

A Japanese is credited also with a further advance in the field of permanent magnet materials. Mishima [32M1, 36M6, 36M7] in 1931, when experimenting with hardenable alloys containing iron, nickel, and aluminum, found alloys having coercive forces as high as 700. Now similar alloys, heat-treated in an improved manner, are made commercially and have a coercive force of 475. They have higher energy products than Remalloy but are mechanically hard and must be ground to final form.

Alnico, the name used in the United States for magnet alloys based on the Fe-Ni-Al system, was improved considerably by the addition of cobalt by Mishima [33M3, 36M6] and Ruder [34R1], and later by the further addition of copper by Horsburgh and Tetley [35H2]. These additions together with the use of an appropriate heat treatment, made it possible to obtain good properties in many sizes and shapes of magnets.

In the years 1935–1938 a number of alloys were discovered which could be formed readily by rolling or drawing and could even be cold drawn to fine wire before the final heat treatment. In some cases good properties were developed during the rolling or drawing process so that aging was not required. Among these alloys should be mentioned especially the Fe-Ni-Cu

[37N2] and Fe-Co-V [40N1] alloys known, respectively, as Magnetoflex (or Cunife) and Vicalloy.

The material now in use (1951) that has the highest energy product $(BH)_m$ and is made in large quantities is known as Alnico 5 in the United States and Alcomax in England. This was reported first by van Urk [40V1] in 1940, the year the British patent [40P2] was issued, and was the result of research carried on at Eindhoven under the direction of von Embden [41J1]. An essential part of the heat treatment of this material is the cooling from a high temperature in a strong magnetic field prior to the final hardening treatment. The Dutch work was inspired by the experiments on heat treating the Mishima alloy in a magnetic field, performed by Oliver and Shedden [38O2] in England, and this in turn was suggested by work on the heat treatment of high permeability alloys carried on at the Bell Telephone Laboratories by Kelsall, Dillinger, and Bozorth [34K1, 34B3]. Heat treatment of Alnico 5 in a magnetic field increases both H_c and B_r, and more than doubles the value of $(BH)_m$, as compared with the values of the same material cooled in the absence of a field.

The discovery that an agglomerate of fine particles (10^{-4} cm or less in diameter) of iron or other material has high coercive force was pointed out by Gottschalk in 1935 [35G6], by Dean in 1941 [41D1], and by Guillaud in 1943 [43G2]. Compacts of iron, with or without admixture of cobalt, are now available commercially in France. A fuller discussion of the physics of such material is given in Chap. 18.

Energy product.—A given volume of magnetic material will produce the highest field in a given air space when the induction B in the material is that for which the energy product BH is a maximum [20E1]. Consider the ring magnet of Fig. 9–3 in which L_s, L_g, A_s, and A_g are the lengths and cross-sectional areas of the specimen and air gap. Since there is no field due to external electric currents,

$$\oint H dL = 0$$

when the integration is made around the closed path, including the gap, determined by the lines of induction. If H and B now are the field strength and induction in the specimen and H_g the field in the gap, and if they are assumed constant, we have

$$HL_s + H_g L_g = 0.$$

Also, since the lines of induction are continuous,

$$H_g A_g = B A_s.$$

These relations combine to give

$$H_g^2 = -BH(V_s/V_g)$$

where V_g and V_s are the volumes of the gap and specimen. Thus H_g is a maximum for given volumes of material and air gap when BH is a maximum. This product is commonly written (BH), and is called the *energy product* because it is proportional to the magnetic energy of unit volume of the material, $BH/8\pi$, and also to the energy stored in unit volume of the gap, $H_g^2/8\pi$. The (BH) vs B curve of Fig. 9-2 shows that the *maximum*

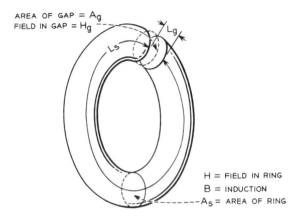

FIG. 9-3. Ring with air gap, defining length of gap (L_g), area of gap (A_g), field in gap (H_g), cross-sectional area of specimen (A_s), and B and H in material.

energy product, $(BH)_m$, of 4.5×10^6 occurs in this specimen when $B = 9000$ and $H = 500$. Such values of B and H, corresponding to $(BH)_m$, may be designated B_0 and H_0.

Leakage flux, traversing the air space outside of the gap, is usually accounted for by adding the factor q to the foregoing equation.

$$qH_g^2 = -(BH)(V_s/V_g).$$

In simple cases q may be calculated [41R1, 43E1], but usually it must be determined experimentally. The problem of efficiency in design is then resolved, as far as the permanent magnet material is concerned, to selecting the material and shaping it so that (BH) in the material is as large as possible.

Demagnetization Curve.—Certain geometrical constructions and empirical mathematical relations are often useful in treating the magnetic data. The demagnetization curve can be simulated by a rectangular hyperbola passing through the points B_r $(H = 0)$ and $-H_c$ $(B = 0)$. Using the general expression for a rectangular hyperbola,

$$(x - x_0)(y - y_0) = c_0^2$$

and making the substitutions $y = B$, $x = H$, $y_0 = 1/b$, $x_0 = -H_c - a/b$, $c_0{}^2 = -a/b^2$, we may write the relation in the form [23W1]

$$\frac{H + H_c}{B} = a + b(H + H_c), \qquad (1)$$

which is similar to the Fröhlich-Kennelly relation

$$\frac{H}{B} = \frac{1}{\mu} = a + bH$$

and differs from it in that the B,H curve is displaced horizontally so that $B = 0$ when $H = -H_c$. Designating the asymptotes $B_s = 1/b$ and $H_s = (a + bH_c)/b = H_c B_s/B_r$, the relation may also be written

$$B = \frac{(H + H_c)B_s}{H + H_s}$$

or

$$\frac{B}{B_r} = \frac{1 + H/H_c}{1 + (H/H_c)(B_r/B_s)}. \qquad (2)$$

The hyperbolic relation given by Eq. (1) may be compared with experiment by plotting $(H + H_c)/B$ vs H, and this is done in Fig. 9-4 using data for Alnico 5. The slope of the straight line is $b = 1/B_s = 0.0743 \times 10^{-3}$, and the intercept at $H + H_c = 0$ is $a = 2.5 \times 10^{-3}$. For this material we then have $B_r = 12\,600$, $H_c = 560$, $B_s = 13\,450$, and $H_s = 598$. The demagnetization curve calculated with these constants is shown as the dotted line of Fig. 9-5 where the lack of perfect correspondence between the parabolic curve and the data is evident.

The energy product (BH) for the hyperbolic curve is a maximum for the point $H = -H_0$, $B = B_0$, where

$$H_0 = H_s - \sqrt{H_s(H_s - H_c)},$$

$$B_0 = B_s - \sqrt{B_s(B_s - B_r)}.$$

The last equation shows that $B_s - B_0$ is the geometrical mean of B_s and $B_s - B_r$.

Values of B_0 and H_0 are such that

$$\frac{H_0}{B_0} = \frac{H_c}{B_r} = \frac{H_s}{B_s},$$

relations that correspond to the simple geometrical construction shown in Fig. 9-5 whereby one diagonal of the rectangle with corners at $B = B_r$ and $H = -H_c$ cuts the demagnetization curve at the point for which $(BH) = (BH)_m$ and, when prolonged, meets the intersection of the asymptotes of the hyperbola. Another geometrical construction appropriate to the parabolic relation is also shown in the figure: let any straight line drawn through the point $B = B_s, H = -H_s$ intersect the line $H = -H_c$

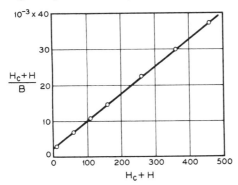

Fig. 9-4. Linear relation between $H + H_c$ and $(H + H_c)/B$ on demagnetization curve.

at a point P and the line $B = 0$ at Q; then a rectangle drawn through these points and the point $H = -H_c, B = 0$ has as its other corner a point P' on the parabolic demagnetization curve.

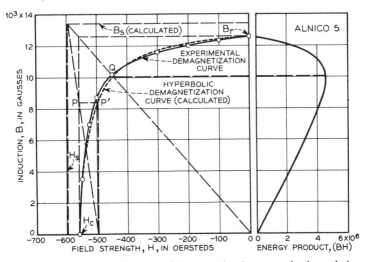

Fig. 9-5. Approximation of demagnetization curve by hyperbola.

The shape of the demagnetization curve between B_r and H_c is fixed by what is often called the *fullness factor*, defined by

$$\gamma = \frac{(BH)_m}{B_r H_c}$$

and for a hyperbolic demagnetization curve given by

$$\gamma = \left[\frac{H_0}{H_c}\right]^2 = \left[\frac{1 - \sqrt{1 - H_c/H_s}}{H_c/H_s}\right]^2 = \left[\frac{1 - \sqrt{1 - B_r/B_s}}{B_r/B_s}\right]^2.$$

Theoretically γ must lie between the limits 0.25 (for a linear demagnetization curve) and 1.0 (for a rectangular loop), and practically it varies from about 0.3 for the cobalt-platinum alloy and 0.4 for KS magnet steel to 0.65 or more for Alnico 5. Values as high as 0.71 have been observed for Vicalloy.

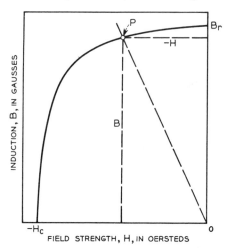

FIG. 9–6. Demagnetization factor of rod is H/B when these quantities are measured with no external applied field.

Equation (2) shows that B/B_r is dependent upon H/H_c alone, provided B_r/B_s is constant. In many of the older materials B_r/B_s does not differ much from 0.5; consequently all of these materials are pretty well described by a single such H/H_c vs B/B_r curve, as was shown by Sanford [27S5]. The theoretical fullness factor for $B_r/B_s = 0.5$ is $\gamma = 0.34$ whereas the actual value for the curve plotted by Sanford is 0.42.

An interesting relation has been observed by Scott [32S7] between the remanent induction B, at the center of a bar magnet, and its dimensional ratio, coercive force, and residual induction. He found that for bars of many materials, having various ratios, m, of length/diameter, the residual induction is determined by the following function:

$$B/B_r = f(m\sqrt{H_c/B_r}).$$

That this is a consequence of the hyperbolic form of the demagnetization curve may be shown in the following way. Refer to Fig. 9–6 and let P be the point on the demagnetization curve that represents the values of B and H at the center of a bar when no external field is applied. The demagnetizing factor of the rod is then

$$\frac{N}{4\pi} = \frac{H}{B},$$

and this for short rods is related to the dimensional ratio approximately by [42B1]

$$N/4\pi = c/m^2$$

in which the constant c has a value lying between 1 and 2. Combining these relations with Eq. (2) gives

$$\frac{B}{B_r} = \frac{1 - \dfrac{cB/B_r}{m^2 H_c/B_r}}{1 - \dfrac{c(B_r/B_s)(B/B_r)}{m^2 H_c/B_r}}$$

or

$$m^2 H_c/B_r = c(B/B_r)(1 - BB_r/B_s^2)/(1 - B/B_r),$$

in which B/B_r depends only on $m^2 H_c/B_r$, provided B_r/B_s is constant. The relation between B/B_r and $m\sqrt{H_c/B_r}$, when $c = 1$ and $B_r/B_s = 0.5$, is plotted as the solid line in Fig. 9–7 where it is compared with Scott's observations. The broken curves in the figure show the relatively small displacements that occur when $c = 1$ and $B_r/B_s = 0.3$, and when $c = 2$ and $B_r/B_s = 0.5$.

Scott notes that for his curve the ratio of ordinate to abscissa is a maximum when the induction corresponds to the maximum energy product of the material. Thus the tangent to the curve, at the knee, drawn through the origin (see Fig. 9–7), locates B_0. This is really an experimental verification of Evershed's rule that a given flux will be carried by the smallest volume of material when (BH) is a maximum.

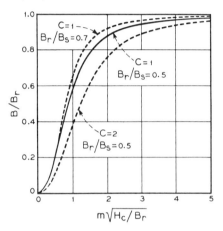

Fig. 9–7. General form of relation existing between B/B_r and H/H_c in a bar of dimensional ratio m.

It can be shown in a straightforward manner that for the hyperbolic form of the demagnetization curve, at the point of tangency

$$\frac{B}{B_r} = \frac{B_0}{B_r} = \frac{1 - \sqrt{1 - B_r/B_s}}{B_r/B_s}.$$

Elenbaas [32E3] was the first to derive Scott's universal curve of B/B_r vs $m\sqrt{H_c/B_r}$. He assumed the demagnetization curve to be elliptical in form.

Reversible Permeability and "Spring-Back."—In addition to the demagnetization curve, other B,H curves are useful in describing the behavior of permanent magnets. The most important of these is the unsymmetrical hysteresis loop, part of which is shown as the line 5–6 in Fig. 9–1, and sometimes known as the *spring-back*. If, for example, an air gap in the magnetic circuit or a negative field causes the remanence to drop to a given value, and then a field is applied in the positive direction, the spring-back curve so obtained depicts the magnitude of the induction as H increases. Several such curves are shown in Figs. 9–2, 40, 58, and 61. In addition the reversible permeability, equal to the initial slope of the spring-back curve, is given for several points on the demagnetization curve. The reversible permeability may change by a factor of about 2 as B decreases from B_r to 0, and it goes through a maximum when B is approximately $B_0/2$. As is to be expected, the materials that have high energy products have relatively low reversible permeabilities.

The spring-back from the induction B_0, corresponding to the maximum energy product, may carry the induction one-half to three-quarters of the way to residual induction as H increases from $-H_0$ to 0. The negative value of field from which the induction springs back to $B = 0$, $H = 0$ is sometimes called the *open-circuit* or *spring-back coercive force*.

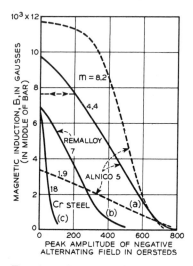

FIG. 9–8. Decrease of remanent induction in bar magnets of various materials and dimensional ratios when subjected to alternating fields of given amplitudes.

Stability.—As already mentioned, the field in the air gap of a permanent magnet should be as constant as possible. It fails, however, to be perfectly constant mainly because of the disturbing effects of temperature and of extraneous magnetic fields. To show the effect of *extraneous magnetic fields*, a straight bar of Alnico 5 was cut to a dimensional ratio of 4.4 and magnetized, and after the applied field was removed B and H in the middle of the bar were found to be 9700 and -460 respectively. A negative alternating field was then applied, increased to a given peak amplitude as indicated by the abscissa of Fig. 9–8, decreased to zero, and B again measured. The decreasing values of B resulting from such fields of increasing amplitudes are shown in curve (a), and similar curves for Remalloy and 3.5 Chrome Steel, each specimen having a dimensional ratio near that for the maximum energy product, are shown as (b) and (c). The broken curves are for Alnico 5 bars having dimensional ratios of 8.2 and 1.9, therefore having inductions well above and below that for $(BH)_m$. The shape of the curve for the longest bar shows some superiority in stability, as may be inferred from the shape of the demagnetization curve.

The horizontal dotted line of Fig. 9–8 illustrates the fact that after an alternating field of a certain amplitude (e.g., $H_m = 200$) has been applied and removed the flux is stabilized at a lower value (e.g., $B = 7600$) against all temporary disturbing fields not exceeding this amplitude. Thus stability against disturbing fields can be attained at the expense of useful flux. The purpose of a keeper is to protect the magnet from extraneous fields when it is not in use.

The induction in a permanent magnet is often permanently changed when it comes momentarily *in contact with iron* or other magnetic material that changes the direction and amount of induction in the magnet. A protective sheath covering the magnet can be used to prevent such a change and thus to maintain the stable field desired in the gap. In one experiment a bar of Alnico 5, $\frac{1}{2}$ in. square and 4.5 in. long, was covered with a sheath

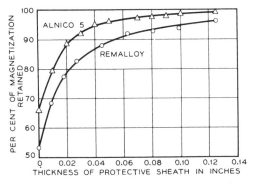

FIG. 9–9. Decrease of remanent induction in bar magnet by contact with iron, when magnet is protected by inert material of thickness shown.

of non-magnetic material of given thickness and then a $\frac{1}{2}$-in. square bar of soft iron was repeatedly brought in contact with the outside of the sheath on all surfaces. The strength of the bar magnet was reduced, the thinner the sheath the more the reduction, up to 34% with the sheath removed entirely. The reduction is plotted against the sheath thickness in Fig. 9–9, which also shows similar data for a somewhat longer bar of Remalloy. For most purposes a sheath $\frac{1}{8}$ in. thick gives sufficient protection and with Alnico 5 this allows only 1% loss under the conditions described. Different kinds of contact with magnetic materials, of course, will require different amounts of protection.

The disturbing effects of *temperature* are more complicated. If a bar is magnetized at room temperature and then heated with no applied field present, the induction will usually decrease as the temperature rises and increase again as it falls, but generally not all of the induction will return. The extent of the recovery depends primarily on the metallurgical stability of the alloy and this varies considerably from one material to another. In a soft magnetic material such as pure iron, heating to 100°C may cause a reduction of about 1% in the flux, and upon cooling the flux will increase again to within 0.01% of its original value [00E1]. Nickel has been heated to within 10–20° of its Curie point so that the remanent induction was reduced to 30% and, upon cooling over 95% of the original induction, was recovered [38A2].

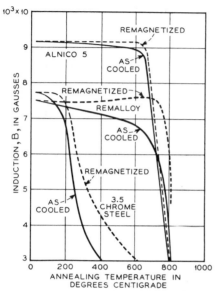

FIG. 9–10. Decrease of remanent induction in bar by annealing at various temperatures.

However, the cheaper magnet materials such as 3.5 Chrome Steel suffer a loss in remanent induction of over 10% when heated to 200°C for a half hour and cooled in air. Curve A of Fig. 9–10 shows the loss for this temperature and for various other temperatures of treatment. The relative losses of 3.5 Chrome Steel, Remalloy, and Alnico 5 are shown graphically by the solid curves of this figure, which indicate the induction remaining at room temperature after magnetizing straight bars having dimensional ratios corresponding to $(BH)_m$ and heating them at the indicated temperature for 30 minutes. The broken lines show the induction after remagnetizing at room temperature. It is apparent that Alnico 5 is quite stable against temperature variations of several hundred degrees.

The loss of induction caused by a cyclic change in temperature can be reduced by "knocking down" the original induction by momentarily applying a reverse field. For example, a bar of 3.5 Chrome Steel was magnetized and the remanent flux reduced 15% by an applied field; then the bar was heated to 145°C and cooled. The flux was then 17% below its original value and was therefore diminished only 2% by the temperature cycle, as compared to the 4% lowering that would have occurred if it had not been stabilized with a reverse field.

The disturbing effect of temperature cycling can be reduced further by repeated cycling over the range of temperature under consideration. Figure 9–11 shows the effects of repeated cycling between 15 and 100°C, in a steel containing 5.5% tungsten [28G2]. The difference between the original and final inductions at 15° may be called *irreversible* and the final change between 15 and 100° *reversible*. The reversible change may be defined by a temperature coefficient.

The *temperature coefficient* of the reversible change of magnetization of permanent magnets usually lies in the range -1 to -5×10^{-4} per °C, often at about 2×10^{-4}. As with highly permeable materials, the coefficient varies not only with the composition and heat treatment but also with the induction as well. For most materials the coefficient is smaller in absolute magnitude when the induction is high, i.e., when the dimensional ratio of the bar is large or the demagnetizing factor small [19G1, 21H2]. In some permanent magnet materials, but not in those used commercially, this coefficient goes through zero and is positive for the upper portion of the demagnetization curve [97A1, 28G2]. Coefficients for a number of alloys are collected in Table 2.

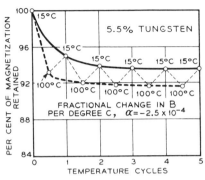

FIG. 9-11. Decrease of remanent induction by repeated cycling between temperatures of 15 and 100°C.

The irreversible change in magnetization that occurs during the heating

TABLE 2. TEMPERATURE COEFFICIENT α OF SOME MATERIALS FOR PERMANENT MAGNETS. $B = B_{t=0}(1 + \alpha t)$

Material	Dimensional Ratio	$\alpha \times 10^4$/°C	Reference
Carbon steel	127	-0.1	
Carbon steel	64	-0.3	[00F1]
Carbon steel	32	-2.5	
Chrome steel	37.0	-2.4	
Chrome steel	12.3	-3.2	[19G1]
Chrome steel	4.0	-4.2	
KS steel	40	-2.0	[21H2]
Remalloy	14	-1.9	*
Alnico 5	11	-2.4	
Alnico 5	4.5	-2.0	*
Alnico 5	1.5	-1.6	
Tool steel		-7.6	
Spring steel, 3 mm		$+7.9$	[28G2]
Spring steel, annealed		-9	
Spring steel, annealed		0	

* Data obtained at the Bell Laboratories.

and cooling of a magnet is generally attributed to metallurgical changes or relief of strains in the material, and takes place even at room temperature, when it is called *aging*. This is a serious problem in materials that have the relatively low coercive force of chrome steel, and such magnets are usually maintained at 100°C or higher for 24 hours before use, in order that the subsequent aging at room temperature may be reduced enough to permit practical use in apparatus.

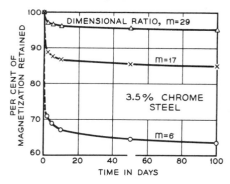

FIG. 9–12. Decrease of remanent induction of chrome steel bars of various dimensional ratios by aging at room temperature.

The aging of recently hardened 3.5 Chrome Steel magnet bars of three dimensional ratios is shown in Fig. 9–12 [37C4]. The change in flux is an exponential function of the time. The demagnetization curves of this material which were measured, respectively, 10 minutes, 1 day, and 316 days after hardening are given in Fig. 9–13. Incidentally, these two figures together

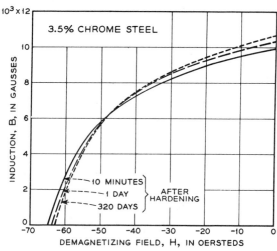

FIG. 9–13. Change of demagnetization with aging at room temperature. Chrome steel.

show that when a bar of large dimensional ratio is magnetized the remanent induction is decreased by aging, but that when the bar is subsequently remagnetized the residual induction is higher than it was before the aging.

In the newer magnetic materials used for permanent magnets, aging at room temperature is not important. At the present time it can be stated as a rule that those materials in which the presence of carbon is essential do age at room temperature and that the others, such as Remalloy and the Alnicos, do not.

An additional cause of loss of remanent magnetism is *mechanical vibration*, usually caused by repeated impacts. This effect again is more important in materials of low coercive force and is practically absent in the alnicos. Severe impacts that produce a 15% loss in tungsten steel will cause only a fraction of a per cent of loss in Alnico 2 [38A1] and a negligible loss in Alnico 5.

Design of Permanent Magnets.—An important problem in efficient design is to select the appropriate material. The basis for such selection has already been discussed in some detail. The next step is to use a minimum of magnet material consistent with the strength of the field desired in a given volume of space. The simple theory of the relations between dimensions and field strength and the properties of the material will first be considered with *leakage neglected*.

In discussing Fig. 9-3 it was pointed out that

$$H_g L_g + H L_s = 0$$

and

$$H_g A_g = B A_s.$$

When dimensions are fixed, these equations give

$$-H = H_g L_g / L_s$$

and

$$B = H_g A_g / A_s$$

or

$$-H = \frac{L_g A_s}{L_s A_g} \cdot B.$$

The point on the demagnetization curve for which $-B/H = (L_s A_g)/(L_g A_s)$ then describes the magnetic state of the material used; in particular the value of B is thereby determined, and the field in the gap is

$$H_g = B A_s / A_g.$$

The values of B and H in the material may easily be determined graphically by drawing a straight line through the origin with the slope $-L_s A_g/(L_g A_s)$ and noting the intersection of this line with the demagnetization curve (see Fig. 9-6). This line is often called the "load line," and its slope is the reciprocal of the demagnetizing factor.

In designing a structure to produce a given field strength in a gap having a given area and length, it is often necessary to concentrate or dilute the flux with a piece of soft iron placed between the permanent magnet and the gap.

Still neglecting leakage, and referring to Fig. 9–3, the equations above may again be used to determine the length and area of the permanent magnet. Then we have

$$L_s = H_g L_g / H$$

$$A_s = H_g A_g / B$$

and the volume of the magnet is

$$V_s = H_g^2 V_g / (BH).$$

In the ideal case B and H should have the coordinates of that point on the demagnetization curve for which the energy product is a maximum.

When the magnetic circuit contains a considerable proportion of high permeability material such as that used as a yoke in Figs. 9–14(a) and (b), it is important that the reluctance of this material, $L/(\mu A)$, be small compared with the reluctance of the gap, L_g/A_g. If the area of the yoke is too small, the induction in it will be high and the permeability correspondingly low, so that in practice A is made large enough so that the reluctance is negligible. Ordinarily annealed iron is suitable as a material for yokes, but often Permendur is used to save space and weight on account of its high permeability at high inductions, and it is invaluable for concentrating the flux near the air gap when H_g approaches the saturation induction of unalloyed iron.

Leakage may be considered in a qualitative way by realizing that the air flux between two points will be proportional to the magnetic potential difference, $\int H dl$, between the points. The leakage will then usually be greatest near the poles of the magnet and around the air gap, but serious leakage in the structure of Fig. 9–14(a) will also occur some distance away from the poles, inside of the structure. This should always be kept in mind in design work and the distance between leakage surfaces should be made as large as possible and their areas small. The design of Fig. 9–14(b) is a poor one because there are large areas at high potential differences. The ideal shape of the structure, considering only the efficiency of the magnetic circuit, often approaches a circle or ellipse in general form, as shown in (c).

The induction inside of the magnet should be B_0 at every point, and consequently its section must be adapted, in some such way as shown by the dotted lines, to take care of the leakage. In an actual magnet of this form, made of Alnico 5 and producing a field of 5500 oersteds in a gap having $L_g = 1.62$ cm and $A_g = 2.85$ cm², the flux in the heaviest section was 240 000 maxwells; this may be compared with a flux of $H_g A_g = 16\,000$

DESIGN OF PERMANENT MAGNETS 361

maxwells in the gap and 79 000 maxwells at the small end of the permanent magnet. Thus in this design about half the leakage occurs near the poles, outside the gap, and half between the heavy neutral section and the small ends of the permanent magnet material.

A design of a "ring pole" magnet is given in Fig. 9–14(d).

Although some simple geometrical shapes are amenable to calculation, estimation of leakage is usually based on the designer's experience with circuits of a similar kind. Cioffi, in unpublished work, has determined

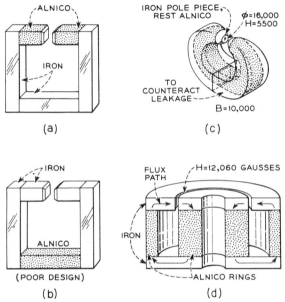

Fig. 9–14. Some simple designs of permanent magnets: (a) good design; (b) poor design; (c) practical magnet with varying section, designed so that $B = 10\,000$ almost everywhere in magnet (unshaded portion is necessary to carry "leakage" flux); (d) practical design of ring-shaped magnet.

leakage in provisional designs by simulating the permanent magnet components with soft iron pieces easily and quickly constructed, and applying a magnetomotive force by means of a coil surrounding the iron, then measuring the useful and total flux per unit of magnetomotive force. It is important that proper measuring instruments be available for measuring flux accurately and in small areas.

The equations used in design may be modified to take account of *leakage*. Dropping the minus sign of H we have

$$HL_s = fH_gL_g$$
$$BA_s = FH_gA_g$$

in which f and F are leakage factors used to facilitate calculation. The factor f takes account of the fact that the flux in the gap is not everywhere perpendicular to the pole face, and therefore the length of the path is somewhat greater than the geometrical value L_g. In practice f is somewhat larger than 1; it is seldom as large as 1.5.

The value of F varies over a much larger range and is often as large as 10; in structures with large air gaps and high fields in the gap it may be 100 or more. This is the ratio of the total flux (measured in the neutral or heavy section of the magnet) to the useful flux (measured in the air gap).

In the magnet of Fig. 9–14(c), $f = 1.006$ and $F = 15.0$, if we consider the useful flux to be that in the cylindrical space bounded by the entire pole face areas. F is naturally much smaller when L_g/L_s is small.

The equations containing f and F may be used to determine L_s and A_s in the manner previously described when leakage was neglected. The "load line" now has a (negative) slope given by

$$\frac{B}{H} = \frac{FA_g L_s}{fA_s L_g}$$

and the volume of magnet required is

$$V_s = FfH_g^2 V_g/(BH).$$

The load line may be drawn with the slope B/H given above, and the point of operation is now determined graphically as the intersection of this with the demagnetization curve.

Model theory [43E1] can often be used to advantage in designing magnet structures, especially if unusually large or small size makes experimental investigation of the actual size difficult. In the use of Alnico 5 it should be borne in mind that the flux path in the magnet is in the same direction as the field present during the special heat treatment it must have to develop its best properties. In directions at right angles the energy product is lower by a factor of 2 or 3.

In designing a *stabilized magnet* one must take account of the fact that the final point of operation will not lie on the demagnetization curve but inside it on a minor loop. Suppose, for example, that the magnet is stabilized by (a) applying a reverse field of 50 oersteds or by (b) reducing the remanent flux by 10%. A minimum of magnet material will then be used when B corresponds to the maximum value of (BH) not for the demagnetization curve but for the curve corresponding to the condition of operation. For (a) the (BH) vs B curve is determined from the locus of points (B,H) that represent the magnetic state of the material after -50 oersteds have been applied and removed as indicated by the path PQR of Fig. 9–15. For (b) the corresponding path is PST. The maximum (BH) products for these curves (a) and (b) are lower than for the normal

(BH) curve, and they occur at somewhat different values of B, depending on the nature of the material and the kind of stabilization required.

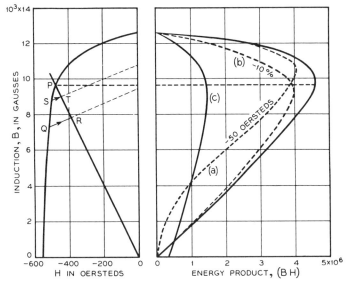

Fig. 9-15. Energy product curves of material stabilized by (a) applying a reverse field of 50 oersteds, by (b) reducing the remanent flux by 10%. Curve (c) shows the useful recoil energy (see text).

In the operation of some kinds of magnets the *load line changes* during use. This occurs, for example, in lifting magnets for which the load line may vary between the limits designated in Fig. 9-16 by the lines OA and OC [40H2, 44E1]. The best design of magnet is obviously not that for which (BH) is a maximum for either the point A or the point C. As the load of the magnet changes, the values of B and H in the magnet itself follow along the minor loop $AC'D$ between A and C'. At the point A all the flux may be considered as leakage since none of it is used in lifting. At C', when the magnet carries its maximum load, the leakage is represented by the ordinate of A'. This is true because the leakage paths are essentially the same, and the leakage flux along these paths is proportional to the magnetomotive force which is reduced in the ratio H_2/H_1 when the load is increased. Edwards and Hoselitz [44E1] thus consider $C'A'$ to be the useful flux, and the product $H_2(C'A')$ to be the useful recoil energy.

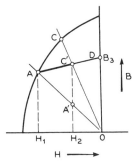

Fig. 9-16. Diagram for determining useful recoil energy.

A little consideration of the geometry will show that for a given point A the useful recoil energy is a maximum when C' is at the middle of the minor loop segment AD, assumed for convenience to be a straight line, and that the recoil energy then is equal to $H_1 B_3$.

The recoil energy curve can then be determined from the demagnetization curve and the minor loop intersections at $H = 0$. Such a curve for Alnico 5 is shown as (c) of Fig. 9-15, and its maximum occurs when the flux density in the "open" position is 9000-10 000, and in the "closed" position it is about 10 500, a value very close to the flux density for maximum energy product for this material. Data for Alnico 2 and a more detailed description of the design of this kind of magnet have been reported by Edwards and Hoselitz [44E1]. Further consideration of variable load has been given in an excellent paper by Cioffi [48C2] and recoil or supplementary energy product curves plotted for Remalloy and Alnicos 2, 5, and 6.

For further study of the design of magnets of various kinds, the reader is referred to papers [35E3, 40H2, 40V1, 44U1] on this subject alone, and to two books [41R1, 43E1] that consider engineering applications in some detail.

PROPERTIES OF ALLOY SYSTEMS

Permanent magnets are almost invariably composed of alloys hardened by heat treatment. They may be divided into three general classes: (1) those made of carbon steel, usually modified and improved by the addition of one or more other elements, and hardened as tool steels are by quenching from some temperature above 700°C; (2) alloys dispersion-hardened by the precipitation of one phase in another or by the formation of a superstructure; this is usually accomplished by heating to a temperature above 1000°C to insure a uniform solid solution and then "baking" at 600–800°C, whereupon one phase (or superstructure), containing no essential carbon, precipitates in a highly dispersed form; and (3) materials formed of very small particles, of the order of magnitude of 10^{-4} cm or less in one dimension.

Since a study of the binary iron-carbon alloys is essential to the understanding of all alloys of class (1), they are discussed here first in some detail even though they are not used in any important way now without substantial additions.

Iron-Carbon Alloys. Constitution.—These alloys have been used as permanent magnets for centuries and have been definitely replaced by improved materials only as recently as about 1910. Consideration of their structure and properties are important now as a foundation for our knowledge of the more complex steels still in general use.

A phase diagram of the system is given in Fig. 9-17. Its general features are the result of the two-phase transformations in pure iron—α,γ and γ,δ—

and the formation of the compound Fe_3C, cementite, containing 6.7% carbon. The eutectic of iron and cementite contains about 4.3% carbon and melts at 1130°C. The α,γ point of pure iron is depressed by the presence of carbon, and a eutectoid is formed at 0.83% carbon and 723°C, the A_1 point. On long-continued heating of high carbon steels, Fe_3C itself breaks down to form iron and graphite. A list of the transformation points and common constituents present in steel is given in Table 3.

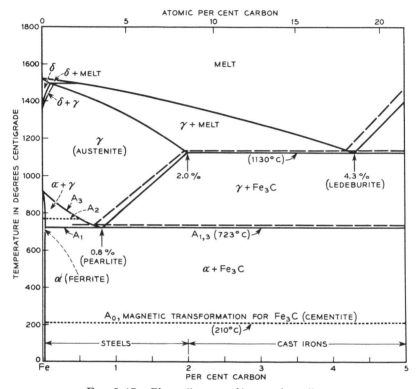

FIG. 9-17. Phase diagram of iron-carbon alloys.

The solubility of carbon in α iron, ferrite, is very small at room temperature (estimated to be 0.006% carbon) and increases to about 0.04% at the A_1 point. Above this temperature, however, carbon is readily dissolved by the γ phase, austenite, and the solubility continues to increase with increasing temperature. On cooling, the amount of carbon held in solution becomes less, and it is the sudden decrease in solubility as the temperature is lowered through the A_1 point that is the foundation for the many and complicated changes that occur during the hardening of steel.

Steel is hardened both mechanically and magnetically by heating to above the A_1 point and cooling rapidly to some lower temperature, or cooling at a controlled rate to room temperature. When cooled rapidly (quenched), the carbon and iron of the original austenite do not have time for appreciable diffusion, and *martensite* is formed. It has a plate-like structure (showing as needles in cross section) and is an unstable phase built on a tetragonal body-centered lattice. Its presence imparts to steel

TABLE 3. CONSTITUENTS AND TRANSFORMATIONS IN CARBON STEELS

CONSTITUENTS

Stable Phases	Unstable Phases	Structural Forms
α, ferrite......	Fe_3C, cementite	Ledeburite, eutectic of austentite and cementite
γ, austenite....	Fe-C, martensite	Pearlite (troostite), alternate layers of ferrite and cementite
δ, delta.......		Bainite, platelike ferrite and granular cementite
Fe_3C, cementite		Martensite, unstable tetragonal phase
C, graphite....		
l, liquid.......		

TRANSFORMATIONS

Symbol	Temperature (°C)	Nature
A_0..........	210	Curie point of Fe_3C
A_1..........	723	γ first forms on heating (A_{c1}) of α saturated with C. Eutectoid forms on cooling (A_{r1})
A_2..........	769	Curie point of α
A_3..........	912 (Pure Fe) to 723 (0.83%C)	α first forms from γ on cooling (A_{r3}) or disappears on heating (A_{c3})
A_4..........	1400 (Pure Fe) to 1490 (0.1% C)	γ first forms from δ on cooling
A_r'..........	600 (Indefinite)	Troostite forms from austenite; cooling rate intermediate
A_r''..........	200 (Indefinite)	Martensite forms from austenite with rapid cooling

its greatest hardness. The temperature at which it forms, A_r'', is indefinite and is lower when the carbon is higher, but it is in the neighborhood of 200°C.

If the cooling is less rapid, or if the material is held near 500°C, *bainite* is formed. It appears to consist of ferrite precipitated along crystallographic planes, with spheroids of carbide included. With slower cooling or by maintaining at about 650°C, *pearlite* is formed of alternate layers of ferrite and cementite. Very fine pearlite which is difficult to resolve under microscope is often called *troostite*; the temperature at which this forms is sometimes designated A_r'. When steel is cooled very slowly, the cementite is spheroidized.

Physical Properties.—The density of steel depends on its composition and heat treatment and the following relations have been suggested by Cross and Hill [28C4] for hot rolled and for annealed alloys.

$\delta = 7.855 - 0.032C$ (hot rolled)

$\delta = 7.860 - 0.04C$ (annealed)

in g/cm^3, C being the carbon content in per cent. Quenching decreases the density up to about 1% carbon. The reported density of cementite is 7.66 [27I1].

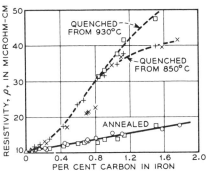

FIG. 9-18. Resistivity of iron-carbon alloys subjected to various heat treatments. ○ and ×, Gumlich [18G2]; △, + and □, Campbell [26C1].

The resistivity is especially sensitive to heat treatment [18G2, 26C1]. As indicated in Fig. 9-18 it is much higher for quenched than for annealed alloys, and at 1.5% carbon the ratio may be as high as 4 or 5 to 1. When the resistivity increases on account of carbon content or heat treatment, its temperature coefficient decreases so that the product remains approximately constant at 0.060–0.065 microhm-cm [18G2].

Magnetic Properties.—The *saturation induction* of iron is decreased by the addition of carbon at a rate faster than it would be by simple dilution; the carbon not only dilutes the iron but reduces the average magnetic moment of the iron atom. The reduction is greater for quenched alloys than for annealed alloys (Fig. 9-19) according to the data reported by Gumlich [18G2] in 1918. His alloys are representative of commercial purity, contain up to 0.2% silicon, 0.6% manganese, and smaller amounts of sulfur and phosphorus. The saturation induction of cementite is about $B_s = 12\,400$ [28S5]; $\sigma_s = 128$ [15°C], $\sigma_0 = 169.3$, $n_0 = 2.015$ per atom of iron [44G1].

As is to be expected the *permeability* of pure iron is depressed considerably by addition of carbon. Data of Gumlich [18G2] and Yensen [24Y1] plotted in Fig. 9-20 show that in the high carbon range the initial permeability lies between 100 and 150 for annealed alloys and is less than half as high when the alloys are quenched. The maximum permeability is depressed much more rapidly by the addition of carbon and is roughly 500 for annealed 0.8% carbon alloy and may be as low as 100 when this alloy is quenched. From these values of μ_0, μ_m, and B_s approximate μ,B curves can be constructed for a given alloy and heat treatment, provided it is noted that the maximum permeability occurs at about $B = 6000$ to 8000.

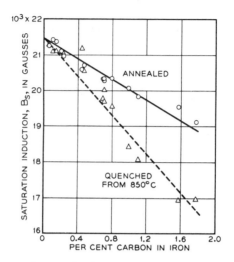

FIG. 9–19. Change of saturation induction of iron-carbon alloys with carbon content and heat treatment. At 1% C (4.7 at. % C) the Bohr magneton numbers corresponding to the straight lines as drawn are $n_0 = 2.0$ and 1.8, respectively.

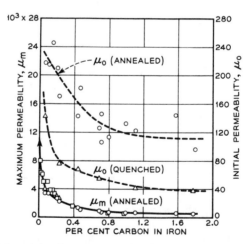

FIG. 9–20. Initial and maximum permeabilities of iron-carbon alloys of various carbon contents and heat treatments.

IRON-CARBON ALLOYS 369

The way in which the *coercive force* and *residual induction* change with carbon content and heat treatment is illustrated [18G2] in Fig. 9–21. Although the results here depicted are somewhat erratic, due partly to the impurity of the specimens which were prepared and measured over 25 years ago, they show that about the best properties for permanent magnets

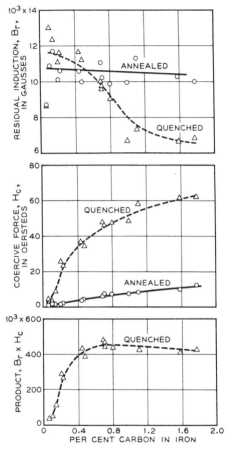

FIG. 9–21. Magnetic properties of carbon steels as dependent on carbon content and heat treatment.

are obtained by quenching an alloy containing 0.8% carbon, and this is the composition that was used for many years in production. The data of Matsushita [22M1] and others indicate that the best quenching temperature lies between 780 and 850°C. Quenching may be carried out with water or oil.

Figure 9–22 shows the effect of *annealing* after quenching at 850°C [22C2, 29K3]. It is apparent that temperatures below 100°C have a definite effect; in fact, aging occurs at room temperature, and at 200°C the effect is marked, as carbide is precipitated from supersaturated solution. Some "overaging," resulting from coagulation of the finely divided precipitate, takes place at 300°C. In general, the coercive force declines as the annealing temperature is increased, and the residual induction rises slowly.

Messkin [29M3] has investigated the effect of *cold rolling* on the coercive force, residual induction, and maximum permeability of a hot rolled sheet containing 0.78% carbon. H_c increases rapidly at first and then more slowly and reaches a value of about 40, considerably less than that attained by quenching. The maximum permeability declines by a factor of about 2.

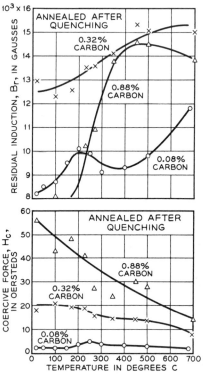

FIG. 9–22. Effect of temperature of anneal of carbon steels previously quenched.

Some experiments on the variation of residual induction with temperature revealed an unusual feature. Bars were magnetized at room temperature and then heated (with no applied field present), and the induction in the material was shown to go through zero at about 200°C, provided the carbon content was more than about 0.4%. This was explained by Smith and Guild [12S1], who performed the experiments, as due to the lamellar structure of alternate plates of ferrite and cementite. Later work [24S2] has confirmed the original explanation. At room temperature the greater hardness of the cementite causes some induction to flow in the iron in a direction opposite to that of the field originally applied. This continues to flow when the cementite loses its magnetism as the temperature approaches its Curie point at 210°C and thus causes a reversal of the original remanence. At higher temperatures the induction behaves normally and decreases gradually toward zero at the Curie point of ferrite, 723–770°, depending on the carbon content.

Additions of Manganese.—Manganese is commonly present in commercial carbon steels, where it is added to facilitate hot rolling of the ingots, and many of the alloys considered in the previous paragraphs have contained 0.2–0.8% of this element. Commercial carbon steel for permanent magnets contains 0.8% manganese [39S5] and is often referred to as "carbon manganese steel."

The effect of this amount of manganese on the phase diagram [39W1] of the Fe-C system is not great. No new phases are present up to 13% manganese and 1.5% carbon. Practically, the eutectoid point S of Fig. 9–17 is displaced toward lower carbon content and lower temperature, and for 1% of manganese this displacement is of the order of 5°C and 0.1% carbon. Theoretically the addition of a third element to an alloy system means that three phases may exist over a range of compositions and temperatures instead of at a single point on the phase diagram, and thus ferrite, austenite, and cementite may exist together in equilibrium over a temperature range of about 10–15°C when the carbon content is 0.8% and the manganese is present to the extent of 1%.

It may be concluded that manganese has little effect on the magnetic properties of carbon steel subjected to a given cycle of heat treatment.

Tungsten Steels.—In attempting to improve the properties of iron-carbon alloys many additions were tried, and in the nineteenth century most success was achieved by the use of tungsten. Tungsten magnets are reported [24H2] to have been made by Remy and Böhler in 1883–1885, and in 1885 Hopkinson [85H1] reported measurements on a French tungsten steel of the kind then "in general use." This has now been supplanted by chrome steel, which has almost identical properties, because the latter is considerably cheaper, and by more expensive alloys with greatly superior qualities.

Tungsten raises the α,γ point and lowers the γ,δ point of iron, and the γ phase ceases to exist when tungsten is present to the extent of about 6%. When carbon is present the eutectoid point (point of double saturation of γ iron with α iron and carbon) occurs at a lower temperature and contains less carbon as the amount of tungsten increases [24O1, 31T2]. Tungsten also lowers the maximum solubility of carbon in austenite. The composition of the magnet steel as now manufactured (0.7 C, 5.5 W) contains more carbon than corresponds to the eutectoid point and so may be classified as hypereutectoid.

Many suggestions have been made regarding the composition of compounds of tungsten with carbon and iron, and in the composition range in which we are interested the compound WC seems to be definitely established. Double carbides of tungsten and iron are often discussed and a reaction

$$m\text{Fe}_3\text{C} + n\text{WC} = m\text{Fe}_3\text{C} \cdot n\text{WC}$$

has been proposed [32M3] to represent a reaction taking place when the carbon content is greater than about 0.4% (5–6% tungsten) and affecting the "spoiling" of the magnetic properties, to be discussed later. Westgren and Phagmén [28W1], however, assigned the formula Fe_4W_2C to the carbide present in high-speed tool steel (0.7 C, 18 W), and Wood [30W2] detected the same carbide in magnet steel. For a critical survey of the structure of this system the reader is referred to the book by Gregg [34G1].

Marie Curie [97C1, 98C1] reported a study of the effect of *composition* on the magnetic properties of tungsten steels of various heat treatments and fixed the optimum at about 5% tungsten and 0.6% carbon, not far from the composition now used. The effect of carbon content on the magnetic properties is given in Fig. 9–23, according to the data of Evershed [25E1] and the very extensive data of Schönert and Hannack [25S3]. The latter measured altogether 1795 specimens containing 5.4 W, 0.3 Mn, and 0.2 Si. Evershed's specimens contained about 6% tungsten.

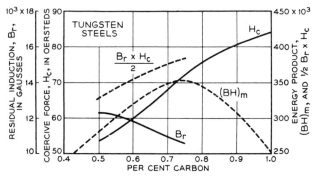

FIG. 9–23. Effect of carbon content on the magnetic properties of tungsten steel containing 5.4–6.0% tungsten, after quenching at 830–850°C.

The impurities likely to be present in commercial magnets, such as manganese, silicon, sulfur, and phosphorus, are usually not harmful to the magnetic quality. Variations in manganese from 0.3 to 0.6% are not objectionable [25S3] and 0.2–0.5% silicon may have a beneficial effect [28S6]. On the contrary even 0.1% of oxygen causes a lowering of the energy product [27E3].

Tungsten steel magnets are *fabricated* by hot rolling or forging and sometimes by casting. When hot rolled it is necessary to reheat to 1200°C between passes to prevent deterioration in the quality of the final product [28S6]. The best magnetic properties are obtained by heating the hot rolled metal for a short time at about 850°C and quenching in water. Sometimes it is necessary to soften the material so that certain mechanical operations can be performed. This softening is accomplished by annealing for an hour at 800–850°C and then cooling in air; such treatment has an undesirable effect on the magnetic properties obtained after the subsequent quenching [26A1]. It is supposed that any such heating, causing the lamellar pearlite to spheroidize, makes it more difficult for the carbon to dissolve in austenite at the quenching temperature and thus makes the carbon less effective in hardening.

In fabrication and heat treatment, care must be taken to preserve a solid mechanical structure. Decarburization is likely to occur, especially if heat treatment at 1250°C is necessary to counteract a previous softening anneal. Such decarburization is enhanced by the presence of even small surface cracks, and the material containing cracks is made still weaker by the decarburization. The drastic water quench often used to develop the best magnetic properties is also the cause of severe strains that may result in breaking or warping if incipient cracks and improper shapes are not eliminated.

The *temperature of quenching* may lie between 750 and 950°C [25E1, 29S2], the preferred temperature being about 850°C. If the quenching temperature is too high or the material is held too long at temperature before quenching, the results are not good [28P1]. When the steel is held as long as 1 hour at the quenching temperature, the "spoiling" occurs when this temperature is somewhat above 800°C, and may be remedied by heating to a considerably higher temperature, 1200 to 1300°C, as was first pointed out by Evershed [25E1]. It is recommended that the material be held at the quenching temperature not longer than 15 minutes in order to prevent spoiling and decarburization.

Deterioration of magnetic quality is often attributed [32M3] to the thermal decomposition of the double carbide or iron and tungsten according to the equation given previously. The mechanism of spoiling cannot, however, be regarded as established. Wood [30W2] and Wood and Wainright [32W4] have used X-rays to investigate tungsten steels and residues from them and have obtained the diffraction pattern of a double carbide (which was assumed to be Fe_4W_2C) as well as that of the simpler carbide WC. No Fe_3C was detected. They found that the spoiling could be correlated with the sharpness of the lines reflected from ferrite and concluded that the good quality of the magnet is associated with high internal strain caused by the presence of W and C atoms. Hoselitz and McCaig [47H4] also relate spoiling to carbide formation.

Tungsten steels age after quenching so that the coercive force decreases and the residual induction increases by several per cent. The change with time at various temperatures is given in Fig. 9–24, according to Gould [29G1], for a steel containing 6.2% tungsten and 0.74% carbon. One recommended practice [34D3] for stabilization has been to accelerate the aging by maintaining for 12 hours at 100°C and, after magnetization, to cycle several times between 0 and 100°C and then to reduce the induction 5% with a reversed field.

Representative demagnetization and energy product curves for tungsten

374   MATERIALS FOR PERMANENT MAGNETS

steel are shown in Fig. 9–25. The broken lines give also the demagnetization curves for this steel in the hot rolled and annealed conditions [25E1].

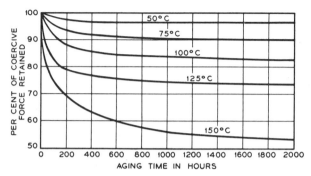

Fig. 9–24. Aging of tungsten magnet steel at various temperatures.

The temperature coefficient of magnetization is about -2 to $-4 \times 10^{-4}/°C$, the density 8.0 g/cm³, the resistivity of good steel 35–40 microhm-cm, and of spoiled steel somewhat lower.

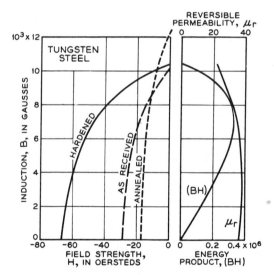

Fig. 9–25. Properties of tungsten magnet steel in various conditions.

Chrome Steels.—These have come into general use only since World War I, when chromium was used as an alloying element because tungsten was then difficult to procure. Long before that time Hopkinson [85H1] had experimented with hardened chrome steel magnets, and Marie Curie [98C1] had studied the effect of chemical composition on coercive force and residual induction. Mars [09M1] again considered their commercial possibilities in 1909, but properties substantially the same as those now well known to the trade were first reported by Gumlich in 1916 [16G1], and the results of his extensive investigation were published in 1922 [22G1].

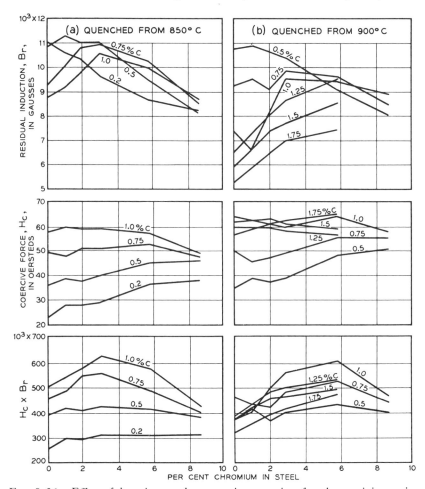

FIG. 9-26. Effect of chromium on the magnetic properties of steels containing various amounts of carbon.

When chromium is added to pure iron the α,γ point is lowered until 7–8% is present (see Fig. 7-20); then it is raised by further addition until, at 11–12%, the γ phase ceases to exist and the γ loop is closed. When carbon is present, however, the γ phase is stable at higher chromium contents. The A_1 point of the iron-carbon system is raised by the presence of chromium, and the carbon content of the eutectoid is lowered; therefore, this point is displaced upward to the left in the diagram as usually drawn

[20M1, 37K2]. The change in some of the critical points with chromium content, on heating and cooling, has been determined by Maurer and Nienhaus [28M4]. Here it may be noted that when high-chrome magnet steel (3.5% Cr, 1% C) is heated it behaves like carbon steel and loses its magnetism (at A_{C2}) before it transforms from α to γ at the A_3 point (A_{C3}), but on cooling it becomes magnetic again as soon as it transforms to the α phase ($A_{R2} = A_{R1}$).

Several carbides have been observed under the microscope and with X-rays [28W2], and the formulas Fe_3C, $(Cr, Fe)_4C$, and $(Cr, Fe)_7C_3$, have been determined or suggested. The solution of the carbides in austenite is slowed up by the presence of chromium, and as in tungsten steel they appear to decompose during prolonged heating; consequently chrome magnet steel must be carefully heated prior to quenching to attain the best magnetic hardness. If "spoiled" by prolonged heating, the magnetic properties may be improved by a brief heating at 1200°C.

The effect of *composition* on B_r, H_c, and B_rH_c is shown in Fig. 9–26, due to Gumlich [22G1]. About the best quality of material is obtained with the composition 3.5% Cr, 1.0% C. There is no substantial gain by going to 6% chromium. The effect of higher carbon contents is to increase H_c, but B_r is lowered so that the product remains practically constant [24H2].

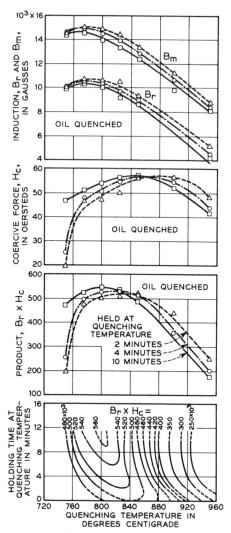

Fig. 9–27. Effect of quenching temperature on the magnetic properties of chrome steels.

Gumlich obtained the best results in 3% chrome steel by quenching from 850°C. The effect of quenching temperature and time have been studied in detail for low chrome steel (1.6% Cr, 1.1% C) by Oberhoffer and Emicke [25O1] and is shown in Fig. 9–27. This material is even more sensitive to these variables than tungsten steel [26S1]. Heat treatment has again been studied by Jellinghaus [49J1].

The good results obtained with the short time of heating before quenching is attributed to the tendency of the carbide to decompose. If desired, the coercive force can be increased at the expense of the residual induction by quenching from a higher temperature. Water quenching produces good magnetic quality, but oil quenching is preferred and causes less warping and cracking. Even in oil, care must be taken to prevent undue warping during the quench.

In order to be machined or cold formed, chrome steel must be softened by heating for a limited time to a temperature just below the A_1 point.

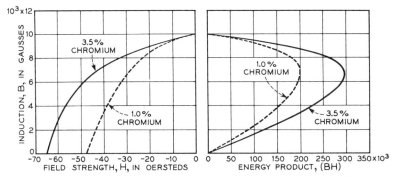

FIG. 9–28. Demagnetization and energy product curves of low and high chrome steel.

This is usually done after the hot rolling, before the quench hardening. Such treatment results in some deterioration in magnetic quality and hence is avoided if mechanical operations are not required. Fabrication is similar to that of tungsten steel.

In *aging* before use in apparatus, magnets of chrome steel are taken through some such cycle of treatments as that proposed by Gumlich [22G1]:

1. Harden.
2. Magnetize.
3. Temper 10 hours at 100°C.
4. Alternate 10 times between 0°C and 100°C.

Deterioration due to vibration may be 2 or 3%.

Representative demagnetization curves for high and low chrome steels are shown in Fig. 9–28. The temperature coefficient varies from -2 to -4×10^{-4} and is said to be increased by the presence of silicon.

The densities [25A1], resistivities [20E2], and mechanical properties

[10M1] of quenched and annealed steels vary considerably, depending on the exact composition and method of treatment. Estimated values are given in the following table for steels containing approximately 1.0% carbon.

TABLE 4. APPROXIMATE DENSITIES AND RESISTIVITIES OF CHROME STEELS

	2% CHROMIUM		3.5–4% CHROMIUM	
	Quenched	Annealed	Quenched	Annealed
Density (g/cm^3)	7.75	7.81	7.7	7.7
Resistivity (microhms-cm)	50–60	20–25	50–60	20–25

Recently the effects of small amounts of various elements on the magnetic properties of 3.5 and 5.5 chrome steels have been studied by Krainer and Raidl [43K1]. $(BH)_m$ is raised by the following: Al, Si, Sn + V, W, Mo, V, Cu, Mn. Ni has no substantial effect. Greiner prepared speech-recording tape and wire from the stainless steel commonly used for cutlery (13% Cr, 0.3% C). A coercive force of 60 was easily obtained.

At the present time chrome steel has the lowest cost of any important commercial magnet steel and is the most economical if high quality is not required.

Chrome-Tungsten Steels.—There is some advantage to be gained [32M3] by having both chromium and tungsten in the same steel. The magnetic properties are somewhat improved and there is less danger of spoiling. Suggested compositions (per cent) are: 6 W, 0.8 Cr, 0.6 C [24P1]; 1.0 W, 1.2 Cr, 1.2 C [09M1]; molybdenum is also used occasionally as an addition to chrome or tungsten steel [39B7]. Trade names of these alloys include *Cobaflux* and *Permanit*.

Molybdenum Steel.—As an alloying element, molybdenum affects the magnetic properties of steel in much the same way that tungsten does, but it acts more strongly so that less of it is required to raise the energy product a given amount.

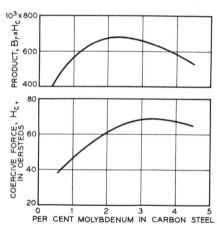

FIG. 9–29. Effect of molybdenum on the magnetic properties of steels containing 1.0–1.3% carbon.

Molybdenum influences A_{C1} little, but depresses A_{R1} strongly [29S3, 13S1] so that a difference between the points of about 200°C exist for a steel containing 2–4% molybdenum. Double carbides are believed to exist, but their compositions are not known.

COBALT STEELS 379

Alloys containing 3.5–4% were investigated by Curie [97C1] who stated that it could be used in place of tungsten but offered no special advantages. Other investigators [09M1] supported this point of view until Stogoff and Messkin [29S3] showed that good properties could be obtained with a molybdenum content as low as 2–2.5%. They found the remanence to be much higher than that reported by previous investigators.

Variation of H_c and $B_r H_c$ with molybdenum content, the carbon being held at about 1–1.3%, is illustrated in Fig. 9–29. The optimum molybdenum is 2–2.5%, and the carbon 0.9–1.0%. The best results are obtained by quenching from 800–825°C in water. Heating before quenching should be limited to 5 or 10 minutes.

The properties reported [29S3] are as follows:

$$H_c = 75 \qquad B_r H_c = 0.8 \times 10^6$$

$$B_r = 10\,500 \qquad (BH)_m = 0.3 \times 10^6$$

The low alloy content of 2.5% is said to facilitate working the alloy. Protection from oxidation is attained by deoxidizing the melt with titanium or vanadium.

Mechanical properties and resistivities of various alloys have been reported by Swinden [13S1]. The temperature coefficient is about -2×10^{-4} per °C. The alloy has not found commercial use.

Cobalt Steels.—The famous "Honda" or "KS" steel was invented by Takagi and Honda [20H2] in 1917 and named in honor of the Japanese industrialist Baron K. Sumitomo. The first alloys made had the approximate composition 0.6 C, 2 Cr, 7 W, 35 Co, and the remainder iron, and so used the elements already known to be good for permanent magnets and introduced, in addition, cobalt. In more recent years a whole series of steels containing various amounts of cobalt up to 40%, as well as the alloying elements W, Cr, Mo, and Mn separately and in combination in various proportions, have been investigated and some of them are made commercially. In England particularly there is a series containing 3, 6, 9, 15, and 35% cobalt, the principal other alloying element being chromium.

The α,γ point of iron (A_3) is raised almost to 1000°C by 35–50% cobalt, and the non-magnetic temperature (A_2) is raised even more so that it coincides with A_3 when the cobalt content is 20% or more. When carbon is present [32V1], the effect of cobalt is to move the eutectoid point also toward higher temperatures and at the same time toward somewhat lower carbon content until with 50% cobalt it is near 850°C and 0.7% carbon. The solubility of carbon in austenite is not greatly altered by cobalt, but when the carbon is thrown out of solution by quenching it tends to form martensite rather than pearlite. Cobalt does not appear to increase temperature hysteresis as chromium and molybdenum do. When carbon and other elements are present, the Curie point (A_2) is considerably lower than that for the pure iron-cobalt alloy, and for KS steel it is in the neighborhood of 750°C.

The efficacy of cobalt in permanent magnets is probably due to the large magnetostriction it exhibits when alloyed with iron. This is effective in increasing the coercive force. Cobalt is also helpful in increasing, rather than decreasing, the saturation magnetization of iron and consequently in maintaining a high remanence; 30% cobalt raises the saturation of iron about 2500 gausses whereas 5% chromium lowers it about 1500. The saturation induction of KS steel is 16 000 to 16 500 [38O1].

Watson [24W2] has made a convenient classification of cobalt steels into three groups that depend on the content of the alloying element chromium

TABLE 5. MAGNETIC CONSTANTS OF COMMERCIAL COBALT STEELS

% Co	B_r	H_c	$(BH)_m \times 10^{-6}$	B_0	H_0
3	7000– 7 500	125–135	0.36–0.38	4000	90
6	7200– 8 000	140–150	0.45–0.50	4500	100
9	7500– 8 100	150–165	0.50–0.55	4700	110
15	8000– 8 500	175–190	0.60–0.65	5200	120
35	9000–10 500	220–260	0.90–1.00	6000	160

and are designated low, medium, and high alloy steels. In each group the coercive force and maximum energy product are closely proportional to the cobalt content. The high alloy steel is not made with cobalt contents over about 15%. Representative values of the constants for some commercial steels have been compiled from various sources and are given in Table 5.

TABLE 6. COMPOSITION AND PROPERTIES OF SOME COBALT ALLOY STEELS

Composition (%)					Properties		Reference
Co	C	Cr	W	Mn	B_r	H_c	
35	0.6	2	7	...	10 600	220	20H2
38	0.9	2	4	0.5	10 000	250	32B9
38	0.9	5.0	4.0	...	9 500	235	35C5
36	0.9	4.5	5.5	0.6	9 700	235	39S5
33	1.2	...	...	4.4	9 300	165	23G1
36	1.1	4.8	...	3.5	9 300	225	23G1
9	0.9	5	1	0.4	7 500	120	39S5
16	1.0	9	...	...	8 000	180	44S1

Of these the 15 and 35 cobalt steels are most used. Although no analyses are available for these specific alloys, the composition of this type of steel has been given [26S2] as 1.0% carbon, 9% chromium, 2.5% molybdenum, in addition to cobalt.

The magnetic properties of certain alloys of known chemical composition are shown in Table 6. These include the cobalt-manganese steels described by Gumlich.

Representative curves of British commercial steels of varying cobalt content are given in Fig. 9–30.

The *heat treatment* of KS steel is relatively simple. After casting or hot rolling or forging, it is heated to 950°C for 5–10 minutes and quenched in

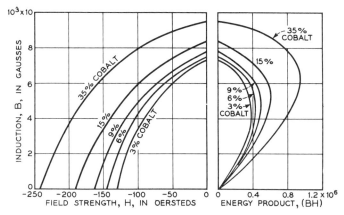

FIG. 9–30. Demagnetization and energy product curves of various cobalt steels.

oil [20H2, 29B2]. If mechanical operations must be carried out cold, the formed steel may be annealed for a short time at 750°C, subjected to the mechanical operations and then hardened as described previously.

Some manufacturers of the low-cobalt steels recommend hardening by the "three-treatment process" [23K1]:

1. Heat in 5–10 minutes to 1150–1200°C and cool in air.
2. Heat to recalescence at about 750°C and cool in air.
3. Heat to 950–1000°C, cool in air until the steel is attracted by a magnet, and then quench in oil.

This treatment is sometimes applied to KS steel, and its significance has been discussed by Schulz, Jenge, and Bauerfeld [26S2] in relation to the formation of austenite, troostite, and martensite. In (1) the carbon is dissolved in austenite and during the cooling it is precipitated in a finely divided form suitable for rapid solution when reheated in (2). In the final stage (3) the carbon redissolves quickly at 950° and when released by quenching causes the formation of martensite which probably contains some complex carbides.

The manufacture of KS steel for compass needles has been described by Gier [30G4]. The ingot was hot rolled at 1000–1100°C and then swaged to 0.10 in. diameter. To counteract the decarburization occurring during the rolling the material was previously carburized for 12 hours at 850°C in a carburizing compound. For hardening, the needles were held for 3 minutes in molten salt at 950°C and then quenched in light oil.

Aging is more important in KS steel than in tungsten steel. The effect of aging temperature and time for several commercial steels have been given by Gould [29G1]. Residual induction is increased by aging. The 15% cobalt steel used for these experiments contained 1.1% carbon, 10.0% chromium, and 1.7% molybdenum.

The recommended field strength for magnetizing the completed magnet is 1500.

Other Steels.—Addition of B, Al, Si, Ti, V, Mn, Cu, Sn and probably many other elements have also been made to carbon steels to improve the properties of permanent magnets [98C1, 09M1, 32M3]. Stogoff and Messkin [28S7], in an investigation of copper alloys, obtained $H_c = 75$ and $B_r = 9000$ in an alloy containing 1% carbon and 5% copper, when quenched from 770°C. Molybdenum and manganese are often used in chrome and tungsten steel, especially when cobalt is also present.

Iron-Cobalt-Molybdenum and Iron-Cobalt-Tungsten Alloys.—An important advancement in the metallurgy of permanent magnet materials followed the prediction by Dean and others [32S6, 33D3, 32K8] that dispersion or precipitation hardening (see Chap. 2, p. 33) would cause magnetic hardening in carbon-free iron alloys in much the same way that mechanical hardening had already been accomplished in Duralumin. Seljesater and Rogers [32S6] were the first to experiment with a number of binary and ternary alloys (including Fe-Co-Mo and Fe-Co-W), and in their investigation they found coercive forces of over 200. Independently Köster exploited the same idea, and he and Tonn have reported the most complete magnetic data on Fe-Co-Mo [32K4, 32K8] and Fe-Co-W [32K5, 32K8] alloys. All of this work was foreshadowed by that of Kroll [29K2] who hardened an iron-nickel alloy by the addition of beryllium and thus increased the coercive force to 20, a figure too low, however, to be of practical importance.

The phase diagrams of Fe-Co-Mo and Fe-Co-W have been discussed in an earlier chapter (Fig. 6-18). According to Köster some of the alloys are hardened during cooling from 1300° by the γ,α phase transformation, others by the precipitation of one phase in another (e.g., Fe_3Mo_2 in α), and still others by both processes. After rapid cooling from 1300°, alloys developed the maximum magnetic hardness after aging at about 700°C. Further work on the Fe-Co-W system has been reported by Rogers [33R3].

Köster and Tonn investigated the magnetic properties of Fe-Co-Mo alloys containing up to 12% Co and 19% Mo, and Fe-Co-W alloys containing 15% Co and 18% W, and 30% Co and 15% W. More magnetic data are now available for the Fe-Co-Mo system because one of these alloys is of commercial importance. Unpublished data of Greiner and Tolman, as well as the earlier data, are shown in Fig. 9–31. The maximum product $B_r H_c$ was observed for the alloy containing 20% Mo and 12% Co, but on account of the similarity of the demagnetization curve to that of KS steel the first alloy in commercial use had the approximate composition of 12% Co and 17% Mo. This is known as *Remalloy* or *Comol*; other trade names are also used.

Typical demagnetization, energy product, and reversible permeability curves of Remalloy are given in Fig. 9–32. The alloy is cast, hot rolled,

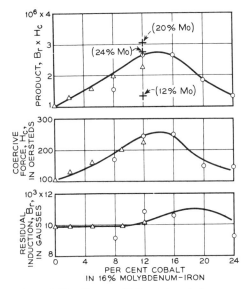

FIG. 9–31. Variation of H_c, B_r and $H_c B_r$ with cobalt content in Remalloy-type magnets.

quenched in air or oil at 1200–1300°C, formed and machined if necessary, then aged for an hour or more at 650–700°C. Variations in the magnetic properties resulting from variations in quenching and aging temperatures are evident from Table 7, taken from the unpublished data of Greiner and Tolman. The alloy is not sensitive to composition or heat treatment except that the carbon should be kept low—an impurity of 0.1% of this element will cause a decrease in B_r of 5% or more and a decrease in H_c of 25%. The saturation induction depends on heat treatment and is over 18 000 after quenching and about 15 000 after aging at 700°C. Remalloy has the following physical properties: density 8.3 g/cm³, resistivity 45 microhm-cm, Rockwell hardness after quenching C27, after aging C55.

In unpublished work E. A. Nesbitt has added chromium, reduced the cobalt content of Remalloy, and obtained a somewhat higher energy product (see Table 16 in this chapter). Reduction of the cobalt content lowers the raw materials cost substantially.

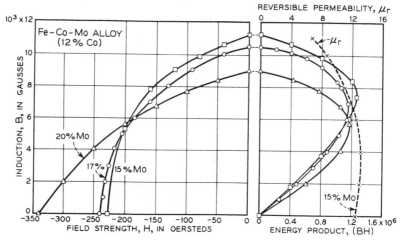

FIG. 9-32. Demagnetization and energy product curves of 3 Fe-Co-Mo alloys.

TABLE 7. DATA FOR FE-CO-MO ALLOYS CONTAINING 72% FE, 12% CO, AND 16% MO

Quenching Temperature	Aging Temperature*	$(B - H)_m$†	B_r	H_c	$B_r H_c \times 10^{-6}$
900	...	15 700	8 000	7	0.06
900	725	14 700	6 800	45	.31
1100	...	18 500	3 000	7	.02
1100	725	15 300	9 500	240	2.3
1300	...	18 600	3 000	7	.02
1300	725	14 800	10 000	260	2.6
1100	675	......	8 500	130	1.1
1200	675	......	11 200	250	2.8
1300	675	......	10 900	255	2.8
1300	...	18 600	3 000	7	.02
1300	610	17 300	11 000	40	.4
1300	725	14 800	10 000	260	2.6
1300	825	15 300	11 000	100	1.1

* Time 1 hour at indicated temperature. Omission of temperature indicates no aging after quenching.
† $H_m = 1700$.

As compared to KS steel, Remalloy has slightly better magnetic properties, is decidedly less costly to manufacture, is less sensitive to aging and variations in temperature (see Fig. 9-8), and has somewhat better mechanical properties—it machines more readily and is not quite so susceptible to cracking and warping.

In a study of the mechanism of hardening of these alloys Rogers [33R3] has compiled data on 20 Fe-Co-W alloys aged at various temperatures. He measured B_r, H_c, density, conductivity, Rockwell hardness, and Young's modulus. Snoek [38S5] determined coercive force and induction at a high field strength as affected by the temperature of aging, of an alloy containing 15 Co and 17 Mo. The papers by Köster contain data on hardness, density, thermal expansion, and resistivity for various alloys.

Seljesater and Rogers [32S6] found that addition of manganese to the Fe-Co-W alloys (26% Co, 20% W, 6% Mn) improved the forgeability markedly.

When mechanical properties are not important and conditions permit the magnet to be ground easily to size, Remalloy usually gives way to one or another of the Fe-Ni-Al alloys with or without additional elements.

Iron-Nickel-Aluminum Alloys.—Interest in the alloys associated with Mishima's name and called "MK" or Alnico 3 has increased rapidly since Mishima's publication [32M1, 36M7] of his results in 1932. The preferred composition (25% Ni, 12% Al, 63% Fe) is not far from that represented by the formula Fe_2NiAl (30% Ni, 14% Al, 56% Fe). Ruder [34R2] carried out an independent investigation of these alloys, showing that they are age-hardenable, and first applied the heat treatments appropriate to such alloys.

In a study of the structure of this system Bradley and Taylor [38B1, 49B7] and Kiuti [41K2] found that they are more complex than supposed from the earlier work of Köster [33K3], and that changes in atomic ordering play an important role in the phase diagram. Figure 9-33 shows a diagram of alloys cooled at 10°C/hr to room temperature, and one for equilibrium at 1300–1350°C. The ordered phases are α' (body-centered FeAl or NiAl structure with Al atoms at the corners), α'' (body-centered Fe_3Al structure), and γ' (face-centered Ni_3Al or Ni_3Fe structure). The α' phase is important in the interpretation of the high coercive force, for at room temperature the good permanent magnets lie in the $\alpha + \alpha'$ area.

Bradley and Taylor interpret the changes in structure during hardening as follows: At 1300°C the alloy is in the α' phase. If it cools slowly, this separates into the two body-centered phases, one having the α structure characteristic of iron, and the other an α' structure having a different composition from the original phase at 1300°C, as indicated by the tie lines in the figure. When the alloy is cooled more rapidly, equilibrium cannot be established, and the distribution of atoms and the distances between them represent a compromise between the conditions corresponding to the original α' phase and the two new phases, having different lattice spacings, into which it would eventually decompose. This makes for an unusually heterogeneous structure and large internal strain.

It was formerly supposed that the internal strain, combined with a high coercive force, was enough to account for the high coercive force on the strain theory (see Chap. 18), but the work of Nesbitt [50N1] (see Chap. 13) has shown that the magnetostriction of some of the alloys of this type is

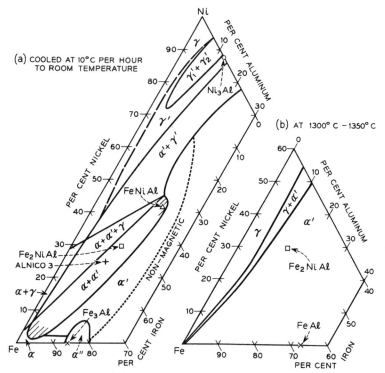

Fig. 9-33. Phase diagrams of Fe-Ni-Al alloys at room temperature (Bradley and Taylor) and at 1300°–1350°C (Kiuti and Bradley).

zero at the same time that the coercive force is 400 oersteds or more. It now appears likely that the heterogeneous structure, produced during the decomposition, causes the formation of particles that are small enough to be single domains and have the high coercive force associated with them, as discussed in Chap. 18.

Following Mishima's report, the *magnetic properties* of the ternary alloys have been investigated repeatedly. Köster [33K2] measured saturation, residual induction, coercive force, and hardness of two alloys (21% Ni, 12.5% Al and 25% Ni, 5% Al) quenched from a high temperature and aged at various lower temperatures; B_r and H_c were maximum for aging at 650–700°C, while Brinell hardness was highest at 500°C—a typical trend for age-hardening alloys. Saturation decreased slowly with increasing aging temperature, indicating precipitation of a component with lower magnetization. Snoek [39S6] showed that, after aging, the saturation of the series of alloys lying between Fe and NiAl was nearly proportional to the iron content. Some data on resistivity and its change with heat treatment have also been reported [38B5, 40H5].

For establishing the best composition and heat treatment, the most comprehensive published work has been by Betteridge [39B8]. For maximum energy product he determined the composition as approximately 27% nickel and 13% aluminum. Heat treatment consisted in quenching from 1100°C into water at 75°C, then aging for 3–4 hours at 650°C. To

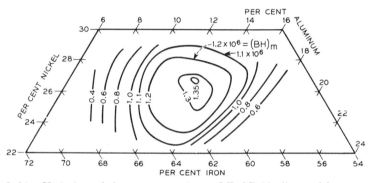

FIG. 9–34. Variation of the energy product of Fe-Ni-Al alloys with composition. Optimum heat treatment was used.

attain the highest $(BH)_m$ for different compositions, each step in the heat treatment must be adjusted; values of $(BH)_m$ for optimum heat treatment for alloys of different composition are shown in Fig. 9–34. At the preferred composition $(BH)_m = 1.35 \times 10^6$, $B_r = 6900$, and $H_c = 520$. An increase in nickel causes a decrease in B_r. The coercive force is highest for 13% aluminum and decreases as the nickel decreases.

The presence of carbon is especially harmful, less than 0.1% causing a severe deterioration (see Fig. 9–35), but it is reported that this amount of carbon can be compensated by an adequate amount of titanium. Silicon and manganese may also be objectionable if more than a few tenths per cent are present.

Commercial alloys, such as *Alnic* or *Alnico 3* [42S1], contain nominally 25% nickel and 12% aluminum and have the following values of the magnetic constants: $B_m = 12\,000$ for $H_m = 2000$, $B_R = 6900$, $H_c = 475$, $(BH)_m = 1.38 \times 10^6$, $\mu_r = 4$ [42S1]. They are brittle and must be ground to size but, under certain conditions, may be drilled with difficulty after being subjected to a heat treatment that causes some diminution in magnetic quality [40R1]. They are practically non-aging and can be heated to 500° without deterioration except that remagnetization may be required. Severe vibration causes about 1% change in flux. The density is 7.1 and the Rockwell hardness C45–C50.

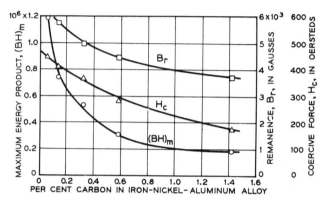

Fig. 9–35. Effect of the carbon content of Fe-Ni-Al alloys on its magnetic properties.

Additions of Cobalt and Copper.—Mishima [36M6] and Ruder [34R1] independently added cobalt to the Fe-Ni-Al alloys and found an increase in H_c but not a very large change in B_r or $(BH)_m$. The alloy with high coercive force (5% Co, 28% Ni, 12% Al) has found commercial use under the name *Alnico 4*—for this material $H_c = 730$, $B_r = 5300$, $(BH)_m = 1.3 \times 10^6$. The cobalt is believed to have little effect on the phase structure of the alloys in the region of interest for permanent magnets.

Mishima found that magnets of good quality could also be made of alloys containing Cu, Cr, Mn, Mo, V, or W. Alloys having copper (without cobalt) also were investigated by Betteridge [39B8] who found an energy product of 1.5×10^6 at the composition 24% Ni, 13% Al, and 3% Cu. His findings have been supported by a later systematic investigation by Lifshitz [41L1]. This alloy is known as *Alni*.

A distinct simplification in heat treatment, together with some improvement in energy product, was effected by the simultaneous addition of both cobalt and copper. Horsburgh and Tetley [35H2] selected the composition 17% Ni, 13% Co, 10% Al, and 6% Cu, and showed that the appropriate cooling rate is such that magnets of large sections, as well as small, may be so treated as to have an energy product of 1.6×10^6. Cobalt increases the quality up to at least 17%, by increasing H_c, B_r remaining practically constant; a content of 10–13% was selected on economic grounds. This alloy is known as *Alnico 2* and by other names as well and has the following magnetic properties: $B_m = 12\,600$ (for the $H_m = 2000$), $B_r = 7400$,

$H_c = 560$, $(BH)_m = 1.6 \times 10^6$, $\mu_r = 4$. In heat treating, it is quenched from 1300°C in water at 90–95°C or in air, then aged for 2–4 hours at 650°C. Energy products of over 1.8×10^6 have been attained, but 1.5×10^6 is more representative of the commercial product. The mechanical properties and density are very similar to those of the ternary alloy. Addition of a few per cent of vanadium to Alnico 2 makes it possible to hot-forge the casting after it has been cooled slowly; magnetic properties are then improved by aging [44F1].

The structure of the ternary system Fe-Co-Al has been investigated in a preliminary way by Köster [33K5] and by Edwards [41E2]. As in the Fe-Ni-Al system, aluminum strongly depresses the Curie point [33K5]. The area in which α and γ are in equilibrium at room temperature extends from the region of 75–90% cobalt on the Fe-Co side and covers the range about 80–95% cobalt on the Co-Al side. Superstructures based on FeAl and FeAl$_3$ are known to exist, but their areas are not yet well defined.

Heat Treatment in Magnetic Field.—In 1938 Oliver and Shedden [38O2] showed that an increase in the residual induction of Alnico 2 was obtained when it was cooled from a high temperature in the presence of a magnetic field. Although the increase was not large—about 20%—this experiment was soon followed by work of great practical importance; less than six months later a patent application [40P2] was made by the Philips Company of Holland, relating to material [41J1] having energy products two or three times that of Alnico 2.

The compositions of the alloys that respond to this kind of treatment are much the same as those of the Mishima alloys, except that the cobalt content is considerably higher, 20–25% as compared to 13 or less for Alnico 2 (see Table 8). The alloy first chosen for commercial use (*Alnico 5*,

TABLE 8. Effect of Heat Treatment in a Magnetic Field on Properties of Some Fe-Co-Ni-Al Alloys (Alnico 5 Type)

Composition				No Magnetic Field During Treatment			Field Present During Treatment			Fullness Factor
% Co	% Ni	% Al	% Cu	H_c	B_r	$(BH)_m \times 10^{-6}$	H_c	B_r	$(BH)_m \times 10^{-6}$	γ
23	16	8.5	...	350	9100	1.2	490	12 700	3.5	0.56
24	13.5	8	1.5	370	9500	1.3	505	13 100	3.8	.57
24	13.5	8	3	535	8300	1.7	600	13 400	4.8	.60
22	14	8.5	...				350	14 000	2.9	.59
20	15	8.5	...				440	13 700	3.1	.48
27	12.5	8	3				520	13 900	3.5	.49
23	15	8.5	3				630	12 900	4.6	.56
23	13.8	8.3	3.5				630	12 800	4.6	.57

Ticonal 3.8) has the composition 13.5% Ni, 24% Co, 8% Al, 3% Cu, 51% Fe. The heat treatment consists in heating the alloy to 1300°C and cooling to room temperature in the presence of a magnetic field of 1000–3000 oersteds, and then aging for several hours at about 625°C. The critical effect of the presence of the field during cooling is shown in Fig. 9–36; by this treatment the demagnetization curve is raised and changed in form so that it has a sharper bend, and the fullness factor, $(BH)_m/B_rH_c$, is unusually high.

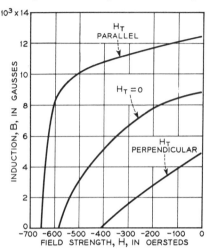

FIG. 9–36. Demagnetization curves of Alnico 5 measured in different directions with respect to the field used during annealing.

Jellinghaus [43J1, 48J4] varied the aluminum content from 6% to 17% in alloys containing 23% Co, 15% Ni, and 3% Cu, and reported changes in the microstructure and magnetic properties. The alloys were composed predominately of the α phase, ferrite. Small amounts of the γ phase, austenite, were found in alloys containing 6% aluminum or less, and a superstructure was found in alloys containing more than 16% aluminum. For alloys having 8–14% aluminum the curves of magnetization vs temperature are not reversible and, therefore, indicate that changes in structure occur with temperature. Conclusive evidence of the nature of these changes is not available. There may be some γ phase present, undetected by X-rays, and its amount would be expected to change with temperature; or the decomposition of α to $\alpha + \alpha'$ may occur as in the ternary Fe-Ni-Al alloys, the change having escaped detection by X-rays. Jellinghaus believes that changes of both kinds occur together.

The saturation induction and Curie points of these alloys are shown in Fig. 9–37. The effect of aluminum content on $(BH)_m$ is shown in Fig. 9–38, and maxima occur in H_c and B_r at the same composition. The magnetic field treatment is effective only in a narrow range of aluminum content centering around 8.5–9%. The properties are not so sensitive to variations in other elements, as Table 8 shows [40P2], but no systematic investigation has been reported. Commercial tolerances of the constituent elements other than aluminum are about ±1%. Contamination with silicon should be kept low. As mentioned in the section on iron-nickel-aluminum alloys, even 0.1% carbon can be objectionable. Microphotographs of fully heat-treated specimens containing, respectively, 0.006 and 0.08%

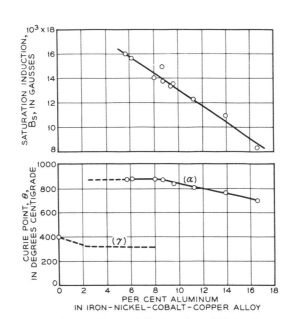

Fig. 9-37. Dependence of saturation and Curie point of Alnico 5 type alloys on the aluminum content (23% Co, 15% Ni, 3% Cu, variable Fe and Al).

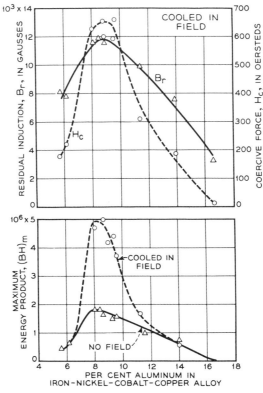

Fig. 9-38. Dependence of energy product of Alnico 5 type alloys on aluminum content and on magnetic anneal.

carbon are shown in Fig. 9-39, for which the author is indebted to E. E. Thomas.

In heat treating, the ingot is heated to 1200–1300°C, then cooled in a field of 1000 oersteds or greater, at a rate not exceeding 10°C/sec, preferably 5°C/sec, when going from 1000 to 700°C. The critical temperature range for the presence of the field is from the Curie point to 800°C, and the field is probably effective at somewhat lower temperatures. The high Curie points of the alloys are therefore important; in Alnico 5 the Curie point

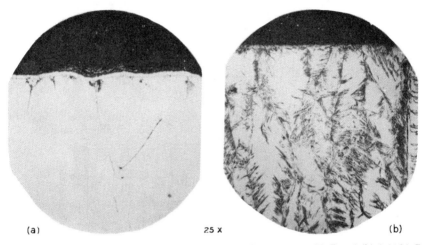

FIG. 9-39. Photomicrographs of Alnico 5 containing (a) 0.006% C and (b) 0.08% C. Magnification 25X.

is 870–900°C, and the action of the field is effective for 100–150°C below this temperature. However, if the Curie point is much below 800°C, this treatment is ineffective.

In a private communication from G. H. Howe he states that the residual induction of the final product increases with the magnitude of the field applied during heat treatment until this is 1000 oersteds, whereas the coercive force attains its highest value when only 100 oersteds are used.

In hardening, the commercial alloy is normally held for 8 hours at 600°C or for a shorter time at 625°C, and this treatment is presumably effective for alloys of somewhat different compositions. The hardening process does not change the residual very much but about doubles the coercive force.

Material that responds to the magnetic heat treatment is anisotropic, and the properties already described obtain only when the field used for measurement is in the same (or opposite) direction as that in which the field was applied during heat treatment. When the two fields are at right angles, the properties are distinctly inferior, the energy product being less by a factor of about 4 to 7 (see Fig. 9-36) [41J1]. This property is important in the use of Alnico 5 in magnets having curved sections (see Fig. 9-14), for in them the field used during heat treatment must be adjusted to the shape of the piece.

As already mentioned, the heat treatment in a field not only increases B_r and H_c but also "squares" the demagnetization curve. The fullness factor $(BH)_m/B_r H_c$ of this material is about 0.60 while for other materials (except Vicalloy) it is usually near 0.4. Associated with the high fullness factor and the high remanence is the high operating induction, B_0, for which (BH) is a maximum. Associated with the high coercive force one expects a low reversible permeability, and for Alnico 5 this is $\mu_r = 3.7$. The mechanical properties of Alnico 5 are similar to those of Alnico 2.

Hoselitz [49M2] showed that a considerable improvement in properties of magnets of the Alnico 5 type was attained if the columnar crystals, formed during the cooling from the melt, are all aligned nearly parallel to the direction of final magnetization. In the usual Alnico 5 some are parallel and some are at 90°, because the melt usually freezes from several surfaces at once. A similar observation was made independently in the United States, where the alloy is called Alnico 5 DG.

The Permanent Magnet Association [45P2, 48O1] in England found that magnetic properties substantially as good as in Alnico 5 could be obtained in the alloy containing only 11% nickel (and 25% Co, 8% Al, 6% Cu). This is called *Alcomax 2*. There is thus a maximum in the energy product near this composition as well as near the Alnico 5 composition, and a minimum at intermediate composition. The treatment used is practically the same in the two alloys. Alcomax 1 contains titanium. Alcomax 3 and Alcomax 4 contain niobium; these alloys are similar to the other Alcomaxes in their preparation and properties (see Appendix 4, Table 3).

The physical basis of the high energy product and high anisotropy of Alnico 5 has not been established, but some relevant experiments and theories may be mentioned here. Anisotropy has been observed in magnetostriction [49H1, 50N1] and in resistivity [49D6] as well as in magnetization. The anisotropy of magnetostriction is that to be expected on domain theory and is discussed in Chap. 13.

Geisler [50G1] has measured the lattice spacings of the phases existing after overaging and has noted a relatively small difference in spacing, or "disregistry," between them—about 0.1% as compared with 1.5% in Cunico.

Nesbitt and Williams [50N2] have obtained powder patterns which show definite domain boundaries that are displaced by application of a field and that outline regions elongated in the direction in which the magnetic field was applied during heat treatment.

Nesbitt has also shown that magnetic anisotropy is well developed after the material has been cooled in a magnetic field to 800°C and then quenched but not aged. The field required for transverse magnetization to half of saturation is then 230 oersteds, which is large compared to the coercive force of 15 oersteds, and this shows that with this treatment the domains are already held firmly in the direction in which the field was applied during heat treatment.

Stoner and Wohlfarth [48S1] and Néel [47N3] have suggested that the material is divided into small regions that are small enough to act as single domains, and that these have the high coercive force associated with material in which domain boundaries are absent and in which domain rotation must occur (see Chap. 18). Néel proposed that the material is composed of needle-shaped particles directed by the field used during heat treatment. It has been suggested by Shubina and Shur [49S7] and by Kittel, Nesbitt, and Shockley [50K1] that the subdivision is caused by thin plates that precipitate in planes controlled by the field present during heat treatment. The latter group suppose that nuclei form at temperatures somewhat below the Curie point, when they are subject to control by the magnetic field, and develop into plates during the aging at 600°C. So far no independent evidence of the existence of such plates has been found, but the experiments of Bradley [49B7] on the related Fe-Ni-Al alloys are suggestive. It is hoped that continued study of single crystals, which have been prepared by E. A. Nesbitt, will aid in the solution of this problem.

Alloys Containing Titanium.—These include the *New Honda* or *New KS* alloys, *Ticonal 2A*, *Alnico 12*, *Alcomax 1*, and *Nipermag*.

Recent investigations of the binary alloys of titanium with iron, cobalt, and nickel [41W2] show the existence of the compounds Fe_2Ti, Co_2Ti, and Ni_3Ti when the titanium content is small, and $FeTi$, $CoTi$, and $NiTi$ when the titanium is considerably higher. At 800°C the solubility of titanium is about 3% in iron and nickel and 7% in cobalt and is substantially less at room temperature.

In the ternary system Fe-Ni-Ti [38V1] the same compounds are observed, and the solid solubility of titanium is greater than in the binaries. Age hardening has been investigated in alloys containing 35, 50, and 75% nickel, and it occurs most noticeably in the lower nickel alloys when 2% or more of titanium is used [40P3]. These alloys are useful because they have high strength at high temperatures [41A3]. In the Fe-Co-Ti system, hardening is also observed [35K7] when more than 2% titanium is present; the alloy having 8% of this element and 55% of cobalt ages at 600°C so that the Brinell hardness increases from 420 to 600. Hardening is also found in the Co-Ni-Ti system, which includes the commercial alloy "*Konel*" (e.g., 17% Co, 73% Ni, 9% Ti) used for vacuum tube filaments and other purposes. No investigations of quaternary or higher systems have been

ALLOYS CONTAINING TITANIUM 395

reported. Most of our information on these systems relates to their magnetic and mechanical properties.

The first important use of titanium in magnetic age-hardened alloys was made by Honda, Masumoto, and Shirakawa. A coercive force of 830 was reported [35K6] for one composition (45% Fe, 28% Co, 16% Ni, 11% Ti) and 920 for another (not stated) [34H1]. These and other data are given in Table 9. The alloys are known as "New KS" or "New Honda" steel. As stated in the original patents [35K6, 36K3], aluminum and other elements may be added, and the demagnetization curve of one such alloy of

TABLE 9. MAGNETIC PROPERTIES OF SOME FE-NI-TI ALLOYS
WITH AND WITHOUT ADDITIONAL ELEMENTS

Composition (%)*				B_r	H_c	Reference	Name
Co	Ni	Ti	Other				
20	11	10	...	9 000	250	35K6	
28	16	11	...	7 500	830	35K6	New KS
...	25	12	4.7 Al	5 200	300	36K3	
...	29	16	4.7 Cu	2 500	490	36K3	
15–36	10–25	8–25	...	6 400	920	34H1	New KS
27	18	6.7	3.7 Al	7 100	780	37N1	
35	18	8	6 Al	5 700	1050	43H5	
35	18	8	6 Al	5 800	950	46R3	Alnico 12
...	32	2	12 Al	5 000	500	36C2	Nipermag
24	14	2.4	7.1 Al	7 900	600	40P2	
24	14	2.4	7.1 Al	11 100	660	40P2†	
22	18	3	7.4 Al	10 500	760	40P2†	
23	16	1.4	7.7 Al	12 500	600	40P2†	
22	18	2.4	7.1 Al	11 000	720	40P2†	
23	15	1.9	6.6 Al	12 800	520	40P2†	

* Remainder iron.
† Heat treated in magnetic field.

specific composition 45% Fe, 27% Co, 18% Ni, 6.7% Ti, 3.7% Al has been published [37N1] and is shown in Fig. 9–40. For this the density is 7.30, and the resistivity is 65 microhm-cm.

Howe [41H3] has described the preparation of the alloy containing 33% Fe, 35% Co, 18% Ni, 6% Al, and 8% Ti. This is the composition of the commercial alloy *Alnico 12*, which is a precipitation-hardenable alloy based on the precipitation of the α and α₂ phases of the Fe-Ni-Al system. Additions of Co, Cu, and Ti go into solution in the Fe- and Ni-Al-rich phases. The melt, to which aluminum and titanium are added last, is poured very hot into sand molds and the castings are stripped as soon as possible and allowed to cool in air. They are then aged for several hours at 650°C. As the work on Ticonal has shown, the presence of a strong magnetic field during the air cooling may be advantageous in alloys of this general type, although they are not improved by the magnetic cooling as much as certain alloys that contain no titanium.

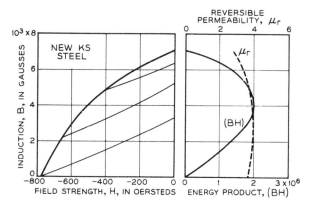

Fig. 9–40. Magnetic properties of New KS or New Honda "steel."

Properties of commercial Alnico 12 are given in Table 9 and Fig. 9–41. The energy product is about 1.75×10^6 [41H3, 46R3].

The patents on New KS and Alnico 12 refer to alloys containing 6% or more of titanium. Use of a considerably smaller amount of this element

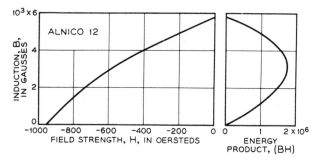

Fig. 9–41. Magnetic properties of Alnico 12.

was made by Catherall [36A3, 36C2] in the alloy called *Nipermag*, containing 32% Ni, 12% Al, 2% Ti, and the rest iron. This has a rather high coercive force (see Table 9) and supports the contention [40P2] that in this type of alloy titanium is effective in obtaining high coercive force but is not necessary for high energy product. Alnico 6 contains 0.5% Ti.

Iron-Nickel-Copper Alloys.—These are important because they include the alloy called *Magnetoflex* or *Cunife* that will stand severe cold reduction and consequently can be formed into thin tape or wire for use in the magnetic recording of speech.

The magnetic hardening of these alloys was first reported by Kussmann and Scharnow [29K1]. They added copper to the alloy containing Fe/Ni = 1/1 and found that the coercive force of specimens in the annealed condition rose rapidly when more than 18% copper was present. At this composition the alloy was homogeneous at high temperatures (γ) and decomposed on cooling into copper-rich (γ_2) and copper-poor (γ_1) com-

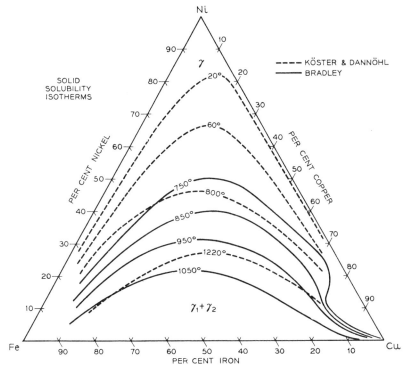

FIG. 9–42. Positions of the γ_1, γ_2 phase boundaries of Fe-Ni-Cu alloys at various temperatures.

ponents, as indicated in the phase diagram of Fig. 9–42. Here the uncertain position of the boundary between γ and $\gamma_1 + \gamma_2$ has been the cause of some discussion, and recent work indicates that this line should not be drawn as smoothly as previously supposed. Results of the most recent investigations are shown in Fig. 9–42. Köster and Dannöhl [35K3] used alloys containing 0.5% manganese, measured resistivity and magnetization at elevated temperatures, and determined the lines from these data. Bradley [41B4] and co-workers used pure alloys and quenched the powdered material from the required temperature, or cooled very slowly when concerned with equilibrium at low temperatures.

In the alloy that has been investigated most (20% Fe, 20% Ni, 60% Cu), there is little doubt that at high temperature (say, 1050°C) it is homogeneous (γ) and that, after cold deformation or heating at 600–700°C, the separation into $\gamma_1 + \gamma_2$ takes place and hardening occurs. Daniel and Lipson [43D1, 44D1] have studied in detail the alloy having the composition represented by $FeNi_3Cu_4$ (12% Fe, 36% Ni, 52% Cu) and found X-ray evidence that during the preliminary stages of its decomposition the structure consists of the original γ phase in which there are periodic changes in the lattice parameter. This structure is detectable after aging the quenched alloy for 15 minutes at 650°C or for a longer time at 550°C, but it is converted entirely to the homogeneous state by maintaining at 800–950°C.

It has been suggested that the large magnetic anisotropy of the rolled sheets is caused by precipitation of γ_2 along definite crystal directions or planes in the oriented crystals, in such a way that the internal strains themselves are anisotropic. The X-ray work of Daniel and Lipson indicates that periodic changes in the parameter of the lattice occur in planes parallel to the cube planes of the decomposing lattice.

A systematic attempt to obtain high coercive force in those alloys was first made by Dahl, Pfaffenberger, and Schwartz [35D3] who reported values

TABLE 10. EFFECT OF TREATMENT ON THE 20% FE, 20% NI, 60% CU ALLOY
(Treatments of first four lines are consecutive [37N2])

Treatment	B_r	H_c	$B_r H_c \times 10^{-6}$	$(BH)_m \times 10^{-6}$
Cast, homogenized, quenched from 1000°C	2000	200	0.4	...
Aged 650°C	3200	400	1.3	...
Cold rolled 50%	4800	400	1.9	...
Cold rolled further to 95%, aged 650°C	5000	440	2.2	...
Cast, homogenized, swaged 95%, quenched from 1070°C, cold swaged to size, aged 600°C (Cunife 1)	5800	600	3.5	2.0
Commercial Cunife 1	5400	550	3.0	1.7

of H_c and B_r for alloys quenched or slowly cooled from 1050°C. The highest H_c was 410 ($B_r = 1400$), for the alloy 10% Fe, 25% Ni, 65% Cu, and the highest $B_r H_c$ was 0.75×10^6 ($\bar{H}_c = 155$, $B_r = 4800$) for the alloy 20% Fe, 40% Ni, 40% Cu. The Brinell hardness of the latter alloy rose only from 120 after quenching to 180 after slow cooling from 1050°C. The important facts that they brought to light are that these alloys can be readily cold worked after cooling slowly, and that, by stretching 30%, the value of $B_r H_c$ for the second alloy is increased to about 1.0×10^6.

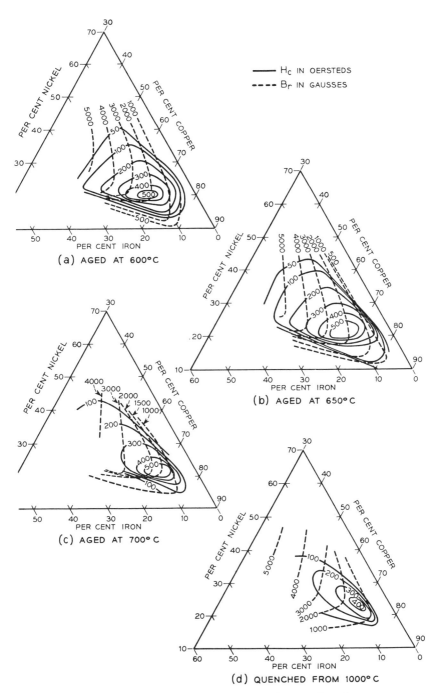

Fig. 9–43. Dependence of H_c and B_r of Fe-Ni-Cu alloys on the aging temperature.

This work was soon followed by that of Neumann, Büchner, and Reinboth [37N2] who investigated in more detail the effect of heat treatment and, especially, of cold deformation. In alloys cast, homogenized, and quenched from 1000°C, they found that aging was effected by heating for an hour at 500–700°C, preferably 650°C. The $B_r H_c$ product was further increased by cold working and still more by subsequent aging. Data for specific treatments are given in Table 10 and Figs. 9–43 and 44.

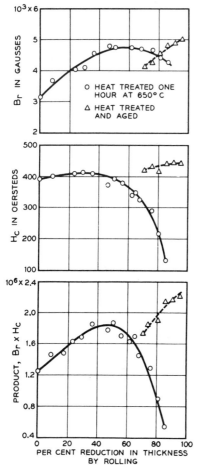

When this alloy has been given a severe cold reduction at some state in its fabrication, distinct anisotropy of magnetic properties is found to exist even after further heat treatment. The data [37N2] of Fig. 9–45 show this effect in sheet that has been reduced 85% in thickness and aged.

Even when these alloys are in the magnetically hard state, they are relatively *soft mechanically* and can be readily worked. Such materials have naturally aroused great interest because they show that magnetic hardness and mechanical hardness are not necessarily concomitant, and they have led to the investigation and discovery of other alloys having similar or superior qualities. The 20% Fe, 20% Ni, 60% Cu alloy is known in Germany under the name "Magnetoflex" and in the United States as "Cunife 1."

Fig. 9–44. Effect of the degree of rolling on the magnetic properties of the alloy containing 20% Fe, 20% Ni, 60% Cu. Circles, after annealing at 650°C and rolling; triangles, after annealing, rolling and again annealing.

Magnetic hardening has also been produced in *iron-rich* alloys containing about 15% nickel and 15% copper. In the first investigation Legat [37L2] determined the phases present in alloys quenched from 1100–1200°C. The best magnetic properties that he attained were in the 70% Fe, 15% Ni, 15% Cu alloy, quenched and then aged at 700°C. The constants are $H_c = 300$, $B_r = 1500$. Dannöhl [38D3] obtained somewhat higher values: $H_c = 330$, $B_r = 3900$, in the alloy containing 12% Ni and 15% Cu, by homogenizing at about 1100°C, quenching, and aging at 600°C. He believes that hardening is by the precipitation of the copper-rich γ_2 phase brought about by raising the temperature until the alloy is all in the γ phase, quenching, then aging at 700°C. When cooled to room temperature, some of the γ may be converted to α, but it is not believed that much hardening is so produced—this is due rather to highly dispersed γ_2.

Alloys of similar composition (70–87% Fe, 10–20% Ni, rest Cu) have been subjected to cold working after quenching [39S17] and used for the recording of speech. Therefore these, as well as the copper-rich alloys,

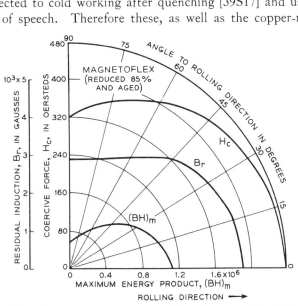

Fig. 9-45. Directional magnetic properties of the alloy containing 20% Fe, 20% Ni, 60% Cu, reduced 85% by rolling and then aged.

have desirable mechanical properties. The highest coercive force reported is 93, and this indicates that precipitation caused by working is less effective magnetically than that produced by aging at elevated temperatures.

Hardening by cold working alone has been produced in the alloy containing 45% Fe, 15% Ni, and 40% Cu. Nesbitt, in unpublished work, has measured a coercive force of 240 and a residual induction of 4400 in wire of the same composition, quenched from 1000°C and then cold drawn from 0.026-in. to 0.006-in. diameter. Before the cold working, the metal is non-magnetic, as may be inferred from the phase diagram (see Fig. 5–66), and the magnetic character is developed during the drawing when it is converted from γ to $\alpha + \gamma_2$. Similar treatment of the alloy containing 60% Fe, 15% Ni, and 25% Cu resulted in material having $H_c = 170$, $B_r = 11\,000$.

Both cobalt and aluminum have been added, separately and together, to alloys like those investigated by Dahl and Neumann. A commercial alloy, *Cunife 2*, has the composition 27.5% Fe, 2.5% Co, 20% Ni and 50% Cu and is processed in the same manner as Cunife 1. The constants are: $H_c = 260$, $B_r = 7300$, $(BH)_m = 0.78 \times 10^6$; it is somewhat stronger than Cunife 1 and is an improvement over that alloy for magnetic recording. Additions of molybdenum have also been investigated.

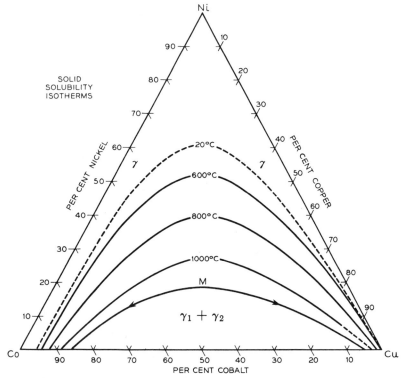

Fig. 9-46. Phase boundaries of the Co-Ni-Cu system at various temperatures.

Cobalt-Nickel-Copper Alloys.—A useful alloy similar to Cunife 1, but containing cobalt instead of iron, has the composition 29% Co, 21% Ni, and 50% Cu and is made under the name *Cunico 1*. Investigation of the structure and magnetic properties of this three-component system was carried out by Dannöhl and Neumann [38D1], and the mechanism of the hardening was studied further by Volk, Dannöhl, and Masing [38V2], as well as Geisler and Newkirk [48G2] (see below). The solid solubility isotherms resemble those of the Fe-Ni-Cu system but are somewhat simpler in form since there is now no α,γ transformation. Solubilities at several temperatures are shown in Fig. 9-46, and the Curie points and saturation intensities of the slowly cooled alloys are given in Figs. 8-25 and 26.

The alloys used by Dannöhl and Neumann contained 0.4-1.1% Mn and about 0.03% Si. The specimens were homogenized for a rather long time at a high temperature, then quenched and aged. They can be cold worked if desired by first cooling slowly from the high temperature, but after working they must again be quenched and aged to develop good properties. After homogenizing for 15 hours at 1140°C and quenching in 10% sodium hydroxide solution, then aging for an hour at various increasing tempera-

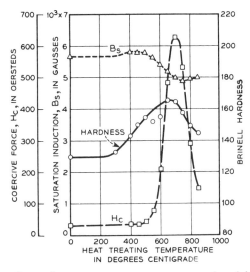

FIG. 9-47. Effect of annealing temperature on some properties of the alloys containing 29% Co, 25% Ni, 45% Cu, 0.7% Mn, after quenching from 1140°C.

tures, the properties of one alloy (29% Co, 25% Ni, 45% Cu) change as shown in Fig. 9-47. Coercive force and Brinell hardness are maximum after aging at 700°C, and the saturation then is somewhat below its value after quenching.

The residual induction and the saturation decrease rather uniformly with composition toward the copper corner. The coercive force is a maximum near the composition 20% Co, 20% Ni, and 60% Cu (Fig. 9-48); the highest $B_r H_c$ occurs at about 30% Co, 25% Ni, and 45% Cu. However, because the fullness factor changes with composition, the energy product $(BH)_m$ is greatest at about 41% Co, 24% Ni, and 35% Cu, the composition chosen for Cunico 2. This alloy is homogenized for 10 hours at 1080°C and aged 12-15 hours at 625°C; if desired, it may be cold worked as described previously. Properties so obtained by Howe (private communication) differ from those recorded by Dannöhl and Neumann, perhaps because the time of aging used by the former is considerably longer (see Table 11).

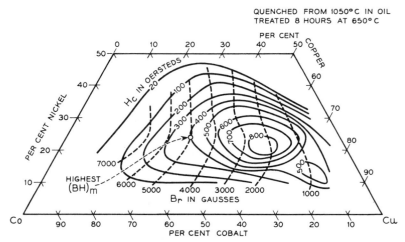

Fig. 9-48. Coercive force and residual induction of some Co-Ni-Cu alloys quenched from 1050°C and aged 8 hours at 650°C.

In their study of the mechanism of hardening, Geisler and Newkirk [48G2, 50G1] showed with the aid of X-rays that when the precipitate first forms it is face-centered tetragonal in structure, and forms on the (100)

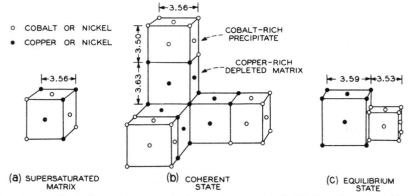

Fig. 9-49. Crystal structures and phase relations in Co-Ni-Cu alloys in which precipitation occurs.

planes of the matrix, being coherent with it. The strained condition of both the matrix and the precipitate is illustrated in Fig. 9-49. During aging the coherency is lost, and the two phases assume their equilibrium structures, which are face-centered cubic with different lattice spacings, one

TABLE 11. SOME PROPERTIES OF CO-NI-CU ALLOYS [38D1], HEAT-TREATED FOR MAXIMUM VALUE OF (BH) PRODUCT

Composition (%)			H_c (oersteds)	B_r (gausses)	$(BH)_m \times 10^{-6}$	B_s (gausses)	μ_r (at B_0)	ρ (microhm-cm)
Co	Ni	Cu						
49	26	25	232	6250	0.7	9750	7	25
45	25	30	335	5900	0.9	9030	5	26
41	24	35	444	5300	1.0	8570	4	26
37	23	40	550	4400	0.9	7570	4	29
(29	21	50)*	660	3400	0.8	5000	3.5	32
(41	24	35)†	450	5300	1.0	...	...	...

* Cunico 1.
† Cunico 2.

with larger and the other with smaller lattice spacing than the original undepleted matrix.

Iron-Cobalt-Vanadium Alloys.—This system includes *Vicalloy*, an alloy that can be cold rolled to thin tape or drawn to fine wire suitable for use in

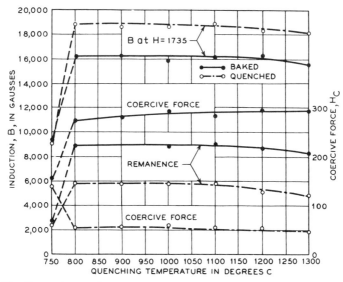

FIG. 9–50. Dependence of magnetic properties of Vicalloy on quenching temperature after quenching, and after quenching and baking.

the magnetic recording of speech [40N1, 42N1, 46N1]. The mechanism of hardening of the alloy is different from that of the materials already discussed and may be referred to as the Vicalloy type of hardening.

The phase structure of these alloys has already been discussed in connection with Vanadium Permendur (49% Fe, 49% Co, 2% V). The transformations are not well established, but it is known that α and γ phases are important and order-disorder transformations occur. Of particular interest for permanent magnets are the alloys containing 50% cobalt and 5 to 15% vanadium.

Nesbitt has shown [46N1] that high coercive force is developed in these alloys in two different ways. According to one method the material, containing preferably 38% Fe, 52% Co, and 10% V, is quenched in oil at 800–1200°C and aged at 600°C for several hours. Before the final aging

TABLE 12. EFFECT OF QUENCHING AND AGING ON VICALLOY
(37% FE, 53% CO, 9.4% V, 0.6% SI)
(Treatments are consecutive, on same specimen)

Heat Treatment	B_r	H_c	$B - H$ for $H_m = 1700$
As cast................................	5050	65	14 500
Aged 600°C, 8 hr..................	6900	295	12 400
Quenched from 1200°C..............	4950	50	16 600
Aged 600°C, 8 hr..................	9000	300	14 400
Furnace-cooled from 1100°C.........	5250	55	16 900
Aged 600°C, 8 hr..................	8650	300	14 100

the material may be hot or cold rolled, or machined. Hardness, after quenching, is about Rockwell C30 and, after aging, C60. Dependence of magnetic properties on quenching temperature is shown in Fig. 9–50, and the effect of aging is given in Table 12. A typical demagnetization curve is included in Fig. 9–55. This material is known as Vicalloy 1 and has an energy product of about 1.0×10^6. Cold working is not essential to developing its quality.

By the use of the other method, a higher coercive force and energy product may be obtained in Vicalloy 2, material containing 10–14% vanadium and subjected to a cold reduction of 50% or more during processing. The alloy is cast, hot worked, cold worked, and aged at about 600°C—no quenching from a high temperature is required. The beneficial effect of cold reduction by various means (drawing, rolling, swaging) is evident from the curves of Fig. 9–51. By such means the energy products of the materials containing 10–11% V are increased by 50–100%, and that of the 14% vanadium alloy is increased from a very small value to about 3×10^6—material containing 14% vanadium is practically non-magnetic if not cold worked. Demagnetization curves of wire that has been cold drawn various amounts are shown in Fig. 9–52; the composition is 34% Fe, 52% Co, 14% V, and 0.3% Mn, and the cast rod was hot swaged to 0.170 in., cold drawn to final size, and aged for several hours at 600°C. In wire that has been cold reduced 98% in area to 0.006-in. diameter, energy products as high as 3.5×10^6 have been attained ($H_c = 525$, $B_r = 10\,000$, $\gamma = 0.67$).

Saturation is reduced by the addition of vanadium and also by aging (Fig. 9-53). Methods of cold work that produce the greatest elongation, such as swaging, rolling between grooved rolls, and drawing, are more effective

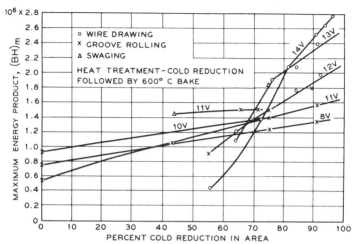

FIG. 9-51. Maximum energy product of some Vicalloys of various compositions and treatments. Cobalt content 52%.

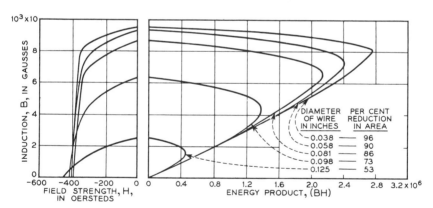

FIG. 9-52. Magnetic properties of Vicalloy II (34% Fe, 52% Co, 14% V) after drawing a wire (originally annealed when 0.187 in. in diameter) by various amounts and baking at 600°C.

than methods that permit spreading, such as rolling between flat rolls. In any case the cold reduction is accompanied by a substantial increase in hardness, for example an increase from Rockwell C30 to C40 during a reduction of 50% in area. Such hardness puts a severe strain on dies and rolls used in fabrication.

Cold reduction also gives rise to directional magnetic properties. The greatest difference is in residual induction, which is about twice as great in the rolling direction as in the cross direction (see Fig. 9–54). The

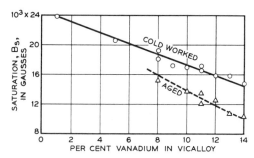

FIG. 9–53. Saturation induction of various Fe-Co-V alloys containing 52% Co, after cold working, and after cold working and aging.

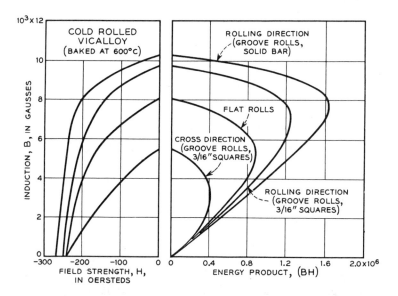

FIG. 9–54. Directional properties of Vicalloy: (1) longitudinal properties of solid bar rolled in grooved rolls and baked at 600°C, (2) after cutting bar into $3/16$-in. squares, (3) properties in sheet rolled with flat rolls and baked, and (4) transverse properties of bar cut in $3/16$-in. squares.

coercive force is less affected. X-ray measurements show an increasing alignment of the crystals with increasing cold work, and this special orientation persists, as expected, during aging. Absence of anisotropy of strain is indicated by the small directional dependence of coercive force.

The mechanism of hardening of Vicalloy may be discussed with the aid of Fig. 6–12. After cold rolling, material containing 50% cobalt and 10% vanadium exists in the α phase, substantially free from any γ. The diagram indicates that during aging at 600°C some α will be converted to γ, and when highly dispersed this will produce hardening. In one respect the hardening is distinctly different from that produced in the usual way, as illustrated by the hardening of the Fe-Co-Mo alloys. In the latter the material is quenched so as to maintain an *unstable*, supersaturated condition at room temperature, then aged to permit an approach to equilibrium

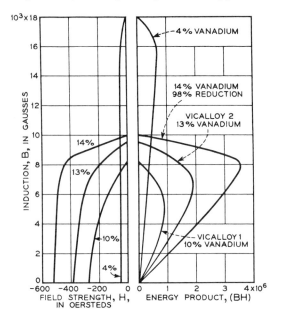

FIG. 9–55. Magnetic properties of Fe-Co-V alloys containing various amounts of vanadium, after hardening.

by the precipitation of a small amount of a second phase. In Vicalloy, on the contrary, after quenching or cold working, the alloy is in the *stable* α phase, and aging permits an approach to the equilibrium stable at an elevated temperature. In the conventional method of hardening the two phases are stable at room temperature and a single phase at high temperature, whereas in Vicalloy the two phases are stable at the higher temperature and a single phase at room temperature. When Vicalloy contains 12% or more of vanadium, it contains both α and γ phases at room temperature before the cold rolling, but an additional amount of α is produced during the cold working, and upon aging some of this is converted again to γ and causes hardening.

After aging, the ductility of the material is lost. As cold worked,

Vicalloy 1 and Vicalloy 2 may be cold formed or machined. A $\frac{1}{8}$-in. wire of Vicalloy 2 can be bent around its own diameter without cracking. The magnetic properties of several Vicalloys are summarized in Fig. 9–55.

Alloys with Noble Metals. Iron-Platinum.—A high coercive force was first attained in these alloys by Graf and Kussmann [35G2]. They found H_c to be a maximum for the alloy FePt (78% Pt) and to have the value 1200 (but the field strength for $B - H = 0$ is $_IH_c = 1700$). Residual induction is 3500, and $(BH)_m = 3.1 \times 10^6$ [37N1].

The transformations that occur in this system are indicated in Fig. 9–56, taken from the work of Isaac and Tammann [36H1], Fallot [34F1, 38F1], and Graf and Kussmann. The last-named authors first determined the Curie points in the composition range 30–70 atomic per cent platinum (60–90% Pt) and reported the transformation temperature of the order-disorder change of FePt to be near 1100°C. Jellinghaus [36J2] established by X-rays the existence of a tetragonal superstructure in FePt containing some rhodium, and this was confirmed by Lipson, Shoenberg, and Stupart [41L2] who showed that the structure was of the AuCu type with $c/a = 1.37$.

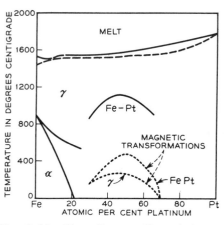

FIG. 9–56. Phase diagram of iron-platinum alloys.

Kussmann [37K3] observed a negative coefficient of thermal expansion in the range 50–60% platinum. A large positive magnetostriction was observed [49A1] at 54% platinum.

Saturation magnetization first *increases* with addition of platinum to iron to about 13 atomic per cent platinum and then decreases and becomes zero at room temperature at about 25 atomic per cent (54% Pt) [35G2, 38F1]. (See Fig. 9–57.) In the region 30–70 atomic per cent platinum, B_s rises again to substantial values (14 000) and once more falls to zero.

The coercive force of FePt depends, as expected, on the heat treatment, and its highest value was obtained by quenching from 1200°C. A demagnetization curve is given in Fig. 9–58 [35G2, 37N1]. Lipson, Shoenberg, and Stupart point out that when the ordered tetragonal phase forms from disordered γ there are three possible orientations that the new structure may assume with respect to the old, and that large strains, responsible for the high coercive force, are set up when the atoms are disturbed from their original positions. Data are available on density [35G2], resistivity, hardness [32N1], and thermal expansion [37K3].

Iron-Palladium.—The original work of Grigorjew on the phase diagram [36H1] has been supplemented in a substantial way by X-ray and magnetic studies. In addition to the γ_2 phase $FePd_3$, reported in the original work

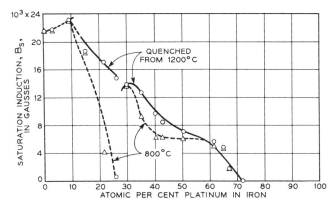

FIG. 9–57. Saturation induction of iron-platinum alloys as quenched or annealed.

and now known to have an ordered cubic structure, Hultgren and Zapffe [39H1] have established with X-rays the existence of an ordered tetragonal phase γ_1 (FePd) having a structure like that of AuCu with $c/a = 0.966$.

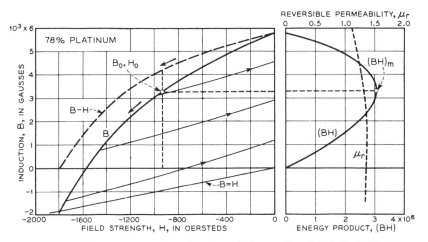

FIG. 9–58. Magnetic properties of an iron-platinum alloy (22% Fe, 78% Pt).

The phases they found at room temperature, after drastic quenching from various temperatures, are shown by the diagram of Fig. 9–59. The general features have been confirmed by others [39A4]. The boundary between the γ and γ_2 fields for $FePd_3$ at equilibrium was determined by Grigorjew to be at about 900°C, in good accord with the later results on the quenched alloys. The X-ray data include measurements of lattice spacings. FePd and FePd$_3$ both increase slightly in volume on ordering (0.5 and 0.3%, respectively).

Fallot [38F1] measured Curie points and saturation over the entire range of compositions (see Chap. 7), and his results are consistent with the investigations of structure reported later by others. Curie points in the Pd-rich alloys depend on heat treatment and show maxima near FePd and FePd$_3$. Jellinghaus [36J2] found that by aging the cast alloy at 500°C the coercive force was increased to a maximum of 260 oersteds.

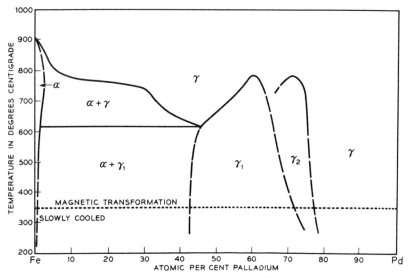

Fig. 9–59. Phase diagram of iron-palladium alloys.

Iron with Other Noble Metals.—In a brief note Zumbusch [35Z1] has reported a coercive force of 5000 oersteds (probably $_IH_c$) in an Fe-Ag alloy. Silver has very slight solid solubility in iron.

Gold also is but slightly soluble in α-iron, about 0.5% at 500°C according to Jette, Bruner, and Foote [34J1], and this accounts for the constancy of the Curie point and the linear decrease in saturation reported by Forrer [36F1] in alloys containing up to 30% gold. Lattice spacings and age hardening have been determined [34J1]. The effect of cold working and annealing on the magnetic properties of the alloy containing 14% iron has been studied by Pan and others [42P1].

Fallot [38F1] has measured Curie points and saturation for alloys of iron with ruthenium, rhodium, osmium, and iridium (see Chap. 7).

Cobalt-Platinum.—Jellinghaus [36J2] found that these alloys have the highest energy product (3.8×10^6) of any of the alloys with noble metals, and for several years they had the highest product known. In an investigation of their structure Gebhardt and Köster [40G3] found that the hexagonal cobalt phase is stable at room temperature until more than 6 atomic per cent platinum (18% Pt) is dissolved, whereupon the structure becomes face-centered cubic (γ). The compound CoPt (77% Pt) is stable from about 30–60 atomic per cent platinum (60–85% Pt), and the high coercive force is attributed to the strains resulting from the transformation of γ to CoPt on cooling, or on aging after quenching at 1000–1200°C. The phase diagram, constructed with the aid of measurements on resistance, thermal expansion, magnetization, and thermal analysis, is reproduced in Fig. 9–60.

The nature of the transformations in CoPt has been studied further by Hultgren and Jaffee [41H4]. X-ray measurements made after quenching from various temperatures confirmed the belief that above 1000°C the lattice is face-centered cubic, and that an ordered tetragonal phase (CoPt) is formed at lower temperatures. In a private communication Newkirk, Martin, and Geisler state that they have confirmed these results, and that they believe the hardening to be associated with transformation of the one phase to the other. In addition, Hultgren and Jaffee reported a body-centered cubic structure stable at still lower temperatures. Gebhardt and Köster observed transformations at 500 and 825°C, but in view of the discovery of the body-centered structure one may question the correctness of their interpretation of these as cubic to disordered tetragonal and disordered to ordered tetragonal transformations. The Curie points of the quenched alloys are also shown in Fig. 9–60.

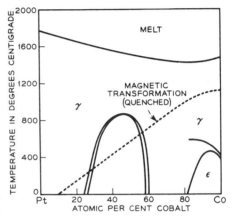

Fig. 9–60. Phase diagram of cobalt-platinum alloys.

Saturation induction of CoPt is greatest after quenching (7200 gauss) and falls upon aging at 700–750°C (to about 4000 gauss). Some small specimens were observed to lose their ferromagnetism after prolonged heating at 750°C.

On account of the large coercive force and low residual induction, the demagnetization curves are quite different, depending on whether B or $B-H$ is used as ordinate. Measurements by Neumann [37N1] are shown in Fig. 9–61. Although his specimen was quenched from 1200°C, the properties correspond closely to those reported by Gebhardt and Köster in a specimen quenched at 1200°C and aged at 700°C. The latter treatment has also been found effective at the Bell Laboratories. Data have been reported [40G3] on density, lattice parameter, resistivity, and hardness. The resistivity of CoPt depends markedly on heat treatment, as would be expected when atomic ordering occurs.

Cobalt-Palladium.—The ϵ,γ transformation (hexagonal to face-centered cubic) in cobalt persists to about 15 atomic per cent palladium (24% Pd). Beyond this the alloys form a continuous series of solid solutions, and the Curie point decreases continually with increasing palladium until it reaches

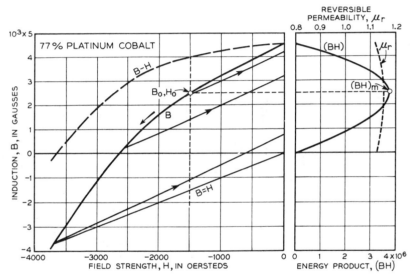

FIG. 9-61. Magnetic properties of a cobalt-platinum alloy (23% Co, 77% Pt).

room temperature at about 92 atomic per cent palladium [35G3] and absolute zero at 100% palladium [30C2]. Saturation magnetization [35G3] at room temperature decreases approximately linearly with atomic per cent palladium toward zero at 92 atomic per cent (95.5% Pd). Plots of resistivity against temperature show breaks only at γ,ϵ transformations and Curie points [36G2].

Constant [30C2] measured the coercive forces of alloys containing 90 and 95 atomic per cent palladium and reported them to be 420 in the hard state at low temperature ($-190°C$). The corresponding residual and saturation inductions at these compositions are low.

Nickle with Noble Metals.—Nickel and platinum form a continuous series of solid solutions, marked only by the formation at about 600°C of ordered structures in alloys near in composition to Ni_3Pt (52% Pt) [37M2,

38K6, 49K4], and to NiPt (77% Pt) [44E3]. The latter structure is tetragonal, like AuCu. The saturation and the Curie point are decreased somewhat on ordering. The Curie point reaches room temperature near 30 atomic per cent (59% Pt) and absolute zero at about 60 atomic per cent platinum (83% Pt). Lattice parameters and resistivities were measured. Values of coercive force have not been reported.

Nickel and palladium form a uniform series of solid solutions over the entire range of compositions, without evidence of superstructure [39H1]. The Curie point falls to room temperature at 85 atomic per cent palladium (92% Pd) and to 0°K at 100% palladium [37M2].

Curie points and saturation of nickel-gold and nickel-ruthenium have been reported by Marian [37M2]. Köster and Dannöhl [36K8] have studied age hardening in nickel-gold alloys and found a coercive force as high as 220, but with a low remanence.

Chromium-Platinum.—Friederich and Kussmann [35F1] showed that these alloys are ferromagnetic in the range of compositions 50–80 atomic per cent platinum (80–95% Pt). The highest saturation induction obtained in quenched alloys is about 1000, and after annealing this rises to about 3000. In an investigation of their structure Gebhardt and Köster [40G3] deduced the portion of the phase diagram reproduced in Fig. 9-62. Apparently the existence of ferromagnetism depends fundamentally on the chromium atoms and is connected with the ordered face-centered phase, the exact nature of which is unknown.

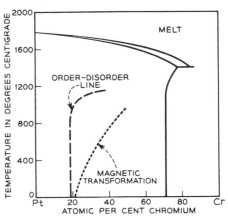

FIG. 9-62. Phase diagram of chromium-platinum alloys.

Manganese-Silver-Aluminum.—Potter [31P3] discovered that the ternary alloy corresponding approximately to the composition $MnAg_5Al$ (8.8% Mn, 86.8% Ag, 4.4% Al) has a coercive force for $B-H = 0$ of $_IH_c = 6000$, a residual magnetization of 40–50 ($B_r = 500$–600) and a saturation of about $B_s = 880$. The field strength at which B goes to zero (H_c) is very small, so that $(BH)_m$ is also small. However, there is some practical [45A1] as well as theoretical interest in a substance with this extremely high value of $_IH_c$ and it is used as *Silmanal*. Best values are obtained by heating for a long time at 250°C. The Curie point is about 360°C. X-ray measurements reveal a simple solid solution with no evidence

of superstructure; the hardening is apparently of the precipitation type.
The alloy is ductile enough to be drawn to a fine wire. It is basically a
member of the Heusler family.

Other Binary Alloys.—In their original investigation of the magnetic
properties of age-hardenable alloys, Seljesater and Rogers [32S6] studied
briefly the alloys of iron with Be, Ti, Mo, and W. The best values of H_c
for the first two of these are well under 100 and, therefore, of little interest.
For Fe-Mo they found $H_c = 220$, $B_r = 7000$, and for Fe-W, $H_c = 150$,
$B_r = 6800$.

TABLE 13. SOME PROPERTIES OF TERNARY ALLOYS OF FE AND CO

Added Element	Extent (%) of α Phase* at 50% Co	Compounds† Reported	High H_c	Corresponding B_r	References
None	...	FeCo	40	10 000	See Chap. 6
Al	50	FeAl$_3$, FeAl, Fe$_2$CoAl	...	...	33K5, 41E2
Be	1	FeBe, CoBe	480	8 000	39K5
Cr	1	FeCr, CoCr	270	10 300	32K6, 35K5, 43N1
Cu	≪1	...	...	...	36J4, 36M9
Mn	17	...	60	10 000	35K5, 45N1
Mo	3	Fe$_3$Mo$_2$, FeMo, CoMo	250	11 000	32K4, 39L2
Ni	25	...	40	10 000	See Chap. 5
Sb	2.5	FeSb$_2$, CoSb	...	...	39G3
Si	6	Fe$_3$Si, FeSi, FeCoSi, CoSi, Co$_2$Si	...	...	35V1
Sn	3	Fe$_2$Sn, FeSn, Co$_2$Sn, CoSn	106	...	35K8
Ta	1	Fe$_2$Ta, FeTa, Co$_5$Ta$_2$	<100	3 000	39K4
Ti	1.5	Fe$_2$Ti, Co$_2$Ti	...	...	32S6, 35K7
V	10	FeV	400	9 500	38K3, 46N1
W	4	Fe$_2$W, Fe$_3$W$_2$, CoW	150	11 500	32K5, 32S6

* At room temperature, second phase appears when third element is present to extent indicated.
† Includes ordered structures.

Other Ternary and Complex Alloys.—Brief reference to other alloy
systems, for which magnetic data exist or which show some promise for
investigation, is made in tabular form only, in Tables 13 to 16. References
there given may be followed for further information.

TABLE 14. SOME PROPERTIES OF TERNARY ALLOYS OF FE AND NI

Added Element	Extent (%) of α Phase* at 50% Ni	Compounds† Reported	High H_c	Corresponding B_r	References
None	...	...	35	7800	...
Ag	...	...	0.7	...	30G2
Al	16	FeAl, NiAl, Fe₂NiAl	475	7000	See Fe-Ni-Al Alloys
Be	...	NiBe	20	...	29K2
Cr	40	FeCr	130	7700	39S4, 45N6
Cu	4	...	440	5000	See Fe-Ni-Cu Alloys
Mn	...	Ni₃Mn, NiMn	55	8000	33K4, 45N2
Mo	13	Fe₃Mo₂, NiMo	...	...	34K5
Si	3	Fe₃Si, FeSi, Ni₃Si, NiSi	...	...	32D1, 38A3, 43G1
Sn	2	Fe₂Sn, FeSn, Ni₃Sn₂, NiSn	80	1500	39S7, 38L7
Ta	>1.5	Fe₂Ta, Ni₃Ta	...	...	BTL, 36G3
Ti	6	Fe₂Ti, Ni₃Ti	...	...	38V1
V	Large	FeV	75	9500	34K7, 45N5
W	8	Fe₃W₂, Ni₆W	...	...	32W1, 32S6

* At room temperature, second phase appears when third element is present to extent indicated.
† Includes ordered structures.

Powder Metallurgy.—Magnets of Alnico 2 (12% Co, 17% Ni, 10% Al, 6% Cu) have been made commercially by mixing powdered ingredients, pressing, and sintering. The magnetic properties of material thus obtained are usually somewhat inferior, and the cost per pound may be several times that of the cast alloys. However, the speicmen is stronger and magnetically more homogeneous, and it may be made to close tolerances without grinding, so that small pieces may be manufactured economically in this manner.

According to Howe [40H3, 40H4, 42H1] sintered Alnico is made as follows. A fine powder is prepared by crushing a cast iron-aluminum alloy (50% Fe, 50% Al), and mixing with it the required amount of the other powdered ingredients except aluminum, care being taken to avoid the presence of oxide. This is pressed into the required form in a die, using 10–30 tons/in.², and then sintered. If it is desired to machine the parts, they may first be sintered at 600°C. With or without such a preliminary treatment they are finally sintered at 1000–1400°C, preferably the latter, in pure dry hydrogen, and cooled at a rate of about 250°C/min. During the sintering, the volume decreases by 2–10% and the dies must, therefore, be designed with this in mind. Magnetic properties are listed in Table 3

TABLE 15. SOME PROPERTIES OF OTHER TERNARY ALLOYS CONTAINING FE, CO, NI OR MN

System	Magnetic Properties Reported	References
Co-Ni-Al	...	41S2
Co-Ni-As	...	12F1
Co-Ni-Cu	$H_c = 710$, $B_r = 3400$, $B_s = 4800$	See p. 402
Co-Ni-Si	...	40F1
Fe-Al-Cu	...	39B9
Fe-Al-Mn	$H_c = 40$	33K8, 36M10
Fe-Al-Si	High μ_0	36M8, 40T3
Fe-Cr-Cu	...	39M2
Fe-Cr-Mn	$H_c = 40$, $B_r = 7000$	33K2, 45N3
Fe-Cr-Mo	$H_c = 300$, $B_r = 7400$	31W1, 34S6*
Fe-Cr-Ti	...	32S6
Fe-Cr-W	$H_c = 210$, $B_r = 5500$	34S6
Fe-Cu-Mo	...	38D4
Fe-Mn-Ti	...	32S6
Fe-Mn-V	$H_c = 60$, $B_r = 11\,400$	*
Fe-Mo-V	$H_c = 240$, $B_r = 8200$	*
Fe-Ti-W	...	32S6
Fe-V-W	...	32K9
Co-Al-Mn	θ	38K8, 38K12
Co-Cu-Mn	$H_c = 40$	38K9
Co-Cu-Mo	...	24D1
Ni-Al-Mn	θ	38K10
Ni-As-Cu	θ	40K3
Ni-Cu-Mo	θ	24D1
Ni-Cu-Sb	θ	41S3
Mn-Ag-Al	$_IH_c = 6000$, $B_r = 600$	31P3

* Unpublished work by E. A. Nesbitt.

of Appendix 5. Magnets of various shapes have been described by Fahlenbrach [49F1].

Steinetz [46S7] has described the preparation of *Permet* (30% Co, 25% Ni, 45% Cu) from powders by sintering above the melting point of copper. It has the following properties: $H_c = 800$, $B_r = 2500$, $(BH)_m = 0.5 \times 10^6$.

Magnetic material of the Fe-Ni-Al type, with additions, may be cast, crushed, and consolidated under pressure with a resin binder. According to Dehler's description [42D1], particles having various sizes up to about 1 mm are mixed with 3–6% of a phenol or polyvinyl resin, pressed at 100°C with 15 000 lb/in.^{2}, and cured at 180°C. Packing factors of 80–90% are obtained. Although B_r and H_c are not far below the corresponding values of the cast material, the demagnetizing curve is rather flat (fullness factor $\gamma = 0.3$) and the energy product low. The material is made commercially under the name of *Tromolit*.

TABLE 16. SOME PROPERTIES OF COMPLEX ALLOYS

System	H_c	B_r	Name or Reference
Fe-Co-Ni-Al	440	7 300	Mishima, Alnicos 1 and 4
Fe-Co-Ni-Cu	260	7 300	Cunife 2
Fe-Co-Ni-Al-Cu	600	12 500	Alnicos 2 and 5
Fe-Co-Ni-Al-Ti	600	12 500	Ticonal
Fe-Co-Ni-Mn	120	7 000	45N4
Fe-Co-Ni-Ti	830	7 500	New Honda
Fe-Co-Cr-Mo	270	9 500	Remalloy 2*‡
Fe-Co-Cr-W	35	5 300	32S6
Fe-Co-Mn-V	100	10 400	*
Fe-Co-Mn-W	...	...	32S6
Fe-Co-Mo-W	...	...	32K8
Fe-Ni-Al-Cu	500	6 300	Alni
Fe-Ni-Al-Mn†	490	9 500	Mishima
Fe-Ni-Al-Ti	1050	6 500	New Honda
Fe-Ni-Cu-Mo	...	...	1040 alloy
Fe-Ni-Cu-V	160	1 500	43N2
Fe-Ni-Mn-Mo	...	...	High μ
Fe-Cr-Mo-W	260	6 600	34S6
Fe-Cr-Ti-W	90	4 500	34S6

* Unpublished work of E. A. Nesbitt.
† Also Fe-Ni-Al with additions of Cr, Cu, Mo, V or W.
‡ $(BH)_m = 1.2 \times 10^6$; 5% Co, 5% Cr, 17% Mo, rest Fe.

Magnetic properties of representative specimens of compacted powders are given in Table 17, where they are compared with specimens of cast Alnico 2 [42S1].

TABLE 17. MAGNETIC PROPERTIES OF SOME MAGNETIC POWDER COMPACTS

Material	Coercive Force H_c (oersteds)	Residual Induction B_r (gausses)	Maximum Energy Product $(BH)_m \times 10^{-6}$
Sintered Alnico 2	520	6900	1.43
Cast Alnico 2	560	7390	1.60
Pressed in resin:			
28% Ni, 13.5% Al	600	3500	0.63
19% Co, 18% Ni, 9% Al, 4% Cu, 4% Ti	800	4200	0.97

Powder Magnets.—Dean and Davis [41D1] have prepared permanent magnets by pressing fine powders formed by electrodeposition of metal on a surface of mercury. In this way a coercive force of over 150, with a residual induction as high as 9000, has been obtained in iron to which a small amount of zinc has been added. Using cobalt, magnets have been made with $H_c = 400$–500 and $B_r = 9000$–11 000, and by mixing iron, nickel, and aluminum, the values, $H_c = 550$, $B_r = 9700$, have been attained. Many other combinations of metals have also been used, including mixtures containing manganese but no iron, cobalt, or nickel. In preparing the powder, iron in ammonium sulfate solution is electrolyzed against a moving agitated surface of mercury containing some zinc. After many hours of deposition the slushy amalgam thus formed is pressed to remove much of the mercury and heated to 260°C in hydrogen to remove the remainder. The presence of zinc (or aluminum) appears to inhibit grain growth of the extremely fine particles and to render the mass non-pyrophoric. The material is then pressed at about 200 000 lb/in.² to form the magnet which is said to be mechanically strong and to have the properties given above.

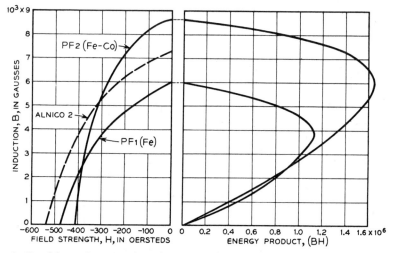

FIG. 9–63. Magnetic properties of magnets of iron powder, with or without cobalt.

Guillaud [43G2] observed that manganese bismuthide, MnBi, in the form of fine particles, had the unusually high coercive force of 12 000 oersteds (for $I = 0$). The theory of this effect, which is to be expected in fine particles of any material having high crystal anisotropy, has been developed by a number of physicists and is discussed in Chap. 18. Magnets of fine iron powder are manufactured in France by reducing iron formate or iron oxalate powder at a temperature of about 400°C, or low enough to prevent the growth of the metal particles, and pressing at about 80 000 lb/in.². Better magnetic properties are obtained when 30–35% of cobalt is added. The dry metallic powder is pyrophoric; therefore it is protected by a liquid such as benzol or acetone during the pressing, but in spite of this a certain amount of oxide is formed. Some formate of calcium or other metal reduceable with difficulty may be mixed with the iron formate to insulate the iron particles subsequently formed and to protect them somewhat from oxidation and sintering [48S15]. Powders can also be made by the carbonyl or the Raney process [47S6].

Demagnetization curves of pressed iron powder (grade PF1) and of iron powder containing cobalt (grade PF2) are shown in Fig. 9–63, taken from a description of these magnets by Steinitz [48S16].

Oxide Magnets.—A novel magnetic material for permanent magnets, described by Kato and Takei [33K7] in 1933, consists of a mixture of oxides of iron and cobalt, subjected to a specific heat treatment. This material has high coercive force and moderately low residual induction, an unusually high resistivity, and a density about half that of iron. It is produced and sold in the United States under the name *Vectolite*, and in France [48S13].

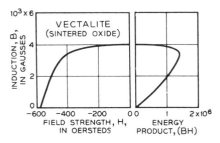

FIG. 9–64. Magnetic properties of the oxide magnet.

The material is fabricated [34H1] from a mixture of magnetite, Fe_3O_4, and cobalt ferrite, $CoO \cdot Fe_2O_3$, by pressing and sintering at a temperature of 1000°C, then magnetizing at about 300°C, and cooling to room temperature in the presence of the field. The oxidizing power of the atmosphere used during heating is apparently of considerable importance [36J3]. Coercive forces of 400–600 oersteds and residual inductions of 3000–5000 gausses are obtained, and a value of $_IH_c$ as high as 1000 has been reported. A demagnetization curve [33K7] is shown in Fig. 9–64. The energy product is limited by the low residual induction.

The magnetic hardening has been investigated in cobalt ferrite and the related ferrites of Cu, Mn, Ni, Mg, and Fe. Takei, Yasuda, and Ishihara [40T2] measured the magnetization of cobalt ferrite cooled rapidly from 850°C in the presence of a longitudinal or transverse field of 4000 oersteds, and in zero field. Differences in permeability were found ($\mu_m = 7, 3$, and 3.5, respectively), but no data on residual induction or coercive force were reported.

The relation of phase transformation to magnetic properties has been studied in some detail in the manganese ferrites by Snoek [36S6] and by Kussmann and Nitka [38K7]. The latter show that ferromagnetism is associated with the well-known cubic spinel structure possessed by $MnO \cdot Fe_2O_3$ and other similar compounds, including Fe_3O_4. Excess manganese favors the formation of tetragonal Mn_3O_4, and excess iron the formation of rhombohedral Fe_2O_3. The latter has a ferromagnetic spinel structure stable above about 400°C and transforms and becomes paramagnetic at room temperature, except when the high temperature form is maintained by rapid cooling. When the manganese and iron contents are intermediate, no change in structure occurs on cooling and hardening does not occur, but increase in manganese oxide beyond about 50% of the total results in an abrupt increase in coercive force [36S6] and a decrease in saturation [38K7]. The change in phase of a certain amount of the material, from cubic form to rhombohedral (precipitation), and the resultant lattice strain, may account for these changes in properties. It is believed that some such kind of hardening takes place in the Japanese cobalt ferrite magnets, but no specific data have been published.

The ferrites have resistivities that are extraordinary in materials that are ferromagnetic. Manganese ferrite may have a saturation of $B_s = 5000$ and at the same time a resistivity of 10^5 ohms-cm at room temperature [36S6]. Ferrites are also important as high permeability materials for use especially at high frequencies and are discussed in this connection in Chap. 8.

CHAPTER 10

MAGNETIC THEORY

FERROMAGNETISM

Ewing's Theory.—Ewing [00E1] was one of the first to attempt to explain ferromagnetic phenomena in terms of the forces between atoms. His theory will be described briefly here, since many physicists today, when thinking about magnetic phenomena, still go back to Ewing's ideas of fifty years ago. He assumed, with Weber, that each atom was a permanent magnet free to turn in any direction about its center. The orientations of the various magnets with respect to the field and to each other were supposed to be due entirely to the mutual magnetic forces. The magnetization curve and hysteresis loop were calculated for a linear group of such magnets and were determined experimentally using models having as many as 130 magnets arranged at the points of a plane square lattice.

The calculations for a linear chain show that as the field (oriented, e.g., at 135° to the line of magnets) is gradually increased in magnitude from zero there is at first a slow continuous rotation of the magnets, then a sudden change in orientation, and finally a further continuous rotation until the magnets lie parallel to the field. The I,H curves calculated for such a group of magnets resemble in general form the actual curves of iron: they show a permeability first increasing and then decreasing, saturation, and hysteresis.

A magnetization curve and hysteresis loops obtained [93E1] with a model of 130 magnets in square array are shown in Fig. 10–1. Experiments with the model showed a variety of other phenomena including rotational hysteresis loss and its reduction to zero in high fields, the effect of strain on magnetization, the existence of hysteresis in the strain vs magnetization diagram, the effect of vibration, and the existence of time lag and accommodation with repeated cycling of the field.

Ewing's general method may be illustrated by calculating the magnetization curve and hysteresis loop for an infinite line of parallel, equally spaced magnets [Fig. 10–2(a)]. It is done most simply by considering first the magnetic potential energy [29M2] of a magnet of moment μ_A and length l, in the field of a similar magnet:

$$W = -\frac{\mu_A^2}{r^3} P_2(\theta) - \frac{\mu_A^2 l^2}{r^5} P_4(\theta) - \cdots . \qquad (1)$$

Here r is the distance between the centers of the magnets, and the $P(\theta)$'s are Legendre functions [23J1] of the angle θ between the direction of the moment of the magnet and the line joining the magnet centers.

$$P_2(\theta) = (1 + 3 \cos 2\theta)/4$$

$$P_4(\theta) = (9 + 20 \cos 2\theta + 35 \cos 4\theta)/64.$$

The potential energy per magnet W_1 for an infinite straight row of magnets can easily be obtained by summing W for all pairs.

$$W_1 = -\frac{2\mu_A{}^2}{r^3}[1.202 P_2(\theta) + 1.038 P_4(\theta)(l/r)^2 + \cdots]. \qquad (2)$$

The behavior of the line when subjected to a field H may be found by adding to W_1 the energy term $-H\mu_A \cos(\theta_0 - \theta)$, where θ_0 is the angle

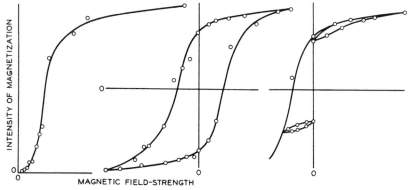

FIG. 10-1. Magnetization curve and hysteresis loops of a Ewing model of 130 pivoted magnets in square array.

between the line of centers and the direction of the field, and finding the value of θ which makes this total energy a minimum for given values of θ_0 and H:

$$\frac{d}{d\theta}[W_1 - H\mu_A \cos(\theta_0 - \theta)] = 0.$$

This gives

$$H = \frac{(d/d\theta)W_1}{\mu_A \sin(\theta_0 - \theta)}.$$

The component of magnetization parallel to H is

$$I = I_s \cos(\theta_0 - \theta),$$

where I_s is the saturation magnetization. By starting with half of the line of magnets pointing in a direction opposite to that of the other half, the initial magnetization is zero, and an unmagnetized or demagnetized material is simulated. Thus a magnetization curve and a hysteresis loop of an assemblage of this sort are obtained by plotting H against I. Such

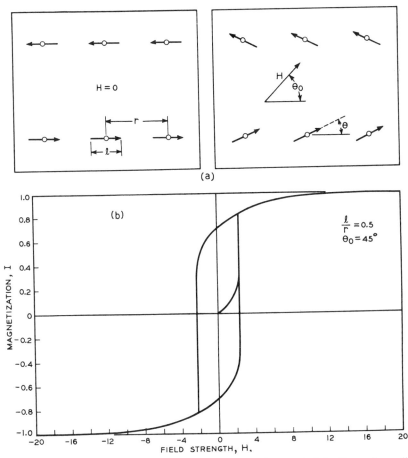

FIG. 10-2. Magnetization curve and hysteresis loop for two infinite lines of equally spaced magnets originally magnetized in opposite directions. The unit of field strength is here $\mu_A r^3/2$ or about 2700 oersteds for iron.

a plot is shown in Fig. 10-2(b), with the scale of H determined by the magnitudes of μ_A and r. The curves are obviously similar to those for real materials.

Limitations of Ewing's Theory.—So far, this calculation is equivalent to what Ewing did more than fifty years ago. But now we know the crystal structure of iron and, in particular, the distances between the atoms. We also know the magnetic moment of each iron atom and know, therefore, the value of μ_A/r^3 which determines the scale of H. Using the appropriate values of $\mu_A = 2.0 \times 10^{-20}$ erg/gauss and $r = 2.5 \times 10^{-8}$ cm, the coercive force H_c for a line of magnets having $l/r = 0.1$ is found to be 4600 oersteds. This calculated value of H_c is affected somewhat by the ratio l/r, but in any case it is found to be of this order of magnitude unless l/r is very close to unity. This magnitude of H is greater, by a factor of over a million, than the lowest value obtained experimentally. Similarly the initial permeability μ_0, according to the model, is about unity, whereas observed values for iron range from 150 to 20 000. Adjustment of l/r to higher values decreases μ_0. Thus, no satisfactory ratio l/r can be found that will give reasonable agreement with experiment, a conclusion reached finally by Ewing himself [22E2].

This calculation of the magnetization curve and hysteresis loop is based on a very much idealized model, and it is difficult to estimate the error to which it may lead. A much better approximation would be to calculate the magnetic potential energy of a group of bar magnets arranged in space in the same way that the iron (or nickel) atoms are arranged in a crystal. This was done by Mahajani [29M2] who showed that application of Eq. (1) with $l = 0$ (but summed to account for the effects of all magnets in the structure) leads to the result that the magnetic potential of the space array is independent of θ; in other words, one orientation of the dipoles is as stable as any other and the magnetization curve would go to saturation in infinitesimal fields no matter in what direction H might be applied. If l is finite, the stable positions of the magnets are parallel to the body diagonals of the cube which is the unit of the crystal structure, and this becomes, therefore, the direction of easy magnetization, a situation which is correct for nickel but decidedly not so for iron. The best correspondence between the action of the model and of iron itself is obtained if the model is made by placing a disk-shaped magnet, magnetized perpendicular to its surface (or a small, circular current of electricity), instead of a bar magnet, at each lattice point of the space array. In the latter case we can explain the direction of easy magnetization in iron and the variation of magnetic energy with direction in the crystal.

In considering Ewing's model it is appropriate to estimate the energy of thermal agitation and to compare it with the magnetic potential energy as calculated from the model. Substituting in Eq. (2) the same values of μ_A and r as were used above, we obtain 10^{-16} erg per atom for the magnetic potential energy in zero field. This is to be compared with the energy of a single molecule at room temperature, 2×10^{-14} erg per atom, as given by the kinetic theory. Thus the energy of thermal agitation is 200 times as great as the calculated magnetic energy. Even at liquid air temperatures the thermal agitation would prevent the atomic magnets from forming stable configurations. Without some additional force the model Ewing used would behave as a paramagnetic rather than a ferromagnetic solid, as was pointed out previously in general terms.

In a real material, however, it is now well established that there are very powerful forces, not contemplated when Ewing made his model and proposed his theory, which maintain the dipole moments of neighboring atoms parallel. These are the electrostatic forces of exchange (see p. 443) which Heisenberg suggested are powerful enough to align the elementary magnets against the strong disordering effect of thermal agitation. Theory accounts only for the order of magnitude of these forces. A rough estimate of the corresponding energy of magnetization is obtained by assuming that it is equal to the energy of thermal agitation at the Curie point, $k\theta/2$. For iron ($\theta = 1043°K$) this gives 7×10^{-14} erg per atom.

The Weiss Theory.—In order to understand how atomic forces give rise to ferromagnetism it is desirable first to review briefly Weiss's theory [07W1] of ferromagnetism, which introduces a so-called "molecular field" that presently will be associated with the nature of these forces. This theory is an extension of the classical theory of paramagnetism which was developed by Langevin. He investigated mathematically the behavior of an ensemble of elementary magnets (atoms), each of moment μ_A, in a field of strength H. The effect of the field is to cause alignment; the thermal agitation, to destroy this alignment. Assuming that the "molecules" are far enough apart so that their mutual forces can be neglected, the energy due to the field is

$$W = -\mu_A H \cos \theta$$

for each dipole oriented so that μ_A makes the angle θ with H. If the ensemble is subject to thermal agitation, the methods of statistical mechanics show that at any temperature, T, the number of dipoles oriented in the solid angle $d\omega$ about the direction θ, is proportional to

$$e^{-W/kT}d\omega = e^{(\mu_A H \cos\theta)/kT}d\omega,$$

according to Boltzmann. For all the molecules the average value of the moment in the direction of the field $\bar{\mu}_A$ is calculated and the ratio of this to the total moment, or the moment when all magnets lie parallel, is found to be

$$\frac{\bar{\mu}_A}{\mu_A} = \text{ctnh} \frac{\mu_A H}{kT} - \frac{kT}{\mu_A H},$$

or

$$\frac{I}{I_0} = \text{ctnh} \frac{\mu_A H}{kT} - \frac{kT}{\mu_A H}.$$

In deriving this, the assumptions are made (1) that the elementary magnets are subject to thermal agitation and momentarily may have any orientation with respect to the direction of the field, and (2) that they are too far apart to influence each other. Quantum theory alters the first of those assumptions by stating that in such an ensemble there will be only a limited number of possible orientations, in the simplest case only two, one parallel and the other antiparallel to the direction of the field. In this case the equation corresponding to Langevin's is

$$\frac{I}{I_0} = \tanh \frac{\mu_A H}{kT}. \tag{3}$$

These two theoretical relations are plotted for variable H and constant T (room temperature) in Fig. 10-3, the constants being those for iron

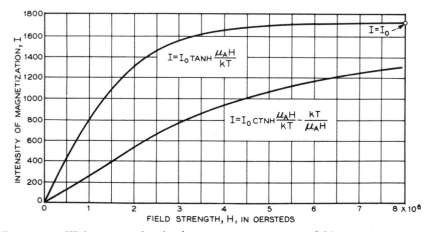

FIG. 10-3. With no mutual action between atoms, enormous fields are necessary to saturate a magnetic material.

($I_0 = 1740$, $\mu_A = 2.04 \times 10^{-20}$ erg/gauss). It is obvious that with the highest fields so far attained in the laboratory (about 300 000 oersteds) the magnetization would attain at room temperature only a small fraction of its final value I_0 if this law were obeyed, and in this range the intensity of magnetization would be sensibly proportional to the field strength:

$$I = \frac{CH}{T},$$

where C is a constant. This relation, known as Curie's Law, is obeyed by some paramagnetic though not by ferromagnetic substances. It is usually written:

$$\chi = \frac{C}{T}.$$

Many more paramagnetic substances obey the similar "Curie-Weiss Law":

$$I = \frac{CH}{T - \theta}. \qquad (4)$$

Weiss pointed out the significance of θ in this equation: it means that the material behaves magnetically as if there were an additional field, NI, aiding the true field H. This equivalence is shown mathematically by putting $\theta = NC$ in Eq. (4) with the result

$$I = \frac{C(H + NI)}{T}.$$

The quantity represented by NI is called the "molecular field," and N is the "molecular field constant." It is interpreted by supposing that the elementary magnet does have an influence on its neighbors, contrary to the assumptions of the simple Langevin theory.

The significance of the molecular field for ferromagnetism is now apparent if we replace the H by $H + NI$ in the more general Eq. (3) and examine the resulting equation:

$$\frac{I}{I_0} = \tanh \frac{\mu_A (H + NI)}{kT}. \qquad (5)$$

This equation is one of the most important in the theory of ferromagnetism. Putting $H = 0$ and

$$\theta = \mu_A N I_0 / k,$$

Eq. (5) reduces to

$$\frac{I}{I_0} = \tanh \frac{I/I_0}{T/\theta}. \qquad (6)$$

This purports to specify the magnetization at zero applied field by a function that is the same for all materials, when the magnetization is expressed as a fraction of its value at absolute zero and the temperature as a fraction of the Curie temperature on the absolute scale. This magnetization vs temperature relation, plotted as the solid line of Fig. 10–4, means that *at all temperatures below θ the intensity of magnetization has a definite value even when no field is applied.*

How is it then that a piece of iron can apparently be unmagnetized at room temperature? The answer, given by Weiss, is that below the Curie point different parts of the iron are magnetized in different directions so that the over-all effect is zero. This is the concept of the *domain*, already discussed in some detail. According to this conception the I of Eq. (5) is that of a domain and is determined experimentally by measuring the magnetization of a specimen when all domains are parallel, i.e., at (technical) saturation ($I = I_s$). Eq. (6) should then be written

$$\frac{I_s}{I_0} = \tanh \frac{I_s/I_0}{T/\theta}. \tag{7}$$

Quantum Theory.—It has already been mentioned that quantum theory alters the Langevin assumption of random distribution and introduces a series of discrete orientations that can be assumed by the elementary magnet in the presence of a field. Another result of the quantum theory is that it supplies a natural unit of magnetic moment, equal to the magnetic moment of a single electron spin. This *Bohr magneton* is

$$\beta = eh/4\pi mc = 9.27 \times 10^{-21} \text{ erg/gauss}$$

with e the charge, m the mass of the electron, h Planck's constant, and c the velocity of light. The magnetic moment of one gram-atom of an element containing one Bohr magneton per atom is

$$N_0 \beta = 5587,$$

N_0 being the number of atoms per gram-atom. The number n_0 of Bohr magnetons per atom can then be determined easily from the saturation moment per gram at 0°K, σ_0, and the atomic weight A:

$$n_0 = \frac{\sigma_0 A}{N_0 \beta}.$$

In general, the magnetic moment* of an atom is due to the resultant of the electron spin and the orbital motion of the components and is

$$\mu_A = Jg\beta$$

where J is a quantum number that may have only half-integral values, $0, \frac{1}{2}, 1, \frac{3}{2}, \cdots$. When the moment is due to electron spin only, $g = 2$; when it is due to orbital motion alone (the spins compensating each other), $g = 1$. In any case,

$$g = \frac{2mc}{e} \cdot \frac{\text{magnetic moment}}{\text{angular momentum}}$$

and its resolution from spin and orbital components is given by the equation on p. 451, where the gyromagnetic ratio is discussed in more detail. In ferromagnetic substances generally almost all of the moment is due to the spin so that g can be taken as about 2. The measured value for iron is about 1.94.

* The expression for the moment, given below, is actually for the maximum magnetic moment resolved parallel to the field. The absolute value of the moment is $g\beta\sqrt{J(J+1)}$.

The number of orientations of the magnetic moment with respect to the magnetic field is $2J + 1$. Taking this into account, the quantum-

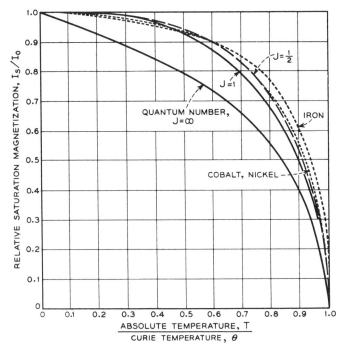

FIG. 10–4. Temperature dependence of the saturation magnetization of iron, cobalt and nickel, as compared with theory.

mechanical analog of the Langevin equation, known as the Brillouin function, may be written [32V3] as follows:

$$\frac{I_s}{I_0} = \frac{2J+1}{2J} \text{ctnh} \frac{(2J+1)a}{2J} - \frac{1}{2J} \text{ctnh} \frac{a}{2J} \qquad (8)$$

where

$$a = \frac{Jg\beta(H + NI_s)}{kT}.$$

In ferromagnetic substances H can usually be neglected in comparison with NI_s. When there is a large number of uncompensated spins per atom (large J), the Brillouin function approaches the classical expression. When $J = \frac{1}{2}$, appropriate for a single electron spin, the tanh formula, already given, is obtained.

Curves for $J = \frac{1}{2}, 1$, and ∞ are plotted in Fig. 10–4, H being neglected in comparison with NI_s. The Curie temperature is

$$\theta = (J + 1)g\beta NI_0/3k. \qquad (9)$$

If we assume that $g = 2$ and $J = \frac{1}{2}$, then

$$\theta = \beta N I_0 / k$$
$$a = \beta N I_s / kT.$$

Below the Curie temperature the data are in pretty good agreement with these assumptions.

Ferromagnetics Above the Curie Point.—All materials that exhibit ferromagnetism are paramagnetic when they are heated above the Curie

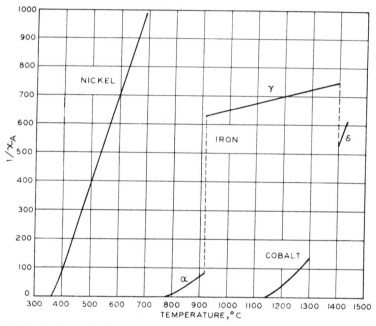

FIG. 10–5. Reciprocal of atomic susceptibility χ_A of iron, cobalt and nickel above their Curie points.

temperature. As the temperature continues to increase, the susceptibility decreases continually according to the Curie-Weiss law for strongly paramagnetic substances, except where there is a change in the phase structure of the material. If the Curie-Weiss law were followed exactly, a straight line would be obtained when $1/\chi$ is plotted against the temperature. The data for iron, cobalt, and nickel (Fig. 10–5) show some variations from the expected linear relations, for reasons not well understood. The breaks in the curve for iron are caused by changes in its crystal structure at the indicated temperatures.

One should be able to calculate from the slope of each of these lines the magnetic moment per atom, and compare this with the moment calculated from the saturation magnetization determined at lower temperatures when the material is ferromagnetic. In paramagnetic materials the measured intensity of magnetization is such a small fraction of saturation that $I \ll I_0$. Then the Brillouin function (8) may be written in series form of which the first term will suffice:

$$\frac{I}{I_0} = \frac{J+1}{3J} a = \frac{(J+1)g\beta(H+NI)}{3kT}.$$

Instead of I/I_0 we may use σ_A/σ_{A0}, the ratio of the measured magnetic moment per gram-atom to the moment that would exist if all of the elementary magnets were aligned parallel. Then the atomic susceptibility is

$$\chi_A = \frac{\sigma_A}{H} = \frac{(J+1)g\beta\sigma_{A0}}{3k(T-\theta)} = \frac{C_A}{T-\theta}, \quad (10)$$

C_A being the Curie constant per gram-atom,

$$C_A = \frac{(J+1)g\beta\sigma_{A0}}{3k}. \quad (11)$$

For nickel the value of C_A, equal to 0.325, is obtained from the slope of the $1/\chi_A$ vs T curve of Fig. 10–5. If we assume that $g = 2$, in close accord with direct measurement, and $J = \frac{1}{2}$, then $\sigma_{A0} = 4850$. This corresponds to 0.87 Bohr magneton per atom, a number much larger than the 0.61 found for the ferromagnetic state. Mott and Jones [36M13] believe that the assumption $J = \frac{1}{2}$ is at fault and suggest, as the best way of explaining the data, that $J = 0$ for some atoms and $J = 1$ for others. If there are 70% of the former and 30% of the latter, theory indicates a paramagnetic Bohr magneton number of 0.81, which is near the observed value 0.87. This would mean that some atoms have no electrons contributing to the paramagnetic susceptibility, and that the others have two electrons each, acting coherently.

Molecular Field Constant, N.—As seen in Fig. 10–4, the curve for $J = \frac{1}{2}$ is in best agreement with the data for variation of saturation magnetization with temperature, below $T = \theta$.

A further indication of the distribution of electrons may be obtained from the Curie points, which are related to the molecular field and J according to Eq. (9). Assuming again that $g = 2$, we can calculate N under the assumption that $J = \frac{1}{2}$ or $J = 1$, with the results shown in Table 1.

In order to select the most satisfactory values of N and J, values of N for nickel derived from various sources are collected in Table 2. Here the most direct determination comes from the magnetocaloric effect [34P2] to be described in more detail in Chap. 15.

No completely satisfactory values of J and N can be selected. Below the Curie point the saturation vs temperature curve is in best agreement with $J = \frac{1}{2}$, while the susceptibility above the Curie point is best explained

TABLE 1. MOLECULAR FIELD CONSTANT, N, AND MOLECULAR FIELD AT 0°K, NI_0
(Calculated from Curie point using the Weiss theory as modified by quantum theory)

Element	θ	N for		$NI_0 \times 10^{-6}$ for	
		$J = \frac{1}{2}$	$J = 1$	$J = \frac{1}{2}$	$J = 1$
Fe	1043	8 950	6 700	15.5	11.6
Co	1393	14 400	10 800	20.7	15.5
Ni	631	18 500	13 800	9.4	3.6

by $J = 1$. As experiments on the magnetocaloric effect show, the molecular field changes with temperature in the vicinity of the Curie point and probably more slowly at other temperatures as well. Further consideration has been given to this question by Smoluchowsky [41S4], who has taken into account the variation of saturation magnetization and molecular

TABLE 2. VALUES OF WEISS MOLECULAR FIELD CONSTANT N DETERMINED
IN VARIOUS WAYS ON VARIOUS ASSUMPTIONS FOR NICKEL

Method	Value of N
Curie point and $J = \frac{1}{2}$	18 500
Curie point and $J = 1$	13 800
Susceptibility above Curie point*	13 400
Magnetocaloric effect, $T > \theta$	11 000
Magnetocaloric effect, $T < \theta$ (300°C)	5 000

* $N = \theta/C$, $C = \rho C_A/A$, where ρ is the density and A the atomic weight; θ is the "ferromagnetic" Curie point.

field constant with temperature. He concludes, for reasons given in detail in the original article, that theory and experiment are more nearly in accord with $J = 1$ than with $J = \frac{1}{2}$, in iron.

A further difficulty in the theory of the molecular field is the implicit assumption that the field has the same value throughout the material. Néel [37N3] has stressed the fact that there will actually be local fluctuations in the strength of the field, and he believes that such fluctuations may account for the apparent differences between the molecular field in the ferromagnetic and paramagnetic states.

Before discussing further the nature of the molecular field it is desirable to review some of our knowledge of the structure of the atoms with which we are concerned.

Atomic Structure of Ferromagnetic Materials. — The structure of an isolated iron atom is shown schematically in Fig. 10–6. The 26 electrons are divided into 4 principal shells, each shell a more or less well-defined region in which the electrons move in their orbits. The first (innermost) shell contains 2 electrons, the next shell 8, the next 14, and the last 2. As the periodic system of the elements is built up from the lightest element, hydrogen, the formation of the innermost shell begins first, and when completed the numbers of electrons in the first four shells are 2, 8, 18, and 32, but the maximum number in each shell is not always reached before the next shell begins to be formed. For example, when formation of the fourth shell begins, the third shell contains only 8 electrons instead of 18; it is the subsequent building up of this third shell that is intimately connected with ferromagnetism. In this shell some electrons will be spinning in one

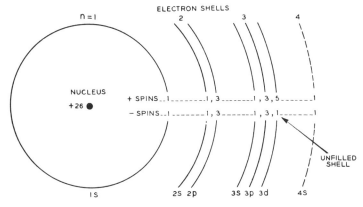

FIG. 10-6. Electron shells in an iron atom, showing location of inner unfilled $3d$ shell responsible for magnetic moment. Electrons of outer $4s$ shell of isolated atom become "free" electrons in a metal.

direction and others in the opposite, and these two senses of the spins may be conveniently referred to as positive and negative. The numbers on the circles show how many electrons with $+$ and $-$ spins are present in each shell in iron, and it will be noticed that all except the third shell contain as many electrons spinning in one direction as in the opposite. The magnetic moments of the electrons in each of these shells mutually compensate one another so that the shell is magnetically neutral and does not have a permanent magnetic moment. In the third shell, however, there are 5 electrons with a positive spin and one with a negative so that 4 electron spins are unbalanced or uncompensated and there is a resultant polarization of the atom as a whole. The existence of a permanent magnetic moment for each atom obviously satisfies one of the requirements for ferromagnetism.

In the free atom the orbital motions of the electrons also contribute to the magnetic moment. When the iron atom becomes part of metallic iron, the electron orbits become too firmly fixed in the solid structure to be influenced very much by a magnetic field. The corresponding orbital moments do not change when the intensity of magnetization changes—this is shown by the gyromagnetic experiments discussed later—and it is supposed that the orbital moments of the electrons in various atoms neutralize one another.

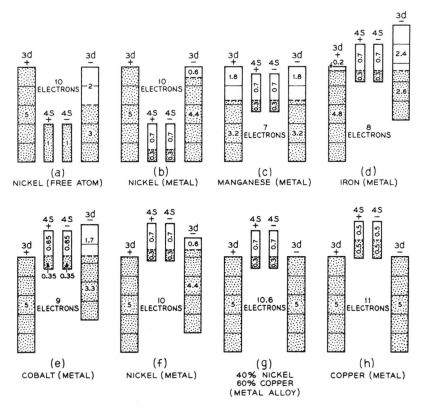

Fig. 10-7. Distribution of electrons among the possible electron positions in a free atom of nickel, and in manganese, iron, cobalt, nickel and copper atoms that form part of a metal.

In the solid structure, neighboring atoms influence the motion and distribution of electrons, particularly in the third shell ($3d$ shell) and the first part of the fourth shell ($4s$ shell). Figure 10-7 illustrates the difference between a free atom and one that is part of a metal. Each of the ten places for electrons in the $3d$ shell is represented by an area which is shaded if that place is occupied. The distribution corresponds in (a) to an isolated atom of nickel, in (b) to a nickel atom in a metal; in the latter situation there is on the average 0.6 electron per atom in the $4s$ shell

ATOMIC STRUCTURE OF FERROMAGNETIC MATERIALS

(these electrons are loosely bound and are the free electrons responsible for electric conduction) and a vacancy or hole of 0.6 electron per atom in the $3d$ shell. In the older atom-model we should have to say that part of the time the $3d$ shell was completely full, whereas part of the time one or more electrons were missing. On the newer approach to the theory of the metallic state, initiated by Bloch [28B2, 36M13], the electrons are treated as traveling waves and their connection with individual atoms is lost. It is the latter picture that is now followed here.

Corresponding to the electron shells, which are spatial configurations, are energy bands, in which the number of electrons is plotted against the energy which they possess.

In nickel, then, as shown in Fig. 10–7(b), the $4s$ bands contain an equal number of electrons with + and − spins, whereas in the $3d$ bands all of the spaces for + spins are filled and there is a vacancy of 0.6 in the − band. The difference between the numbers of + and − spins is equal to the net magnetic moment per atom and is determined directly from the saturation magnetization at 0°K.

In Fig. 10–7(f) the diagram for nickel is repeated, this time with the tops of the unfilled positions on the same level to bring out an analogy with the filling of vessels with water. Diagrams for manganese, iron, cobalt, nickel, and copper (and a Ni-Cu alloy) are shown in parts (c) to (h). In each case the 18 electrons in closed shells are not shown. In iron the situation is somewhat different from that in nickel; neither the $3d+$ nor the $3d-$ shell is filled. This is deduced from the assumed relative constancy of the number of electrons in $4s$, from the excess of holes in $3d+$ over those in $3d-$ ($n_0 = 2.2$), and from the total number, 26, of extranuclear electrons.

The distribution in space of electrons belonging to the $3d$ and $4s$ shells is known approximately and is depicted for iron in Fig. 10–8. In (a) the ordinate shows the number of electrons there are at various distances from the nucleus. The $3d$ shell is thus seen to be a rather dense ring of electrons, as contrasted with the $4s$ shell which extends farther from the nucleus, so far that in the solid the shells of neighboring atoms overlap considerably. Half of the internuclear distance in solid iron is indicated by the line R.

In Fig. 10–8(b) the number of electrons having energy between E and $E + dE$ is plotted against the energy E; this band representation is similar to that of Fig. 10–7, but now the squares and rectangles are replaced by the more appropriate curved surfaces. If (b) is turned $90°$ relative to (a) the two pairs of curves bear some resemblance to each other. This is so because the energy of binding is generally less at greater distances from the nucleus. The $3d+$ level is represented as lower in energy than the $3d-$ since one of these bands is preferred.

The line "Fe" in Fig. 10-8(b) represents the limit to which the $3d$ and $4s$ shells are filled in iron; neither $3d+$ nor $3d-$ is completely full. The lowest energy levels are filled first, and the picture is drawn so that the

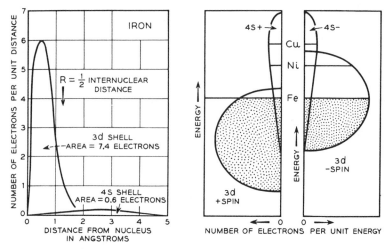

FIG. 10-8. Distribution of electrons in the $3d$ and $4s$ shells in iron, and the filling of the corresponding energy bands in iron, nickel and copper.

analogy with the filling of connected vessels with water is apparent. In cobalt and nickel the extra one and two electrons completely fill $3d+$ but not $3d-$, as indicated by the line "Ni" for nickel. On account of the

TABLE 3. NUMBER OF ELECTRONS AND VACANCIES (HOLES) IN VARIOUS SHELLS IN METAL ATOMS NEAR IRON IN THE PERIODIC TABLE

Element	Number of Electrons in Following Shells:				Total	Holes in:		Excess Holes in $3d-$ over $3d+$
	$3d+$	$3d-$	$4s+$	$4s-$		$3d+$	$3d-$	
Cr	2.7	2.7	0.3	0.3	6	2.3	2.3	0
Mn	3.2	3.2	.3	.3	7	1.8	1.8	0
Fe	4.8	2.6	.3	.3	8	0.2	2.4	2.22
Co	5	3.3	.35	.35	9	0	1.7	1.71
Ni	5	4.4	.3	.3	10	0	0.6	0.60
Cu	5	5	.5	.5	11	0	0	0

shapes of the bands it can be seen that the additional electrons do not alter greatly the number in $4s$, and from the saturation intensity of nickel we estimate this number as 0.6. In copper the additional electron is sufficient to fill both $3d$ shells with one electron to spare, and this electron must go into the $4s$ shell which then becomes half full as shown by the line "Cu," and also by (h) of Fig. 10-7. The diagram does not show changes in the relative levels of the $3d+$ and $3d-$ bands that occur in going from one element to another; when both $3d$ bands are filled, as in copper, these

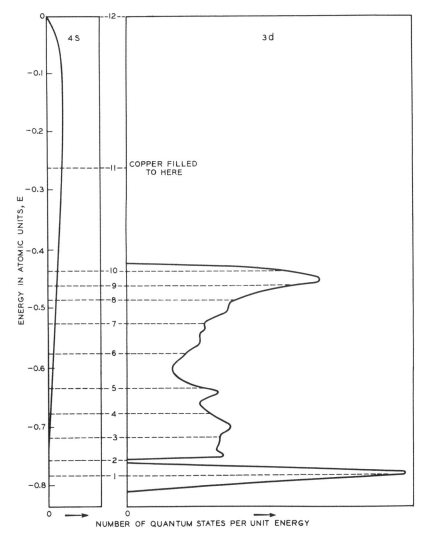

Fig. 10–9. Energy levels in the $3d$ and $4s$ bands of copper, as calculated by Slater and Krutter.

levels are the same. The numbers of electrons and "holes" in metals near iron in the periodic table are given in Table 3.

A more accurate estimation of the form of the $3d$ and $4s$ bands for copper is given in Fig. 10–9 derived by Slater [36S5] from the work of Krutter [35K4].

440 MAGNETIC THEORY

An especially simple and interesting illustration of the atom model described is afforded by the alloys of nickel and copper, as was first pointed out by Stoner [33S3]. The substitution of one copper for one nickel atom in the lattice is equivalent to adding one electron to the alloy. This electron seeks the place of lowest energy in the alloy and finds it in the

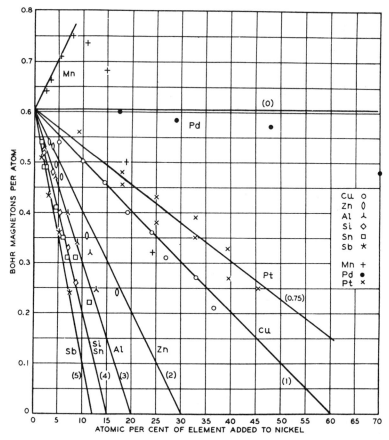

Fig. 10–10. The saturation magnetization of nickel as affected by the addition of other elements having 1, 2, 3, ... electrons in the outmost shell.

$3d-$ shell of a nickel atom rather than in the copper atom to which it originally belonged. This lowers the magnetic saturation of the alloy by one Bohr unit, since the added electron in the $3d-$ band just neutralizes the moment of one in the $3d+$ band. Addition of more copper to nickel decreases the average moment until the empty spaces in the $3d-$ band are just full; this occurs when 60% of the atoms are copper (Fig. 10–7g) and then the magnetic saturation at 0°K will be just zero. This is the

ATOMIC STRUCTURE OF FERROMAGNETIC MATERIALS 441

explanation given by Stoner [33S3], of the experimental results of Sadron [32S3] and Marian [37M2] shown in Fig. 10–10. There are shown also the saturation moments for other alloys of nickel; it is evident that zinc with two $4s$ electrons fills up the $3d$ band twice as fast as copper, aluminum three times as fast, silicon and tin four times, and antimony five, in good accord with theory. In each of these cases the added atoms have filled up $3d$ bands, losing their more loosely bound $4s$ electrons when there are available places of lower energy. The data for palladium indicate that this element has the same number of outer electrons as nickel, as might be expected since palladium lies directly below nickel in the periodic table. When the similar but heavier platinum is added to nickel, the decrease in

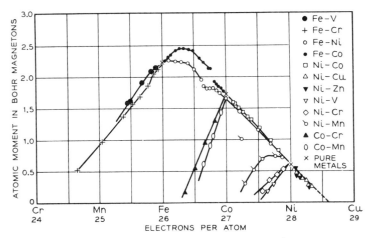

FIG. 10–11. Saturation magnetization as dependent on electron concentration. Data by Peschard [25P1], Weiss, Forrer, and Birch [29W3], Forrer [30F4], Sadron [32S3], Fallot [36F1, 38F1], Farcas [37F5], Marian [37M2], and Guillaud [44G1].

average atomic moment indicates that some of the outer electrons of platinum go into the $3d$ band of nickel, but that they do not fill this level as rapidly as the outer electrons of copper do when this element is added.

Variation of atomic moment with the number of electrons in the $3d$ and $4s$ shells is shown graphically in another way in Fig. 10–11, patterned after figures of Slater [37S8] and Shockley [39S1]. According to the theory just described, one expects that in alloys the average moment per atom will depend only on the number of electrons per atom, and this expectation is here shown to be reasonably well fulfilled when the alloying atoms are only one or two atomic numbers apart, as in the series Ni-Cu and Fe-Ni. Large departures are noted when the atomic numbers of the component atoms differ by 4 or 5, as in Cr-Ni and V-Ni.

Pauling [38P1] concludes from the form of Fig. 10–11 that the $3d$ band is broken into two parts, the upper one capable of holding 2.4 and the lower one 2.6 electrons of each spin, per atom. If we start with the $3d$ bands full and decrease the electron concentration, the upper band of one spin will empty first, the magnetic moment thereby increasing, and then the upper band of opposite sign will begin to empty and the magnetic moment consequently begin to decrease. The size of the bands is designed to explain the maximum moment of about 2.4 Bohr magnetons per atom that occurs in the iron-cobalt alloys.

Electron shells that are completely filled behave more like hard elastic spheres than those which are only partially filled. In solid copper with a complete $3d$ shell and a $4s$ shell just begun, the $4s$ electrons "overlap" those of neighboring atoms so much that their connection with any one atom is practically lost; the $3d$ shells, on the contrary, have very little overlap with neighboring atoms. In the ferromagnetic metals the $3d$ shells are incomplete and their overlap is greater than in copper; this affects the interaction responsible for the Weiss molecular field, to be given in a later section. But copper would not be ferromagnetic even if the interaction were large, because the completed shell means that the saturation magnetization is zero. In reality, copper is diamagnetic.

Collective Electron Ferromagnetism.—The band picture just described has been used by Stoner [33S3, 38S8] in developing a quantitative theory of the variation of saturation magnetization with temperature below the Curie point and of susceptibility with temperature above the Curie point. The theory is based on the general form of the energy bands worked out by Mott and Jones [35M1, 36M13] and follows the work of Slater [36S5, 36S12]. He assumes that the exchange interaction can be represented by a molecular field that is proportional to the magnetization, as it is in the Weiss theory. In the band model (Fig. 10–8) this means that the $+$ and $-$ bands are displaced vertically by an amount proportional to the difference in the number of electrons in the bands. He assumes also that the shape of the band at the top is parabolic in form. The problem is then to find the distribution of electrons with minimum total energy as the temperature is increased from absolute zero to the Curie point and above. The Fermi energy and spread in energy due to the thermal motion of electrons, which is associated with the increase in temperature, are balanced against the energy of exchange interaction until equilibrium occurs. Stoner uses $k\theta'$ as a (parametral) measure of the exchange energy, and ϵ_0 as the energy of the top of the unfilled band when the exchange is zero.

The variation of magnetization (I_s/I_0) with temperature (or T/θ) is shown in Fig. 10–12 as calculated, with some difficulty, for various values of $k\theta'/\epsilon_0$. The value $k\theta'/\epsilon_0 = \infty$ corresponds to the modified Weiss law for $J = \frac{1}{2}$. The curves at the right show the variation of the reciprocal mass susceptibility $(1/\chi)$ with temperature above the Curie point; in accordance with experiment, these lines (except the uppermost) are slightly concave upward.

For further consideration of the collective electron theory reference may be made to the original papers by Stoner and by Wohlfarth [49W5, 49W6] and to reviews by Van Vleck [45V1] and Stoner [48S3].

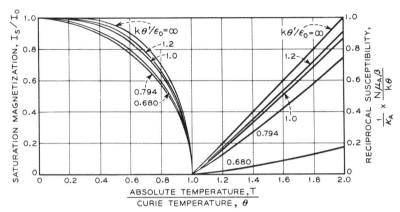

Fig. 10–12. Saturation magnetization as dependent on temperature up to the Curie point ($T/\theta = 1$), as calculated by Stoner using his "itinerate electron" model. Parameter $k\theta'/\epsilon_0$ depends on exchange integral and Fermi energy. Above $T = \theta$, ordinate is inversely proportional to atomic susceptibility.

Interpretation of the Molecular Field.—It was shown by Heisenberg [28H1] that the molecular field postulated by Weiss can be explained in terms of the quantum mechanical forces of exchange acting between electrons in neighboring atoms. Imagine two atoms some distance apart, each atom having a magnetic moment of one Bohr magneton due to the spin moment of one electron. A force of interaction has been shown to exist between them, in addition to the well-known electrostatic and (much weaker) magnetic forces. It is known that, as one would expect, such forces are negligible when the atoms are two or three times as far apart as they are in crystals. It is supposed also, on the basis of rough calculations by Bethe [33B2], that as two atoms are brought near to each other from a distance these forces cause the electron spins in the two atoms to become parallel (positive interaction). As the atoms are brought still nearer together the spin-moments are held parallel more firmly until at a certain distance the force diminishes and then becomes zero, and with still closer approach the spins set themselves antiparallel with relatively strong forces (negative interaction, antiferromagnetism). In the curve of Fig. 10–13 the energies corresponding to these forces are shown in schematic form as a function of the distances between atoms.

Bethe's curve was drawn originally for atoms with definite shell radii and varying internuclear distances. It may equally well be used for a series of elements if we take account of the different radii of the shell in which the magnetic moment resides. The criterion of overlapping or interaction for the metals of the iron group is the radius R of the atom (half the internuclear distance in the crystal) divided by the radius r of the $3d$ shell. In Fig. 10–13 this ratio R/r has been used as abscissa and the elements iron, cobalt, and nickel have been given appropriate positions on the curve. The ferromagnetism of gadolinium, discovered in recent years, is apparently associated with a large R/r and small interaction as compared to nickel, and it is placed on the curve accordingly. Slater [30S3] has shown that the ratio R/r is larger in the ferromagnetic elements than in other elements having incomplete inner shells, and that the point at which the curve crosses from the nonferromagnetic to the ferromagnetic region

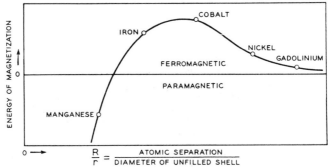

FIG. 10–13. Bethe's curve relating the exchange energy of magnetization to the distance between atom centers, with fixed diameter of active shell.

is near $R/r = 1.5$. Values of $2R$, $2r$, and R/r, as calculated by Slater for some of the elements with incomplete inner shells, are given in Table 4. More recent calculations of $3d$ electron shells [38M3] indicate that his values of r are somewhat too large, but his conclusion that R/r exceeds a certain value in ferromagnetic materials remains valid. Néel [37N3] has used $R - r$ instead of R/r as abscissa.

The positions of the various elements on the curve of Fig. 10–13 should not be regarded as definite, since the energy of interaction varies with the kind of structure and particularly with the number of neighbors, and in reality a different curve should be drawn for each structure.

The energy of interaction J_1—the positive ordinate of Fig. 10–13—has been related [30S2, 40B3] to the Curie temperature θ, on the absolute scale, and to the Weiss molecular field constant N, by the equation:

$$J_1 = \beta N I_0/z = k\theta/z$$

in which z is the number of nearest neighbors (with which interaction occurs) and β is the Bohr magneton. The general form of the interaction

TABLE 4. INTERNUCLEAR DISTANCES (2R), AND DIAMETERS (2r) OF INCOMPLETE INNER SHELLS, OF SOME ATOMS, IN ÅNGSTROMS [SLATER, 1930]

Atom	$2R$	Inner Shell $2r$	Ratio R/r	Incomplete Inner Shell	Curie Temperature θ, °K
Mn	2.52	1.71	1.47	$3d$	
Fe	2.50	1.58	1.63	$3d$	1040
Co	2.51	1.38	1.82	$3d$	1400
Ni	2.50	1.27	1.97	$3d$	630
Cu-Mn	2.58	1.44	1.79	$3d$	600
Mo	2.72	2.94	0.92	$4d$	
Ru	2.64	2.33	1.13	$4d$	
Rh	2.70	2.11	1.28	$4d$	
Pd	2.73	1.93	1.41	$4d$	
Gd*	3.35	1.08	3.1	$4f$	290
W	2.73	3.44	0.79	$5d$	
Os	2.71	2.72	1.02	$5d$	
Ir	2.70	2.47	1.09	$5d$	
Pt	2.77	2.25	1.23	$5d$	

* Calculated using Slater's formula.

curve is substantiated in a qualitative manner by the observed variation of the Curie points of the iron-nickel alloys, shown in Fig. 10–14. The maximum in the curve near 70% nickel apparently corresponds to the maximum of the interaction curve of Fig. 10–13. In alloys of higher nickel content the curve indicates that the Curie point should be increased if the material is compressed. The opposite should be true of the face-centered alloys having less than this amount of nickel. These predictions are borne out by the fact, observed by Steinberger [33S2], that under a hydrostatic pressure of 10 000 atm the 30% nickel alloy becomes practically non-ferromagnetic at room temperature (permeability is independent of field strength and equal to 1.7). On the contrary, the effect of the pressure on the phase equilibrium is unknown so that the data might be explained also by a change of phase brought about by the change of pressure. More data are needed to clarify the theory.

P. R. Weiss [48W2], working with Van Vleck [39W3], has extended the Bethe-Peierls procedure for calculation of order-disorder phenomena to the quantum-mechanical calculation of the molecular field for various 1-, 2-, and 3-dimensional arrays. He takes into account specifically the interaction of each atom with its nearest neighbors and by approximation the effect of atoms at greater distances. One result of this work, which is in agreement with the earlier work of Bloch [30B3], is that a 3-dimensional structure appears to be necessary for ferromagnetism. This is contrary to the works of Kramers and Wannier [41K3] and Onsager [44O1], which are based on the less rigorous Ising model [25I1] and lead to the conclusion that a 2-dimensional array may be ferromagnetic. Another result of Weiss is that a simple cubic lattice may be ferromagnetic, in contradiction

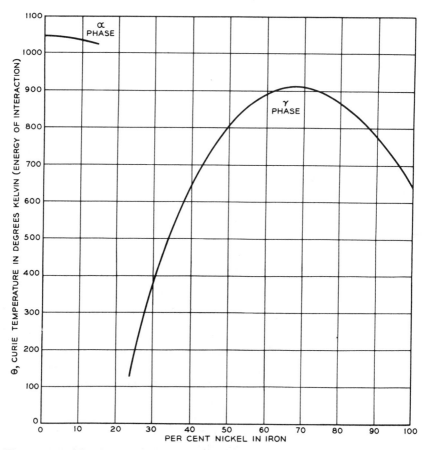

FIG. 10–14. The Curie temperatures of iron-nickel alloys, showing maximum corresponding to the maximum of Bethe's curve.

to the result of Heisenberg [28H1] which indicated that eight nearest neighbors were required. Results of the calculations are expressed in terms of the interaction energy J_1 and the Curie temperature θ for the various structures. Values of $k\theta/J_1$ are given in Table 5. It is to be noted that $k\theta/J_1$ is not equal to z, as assumed in the first approximation.

An excellent summary of the present status of the calculation of the molecular field is given by Van Vleck [45V1].

TABLE 5. VALUES OF $k\theta/J_1$ FOR VARIOUS ARRAYS OF DIPOLES, ACCORDING TO QUANTUM-MECHANICAL CALCULATIONS OF P. R. WEISS [48W2]

Comparison is made with earlier, more approximate calculations of Heisenberg. In the table, z is the number of nearest neighbors, (a) refers to Stoner's method of estimating J_1 and θ [30S2], using $Jg = 1$, and (b) to Heisenberg's more detailed method [28H1].

		Values of $k\theta/J_1$		
Array	z	(a)	(b)	Weiss
Linear chain.........	2	1	0	0
Square net...........	4	2	0	0
Hexagonal net.......	6	3	0	0
Simple cubic lattice...	6	3	0	1.85
Body centered lattice..	8	4	2	2.90 ($J = \tfrac{1}{2}$)
				6.66 ($J = 1$)

Thermal Expansion Near $T = \theta$.—There is an anomalous expansion of the high-nickel alloys (due to loss of magnetism) as the alloy is heated through the Curie point, a contraction of the low-nickel alloys, and no

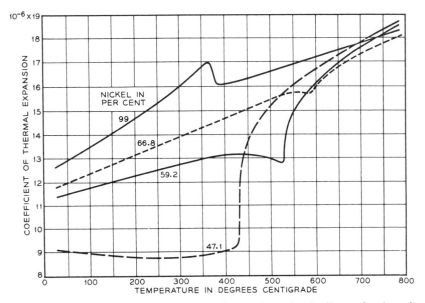

FIG. 10–15. The expansion coefficients of some iron-nickel alloys, showing the magnetic anomaly and its change in sign at about 70% nickel [17C1].

anomaly in the alloys having about 70% nickel, as indicated by the data of Fig. 10–15. Bethe's curve represents the change of interaction energy with volume as a material is expanded or contracted, and it is to be expected that there will be a reciprocal effect, a change in volume as the material passes through the Curie temperature (see Chap. 14). Theoretical analysis [35K2, 36D8, 39B5, 39S1] shows that the effect to be expected does agree in sign with experiment, and that the anomalous expansion or contraction depends on the slope of Bethe's interaction curve. In the case of Invar (about 36% Ni) the slope is positive and unusually large, and theory indicates a large magnetic contraction. This will oppose the normal thermal expansion so that the net expansion is almost zero below the Curie point, which is 250–300°C.

Iron lies to the left of the maximum in the Bethe interaction curve (Fig. 10–13), as indicated by its expansion curve. Calculations by Kornetski [35K2] indicate that the interaction energy doubles for a 2% increase in lattice constant. The behavior of cobalt, nickel, and alloys of cobalt-nickel and of nickel-copper indicates that all of these substances should lie to the right of the maximum. It should be expected that iron-cobalt, like iron-nickel, alloys should lie in the region including the maximum. This is not observed; instead, the Curie point continually decreases as iron or nickel is added to cobalt. In iron-cobalt alloys the change of Curie point with composition is obscured by a change of phase so that no easy test of the theory is possible, but the "virtual" Curie points, evaluated by Forrer [30F2] by extrapolation of the I_s vs T curve of the ferromagnetic phase, do show the expected maximum.

$T^{3/2}$ **Law.**—Bloch [30B3] has made a calculation of the variation of saturation magnetization with temperature at low temperatures, using the Heisenberg model and considering exchange forces between atoms. These can be calculated more rigorously if the spins of almost all the atoms are parallel and the remainder, with reversed spins, are not numerous enough to interact with each other. Calculation is made for material having one active electron spin per atom, the atoms being arranged at the points of a cubic lattice. The expression derived is

$$I_s/I_0 = 1 - A(kT/J_1)^{3/2},$$

J_1 being the exchange interaction and A a constant that is 0.1174 for simple cubic lattices, 0.0587 for body-centered, and 0.0294 for face-centered lattices. Opechowski [37O5] has extended the calculation to include terms of higher power.

In comparing with experiment we may, with some license, make use of the relation $J_1 = k\theta/z$, based on the atom model and taking account of nearest neighbor interactions only. We then have, for a body-centered lattice such as iron,

$$I_s/I_0 = 1 - 0.863(T/\theta)^{3/2}.$$

Experimentally one finds a tolerably good agreement with the $\frac{3}{2}$ power of T/θ, but very poor agreement with the coefficient, which empirically is about 0.11 instead of 0.86 (see Fig. 14–10). This discrepancy is somewhat alleviated if one uses the formula of Moller [33M2], as shown by Kittel [49K1].

Gyromagnetic Effects.—In the discussion of the structure of ferromagnetic atoms, use was made of the concept of electron spin. This section will give some of the evidence for the existence of this spin, its experimental determination, and its relation to magnetic phenomena.

In principle, the ratio of the moment of momentum to magnetic moment may be determined as illustrated in Fig. 10–16. An electron of

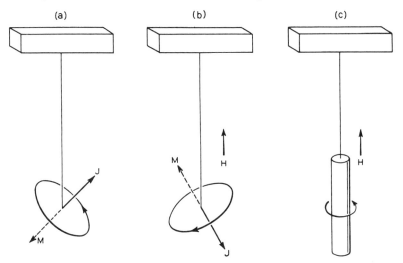

Fig. 10–16. The magnetic moment M and the angular momentum J of an orbital electron. Change in direction of M causes a corresponding change in J, the gyromagnetic ratio J/M remaining constant.

mass m and negative charge e revolves about its nucleus f times per second in an orbit of radius r. The magnetic moment due to the circulating current is at right angles to the plane of the orbit and is

$$M_0 = ef\pi r^2/c.$$

The moment of momentum is in the opposite direction and its magnitude is

$$J_0 = 2mf\pi r^2.$$

The ratio of the moments for this orbital motion is then

$$\rho_0 = \frac{J_0}{M_0} = \frac{2mc}{e}.$$

Imagine now that the atom is suspended in space as shown in (a). If a strong magnetic field is applied, the vector M representing the magnetic moment will rotate around the axis of the suspension, and J will rotate with it, as the electron precesses. As long as there is no external force or friction the angle between M and the axis will not change but only the speed of its rotation will vary. On the contrary, if there is an exchange of energy with other atoms as there is in a real material subject to temperature agitation, then M approaches parallelism with H as shown in (b), and the components of M and J parallel to the axis change in the same ratio. Consequently, the change in the magnetic moment about the axis of the suspension may be said to cause a change in the moment of momentum about the same axis. As a result of the concerted action of all of the atoms composing a rod (c), and the recoil of the rod as a whole, the suspension is subject to a torque equal to the (negative) time rate of change of the moments of momentum of the constituent electrons:

$$L = -dJ/dt.$$

Thus a rod suspended as shown in Fig. 10–16(c) may be magnetized a known amount, its resulting rotation measured, and its gyromagnetic ratio J/M so determined (Einstein-de Haas effect [15E1]). The same ratio may be found also by measuring the magnetic moment M caused by rotating a similar rod with a known angular acceleration (Barnett effect [15B1]).

The existence of a magnetic moment and an angular momentum associated with an electron apart from its orbital motion in the atom, was suggested by a number of people, including O. W. Richardson [23R1] in 1921. It was applied in 1925 by Goudsmit and Uhlenbeck [26G1] in order to explain the structure of atomic spectra, and thereupon became a well-established concept. The magnetic moment assigned to this spin of the electron about its own center was equal to one Bohr magneton which by definition is that of the smallest electron orbit on the Bohr theory:

$$\beta = \frac{eh}{4\pi mc} = 9.27 \times 10^{-21} \text{ erg/gauss}.$$

The unit of angular momentum was taken as one-half of that for the smallest Bohr orbit or as

$$J_s = \frac{h}{4\pi}.$$

The gyromagnetic ratio for the spin motion, denoted by ρ_s, is

$$\rho_s = \frac{J_s}{\beta} = \frac{mc}{e} = \frac{\rho_0}{2},$$

and is thus half the ratio for the orbital motion of the electron. Dirac has shown that these results are consequences of relativistic quantum theory.

In general the gyromagnetic ratio J/M is

$$\frac{J}{M} = \rho = \frac{mc}{e} \cdot \frac{2}{g}$$

where g is closely related to the Landé splitting factor. For spin moment, $g = 2$; for orbital moment, $g = 1$. When the moment of an atom is the resultant of finite spin and orbital moments, g may be found in terms of the quantum numbers, s and l, expressing the angular momenta of the spin and orbital components

$$g = \frac{3}{2} + \frac{s(s+1) - l(l+1)}{2j(j+1)}.$$

Here s may have any of the half-integral values $0, \frac{1}{2}, 1, \frac{3}{2} \cdots$ and l any of the integral values $0, 1, 2 \cdots$, while the number j, representing the angular momentum of the resultant, may be any positive number equal to the sum or difference of s and l. (The actual value of the resultant angular momentum is

$$J = (h/2\pi)[j(j+1)]^{\frac{1}{2}},$$

and that of the magnetic moment is

$$M = (eh/4\pi mc)g[j(j+1)]^{\frac{1}{2}},$$

but the components parallel to the applied field are $jh/2\pi$ and $gjeh/(4\pi mc)$, respectively.) For some values of s, l, and j, e.g., 4, 2, and 2, g is greater than 2, and for some values it is less than 1.

The sign as well as the magnitude of the rotation is of importance. All experiments are consistent with the idea that the magnetic moment is due to the spinning or circulation of negative electrons rather than of positive charges.

The results to be described in the following pages show that in ferromagnetic materials generally the value of g has nearly the value 2 and not at all the value 1; hence we conclude that ferromagnetic processes are concerned primarily with the spins of the electrons and not their orbital motions. When a *change in magnetization* takes place we, therefore, attribute it chiefly to a change in the *direction of spin* of some of the electrons and believe that the orientations of the orbits are disturbed but slightly. In some paramagnetic materials, on the contrary, the reorientation of orbits plays an important part.

Gyromagnetic Experiments.—The first gyromagnetic experiment to be performed successfully was magnetization by rotation. After an unsuccessful trial by Perry in 1890 [35B5], the experiment was considered independently by Barnett [15B1] who in 1914 obtained the result, then inexplicable, that g was approximately twice the classical value 1.

Richardson [08R1] in 1907, was the first to propose rotation by magnetization, and Einstein and de Haas [15E1] performed the experiment in 1915. It was repeated in 1918 by Stewart [18S1] who, for the first time, obtained a result consistent with Barnett's. It has been confirmed since by a number of others.

In recent years the method most often used (rotation by magnetization) is that suggested by Chattock and Bates [23C1] and developed by Sucksmith and Bates [23S2, 25S1]. As modified by Barnett [35B5], it is shown diagrammatically in Fig. 10–17. A rod of the material under investigation (the "rotor," M) is wound with a magnetizing coil and suspended by a fine quartz fibre in a second (induction) coil A. The leads from the latter are connected in series with an adjustable resistance R and a third coil B, inside of which is a small permanent magnet (moment m) mounted below

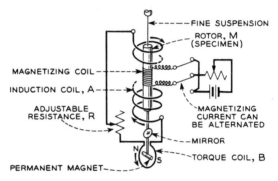

FIG. 10–17. Diagram of the method of determining the gyromagnetic ratio, ρ.

the rotor and connected rigidly to it. A change in the magnetic moment of the rotor is produced by changing the current in the magnetizing coil. This causes a gyromagnetic rotation of the rotor and at the same time induces a voltage in coils A and B. R is adjusted so that the current flowing is of such strength that the field produced by it in B acts on the permanent magnet to annul the gyromagnetic torque of the rotor. The magnetizing current is alternated with a period equal to the natural period of rotation of the rotor assembly and the final deflection δ noted for various values of R. R is plotted against δ and its value, R_0, determined for zero deflection by interpolation.

Let

$$L_A = -dJ/dt$$

be the torque caused by the gyromagnetic effect. The current induced in coils A and B by a change in the moment M of the rotor is

$$i = E/R = (dM/dt)(K_A/R),$$

where K_A is a constant of coil A. This current produces a torque on the magnet m in B:

$$L_B = miK_B,$$

K_B being a constant of coil B. When $R = R_0$, $L_A = -L_B$ and

$$\rho = \frac{dJ}{dM} = \frac{mK_AK_B}{R_0}.$$

The value of ρ is calculated by this formula after finding the values of the coil constants, the resistance R_0, and the moment of the permanent magnet.

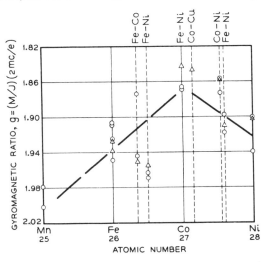

Fig. 10-18. Dependence of Landé factor, $g = 2/(\rho e/mc)$, on electron concentration, as determined by Barnett.

Barnett has taken great care to eliminate various errors caused mainly by the presence of undesirable fields such as the earth's and by asymmetry and magnetostriction of the rotor.

Experimental Values of g.—The results of gyromagnetic experiments are given preferably in terms of g:

$$g = (M/J)(2mc/e),$$

and are collected in Table 6 and Fig. 10-18. Here a g value of 2 means that electron spin only is operative; the ratio would be 1 if change in orbit orientation were the only effect. The apparent slight difference of most of the values from 2 indicates that there is some small but definite change in orbit-orientation in ferromagnetic materials when they are magnetized, and calculations by Brooks [40B5] agree in order of magnitude with the effect observed. In the weakly ferromagnetic pyrrhotite (FeS)

TABLE 6. HISTORICAL TABULATION OF $g = 2/(\rho e/mc)$
FOR FERROMAGNETIC SUBSTANCES

Date	Investigators	Material	g
1914	Barnett* [15B1]	Steel	1.98
1915	Barnett* [15B1]	Steel	2.11
1915	Einstein & de Haas [15E1]	Iron	1.0
1917	Barnett* [17B1]	Iron, nickel, and cobalt	>1<2
1918	J. Q. Stewart [18S1]	{ Iron	1.96
		{ Nickel	2.13
1919	E. Beck [19B1]	{ Iron	1.89
		{ Nickel	1.75
1919	Arvidsson [20A1]	Iron	2.13
1922	Claassen [22C3]	Soft iron to glass-hard steel	1.77 to 0.83
1923	Chattock & Bates [23C1]	{ Iron	1.98
		{ Nickel	1.98
1920–25	Barnett & Barnett* [25B1]	Numerous	
1923	Sucksmith & Bates [23S2]	Fe, Ni, Heusler alloy	2.00
1925	Sucksmith [25S1]	{ Cobalt	1.96
		{ Magnetite	2.04
1935	Sucksmith [35S3]	Cu-Ni alloy	1.90
1932–35	Coeterier & Scherrer [32C3, 35C1]	{ Iron	2.00 −
		{ Pyrrhotite	0.62 −
1935	Ray-Chaudhuri [35R1]	Iron Oxides	1.92 to 1.96 −
1931–38	Barnett [31B2, 34B4, 40B6, 44B1]	Numerous	
1938	Kikoin & Goobar [40K6]	Iron	1.92+
1949	Barnett & Giamboni [49B8]	Permalloy	1.98
1949	Meyer [49M3]	{ Iron	2.01
		{ Nickel	2.00
1949	Meyer [49M4]	{ FeNi	1.99
		{ Fe_3Ni	2.07
		{ FeCo	2.00
		{ Fe-Co-Ni	2.02

* By Barnett effect, magnetization by rotation; others by Einstein-de Haas effect, rotation by magnetization. Measurements on Fe-Se and Mn-Se have also been reported [39G2].

the experimental value 0.62 has been shown by Inglis (34I2) to be in harmony with the theoretical value 0.67, for a possible state of the iron atom ($s = -\frac{1}{2}, l = 2, j = \frac{3}{2}$) in which orbital moment is of importance.

Since the most accurate determinations of g have been made by Barnett, his extensive data are shown separately in Fig. 10–18. Distribution of the points marked with circles (Barnett effect) and with triangles (Einstein-de Haas effect) shows that there is no systematic difference in the results obtained by these methods.

DIAMAGNETISM 455

Gorter [41G1] has shown that the materials investigated may be divided into four groups, centered around manganese (Heusler alloy), iron, cobalt, and nickel, which have characteristic values of g (namely, 1.98, 1.94, 1.85, and 1.90 respectively). This idea is extended in Fig. 10–18, where g is plotted against atomic number, alloys taking appropriate intermediate positions. There is a definite trend indicating that g is determined by the number of extra-nuclear electrons in the atom. The trend is parallel to the g values calculated by Gorter and Kahn [40G5] from the deviation of susceptibility of salts of Mn^{++}, Fe^{++}, Co^{++}, and Ni^{++} from the spin-only value, although the latter g values are less than those observed in the corresponding metals.

Gyromagnetic ratios for paramagnetic materials have been determined by Sucksmith [31S3, 32S4, 35S3] and are given in Table 7. The departure from the values 1 and 2 show that changes in both spin and orbital moments occur during magnetization. In the last column are added theoretical values deduced from spectroscopic data.

TABLE 7. VALUES OF g FOR SOME PARAMAGNETIC SUBSTANCES
(SUCKSMITH)

Substance	g Value obs.	calc.
Nd_2O_3	0.78	0.76
Gd_2O_3	2.12	2.00
Dy_2O_3	1.36	1.33
Eu_2O_3	>4.5	6.56
$FeSO_4$	1.89	<2.00
$CoCl_2$-$CoSO_4$	1.54	<2.00
$CrCl_2$	1.95	<2.00
$MnCO_3$-$MnSO_4$	1.99	2.00
Ni-Cu (56% Ni)	1.9	2.00

DIAMAGNETISM

Faraday showed in 1845 that all substances may be classified as diamagnetic or paramagnetic. Now ferromagnetic substances are not usually thought of as paramagnetic but are considered to be in a separate classification. Faraday distinguished between the paramagnetism and diamagnetism of feebly magnetic substances by suspending them in a strong inhomogeneous field and noting whether they were drawn into or repelled from the strongest part of the field. The force with which a diamagnetic material is repelled by a field is proportional to the strength H and gradient dH/dx of the field and to the (volume) susceptibility κ and volume v of the material:

$$f = \kappa v H dH/dx.$$

The volume susceptibilities of solid diamagnetic substances are usually about -1×10^{-6} or -2×10^{-6}, and are normally independent of field strength. Bismuth has the exceptionally large value of -13×10^{-6}, but even in this substance the numerical value of the susceptibility is millions of times less than that of the ferromagnetic metals. Bismuth, gallium, tin, and graphite are exceptional in that their susceptibilities vary with the field at low temperatures ($<50°$K) [36S13, 49S4].

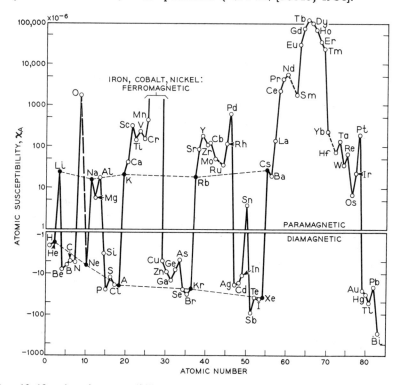

Fig. 10-19. Atomic susceptibilities of the elements at room temperature. Dotted lines connect alkali atoms and rare gases.

While diamagnetic and paramagnetic properties are not often of technical use (they are sometimes used in ore separation), a study of their behavior is very important for our knowledge of the nature of matter. Diamagnetism is an atomic property and usually occurs when the atom has a symmetrical electronic structure and no permanent magnetic moment. In discussing diamagnetism it is appropriate to use the term *atomic susceptibility* defined as $\chi_A = \kappa A/d$, A being the atomic weight and d the density. Then $\chi_A H$ is the magnetic moment of one gram atomic weight, and when divided by Avogadro's number is the average magnetic moment per atom resolved parallel to the magnetic field. Similarly $\chi_M = \kappa M/d$ is the molecular susceptibility, M being the molecular weight.

As Curie [95C1] showed in 1895 in his classical researches, diamagnetism is usually independent of temperature and is not disturbed markedly by

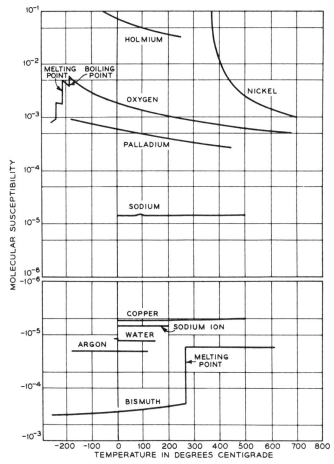

FIG. 10–20. Variation of atomic or molecular susceptibility with temperature, for some representative materials.

atomic collisions or the positions of near-by atoms. However, some change in the diamagnetic susceptibility is generally noted when a substance changes from the solid to the liquid state.

Following the work of Curie, Honda [10H2] and Owen [12O1] reported measurements of the susceptibilities of many elements. Figure 10–19 shows the atomic susceptibilities as dependent on atomic number according to more recent data summarized by Selwood [43S2]. Here it is apparent that the rare-earth metals are strongly paramagnetic, that the metals of the "transition groups" containing palladium and platinum are rather strongly paramagnetic, and that most good metallic conductors of electricity are weakly paramagnetic. This implies what is, in fact, the case, that in a metal the conducting electrons are paramagnetic. Some of the metals and most of the non-metals are weakly paramagnetic or diamagnetic. Antimony and bismuth have moderately strong diamagnetism. Most ionic and molecular compounds are also diamagnetic because their electrons are paired and their permanent moments tend to cancel each other; however, gaseous oxygen, O_2, and nitric oxide, NO, are paramagnetic. Figure 10–20 shows the atomic or molecular susceptibilities of some typical materials and the way in which they change with temperature.

Theory of Diamagnetism.—Our understanding of the origin of diamagnetism is due largely to Langevin [05L1] who, extending the early ideas of Weber, published his celebrated paper on magnetism and electron theory in 1905. He considered an atom with a single electron of charge e and mass m traveling with velocity v in a circular orbit of radius r. This "current" gives rise to a magnetic moment M, proportional to the current and the area of the orbit:

$$M = \frac{ve/c}{2\pi r} \cdot \pi r^2 = \frac{ver}{2c}.$$

Application of a magnetic field H will create an emf in the orbit, as a result of the flux threading it, and this will cause the electron to change its velocity by an amount

$$\Delta v = \frac{eHr}{2mc}$$

in such a direction as to change the moment M by the amount

$$\Delta M = \frac{er\Delta v}{2c} = \frac{e^2 H r^2}{4mc^2}.$$

The susceptibility of this "atom" is, therefore, $-\Delta M/H$, and the atomic susceptibility, or susceptibility of one gram atomic weight, is

$$\chi_A = -\frac{N_0 e^2 r^2}{4mc^2}.$$

If an atom contains many electron orbits, oriented in all directions with respect to the field, the expression for atomic susceptibility becomes

$$\chi_A = -\frac{N_0 e^2 \sum \overline{r^2}}{6mc^2} = -2.83 \times 10^{10} \sum \overline{r^2}$$

in which the mean square orbital radius is summed over all the orbits in the atom.

The foregoing equation, known as Langevin's equation of diamagnetism, is applicable to all atoms, whether they be diamagnetic or paramagnetic. If an atom has a permanent moment and the material is consequently paramagnetic, it is still true that its susceptibility has the diamagnetic component given by this equation.

Knowing the number of electrons in the atom, one may calculate from the diamagnetic susceptibility the average (rms) radius of an atom and compare this with the radius determined by other means. Values thus obtained are consistent with our other knowledge of atomic structure.

In quantum mechanics the Langevin formula still holds when the field of the nucleus is spherically symmetrical, as it is in atoms or ions. The meaning of the orbital radius r, however, is somewhat different, since the electrons are believed to be distributed in less localized orbits. Calculation of diamagnetic susceptibilities by wave mechanics has been attempted for a number of the simple atoms and molecules and has been moderately successful [32V3] for the rare monatomic gases He, Ne, A, Kr, and Xe, and molecular hydrogen, H_2. For helium, theory (Slater [28S9] and Stoner [29S6]) indicates an atomic susceptibility of -1.65 to -1.90×10^6, depending on the method of calculation; Havens [33H4] measured value is -1.91. For the other elements both theoretical and measured susceptibilities vary over a greater range.

Positive ions of the alkali and alkaline earth metals, and of the halogens, have the rare gas structure and can also be treated theoretically. The lack of precision in the calculation arises in the approximate nature of the wave functions and the effective molecular charge assumed.

Simple Salts and Their Solutions.—Most salts are diamagnetic because the ions of which they are composed (e.g., Na^+ and Cl^-) have the completed electron shells characteristic of the rare gases and consequently have no permanent magnetic moment. Measurements are made either on solid salts or on solutions, and the susceptibilities of the separate ions can be deduced from a series of such measurements. There is some ambiguity in separating the susceptibility of the salt into its component parts, and there is also some change with concentration and state of aggregation, but each ion has a value of χ_A that is approximately constant. Values of the separate ions, deduced by Miss Trew [41T1] and others, are plotted in Fig. 10–21 against the number of extra-nuclear electrons in the ion. The increase of χ_A with the number of electrons n is to be expected according to simple theory; the relative large increase between $n = 10$ and $n = 18$ is shown also by ionic radii deduced quite independently from investigations of crystal structure and of the refraction of light. In ions having the same number of electrons but a different nuclear charge, as in the series

I⁻, Xe, Cs⁺, and Ba⁺⁺, the diamagnetic susceptibility decreases numerically with increasing charge because the larger charge draws the electrons closer to the nucleus and thus diminishes the value of $\sum \overline{r^2}$ of Langevin's equation.

The principle of additivity of the diamagnetism of ions may be extended to include also the more complex ions NO_3^-, SO_4^{---}, NH_4^+, and others.

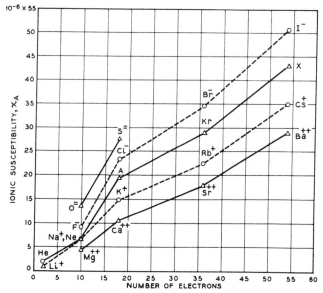

FIG. 10–21. Ionic susceptibilities of similar structures increase with number of electrons in ion.

Values for these are less certain than those for the simple ions already considered.

Other Compounds.—In his early comprehensive experiments (1908–1913) Pascal [10P2] found that the molecular susceptibility of many organic compounds is the sum of the atomic susceptibilities of the atoms composing them, plus additional terms characteristic of the various bonds occurring in the molecule. The atomic susceptibilities and constants of some elements and bonds are given in Table 8. For example, the calculated value of $-x_M \times 10^6$ for ethyl bromide, C_2H_5Br, is $(2 \times 6.00) + (5 \times 2.93) + 30.6 - 4.1 = 53.1$, and is close to the observed value of 53.3. Pascal's constants have been extended to many more complicated organic structures and groups of atoms. His work was also the basis of the additivity relations of inorganic ionic compounds discussed in previous paragraphs.

While the Langevin theory is not very illuminating when applied to a simple diatomic molecule like H_2, quantum mechanics can be called upon, provided the wave functions are known with sufficient accuracy. In the expression for χ_M there is a term, always positive, added to the negative term of the Langevin equation. Van Vleck and Frank [29V1] have used the wave functions of molecular hydrogen to calculate both terms of the

TABLE 8. CONTRIBUTION OF ATOMS AND BONDS TO SUSCEPTIBILITY OF ORGANIC MOLECULES (PASCAL)

Element	$-\chi_A \times 10^6$	Element	$-\chi_A \times 10^6$	Bond	$-\chi_B \times 10^6$
H	2.93	Cl	20.1	C—C	0
C	6.00	Br	30.6	C=C	−5.5
N (chain)	5.57	I	44.6	C≡C	−0.8
N (amine)	2.11	Na	9.2	N=N	−1.8
O (alcohol)	4.61	K	18.5	C—Cl	−3.1
S	15.0	Zn	13.5	C—Br	−4.1
P	26.3	Si	20	Benzene ring	1.4

expression and found them to be -4.66 and $+0.51 \times 10^{-6}$. The net result of -4.15×10^{-6} may be compared favorably with the experimental values [33H4] ranging from about -3.9 to -4.0×10^{-6}.

Calculation of heavier diatomic molecules cannot be carried out so rigorously, and the approximate wave functions generally used for these substances give too high a value of χ_M. Our knowledge of gases is also limited by the fact that measurements are difficult to make with accuracy.

The magnetic properties of water are interesting for three reasons: (1) it is often used for calibration of apparatus; (2) its susceptibility must be subtracted from that of a solution to obtain the value for the solute; and (3) the variation of its susceptibility with temperature has been measured accurately and this gives us a conception of the amount and nature of the temperature variation of susceptibility of diamagnetic substances generally. The value of χ_M is 0.7218×10^{-6} at 20°C and increases about 0.013%/°C according to Auer [33A6]. The opinion of various investigators is that this change is caused by depolymerization of the water with increase in temperature. Experience with water supports the general conclusion that diamagnetism is independent of temperature except for the slight change in the effect of surrounding molecules on the electron distribution of the atom or molecule under consideration. Such changes occur, of course, during melting and vaporization (Fig. 10–20) and when molecules combine to form groups, as in polymerization.

PARAMAGNETISM

The paramagnetism of the materials examined by Curie was dependent on temperature, the susceptibility decreasing with increasing temperature. Later it was found that there was a large class of paramagnetic substances for which the susceptibility was practically independent of temperature (Fig. 10–20); this class includes many of the metallic elements when, and only when, they are in the solid conducting state, and their paramagnetism is caused by the conduction electrons. This kind of paramagnetism is referred to as *weak paramagnetism*, whereas the former class has *strong paramagnetism* and is caused by a permanent magnetic moment of the component atoms or molecules.

The origin of the two kinds of paramagnetism is shown schematically in Fig. 10–22, where the curved lines represent electron shells in an atom. The outer line is broken to indicate that electrons in this shell are loosely bound and become conduction electrons in a solid metal or valence electrons in a compound. In a metal the spins of a portion of these electrons can be

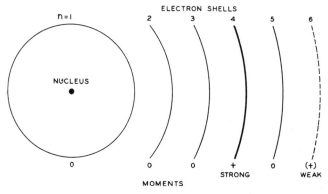

Fig. 10–22. Paramagnetism occurs when inner shell (heavy line) is incomplete and therefore has permanent moment, or when outer electrons (broken line) become conduction electrons in metal (weak paramagnetism).

changed by an applied field in a way that can be explained by quantum mechanics, and they therefore give rise to weak paramagnetism as shown, for example, by metallic sodium. Most of the other shells have their full complement of electrons, and the magnetic moments due to their orbital and spin motions are self-neutralizing so that they contribute only diamagnetism to the atom. However, some of the shells in some atoms are incomplete and therefore have a resultant moment that is large compared with the spin of the conduction electrons or the diamagnetic moment of the closed shells. Such incomplete shells occur notably in the iron group and rare earth group of elements, also in the palladium and platinum groups, and all of these show strong paramagnetism. In the extreme cases of the rare earth elements dysprosium and holmium, the fourth shell (Fig. 10–22) contains electrons whose spin and orbital moments combine to give the largest magnetic moments of any atoms.

The Langevin theory of paramagnetism and the quantum-theory

modifications have already been discussed on p. 427 in relation to ferromagnetism. There it was shown that according to the classical Langevin theory

$$\frac{\bar{\mu}_A}{\mu_A} = \operatorname{ctnh}\frac{\mu_A H}{kT} - \frac{kT}{\mu_A H},$$

μ_A being the magnetic moment per atom and $\bar{\mu}_A$ its average component in the direction of the field. From this it follows that, when $\mu_A H \ll kT$, as it is in paramagnetic materials, the atomic susceptibility is

$$\chi_A = \frac{N_0 \bar{\mu}_A}{H} = \frac{N_0 \mu_A{}^2}{3kT},$$

N_0 being the number of atoms per gram-atom. The number of Bohr magnetons per atom, as calculated from this expression, is known as μ_{eff} and is

$$\mu_{\text{eff}} = \frac{\mu_A}{\beta} = \left(\frac{3k\chi_A T}{N_0 \beta^2}\right)^{\!1/2} = 2.83\,(\chi_A T)^{1/2} = 2.83\sqrt{C_A}.$$

This quantity is often used to express the results of experiment, even though the classical theory on which it is based does not apply.

The effective moment of an atom may change with temperature or field strength (in a calculable way), and Curie's law will not then be obeyed. Under certain conditions χ_A will be of the form

$$\chi_A = \frac{N_0 \mu_A{}^2}{3kT} + N_0 \alpha,$$

the atomic moment being $\mu_A = jg\beta$ and the constant α being calculable by means of wave mechanics. Such an expression, with the constant α included, accounts quantitatively for the behavior of some of the rare earth ions and is applicable in principle in other cases where, however, it cannot often be worked out in detail.

Rare Earths.—The ions of the rare earth elements cerium (58) to ytterbium (70) are good examples of strongly paramagnetic substances. The permanent magnetic moment resides in the $4f$ subshell of the fourth electron shell in the usually trivalent ions. This inner shell is protected by the outer shells from the influence of neighboring atoms, and hence the assumptions underlying the theory are fairly well satisfied.

Ionic susceptibilities are deduced from measurements on solutions and on solid salts and oxides by subtracting the contributions of water and other ions, and they are plotted in Fig. 10–23 from data summarized by Selwood [43S2]. The susceptibilities decrease with increasing temperature, as expected, except that χ_A for the samarium ion Sm^{+++} is almost constant above room temperature and that for europium shows also some unusual behavior with temperature.

Calculation of the susceptibilities of the rare earth ions by quantum mechanics is one of the important successes of the Van Vleck theory [32V3], based on the earlier work of Hund [25H1]. Comparison with experiment is made in the figure. Both terms of the last equations given previously, those dependent on and independent of temperature, were derived quantitatively from theory, and the unusual change of χ_A with temperature for Sm and Eu were satisfactorily explained.

Rare earths in metallic form have large susceptibilities, and gadolinium when cooled below 16°C is ferromagnetic. Values of $\chi_A \times 10^6$ are more

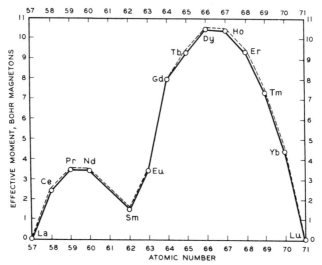

FIG. 10–23. Effective moments of the trivalent ions of the rare earth elements. Broken line is according to Van Vleck theory.

than 100 000 for ytterbium and dysprosium at room temperature. Important deviations from Curie's law are observed, and a law of the Curie-Weiss form,

$$\chi_A = C/(T - \theta)$$

is closely followed. For gadolinium, as for the other ferromagnetic elements, θ is approximately the Curie point, but the existence of θ in the Curie-Weiss equation does not mean necessarily that the material will be ferromagnetic below the temperature $T = \theta$; rather, the simple law breaks down before this point is reached.

Ions of the Iron Group.—Strong paramagnetism occurs also in the ions of the elements lying between (but not including) scandium and copper and is most marked in the ions of manganese, iron, and cobalt. The moments of the ions are generally less than those of rare earth ions, because the incomplete shell responsible for the paramagnetism now has a capacity of 10 instead of 14 electrons.

In calculating the moments from quantum theory the method used for the rare earths is not at all successful, as shown by the dotted line of Fig. 10–24. Results of calculations made with other assumptions are shown by the other lines, (a) and (b), and it is seen that the observed values lie between these curves. Stoner [36S11] has indicated the reason for this. The shell responsible for the moment is the outermost shell of the ion and

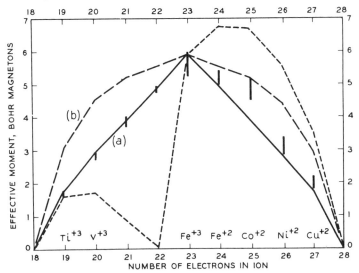

FIG. 10–24. Effective moments of ions of iron group. Vertical lines indicate range of observed values, connecting lines the theoretical values based on different assumptions.

thus it is subject to the influence of surrounding ions. These affect the orbital motions in a way difficult to calculate, but if the effect is severe only the spin moments will be influenced by the magnetic field and curve (a) results. On the contrary, if the interaction between neighboring ions is weak, the orbital motions will be influenced by the field and curve (b) is calculated. When the influence of surrounding ions on the unfilled shell is moderate, the actual ionic moments may be expected to lie between the two theoretical limits, as in fact they do.

Paramagnetism at Low Temperatures.—At ordinary temperatures the Langevin equation may be represented with sufficient accuracy by the first term of the series expansion, as noted (p. 433). However, at low temperatures and in strong fields $\mu_A H/kT$ may be so large that the complete equation, or its quantum equivalent, must be used. Under these conditions the susceptibility is no longer independent of the field strength, and the magnetic moment of the material approaches a limiting value corresponding to alignment of the molecular magnets parallel to the field.

The first experimental test of this phenomenon, by Woltjer and Onnes [23W3], was carried out at temperatures down to 1.3°K and field strengths up to 22 000 oersteds, and the results confirmed the theory in a satisfactory manner. A slightly different relation was predicted by quantum theory, and this agrees with experiment entirely within the limits of accuracy of the experiments. For the Gd^{+++} ion, $J = \frac{7}{2}$. The theoretical curves for $J = \frac{1}{2}$, $J = 1$, $J = \frac{7}{2}$, and $J = \infty$ (classical) are shown in Fig. 10–25,

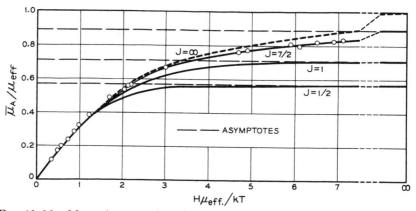

FIG. 10–25. Magnetic properties of gadolinium sulfate at low temperatures; μ_{ef} is the atomic moment in Bohr magnetons and $\bar{\mu}$ its average component in the direction of the field.

where $\bar{\mu}_A/\mu_{eff}$ is plotted against $H\mu_{eff}/kT$. Experimental points lie close to the curve for $J = \frac{7}{2}$. For further discussion of this point the reader is referred to Giauque [27G1] and Van Vleck [32V3].

The magnetic behavior of certain paramagnetic salts is important in the attainment of very low temperatures. A temperature of 0.0015° above absolute zero was observed by de Klerk, Steenland and Gorter [50D3].

Paramagnetic Gases.—Most monatomic gases, such as those in the eighth column of the periodic table, are diamagnetic because they have complete electron shells and, therefore, no residual magnetic moments. Metal vapors, however, have monatomic molecules in which the single electron in the outer shell should give rise to paramagnetism. Measurements on such vapors are difficult, but potassium is reported [28G3] to be paramagnetic and to obey Curie's law over a limited range of temperatures. It has a susceptibility, according to the rather inaccurate measurements, corresponding to a single, uncompensated electron ($\chi_A = 0.375/T$).

Molecules containing more than one atom are generally diamagnetic because they contain an even number of electrons whose moments neutralize each other. There are two well-known exceptions to this rule, oxygen (O_2) and nitric oxide (NO). The former has an even number of electrons that remain uncompensated in a way not understood; the latter has an odd number of electrons. The paramagnetism of many other "odd" molecules has been studied to a lesser extent.

Oxygen [29W6] follows Curie's law except at high pressures [39K9] when the Curie-Weiss law is obeyed, with the constant θ small and negative. Liquid oxygen also follows the Curie-Weiss law, and the susceptibility increases with decreasing temperature through the solidification point.

The molecular susceptibility of nitric oxide has been measured [31S8] from 20 to $-160°C$ (below its normal boiling point), and the effective Bohr magneton number, μ_{eff}, was found to vary by about 15% over this range of temperature. Accurate calculation of μ_{eff} and its variation with temperature has been carried out by Van Vleck [32V3], who based his work on the electronic structure derived from spectroscopic data. The close agreement found between calculated and observed values is a tribute to the adequacy of the theory.

Paramagnetism of Free Electrons.—The weak temperature-independent paramagnetism of many of the metals (Fig. 10–20) was inexplicable on the Langevin theory, according to which all paramagnetism should decrease with increase in temperature. According to modern theory, as first shown by Pauli [26P2], the conduction electrons of a metal have spin moments which are only occasionally affected by a magnetic field. In the simplest case with one free electron per atom, quantum theory predicts an atomic susceptibility of $1.88 \times 10^{-6} (A/d)^{2/3}$, where A is the atomic weight and d the density in g/cm^3. There is also a smaller diamagnetism predicted by quantum theory, equal to one-third the paramagnetic contribution.

In actual metals the electrons are not wholly free but are partially bound to the atoms and are influenced by the so-called correlation and exchange forces, quantum mechanical in nature. Also, the ionic core of the atom has an important diamagnetic contribution to the susceptibility. These various components have been calculated for sodium by Sampson and Seitz [40S5] and when summed they come close to the observed susceptibility. For other elements the calculations are less satisfactory.

The paramagnetic contributions of the electrons will vary with the nature of the metal, and the diamagnetism of the ionic cores will be different for different atoms, and it is easy to see qualitatively that these and other factors will cause some metals to be weakly paramagnetic and others to be weakly diamagnetic. The order of magnitude of the susceptibility is explained by theory, and more exact calculations must await further quantitative development of atomic theory.

Molecular Beams.—According to the quantum theory, the magnetic moment of a molecule may have only a limited number of orientations with respect to the direction of a magnetic field in which it is placed. A direct test of this was made successfully by Stern and Gerlach [24G2] in 1921. In this experiment silver atoms, issuing from an oven containing melted silver (Fig. 10–26), are formed into a beam by using small slits, and then pass into the non-uniform field of an electromagnet. In the field the (monatomic) molecules are oriented so that the magnetic moment of each is either parallel or antiparallel to the direction of the field. Those oriented parallel to the field are deflected in the direction of the stronger field by a force proportional to the gradient (downward in the diagram); those oriented in the opposite direction are deflected toward the weaker field (upward). Thus the beam is split into two components, as detected by the image formed on the screen, and from the separation of the images

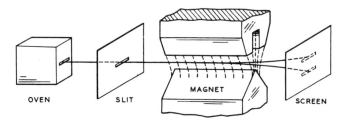

Fig. 10–26. Diagram of Stern-Gerlach experiment for determining magnetic moments of gaseous atoms. Beam of vaporized metal is split by strong inhomogeneous magnetic field.

the atomic moment can be estimated. It is necessary to know the velocity of the beam (calculated from the temperature of the oven) and the gradient of the field (about 200 000 oersteds/cm in some experiments).

Later experiments [25G1], designed for more precise determination of the atomic moment, show that for the alkali metals and Cu, Ag, and Au the moments have the theoretical value of one Bohr magneton, within the experimental error of about 1–4%. The elements Zn, Cd, Hg, Sn, Pn, etc., have zero atomic moment (Leu [27L1]) as expected, because the two electrons in the outer shell are mutually compensating. Other elements, e.g., thallium, are found to have the non-integral number of Bohr magnetons derived by quantum theory from spectroscopic data.

Nuclear Moments.—The nucleus of an atom may itself have a magnetic moment. This is always much smaller than a Bohr magneton and is detected most readily when the outer electronic structure is balanced magnetically so that its net moment is zero. Nuclear moments do not have any important influence on the magnetic properties of materials, but they are very important for our knowledge of nuclear structure, and techniques for their determination have been developed almost continually since 1931. Here the particular method known as the "molecular beam resonance method" and used first in 1938 by Rabi and his colleagues [39R3, 46K4] will be outlined briefly.

As in other experiments a molecular beam is provided and defined by slits as shown in Fig. 10–27. When the molecules enter the field, the vector representing the magnetic moment precesses around the direction of the field. About half of the molecules are deflected downward by the field gradient produced by magnet A, and upward by the gradient of magnet B, and arrive at the point D where they are detected by a suitable device. Molecules of different velocities are focused at the same point after traversing other paths, provided no change occurs in the magnetic moment along the path.

A change in the vertical component of the moment is produced in the uniform field H of magnet C by superposing upon the field a rather weak

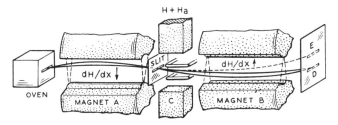

FIG. 10–27. Rabi method of determining nuclear moments by "radio-frequency spectrometry." Beam of atoms is bent by magnets as shown, and is also acted on by a high-frequency magnetic field generated in the copper tubes shown.

alternating field, of frequency f, produced by current in the copper tubes shown. The disturbing action of this field may best be explained by analogy with a spinning top precessing about a vertical axis under the influence of gravity. If a disturbing force is produced, as by oscillating the point of support, the top will become unstable, provided the frequency of oscillation is equal to the frequency of precession. In the case of the molecules, instability is produced when their frequency of precession (proportional to the field strength H) and the frequency of oscillation of the weaker field are equal, and unstable molecules then deviate from their previous paths as indicated by the dotted line E. H is varied slowly until the number of molecules arriving at D is a minimum, and H and f are noted.

The ratio of the magnetic moment to the angular momentum is expressed in terms of the Landé factor g:

$$g = (4\pi Mc/e)f/H = 0.001312 f/H,$$

M being now the mass of the proton. For the Li^7 atom experiment gives $f = 5.61 \times 10^6$ cps, $H = 3400$ oersteds, and hence $g = 2.167$. The magnetic moment of the Li^7 nucleus, for which the spin is $\frac{3}{2}$, is

$$M_n = \tfrac{3}{2}gB_n$$

or 3.25 nuclear magnetons. The nuclear magneton B_n is only $\frac{1}{1840}$ times the Bohr magneton β:

$$B_n = \frac{eh}{4\pi Mc} = \frac{\beta m}{M} = \frac{\beta}{1840}.$$

Thus the nuclear magnetic moment in this case is less than 0.1% of the moment of the iron atom, and most nuclear moments are even less.

More recently other resonance methods have been used for determining nuclear moments, and these are applicable to solids and liquids because they do not require molecular beams. They have been described by Purcell, Torrey, and Pound [46P1], Bloch, Hansen, and Packard [46B6], and others.

Antiferromagnetism.—There is a class of materials, paramagnetic as judged by their small positive susceptibility, whose behavior can be interpreted in terms of negative exchange interaction. Their molecular

TABLE 9. CURIE POINTS, θ, OF SOME ANTIFERROMAGNETIC SUBSTANCES

Substance	θ, °K	Substance	θ, °K
VCl$_3$	30	MnF$_2$	72
Cr$_2$O$_3$	320	FeO	198
CrSb	700	Fe$_2$O$_3$	950
CrCl$_2$	70	FeF$_2$	79
MnO	116	FeCl$_2$	24
MnO$_2$	90	CoO	271
MnS	165	CoCl$_2$	25
MnSe	160	NiO	520
MnTe	307	NiCl$_2$	50

For more recent data see Nagamiya et al., *Advances in Physics* **4**, 1–112 (1955).

susceptibilities, of the order of 10^{-3}, increase with increasing temperature to a critical temperature, often called the Curie point (or λ-point), above which the susceptibility decreases according to the Curie-Weiss law. Table 9 shows the Curie points of a number of compounds investigated by Squire, Bizette, and Tsai [38S9, 39S15, 43B5], Millar [28M7, 29M6], Foëx and Graff [39F2, 48F3], Starr, Bitter, and Kaufmann [40S8], Bizette [46B4], Foëx and Blanchetais (49F2] and Ellefson and Taylor [34E4].

Susceptibility per gram is plotted against temperature for MnSe and MnTe in Fig. 10–28, according to Squire [39S15]. Hysteresis is observed below the Curie point, as illustrated in Fig. 10–29, but disappears at still lower temperatures [39S15], and susceptibility is somewhat field-dependent.

ANTIFERROMAGNETISM

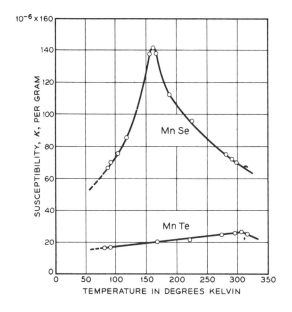

FIG. 10–28. Magnetic susceptibility per gram as dependent on temperature, for antiferromagnetic substances MnSe and MnTe.

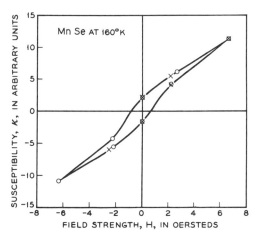

FIG. 10–29. Hysteresis loop of antiferromagnetic substance MnSe, just below Curie point.

The hysteresis observed at temperatures below 1°K in iron alum and other materials used in magnetic cooling experiments may be attributed to antiferromagnetism, as suggested by C. J. Gorter in private conversation with the author. Measurements of specific heat near the Curie point have been made by Kelley [39K8] and others.

The theory has been developed by Néel [32N2, 36N3], Bitter [37B6], and Van Vleck [41V1] and has been reviewed by the latter [45V1]. A more recent development, with special application to the ferrites, has been made by Néel [48N1]. The theory is based on the idea that the exchange forces will maintain neighboring atoms antiparallel when the temperature is low enough. Analysis shows that a sharply defined Curie point is to be ex-

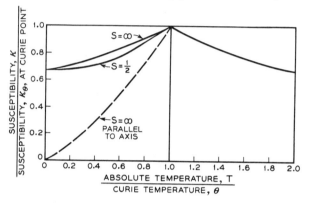

FIG. 10–30. Theoretical susceptibility of antiferromagnetic material as dependent on temperature, for two atom states. Scales are adjusted so that the relative susceptibility is one at the Curie point θ. $\kappa/\kappa_\theta = 1$ when $T < \theta$, perpendicular to axis.

pected, and that below this temperature the susceptibility of a polycrystalline material should decrease (as observation shows that it does) and approach a limit at 0°K of $\frac{2}{3}$ of the susceptibility at the Curie point (see Fig. 10–30). Experimentally the limits approached are from 0.3 to 0.85 instead of 0.67. As Stout and Griffel [49S6] have shown, the susceptibility of single crystals of tetragonal MnF_2 is highly anisotropic below 70°K and is much higher perpendicular to the tetragonal axis than parallel to it.

When atomic magnetic moments are placed at the points of a lattice such as the body-centered cubic lattice, which may be resolved into two distinct sublattices A and B, then one may imagine all of the A atoms to be oriented in one direction and the B atoms in the opposite, so that nearest neighbors are always antiparallel. Then one can speak of a domain as the region over which this kind of ordering extends. In one domain the atomic moments might be parallel to [100] and [$\bar{1}$00] directions, whereas in a neighboring domain they would be parallel to [010] and [0$\bar{1}$0] directions as indicated in Fig. 10–31. Néel reports that, according to theory, the Bloch walls separating them have about the same thickness as in ferromagnetic domains. Antiferromagnetic domains, not so far observed, differ from ferromagnetic domains in that each has a net moment of zero in zero applied field. They are not redirected by an external field; consequently, the high susceptibilities of ferromagnetic materials do not occur. It is possible, however, that the direction of the line along which the local magnetization is directed may be changed in orientation at will by applying a mechanical force.

A direct test of antiferromagnetism has been carried out using neutron diffraction. The scattering caused by the magnetic coupling of the neutron with the magnetic atom has been shown by Halpern and Johnson [39H6] to be of the same order of magnitude as the scattering from the nucleus. At the suggestion of Smart, experiments were made by Shull [49S3] at 80°K, well below the Curie point of MnO, and it was found that at this temperature the regular antiparallel arrangement of the magnetic Mn atoms caused additional diffraction lines to appear. Similar results were found for α Fe$_2$O$_3$.

Some of the properties of domains of this kind may be calculated using molecular field theory. Consider first that the only interaction is between nearest neighbors, and that this is negative so

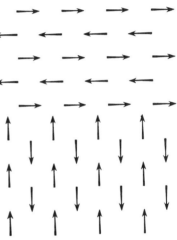

FIG. 10–31. Possible domain structure of antiferromagnetic material.

that neighboring atoms are maintained antiparallel at sufficiently low temperatures. Then the Curie point will be the same as in a ferromagnetic material with the same energy of interaction; and above the Curie point

$$I = \frac{CH}{T - \theta},$$

$$\theta = CN,$$

N being the molecular field constant.

When the next-nearest neighbors exert an appreciable influence, the Curie point will still be determined by the total interaction energy. If we write a molecular field constant N_1 for interaction between nearest neighbors and N_2 for interaction between next-nearest neighbors, the Curie point is

$$\theta = C(N_1 + N_2).$$

Above the Curie point the induced moments are parallel, and the stronger interaction between nearest neighbors is opposed by the weaker interaction between next-nearest neighbors. Then in the Curie-Weiss law,

$$I = \frac{C(H + NI)}{T} = \frac{CH}{T - \theta'};$$

the molecular field constant N is equal to $-(N_1 - N_2)$, and θ' is therefore $-C(N_1 - N_2)$. The quantities C, θ', and θ are determined by experiment, so that N_1 and N_2 can be evaluated:

$$N_2 + N_1 = \theta/C$$
$$N_2 - N_1 = \theta'/C$$

and

$$N_2 = (\theta + \theta')/2C$$
$$N_1 = (\theta - \theta')/2C.$$

The data for MnF_2 give $N_2/N_1 = 0.22$.

At the Curie point the susceptibility is

$$\chi = C/(\theta - \theta') = 1/2N_1.$$

At absolute zero, when a magnetizing field is applied along the line parallel to which the spins are directed, the susceptibility $\chi_\parallel$ is zero, because no torque acts on the spins. The susceptibility, $\chi_\perp$, at right angles to this direction can be evaluated by adding vectorially the applied and molecular fields and calculating the change in magnetization produced thereby. The numerical value,

$$\chi_\perp = 1/2N_1$$

is the same as the susceptibility at the Curie point. The theoretical curves for $\chi_\parallel$ and $\chi_\perp$ are shown in Fig. 10–30. The theoretical curve for χ for an assemblage of randomly directed domains is also known. It is easy to see that $\chi/\chi_\perp = \frac{2}{3}$ at $0°K$.

P. W. Anderson [50A2] has pointed out that this simple form of the molecular field theory is not applicable to NaCl-like structures such as MnO because these do not allow a division into two sublattices, but require four in order that neighboring atoms will be in different sublattices. The molecular field theory is applied with some modification to

ANTIFERROMAGNETISM   475

this case. One of the results is that $\theta/\theta' = 3$, in good accord with experiment for a number of substances.

Anderson also points out that the results of Shull and Smart [49S3], showing that next-nearest neighbors are antiparallel, is hard to understand on the basis of ordinary exchange coupling acting through the intermediate oxygen atom. He explains this in terms of "super-exchange," as proposed by Kramers [34K15, 40K7]. According to this, exchange coupling can bridge the gap between two magnetic atoms when they are separated by atoms with unfilled shells. Anderson finds that such coupling has a certain directional effect, and in oxygen is stronger when the adjacent atoms are on opposite sides rather than in directions at 90° from each other.

A theory of antiferromagnetism was also developed by L. Landau in 1933 (*Sov. Phys.* **4**, 675).

CHAPTER 11

THE MAGNETIZATION CURVE AND THE DOMAIN THEORY

Three Parts of Magnetization Curve.—A typical magnetization curve, showing the relation between B and H in a specimen initially unmagnetized, may be divided into three main parts. Using the usual metaphor, we say that the natural divisions are at the instep and the knee of the curve, as indicated by the horizontal broken lines in Fig. 11-1. In the *first region* (toe to instep) the curve starts from the origin with finite slope $dB/dH = \mu_0$ and rises so that it is concave upward, following usually the Rayleigh relation

$$\mu = \mu_0 + \nu H$$

or

$$B = \mu_0 H + \nu H^2$$

wherein ν is $d\mu/dH$ and has a constant value. This portion of the curve is usually said to be reversible because the curve is retraced approximately (but not exactly) when the field strength is diminished; that is, $dB/dH \approx \mu_r$.

The *second part* of the magnetization curve has the greatest slope and is irreversible in that the path followed when H is decreased is quite different from the upward curve. Here $dB/dH \gg \mu_r$, sometimes by a factor of 100 as dB/dH attains values of 10^6. Above the knee the *third part* has a smaller slope that approaches one when B is plotted against H, and zero when I or $B-H$ is the ordinate, as H increases indefinitely. The curve conforms more or less to the Frölich-Kennelly relation

$$\frac{1}{\mu} = a + bH.$$

This section of the curve is also reversible to a considerable extent and $dB/dH \approx \mu_r$.

The second and third sections of the magnetization curve just described correspond to two sections of the hysteresis loop. As indicated in Fig. 11-4, one part of the loop extends approximately from remanence along the steepest part of the loop to the point of maximum curvature; the other part, from that point to saturation and back to remanence.

General Description of Domain Theory.—Many of the phenomena of the magnetization curve and hysteresis loop can be described to advantage in terms of the domain theory. The theory that a ferromagnetic material is composed of many regions, each magnetized to saturation in some direction, was first stated by Weiss [07W1] in 1907, and the description he gave at that time is accepted today with some modification. The existence of such domains is now plainly evident from the study of powder patterns, as described later in this chapter. In the unmagnetized state the directions

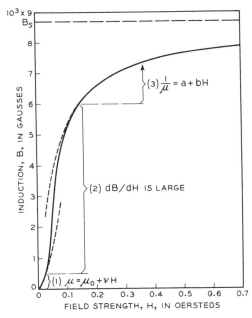

FIG. 11-1. Three sections of magnetization curve separated by the "knee" and the "instep."

in which the domains are saturated are either distributed at random or in some way such that the resultant magnetization of the specimen as a whole is zero. Application of a field changes only the direction of magnetization in a given volume, not the magnitude. The three main parts of the magnetization curve are distinguished from each other by the nature of this change in direction, which may take place in any of several ways.

Two important problems in ferromagnetism are to explain (1) the manner in which the magnetization is changed by the presence of an applied field and (2) the anisotropic character of the magnetic crystal which determines the direction in which the domain is spontaneously saturated. Further matters of interest are the size of the domain and the circumstances that determine its boundaries.

In the absence of an external field the direction of magnetization in a domain is affected by the crystal structure or by strain. The magnetization curves of single crystals of iron and of nickel, taken in the three principal directions in each crystal and reproduced in Fig. 11-2, show that in iron the magnetization proceeds most readily in a direction parallel to

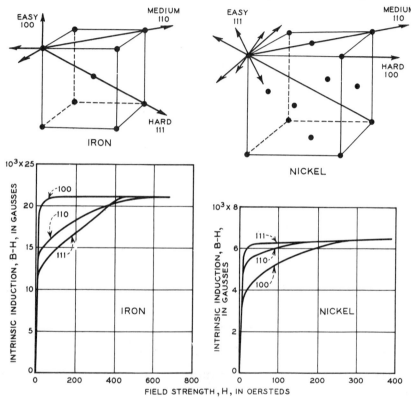

FIG. 11-2. Magnetic properties and crystal structures of single crystals of iron and nickel.

a cubic axis and in nickel parallel to the body diagonal of the cube, a direction as far removed as possible from an axis. It is concluded that in a domain in a crystal of iron not subjected to strain or external magnetic field the magnetization is always parallel to one of the crystal axes ([100] directions). Since there are six such directions that are equivalent, a demagnetized crystal of iron will normally have one-sixth of its domains oriented in each of these directions of easy magnetization. Nickel has eight equivalent directions ([111] directions) each equally inclined to the three crystal axes. In each domain the intensity of magnetization is equal

GENERAL DESCRIPTION OF DOMAIN THEORY 479

to the saturation intensity of the material measured at the given temperature.

The magnetic moment of any one domain is specified by the magnitude and direction of its magnetization and by its volume. Changes in the magnitude of the spontaneous magnetization may be effected by change in temperature or in some slight measure by application of the highest fields

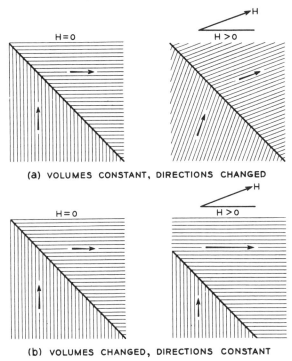

Fig. 11-3. Two mechanisms of change of magnetization: (a) domain rotation and (b) moving boundary.

attainable. Ordinarily at constant temperature the moment of a domain, and therefore the magnetization of the ferromagnetic body of which it is a part, is changed by:

(1) A change in the direction of magnetization of the domain—"rotation"; or
(2) A change in the volume of the domain—"moving boundary."

These processes are shown schematically in Fig. 11-3. R. Becker [32B8, 33B3] has discussed the nature of these changes and shown that the *moving boundary*, proposed also by Bloch [32B2], is particularly important for

480 THE MAGNETIZATION CURVE AND THE DOMAIN THEORY

changes occurring in low and medium fields. The magnetization of a body of volume V, originally unmagnetized, may be written

$$I = \frac{\sum_m \delta(I_m V_m)}{\sum V_m} = \frac{I_s \sum (V_m \sin\theta\, \delta\theta + \cos\theta\, \delta V)}{V}$$

where I_m and V_m refer to magnetization and volume of the mth domain, and θ is the angle between the direction of the magnetization of the domain

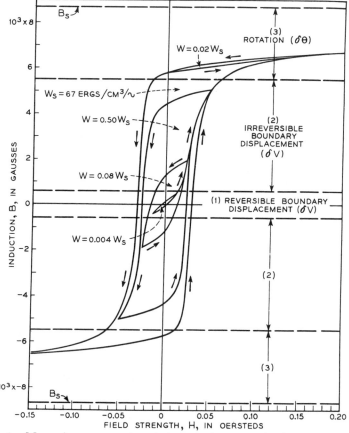

FIG. 11–4. Magnetic processes in relation to the magnetization curve and the hysteresis loops. Material, 4–79 Permalloy.

and the direction of the field. The terms containing $\delta\theta$ and δV refer respectively to processes (1) and (2) above.

Processes may be classified also as irreversible or reversible according to whether the energy dissipated in heat is a relatively large or small fraction of the potential energy.

Considering these classifications the three main parts of the magnetization curve may be identified with the following processes:

(1) Reversible boundary displacement.
(2) Irreversible boundary displacement.
(3) Reversible rotation.

A fourth process, irreversible rotation, may' occur in a rotating field.

Figure 11-4 shows the three important processes in relation to the magnetization curve and to major and minor hysteresis loops, and (by area of loops) to the corresponding energy losses due to hysteresis, for a specimen of a high permeability alloy (4-79 Molybdenum Permalloy).

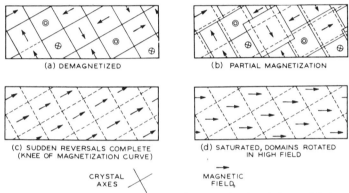

FIG. 11-5. Change of domain structure with magnetization (schematic), in material like iron.

Figure 11-5(a) is a highly schematic representation of the domains in a part of a single crystal that itself may be a portion of a polycrystalline specimen of iron. The arrows, circles, and crosses represent the directions of local magnetization; the sides of the squares, the directions of the crystal axes. Figures 11-5(b) and (c) represent the states of magnetization corresponding to the instep and knee of the curve, attained by reversible and irreversible boundary displacement, and (d) represents saturation attained in high fields by the process of rotation.

Stability of Domain Orientation.—As mentioned previously, the direction of magnetization in a domain is determined by the crystal structure, provided strain and magnetic field are absent, but when the latter are present either of them may be controlling. The effect of crystal orientation on ease of magnetization has already been shown in Fig. 11-2, and the importance of tensile stress is illustrated in Fig. 11-6 for two substances oppositely affected by tension.

The relative importance of the three factors—crystal structure, stress, and field—may be derived from the following expressions for the potential energy, which are discussed in more detail in later chapters:

$$E_k = K(\alpha_1{}^2\alpha_2{}^2 + \alpha_2{}^2\alpha_3{}^2 + \alpha_3{}^2\alpha_1{}^2) \tag{1}$$

$$E_\sigma = \tfrac{3}{2}\lambda_s \sigma \sin^2 \theta \tag{2}$$

$$E_H = -HI_s \cos \phi. \tag{3}$$

In (1), E_k is the magnetic crystal energy density of a domain oriented so that its "direction," i.e., the direction of its magnetization, has the direction-cosines α_1, α_2, and α_3 with respect to the three axes of the cubic crystal.

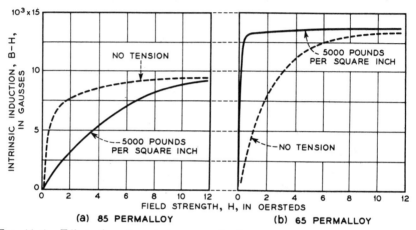

FIG. 11–6. Effect of tension on the magnetization of materials having negative or positive magnetostriction, (a) and (b), respectively.

K is the crystal anisotropy constant, the value of which is about 420 000 ergs/cm³ for iron and $-45\,000$ for nickel. In (2), E_σ is the magnetic strain energy, λ_s is the magnetostriction expansion occurring between the demagnetized state and saturation, σ the tensile stress to which the domain is subject, and θ the angle between the magnetization and the tension. In (3), E_H is the energy of the domain due to the presence of a field, I_s is the saturation magnetization, H the field strength, and ϕ the angle between I_s and H. The way in which the energy changes with the orientation of the domain is represented in Fig. 11–7 in which the stable positions, corresponding to minimum energy, are indicated. The figures correspond to Eqs. (1) to (3) except for additive constants.

The foregoing equations are quantitative expressions of rules that are often used in a qualitative way in applying domain theory to a new problem. For example, Eq. (2) states that for a material like nickel with negative magnetostriction (λ_s negative), application of tension (σ positive) to an unmagnetized specimen will cause the domains to become oriented more nearly at right angles (θ larger) to the line of tension. In the limiting case of large tensions the domains will thus be lined up at right angles to the direction of tension, the directions of magnetization occupying various all orientations in the plane perpendicular to the axis.

Equation (1) shows that for a material with a relatively large (positive or negative) value of the anisotropy constant K there will be only a rela-

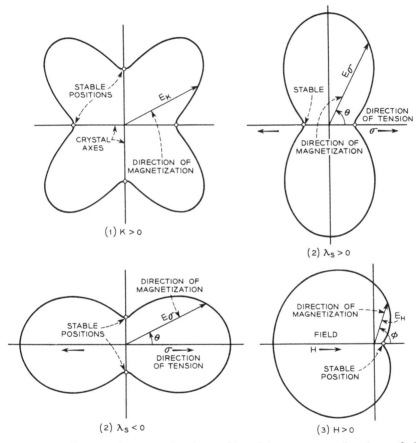

Fig. 11-7. Energy diagrams, showing stable minimum-energy directions of the magnetization under different conditions of field and stress.

tively slow approach to saturation with increasing field strength. The domains are held firmly parallel to the crystal directions $\langle 100 \rangle$ when K is positive, and $\langle 111 \rangle$ when K is negative. Energy E_H with a high negative value in the direction of H is then necessary before the domains will be drawn from the stable crystal directions to become parallel to H. This means practically that to attain a relatively high permeability at an induc-

484   THE MAGNETIZATION CURVE AND THE DOMAIN THEORY

tion of three-quarters or more of saturation a polycrystalline material must either have a small value of K or must have crystals oriented in a special way so that there is a preponderance of easy directions of magnetization in the crystals parallel to the specified direction in the specimen.

Further description of the domain theory is given in Chap. 18, where it is applied quantitatively to various magnetic phenomena.

Approach to Saturation.—Several empirical equations have been proposed to represent the course of the magnetization curve in field strengths considerably greater than the coercive force. Probably the first of these was Lamont's Law [1867L1] which states that the susceptibility κ is proportional to the difference between the intensity of magnetization and its saturation value:

$$\kappa = C(I_s - I).$$

This is equivalent to Frölich's relation [81F1]

$$I = \frac{H}{a_1 + b_1 H}$$

with $a_1 = 1/CI_s$ and $b_1 = 1/I_s$. When H is not too high, so that B may be replaced by $4\pi I$, this is equivalent to

$$\frac{1}{\mu} = a' + b'H,$$

a relation proposed by Kennelly [91K1] and one that is particularly useful because it shows a linear relation between reluctivity $(1/\mu)$ and field strength. In the highest fields this should be written, as Kennelly acknowledged,

$$\frac{1}{\mu - 1} = a + bH$$

and may then be used to determine B_s, the saturation value of $B-H$, since

$$b = d\left(\frac{1}{\mu - 1}\right)\bigg/ dH = 1/B_s.$$

The Frölich-Kennelly relation may be rewritten

$$I = I_s\left[1 - \frac{4\pi a I_s}{H} + \left(\frac{4\pi a I_s}{H}\right)^2 \cdots\right]$$

to show its relation to the equation

$$I = I_s(1 - A/H)$$

used by Weiss [10W1] for extrapolation to determine the value of I_s. In these equations A or a may be termed the coefficient of magnetic hardness.

A plot of $1/\mu$ vs H is reproduced in Fig. 11-8 for 4-79 Molybdenum Permalloy, an example of a high permeability material. Similar plots have been published [16B1] for iron, cobalt, nickel, 4% silicon-iron, and other materials. There is a break in the curve for iron at an induction of about 18 000 gausses and a similar one for the iron-silicon alloy at about $B = 16\,000$. As pointed out by Ball [16B1], whose data are shown here, breaks of the kind observed may be expected when the material contains

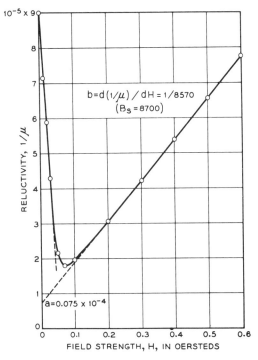

FIG. 11-8. Comparison of Frölich-Kennelly relation (straight line) with experiment. Material is 4-79 Permalloy.

two or more components which have different saturation values that are approached at different rates. However, this is not the only cause of such a sudden change in slope, for it has been observed that in the [111] direction in a single crystal of iron the B,H curve has a sudden break as it nears saturation (Fig. 11-2), and such breaks for crystals oriented in certain ways in a polycrystalline specimen may integrate to produce the shape of curve observed [37S4]. Data for iron [37S4] show that there is a point of inflection in the $\mu-1$ vs $B-H$ curve for this material as saturation is approached (Fig. 3-6).

All of the empirical relations already given imply that the curve relating

$\mu-1$ and $B-H$ approach saturation linearly. For example the Fröhlich-Kennelly relation may be transformed to

$$\mu - 1 = (1/a) - (b/a)(B - H)$$

in which this linear relation is evident. The form used by Weiss may be expressed

$$\mu - 1 = (1/A)(B - H) - (B_s/A)(B - H)^2$$

This latter equation departs from linearity by being *concave downward* as may be inferred from the negative second derivative, $d^2\mu/dB^2$. As a matter of experimental fact all μ,B curves depart from linearity by being *concave upward* at high values of B and concave downward only when B is three-quarters or less of saturation. Thus the value of B_s cannot be determined by extrapolation over any considerable range in B because the curvature of the μ,B curve for a new material is not predictable. In iron the curve is sufficiently linear only when $B-H$ is more than 99% of B_s [37S4].

Polley [39P4] and Becker and Döring [39B5] have used the more general equation

$$I = I_s - a_0/H - b_0/H^2 - c_0/H^3 \cdots + \kappa_0 H$$

which, though it takes no account of the part of the curve that is concave upward, nevertheless is the most satisfactory representation of the facts in sufficiently high fields, and (omitting $c_0, \cdots$) when expressed in the form

$$\left(\frac{dI}{dH} - \kappa_0\right)H^3 = a_0 H + 2b_0$$

gives good agreement with the data for nickel, as shown in Fig. 11-9 by the linearity of the curve obtained by plotting the left side of the equation against H. From such curves relating to data taken at temperatures from $-253°C$ to $+135°C$ Polley has determined the way in which κ_0 and a_0 vary with temperature; this variation is indicated in Fig. 11-10.

The term $\kappa_0 H$ is explained as the increase in spontaneous magnetization caused by the increased alignment of spins in the high field. The value of κ_0 determined by Polley is much larger than that predicted by the Weiss theory, but Holstein and Primakoff [40H7] have calculated its value, using the theory of spin waves, and find good agreement with experiment.

The term in b_0 is attributed to the magnetic crystal forces, which oppose the action of the applied field. In a polycrystalline material with crystals oriented at random the approach to saturation (reversible rotation) may be calculated [31A4, 32G3, 39B5, 41H2] by averaging the magnetization curves for all crystal orientations. The result is

$$I = I_s - \frac{8K^2}{105 I_s H^2} - \frac{192 K^3}{5005 I_s^2 H^3} \cdots$$

according to which the first two constants in the previous equation are

$$a_0 = 0$$
$$b_0 = 0.0762 K^2 / I_s.$$

As Czerlinski [32C6] and Polley [39P4] have shown, values of K for iron and nickel determined from this equation (or its derivative) agree well

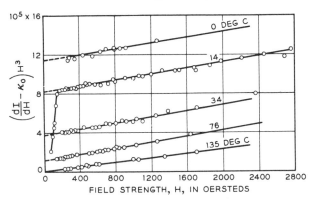

FIG. 11-9. Test of Polley's relation at various temperatures.

with those determined directly from the properties of single crystals, and the validity of this physical interpretation of the constant b_0 is thus established. Holstein and Primakoff, and Néel [48N4], have shown that a

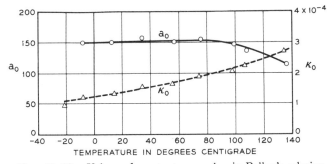

FIG. 11-10. Values of constants a_0 and κ_0 in Polley's relation.

different coefficient of the b_0 term is to be expected in high fields when one takes into account the magnetic interaction between crystals. Néel derives the expression

$$I = I_s - \frac{8K^2}{105 I_s (H + 4\pi I_s/3)^2}.$$

In moderately high fields the second term in parentheses reduces the fraction by a factor of about 2. In an experimental test of this equation

Néel finds that his own data on well-annealed nickel differ from those of Polley in such a way that the derived value of K agrees well with that determined directly from single crystals, about 45 000.

In materials subjected to severe internal strain the equations of p. 482 are applicable, and if $\lambda_s \sigma \gg K$ the strain will predominate. Letting σ_i represent the rms value of the stresses oriented at random in the material, Becker and Polley [40B2] have derived

$$I = I_s - \frac{3}{5} \frac{\lambda^2 \sigma_i^2}{I_s^2 H^2},$$

thus the combined contributions of strain and crystal orientation may be expressed

$$b_0 = \frac{8K^2}{105 I_s} + \frac{3\lambda^2 \sigma_i^2}{5 I_s}.$$

In hammered, hard nickel these authors found b_0 to be about ten times as large as in the same material annealed, and concluded that the second term for b_0 is then predominant. From their measurements we have

$$b_0 = 2.5 \times 10^6 \quad \text{and} \quad \sigma_i = 16 \text{ kg/mm}^2,$$

a value of reasonable magnitude.

Buhl [49B9] has compared the value of σ_i calculated from the approach to saturation with that derived from the initial permeability. According to domain theory (see Chaps. 13 and 18), the initial permeability in a highly strained material is

$$\mu_0 - 1 = \frac{8\pi I_s^2}{9\lambda_s \sigma_i}.$$

Pure carbonyl nickel was severely deformed, and b_0 and μ_0 were measured. Values of the internal stress σ_i, calculated from the two kinds of measurements, differed by no more than a factor of 2.

The term in a_0/H has been explained by Néel [48N4] as due to the nonmagnetic cavities or inclusions which are always present in materials. As a result of these, the intensity of magnetization varies from place to place and the effect of the field is quite different from that to be expected for a perfect crystal. For the rather complicated calculation the reader is referred to the original article. The result shows proportionality with $1/H$, and the absolute value of a_0 is derived as a function of the amount of cavities or inclusions. Experiments were also performed, in collaboration with Lorin, with sintered material with variable density and a calculable proportion of cavities. These show the expected linear relation between the magnetization per gram and $1/H$.

Other relations between B and H have been proposed empirically to represent the approach to saturation over a wide range of field strengths. An extensive list of formulas has been compiled by von Auwers [34A2], and the formulas include logarithmic, exponential, and arctan functions and use as many as five constants.

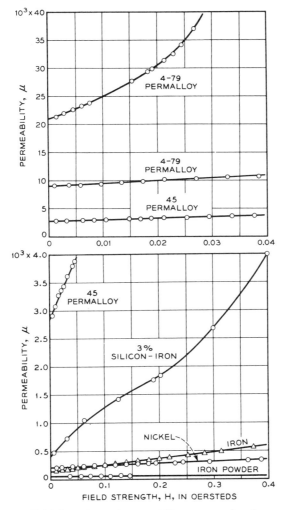

FIG. 11-11. Initial portions of permeability curves of various materials.

Initial Portion of Curve.—Rayleigh [87R1, 24J1] showed that, when the magnetization is made smaller and smaller, the permeability of iron approaches a constant value much greater than one. He also found that when the magnetization is sufficiently low the law

$$\mu = \mu_0 + \nu H$$

is valid. Here μ is the normal permeability in the field H, μ_0 the initial permeability, and ν a constant equal to $d\mu/dH$. This relation is now universally used for extrapolating the μ,H curve to zero field strength to determine μ_0. For most materials this curve is a straight line when B is less than about $B_s/10$, as shown in Fig. 11–11. However, for a few materials, including commercial silicon steel, the line seems to depart from linearity at small inductions and to bend downward to approach a value

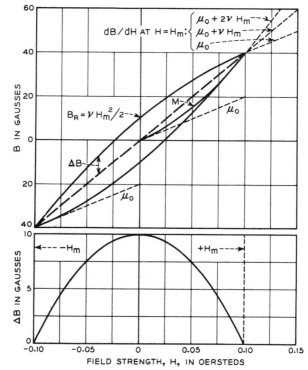

FIG. 11–12. Theoretical form of Rayleigh hysteresis loop at low inductions.

of μ_0 considerably lower than that expected from measurements at higher inductions. One curve of this sort is included in Fig. 11–11 and is to be regarded as the exception rather than the rule among magnetic materials.

The relation described above may be written

$$B = \mu_0 H + \nu H^2$$

and is represented by the line M of Fig. 11–12.

Rayleigh suggested further that the sides of a hysteresis loop with tips at $H = \pm H_m$ and $B = \pm B_m$ may be represented by parabolic curves which, as they leave the tips, have initial slopes of μ_0, the same as the initial slope

INITIAL PORTION OF CURVE 491

of the magnetization curve. Under these assumptions it follows that the equations for the upper and lower branches of the loop are respectively

$$B = \mu H \pm \frac{\mu - \mu_0}{2H_m}(H_m^2 - H^2)$$

where $\mu = B_m/H_m$. Since

$$\mu = \mu_0 + \nu H_m,$$

we have

$$B = (\mu_0 + \nu H_m)H \pm (\nu/2)(H_m^2 - H^2).$$

A Rayleigh loop of this kind is drawn in Fig. 11-12.

The slope of the lower branch of the loop is

$$dB/dH = \mu_0 + \nu H_m + \nu H,$$

and this may be compared with the slope of the normal magnetization curve

$$dB/dH = \mu_0 + 2\nu H.$$

Rates of change of slope (d^2B/dH^2) for the hysteresis loop and magnetization curve differ by a factor of 2 and are respectively equal to the constants ν and 2ν. Values of dB/dH for some points on the hysteresis loop and the normal magnetization curve are given in the Table 1.

TABLE 1. SLOPES OF RAYLEIGH LOOP AND MAGNETIZATION CURVE AT VARIOUS VALUES OF H

	$H = -H_m$	$H = 0$	$H = +H_m$
Hysteresis loop: lower branch..	μ_0	$\mu = \mu_0 + \nu H_m$	$\mu + \nu H_m = \mu_0 + 2\nu H_m$
Magnetization curve.........	...	μ_0	$\mu + \nu H_m = \mu_0 + 2\nu H_m$

The residual induction is

$$B_r = (\nu/2)H_m^2,$$

and the coercive force is

$$H_c = \sqrt{H_m^2 + \mu^2/\nu^2} - \mu/\nu$$

or approximately, when $H_m \ll \mu/\nu$,

$$H_c = \nu H_m^2/(2\mu).$$

The hysteresis loss W_h in ergs/cm^3 per cycle is $1/4\pi$ times the area of the B,H loop:

$$W_h = (1/3\pi)\nu H_m^3 = (\nu/3\pi)B_m^3/\mu^3.$$

During traversal of the loop from one end to the other, that part of the change in induction given by

$$B_{\text{rev}} = \mu_0(2H_m)$$

may be referred to as the reversible change in induction. The remainder,

$$B_{\text{irr}} = \tfrac{1}{2}\nu(2H_m)^2 = 2\nu H_m{}^2,$$

may be called the irreversible change in B. The ratio,

$$B_{\text{irr}}/B_{\text{rev}} = \nu H_m/\mu_0,$$

increases linearly with field.

In dealing with alternating fields the expression for B in terms of H may be written so that the time t is the independent variable. If we write

$$H = H_m \cos \omega t,$$

it follows that

$$B = \mu H_m \cos \omega t \pm (\nu/2) H_m{}^2 \sin^2 \omega t$$

or

$$B = B_m \cos \omega t \pm B_r \sin^2 \omega t.$$

Expansion of $\sin^2 \omega t$ in Fourier series yields

$$B = B_m \cos \omega t \pm B_r \left(\frac{8}{3\pi} \sin \omega t - \frac{8}{15\pi} \sin 3\omega t - \frac{8}{105\pi} \sin 5\omega t \cdots\right).$$

Neglecting the higher terms we have the relations

$$B = B_m \cos \omega t + 0.85 B_r \sin \omega t$$

$$H = H_m \cos \omega t$$

which describe a loop having maximum amplitude of induction $B_m{}' = (B_m{}^2 + 0.85 B_r{}^2)^{\frac{1}{2}}$ and of field strength $H_m{}' = H_m$, having an apparent remanence $B_r{}' = (8/3\pi) B_r$, and having an area equal to the true loop for which the amplitudes are B_m and H_m.

The Fourier expansion shows that the amplitude of the third harmonic component of B, relative to the fundamental, is

$$B_3/B_m = (8/15\pi)(B_r/B_m) = 4\nu H_m/(15\pi\mu).$$

The loss angle ϵ, by which the induction lags behind the field strength, is given by

$$\tan \epsilon = (8/3\pi)(B_r/B_m) = \frac{8}{3\pi} \cdot \frac{\nu}{2\mu^2} \cdot B_m.$$

In the domain theory the initial permeability is a measure of the ease with which a domain boundary may be moved. Quantitative relations between μ_0 and other quantities such as internal stress and magnetostriction will be discussed in Chap. 18.

Experimental Test of Rayleigh's Law.—Since the experiments of Rayleigh the most careful measurements on the μ,H curve and the size and shape of the hysteresis loop at low inductions have been made by Ellwood [35E2]. He used a specimen of compressed iron powder having $\mu_0 = 32$ and $\nu = 1.0$ and made measurements of μ at inductions as low as $B = 0.0002$ [34E2] and measured loops having $B_m = 2$ to 100. Rayleigh's expressions for $d\mu/dH$, B_r, and W_h were satisfied within the experimental error by the same value of ν. The shape of the hysteresis loops was somewhat different from the Rayleigh loop described previously. To show this in convenient fashion Ellwood plotted $B - \mu_m H$, the vertical distance between the point under consideration and the line joining the tips of the loop. This magnitude, ΔB, is shown schematically in Fig. 11-12, and the data for three

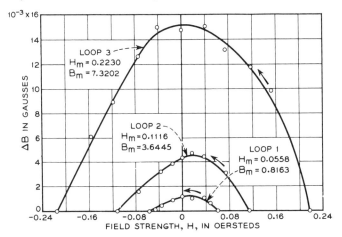

FIG. 11-13. Relation between ΔB and H observed at low inductions.

loops are recorded in Fig. 11-13. The skewing of the loops is particularly apparent for the lower inductions; for Rayleigh's loops $\Delta B = (\nu/2)(H_m{}^2 - H^2)$ and the loop is symmetrical around $H = 0$. The tangent to the ΔB vs H curve at $\pm H_m$ for a symmetrical loop is

$$\tan \theta = \nu H_m$$

while the observed θ's on the descending branch are larger at $+H_m$ and smaller at $-H_m$.

Values of μ_0 and ν for some typical and for some unusual materials are given in Table 2.

The normal increase of μ_0 with temperature is accompanied by a corresponding increase in ν [26W1, 38K15]. The latter constant is also affected markedly by stress and usually but not always increases or decreases in the same direction as μ_0.

A theory of the physical basis of the Rayleigh form of loop, based on an ensemble of domains having a statistical distribution of coercive forces

494 THE MAGNETIZATION CURVE AND THE DOMAIN THEORY

TABLE 2. CONSTANTS FOR SOME MATERIALS AT LOW INDUCTIONS

Material	μ_0	$\nu = d\mu/dH$	Calc. $W_h \times 10^6$ ($B = 1$)
Iron	200	2 000	27.0 ergs/cm^3
Iron (unannealed)	80	60	13.0
Pressed Iron Powder	30	1.0	3.9
Cobalt	70	10	.3
Nickel	220	250	2.5
2-81 Mo Permalloy Powder	120	0.5	.03
45 Permalloy	2 300	16 000	.14
4-79 Mo Permalloy	20 000	350 000	.005
Supermalloy	100 000	12×10^6	.001
45-25 Perminvar	400	0.1	.0002

(caused in turn by variations in internal strains), has been proposed by Preisach [35P2, 39B5], Néel [42N2], and Kondorsky [*J. Phys. USSR* 6, 93 (1942)].

Rectangular Hysteresis Loops.—In some materials the steep portion of the hysteresis loop becomes simply a vertical line—the loop becomes a rectangle, or nearly so, with lines extending horizontally from the tips of the loop.

Such a loop for 69 Permalloy, under a tension of 11 kg/mm^2, is illustrated in Fig. 11–14, and here the coercive force determined in the usual manner has the value indicated by H_s, sometimes called the "starting field" because that is the field necessary to start the change in magnetization along the steep part of the loop. A different value of H for which $B = 0$ is obtained if the field is reduced from a high positive value to zero, and then a negative field equal to or greater than H_s is applied for a short time only. B will then decrease slightly from its value B_r but will not be reduced to zero until the negative field is made equal to H_0, called the "critical field"; the path then followed is indicated in the same figure.

When the maximum field strength of the hysteresis loop is not sufficient to saturate the material, the steep sides of the loop coincide with the inner loop of Fig. 11–14, as illustrated in Fig. 11–15 for two loops having different values of B_m.

Acting on a suggestion of Langmuir, Sixtus and Tonks [31S5] showed that when the magnetization changes along the steep part of a rectangular loop a wave of magnetization moves along the specimen with a measurable speed, the material being magnetized practically to saturation in one direction on one side of the wave-front and in the opposite direction on the other side. In one experiment the wave was stopped before it enveloped the whole wire, and the nature of the boundary between the two regions was examined.

The 68 Permalloy used for Figs. 11–14 and 15 is typical of a material

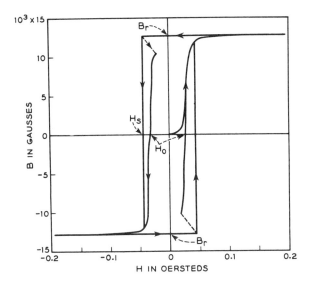

FIG. 11–14. Definition of starting field H_s and critical field H_0 of material having hysteresis loop with straight sides. Inner loop is observed when H_s is applied for short time only (see dotted arrow).

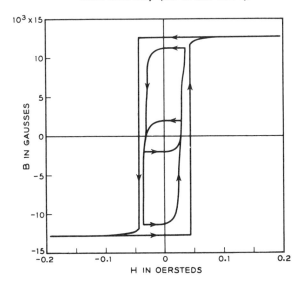

FIG. 11–15. Hysteresis loops of 68 Permalloy at low and high maximum inductions.

having positive magnetostriction and low anisotropy constant. The domains are then easily oriented by tension so that in the absence of an applied field they lie parallel to the line of tension, half in each of the two opposite directions. When a field is applied, half of the domains make 180° reversals by displacement of the domain boundaries until the material is saturated, as shown by the initial magnetization curve in Fig. 11–14. Since domains directed parallel and antiparallel to the field have great stability, the material will remain saturated when a strong positive field has been reduced to zero, and even when some negative field has been applied. When a nucleus of antiparallel magnetization has been formed by a 180° reversal at some place in the material, this region will immediately influence the adjacent region magnetically [31B4], and as a result the wave of reversal is propagated in both directions from the nucleus. The critical

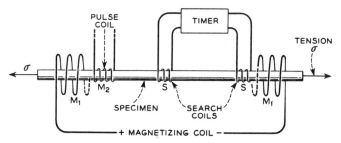

FIG. 11–16. Arrangement for measuring the velocity of large domain boundaries.

field H_0 is the least field strength that will cause motion of the boundary between large domains, oppositely directed, and is useful in interpreting certain magnetic processes.

Observations of these large discontinuities were first made in nickel by Forrer [26F1] and later in Permalloy under tension by Preisach [29P1, 32P1]. Similar properties were observed by Bozorth, Dillinger, and Kelsall [34B3] in material heat-treated in a magnetic field. Measurements of the speed and nature of the propagation were reported in a series of articles by Sixtus and Tonks appearing during the period 1931–1935 [35S6] and in later articles by Sixtus. Additional theoretical treatment of nucleation and the properties of the boundary between domains has been made by Döring [38D2] and by Dijkstra and Snoek [49D5]. The movement of the boundary is limited by eddy currents and perhaps by spin-relaxation phenomena. (Williams, Shockley, Kittel [50W4, 50S8, 51K1].)

The speed of propagation was measured in the manner illustrated in Fig. 11–16 [38S4]. The wire specimen was placed in the magnetizing coil M_1 and subjected to a given tension. Search coils S were placed a suitable distance apart and connected to an electronic device for measuring the time interval between the voltage pulses produced in them by the wave of induction reversal. The wave was started by a current pulse in an additional coil M_2 placed near one end of the specimen. The change of H_0 and H_s and the velocity of propagation with tension are shown for two different materials [38S4] in Figs. 11–17 and 18. In the hard-drawn, iron-nickel wire containing 14% nickel (Fig. 11–17) both H_0 and H_s decrease with increasing tension σ. The velocity of propagation is zero at $H = H_0$, increases linearly with H, and is always higher for higher values of σ. In annealed 78.5 Permalloy (Fig. 11–18), on the

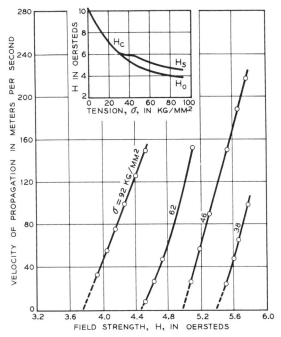

FIG. 11–17. Velocity of propagation of magnetization in a hard wire of a 14% nickel-iron alloy under tension.

contrary, H_s increases with increasing σ, the velocity for a given H decreases with increasing σ, and the velocity for given tension increases parabolically instead of linearly with field strength.

Rectangular loops are observed also when certain materials are heat-treated in a longitudinal magnetic field. The iron-nickel alloys whose properties are modified by this treatment have nickel contents between 50 and 90%; these alloys, and those containing cobalt also, are discussed in Chaps. 5 and 6. They exhibit the same phenomena of wave propagation and have characteristic values of H_0 and H_s, but so far these qualities have been studied but little in a quantitative way [31B4]. In 65 Permalloy (0.006-in. thick), heat-treated in this manner, the time necessary for the flux change to occur may exceed a minute when the field strength just exceeds the starting field H_s.

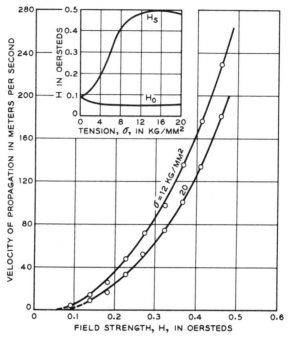

Fig. 11-18. Velocity of propagation of magnetization in a wire of annealed 78 Permalloy under tension.

Constricted Loops.—It has already been pointed out that the Perminvars, important for constancy of permeability in weak and moderate fields, have hysteresis loops that are unusually narrow near the origin—they are constricted or "wasp waisted." In the original investigation [28E1] it was noted that the heat treatment (i.e., "baking") that increased the constancy of permeability in weak fields also enhanced the constricted character of the loop. Figure 11-19(b) shows a constricted loop for 65 Permalloy that has been baked for 24 hours at 425°C, and, for comparison, loops with three other heat treatments are given in (a), (c), and (d)

Materials that have constricted loops when baked have normal unconstricted loops when cooled rapidly. The initial and maximum permeabilities, as well as coercive force and remanence of such materials, are sensitive to heat treatment, but measurement shows no change in the saturation induction. Alloys that respond to baking also respond to heat treatment in a magnetic field. Although the correspondence is not exact, in general it may be said that the iron-cobalt-nickel alloys that respond to these treatments are those represented by the Perminvar region [35E1] of Fig. 5-79. Presumably atomic ordering is involved.

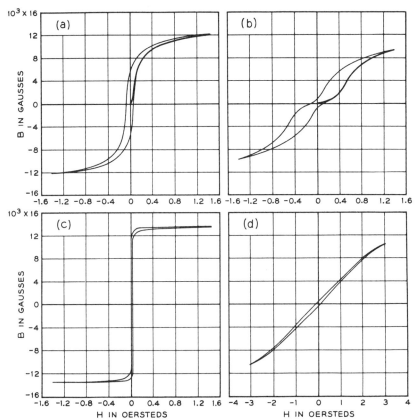

FIG. 11-19. Hysteresis loops of 65 Permalloy heat-treated in various ways: (a) annealed at 1000°C; (b) baked at 425°C for 24 hours, (c) heat-treated in a longitudinal field; (d) heat-treated in a transverse field.

Coercive Force and Residual Induction.—In weak fields, as we have just seen, both the coercive force and the residual induction are proportional to the square of the maximum field strength $H_m{}^2$. In strong fields they approach limiting values called the *coercivity* and the *retentivity*, respectively. The whole H_c vs H_m and B_r vs H_m curves resemble somewhat the normal magnetization curve, as shown by data for 4-79 Molybdenum Permalloy plotted in Fig. 11-20; they differ from the magnetization curve in that they start from the origin with zero slope.

The upper portions of the curves are represented fairly well [20S1] by the equations

$$H_m/H_c = c_1 + c_2 H_m$$

$$H_m/B_r = d_1 + d_2 H_m$$

in which c_1, c_2, d_1, and d_2 are constants, c_2 and d_2 being the reciprocals of

500   THE MAGNETIZATION CURVE AND THE DOMAIN THEORY

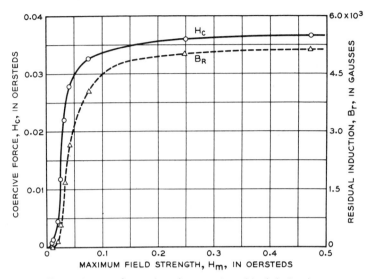

FIG. 11–20. Dependence of coercive force and residual induction on maximum field strength.

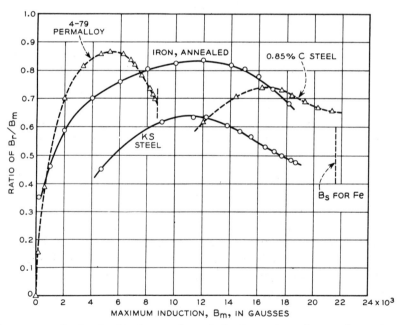

FIG. 11–21. Ratio of residual to maximum induction as dependent on the maximum induction of various materials.

INTERPRETATION OF RETENTIVITY BY DOMAIN THEORY   501

the coercivity and retentivity, respectively. These relations may be used for interpolation and limited extrapolation. Their resemblance to the Kennelly reluctivity relation is obvious.

Values of c_1 and of the coercivity $H_{c\infty} = 1/c_2$ are given in Table 3 for four materials having widely different coercivities. The equation relating H_c to H_m requires that, when the coercive force H_c is 95% of the coercivity

TABLE 3. VALUES OF CONSTANTS IN RELATION $H_m/H_c = c_1 + c_2 H_m$, AND CALCULATED FIELD REQUIRED FOR H_c TO BE 0.95 OF ITS LIMITING VALUE

Material	c_1	$1/c_2 = H_{c\infty}$	H_m for $H_c = 0.95 H_{c\infty}$
4-79 Mo Permalloy	0.3	0.037	0.2
Iron	.25	.73	3.7
Mild Steel	.28	4.1	23.0
Honda Steel [20S1]	.4	220.0	1750.0

$H_{c\infty}$, the value of H_c is approximately $20 c_1 H_{c\infty}$. Values of H_m required to make $H_c = 0.95 H_{c\infty}$ are calculated from the constants and shown in the last column.

Although B_r increases continually with increase in H_m, the ratio B_r/B_m passes through a maximum and then decreases to its limiting value for large values of H_m. Thus it may be said that the residual induction approaches its limit more rapidly than the normal induction approaches its limit, B_s. Data for several materials are plotted in Fig. 11–21.

Both B_r and H_c decrease as the temperature increases [10T1, 10H1, 32K2], and apparently also the ratio B_r/B_s [10T1].

The retentivity may vary over wide limits depending upon a number of factors, among which may be mentioned composition, fabrication, heat treatment, stress, and temperature.

In the first part of Table 4 are shown some relevant data for materials that may be classified as "normal." It is apparent that the ratio of retentivity to saturation induction, B_r/B_s, for these materials does not vary too much from the value 0.5. In the lower part of the table, on the contrary, are listed some materials and treatments that show extreme values of B_r/B_s; 65 Permalloy is chosen as an example of a material showing unusual variation, and for it the limiting values 0 and 1 are approached closely. A variation in the rate of cooling of this material during the heat treatment was sufficient to change the ratio of B_r/B_s from 0.41 to 0.07. Heat treating in a magnetic field resulted in the ratio 0.90 or 0.04, depending upon whether this field was parallel or at right angles to the field applied during measurement. The effect of tension was to give the extreme ratio of 0.98, and presumably pressure results in a low ratio.

Interpretation of Retentivity by Domain Theory.—The different values of retentivity observed are readily interpreted in a qualitative way by means of domain theory. In a magnetic material the initial or demagnetized state may be represented by a series of vectors emanating from a point, the direction of each vector representing the direction of magnetization of a

TABLE 4. RETENTIVITY OF VARIOUS MATERIALS

Material	Treatment	B_r	B_s	B_r/B_s	Theoretical B_r/B_s
Iron..........	Cold drawn	8000–11 000	21 600	0.4–0.5	0.5
	Annealed	6000–14 000	21 600	0.3–0.7	.5
Nickel.........	Cold drawn	2900– 3 900	6 100	0.5–0.65	...
	Annealed	2000– 4 000	6 100	0.3–0.65	.5
4 Si-Fe.........	Annealed	6000– 8 000	19 800	0.3–0.4	.5
	Cold rolled Annealed	14 000	20 200	0.7	...
45 Permalloy....	Annealed	7500– 9 500	16 000	0.45–0.6	.5
	Cold rolled 95% Annealed	7 000	16 000	0.45	.5
4-79 Molybdenum Permalloy	Fast cool	3800– 5 100	8 700	0.45–0.65	.5
65 Permalloy....	Hard	6 500	14 400	0.4–0.5	.5
	Quenched	5 900	14 400	.41	.5
	Fast cool	4 500	14 400	.31	...
	Slow cool	1 600	14 400	.11	...
	Baked	1 000	14 400	.07	0
	Under tension	14 100	14 400	.98	1.0
	Annealed in longitudinal field	13 000	14 400	.90	1.0
	Annealed in transverse field	600	14 400	.04	0

group of domains. If there is no preferred direction of magnetization in the specimen as a whole the ends of the vectors will be uniformly distributed over the surface of a sphere as indicated in (A) of Fig. 11–22(a). In an unannealed material having high internal strains oriented at random, a high field will cause saturation (B), and upon reduction of the field to zero the domains will assume orientations determined by the directions of the local strains and will lie parallel or antiparallel to their original directions. As shown at (C) they will all have a component of magnetization parallel to the direction in which the specimen was saturated. Under these conditions the retentivity will be half of saturation:

$$\frac{B_r}{B_s} = \frac{1}{2\pi} \int_0^{\pi/2} 2\pi \sin \theta \cos \theta \, d\theta = \tfrac{1}{2}.$$

Here θ is the angle between the local magnetization and the field.

INTERPRETATION OF RETENTIVITY BY DOMAIN THEORY   503

In the usual annealed material having random crystal orientation the same vector diagrams (a) are appropriate; the stable positions are now determined by crystal anisotropy.

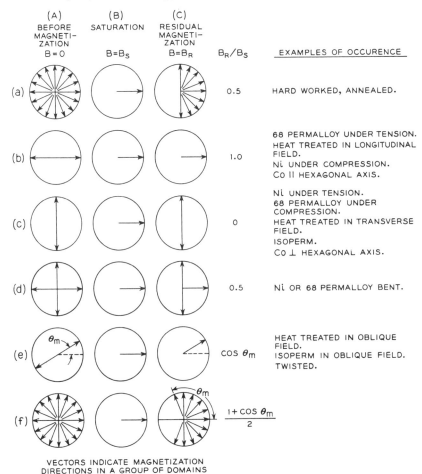

FIG. 11-22. Schematic representation of domain distributions in various stages of magnetization, for various types of materials.

When *tension* is applied to a material having positive magnetostriction (e.g., 68 Permalloy), the domains are oriented parallel and antiparallel to the magnetic field as shown in (A) of Fig. 11-22(b) (see Fig. 13-23). After magnetization (B) and reduction of the field strength to zero (C), the retentivity is equal to saturation in the ideal case, and by experiment values of B_r/B_s of 0.99 or more have been observed [38K4]. Similar diagrams apply when tension is applied to the Permalloys containing

60-80% nickel; to nickel and the high nickel Permalloys under compression; and to some alloys heat-treated in a longitudinal magnetic field.

In nickel under tension, and in Permalloy heat-treated in a transverse field, domains are oriented originally at right angles to the axis of tension and are changed by the applied field as indicated in Fig. 11-22(c). Here the retentivity approaches zero as a limit, values of B_r/B_s as low as 0.04 having been observed. A similar orientation of domains occurs in the Isoperms made by special treatment of certain Fe-Ni or Fe-Ni-Cu alloys. In the Fe-Ni Isoperm the domains are oriented in the rolling plane of the sheet at right angles to the direction of rolling [35S2]; in the Fe-Ni-Cu Isoperms they are perpendicular to the rolling plane [34K4].

When a magnetostrictively negative specimen of strain-sensitive material is *bent*, the part under compression will be represented by (b) of Fig. 11-22, the part under tension by (c), and the specimen as a whole by (d). At remanence the domains are oriented so that approximately half are parallel and half at 90° to the direction in which the field has been applied. A ratio B_r/B_s of 0.5 has been observed in bent nickel [26F1, 31K6]. A similar behavior may be expected in materials having positive magnetostriction. Strain-sensitive materials when *twisted* generally have a stable domain orientation inclined to the axis of the specimen and may be represented by Fig. 11-22(e). B_r/B_s may be expected to approximate 0.7 when $\theta = 45°$.

When crystal anisotropy and strain are both weak, the retentivity is sometimes much lower than the theoretical value $0.5B_s$, characteristic of any "normal" isotropic material. Three ways will be considered in which the low retentivity might be explained. It is believed that the true explanation of Perminvar lies in the last mechanism to be described [48B2].

The *first* way of interpreting a low retentivity is purely formal, and no mechanism corresponding to it has been suggested. According to this, there are stable configurations corresponding to values of θ greater than 90°, as shown in (C) of Fig. 11-22(f). Then the ratio B_r/B_s is

$$\frac{B_r}{B_s} = \frac{\int_0^{\theta_m} 2\pi \sin\theta \cos\theta \, d\theta}{\int_0^{\theta_m} 2\pi \sin\theta \, d\theta} = \frac{1 + \cos\theta_m}{2},$$

(in which θ_m represents the maximum value of θ) and may vary from 0.5 to 0 as θ_m varies from 90 to 180°. Values of θ_m greater than 90° may be expected when a specimen is subjected to a demagnetizing field or other field opposite in direction to that originally applied, but it is improbable that domains can be so oriented in a closed magnetic circuit and therefore at residual induction.

The *second* possible way to account for values of B_r/B_s of less than 0.5 ($\theta_m > 90°$) involves the moving boundary as an important step in the change in magnetization from saturation to remanence as well as from the demagnetized state to saturation. In the upper part of Fig. 11-23 are diagrams for two domains in a material, such as 68 Permalloy, having a stable orientation fixed by the angle $\theta = 60°$. The directions of magnetization in these domains are originally oppositely directed as shown in (A) and are both parallel to one of the cubic axes of the crystal, the directions of the other two axes being less stable because the microstresses, resulting

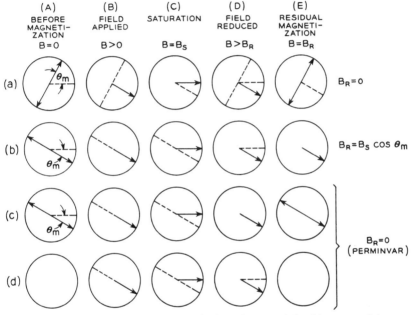

FIG. 11-23. Proposed domain distributions for materials of low retentivity.

from the heat treatment, favor the first axis. Upon application of a field in the direction $\theta = 0$ a second axis becomes more stable and both domains assume the orientation (B) by movement of 90° boundaries. A stronger field causes saturation (C) and reduction of the field results in condition (D), the same as (B). When the field strength is reduced to zero, the stable position is again parallel to the first axis, and since both directions along this axis are equally stable and are both inclined 90° to the direction shown in (D), the domains will go half one way and half the other by displacement of 90° boundaries. As pictured in (E) the retentivity is zero. On the contrary, if θ is small, say 30°, the same argument leads to $B_r = B_s \cos \theta$: the transition from (A) to (B) takes place by displacement of a 180° boundary between the two domains.

There is a lower limit to the ratio B_r/B_s for the mechanism just described. For the calculation of this limit let θ be the angle between the magnetic field and the nearest cubic axis of the crystal in which the domain lies; determine the average value of $\cos\theta$ over all possible values of θ. For each orientation of the field with respect to the crystal axis there are three possible directions for the local stress, one for each of the axes, and for two of these directions the retentivity will be zero as explained in the upper diagrams of Fig. 11-23. Then the quantity to be evaluated is $\overline{(\cos\theta)}/3$, with the average to be taken over that portion of the surface of a unit sphere that lies between the crystallographic directions [100], [110], and [111]. This average is

$$\overline{\cos\theta} = \frac{\int_{\phi=0}^{\pi/4}\int_{\theta=0}^{\operatorname{ctn}^{-1}\cos\phi} \sin\theta\cos\theta\,d\theta d\phi}{\iint \sin\theta\,d\theta d\phi} = \frac{3}{\sqrt{2}} - \frac{3\sqrt{2}}{\pi}\tan^{-1}\sqrt{2}$$

and is approximately 0.834. B_r/B_s is then $0.834/3 = 0.278$. Although this is considerably lower than 0.5, it is still not nearly low enough to account for the value 0.07 observed for 68 Permalloy heat-treated for a long time at 450°C.

The *third* method by which retentivity may be reduced in theory below the value 0.5 takes into account the lack of domain structure in material having low crystal anisotropy [48B2]. Experiments on powder patterns, which will be described later, indicate that the conventional domain structure, such as that observed in iron, does not exist in the Permalloys in which the anisotropy is small. This may be connected with the large width of the domain walls and the low wall energy which are associated with low anistropy and low magnetostriction [48B2] (see Chap. 18). Kittel [49K1] has suggested that under these circumstances flux-closure is the predominant force, and that the flux lines in large areas will bend and form closed paths when the material is unmagnetized. When a magnetic field is applied, these lines will be broken, but when the field is removed, the lines will reform into much the same pattern as that which existed before the field was applied. The remanence would thus approach zero.

It is significant that low retentivity is possessed by those iron-nickel alloys that have very small crystal anisotropy and have been heated for a long time at a low temperature (450°C) to minimize internal strains. Low anisotropy is also known [39T2] to exist in the iron-rich Fe-Si alloys, and an unpublished experiment of Boothby shows that an abnormally low retentivity is possessed by an alloy containing 6.5% silicon, cooled slowly from 1100°C. The ratio $B_r/B_s = 2000/18\,500 = 0.11$ and is in contrast to the high retentivity and permeability obtained when the same specimen is heat-treated in a longitudinal field.

A further discussion of coercive force and domain theory is given in Chap. 18.

Hysteresis.—A representative series of normal hysteresis loops is shown in Fig. 11–24. When the maximum field strength H_m for the loop is small,

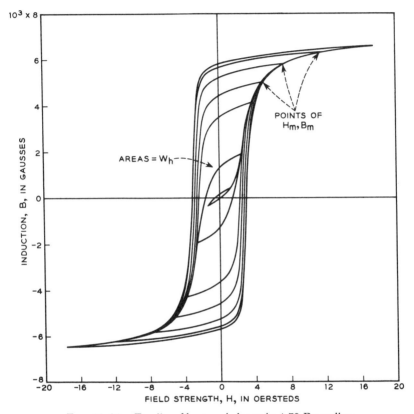

FIG. 11–24. Family of hysteresis loops in 4-79 Permalloy.

the loop is lens-shaped, its sides parabolic. When H_m is increased, the lens becomes thicker in proportion to its length; when H_m is increased further, the horizontal width of the loop tends to become more nearly the same at every value of B and at the same time the loop becomes somewhat S-shaped; when H_m is still greater, the shape of the loop changes only by the addition of "tails" in the first and third quadrants, the tails following closely the normal magnetization curves measured in the two opposite directions.

The energy loss associated with the hysteresis loop is readily determined

508 THE MAGNETIZATION CURVE AND THE DOMAIN THEORY

from the area of the loop (Warburg's law [81W1]) and is

$$W_h = (1/4\pi) \oint H dB = \oint H dI \text{ ergs/cm}^3 \text{ per cycle}$$

when B and H are in gausses and oersteds. When W_h is about 10 000, as it is when soft iron is subjected to a large cyclic field, the energy released is sufficient to raise the temperature of the specimen about 0.0003°C per cycle.

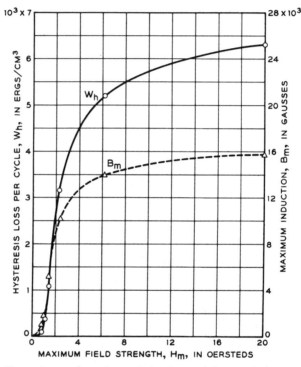

FIG. 11-25. Dependence of maximum induction and hysteresis loss on maximum field strength.

In Fig. 11-25 the hysteresis loss is plotted against the maximum field strength in each cycle for a specimen of soft iron having a maximum permeability of 9000 and coercivity of 0.75. Although the W_h, H_m and B_m, H_m curves rise most rapidly at about the same field strength (near $H_m = H_c$), the W_h curve continues to rise thereafter more than the B_m curve.

In Fig. 11-26 hysteresis loss is plotted against maximum induction for the same specimen over a large range of values of B_m and W_h. At the lowest values W_h increases as B_m^3, as already discussed in connection with the Rayleigh loop. For somewhat higher values of B_m, Rayleigh's law,

$$W_h = (1/3\pi)\nu H_m^3 = (1/3\pi)\nu B_m^3/\mu^3$$

is still valid, although W_h is not proportional to B_m^3 but more nearly proportional to B_m^2 because of the way in which μ changes with B_m. In the specimen under discussion Rayleigh's law begins to fail in the neighborhood of $B_m = 200$ and is off by a factor of 2 at $B_m = 1000$.

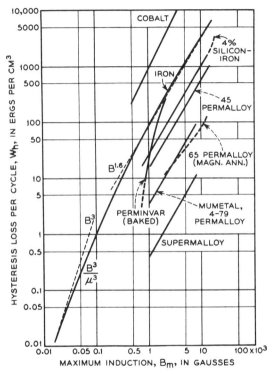

Fig. 11–26. Hysteresis loss of some common materials at various maximum inductions. Note comparison with Steinmetz and Rayleigh laws.

For values of B_m ranging from 500 to 15 000 in iron the hysteresis is given rather accurately by the law of Steinmetz [91S1, 92S1],

$$W_h = \eta B^{1.6},$$

which is represented by the broken line in the figure. This relation is much used (in cgs units) for interpolation and limited extrapolation, and is appropriate for a variety of materials. Caution must be exercised, however, in applying it to a new material for which its validity has not been tested. Data are shown for a variety of materials in the same figure, and for nickel in Fig. 11–31.

Some of the curves in Fig. 11–26 increase in slope when B_m exceeds a certain value, while others are straight at the highest values of B_m attained. Many observers [11W1, 16B1, 23J1, 26W2, 33S7] have reported values of the exponent ranging from 1.6 upward to 2.5 or 3.0 at inductions of 15 000 to 18 000 gausses, and some have observed a decreasing exponent when B exceeds 15 000. The reason for the increased slope has been investigated by Ball [16B1], who concluded that at least one reason for the upward turn is lack of uniformity in the specimen, and proved his point by making an

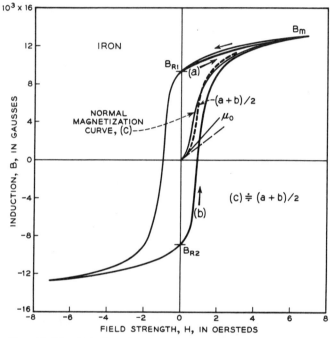

Fig. 11–27. Relation of magnetization curve to hysteresis loop.

artificially non-uniform ring; he measured iron and cobalt specimens separately and as a single composite ring and observed the rise in the curve for the composite ring at high inductions. Some of the increased slope in the older curves for commercial silicon-iron was shown to be due to scale left on the surface of the specimens after annealing. More data on hysteresis loss at high inductions in pure homogeneous material is necessary to establish the normal course of the W_h, B_m curve for inductions ranging from 15 000 to saturation.

As the curves of Fig. 5–88 show, the Steinmetz law fails for materials having the characteristic constricted loops of Perminvar; it also fails for materials having rectangular loops as a result of heat treatment in a longitudinal magnetic field. Data for some permanent magnets have been reported by Jellinghaus [49J2].

Other empirical relations involving W_h have been proposed and found limited use. One of these [10R1], valid over a limited range, is

$$W_h = aB_m + bB_m^2.$$

Another relation gives the hysteresis loss in terms of B_m and H_c; if the loop has the same horizontal width at every value of B,

$$W_h = (2H_c)(2B_m)/4\pi = H_c B_m/\pi,$$

or, more generally, one may write

$$W_h = kH_c B_m/\pi.$$

It has been found [22A1] that the value of k varies approximately linearly with B_m over the range of B_m used in power transformers, and for iron, silicon-iron, cast iron, and cobalt it is given approximately by

$$k = 0.67 + 0.36 \times 10^{-4} B_m$$

so that it has the convenient value of 1 in the neighborhood of $B_m = 10\,000$.

The error function (probability integral) [42W4] and hyperbolic tangent may be used to simulate magnetization curves and hysteresis loops:

$$B = B_s \operatorname{erf} \frac{H \pm H_c}{H_c}$$

$$B = B_s \tanh \frac{H \pm H_c}{H_c},$$

the minus sign being used for the ascending branch, the plus sign for the descending portion of the loop. Still other mathematical expressions have been used and referred to by Sequenz [35S7].

Elenbaas [32E2] and Kühlewein [32K7] have pointed out an interesting relationship between certain parts of the hysteresis loops and the normal magnetization curve. As indicated in Fig. 11-27, B,H curves are drawn from the two remanence points to saturation in one direction [curves (a) and (b)]. For a given value of H the average value of the B's for these two curves is equal approximately to B for the normal magnetization curve (c). That is,

$$c = (a + b)/2.$$

This relation is quite accurate for a variety of materials; it is not to be relied upon in a quantitative way, especially when the hysteresis loops are abnormal in shape.

512 THE MAGNETIZATION CURVE AND THE DOMAIN THEORY

Hysteresis is affected by many factors, among the most important of which the following may be listed.

Gross composition   Heat treatment
Impurities   Temperature
Fabrication   Stress

Figure 11-28 shows the large effects of *heat treatment* and *impurities*. Commercial ingot iron, cold rolled 50% after anneal, has the loop shown

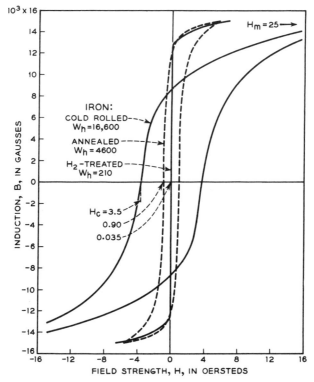

FIG. 11-28. Effect of treatment of specimen on the hysteresis of iron.

with $W_h = 16\,600$ for $B_m = 15\,000$. After annealing in the usual way at 900°C, W_h is reduced by a factor of 3.6 to 4600. After purification accomplished by heat treating in pure hydrogen at 1450°C for 24 hours, W_h is 210, a further reduction by a factor of 22. The effect of *gross composition* is well illustrated in the Permalloy series—the hysteresis losses for nickel and for 78.5 Permalloy are, respectively, 10 000 and 300 for $B_m = 5000$.

The effect of *fabrication* is first of all to produce internal strains which may, however, be eliminated by annealing. A further result of fabrication is often a special crystal orientation that remains after the usual annealing.

Such orientation is well known in silicon-steel strip that has been annealed after controlled cold rolling, and a permeability curve for this material is shown in Fig. 4–20 where it is compared with the same alloy fabricated by

TABLE 5. HYSTERESIS LOSSES (ERGS/CM^3) IN VARIOUS MATERIALS, AT SPECIFIED MAXIMUM INDUCTION, B_m
(The coefficient η is that used in the Steinmetz equation $W_h = \eta B^x$, x being taken here as 1.6.)

Material	Composition (%)	Treatment	B_m	W_h	$\eta \times 10^6$
Iron.........	99.9 Fe	Cold rolled 50%	10 000	10 000	4 000
Iron.........	99.9 Fe	Annealed 900°C	10 000	3 000	1 200
Iron.........	99.9 Fe	Annealed 1400°, H$_2$	14 000	190	45
Silicon-iron.....	4 Si, 96 Fe	Hot rolled, annealed 800°	10 000	1 300	500
Silicon-iron.....	3 Si, 97 Fe	Cold rolled, annealed 1200°	10 000	300	120
45 Permalloy...	45 Ni, 55 Fe	Annealed 950°	10 000	700	280
45 Permalloy...	45 Ni, 55 Fe	Annealed 1200°, H$_2$	10 000	100	40
Hipernik.......	50 Ni, 50 Fe	Annealed 1200°, H$_2$	10 000	100	40
4-79 Molybdenum Permalloy	4 Mo, 79 Ni, 17 Fe	Annealed 1100°	5 000	40	15
4-79 Molybdenum Permalloy	4 Mo, 79 Ni, 17 Fe	Annealed 1350°, H$_2$	5 000	8	3
Supermalloy....	5 Mo, 79 Ni, 16 Fe	Annealed 1300°, H$_2$	5 000	4	1.5
Mumetal.......	76 Ni, 3 Cu, 2 Cr, 19 Fe	Annealed 1050°	5 000	50	20
Perminvar	45 Ni, 25 Co, 30 Fe	425°, 24 hr	600	0	...
			1 000	15	...
			10 000	4 000	...
Chrome steel...	3 Cr, 1 C, 0.4 Mn, 96 Fe	825°, quenched	14 000	0.25 $\times$ 10^6	60 000
KS Steel.......	36 Co, 7 W, 3 Cr, 1C, 0.5 Mn, 52 Fe	950°, quenched	12 500	0.9 $\times$ 10^6	250 000
Alnico 5........	24 Co, 14 Ni, 8 Al, 3 Cu, 51 Fe	Annealed 1300°, cooled in field, baked 600°	14 000	2.5 $\times$ 10^6	600 000

hot rolling. The ratio of the hysteresis losses for these two materials at $B_m = 10\,000$ varies considerably, but it is sometimes as high as 5 to 1. Since H_c and B_r almost invariably decrease with increasing *temperature*, when no phase transformations occur, it follows that hysteresis also decreases. This is borne out by the data discussed in other chapters. The effect of *stress* on hysteresis is referred to in Chap. 13.

Data on hysteresis loss in various materials are summarized in Table 5. The large range of values should be noted, extending from 4 ergs/cm^3 per cycle at $B_m = 5000$ for Supermalloy to 2.5×10^6 for Alnico 5. Attention should be paid also to the unusually low losses in Perminvar already discussed in connection with Figs. 5–87 and 88.

Hysteresis is often measured in watts/kg referred to a field alternating f times per second. For convenience the relations between these units and W_h expressed in ergs/cm^3 per cycle is given below:

$$\text{No. of watts/kg} = \frac{10^{-4} f W_h}{\delta}$$

$$\text{No. of watts/lb} = \frac{10^{-4} f W_h}{2.205 \delta}.$$

Here δ is the density in g/cm^3. The number of equivalent ergs/cm^3 per cycle in one unit for iron and for 4% silicon-iron is given in Table 6.

TABLE 6. UNITS OF HYSTERESIS LOSS

Unit	f	Equivalent ergs/cm^3/cycle	
		Fe	4 Si-Fe
1 watt/kg.......	60	1312	1267
	50	1574	1520
1 watt/lb.......	60	2893	2793
	50	3472	3352

Rotational Hysteresis.—When a disk of magnetic material is placed in a uniform magnetic field directed parallel to a diameter, the induced magnetization is generally not parallel to the field. As a result there is a torque acting on the disk, of magnitude

$$L = -HI \sin \theta.$$

Here θ is the angle between H and I, and the negative sign indicates that the torque opposes the motion. When the disk is turned slowly through 360°, the energy consumed is called the rotational hysteresis loss per cycle and is

$$W_r = -\int_0^{2\pi} L d\phi.$$

When L is constant with the azimuth angle ϕ, the loss is simply

$$W_r = -2\pi L.$$

Determinations of W_r are usually made by measuring L in a torque magnetometer [38B3, 37W3] and can presumably be made by measuring the increase in temperature due to the heat evolved, provided the eddy-current loss is known to be negligible. L has also been determined by measuring the magnitude and direction of I with a ballistic galvanometer, and determining the product $HI \sin \theta$ [15G1]. The torque may be determined equally well [32H2] from the relation

$$L = -H_a I \sin \theta_a$$

where H_a is the magnitude of the applied field, uncorrected for demagnetizing force, and θ_a the angle between H_a and I. The equivalence of the two equations for L can be shown by a simple analysis, assuming only that the demagnetizing field is antiparallel to the magnetization (see p. 591).

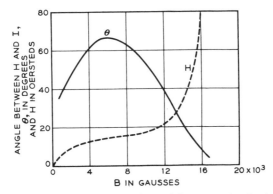

Fig. 11-29. Rotational hysteresis: angle of lag of the magnetization behind the field, for various inductions, in magnet steel having a coercive force of 14.

Consequently $I \sin \theta_a$ may be determined by measuring the flux in a coil placed perpendicular to H_a, and L may be obtained by multiplying by H_a [38I2].

The angle θ by which I lags behind H has been determined by Gans and Loyarte [15G1] for a disk of steel having a coercive force of 14 oersteds. As shown in Fig. 11-29 the angle has a maximum value at about $\frac{1}{3}$ or $\frac{1}{4}$ of saturation and is zero at $B = 0$ and $B = B_s$. The rotational hysteresis loss W_r also passes through a maximum, and this occurs at a much higher induction, just beyond the knee of the magnetization curve. After this, W_r declines toward zero at saturation as indicated in Fig. 11-30. The general shape of the W_r,B curve was first established by Baily [96B1], after it was predicted as a consequence of the Ewing theory.

As compared with the usual or alternating hysteresis, W_h, the rotating hysteresis is the larger at low inductions and the smaller at high inductions. This fact is illustrated for nickel [08W1] in Fig. 11-31 and for a commercial

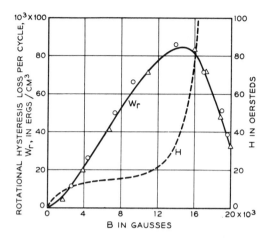

Fig. 11–30. Rotational hysteresis loss at various inductions, in steel.

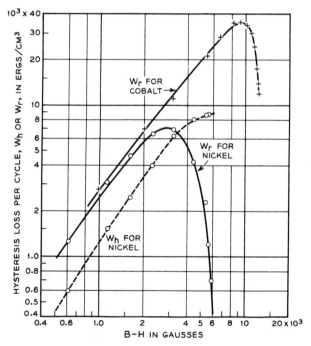

Fig. 11–31. Comparison of rotational (W_r) and alternating (W_h) hysteresis loss.

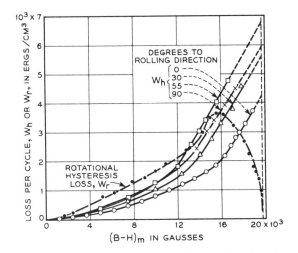

Fig. 11–32. Alternating and rotational hysteresis loss in grain-oriented silicon-iron sheet.

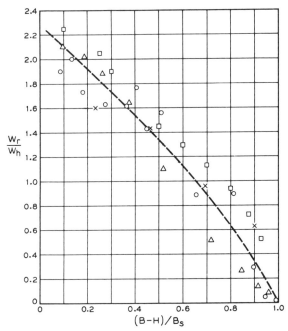

Fig. 11–33. Ratio of rotational to alternating hysteresis loss in various materials.

iron-silicon alloy (3.1 Si) in Fig. 11–32 [39B6]. The latter alloy is highly directional in its magnetic properties, and in such a material it is difficult to determine $\int_0^{2\pi} L d\theta$ because L varies over a wide range of values during the cycle. Brailsford [38B3] has overcome this difficulty nicely by superposing two or three similar disks oriented relative to each other so that the high positive and high negative values of L neutralize each other to a high degree, the remaining L varying over a range small enough to permit accurate integration of the L,θ curve. Using this method he has determined W_r [39B6] for iron and for iron-silicon alloys containing 1.9, 3.1, and 3.7% silicon. Data for W_h for the same specimens were determined at the same time.

The ratio W_r/W_h for several materials is plotted against B/B_s in Fig. 11–33. Although the points are rather scattered, it appears that the ratio has a finite value at $B = 0$, is unity in the neighborhood of $B/B_s = 0.6$ to 0.8, and goes to zero at saturation. The dependence of the ratio on such things as crystal anisotropy and strain has apparently not been investigated. Data for cobalt [01B1] of unknown analysis are included in Fig. 11–31.

Distribution of Heat-loss over Magnetic Cycle.—Warburg's law gives accurately the hysteresis loss for a complete cycle of magnetization but gives no indication of the amount of heat generated in any portion of the cycle. The latter problem was attacked first by Adelsberger [27A1] who found the expected heating when traversing most of the loop and, unexpectedly, a cooling during part of the cycle.

Determination of the heat produced during a small portion of a magnetic cycle requires careful measurement with special technique. The rise in temperature during a complete cycle is about 0.0003°C in annealed iron; consequently perceptible measurement of about 10^{-6} °C is necessary for detailed examination of the loss in energy. This is accomplished by using thermocouples of many junctions, junctions of only one kind being in contact with the specimen. Thermal insulation and control of ambient temperature are important, and some scheme for amplifying the galvanometer deflections is usually employed for measuring the thermocouple emf's.

A representative curve, obtained by Okamura [36O2] for partially annealed iron and showing the double heating and cooling typical of a high permeability material, is reproduced in Fig. 11–34. Upon reducing the field strength from a high value, corresponding to a magnetization of over 90% of saturation, a cooling is observed first. At lower inductions this changes to marked heating, then to cooling when the induction reverses sign, and finally heating occurs again when magnetization increases toward saturation in the opposite direction. A similar course is followed during the following half cycle, as shown, but the net heating over the com-

DISTRIBUTION OF HEAT-LOSS OVER MAGNETIC CYCLE 519

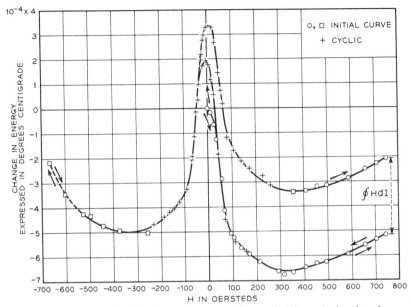

FIG. 11-34. Change in the thermal energy of annealed iron during its change in magnetization.

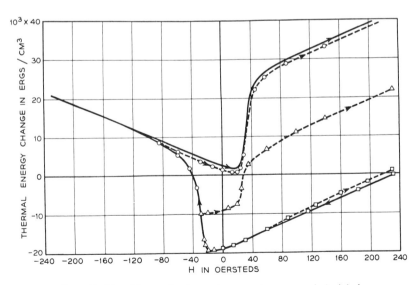

FIG. 11-35. Change in thermal energy in unannealed nickel.

plete cycle remains positive and equal to $\oint HdI$, in accordance with Warburg's law.

A different behavior is observed for unannealed nickel, as illustrated by the curve of Fig. 11-35, measured by Miss Townsend [35T2]. The cycle consists simply of cooling and subsequent heating. In contrast to the curve for soft iron, the greatest drop in temperature is as large as the net rise in temperature during the half cycle.

It is apparent that for each kind of material the heating and cooling over part of the cycle are reversible, and over part irreversible. The reversible changes occur at high inductions, whereas the irreversible changes occur at low inductions, when the field strength is near the coercive force. Okamura has attempted to separate the reversible from the irreversible portion, and

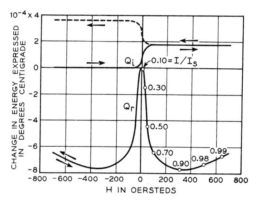

FIG. 11-36. Attempted separation of thermal energy into reversible and irreversible components.

the result of his separation for soft iron is given in Fig. 11-36. The curve for the reversible heat is traced and retraced for each cycle, while that for the irreversible heating never lowers, but rises by $\oint HdI$ for each full cycle. The irreversible heat Q_i is identified with the hysteresis loss; the reversible heat Q_r, with the magnetocaloric effect, which is discussed in more detail in Chap. 15.

It is interesting to see over what part of the loop Q_i changes most rapidly. If we accept Okamura's method of separation of Q_i from Q_r, we find (Fig. 11-37) that the change is most rapid when $B-H$ is changing most rapidly with H, and that the change in Q_i occurs over a smaller range in H than does the change in $B-H$. This is in accordance with our previous information regarding magnetic processes—the irreversible changes (boundary displacements) occur along the steep portions of the magnetization curve and hysteresis loop, and the reversible changes (domain rotations) occur above the knee of the curve.

The equation for the magnetocaloric effect relates the (reversible) heat evolved to the change of magnetization with temperature at constant field strength:

$$\Delta Q = - T(\partial I/\partial T)_H \Delta H \text{ ergs/cm}^3.$$

When the magnetization is small or of intermediate value its change with temperature, $\partial I/\partial T$, is positive (e.g., see Fig. 14–2); $\Delta Q/\Delta H$ is therefore

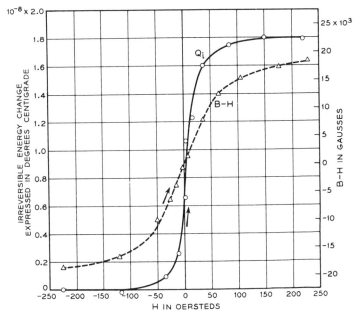

FIG. 11–37. Relation between irreversible thermal loss and magnetic induction in annealed iron.

negative and an increase of H causes cooling proportional to the absolute temperature T and inversely proportional to the density d. When the magnetization is near saturation, $\partial I/\partial T$ is negative and application of a field produces heating; release of the field, cooling. In the highest fields when saturation is already attained, $\partial I/\partial T$ is the change in spontaneous magnetization I_s with temperature and is constant to a first approximation; the heating is then linear with H.

This is the qualitative explanation [3602] of the lower curve of Fig. 11–36, and Okamura has measured $\partial I/\partial T$ and found agreement between the Q_r calculated therefrom and the Q_r observed.

When the magnetic state of a body is changed, the work done on the material, the change in magnetic energy, is $\int H dI$, and during a complete magnetic cycle the net amount of work done is equal to the heat liberated, $Q_i = \oint H dI$. When the magnetic state of hard nickel, represented by the tip of a hysteresis loop, is changed by decreasing the field, it is surprising

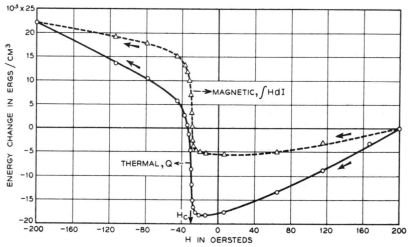

FIG. 11-38. Relation between thermal energy and magnetic energy in unannealed nickel.

to find that the change in the magnetic energy is far less than the heat liberated. As shown in Fig. 11-38 [38H1],

$$\left| \int H dI \right| < |Q|.$$

In other words, the material cools faster than it would if all of the decrease in magnetic energy were used up in the cooling. This must mean that there is an increase in the potential energy of the material other than the potential energy of magnetization. An indication of a change in potential energy of this sort was first observed by Constant [28C3] in magnet steel and amply confirmed by Hardy and Quimby [38H1]. More recently Bates and his colleagues [41B3, 43B1, 47B3, 48B5, 48B6, 48B7, 49B13] have emphasized the importance of this phenomenon and have plotted

$$\Delta = \int H dI - Q$$

for a variety of materials. Some of their curves, obtained by plotting Δ vs $B-H$, are given in Fig. 11-39. Other data on the evolution of heat

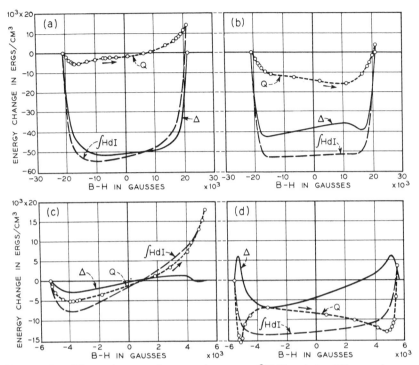

FIG. 11–39. Thermal energy Q, magnetic energy $\int H dI$, and the difference between them Δ, in several materials: (a) unannealed iron, (b) annealed iron, (c) unannealed nickel, (d) annealed nickel.

have been reported by Ellwood [30E5] and Honda, Okubo, and Hirone [29H4].

The physical processes involved in the evolution and absorption of heat have been analyzed and discussed by Stoner and Rhodes [49S8]. The reversible processes are:

(1) the increase in the magnetization above the "spontaneous" or "intrinsic" value I_s, as discussed above;
(2) the rotation of the domains against the forces of crystal anisotropy, occurring mainly in considerably lower fields; and
(3) reversible boundary displacement observed in weak fields.

The irreversible processes are:

(4) the sudden movement of domain walls (as observed in the Barkhausen effect); and, in some cases,
(5) sudden changes in domain orientation (as in single domain particles).

The calculation of the increase in magnetization according to (1) can be carried out as described above, using the magnetocaloric equation in which the change of magnetization with temperature, dI/dT, is taken from the known variation of saturation magnetization with temperature (Chap. 14). Calculation of the rotational process (2) involves the change of the crystal anisotropy with temperature, and this can be estimated from the data available (Chap. 12). At present the other processes are not known well enough to permit theoretical analysis. Process (1) results in a heating as the field is increased, process (2) results in cooling, and the relative magnitudes of the two effects determines whether there is a net heating or cooling. Differences in the shapes of the experimental curves are attributed to the relative importance of these two processes.

In comparing the theory with the data Stoner and Rhodes have calculated the heating resulting from the change in intrinsic magnetization, because this can be done with considerable accuracy, and then considered the remainder in terms of the other processes. In iron this remainder is roughly constant in high fields and is about one-half of the amount calculated for the rotational process, as estimated from the data of Hardy and Quimby [38H1]. It begins to deviate strongly from this value only when the field is so small that irreversible processes are known to occur. In cobalt the rotation effect predominates, a fact associated with the high Curie point and high crystal anisotropy.

Separation of the effects of the reversible and irreversible is highly desirable, and a method has been outlined for accomplishing this in a much more quantitative manner than that used by Okamura. Such a separation would contribute to our knowledge of the irreversible processes.

Barkhausen Effect.—In 1919 Barkhausen [19B2] discovered the effect known by his name and interpreted the experiment as demonstrating the irregularities in the magnetization of iron. This effect may be demonstrated simply as shown by the right side of Fig. 11–40, which is self-explanatory. The noise heard in the phones, or in a loud-speaker, persists only during a change in magnetization on the steep part of the magnetization curve or hysteresis loop. It is thus identified with irreversible changes in magnetization that occur as distinct events.

Oscillographic records [29B1] of the discontinuities are reproduced in Fig. 11–41. Here the impulses occur in both directions from the zero line only because two search coils, oppositely wound, were connected in series to the amplifier.

As pointed out by Tyndall [24T1], the change in magnetic moment for a single discontinuity may be determined from the area under the impulse on the oscillographic record. In a specimen of silicon-steel he found discontinuities corresponding to a change in moment as large as 0.008 unit, equivalent to the complete reversal of magnetization in a volume of 4×10^{-6} cm^3. Both he and Preisach [29P1] measured the sizes of many individual discontinuities, and the latter found that on the steepest part of the hysteresis loop the discontinuous change so determined was equal to the total change as measured in the usual way with a ballistic galvanometer, while on the flatter portions of the loop the discontinuous part was

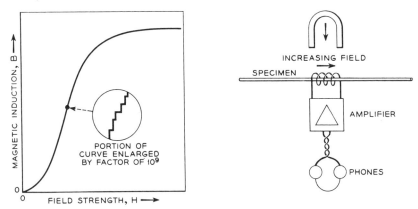

FIG. 11–40. Barkhausen effect and method of observing it.

negligibly small. The same result was obtained by Bozorth [29B1] who, with a rectifier, suppressed the impulses arising from one of the two coils used to produce the record of Fig. 11–41 and integrated electrically those from the other coil. When the field was changed rapidly, the impulses from the two coils overlapped so that the change in magnetization measured

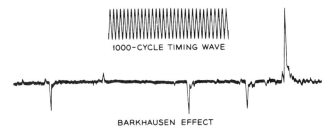

FIG. 11–41. Oscillograph records of Barkhausen effect, and timing line of 60 cps. Impulses are normally in one direction only, but are in both directions here because there are two opposing pick-up coils.

in this way was too small; when the field was varied more and more slowly, the observed discontinuous change occurring on the steep part of the hysteresis loop approached as a limit the total change observed with the ballistic galvanometer. Data are given in Fig. 11–42.

The average size of the Barkhausen discontinuity has been the subject of investigation by Bozorth and Dillinger [30B1]. The theory of their determination is as follows. Assume that the magnetic moment of a small volume, v, is initially vI_s, directed parallel to the axis of the specimen and surrounding coil, and that it changes suddenly to vI_s antiparallel to the original orientation. The form of the voltage impulse in the coil and

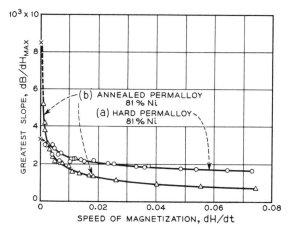

FIG. 11–42. When the field strength is changed more and more slowly the measured change in the discontinuous magnetization approaches the total change in magnetization, at the steepest part of the loop.

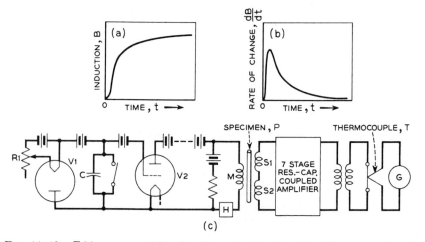

FIG. 11–43. Eddy currents delay the change in induction (a). The area under the Barkhausen impulses is proportional to the change in induction (b). Circuit used in measuring the power loss associated with the Barkhausen effect (c).

the current impulse at the output of the amplifier will depend on the decay of the eddy currents induced by the changing magnetization. The change of induction with time may be represented by Fig. 11–43(a). If A is the cross-sectional area of the specimen and V the volume enclosed by the coil, the momentary current will be

$$i_1 = CAdB/dt = CA(v/V)2B_s f(t)$$

in which C is a proportionality constant and $f(t)$ a function resembling that shown in Fig. 11–43(b) and satisfying the relation

$$\int_{-\infty}^{+\infty} f(t)\, dt = 1.$$

The area under the i_1, t curve then is

$$\int_{-\infty}^{+\infty} i_1\, dt = CA(v/V)2B_s$$

and is a measure of the size of the discontinuity v.

When a succession of discontinuities occurs at the average rate of n per second, the average current in the amplifier output is

$$\bar{i} = CA(\overline{dB/dt}) = CAn(\bar{v}/V)2B_s.$$

The mean square current is

$$\overline{i^2} = n\int_{-\infty}^{+\infty} i_1^2\, dt = 4nC^2 A^2 B_s^2 (\overline{v^2}/V^2) \int_{-\infty}^{+\infty} f^2(t)\, dt.$$

Combining these equations gives

$$\frac{\overline{v^2}}{\bar{v}} = \frac{V\overline{i^2}}{2C^2 A^2 B_s (dB/dt)\int_{-\infty}^{+\infty} f^2(t)\, dt}.$$

From the theory of eddy-current decay in a cylinder we have [30B1]

$$\int_{-\infty}^{+\infty} f^2(t)\, dt = \frac{2\rho}{A\mu_r}$$

in terms of the resistivity ρ and reversible permeability μ_r. Using also the relations $V = Al$ and $\overline{v^2} = 0.7\, \bar{v}^2$, we get finally

$$\bar{v} = \frac{0.2l\mu_r \overline{i^2}}{C^2 \rho B_s dB/dt}.$$

Here the numerical factor 0.2 is known only approximately. The length of the specimen in which discontinuities affect the coil, denoted by l, is not simply the part between the ends of the coil but depends on μ_r, especially when μ_r is large and is determined experimentally. This equation shows that the magnitude of the Barkhausen noise, which may be defined as $\overline{i^2}$, not only is proportional to $\bar{v}$, B_s, and dB/dt, but depends on μ_r/ρ and the effective length of the specimen:

$$\overline{i^2} = \frac{C^2 \bar{v} B_s dB/dt}{0.2 l \mu_r/\rho}.$$

When a given search coil is used around a single cylinder or wire, the noise is independent of the diameter of the specimen, to a first approximation. If the specimen is a bundle of wires, the noise is proportional to the number in the bundle; consequently for given coil and volume of material the noise is enhanced by sub-dividing the specimen longitudinally so as to increase the rate of decay of eddy currents without decreasing dB/dt.

Determinations of $\bar{v}$ for various materials have been made using the relation given above. The diagram of Fig. 11–43(c) shows the circuit used. The specimen P is subjected to a slowly changing magnetic field regulated by the resistance R_1, and the two search coils S_1 and S_2 surround the specimen and lead through an amplifier and transformer to a thermocouple T in which the energy dissipated is measured by the deflection of the galvanometer G. In addition to $\overline{i^2}$ so measured, dB/dH, μ_r, and l were determined for each specimen as dependent on field strength for one half of the hysteresis cycle. Values of dB/dH and $\overline{i^2}$, recorded photographically, are shown for annealed iron in Figs. 11–44(a) and (b); actual values of these quantities and μ_r and of the desired quantity $\bar{v}$ are shown in (c).

In Fig. 11–44(d) the relation of $\bar{v}$ to the hysteresis loop is apparent; in iron the largest discontinuities occur at the steepest part of the loop, and it may be significant that this is where most of the hysteresis loss occurs, as indicated in Fig. 11–37. On the contrary, the largest discontinuities in 50 Permalloy occur definitely off the steep part of the loop and near the point of maximum curvature. The fluctuations are apparent in Fig. 11–45(a) showing dB/dH as recorded for two successive loops without the aid of an amplifier, and here it is evident that the largest fluctuations occur well beyond the place where dB/dH is a maximum. Quantitative results are given in Fig. 11–45(b). It would be interesting to know whether or not the heat of hysteresis is greatest at this same point on the loop for 50 Permalloy.

This method of determining the average size of Barkhausen discontinuities has been used by Förster and Wetzel [41F1] who have obtained values of $\bar{v}$ in substantial agreement with those already mentioned. Similar results have also been obtained by Tebble, Skidmore and Corner [50T1], using improved technique and analysis.

In the domain theory it is important to know not only the magnitude of the change in magnetization but also the orientations before and after the change. One method of approach to this problem is a study of the *trans-*

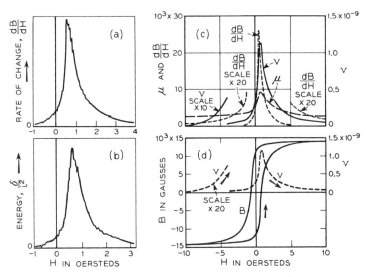

Fig. 11–44. Photographic records of (a) dB/dH and (b) the instantaneous power loss associated with the Barkhausen effect in iron. Data so obtained are used (c) to determine the average volume v of the Barkhausen discontinuity, shown in (d) in relation to the hysteresis loop.

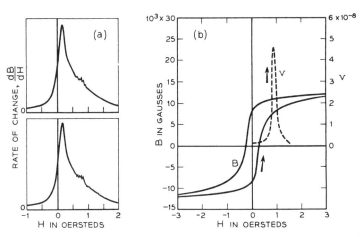

Fig. 11–45. Volume of Barkhausen discontinuities in 50 Permalloy, in relation to the hysteresis loop.

verse Barkhausen effect [32B4] in which is measured the ratio of the changes in magnetic moment parallel and at right angles to the direction of the magnetic field. The specimen used in these experiments is in the form of a tube that may be magnetized axially in the usual way by putting it in a solenoid, or circumferentially by passing a current through a wire lying along the axis. The search coils are placed coaxially in the way shown in Fig. 11-43 for the longitudinal effect, and the theory of the measurements is the same. The effect of the geometry of the specimen on the rate of decay of eddy currents is the same for both effects.

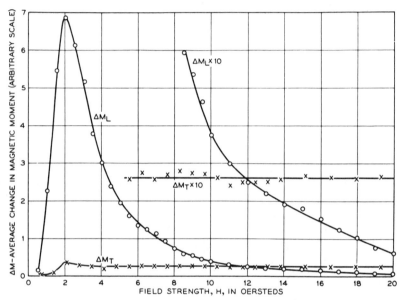

FIG. 11–46. Relative sizes of the longitudinal and transverse Barkhausen effect in iron.

Data for annealed iron are shown in Fig. 11-46. The transverse change in moment is relatively small for low inductions and it is only when $B = 15\,000$ ($H = 12$) that the two are equal. In higher fields the change in transverse moment becomes several times as large as the longitudinal change. In unannealed iron the ratio of transverse to longitudinal change at any specified induction is even larger than for annealed iron. The results show clearly that the change in the direction of magnetization sometimes makes a large angle with the axis of the specimen along which the field changes. At low inductions the most plausible interpretation of the data is that there exists a preponderance of 180° reversals with initial and final positions inclined but slightly to the axis.

An individual rather than statistical study of the change in the direction of magnetization in domains has been made by McKeehan, Beck, and Clash [32B5, 34M5, 35C4]. In their experiments a single crystal of iron containing 3.2% silicon was rotated in a constant field capable of magnetizing the disk to about 0.8 of saturation. Coils around the disk picked up components of voltage 90° to each other, and these impulses after amplification were applied to the two pairs of plates of a cathode ray oscillograph. Each discontinuity in the disk thus registered on the screen as a line extending outward from the center, and its direction indicated the direction of change in moment of the domain in the disk.

The results showed a preponderance of 180° reversals along [100] directions with a considerable amount of variation from this angle. Changes in the direction of magnetization by 90° were seldom noted. It was also observed that many of the largest impulses gave zigzag lines on the oscillo-

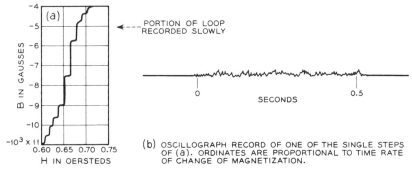

FIG. 11-47. Large discontinuities in a specimen of iron (a), and oscillographic record (b) of one large discontinuity showing that it is composed of many small ones.

graph, lines that changed somewhat in direction during the impulse; this indicates that the reversal of one domain sets off the change of another closely coupled to it magnetically, in the same way that domains are set off in the large Barkhausen discontinuities associated with loops having vertical sides. Similar observations have been made by Förster and Wetzel [41F1]. Proof that the large Barkhausen jumps do contain many such small Barkhausen units is given in Fig. 11-47 [31B4]. Here the largest single jump in part (a) is recorded by a rapid oscillograph giving the record shown in (b), and one notes the many irregularities occurring between two quiet periods. Even the large jumps are not noticeable in hysteresis loops measured in the usual way; they are only evident when H is changed slowly and B is recorded accurately.

The number of domains that are associated and form what may be called a "cluster" vary over wide limits. In materials having rectangular hysteresis loops the cluster sometimes constitutes the whole specimen; in other materials having loops with no large value of dB/dH one may expect the domains to act as independent units and in many cases they are known to do so. Usually, however, there are considerable fluctuations in the rate of change of B that involve large numbers of domains. The corresponding irregularities in the magnetization curve have been noticed, especially when using automatic recording with a relatively fast galvanometer and slow change in field strength, and have been reported by Forrer and Martak [26F1, 32F1] and Bozorth and Dillinger [31B4]. To show the nature and size of these fluctuations two ring specimens of 45 Permalloy were each wound with primary and secondary turns, the secondaries connected in series opposition to a recording fluxmeter, and H changed slowly. If the

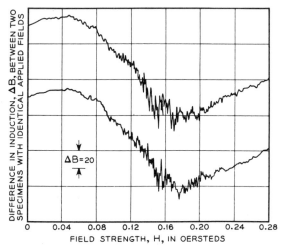

Fig. 11-48. Large irregularities of magnetization shown in records of induction in two coils connected in opposition. These are non-reproducible from cycle to cycle. Material is 4-79 Permalloy.

specimens had been identical in magnetic characteristics, the traces on the recording paper, reproduced in Fig. 11-48, would have been horizontal straight lines.

It is apparent from the figure that (1) as may be expected the specimens do not have exactly the same B,H curve, reproducible differences of about 100 gausses being apparent in each cycle; and that (2) there are fluctuations from cycle to cycle as large as 30–40 gausses, corresponding to the reversal of a volume of the order of 10^{-4} cm³. The latter, therefore, are due to the clustering of many thousands of Barkhausen discontinuities of the usual size.

Powder Patterns.—For many decades iron filings have been used to portray the directions of lines of magnetic force in air and to detect flaws or inhomogeneities in magnetic materials. In 1931 it occurred to von Hámos and Thiessen [31H3, 32H3] to use fine magnetic powder to detect the local inhomogeneities in magnetization that the domain theory predicts. Independently Bitter [31B5] applied a suspension of siderac (Fe_2O_3), having particles about 10^{-4} cm in diameter, to a polished magnetized surface and observed under the microscope that the powder formed parallel lines regularly spaced about 0.1 mm apart and approximately perpendicular to the direction of magnetization.

The technique and interpretation of such patterns constituted then the subject of study by a number of workers [32A3, 33S6, 34K12, 34M6, 37E3, 37S5]. The preparation of colloid for these studies has been described in some detail by Elmore [38E4, 49W1] who recommends a suspension of magnetite, ground to colloidal dimensions, peptized with hydrochloric acid and protected by 1% soap. A more stable colloid has been prepared by Baker, Winslow, and Bittrich of the Bell Laboratories by precipitating Fe_3O_4 from a solution of $FeCl_2$ and $FeCl_3$ with NaOH and stabilizing the colloidal suspension with dodecyl amine.* Electrolytic polishing [42E2, 49W1] overcomes the objectionable mechanical polishing which disturbs the surface.

A notable advance was made by McKeehan and Elmore [34M3, 34M7, 36E3] who first observed a well-defined pattern on a demagnetized single crystal. Figure 11-49 shows such a pattern (b) and also those observed when the magnetization is directed (a) into, or (c) out of, the same portion of the surface as that shown in (b). The suspension used for the experiments was a true colloid of Fe_2O_3 particles small enough to show Brownian movement, and a change in magnetization of the magnetic specimen was accompanied by a movement of the lines immediately visible to the eye. See also Akulov and Dekhtiar, *Ann. Physik* **15**, 750 (1932).

More recent work by Williams, Bozorth, and Shockley [49W1, 49W2, 49B1, 49B3] has made visible for the first time the domain boundaries characteristic of unstrained iron and has improved considerably our knowledge of the processes of magnetization. They used single crystals containing 3.8% silicon and having surfaces cut nearly parallel to crystallographic planes. The specimens were annealed and polished carefully, first mechanically and then electrolytically. After mechanical polishing the powder pattern on a surface almost parallel to (100) is the "maze" pattern of Fig. 11-50(a), similar to that of Fig. 11-49. After electrolytically

---

* The stabilizing solution is made by adding 2 g of cocoanut oil amine (mostly dodecyl amine), as obtained from Armour and Co., to about 10 cm³ of normal hydrochloric acid to bring to a pH of 7. The solution is then diluted with distilled water to 50 cm³ and 20 cm³ of the Fe_3O_4 slurry added. After thorough mixing, the whole system is brought to 150 cm³ with distilled water and stirred vigorously at 6000 rpm. It is finally diluted again to a total of 600 cm³ and is ready for use. The particles are about 1000 A in diameter, as determined by Heidenreich, and are charged positively with respect to the solution.

polishing and reapplying the powder to the same area, the result is the "tree" pattern of Fig. 11–50(b). It is evident from this and other experiments that the maze pattern is characteristic of a strained surface and that the tree pattern shows the domain boundaries of strain-free material.

The directions of magnetization in the domains can be determined in several ways, using techniques described in the original paper. The result

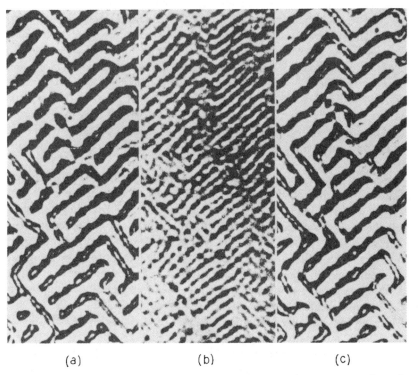

(a)   (b)   (c)

FIG. 11–49. "Maze" pattern of fine magnetic powder on the polished surface of an iron crystal, caused by domain structure. In the middle section (b) the magnetization is zero, in (a) and (c) the field is directed inwards and outwards from the surface, respectively. Magnification about 200 times.

for a portion of one tree pattern is shown in Fig. 11–50(c). The local magnetization in unmagnetized material is always parallel to one of the crystal axes, and the boundaries separate domains magnetized at 90° or at 180° to each other.

Visible movement of domain boundaries takes place upon application of field or stress. The effect of uniform tension is shown in Fig. 11–51. In this material tension increases the magnetization in the direction of the tension, and the mechanism by which this is accomplished is here apparent:

POWDER PATTERNS 535

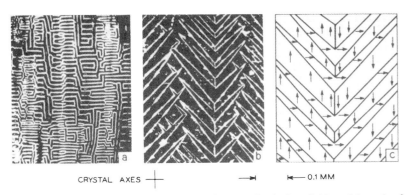

FIG. 11-50. Powder patterns obtained after mechanical polishing (a) and after electrolytic polishing (b) of the same surface area. Specimen unmagnetized.

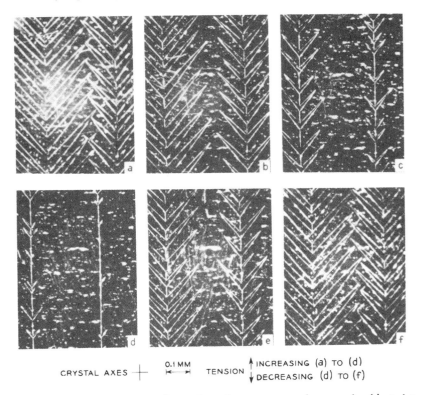

FIG. 11-51. Powder patterns observed on the same area of a crystal subjected to increasing tension (a to d) and then decreasing tension (d to f). Tension favors growth of domains magnetized parallel to tension axis.

536 THE MAGNETIZATION CURVE AND THE DOMAIN THEORY

domains oriented parallel to the axis of tension are enlarged by displacement of domain boundaries, at the expense of domains oriented at right angles, so that the latter domains disappear almost entirely when the tension is sufficiently large. With release of tension the original kind of tree pattern forms, but the details of the pattern are not the same. This shows that the boundaries are not fixed to the structure of the crystal in the way that they are in the maze pattern, where the local stresses always cause the return of the powder lines to the same places after they have been disturbed temporarily by field or uniform stress.

The tree pattern occurs only when the surface is slightly inclined to a (100) plane. When the surface is accurately parallel, the domains appear

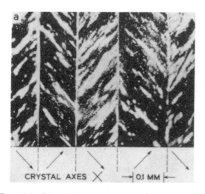
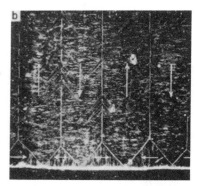

FIG. 11-52. Patterns on surfaces accurately parallel to (100) crystal planes. The long domain boundaries are parallel to (001) planes, and triangular domains are "domains of closure" for the flux of the larger domains.

to be thin sheets. Two patterns on such surfaces are shown in Fig. 11-52. In (a) a concentrated colloid was used, and a number of experiments show that this causes the appearance of striations aligned at right angles to the direction of magnetization. Adjacent domains are here at 90° to each other. In (b) adjacent domains differ by 180° in the middle of the specimen, whereas near the edges the structure shows "domains of closure" which close the flux paths from the larger domains without the formation of poles.

The patterns of Fig. 11-53 show the movement of domain boundaries caused by a magnetic field. As indicated schematically at the right, the field causes the irregular boundary to be displaced downward; this displacement permits the domains oriented favorably with respect to the field (those at the top) to grow at the expense of the others.

When the surfaces are not parallel or nearly parallel to simple crystallographic planes, the patterns are likely to be more complicated. Figure 11-54 shows two examples of such patterns. Although the simple patterns

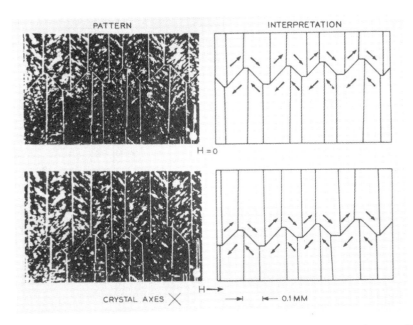

FIG. 11–53. Domain structure and wall movement as specimen (with surface parallel to a (100 plane) is magnetized parallel to a [011] direction. Arrows show directions of domain magnetization.

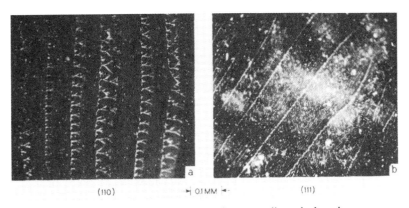

FIG. 11–54. Powder patterns showing rather complicated domain structure on (110) (partially magnetized) and (111) surfaces of iron crystal surfaces (3.8% Si).

are well understood, it has not yet been possible to understand in detail the more elaborate ones. It is believed, however, that the basic principles that apply to the simple ones are also applicable to the more complex ones. These principles are discussed in Chap. 18.

Experiments on cobalt have also been instructive. Bitter [32B3] observed two types of patterns on polycrystalline material, and Elmore [38E3, 38E7], working with single crystals, found the hexagonal lace-like patterns on surfaces parallel to the hexagonal planes (00·1) perpendicular to the crystal axis, and the straight line patterns on prism planes, as shown in the photographs of Fig. 11-55, taken by H. J. Williams. These patterns are in accord with the magnetic properties of cobalt, known to have a direction of easy magnetization parallel to the crystal axis. The domains are then expected to be long in the direction of the axis and packed together

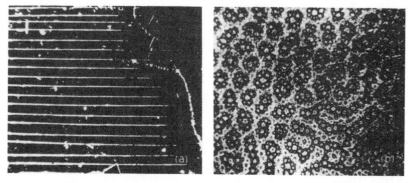

Fig. 11-55. Powder patterns of cobalt on surfaces parallel (a) and perpendicular (b) to the hexagonal crystal axis, showing the rod-like (or plate-like) domain structure.

like a bundle of needles (or sheets). The boundaries of such domains thus correspond to the patterns. Moving pictures of the patterns taken with slowly changing field strength show sudden displacements of the boundaries corresponding to jumps much larger than those usually attributed to the Barkhausen effect.

Germer [42G1] has measured the strength of magnetic fields close to the surface of an unmagnetized cobalt crystal and found that near a hexagonal face it is of the order of 10^4 oersteds and falls off with distance from the surface so that it is relatively weak at 0.01 mm. The fields near prism faces are weaker and fall off more slowly with distance, in the way that one would expect if the domains are needle-like as assumed.

Patterns observed more recently by Williams, Nesbitt and Bozorth, and references to earlier brief notes by Williams (summarized in [49W1]), are given in the report of the conference held in Grenoble in July 1950 [51B1].

Incremental and Reversible Permeability.—When a magnetic field acting on a specimen is maintained constant and an additional field is alternated cyclically between two limiting values, the fields and inductions thus produced are said to be superposed, and the cyclically changing induction ΔB, divided by the cyclically changing field strength ΔH, is the *incremental*

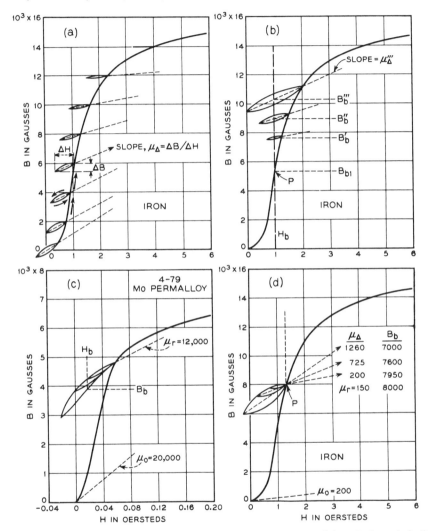

Fig. 11-56. Minor hysteresis loops taken under various conditions in iron (a,b,d) and 4-79 Permalloy (c).

permeability μ_Δ. A normal magnetization curve and superposed hysteresis loops of constant ΔH are shown in Fig. 11-56(a).

The point midway between the tips of a superposed loop is designated H_b, B_b, and these quantities are called the biasing field strength and biasing induction respectively. Figure 11–56(b) shows three loops having the same biasing field; (c) and (d) show superposed loops at constant maximum field strength, both H_b and B_b varying with amplitude.

The *reversible permeability* μ_r, for values of H_b and B_b lying on the normal magnetization curve, may be determined in either of two ways illustrated in Fig. 11–56(b) and (d). As indicated in (b), μ_Δ is determined for several values of ΔH (H_b remaining constant), ΔH *increasing* for each succeeding determination, and μ_r obtained by extrapolating μ_Δ to $\Delta H = 0$. Alternatively, as in (c) and (d), the steady field is applied in one direction, and the alternating field is applied first in the opposite direction and then cycled between limits, and μ_r is obtained again by extrapolating μ_Δ to $\Delta H = 0$. Using either method, μ_r is determined for a point H_b, B_b on the normal magnetization curve and may be determined in a similar manner for any point on any B,H curve. Measurements may be made by the usual method employing the ballistic galvanometer, but they are usually made more easily using three separate windings on a ring specimen, one producing the biasing field (H_b), another connected to a ballistic galvanometer (for B_b), and the third leading to an inductance bridge (for μ_Δ). In using the bridge, care must be taken to avoid eddy-current shielding in the specimen and to prevent the influencing of the bridge measurement by too low an impedance in the circuit containing the biasing winding. Account must be taken also of accommodation, for the first loop traced will have a somewhat different value of B_b than the loop obtained after repeated cycling of ΔH.

Incremental permeability is of technical importance in the design of "output" transformers and loading coils used in communication. In the output transformer the primary winding carries the same current as the vacuum tube to which it is connected; therefore the core is magnetically biased and the signal to be transmitted causes a superposed alternation in induction of relatively small amplitude. The desirable property of the material is thus a high reversible or incremental permeability with high biasing field. Both 45 Permalloy and Vanadium Permendur are good in this respect.

Loading coils supply inductance in communication lines carrying both telephone and telegraph currents. The coil should have an inductance independent of current so that the impedance and therefore the strength of the telephone signal should not vary when the telegraph current goes on and off. Such a variation is known as the "flutter effect" [21F1, 29D1] and is due to the variation of incremental permeability with biasing field strength. This variation can be reduced by interposing an air gap in the magnetic circuit, thus reducing the magnitude of the biasing field strength at the expense of the superposed permeability.

The way in which the incremental and reversible permeabilities vary with the biasing field strength H_b is shown in Fig. 11–57 for the normal magnetization curve for 45 Permalloy. Both μ_Δ and μ_r decrease continually as H_b increases and approach the limit one.

Small hysteresis loops superposed on a large one are shown in Fig. 11–58. The alternating amplitude of B is held constant at $\Delta B = 500$ or 1000;

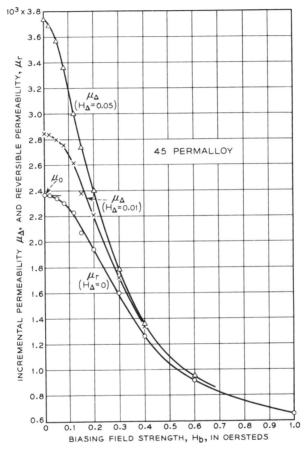

Fig. 11–57. Incremental and reversible permeabilities of 45 Permalloy measured with various biasing fields and incremental field strengths.

hence the length of the loop is an inverse measure of the incremental permeability, which is seen to be a maximum in the second (or fourth) quadrant for this material, Molybdenum Permalloy.

When μ_r is plotted against H_b, as one traverses a hysteresis loop of large amplitude, we have the characteristic shape of the "butterfly" loop formed by the solid line of Fig. 11–59. Hysteresis is obvious, and its importance

542 THE MAGNETIZATION CURVE AND THE DOMAIN THEORY

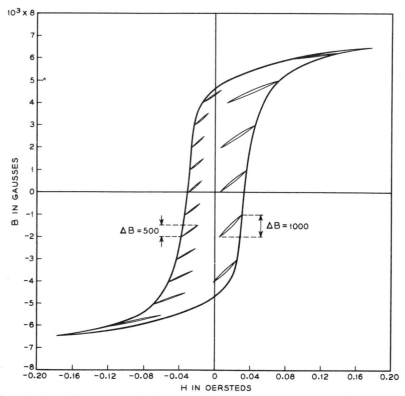

Fig. 11–58. Minor and major hysteresis loops of 4-79 Permalloy. For the minor loops ΔB is constant at 500 (left side) or 1000 (right side) gausses.

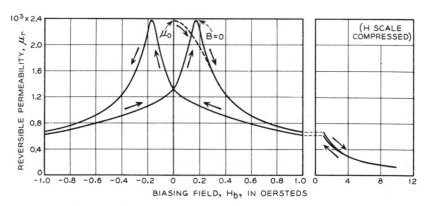

Fig. 11–59. Reversible permeabllity vs biasing field. Material 45 Permalloy.

in output transformers is apparent. Goldschmidt [30G3] has plotted μ_0/μ_r, instead of μ_r, against H_b for such loops and found that two sides of the loop are straight (Fig. 11–60); this recalls the more familiar Frölich-Kennelly reluctivity relation in which $1/\mu$ is approximately linear with H.

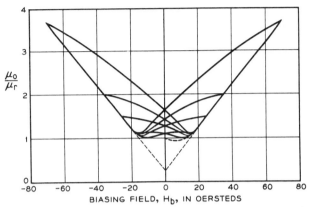

Fig. 11–60. Reciprocal of reversible permeability plotted against biasing field.

The loop of Fig. 11–59 refers to 45 Permalloy, and shows that for this material μ_r is a maximum at $B = 0$ on the hysteresis loop ($H = H_c$), and also that μ_r at this point is equal to μ_0.

Reversible Permeability vs Induction.—It was suggested by Gans [10G1] that μ_r is determined by B_b alone; consequently μ_r is often plotted against B_b as in Fig. 11–61 for 45 Permalloy. Careful measurements show that μ_r is not a single-valued function of B_b, but that for some materials, including 45 Permalloy and iron [29S1], the departure from this simplification is not large. In many cases, however, the relation is far from being so simple. In Fig. 11–62 are loops that show almost as much deviation from a single curve as the loops obtained by plotting μ_r vs H_b. It is also apparent here that the maximum value of μ_r for the loop is not generally equal to μ_0.

Materials like Perminvar that have hysteresis loops of unusual shapes have also peculiar reversible permeability curves. One of these has been shown in Fig. 5–98.

It may be significant that there are only small departures from Gans's rule when the material has a "normal" domain structure and changes in magnetization take place as illustrated in Fig. 11–5. When the anisotropy constant is small and the domain structure sensitive to strain and heat treatment, we may expect the unusual magnetic properties found in the Perminvars and 68 Permalloy annealed in various ways. The reversible permeability is affected by the same factors.

The reversible permeability is a measure of the firmness with which the domains are held in position by the biasing field. The difficulty in its quantitative theoretical evaluation is that the stabilities of the various domains differ widely from each other. When $B = 0$ and $\mu_r = \mu_0$, the problem is somewhat simplified and is discussed in Chap. 18. When the

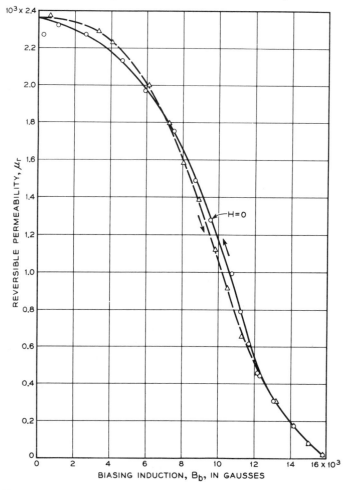

FIG. 11-61. Reversible permeability vs induction for 45 Permalloy.

biasing induction is near saturation, the reversible permeability is substantially equal to dB/dH and then it can be calculated in the same manner as the approach of magnetization to saturation, assuming that the domains are rotated against the forces of crystal anisotropy or strain. But at intermediate inductions the problem is difficult because the domain configurations are complicated. That the configurations in some materials are substantially different for increasing and decreasing inductions, at the same value of the induction, is shown by the fact that the corresponding values of μ_r are different, as pointed out in Fig. 11-62.

There are wide differences in the reversible permeability curves for different materials, as suggested by the wide range of values known to exist for μ_0 and B_s. Gans [10G1] found that many such curves could be brought

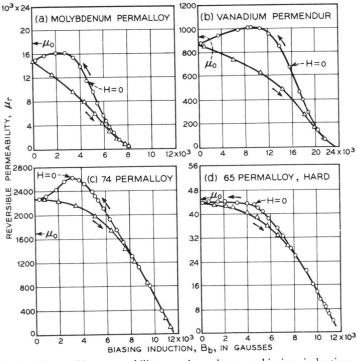

FIG. 11-62. Reversible permeability as dependent on biasing induction in (a) Molybdenum Permalloy, (b) Vanadium Permendur, (c) Permalloy containing 74% nickel and (d) hard Permalloy containing 65% nickel.

close to coincidence by plotting κ_r/κ_0, or μ_r/μ_0 against I/I_s. How well this "law of corresponding states" is in accord with the facts is shown in Fig. 11-63, in which are plotted data for several common and useful materials. The relation between κ_r (or μ_r) and I (or B) suggested by Gans [11G1] may be expressed in parametral form as follows:

$$(B - H)/B_s = \operatorname{ctnh} x - 1/x = L(x)$$

$$(\mu_r - 1)/(\mu_0 - 1) = 3(1/x^2 - 1/\sinh^2 x) = 3L'(x)$$

and is drawn as curve (a) in the figure. Brown [38B4] has derived this relation making certain assumptions regarding the nature of the domains, the principal one being that they have no direction of easy magnetization. On the contrary, if the directions of the domains are restricted so that they always lie parallel or antiparallel to the field, theory gives curve (b); and if they lie parallel or antiparallel to a single crystallographic direction in crystals that are oriented at random in the material, we have curve (c). The theory of the last curve is designed for cobalt, and the data show fair

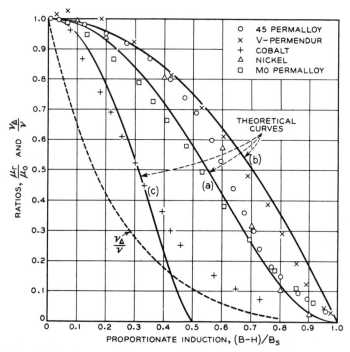

FIG. 11–63. Reduced reversible permeability vs reduced biasing induction for several materials, in comparison with the theory: (a) Gans' relation, (b) isotropic domains, (c) anistropic domains (Brown). Broken curve shows the approximate change in $d\mu/dH$ (for weak fields) as dependent on biasing induction, observed for a variety of materials.

agreement when B is less than half of B_s. Most of the experimental points for other materials lie in the region between the theoretical curves (a) and (b), or near them.

Incremental Permeability vs Amplitude.—When H_Δ, the amplitude of the alternating field, is increased the incremental permeability is also generally increased. For small amplitudes this increase is linear and by analogy with Rayleigh's law,

$$\mu = \mu_0 + \nu H,$$

we may write

$$\mu_\Delta = \mu_r + \nu_\Delta H_\Delta$$

in which ν_Δ is $d\mu_\Delta/dH_\Delta$ and has a constant value when H_Δ, equal by definition to $\Delta H/2$, is small. Similarly the hysteresis losses for small, superposed loops may be calculated from the relation

$$W_\Delta = \nu_\Delta H_\Delta^3/(3\pi).$$

Ebinger [30E4] has measured ν_Δ for a variety of materials and plotted ν_Δ/ν vs $(B-H)/B_s$. He finds that the points lie reasonably close to a

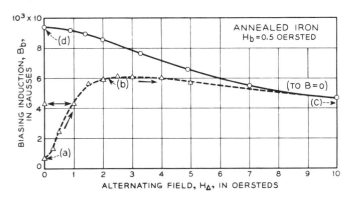

Fig. 11-64. Dependence of biasing induction, at constant biasing field, on a superposed alternating field.

single curve that drops rapidly from the initial point and approaches asymptotically the axis of abscissa. His curve is reproduced as a broken line in Fig. 11-63.

Curves showing μ_Δ vs H_Δ for various values of H_b have often been determined and are similar in general form to μ vs H curves for the normal magnetization curve. Empirical methods for estimating μ_Δ for relatively large values of ΔB have been suggested by Spooner [23S3].

As already indicated in Fig. 11-56 the superposition of a small alternating field on a steady biasing field is attended by an increase in the value of the biasing induction B_b. Figure 11-64 shows the relation between B_b and H_Δ for one value of the biasing field applied to annealed iron. As the amplitude H_Δ of the alternating field increases indefinitely, the biasing induction rises from the initial point (a), goes through a maximum (b), and then decreases toward zero (c). As H_Δ is then gradually decreased to zero, the induction rises continually to a definite value at (d). The points marked (a), (b), (c), and (d) in the figure are shown again in Fig. 11-65 where the point (a) is shown on the normal magnetization curve. The point (d) lies

548　THE MAGNETIZATION CURVE AND THE DOMAIN THEORY

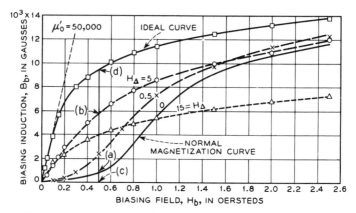

Fig. 11–65. Dependence of biasing induction on biasing field for material in different magnetic states. Points (a), (b), (c), (d) correspond to similar points on previous figure.

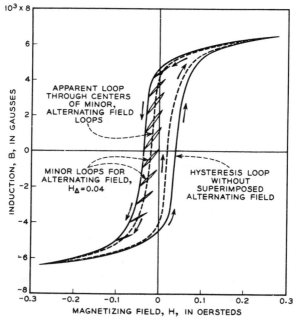

Fig. 11–66. Apparent reduction of hysteresis by superposition of alternating field.

on the "ideal" or "anhysteretic" magnetization curve, which is the locus of points such as (d) for various values of the biasing field H_b [15S1, 20G2] and is drawn as the uppermost curve in the figure. Its slope is a maximum at the origin where it is equal approximately to the maximum slope of the symmetrical hysteresis loop of high H_m. Other curves in this figure are for induction vs biasing field strength with an alternating field of constant (designated) amplitude superposed. The data of the last two figures refer to annealed iron 0.125 in. thick, the superposed fields being 60 cps; here eddy-current shielding is not always negligible. Other data have been reported by Ashworth [38A2] and Procopiu [30P2].

Hysteresis with Superposed Fields.—It is sometimes stated that the hysteresis loss of a magnetic material may be reduced by superposing an alternating field. This reduction is apparent and not real, as may be seen from Fig. 11–66. When no alternating field is present, the major hysteresis loop for a specimen of Molybdenum Permalloy is as shown by the solid line. When an alternating field of amplitude $H_\Delta = 0.04$ is superposed, the average B for a given constant applied H (H_b) will be a point on the broken line passing through the centers of the minor loops, and the apparent loop thus traced is found to have only 70% of the area of the large loop. But when the actual path of the B,H curve is considered in detail, it is realized that when the alternating field is present the total hysteresis loss is equal to that of the major loop *plus* the sum of all the minor loops being traced f times per second.

Hysteresis losses in loops of small amplitude may be calculated by the modified Rayleigh relation, provided $d\mu_\Delta/dH_\Delta$ is known, as indicated above. Losses in minor loops in silicon transformer sheet have been studied in some detail by Ball [15B2], Chubb and Spooner [15C1], Sidhu [33S7], and others. As examples of the results, Figs. 11–67 and 68 [15C1] show loops of small amplitude ($B_\Delta = 1000$) and various biasing indications ($B_b = 0$ to 12 000), and of various amplitudes ($B_\Delta = 1000$ to 7000) and a fixed biasing induction ($B_b = 4000$). Figure 11–69 summarizes their results for a large range of values of amplitude and bias. The material used in obtaining these data is not representative of the modern commercial product.

For a fixed value of B_Δ, the ratio of the hysteresis loss for a given B_b to that for $B_b = 0$ is sometimes called the *displacement factor*. Figure 11–69 shows that for silicon-iron this factor always increases with B_b, and such an increase has been confirmed by others. Sidhu [33S7] has measured superposed hysteresis loss in a 50% nickel-iron alloy and found that the displacement factor goes through a minimum of about 0.2 when B_b is about 3000, then rises to 1 at about $B_b = 7500$ and to 2 or more when $B_b = 10\,000$. He has proposed the relation

$$\log W_h \log B_m = A + C \log B_m$$

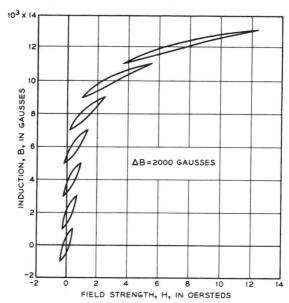

Fig. 11-67. Minor hysteresis loops in silicon iron transformer sheet commercially available in 1915.

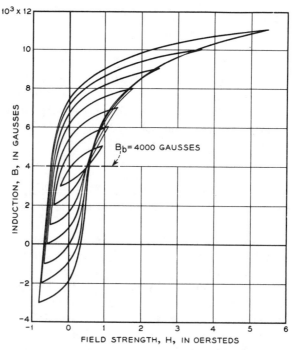

Fig. 11-68. Loops of various amplitudes, with constant biasing induction. Silicon-iron.

HYSTERESIS WITH SUPERPOSED FIELDS

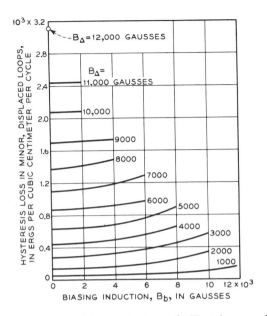

FIG. 11-69. Dependence of hysteresis loss of silicon-iron on biasing induction.

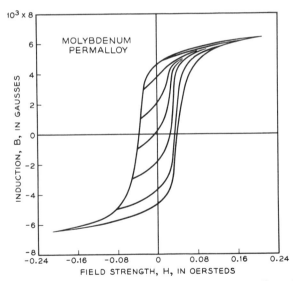

FIG. 11-70. Various hysteresis loops of Molybdenum Permalloy.

in which the constants A and C depend on B_Δ, and B_m is the maximum value of B in the loop (equal to $B_b + B_\Delta$).

Loops of various amplitudes and bias are shown in Fig. 11-70 for Molybdenum Permalloy, which has high permeability and a coercive force of 0.04.

Superposed Non-parallel Fields.—Fields at right angles to each other may be superposed conveniently by using a tube of the material under investigation and producing an axial field with a solenoid and a circumferential field with a current carried by a conducting wire placed on the

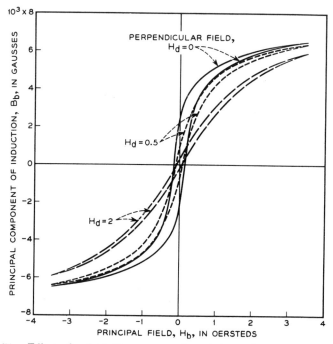

Fig. 11-71. Effect of a field H_d, superposed at right angles to the principal field H_b, on the hysteresis loop measured parallel to the principal field.

axis of the tube. A less elegant experiment may be made by passing the current through a solid rod of the magnetic material. Both alternating and constant fields have been used in such experiments, the first of which was reported in 1891 by Fenzi [91F1] who showed that the hysteresis measured in the usual way was greatly reduced when an alternating current was passed through the rod under test.

Results of tests with superposed alternating or constant fields, produced in a thin-walled partially annealed Permalloy tube, are given in Fig. 11-71. Here curve 0 is the hysteresis loop taken in the usual way with no superposed field. For curves 0.5 and 2 the constant superposed fields H_d were circumferential and had magnitudes of 0.5 and 2 oersteds respectively. The induction measured, B_p, was always that component of the total induction that was parallel to the principal field H_b. Extensive measurements on perpendicular fields and fields inclined at various angles have been reported by Sugiura [31S6].

During the experiments of Fig. 11-71 the actual field in the material is changing in both magnitude and direction. If there were no rotational hysteresis, curves 0.5 and 2 could be calculated from curve 0 by simply adding the superposed fields vectorially and estimating the component of

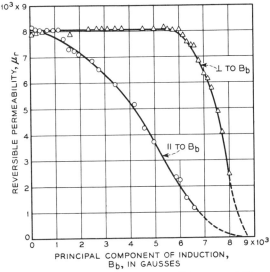

Fig. 11-72. Effect of biasing induction on reversible permeability when it is measured parallel or perpendicular to the direction of measurement.

induction measured by the search coil. In fact curve 2 can be calculated from curve 0 with an accuracy limited by the width of the hysteresis loop. For example, consider the point $H_b = 2$ on curve 2. When $H_d = 2$ is applied, the total field is $2\sqrt{2}$ and is inclined 45° to the axis of the tube. According to curve 0 this field should produce an induction of 6300 parallel to the resultant field, and the component parallel to H_b is 6300 sin 45° = 4450, a value corresponding closely to an observed point on the ascending branch of curve 2 at $H_b = 2$.

The area of the B_b, H_b loop is reduced by either alternating or constant superposed H_d. When H_d is alternating, the total hysteresis loss in the material includes also that caused by the alternating H_d even when H_b is not changing. When H_d is constant there is also some rotational hysteresis loss not included in the loss measured by the area of the B_b, H_b loop.

An interesting case of superposition of fields occurs when a weak alternating field is applied at right angles to a constant field. The small alternating induction determines a kind of incremental permeability, and by extrapolation one obtains the reversible permeability for perpendicularly superposed induction. Figure 11–72 shows that the reversible permeability of Molybdenum Permalloy thus determined (upper curve) is larger than the ordinary reversible permeability (lower curve) except at the end points where the values must coincide. Some measurements have also been reported briefly by Webb [38W2] on perpendicular incremental permeability and the way in which it is affected by twisting the specimen.

CHAPTER 12

MAGNETIC PROPERTIES OF CRYSTALS

It is well known that single crystals of ferromagnetic materials are magnetically anisotropic, even when the crystals have cubic symmetry. In iron the ease of magnetization is much greater in the directions of the cubic axes, $\langle 100 \rangle$, than in directions farthest removed from them, $\langle 111 \rangle$. In nickel the reverse is true, as illustrated in Fig. 11–2. This figure shows also the crystal structures of iron and nickel.

In hexagonal cobalt (Fig. 12–1) the variation of magnetizability with crystallographic direction is even greater than in iron or nickel. At room

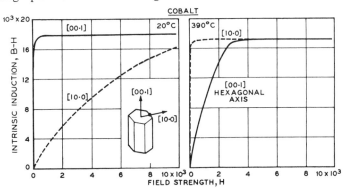

FIG. 12–1. Magnetization curves of cobalt in the directions of easy and hard magnetization at room temperature and at 390° C.

temperature the hexagonal axis is the direction of easy magnetization; at higher temperatures (above 275°C) this is the direction of most difficult magnetization.

In the investigation of crystal anisotropy an important problem has been to obtain crystals of sufficient size for measurement. Early work by Weiss [96W2, 05W2] was carried out on natural crystals of magnetite (Fe_3O_4) and pyrrhotite (approx. FeS). In later work, single crystals of various metallic substances have been prepared (1) by slow cooling of the melt [14F1, 25B4, 37C2] and (2) by the strain-anneal method. According to the latter, worked out for iron by Edwards and Pfeil [24E2], the decarbonized polycrystalline material is stretched a few per cent, usually

555

2–5%, and annealed at a definite temperature, about 850–900°C. "Exaggerated" grain growth occurs for a critical combination of extension and annealing. (The critical values may be different for each lot of material.) This method has been applied especially to iron, which is not amenable to method (1) since it has a transformation point in the solid state. However, slow cooling of the melt has been used for cobalt which has a transformation, associated with which is a relatively small heat effect, at about 400°C.

In the earlier measurements of Weiss and his colleagues, and in Japanese work, the anisotropy was determined either by forming the crystal into ellipsoids (so that the demagnetizing factor is known) and measuring B and H in the usual way in different directions, or by forming a disk and measuring the torque in a strong magnetic field, as described below.

In the more recent work of Williams and Bozorth [37W5, 39W2], single crystals of many ferromagnetic materials were prepared and oriented as described by Cioffi and Boothby [39C1, 37C2], and Walker, Williams, and Bozorth [49W4], by slow cooling of the melt. Pure materials were selected and placed in a refractory container of high purity alumina. Melting took place in an atmosphere of pure dry hydrogen, and freezing was carried out by lowering the temperature slowly (about 2°C/hr in some cases) until solidification was complete. Alloys were homogenized by maintaining just below the melting point for 24 hours or more, depending on the difference in temperature between the solidus and liquidus lines on the phase diagram. Elements and alloys so prepared included Co, Ni, Fe-Ni, Co-Ni, Fe-Si, Fe-Al, Ni-Cu, and Fe-Co-Ni. Some of the crystals weighed as much as a pound.

Specimens were cut from these crystals in the form of hollow parallelograms or "picture frames," each side being aligned by X-rays parallel to a desired crystallographic direction. The purpose of this form is to provide a closed magnetic circuit so that the uncertainty of a demagnetizing factor can be obviated. In a cubic crystal a parallelogram can always be constructed so that each side is parallel to a direction of the form $\langle hkl \rangle$, the sides being, for example, $[hkl]$, $[hlk]$, $[\bar{h}\bar{k}l]$, $[\bar{h}l\bar{k}]$ (see Fig. 12–2a). Specimens cut from an iron-silicon alloy as shown in the figure have sides parallel to $\langle 100 \rangle$, $\langle 110 \rangle$, and $\langle 111 \rangle$, respectively. Such specimens were wound with primary and secondary windings for measurement by the usual method using a ballistic galvanometer [37W5].

Other specimens were formed into short cylinders (disks), and measurements of torque were made as shown in Fig. 12–3. The specimens are mounted in a carriage held by two torsion fibers (of phosphor bronze) that are fastened to a rigid support at the top and to a circular scale S at the bottom, as described by Tarasov [39T4]. When the field is excited in the electromagnet, the crystal tends to turn so that the direction of easy magnetization will be parallel to the field. The torque so produced is balanced by turning the bottom of the lower fiber until the crystal regains its original orientation as determined by reflection of a light beam from the

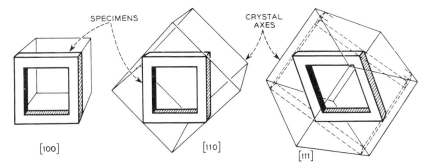

FIG. 12-2. Method of cutting magnetic specimens for determining magnetization in [100], [110] and [111] directions.

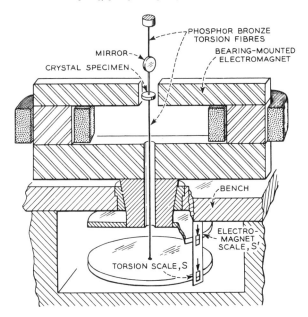

FIG. 12-3. Torque magnetometer. Electromagnet can be rotated so that the field is in any desired direction in the crystal specimen. The torque induced in the crystal is balanced by twisting the fibre a measured amount by means of the scale S at the bottom.

mirror. The scale reading S_2 is then compared with the original reading S_1 (with $H = 0$), and $S_2 - S_1$ is a measure of the torque. The orientation of the crystal axes with respect to the applied field is varied by turning the electromagnet, which is mounted on a heavy bearing, and noting its position on a suitable scale, S'. One then plots the torque against the crystal orientation and deduces from the curve the crystal anisotropy constant in the manner to be described under "Values of Constants."

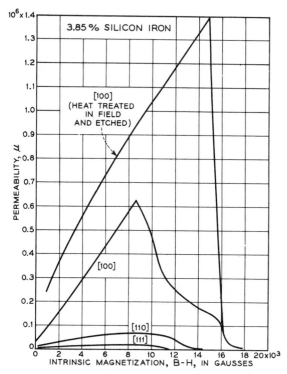

Fig. 12–4. Permeability vs intrinsic induction in the three principal directions in single crystals containing 3.8% silicon. Top curve, after heat treatment at 1300° C and cooling in a magnetic field.

Williams [37W3] has described a simpler apparatus and has discussed the various magnetic properties that it can be used to measure, such as magnetization, hysteresis, coercive force, and remanence, in addition to torque.

Permeability.—The permeability of single crystals has been difficult to measure with accuracy, because of the demagnetizing factors of the straight specimens employed, until the "picture frame" specimen was employed by Williams (see Fig. 12–2). The lack of demagnetizing field in these specimens permits accurate measurement of both maximum and initial permeability. The μ vs $B-H$ curves for crystals of 3.8% silicon-iron are reproduced in Fig. 12–4. The three lower curves were obtained [37W5] after the specimens, cut from the same crystal, were annealed together in pure hydrogen at 900°C. After subsequent annealing of the [100] specimen at 1300°C and then in a magnetic field, and etching away its surface, still higher permeability was measured, as shown by the upper curve. Measurement of a specimen of pure iron showed that $\mu_m = 1\,400\,000$ in the [100] direction [37B1].

B vs H curves for silicon-iron in lower fields (0–2 oersteds) are given in Fig. 4–23, and μ vs H curves for still lower fields, permitting extrapolation to μ_0, are shown in Fig. 4–27.

In these specimens it is noted that μ_0 depends on crystal direction. However, theory indicates that μ_0 is the same in all directions in a material with domains distributed equally among the six directions of easy magnetization. In order to explain Williams' results, Kondorski [38K13] and Becker and Döring [39B5], following a proposal by Bozorth [37B1], suggest that in the [100] specimen the domains are oriented along the easy directions that are nearest to the sides of the rectangular specimen. Thus there would be no domains oriented perpendicular to the side (and at 90° to the applied field), and there would be 3 times the normal or random number of domains oriented parallel or antiparallel to the field. If we suppose that 180° boundaries are responsible for initial permeability (no magnetostriction is then involved), the permeability should then have 3 times the normal value characteristic of a material in which all $\langle 100 \rangle$ directions are occupied with domains. Similarly the [110] specimen would have 1.5 times the normal value, and the [111] crystal would have just the normal value. The initial permeabilities should then be in the ratio 6 : 3 : 2 for the [100], [110], and [111] directions, as observed.

The same [100] specimen as that used by Williams for measurement of μ_0 was examined by Williams and Shockley [49W2] by powder pattern technique. This shows quite clearly that the magnetization in this specimen does proceed in fact by movement of a 180° boundary. The [110] specimen shows also that movements of 180° boundaries are responsible for a large part of the change in magnetization.

By measuring the torque in weak fields Shoenberg and Wilson [46S6] were able to show that the domains were not distributed equally among the various $\langle 100 \rangle$ directions.

Honda and Nishina [36H6] measured μ_0 for iron crystals in the form of rods. In the [100], [110], and [111] directions they are in the ratio 2.7 : 1.7 : 1 and not quite in the ratio 6 : 3 : 2 = 3 : 1.5 : 1 found by Williams in silicon-iron. Later Shimizu [41S9] observed the ratio 1.8 : 1.3 : 1.

Permeability of nickel in different directions is shown in Fig. 12–5. The permeability at low fields has been measured by Williams, and his (unpublished) data show that $\mu_0 = 100$ in the [100], [110], and [111] directions (Fig. 12–6). This is contrary to a brief report by Okamura

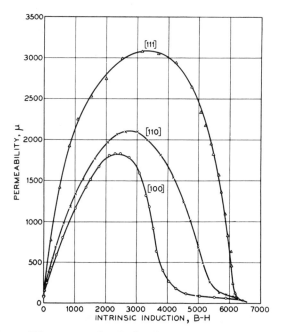

Fig. 12-5. Permeability curves in single nickel crystals. (Saturation induction was not determined with precision.)

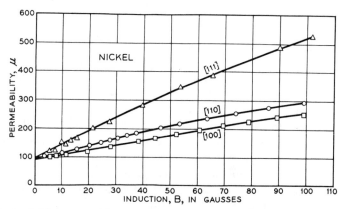

Fig. 12-6. Initial permeability curves, showing isotropy of initial permeability in nickel crystals.

[39O3]. Williams has observed an absence of anistropy of μ_0 in Permalloys containing 35, 76, or 81% nickel. Attempts to observe the domain structure of these materials by the powder pattern technique has not been successful.

The initial susceptibilities of single crystals of cobalt and Mn_2Sb have been measured by Guillaud, Bertrand, and Vautier [48G1, 49G7]. For a direction perpendicular to the hexagonal axis of cobalt (a direction of hard magnetization) it is to be expected that magnetization will take place only by rotation against the forces of crystal anisotropy. Comparison of theory and experiment is given below, under "Calculation of Magnetization Curves."

Remanence and Coercive Force.—Kaya [33K10] measured the hysteresis loops of single-crystal iron rods about 2 mm in diameter and 15-30 cm

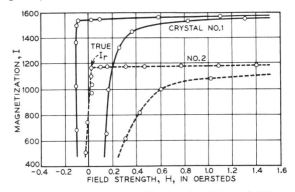

FIG. 12-7. Hysteresis loops of two single crystals of iron of different orientations.

long. It was difficult to determine the residual magnetization directly because the demagnetizing fields were rather uncertain and the applied field for $H = 0$ was therefore not known accurately. However, each loop showed a characteristic sharp corner near the point $H = 0$, as shown in Fig. 12-7, and the value of I at this point was assumed to be I_r. As so determined, I_r was found to be given by the relation

$$I_r = I_s/(\alpha_1 + \alpha_2 + \alpha_3),$$

the α's being the direction cosines of the rod axis with respect to the three crystal axes. This is shown in Fig. 12-8. Additional data by Kaya and Takaki [36K7] supported this relation and are included in the figure.

One expects, on the basis of simple domain theory, that when the field is removed from a saturated crystal the domains will lie along the direction of easy magnetization that is nearest to the direction in which the field has been applied. However, this would mean generally that there would be a component of magnetization at right angles to the field direction, and in long thin specimens there would be large demagnetizing forces opposing such magnetization. If the demagnetizing force is controlling, so that the normal component of magnetization is essentially zero, then one expects the domains to be distributed in a calculable way among the three directions of easy magnetization (in iron) that are nearest to the field direction. Gorter [33G7] pointed out that under these assumptions the retentivity in a cubic crystal is just that observed by Kaya. This can be proved in a straightforward manner. The extreme values of I_r/I_s to be expected are 1 and $1/\sqrt{3}$ (for the [100] and [111] directions).

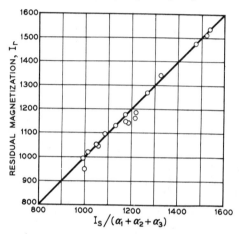

Fig. 12-8. Observed remanence plotted against the calculated remanence in single crystals of iron.

The theoretical retentivity in a specimen with no demagnetizing field in any direction, calculated on the assumption that one-third of the domains are magnetized along each of the three directions of easy magnetization (in iron) that lie nearest to the field-direction, is found to be

$$I_r = I_s/(3\alpha_1),$$

α_1 being the cosine of the smallest angle between H and [100]. The value of I_r/I_s then lies between $1/\sqrt{3}$ and $\frac{1}{3}$ and never far from $\frac{1}{2}$. The theory for a disk-shaped specimen, with the field lying in the plane of the disk, has been given by Vonsovskii [40V2].

Some data on the dependence of *coercive force* on crystal orientation have been reported by Kaya [33K10], Shur and Jaanus [37J1, 41S10, 46S5], Sixtus [37S9], and Williams [37W5]. A theoretical treatment has been given by Vonsovskii [40V2]. The lowest value of coercive force appears to be in the direction of easiest magnetization. Further conclusions do not seem to be justified at present.

Crystal Anisotropy Energy.—In the quantitative evaluation of magnetic crystal anisotropy it is convenient to use the crystal anisotropy constants, K_1, K_2, $\cdots$. These are defined by first expressing the free energy of crystal anisotropy as a function of the direction cosines, α_1, α_2, α_3, of the magnetization vector with respect to the crystallographic axes. In a cubic crystal the expression, written in ascending powers of the α's, is

$$E_k = K_0 + K_1(\alpha_1^2\alpha_2^2 + \alpha_2^2\alpha_3^2 + \alpha_3^2\alpha_1^2) + K_2\alpha_1^2\alpha_2^2\alpha_3^2 + \cdots.$$

The units commonly used are ergs/cm^3. Terms in odd powers are absent because they are inconsistent with the cubic symmetry. The quadratic term is absent because

$$\alpha_1^2 + \alpha_2^2 + \alpha_3^2 = 1,$$

and terms in α_i^4 are omitted because of the relation

$$(\alpha_1^2 + \alpha_2^2 + \alpha_3^2)^2 = 1 = \alpha_1^4 + \alpha_2^4 + \alpha_3^4 + 2(\alpha_1^2\alpha_2^2 + \alpha_2^2\alpha_3^2 + \alpha_3^2\alpha_1^2).$$

Higher terms have never been found necessary to express experimental results, and usually the term in K_2 is redundant. An expression of this kind has been used by Mahajani [29M2], Akulov [29A3], Gans [32G3], and others.

The difference in the crystal anisotropy energy density between two different crystal directions can be determined by measuring the difference in energy necessary to magnetize a crystal in these two directions. Thus let

$$A_{hkl} = \int_0^{I_s} H dI$$

be the work necessary to magnetize a specimen so that the magnetization is everywhere parallel to [hkl]. If the work is done against the crystal forces alone, $E_k = A$ for a given crystal direction; however, we know that there will be other forces, such as those of magnetostriction, which will contribute to $\int H dI$. When the work done against these forces is the same in various crystallographic directions, then the difference in A for any two directions is equal to the difference in E_k for these directions. Practically, the determination of K_1 and K_2 is not limited by such a condition.

In particular let the two directions, along which A is measured, be [110] and [100]; then the corresponding direction cosines are $(1/\sqrt{2}, 1/\sqrt{2}, 0)$ and $(1, 0, 0)$, and

$$A_{110} - A_{100} = E_{110} - E_{100} = K_1/4,$$

the E's representing the values of E_k in the directions indicated. Similarly,
$$A_{111} - A_{100} = K_1/3 + K_2/27$$
and, conversely,
$$K_0 = A_{100}$$
$$K_1 = 4(A_{110} - A_{100})$$
$$K_2 = 27(A_{111} - A_{100}) - 36(A_{110} - A_{100}).$$

K_1 can then be determined by measuring the area between the magnetization curves for the [110] and [100] directions, such as those of Fig. 11-2, and K_2 by using also the curve for [111]. It is evident that K_1 is positive for iron and negative for nickel.

For crystals of hexagonal symmetry, such as cobalt, it is more convenient to use the sine instead of the cosine of the angle between the magnetization and the hexagonal axis. Letting this angle be ϕ, we have

$$E_k = K_0 + K_1 \sin^2 \phi + K_2 \sin^4 \phi + \cdots.$$

Higher terms and terms depending on the orientation in the (00·1) plane have so far been found unnecessary except perhaps for pyrrhotite. The energy required to magnetize to saturation in the direction of the axis [00·1] is

$$E_0 = K_0$$

and at right angles to this axis, parallel to [10·0] or [11·0] or intermediate directions, is

$$E_{90} = K_0 + K_1 + K_2$$

per unit volume of material. In cobalt the sign of $K_1 + K_2$ changes at about 275°C, and the direction of easy magnetization changes then from [00·1] to [10·0], as will be discussed.

The crystal anisotropy of a tetragonal crystal, Mn$_2$Sb, has been studied by Guillaud [43G2]. The expression for the energy is

$$E_k = K_0 + K_1 \sin^2 \phi + K_2 \sin^4 \phi + K_3 \cos^2 \alpha \cos^2 \beta,$$

ϕ being the angle between the magnetization and the tetragonal axis, [001], and α and β the angles with the other two axes. We then have

$$E_{001} = K_0$$
$$E_{100} = E_{010} = K_0 + K_1 + K_2$$
$$E_{110} = K_0 + K_1 + K_2 + K_3/4.$$

The K's can be evaluated by use of the relations

$$E_{100} - E_{001} = K_1 + K_2$$
$$E_{110} - E_{100} = K_3/4.$$

K_1 and K_2 can be determined separately in crystals of this symmetry, as in hexagonal crystals, from the shape of the I vs H curve in a direction of difficult magnetization, e.g., [100] in Mn$_2$Sb.

Torque Curves.—As mentioned previously, crystal anisotropy can be studied by measuring the torque that exists when an anisotropic crystal

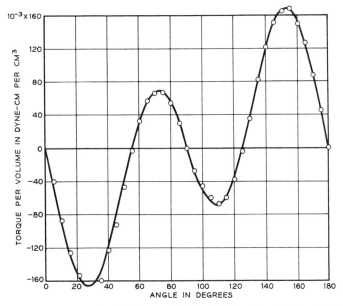

FIG. 12-9. Torque curve of a single crystal disk of silicon-iron in the (110) plane, taken with an applied field of 5800 oersteds. The line is according to theory with $K_1 = 287\,000$, $K_2 = 100\,000$.

is placed in a uniform magnetic field, and anisotropy constants are readily determined by measurement of the torque. Figure 12-9 shows a typical torque curve for a disk cut from a single crystal of iron (containing 3.85% Si), with the plane of the disk lying parallel to the (110) plane.

The torque acting on each unit volume of a crystal is equal to the rate of change of energy density with angle:

$$L = -dE/d\theta,$$

θ denoting the angle between the *direction of magnetization* and a crystallographic axis. This relation is not immediately useful in determining the anisotropy constants, since the direction of magnetization is undetermined; however, as H is increased indefinitely the direction of I approaches the direction of H and hence θ is definitely known for the torque obtained by extrapolating the observed torque to $H = \infty$. A typical plot of L vs H, showing approach to a finite limit at $H = \infty$, is shown in Fig. 12–10(a); (b) shows that the L vs $1/H$ curve is a convenient method of plotting for extrapolation, a procedure frequently used [36S15, 39T4, 41B5, 48K10].

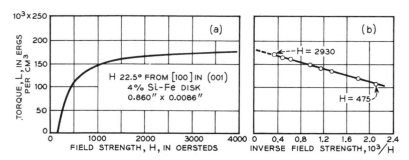

Fig. 12–10. Torque on a single crystal of silicon-iron, plotted against: (a) the applied field and (b) the reciprocal of the applied field.

In a cubic crystal magnetized parallel to the (100) plane the torque in infinite field is

$$L_{100} = -K (\sin 4\theta)/2,$$

and in the (110) plane

$$L_{110} = -K_1(2 \sin 2\theta + 3 \sin 4\theta)/8$$
$$+ K_2 (\sin 2\theta - 4 \sin 4\theta - 3 \sin 6\theta)/64,$$

θ being the angle between I_s and the [001] direction. The slope of the L_{100} vs θ curve at $\theta = 0$ is then

$$(dL_{100}/d\theta)_{\theta=0} = -2K_1(\cos 4\theta)_{\theta=0} = -2K_1,$$

and the maxima and minima of the curve, at $\theta = 22.5°, 67.5°, \cdots$, are

$$(L_{100})_m = \pm K_1/2.$$

Thus K_1 is readily determined by measuring the maximum value of L in the (100) plane.

In the (110) plane, in which both K_1 and K_2 are involved, the constants can be adjusted by trial and error to give the best fit to the torque data; this is easily accomplished since the K_2 term is relatively small. When $K_2 = 0$, the first minimum and maximum occur at $\theta = 25°31'$ and $70°20'$, when L has the values $-0.561K_1$ and $+0.210K_1$, respectively.

Values of the K's for various materials will now be summarized. A more detailed discussion of torque curves is given in a later section.

Values of Constants.—Anisotropy constants of various materials, including elements, alloys, and compounds, are here collected from various reports in the literature and presented in tabular and graphical form. A detailed summary of results obtained before 1934 has been given by McKeehan [34M1].

Iron.—Crystal anisotropy was observed by Beck [18B1] in 1918, in crystals prepared from the melt by the thermit process and containing

TABLE 1. ANISOTROPY CONSTANTS (ERGS/CM^3) OF IRON AND NICKEL OBTAINED BY VARIOUS AUTHORS, AT ROOM TEMPERATURE

$K_1 \times 10^{-3}$	$K_2 \times 10^{-3}$	Data by:	Constants by:
Iron			
478	...⎫	Honda and Kaya [26H2]	Gans and Czerlinski [33G8]
428	...⎭		
400	290⎫	Honda et al. [28H3]	⎰Gans [32G3]
410	140⎭		⎱Piety [36P3]
410	...	Czerlinski [32C6]	Czerlinski [32C4]
427	−170	Piety [36P3]	Piety [36P3]
442	140⎫	Piety [36P3]	McKeehan [37M1]
421	150⎭		
470	...⎫	Tarasov [39T2]	Tarasov [39T2]
525	...⎭		
Nickel			
−51	...	Kaya [28K2]	Gans and Czerlinski [32G3]
−58	...⎫		
−40	...⎬	Sucksmith et al. [28S10]	Gans and Czerlinski [32G3]
−47	...⎭		
−50	...	Czerlinski [32C4]	Czerlinski [32C4]
−34	+50	Honda et al. [35H4]	McKeehan [37M1]
−47	...	Brükhatov and Kirensky [37B5]	Brükhatov and Kirensky [37B5]
−48	...	Williams and Bozorth [39W2]	Williams and Bozorth [39W2]
−50	...	Polley [39P4]	Polley [39P4]
−59	...	Williams and Bozorth	(unpublished)

1.6% silicon. He drew magnetization curves, and a value of K_1 has been derived from them by McKeehan [34M1]. More accurate and extensive measurements were made in the period 1925–1930 by a number of workers, especially Webster [25W4], Gerlach [26G2, 30G7], Honda, Kaya, and Masumoto [26H2, 28H3], Dussler [28D1], Gries and Esser [29G2], Sizoo [29S7], and Foster and Bozorth [29F1, 30F3]. Among these Honda, Masumoto, and Kaya [28H3] and Webster [25W4] should be especially mentioned. From their work, and that of Piety [36P3], good values of K_1 and approximate values of K_2 have been derived by Gans and Czerlinski [32G3, 33G8] and Piety [36P3], and are given in Table 1.

Variation of K_1 and K_2 with temperature is given in Fig. 12–11 [28H3, 37B1]. It is to be noted that K_1 becomes very small at temperatures (e.g., 500°C) at which the saturation magnetization is still a large fraction of its value at 0°K.

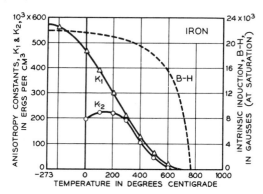

FIG. 12–11. Crystal anisotropy constants of iron at various temperatures.

Cobalt.—Single crystals were prepared and studied by Kaya in 1928 [28K3]. Values of K_1 and K_2, derived [37B1] from the more recent data of Honda and Masumoto [31H2], are shown graphically in Fig. 12–12 as

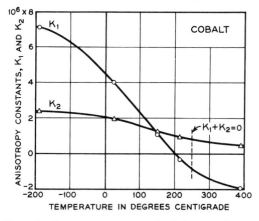

FIG. 12–12. Crystal anisotropy constants of cobalt at various temperatures.

dependent on temperature. At room temperature the best values are $K_0 = 40 \times 10^6$, $K_1 = 3.98 \times 10^6$, $K_2 = 1.98 \times 10^6$, $K_1 + K_2 = 5.96 \times 10^6$. The temperature at which $K_1 + K_2$ changes sign is about 250°C. At about 480°C cobalt becomes cubic, and above this temperature no data on anisotropy are available.

Nickel.—Data for nickel have been reported by Kaya [28K2], Sucksmith, Potter, and Broadway [28S10], Honda, Masumoto, and Shirakawa [35H4],

Kleis [36K2], and Williams and Bozorth [39W2, 39W4]. At room temperature the constants derived from the work of Honda, Masumoto, and Shirakawa are $K_1 = -35\,000$, $K_2 = 46\,000$. The older works of Kaya and Suchsmith, Potter and Broadway, and the newer results of Brükhatov and Kirensky [37B5, 38B7], yield values of K_1 lying between 40 000 and 58 000. Data are given in Table 1.

The work of Brükhatov and Kirensky, supplemented by the work of Williams and Bozorth [39W4] at lower temperature (20°K), has shown that the anisotropy constant is given closely by the empirical relation,

$$K_1 = K_{10}e^{-\alpha T^2}$$

in which T is the absolute temperature and α and K_{10} are constants. Some later results [39W2] indicate that $K_{10} = -750\,000$. Data are

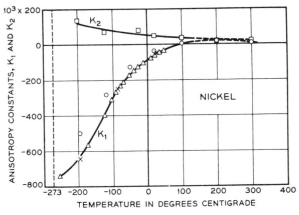

FIG. 12–13. Crystal anisotropy constants of nickel at various temperatures.

shown in Fig. 12–13. This relation does not hold above 100°C, according to the data of Honda, Masumoto, and Shirakawa, for at this temperature the direction of easy magnetization is [110] rather than [111]. Their data indicate that K_1 becomes positive at a slightly higher temperature. It has been shown [36B1] that in this case K_2 must be taken into account. Provisional values of K_2, calculated from the data of Honda, Masumoto, and Shirakawa, are also plotted in Fig. 12–13.

Variation of K_1 with temperature has also been reported by Polley [39P4]; his results were obtained by measuring the rate of approach of the I,H curve of polycrystalline nickel to saturation. Values of K_1 at room temperature were similarly determined by Czerlinski [32C4].

Iron-Nickel Alloys.—It was first pointed out by Lichtenberger [32L1] that the anisotropy constant K_1 of the iron-nickel alloys changes sign near 70% nickel. The change of sign was confirmed by Burgers and Snoek's

[35B4] measurement of a single crystal of a 50% nickel alloy, which was found to have a positive K_1. Kleis [36K2] determined the critical composition as 76% nickel. McKeehan and Grabbe [39M1, 40G1] showed that the constant and therefore the critical composition depend on atomic ordering, which is known to exist in this region of composition. In the disordered state the crossing point came at 75%; in the ordered state,

TABLE 2. VALUES OF K_1 AND K_2 FOR VARIOUS FE-CO, FE-NI, CO-NI AND FE-CO-NI ALLOYS
(Taken from summary by McKeehan [37M1], based on data by Shih [34S12, 36S14], Kleis [36K2], and McKeehan.)

Composition (%)			20°C		200°C	
Fe	Co	Ni	$K_1 \times 10^{-3}$	$K_2 \times 10^{-3}$	$K_1 \times 10^{-3}$	$K_2 \times 10^{-3}$
100	..	...	420	150	300	22
70	30	...	102	160	...	...
60	40	...	45	−110	...	...
50	50	...	−68	−390	...	...
30	70	...	−433	50	...	...
50	..	50	33	−180	25	−80
35	..	65	15	−70	12	−40
30	..	70	7	−17	2	−4
10	..	90	−7	−23	−2	−10
...	..	100	−34	53	5	20
...	65	35	−258	150	...	...
...	50	50	−108	−40	...	...
...	40	60	−74	40	...	...
...	20	80	−4	8	...	...
...	10	90	16	−40	...	...
...	3	97	−10	9	...	...
50	10	40	61	−160	19	4
25	25	50	4	16	4	2
20	15	65	9	−110	−1	−18
15	25	60	−26	34	−10	−45
10	40	50	−72	−4	−54	41
10	30	60	−38	−80	−17	−50
10	20	70	−29	17	−25	70
10	10	80	−2	−39	−2	−20

near 70% nickel. In both states the curves of K_1 vs composition are rather flat, tending to lie parallel to the axis of composition. The course of the whole K_1 vs composition curve from 35% to 100% nickel, for one heat treatment (slow cooling), is given in Fig. 12–14 [39W2].

Kleis observed that, between 50 and 65% nickel, the [110] direction and not [100] or [111] is that of most difficult magnetization. Therefore, the term in K_2 is important and has the opposite sign from K_1. Values of K_1 and K_2 at 20° and 200°C are included in Table 2.

Values of K_1 for alloys containing 0–16% nickel were determined by Tarasov [39T5] with specimens cut from cold-rolled polycrystalline sheets, using the very doubtful assumptions that the amount of special orientation

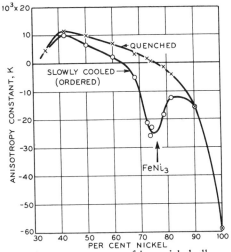

FIG. 12-14. Crystal anisotropy constant of iron-nickel alloys after quenching and after cooling slowly to effect atomic ordering. (1952)

in the sheets does not vary with composition. The constants so determined vary linearly with composition and appear to approach the value $K_1 = 0$ at 25% nickel.

Iron-Cobalt-Nickel Alloys.—These alloys, including the binary iron-cobalt and cobalt-nickel series, have been investigated by Shih [34S12, 36S14], and the results have been extended and summarized by McKeehan [37M1]. Values of K_1 for 20°C are given in Fig. 12-15, taken from McKeehan, and include the iron-nickel alloys already considered. In the original and in Table 2 are data for 200°C as well. The points of zero anisotropy are near 70% nickel, 30% iron in the iron-nickel alloys; near 45% cobalt, 55% iron in the iron-cobalt alloys; and near the nickel end of the cobalt-nickel alloys. The course of the boundary $K_1 = 0$ for ternary alloys is rather uncertain, but it is believed to pass near the selected Perminvar composition (30% Fe, 25% Co, 45% Ni) when the usual annealing temperature of this alloy (400–600°C) is used. A constant $K_1 = -2000$ has been determined at the Bell Laboratories, using a single crystal grown from the melt.

Iron-Silicon Alloys.—Ruder [25R1] showed in 1925 that [100] was a direction of easy magnetization. In a systematic investigation by Tarasov [39T2] (see Fig. 12-16) of alloys containing up to 7.5% silicon, he found K_1 to decrease continually toward zero which, by extrapolation, seems to occur at about 11–12% (20–22 atomic per cent) silicon. (The magnetostriction is zero at about 7.5%.) A disturbing factor in these measure-

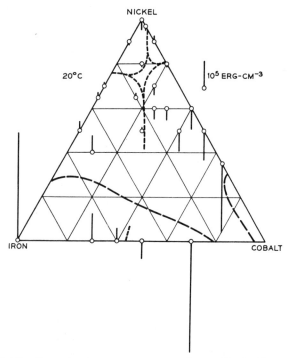

FIG. 12-15. Crystal anisotropy constants of some Fe-Co-Ni alloys.

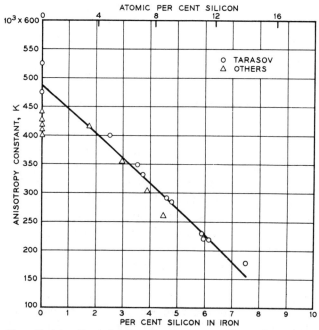

FIG. 12-16. Crystal anisotropy constants of iron-silicon alloys.

ments is the high value, $K_1 = 527\,000$, found for iron; others (see p. 567) have found 400 000–480 000. The curve for iron-silicon alloys lies close to, but somewhat above, the several points measured by Beck [18B1],

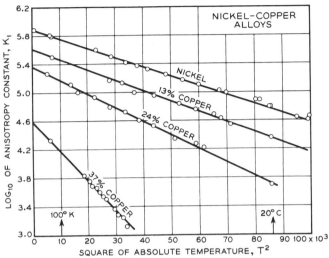

Fig. 12–17. Variation of crystal anisotropy of nickel-copper alloys with temperature and composition.

Williams [37W5], Schlechtweg [36S15], and Tarasov and Bitter [37T1], as shown in the figure.

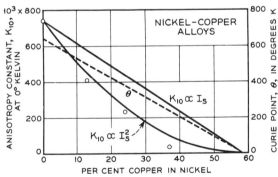

Fig. 12–18. The observed crystal anisotropy constants of nickel-copper alloys at 0° K are nearly proportional to the square of the saturation magnetization at 0°K.

Nickel Alloys.—Nickel-copper alloys are of interest because they exhibit linear decrease of saturation moment and Curie point with increasing copper content. Three alloys (13, 24, 37% copper) were investigated by Williams and Bozorth [39W2] at temperatures from −190 to 20°C. The data lay on straight lines when log K_1 was plotted against T^2, and they showed that the relation of Brükhatov and Kirensky was obeyed. Results for the alloys and pure nickel are shown in Fig. 12–17. Derived values of K_{10}, the anisotropy constant at 0°K, are plotted vs composition in Fig. 12–18. For convenience the Curie point is also indicated, taken from the data of Krupkowski [29K5]. In a later section comparison will be made with Van Vleck's theory [37V1, 47V1].

Puzei [49P2] has measured K_1 in Ni-Cu and Ni-Mo alloys.

Heusler Alloy.—This material is similar to nickel in that the direction of easy magnetization is [111]. The anisotropy constant can be derived from Potter's [29P3] data and has the value −900 000 at room temperature.

Various Iron Alloys.—Zaimovski [35Z2] and Messkin and Somin [36M10] have reported values of K_1 for alloys of iron with Al, Cr, Ti, Mo, and W. These are derived from measurements of coercive force and the shape of the magnetization curve. In some of the alloys, at least, two phases are present, and the values of K_1 derived from the data are not to be relied upon.

Snoek and Went [47S2] have also measured the effect of various elements on the anisotropy of iron alloyed with Al, Co, Cr, Mn, Mo, Ni, Si, Sn, V, W, and Zn. The anisotropy energy was evaluated from the area between the ideal magnetization curve and the line for saturation magnetization. They found K to pass through zero for iron-aluminum, iron-silicon, and iron-cobalt alloys only, in the first-mentioned alloys at about 12% aluminum. They also investigated the ternary Fe-Si-Al alloys and agree with Zaimovsky and Selissky [41Z2] in finding practically zero anisotropy and zero magnetostriction in the alloy Sendust (see Fig. 4–32). In the Fe-Si-Al system the anisotropy constant appears to be positive when the aluminum is less than 15% and the cobalt less than about 35%.

Compounds.—Anisotropy was observed years ago in pyrrhotite (FeS to Fe_6S_7) and magnetite (Fe_3O_4). In the former Weiss [05W2, 29W2] found that magnetization was easiest in the (00·1) plane of the hexagonal crystal, and its behavior is thus similar to that of cobalt at temperatures above 250°C. Magnetization parallel to the hexagonal axis is so difficult that it is estimated, by extrapolation, that a field of 150 000–200 000 would be necessary to saturate the crystal; the energy of magnetization in this direction is then about 5×10^6 ergs/cm^3. The difference in energy for magnetization in the two principal directions lying in the hexagonal (00·1) plane is 200 000–250 000. Kaya and Miyahara (39K1) have found, contrary to the work of Weiss, that in this plane the magnetic properties show perfect hexagonal symmetry.

The constant for cubic magnetite, derived from the work of Weiss [96W2, 29W2], is $K_1 = -85\,000$ to $-100\,000$ at room temperature. Bickford has derived values of K_1 from ferromagnetic resonance experiments (Chap. 17), and his results are shown in Fig. 12–19 for temperatures lying between −160 and +20°C. K_1 passes through zero at −143°C.

Recently Guillaud [43G2] has measured the anisotropy of MnBi, MnSb, and Mn$_2$Sb. The first is hexagonal and is most easily magnetized along the hexagonal axis until the temperature is reduced below 85°K. The

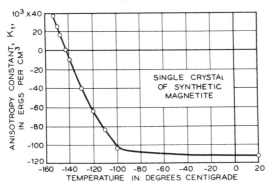

FIG. 12–19. Crystal anisotropy of magnetite, determined from ferromagnetic resonance experiments.

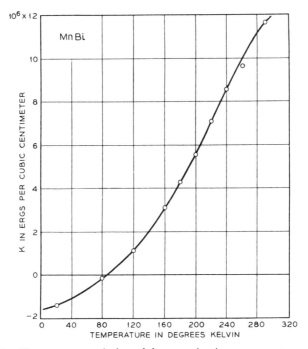

FIG. 12–20. Temperature variation of the crystal anisotropy constant of MnBi.

energy necessary to change the magnetization from [00·1] to [11·0] is $K_1 + K_2$, as shown previously, and this energy is plotted against temperature in Fig. 12–20. At 20°C Guillaud evaluates $K_1 = 8.9 \times 10^6$

and $K_2 = 2.7 \times 10^6$ ergs/cm³. At room temperature the anisotropy is about twice that of cobalt.

Although MnSb has not been subjected to a detailed study, there is evidence that the hexagonal axis is here also the direction of easy magnetization. The value of $K_1 + K_2$ is constant at 1.0×10^6 from room temperature to 20°K, remarkably insensitive to temperature. Its relation to the crystal structure, particularly the axial ratio, has not been investigated.

The compound Mn_2Sb has tetragonal symmetry, and at room temperature the tetragonal axis [001] is the direction of easy magnetization,

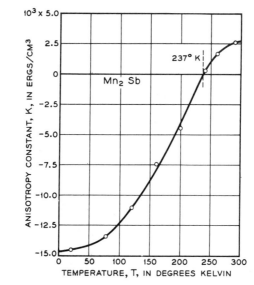

FIG. 12-21. Temperature variation of the crystal anisotropy constant of Mn_2Sb.

whereas at lower temperatures it is the direction of difficult magnetization. The asymmetry in the (001) plane is slight but detectable—variations in magnetization are about 3%, at most, in a constant field. There is some evidence that the symmetry becomes orthorhombic at 77°K. Evaluation of anisotropy constants may be made with the same equation as that used with hexagonal crystals, and $K_1 + K_2$ will then represent the difference in energy for the [001] and [100] directions. Values of $K_1 + K_2$ at various temperatures are given in Fig. 12-21. Evaluation of the separate values was made at 17°C: $K_1 = 181\,000$, $K_2 = 79\,000$.

It has been reported [35M2] that FeBe has high crystal anisotropy, but no quantitative measurements have been given.

Calculation of Magnetization Curves.—In 1930 Webster [30W3] and, in 1931 independently, Heisenberg [31H6] improved upon our concept of the domain theory. They proposed, in accordance with present-day ideas already described in this book, that in the unmagnetized state a ferromagnetic material is composed of many domains, each of which is magnetized to saturation in one of the directions of easy magnetization already known from experiments on single crystals. Since there are always several directions of easy magnetization (e.g., 6 in iron), the distribution of domain directions among them can always be such as to leave a net magnetization of zero.

Heisenberg proposed also that when the field is strong enough the magnetization of each domain will be turned from the direction of easy magnetization in a calculable way, provided one knows how the energy of anisotropy varies with the angle through which the magnetization has turned. The theory was applied to the calculation of the I vs H curve of cobalt (2 directions of easy magnetization) and was extended by Akulov [31A3] and by Gans and Czerlinski [32G3, 33G8] to the calculation of I vs H curves of nickel and iron, as will be shown. In carrying out these calculations we neglect for the moment the

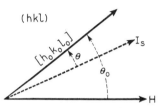

FIG. 12-22. Angles between vectors used in calculating magnetization curves of single crystals.

effects of strains and all factors which influence the displacement of boundaries between domains, and we assume that each domain has already been oriented by weak fields so that it is magnetized in that direction of easy magnetization which is nearest to the direction of the field. The calculation thus refers only to the *upper part* of the magnetization curve.

Consider a specimen of a single crystal subjected to a field H that makes a given angle, θ_0, with a direction of easy magnetization, $[h_0k_0l_0]$, as in Fig. 12-22. As H increases, the magnetization of the domains, I_s, will deviate from $[h_0k_0l_0]$ and approach H. It is assumed here that I_s lies in the same plane as H and $[h_0k_0l_0]$, namely, (hkl). The angle of deviation θ may be calculated by minimizing the total energy, composed of crystal anisotropy energy E_k and energy of interaction of magnetization and field $-HI_s \cos(\theta_0 - \theta)$. This energy,

$$E = E_k - HI_s \cos(\theta_0 - \theta)$$

is a minimum when $dE/d\theta = 0$; therefore when

$$H = \frac{dE_k/d\theta}{I_s \sin(\theta_0 - \theta)}.$$

At this time

$$I = I_s \cos(\theta_0 - \theta).$$

Knowing the dependence of E_k on θ, one can calculate I as a function of H, using θ as parameter.

In a cubic crystal with known K_1, let $(hkl) = (001)$ and $[h_0 k_0 l_0] = [100]$, and let $\theta_0 = 45°$, so that H is directed along the [110] direction. Then

$$E_k = K_1(\alpha_1^2\alpha_2^2 + \alpha_2^2\alpha_3^2 + \alpha_3^2\alpha_1^2) + K_2\alpha_1^2\alpha_2^2\alpha_3^2$$

$$\alpha_1 = \cos\theta, \qquad \alpha_2 = \sin\theta, \qquad \alpha_3 = 0$$

$$E_k = (K_1/8)(1 - \cos 4\theta)$$

$$dE_k/d\theta = (K_1/2)\sin 4\theta$$

and

$$H = \frac{K_1 \sin 4\theta}{2I_s \sin(45° - \theta)}$$

$$I = I_s \cos(45° - \theta).$$

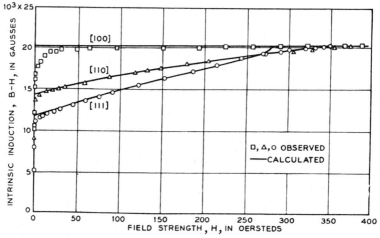

Fig. 12–23. Calculated and observed magnetization curves of 3.8% silicon-iron

The I vs H curve is then easily plotted, with θ as parameter, from $I = I_s/\sqrt{2}$ to I_s. The field necessary to magnetize to saturation in the [110] direction is obtained by putting $\theta = 45°$, when

$$H_s = 2K/I_s.$$

Similarly, when H is parallel to [111], the I,H curve may be calculated by putting $(hkl) = (011)$ and $\theta_0 = 54°44'$ (or $\sin\theta_0 = 1/\sqrt{3}$), and using the resulting equations

$$H = \frac{8K_1(2\sin 2\theta + 3\sin 4\theta) + K_2(2\sin 2\theta + 4\sin 4\theta - 3\sin 6\theta)}{64 I_s \sin(55° - \theta)}$$

$$I = I_s \cos(55° - \theta).$$

Results of calculations for both [110] and [111] are compared with the experiments of Williams [37W5] in Fig. 12–23 for iron containing 3.8% silicon. Values of $dE_k/d\theta = -L$ for various crystallographic planes (hkl) are given in Table 3.

Expressions for $dE_k/d\theta$ for various planes have been given in terms of crystallographic indices by Bozorth [36B1] and in terms of direction cosines by Tarasov and Bitter [37T1]. Relevant calculations have also been made by Schlechtweg [36S15].

TABLE 3. VALUES OF $dE_k/d\theta$ AS $f(\theta)$
(θ is measured in (hkl) plane from direction $[h_0k_0l_0]$)

(hkl)	$[h_0k_0l_0]$	$dE_k/d\theta$
100	001	$K_1(\sin 4\theta)/2$
100	011	$K_1(-\sin 4\theta)/2$
110	001	$K_1(2\sin 2\theta + 3\sin 4\theta)/8 + K_2(\sin 2\theta + 4\sin 4\theta - 3\sin 6\theta)/64$
110	$1\bar{1}0$	$K_1(-2\sin 2\theta + 3\sin 4\theta)/8$ $+ K_2(-\sin 2\theta + 4\sin 4\theta + 3\sin 6\theta)/64$
110	$1\bar{1}1$	$K_1(-2\sin 2\theta - 7\sin 4\theta)/24 + K_1(\cos 2\theta - \cos 4\theta)/3\sqrt{2})$ $+ K_2(-3\sin 2\theta - 28\sin 4\theta - 23\sin 6\theta)/576$ $+ K_2(3\cos 2\theta - 8\cos 4\theta + 5\cos 6\theta)/(144\sqrt{2})$
111	$1\bar{1}0$	$K_2(\sin 6\theta)/18$
111	$11\bar{2}$	$K_2(-\sin 6\theta)/18$
211	$01\bar{1}$	$K_1(2\sin 2\theta - 7\sin 4\theta)/24$ $- K_2(-13\sin 2\theta + 20\sin 4\theta - 25\sin 6\theta)/576$
211	$1\bar{1}\bar{1}$	$K_1(-2\sin 2\theta - 7\sin 4\theta)/24$ $- K_2(13\sin 2\theta + 20\sin 4\theta + 25\sin 6\theta)/576$

It is instructive to plot E_k as dependent on the direction of magnetization. This is done [36B1] in Fig. 12–24 for a cubic crystal in which K_1 is the only important constant. The difference between iron (positive K_1) and nickel (negative K_1) has been shown remarkably well by Bitter [37B4] by the use of space models (see Fig. 12–25), the surfaces of which represent the energy E_k; the minima for iron at the face centers and for nickel at the "corners" are apparent.

The effect of the field on the total energy, caused by the addition of the term $-HI_s \cos \theta$, is shown graphically in Fig. 12–26 [36B1]. The particular case chosen is for $K_1 = 33\,000$, $K_2 = -180\,000$, and has [110] as the direction of most difficult magnetization, as observed by Kleis [36K2] for the 50% nickel-iron alloy.

A procedure for the graphical determination of the change of magnetization with field in a cubic crystal, for any given direction of field, has been described [41B5]. In the general case the magnetization I_s does not lie in the same plane as H and the direction of easy magnetization. A method of calculation has also been given by Schlechtweg [36S15].

As mentioned previously, [110] rather than [100] or [111] is sometimes

580 MAGNETIC PROPERTIES OF CRYSTALS

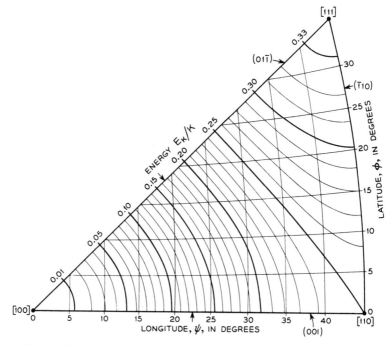

FIG. 12-24. Lines of equal crystal energy on stereographic projection of a cubic crystal, with scales of lattitude ϕ and longitude ψ.

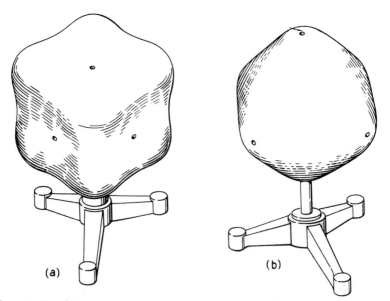

FIG. 12-25. Space variation of crystal energy in crystals having directions of easy magnetization parallel to (a) [100] directions and (b) [111] directions.

the direction of easy magnetization, and sometimes the direction of difficult magnetization. The relations between K_1 and K_2 for the different possible orders of ease of magnetization in the three principal directions are given in Table 4 [36B1].

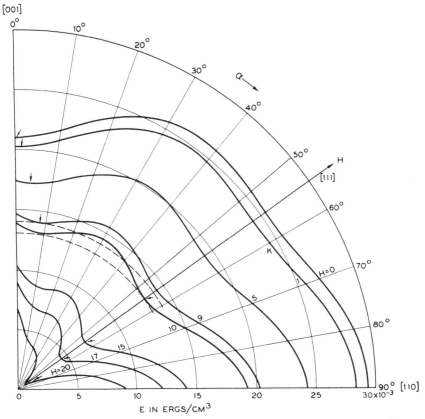

FIG. 12–26. Crystal energy as a function of field strength and direction of magnetization in the (110) plane, for $K_0 = 21\,000$, $K_1 = 33\,000$, $K_2 = -180\,000$.

The rate of *approach to saturation* obviously varies with crystallographic direction, and for a given direction the change of I with H can be calculated as described previously. The approach to saturation in a polycrystalline material, in which the crystals are oriented at random, has been calculated by Akulov [31A4], Gans [32G3], and Becker and Döring [39B5] with similar results. These references should be consulted for the details of the calculation. In high fields the expression is

$$\frac{I}{I_s} = 1 - \frac{8K_1^2}{105 I_s^2 H^2}.$$

TABLE 4. DIRECTIONS OF EASIEST, INTERMEDIATE AND HARDEST MAGNETIZATION

K_1 K_2	$+$ $+\infty$ to $-9K_1/4$	$+$ $-9K_1/4$ to $-9K_1$	$+$ $-9K_1$ to $-\infty$	$-$ $-\infty$ to $9\|K_1\|/4$	$-$ $9\|K_1\|/4$ to $9\|K_1\|$	$-$ $9\|K_1\|$ to $+\infty$
Easiest	[100]	[100]	[111]	[111]	[110]	[110]
Intermediate	[110]	[111]	[100]	[110]	[111]	[100]
Hardest	[111]	[110]	[110]	[100]	[100]	[111]

Higher terms involving K_2 have also been evaluated [32G3, 39B5]. The use of this relation has already been discussed in Chap. 11. Experiments indicate that the properties in high fields are affected by other factors in addition to crystal anisotropy. When these are taken into account the foregoing expression can be used to evaluate K_1 from measurements on polycrystalline material, provided random crystal orientation obtains. Anisotropy constants of iron and of nickel at various temperatures have been thus determined by Czerlinski [32C4] and Polley [39P4].

In *lower fields*, below those corresponding to residual induction, when domain boundary movements predominate, adequate theory is not available for the calculation of magnetization curves. However, Heisenberg [31H6] has proposed a method whereby the distribution of the magnetization of domains among the various possible directions of easy magnetization may be calculated under certain simplifying assumptions. This carries with it the corollary, as pointed out by Bozorth [32B10], that the component of magnetization perpendicular to the field may be calculated as a function of the component parallel to the field.

Heisenberg's assumption is that the domain directions will be distributed at random among the various easy directions, with the restriction that the parallel component of magnetization parallel to the field will have some given value. Then for a given value of I (parallel component) the most probable distribution of the domains can be calculated and I_n (normal component) can be determined from them. The assumption is open to the criticism that the actual distribution will not be expected to be quite random because local internal strains will give a preference for some directions over others.

On this basis Bozorth [32B10] has calculated I_n vs I curves for single crystal disks of iron and compared them with the experimental results of Honda and Kaya [26H2]. In order to effect the comparison it has been necessary to correct the experimental results for the demagnetizing field, which will act on I_n and tend to diminish it, and cause the true H to deviate from the applied field (except in the [100], [110], and [111] directions). In Fig. 12–27 are shown the calculated, observed, and corrected curves for a true H acting 40° from [010] in (001). Calculations apply only to values of I below $I_s \cos 40°$; above this value the rotation process is assumed to apply. Comparisons for other directions are given in the original article.

In a uniaxial crystal such as cobalt, in which the direction of easy magnetization is parallel to the axis, magnetization at right angles to the axis proceeds by domain rotation alone and the magnetization curve can

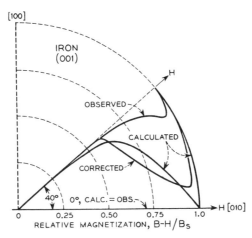

FIG. 12–27. Polar diagram of magnetization in a (001) plane in iron, when the field is applied parallel to [010] or 40° therefrom. Each curve is the path traced by the end of the magnetization vector as the field is increased from zero.

be calculated from the crystal anisotropy, as Heisenberg [31H6] has shown. Following the general method for rotation processes already outlined, values of I and H are calculated from the relations:

$$I = I_s \sin \varphi$$

$$H = (2K_1/I_s) \sin \varphi \, [1 + 2(K_2/K_1) \sin^2 \varphi],$$

the parameter φ being the angle between the spontaneous magnetization and the axis. The results for cobalt are in good agreement with experiment.

The expression for the initial permeability for such a rotational process is

$$\mu_0 = 1 + 4\pi I_s^2/(2K_1),$$

and Guillaud, Bertrand, and Vautier have compared the theory with the data for cobalt [48G1] and Mn_2Sb [49G7] at various temperatures. At room temperature the calculated and observed values of μ_0 for cobalt were 3.4 and 2.9; better agreement was obtained at $-196°C$. In Mn_2Sb the tetragonal axis is the direction of easy magnetization down to $-33°C$, when K_1 changes sign. In this material there is rather good agreement

584   MAGNETIC PROPERTIES OF CRYSTALS

between theory and experiment until K_1 becomes very small. Measurements of μ_0 and H_c were also made parallel to the axis.

Calculation of Torque Curves.—As an example, consider the calculation of the torque L as a function of the field H when the latter is directed 22.5° from a [001] direction in the (100) plane of a cubic crystal. In

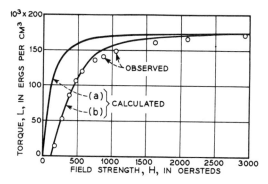

FIG. 12-28. Torque curve for silicon-iron, calculated (a) from simple theory, (b) with due regard for demagnetizing field. Observed points are shown.

accordance with domain theory, each domain has the intensity of magnetization I_s, and in a field is acted on by the torque

$$L = -dE_k/d\theta$$

per unit of volume, as already described. As in the previous section, conditions for minimum energy give

$$H = \frac{dE_k/d\theta}{I_s \sin(\theta_0 - \theta)}$$

with the notation of Fig. 12-22.

In the (100) plane, evaluation of $dE_k/d\theta$ gives

$$L = -(K_1/2) \sin 4\theta$$

and

$$H = -\frac{L}{I_s \sin(\theta_0 - \theta)}$$

and therefore one can plot L vs H using θ as a parameter and $\theta_0 = 22.5°$.

The theoretical torque curve, calculated assuming that the applied field is the actual field, is compared in Fig. 12-28 with the data [41B5] for iron containing 3.8% silicon. The departure of the points from the theoretical curve (a) at low H's is marked. A still more pronounced departure is observed when $\theta_0 = 40°$, as shown by the theoretical curve ($m = \infty$) and observed points of Fig. 12-29. The departures are attributed primarily to the fact that the experimental disks have demagnetizing factors that affect the situation in the following way:

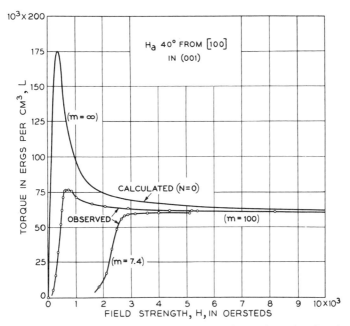

FIG. 12-29. Torque curve for silicon-iron disks of two dimensional ratios. Theoretical curve is for infinite dimensional ratio.

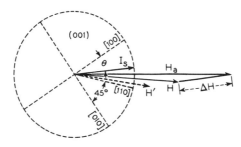

FIG. 12-30. Orientations of vectors I_s and $H = H_a - \Delta H$ in (001) plane, to illustrate effect of demagnetizing field ΔH on the direction of the field and on the magnitude of the torque.

In Fig. 12-30 let H_a represent the direction and magnitude of the applied field that has turned I_s through the angle θ from its original [100] position. The magnetization of the disk gives rise to a demagnetizing field ΔH, directed oppositely to I_s, of magnitude NI_s, N being the demagnetizing factor. This field adds vectorially to H_a to give the true field H. Now H

will be nearer to the [110] direction than will H_a; consequently, if ΔH or the angle between H_a and [100] is large enough, the true field will lie almost parallel to [110] and some of the domains will spring to another equally stable position lying between [110] and [010]. These domains will have torques opposite in sign to those lying on the [100] side of [110], and the resultant of the torques of all domains will be considerably lowered. The thicker the crystal or the smaller the diameter, the greater will be ΔH and the smaller the torque in all except very intense fields. The intensity of the applied field at which the torque becomes appreciable will obviously be about equal to the demagnetizing field for saturation magnetization, and this is shown by the data of Fig. 12–29, the demagnetizing field of the thickest disk being about 2000 oersteds. When the curve (b) of Fig. 12–28 is calculated with due regard for the demagnetizing field, agreement with experiment is found.

Torque curves for other orientations can be calculated following the same procedure. Incidentally, it may be noted that the torque is the product of the field strength and the component of I, namely, I_n, that is normal to the direction of H:

$$L = -HI_n.$$

This follows from fundamental principles and is consistent with the foregoing because $I_n = I_s \sin(\theta_0 - \theta)$.

TABLE 5. DIRECTIONS OF EASY MAGNETIZATION IN ROLLED SHEETS, IN TERMS OF THE ANGLE BETWEEN THE EASY DIRECTION AND THE DIRECTION OF ROLLING
(Composition of material is in per cent. Data are from several sources [31D3, 37B1, 40C3].)

Material	Cold Rolled (degrees)	Recrystallized (degrees)
Fe	45	0, 90
3.5 Si-Fe, hot rolled		0
3.5 Si-Fe		0, 90
4.5 Mo-Fe	45	0
78 Ni-Fe	0	0 or 90
50 Ni-Fe	0	90
50 Ni-Fe, rolled 95%	0	0, 90
93 Ni-Fe	0	45
Ni	0, 90	45
49 Fe, 49 Co, 2 V	90	0

Anisotropy in Polycrystalline Sheet.—Dahl and Pfaffenberger [31D3] showed in 1931 that almost all magnetic materials, both before and after annealing, have some magnetic anisotropy. Directions of easy magnetization for some common materials in sheet form in relation to the direction of rolling of the sheet are given in Table 5.

A relation between the torque curves of these materials and the orientation of the crystals composing them was worked out by Akulov and Brüchatov [32A5], and later in more detail by others [36B1, 40M1]. Sixtus [35S11] and Tarasov [38T2, 39T6] studied silicon-iron sheet in the hot- or cold-rolled or annealed condition, and they found good correlation between the torque curves and the crystal orientations as determined by other means. Tarasov's curves of cold-rolled silicon-iron, subsequently annealed at various temperatures, are shown in Fig. 12–31. The curve labeled "cold rolled" shows that directions of *difficult magnetization* occur at 0° and 90° to the rolling direction (the slope of the curve is positive and large and the material has positive K). After annealing at 900° these are directions of *easy magnetization*. In general, the curves show a gradual transition from the curve for the cold rolled material to that for the annealed (recrystallized) material. The latter corresponds to a single crystal oriented [35B9] with its (110) plane in the rolling plane and the [001] and [110] directions parallel, respectively, to rolling and cross directions, with the torque amplitude reduced by variations or "scattering" of the actual crystals from this ideal position.

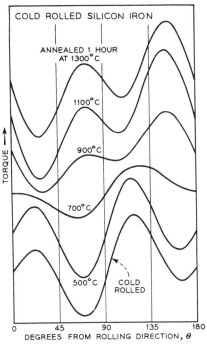

Fig. 12–31. Torque curves of silicon-iron, reduced 40% by cold rolling and annealed for one hour at various temperatures.

In silicon-iron rolled and annealed according to the technique of Goss [35G5] the crystals are aligned with high accuracy and the torque curve is close to that of a single crystal in both shape and amplitude. A torque curve of such material and, for comparison, a calculated curve for a single crystal in the (110) plane are given in Fig. 12–32.

The relation of magnetic properties to crystal orientation in *iron-nickel* alloys has been intensively investigated, especially in connection with the material Isoperm in which special crystal and domain orientations are produced by controlled cold rolling and annealing.

Through the work of Snoek [35S12] and Pawlek [35P3] it has been established that the crystals in the face-centered iron-nickel alloys are aligned by severe cold rolling and subsequent recrystallization, so that a (100) plane lies in the rolling plane and a [001] direction in the rolling direction ("cube texture"). Conradt, Dahl, and Sixtus [40C3] have found the torque curve of an alloy of 93% nickel, rolled 99% and recrystallized at 950°C (see Fig. 12–33), to be that expected from the texture.

When the recrystallized material is rolled further in the original direction (or at 90°) the cube texture persists, according to X-ray tests, but the magnetic anisotropy is such that the rolling direction is a direction of

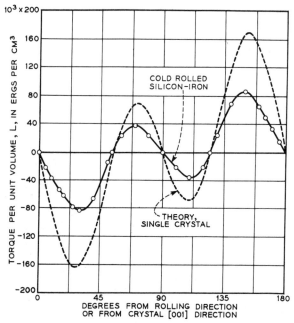

Fig. 12–32. Comparison of torque curve calculated for a single crystal and observed for grain-oriented polycrystalline material.

difficult magnetization and the cross direction is the direction of easy magnetization, as pointed out by Six, Snoek, and Burgers [34S5]. The domain structures caused by the various treatments are illustrated in Fig. 12–34.

After the material has been so recrystallized and rolled, the torque curve is of the form

$$L = A \sin 2\theta + B \sin 4\theta.$$

Rathenau and Snoek [41R2] assumed that the 2θ term was caused by the rolling and the 4θ term was due entirely to the crystal orientations. By placing two disks together so that the rolling direction of one coincided with the cross direction of the other, it was easy to determine the coefficient

ANISOTROPY IN POLYCRYSTALLINE SHEET 589

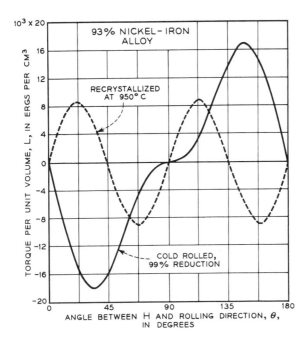

Fig. 12–33. Dependence of torque curves of a 93% nickel-iron alloy on heat treatment. Solid curve, cold-rolled with 99% reduction in thickness: direction of easy magnetization is parallel to the direction of rolling. Dotted curve, same material recrystallized, with direction of hard magnetization parallel to the rolling direction.

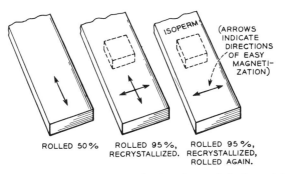

Fig. 12–34. Directions of easy magnetization (arrows) in iron-nickel alloy with various treatments. Development of Isoperm.

of sin 4θ and thus the anisotropy constant. Similar measurements were made on the same material after relieving most of the strain by annealing at 600°C. Values of K_1 obtained in both cases (see Fig. 12-35) differ somewhat from those found with single crystals (see Fig. 12-14).

The nature of the anisotropy caused by rolling has been the subject of some investigation. Conradt, Dahl, and Sixtus [40C3] started with annealed material with random crystal orientation, then cold rolled it by various amounts. Compositions ranged from 40–100% nickel. Rolling of about 50% produced the greatest anisotropy, and this was such that the rolling direction was a direction of *easy* magnetization, as shown by the torque curve of Fig. 12-33, and contrary to the result of rolling the material with cube texture. The constant of anisotropy of rolling (coefficient A in the foregoing equation) then attains its highest value at about 75% nickel, and there has a value of 70 000, much larger than the crystal anisotropy constant. It disappears almost completely after annealing at about 500°C. It does not change sign when the magnetostriction does, at 81% nickel, and therefore it is unlikely that the anisotropy is caused by rolling strains. In the opinion of Conradt, Dahl, and Sixtus [40C3, 42C4], based on additional data, it is associated with the formation of superstructure in a way not understood in detail.

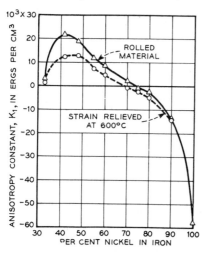

FIG. 12-35. Cubic anisotropy of iron-nickel alloys in cube position after annealing and cold rolling (crosses) and after subsequent annealing at 600° C (circles).

Rotational Hysteresis.—Since the torque is the (negative) rate of change of energy with angle, the energy can be obtained by integration of the torque curve (L vs θ). If the energy of anisotropy is the only energy involved, the integral of the L vs θ curve from 0 to 360° is zero. Expressing the torque caused by the anisotropy as L_a,

$$\int_0^{2\pi} L_a\, d\theta = 0.$$

If other energy, such as rotational hysteresis, is present, this is measured by

$$W_r = -\int L\, d\theta.$$

Here θ may be the angle between the magnetization and the field in the specimen H, or it may be the angle between the magnetization and the direction of the applied field H_a. In vector notation [38I2]

$$L = I \times H$$

and since

$$H = H_a - NI,$$

N being the demagnetizing factor, then

$$L = I \times H_a - N(I \times I) = I \times H_a.$$

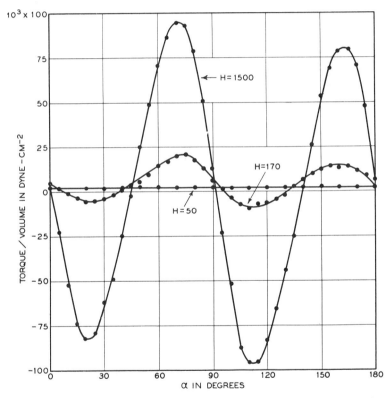

FIG. 12-36. Torque curves of hard-rolled iron, showing rotational hysteresis at the lower field strengths.

Therefore L is equal to both $I \times H$ and $I \times H_a$, and W_r can be evaluated by integrating the curve obtained by plotting L against either H or H_a.

Rotational hysteresis loss is evident in the torque curves of Fig. 12-36, taken by Williams [37W3] on a specimen of hard rolled iron. Rotational hysteresis is easily derived from these curves for $H_a = 50$, when anisotropy is practically zero. At $H_a = 170$, anisotropy is evident, but the curve is obviously unsymmetrical and the area under it has a net negative value. At $H_a = 1500$, hysteresis is undetectable.

Brailsford has measured rotational hysteresis in silicon-iron sheets at various field strengths. In order to measure more accurately the net area under the torque curve, when anisotropy is large, he has used three disks placed on top of each other so that the directions of easy magnetization are staggered by 120°. The large positive and negative values of L are thus much reduced, and the net area can be determined more accurately. His results have already been discussed in Chap 11 (Fig. 11–32).

Origin of Anisotropy.—So far in this chapter the discussion of anisotropy has been on a purely empirical basis, the expressions for it having been developed in the form of series that are consistent with the crystal symmetry. The explanation of anisotropy in terms of atomic and crystal structure has been considered in a number of papers and will now be reviewed briefly.

The quantum forces of exchange that are responsible for the molecular field are isotropic in character and do not account for variation of properties with direction. One of the first proposals regarding the origin of anisotropy was made by Mahajani [26M5, 29M2] who considered the magnetic interaction between magnets of various kinds, located at lattice points of the crystal. The magnetic action of a cubic lattice of dipole moments, always aligned parallel to each other, is purely isotropic. On the contrary, if bar-shaped magnets are located at the points of a body-centered lattice such as iron, they will have stable positions when they are parallel to a cube diagonal, [111]. If the magnets are formed by current loops or are small, flat, disk-shaped permanent magnets magnetized perpendicular to the surface, they are stable when parallel to [100]. Agreement with experiments is thus obtained if the atomic magnets in iron are assumed to be disk-shaped. Similarly it has been shown [37B8, 37M8] that one can explain qualitatively the direction of easy magnetization in nickel, [111], if one assumes that here the atomic magnets are bar-shaped (elongated in the direction of magnetization).

In trying to explain the anistropy of cubic crystals with a more realistic atom model it has been shown (e.g., by Van Vleck [37V1]) that interaction of a purely magnetic nature is quantitatively deficient by a factor of approximately 1000. It has been proposed by Heisenberg [30H2], Powell [30P3] and Fowler [31F2], Bloch and Gentile [31B6] and others that spin-orbit coupling must be invoked. In the solid lattice the electron orbits are restricted by the electrostatic fields of neighboring atoms so that they cannot be freely oriented by a magnetic field. There is also strong electrostatic interaction between the spin and orbital motion of a single electron. Consequently, when the external magnetic field orients the spin of one electron, it reacts on its own orbit, which reacts on the orbits of the neighboring atoms, which, in turn, influence the spins in these orbits. The directional character of the orbits is thus communicated to the spins.

The most tangible and complete quantitative investigation of this model has been that of Van Vleck [37V1]. The results lead to an anisotropy that is of about the right order of magnitude, and the general character of the temperature dependence agrees with experiment in that it decreases with increasing temperature more rapidly than the saturation intensity of magnetization.

Van Vleck considered two models in his calculation. According to the first or *dipole* model, each magnet has only a dipole moment, and then anisotropy will exist only if all the dipoles are not perfectly parallel. According to the second model a *quadrupole* moment is attributed to each atom; in this case anisotropy will occur even when all magnets are parallel, as at absolute zero. The reader is referred to the original paper for a description of the calculations.

When a metal such as nickel is alloyed with a nonmagnetic metal, copper, theory indicates that according to the first model K_1 should decrease, remaining proportional to the saturation magnetization at 0°K, namely, I_0. According to the second model, one expects $K_1 \propto I_0^2$. The experiments of Williams and Bozorth [39W2], described previously, are more nearly in accord with the second or quadrupole model, as shown in Fig. 12-18.

A further investigation, using the electron band model, was carried out by Brooks [40B5]. The calculated anisotropy is found here also to be in general agreement with experiment. In addition the signs of the anisotropy constants K_1 in iron and nickel are accounted for, and the existence of the change in sign of K_1 in the face-centered, iron-nickel alloys is explained in a reasonably satisfactory (qualitative) way. As iron is added to nickel and more holes appear in the d-band, the increased number of unbalanced spins tends theoretically to make K_1 more positive (the absolute value of the negative K_1 becomes less) and finally greater than zero when enough iron is added (experimentally, 30%).

Calculations on spin-orbit coupling in cubic crystals have also been reported by van Peype [38P2].

When Van Vleck applied his method of calculation to a non-cubic crystal such as *hexagonal cobalt*, he found that the anisotropy calculated for the dipole model could be of the same order as that observed. An anisotropy 100 times that observed for cobalt could be accounted for in a hexagonal structure with atomic distances varying in an extreme way with crystal direction. In hexagonal cobalt the distances depart only slightly from those that exist in the cubic form and so extreme values of K are not expected. In some of the hexagonal and tetragonal compounds of manganese examined by Guillaud the departure from cubic symmetry is larger and the observed anisotropies are correspondingly greater.

Variation of the anisotropy of hexagonal cobalt with temperature has been calculated by Vonsovsky [40V3], using the particular model of Bloch and Gentile. His results are consistent with the fact that the sign of the anisotropy reverses at elevated temperatures (experimentally, 250°C). A critical review of the origin of anisotropy has been given by Van Vleck [47V1].

The anisotropy of magnetostriction is discussed in the following chapter. Magnetic anisotropy of paramagnetic and diamagnetic crystals has been much investigated but will not be discussed here. A brief reviewing article, by Lonsdale [38L4], refers to the discovery of the effect by Plücker in 1847 and to the following investigations of Faraday.

CHAPTER 13

STRESS AND MAGNETOSTRICTION

STRESS

The magnetic properties of most ferromagnetic materials change with the application of stress to such an extent that stress may be ranked with field strength and temperature as one of the primary factors affecting magnetic change. In some materials a tension of 10 kg/mm² (14 200 lb/in.²) will increase the permeability in low fields by a factor of 100; in others the permeability is decreased by tension and in still others (e.g., iron) the permeability in low fields is increased and that in higher fields decreased. In all materials the saturation induction is unaffected by a stress within the elastic limit, and it is affected by stresses large enough to produce plastic flow only when a change of phase or state of atomic ordering occurs in the material.

Application of a stress beyond the elastic limit generally decreases the permeability of annealed materials. The presence of internal strains, caused by mechanical overstrain or precipitation hardening, makes the material harder magnetically as well as mechanically.

In discussing the effect of a unidirectional stress it is convenient to divide materials into two classes, which have:

(1) Positive magnetostriction, or
(2) Negative magnetostriction.

In materials of class (1) the magnetization is *increased* by tension (except at $I = 0$ or I_s), and the material *expands* when magnetized; in class (2) the magnetization is *decreased* by tension and the material *contracts* when magnetized. The quantitative relation between magnetostriction and stress is discussed below under "Domain Theory of Effect of Stress," and in Chap. 18. Early work on the strain-sensitivity of Permalloy, and its relation to magnetostriction, was reported by Buckley and McKeehan [25B2].

Permalloy containing 68% nickel is a good example of a material having positive magnetostriction and high strain sensitivity. Figure 13-1 depicts magnetization curves for this material with tensions (σ) of 0 and 2 kg/mm² (2800 lb/in.²) and shows the rapid rise of induction with field strength ($dB/dH = 1\,000\,000$) to the saturation value when stress is present. By a stress of 2 kg/mm² the maximum permeability is increased sevenfold, and μ at $H = 0.25$ is increased 20 times. The same figure shows the corresponding *decrease* in permeability with tension in nickel, a typical example of a material having negative magnetostriction.

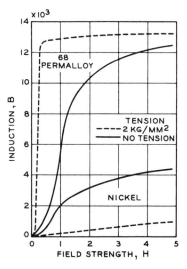

FIG. 13-1. Effect of tension on the magnetization curves of 68 Permalloy (positive magnetostriction) and of nickel (negative magnetostriction). Tension held constant during magnetization.

Hysteresis loops of nickel and 68 Permalloy are shown in Fig. 13-2. The effect of stress is reversed in sign over a small portion of the loop just before the induction goes to zero. When the tension in a positively magnetostrictive material is sufficiently great, the sides of the loop become vertical. A similar effect occurs when linear pressure is applied to a material having negative magnetostriction.

Strain Beyond Elastic Limit.—When increasing tension is applied to a material having positive magnetostriction, the induction at constant field strength (applied after the stress is fixed) increases until the elastic limit is reached and then usually decreases continuously to a low value as the material is hardened by plastic deformation. The data of Fig. 13-3 refer to the same specimen as that of Fig. 13-1, but now the tension is increased from zero to about twice the elastic limit; the induction in a field of 0.1 oersted increases until it is almost equal to the saturation density (13 300) and then decreases when plastic flow begins.

When the material, like nickel, is magnetostrictively negative, the point on the curve at which plastic flow begins is not readily determined. Data illustrating this point are given in Fig. 13-4, which includes also the data for iron. The elastic limit for magnetostrictively negative materials is more easily located when one plots the induction for a given field strength determined after the stress has been applied *and removed;* here an overstrain of 1% is easily discernible and under favorable circumstances considerably less than 0.1% can be detected. Curves obtained in this way for iron and nickel are shown by the broken lines of Fig. 13-4.

The effects of a plastic strain on the initial and maximum permeabilities and on the coercive force correspond to the phenomena already described; μ_0 and μ_m are decreased and H_c increased progressively, the more the stress exceeds the elastic limit. Figure 13-5 shows how μ_m is affected by a tension, σ, applied to a material having positive magnetostriction (68 Permalloy); with increasing stress, μ_m increases up to the elastic limit and the

STRAIN BEYOND ELASTIC LIMIT 597

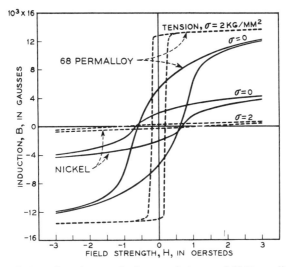

FIG. 13–2. Effect of tension on the hysteresis loops of 68 Permalloy and of nickel. Maximum field strength, 20 oersteds.

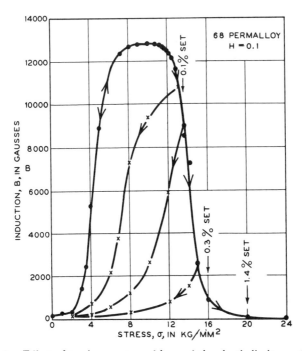

FIG. 13–3. Effect of tension, up to and beyond the elastic limit, on the induction of 68 Permalloy in at field strength of 0.1 oersted.

μ_m vs σ curve can be traversed in both directions as long as this limit is not exceeded. When σ is increased beyond the elastic limit, the permeability declines, the more the higher the stress. When σ is decreased after having attained a maximum stress σ_m which is above the elastic limit, the permeability decreases continually; and when σ is again increased, the permeability increases until $\sigma = \sigma_m$, after which it again declines.

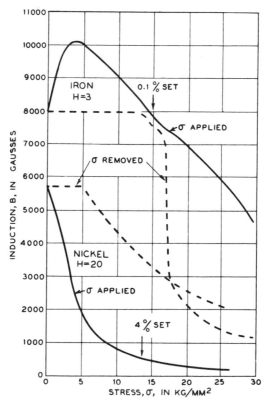

FIG. 13-4. Effect of tension on the induction of iron and of nickel at field strengths of 3 and 20 oersteds, respectively.

The effect of tension on the initial permeability of nickel has been especially important in the development of the domain theory, and in this connection Kersten [31K6] has compiled data for tensions both below and above the elastic limit. His data are given in Fig. 13-6 which shows the agreement between experiment and theory, the latter requiring that $\mu_0 - 1$ be inversely proportional to tension. A fuller discussion of the relation with theory is given in a later section, "Internal Strains," and in Chap. 18.

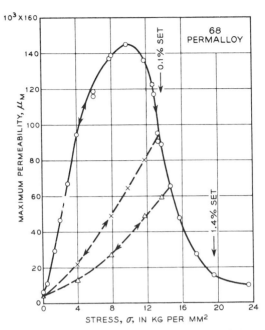

FIG. 13–5. Effect of tension on the maximum permeability of 68 Permalloy.

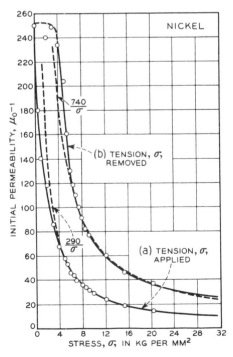

FIG. 13–6. Effect of tension on the initial permeability of nickel.

Stress Within the Elastic Limit.—The effects of stress on magnetic properties are usually described by magnetization curves recorded when the specimen is subjected to a tension (σ) which is maintained constant while the field strength is varied. The value of B thus obtained with a given σ and H is not the same as that observed when H is applied first and σ subsequently; this is illustrated in Fig. 13–7, for nickel, in which the two processes are designated (σ,H) and (H,σ), respectively. The (H,σ) curve lies above the (σ,H) curve for materials having either positive or negative magnetostriction.

When H is applied first and maintained constant, successive application and release of σ cause unequal changes in B until the process has been

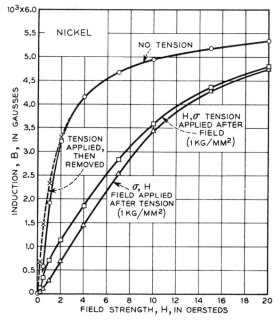

Fig. 13–7. Effects of applying stress (0.1 kg/mm²) and field in different orders, in nickel.

repeated a number of times; thus for the "cyclic state" the change in B produced by the stress is often considerably less than for the first application of the same stress. An extreme case of this effect is illustrated in Fig. 13–8 for 68 Permalloy which has been subjected to a definite succession of field strengths before the stress was applied. After the stress has been applied and removed a number of times, in a steady field, the induction approaches a limit that may be quite different from the original value obtained by application of the field alone. The dotted line in Fig. 13–7 shows this limiting value of induction for nickel after repeated application and removal of a tension of 1 kg/mm². The displacement upward from the normal magnetization curve, caused by such a procedure, is most noticeable at low inductions.

In the following description of the various phenomena associated with elastic stress, nickel and 68 Permalloy will often be used [45B1] as examples of materials with negative and positive magnetostriction, respectively. Although nickel has the greater magnetostrictive change in length when magnetized to saturation, it is not as sensitive to changes in stress as 68 Permalloy. This difference may be attributed to the relatively low magnetic anisotropy of 68 Permalloy as compared to nickel and should be borne in mind in considering the following data. Iron occasionally is used

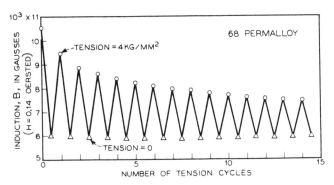

FIG. 13-8. Change of induction caused by repeated application and removal of a tension of 4 kg/mm² in 68 Permalloy.

as an example of a material having sometimes positive, sometimes negative magnetostriction, depending on the field strength and on the previous condition of magnetization.

Magnetization curves of nickel and of 68 Permalloy, under constant tensions, have been shown in Fig. 13-1. In taking these curves the specimen was demagnetized with the tension on and the tension kept constant during the measurements of B and H. Tension causes the magnetic induction to increase in 68 Permalloy and decrease in nickel, and as the stress is increased from zero the largest changes in induction occur, first, near the steepest part of the magnetization curve, and with higher stresses nearer the knee of the curve for nickel and nearer the origin for 68 Permalloy. When the tension is high, the magnitude of the change in 68 Permalloy is large and is limited only by the saturation magnetization of the material and the ability of the material to maintain the stress. As the field strength is increased from zero to a high value, the change of induction caused by the presence of a constant stress generally increases from zero, passes through a maximum and approaches zero again as a limit.

Figure 13–9 shows that the behavior of annealed iron is more complicated. Low tensions cause an increase in induction when this lies below the knee of the curve, a decrease for higher inductions; high tensions cause an increase in B for points near the origin of the curve and a decrease for other points (the Villari reversal [1865V1]). Below the knee of the magnetization curve ($B = 13\,000$), the point at which the change of induction with tension, $dB/d\sigma$, changes sign varies from very small values near the knee

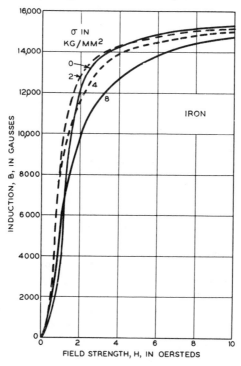

Fig. 13–9. Magnetization curves of iron subject to various tensions.

to about $\sigma = 2$ kg/mm² at $B = 10\,000$ and about $\sigma = 8$ at $B = 1000$–2000. The behavior at low inductions is indicated in Fig. 13–10 in which the curve for $\sigma = 4$ crosses that for $\sigma = 8$ at $H = 0.6$, $B = 2000$.

The effect of tension on the initial permeability and on permeabilities at low field strengths is shown for a commercial high-permeability alloy in Fig. 13–11. The specimen of 4-79 Molybdenum Permalloy was heated to 1050°C and cooled in the usual way. The initial permeability decreases continually with increasing tension but comes back to its original value upon release of the tension if the elastic limit (about 15 kg/mm²) has not been exceeded. This is shown by the broken line. For field strengths of 0.01 oersted and somewhat higher each μ vs σ curve goes through a pro-

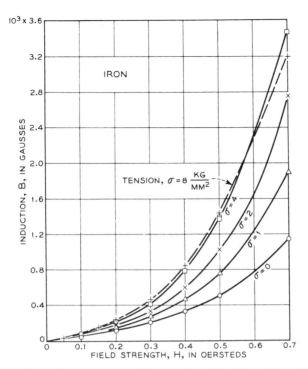

FIG. 13-10. Magnetization curves of iron at low inductions with various tensions.

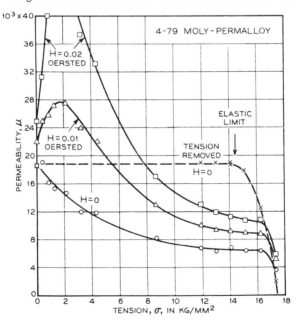

FIG. 13-11. Effect of tension on the permeability of 4-79 Permalloy at various field strengths.

nounced maximum and then falls below the points for $\sigma = 0$. In stronger fields the maximum is less pronounced and as saturation is approached $d\mu/d\sigma$ approaches zero, as is to be expected. Study of Figs. 13-9 to 11 will bring out several points of similarity and of difference in the behavior of this alloy and iron.

It has already been mentioned that 68 Permalloy is especially sensitive to stress and that there are marked differences between the magnetization

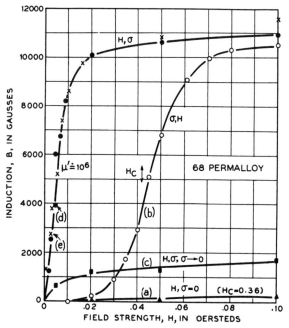

Fig. 13-12. Various kinds of magnetization curves of 68 Permalloy: (a) normal curve (no tension); (b) $\sigma(4 \text{ kg/mm}^2)$ applied first and then H; (c) H applied, σ applied and removed several times; (d) H applied first and then σ (solid circles); (e) σ applied, then H, then a high alternating field superposed and reduced to zero (crosses).

curves when σ and H are applied in different orders. Figure 13-12 shows the (σ,H) and (H,σ) curves for low fields when $\sigma = 4 \text{ kg/mm}^2$. The maximum slope of the (H,σ) curve is about $B/H = 10^6$, and both this high slope and the shape of the curve bring to mind the ideal magnetization curve, which for comparison has been determined on this same specimen. As nearly as one can tell from the data, these two curves coincide.

As pointed out years ago by Ewing [00E1], there is often marked hysteresis in the relation between stress and magnetization with constant field strength. As illustrations of the complicated interrelations between σ, B, and H, Fig. 13-13 shows a B vs H loop of 68 Permalloy taken with a stress

$\sigma = 4$ kg/mm continually maintained (broken line) and a loop with the same stress released and then applied at various constant values of H (solid line). Figure 13–14 shows a similar loop taken with no stress

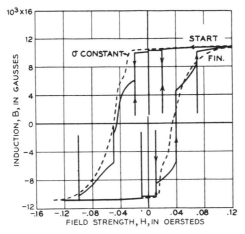

FIG. 13–13. Hysteresis loop of 68 Permalloy under tension: tension held constant at 4 kg/mm², except released and reapplied at certain field strengths.

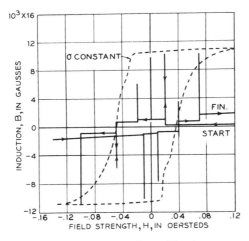

FIG. 13–14. Hysteresis loop of 68 Permalloy. Tension zero except when applied and removed at certain field strengths.

applied except at certain values of H at which it is applied and immediately released.

Hysteresis loops of nickel and of iron measured with constant tensions are reproduced in Figs. 13–15 and 16, respectively. The crossing points of loops taken with different values of stress are usually in the second and

606   STRESS AND MAGNETOSTRICTION

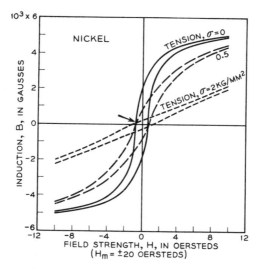

Fig. 13–15.   Effect of tension on the hysteresis loop of nickel.

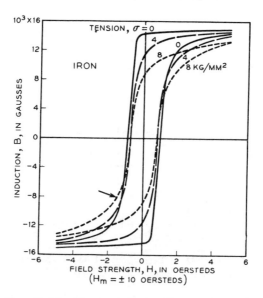

Fig. 13–16.   Hysteresis loops of iron under tension.

fourth quadrants, as shown for nickel; the behavior of iron is abnormal in this respect and is associated with the Villari reversal. Figure 13-17 shows that 68 Permalloy (positive magnetostriction) may be classed as normal; it shows also the "squaring" of the loop exhibited generally by magnetostrictively positive materials under tension and magnetostrictively negative materials under compression.

The increase in permeability that occurs in a material subject to alternating stresses at a constant low field strength, illustrated in Figs. 13-7 and 12 (broken lines), has its counterpart in the reduction in the area of the hysteresis loop. Figures 13-18 and 19 for 68 Permalloy and nickel, respec-

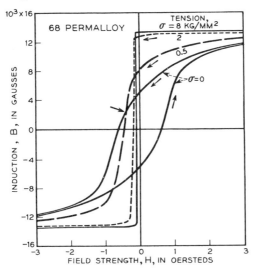

FIG. 13-17. Parts of hysteresis loops of 68 Permalloy under tension. Maximum field strength 5 oersteds.

tively, show that the hysteresis is reduced to 5–30% of the normal amount by twice applying and removing a tension considerably less than the elastic limit.

Kersten [31K6] has recorded hysteresis loops of wires bent beyond the elastic limit and then held straight under elastic stress by placing in a capillary tube. In this way part of the material is put under elastic tension and part under compression. In one series of experiments wires 0.5 mm in diameter were bent plastically until their radii of curvature were 70–100 mm when unrestrained by external forces. The hysteresis loops obtained after placing these wires in the capillary are reproduced in Fig. 13-20. It should be noted that the induction changes precipitously when it is about half of saturation and that thereafter the approach to saturation starts linearly.

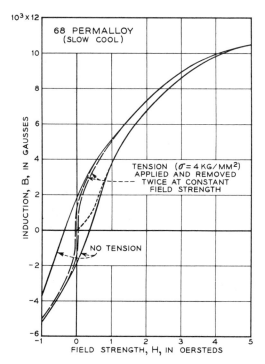

Fig. 13–18. Inner loop, stress applied and removed twice at constant field; outer loop and magnetization curve, no stress. Material, 68 Permalloy annealed and slowly cooled.

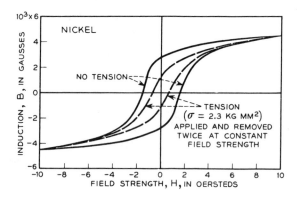

Fig. 13–19. Same as in previous figure. Material, nickel.

STRESS WITHIN THE ELASTIC LIMIT 609

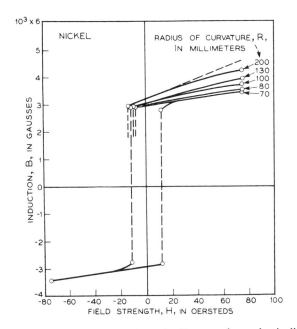

Fig. 13–20. Loops of nickel wire 0.5 mm in diameter, bent plastically with various radii of curvature, then held straight during measurement.

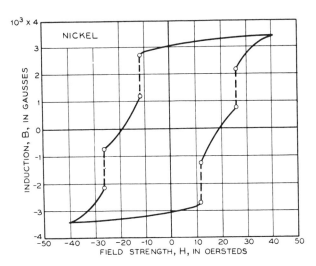

Fig. 13–21. Hysteresis loop of nickel after plastic and elastic strain (Forrer).

A wide variety of shapes of loops obtained in material strained in this manner was previously reported by Forrer [26F1]. Some of them showed two large discontinuities on each side as indicated in Fig. 13-21. Kersten has made clear in a qualitative way the occurrence of this kind of discontinuity in hysteresis loops. Figure 13-22(a) represents a wire bent so that at no point does the stress exceed the elastic limit, and the inclined straight line indicates the compressional and tensional values of the stress at points along a diameter. When the radius of curvature is smaller, as in (b), the stresses at and near the edge of the wire are sufficient to cause plastic flow, and their distribution is as shown. When the wire is now allowed to become unrestrained, the stresses redistribute themselves as shown in (c), and if the wire is held straight in a capillary tube the stresses are as in (d) or (e) depending on the amount of plastic strain. In (d) there are two regions of (unequal) compression and two of tension, and the hysteresis loop for this state will therefore be compounded of the loops of nickel in these four states. Since nickel in compression, like 68 Permalloy in tension, has a loop with a vertical portion occurring at a field strength that depends on the magnitude of the stress, the composite loop will contain two vertical portions on each side (Fig. 13-21). This and other more complicated loops obtained by various combinations of bending, stretching, and twisting have been observed in nickel [26F1], in the Permalloys, and in the iron-nickel alloy containing 8% nickel [32P1, 29P1].

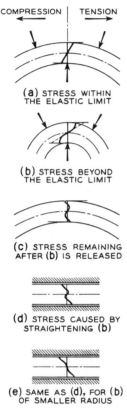

Fig. 13-22. Explanation of Forrer loops (see text).

Domain Theory of Effect of Stress.—Since a magnetic material changes its shape upon magnetization (magnetostriction), it will change its magnetic properties when strained. When the changes in stress and magnetostriction are small and reversible, there is a thermodynamic relation between the change of induction B, with stress (tension) σ, and the length l, as affected by magnetizing field strength H:

$$\frac{1}{l}\frac{\partial l}{\partial H} = \frac{1}{4\pi}\frac{\partial B}{\partial \sigma}.$$

This relation is derived in Chap. 15. We see from it that the magnetization will be increased by tension if the magnetostriction is positive and increases continually with field, that B is decreased if $\Delta l/l$ is negative.

The dependence of magnetization on stress may also be described in terms of the energy associated with the stress and the direction of spontaneous magnetization I_s in a domain. As already stated in Chap. 11 and derived in Chap. 18 the magnetic strain energy density is

$$E_\sigma = \tfrac{3}{2}\lambda_s \sigma \sin^2 \theta,$$

where λ_s is the magnetostrictive expansion at saturation, and θ the angle between the saturation magnetization I_s and the tension σ. This expres-

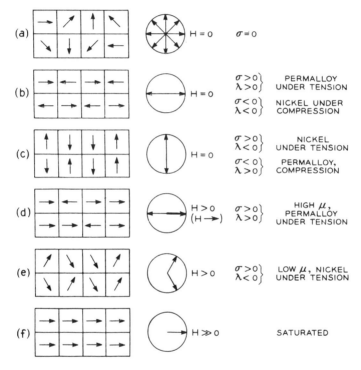

FIG. 13–23. Effect of stress on the domain structures of materials of positive or negative magnetostriction.

sion shows that when λ_s and σ are positive, as when tension is applied to iron, the energy is a minimum for $\theta = 0$, and therefore that the domains are stable when I_s is parallel to σ. Consequently I for such a material is generally increased by tension. When $\lambda_s \sigma$ is negative, as in nickel under tension, the domains will tend to be oriented with I_s perpendicular to the axis of tension, and magnetization will be decreased by tension.

Energy diagrams showing stable domain orientations have already been given in Fig. 11–7. Diagrams illustrating the effect of tension and compression on materials of various kinds are given in Fig. 13–23, which is self-explanatory. Such diagrams facilitate greatly the description of the domain structure of a material and aid in predicting its behavior.

An early and instructive application of the domain theory of stress was made by Becker and Kersten [30B4] in 1930. Increasing *tension* was applied to *nickel* and B vs H curves taken with H parallel to σ. As already shown in Fig. 13–1 the permeability is decreased and the lower part of the curve soon becomes flat as σ increases. When σ is sufficiently high (large compared with internal stresses), one should be able to calculate the course of the initial portion of the B vs H curve. Following Becker and Kersten we write the energies associated with stress and magnetic field:

$$E_\sigma = \tfrac{3}{2}\lambda_s \sigma \sin^2 \theta$$

$$E_H = -HI_s \cos \theta.$$

The sum of these energies is a minimum when

$$(d/d\theta)(E_\sigma + E_H) = 0.$$

Since $I = I_s \cos \theta$, we find

$$\frac{I}{H} = \frac{I_s^2}{3(-\lambda_s)\sigma}$$

or

$$\mu_0 - 1 = \frac{4\pi I_s^2}{3(-\lambda_s)\sigma}.$$

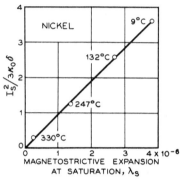

FIG. 13–24. Relation between the magnetostriction and the quantity $4\pi I_s^2/3(\mu_0 - 1)\sigma$.

Using μ_0 as a measure of the slope of the B vs H curve in low fields, curve (a) of Fig. 13–6 was plotted [31K6]. When known values of I_s and λ_s are used, and σ expressed in kg/mm^2,

$$\mu_0 - 1 = \frac{290}{\sigma}.$$

Agreement between theory and experiment was an important confirmation of the validity of domain theory.

The relation between μ_0, λ_s, and σ has been tested further in nickel by Scharff [35S9] over a range of temperature and stress. She plotted $4\pi I_s^2/3(\mu_0 - 1)\sigma$ against λ_s and obtained the straight line of Fig. 13–24, with the proportionality constant close to one. The experimental values of λ_s used were those of Nagaoka and Honda [02N1]. More recent data of Döring [36D4] show that, as the temperature varies, λ_s remains proportional to I_s^2. Then one should expect, since $I_s^2/\lambda_s = $ constant,

$$\mu_0 - 1 \propto \frac{1}{\sigma}$$

EFFECT OF VERY SMALL STRESSES

as the temperature is changed. Scharff, however, found $(\mu_0 - 1)\sigma$ to vary with temperature over a three fold range. Recently Döring [47D1] has shown that a variation of $(\mu_0 - 1)\sigma$ with temperature might be expected. Neither the theory nor the experiments are final: it is not certain that the theory can be used at temperatures near the Curie point, and measurements of μ_0, σ, and λ_s have not been made on the same specimen.

Effect of Very Small Stresses.—When a small stress, σ, is repeatedly applied to and removed from a magnetic material subjected to a steady field

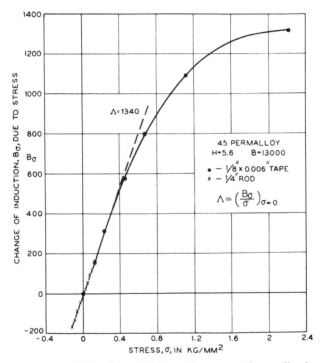

Fig. 13–25. Change of induction of 45 Permalloy caused by application of tension or compression. Biasing field, $H = 5.6$.

("polarized"), the change in induction so produced, B_σ, is proportional to σ. As may be noted in Fig. 13–25, the B_σ vs σ curve goes through the origin without change in slope, and as σ increases the curve approaches a limiting value of B_σ. The data of this figure refer to 45 Permalloy, to which was applied a constant polarizing field $H_0 = 5.6$. Data of this kind may also be plotted, as in Fig. 13–26, with B_σ/σ as ordinate. Here the curves refer to 68 Permalloy, and when the polarizing field is small they rise characteristically from a finite ordinate at $\sigma = 0$ to a maximum and then drop asymptotically toward zero. When H_0 corresponds to a point

614   STRESS AND MAGNETOSTRICTION

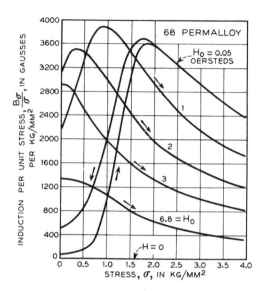

Fig. 13–26. Change in induction per unit stress for various magnitudes of stress. Data used to extrapolate to zero stress to determine Λ.

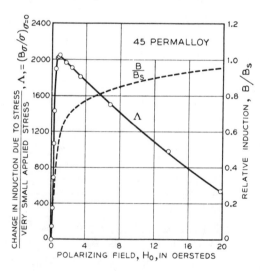

Fig. 13–27. Stress sensitivity at zero stress, Λ, of 45 Permalloy as dependent on biasing field strength, and in relation to induction.

on the magnetization curve well beyond the knee, the B_σ/σ curve falls continually from its value for $\sigma = 0$. Hysteresis is especially apparent in low fields.

A significant fundamental constant, and one of practical importance in the design of magnetic microphones, is the change in induction with stress for very small stresses. This is designated $\Lambda = (B_\sigma/\sigma)_{\sigma=0}$. For 45 Permalloy the dependence of Λ on the polarizing field strength is given by the solid curve of Fig. 13-27 [45B1]. The curve rises rapidly to a maximum, occurring at a field strength somewhat greater than that corresponding to maximum permeability, and then falls towards zero. When Λ is plotted against the polarizing induction B_0, as in Fig. 13-28, it is apparent that its maximum value Λ_m occurs when B_0 is somewhat greater than half of the saturation B_s. Values of the reversible permeability are also given in this figure. Figure 13-29 shows the hysteresis in the Λ vs H_0 curve; here the value of H_0 for $\Lambda = 0$ is somewhat less than the coercive force as may be inferred from the position of the intersections of the various loops of Figs. 13-15 and 17.

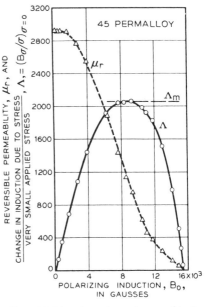

FIG. 13-28. Λ as dependent on biasing induction; also, reversible permeability.

The constant Λ_m, the maximum value of Λ obtained by plotting it against H_0 or B_0, has been determined [45B1] for ten iron-nickel alloys and the results are given in Fig. 13-30. The cross-over near 81% nickel is obviously associated with the vanishing magnetostriction of this alloy. The maximum near 60% nickel is apparently associated with the low magnetic anisotropy of alloys near this in composition. The minimum near 90% nickel is due to a combination of factors—the changes of saturation magnetization, of magnetostriction, and of anisotropy with composition.

Domain theory has been used by Bozorth and Williams [45B1] to calculate both Λ as a function of B_0 for a given material, and Λ_m as dependent on the nickel content of Fe-Ni alloys, as follows:

Assume in accordance with Fig. 13-31 that a single domain is orientated so that its magnetization I_s makes the angle θ with the axis of the specimen along which H and σ are applied. The domain has been rotated into this

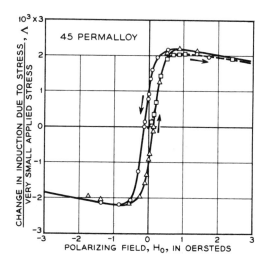

Fig. 13–29. Hysteresis in Λ in 45 Permalloy.

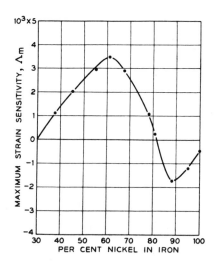

Fig. 13–30. Maximum stress-sensitivity, Λ_m (at biasing field for highest Λ), as dependent on nickel content.

position by application of a field H_0 which has turned it by the angle ϕ from its original orientation, which was parallel to the direction of easy magnetization (e.g., [100]) in the single crystal of which the domain is a part. The value of H_0 is calculated from the energy of magnetic anisot-

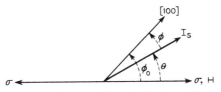

FIG. 13–31. Vectors and angles used in calculating effect of small stresses on magnetization.

ropy of the crystal E_k and the magnetic energy due to the applied field, by methods already described in Chapter 12. When the vectors H and I_s lie in a cube plane, (100), the crystal energy reduces to

$$E_k = K(1 - \cos 4\phi)/8,$$

(K being the anisotropy constant) and the energy due to the field is

$$E_H = -HI \cos \theta.$$

When these energies are added together, they have a minimum value when $H = H_0$ and $I = I_0$, given by

$$H_0 = \frac{2\pi K \sin 4\phi}{B_s \sin \theta}$$

$$I_0 = I_s \cos \theta.$$

To find the effect of a small stress on the orientation of the domain, we write the expression for the domain energy which is now composed of three parts: E_k determined by the crystal anisotropy, E_H by the magnetic field, and E_σ by the strain:

$$E_\sigma = -\tfrac{3}{2}\lambda_s \sigma \cos^2 \theta.$$

This shows that the energy is proportional to the stress and to the fractional change in length λ_s which occurs during magnetization to saturation. Also, the $\cos^2$ term indicates the twofold symmetry corresponding to stress along a line. The total energy is then

$$E = E_k + E_H + E_\sigma,$$

and to determine the orientation at equilibrium, we put

$$dE/d\phi = 0.$$

This gives

$$\Lambda = \left(\frac{dB}{d\sigma}\right)_{\sigma=0} = \frac{3\lambda_s B_s}{2K} (B_0/B_s)(1 - B_0^2/B_s^2) \cdot f$$

when B is substituted for $4\pi I$. Here f is a small factor that varies from 1.37 to 1.60, depending on the original crystal orientation. Putting $f \approx 1.5$, we then have

$$\Lambda = 2.2(\lambda_s B_s/K)(B_0/B_s)(1 - B_0^2/B_s^2).$$

When K is in ergs/cm^3 and B in gausses, Λ is in gausses cm^{-2} dyne^{-1}. When σ is in kg/mm^2, Λ is changed by a factor of nearly 10^8. In a given material, Λ is a maximum when the induction is somewhat over half of saturation, namely, $(\sqrt{3}/3)B_s = 0.58B_s$, and the value of Λ at its maximum is

$$\Lambda_m = 0.77\lambda_s B_s/K.$$

For 45 Permalloy the values of the constants $B_s = 15\,000$ and $\lambda_s = 25 \cdot 10^{-6}$ are known to be approximately correct, and $K = 15\,000$ ergs/cm^3

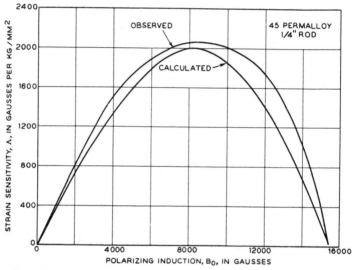

Fig. 13-32. Comparison of strain-sensitivity for small stresses, Λ, as observed and as calculated, as dependent on biasing induction.

is known with less certainty. Using these values in the foregoing expression the change of induction with small stresses can be calculated as a function of the polarizing induction. The curve showing this relation is given in Fig. 13-32 where it is compared with the observations. Fair agreement is obtained for all flux densities.

The assumption made in the derivation of the expressions for Λ, that the magnetization is changed by rotation of the domain vector, is not valid at low inductions. Nevertheless, Λ at the lowest induction can be calculated with some accuracy from the fundamental constants λ_s and K, using the equation already given.

The foregoing relation indicates that for maximum effect of stress on induction the most desirable material is one for which λ_s and B_s are high and K is low. If K is unusually low or if the material is hard worked so that the internal stresses σ_i are high, the resistance of a domain to a force trying to change its magnetization will depend not on the crystal anisotropy constant K but on the product of magnetostriction and internal stress, $\lambda_s\sigma_i$. Under these conditions the foregoing equation does not apply. In general, the controlling factor will be K or $\lambda_s\sigma_i$, whichever is larger.

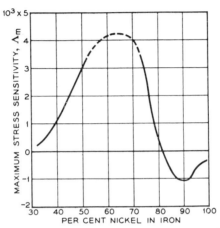

Fig. 13-33. Maximum strain sensitivity at low stresses, as calculated from λ_s, B_s, and K. Note general similarity to observed curve of Fig. 13-30.

The expression $\Lambda_m = 0.77 \cdot 10^8 \lambda_s B_s / K$ is plotted in Fig. 13-33 for the various iron-nickel alloys tested, by use of constants given elsewhere in this book. Comparison of this calculated curve with the observations, as given in Fig. 13-30, shows a striking similarity in the general features of the curves, and the agreement as to magnitude is not bad. The theoretical curve approaches $\Lambda_m = \infty$ when $K = 0$ and the curve in the figure has been arbitrarily reduced in this region (55-70% nickel) since the simple assumptions of the theory no longer hold; Λ is limited by internal strain σ_i instead of by the crystal forces measured by K, no matter how perfect the anneal. If we assume that the internal strain σ_i/E (E is Young's modulus) is equal to the magnetostrictive strain λ_s, we can calculate the highest expected xalue of Λ_m to be attained in 65 Permalloy, provided it has been perfectly annealed. This value is about six times that observed.

The effect of stress on boundary movement in low fields, where Rayleigh's law is valid, has been considered by Brown [49B2], who obtained an expression that agrees well with experiment.

Internal Strains.—When a metal is strained beyond the elastic limit, many changes occur in its structure and properties. Drastic changes occur in many of the mechanical properties such as hardness and in the magnetic properties such as permeability and coercive force. This is illustrated in Figs. 13-34 and 35, which refer to swaged wires of iron and 4-79 Molyb-

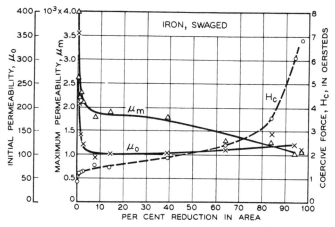

FIG. 13-34. Magnetic properties of iron as affected by reduction in area by swaging.

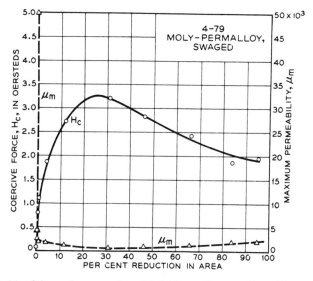

FIG. 13-35. Magnetic properties of 4-79 Permalloy as affected by swaging.

denum Permalloy. Such qualities depend primarily on the spatial fluctuations in atomic spacings, the deviations of interatomic distances from those of the ideal lattice. The most direct test and measure of such strains are furnished by X-ray analysis; the root-mean-square lattice fluctuations can be determined from the width of X-ray reflections. As an example, Fig. 13–36 shows the strains so determined by Haworth [37H1] in 70 Permalloy after severely cold working and annealing at various temperatures.

We have no well-established concept of the geometrical nature of internal strains. The most definite and successful theory has been stated by Taylor [34T1], Orowan [34O1], and Polanyi [34P1] and describes the formation and movement of *dislocations*. Some possible kinds of dislocations are illustrated in Figs. 13–37 and 38. Through some accident of growth or deformation or because of the presence of foreign material there will be an excess or deficiency of atoms in the otherwise regular lattice. In Fig. 13–37 is pictured a cross section of a crystal in which the dislocation D extends along a line perpendicular to the section. Some atoms will be forced nearer to each other, some farther apart. As indicated in (b), (c),

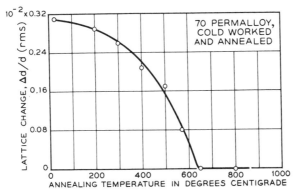

FIG. 13–36. Change in width of X-ray line reflecting from 70 Permalloy, during annealing after cold working.

and (d), dislocations may move when influenced by mechanical forces or by temperature and may disappear when they reach the surface or when two dislocations of opposite kind neutralize each other. This may occur in the structure of Fig. 13–38 if the two terminated planes of atoms, A, move to fill the vacant place shown at the middle of the diagram.

In studying the relation of ferromagnetism to internal strain the dislocation model has been used only by Brown [41B2], and so far no way has been found to extend it in a quantitative way. In most of the work on ferromagnetism it is assumed only that the direction and magnitude of strain have different values at different places in the material. The direction of the strain in any small region will determine the direction of easy magnetization there, provided the crystal anisotropy does not overwhelm the strain, and the magnitude of the strain will determine the magnetic force necessary to change the direction of magnetization.

Kersten [31K1, 32K11, 33K11] has worked out several relations between

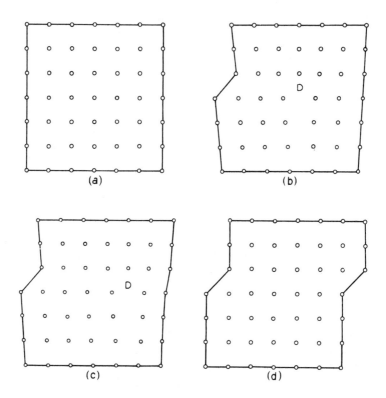

Fig. 13-37. Possible dislocations, D, and their movement.

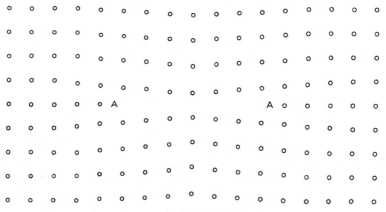

Fig. 13-38. Double dislocation A-A.

internal strain and magnetic quantities, for example, initial permeability, reversible permeability at remanence, change of remanence with tension, and the area $\int_{I_r}^{I_s} H dI$, between the descending branch of the hysteresis loop and the line for saturation, $I = I_s$. These relations will now be derived, following Kersten, and compared with experiment. In all cases considered here, it is assumed that the internal strains are large and that the process of magnetization is entirely domain rotation. Internal strains are also closely related to coercive force and change of Young's modulus with magnetization—these relations are discussed later in this chapter and in Chap. 18.

The initial permeability μ_0 is calculated with the aid of Fig. 13-39, in which θ_0 represents the angle between the field H and the internal stress σ_i in one domain, and θ is the angle by which H rotates I_s from its stable position parallel to σ_i. The energy density associated with the strain is

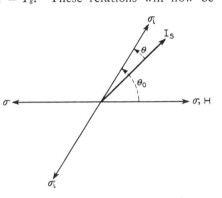

FIG. 13-39. Vectors and angles for calculating initial permeability, or effect of applied stress, σ, in material having internal stress, σ_i.

$$E_\sigma = \tfrac{3}{2}\lambda_s \sigma_i \sin^2 \theta,$$

λ_s being the magnetostriction at saturation, and the energy associated with the field is

$$E_H = -HI_s \cos(\theta_0 - \theta).$$

With a given field strength H the equilibrium position of the vector I_s is that for which the total energy is a minimum. Then

$$d(E_\sigma + E_H)/d\theta = 0$$

or

$$3\lambda_s \sigma_i \sin\theta \cos\theta = HI_s \sin(\theta_0 - \theta).$$

When H is small, θ is small, and this relation reduces to

$$\theta = \frac{HI_s \sin \theta_0}{3\lambda_s \sigma_i}.$$

The component of magnetization parallel to H is

$$I = I_s \cos(\theta_0 - \theta),$$

and the change of magnetization with field is

$$\frac{dI}{dH} = \frac{dI}{d\theta} \cdot \frac{d\theta}{dH} = \frac{I_s^2 \sin^2 \theta}{3\lambda_s \sigma_i}.$$

This is the initial susceptibility, $\kappa_0 = (\mu_0 - 1)/4\pi$. If the internal strains are oriented at random, the susceptibility of the material will be obtained by using the average value of $\sin^2 \theta$ for all values of θ between 0 and 180°, or $\frac{2}{3}$. Then

$$\mu_0 - 1 = \frac{8\pi I_s^2}{9\lambda_s \sigma_i}.$$

Values of σ_i calculated from this formula can be compared with values determined by other methods, either other magnetic methods or quite different methods such as measurements of X-ray line breadths. One finds rather close agreement between the results of the various magnetic methods and fair agreement between the results of magnetic and X-ray methods. Close agreement is not to be expected in the latter case because the quantities depend in a different way on the distribution of strains present. The value of σ_i calculated from μ_0 can be compared with σ_i calculated from Haworth's [37H1] X-ray measurements on severely cold worked 70 Permalloy. The measured root-mean-square X-ray strain of 3×10^{-3}, multiplied by Young's modulus, 1.8×10^{12} dynes/cm^2, gives $\sigma_i = 5 \times 10^9$ dynes/cm^2. Using this value in Kersten's formula with appropriate values of I_s and λ_s, the calculated value of μ_0 is found to be 68. This compares very well with the measurement [41W1] of 54 on a different specimen of almost the same composition.

Another magnetic method of estimating internal strain involved the *change of remanence with tension* when the tension is very small. In Fig. 13–39 replace the vector H by the vector σ, and consider the rotation of magnetization caused by the applied stress σ. The energies that are changed by the rotation are:

$$E_{\sigma_i} = \tfrac{3}{2}\lambda_s \sigma_i \sin^2 \theta$$

$$E_\sigma = \tfrac{3}{2}\lambda_s \sigma \sin^2 (\theta_0 - \theta).$$

The minimum in the sum of these energies occurs when

$$\theta = (\sigma/\sigma_i) \sin \theta_0 \cos \theta_0,$$

provided θ is small. The remanence is

$$I_r = I_s \cos (\theta_0 - \theta),$$

and the change of remanence with tension is

$$dI_r/d\sigma = (I_s/\sigma_i) \sin^2 \theta_0 \cos \theta_0.$$

For a random distribution of internal strains the average of the trigonometric term is $\frac{1}{4}$; therefore, for the material as a whole,

$$\left(\frac{dI_r}{d\sigma}\right)_{\sigma=0} = \frac{I_s}{4\sigma_i}.$$

Kersten [33K11] has compared the values of σ_i calculated from observations of μ_0 and of $dI_r/d\sigma$, and the results, shown in Fig. 13–40, are in good agreement. Allowance must be made for the fact that deviation from perfect randomness of distribution will affect the quantities in a different way. Förster and Stambke [41F2] have compared μ_0 and $(dI_r/d\sigma)_0$ measured on the same specimens and have found the ratio to have the constant value required by theory, when the value of σ_i varied by about a factor of 5.

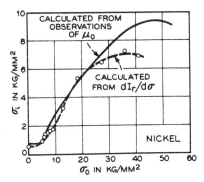

FIG. 13–40. Internal stress, σ_i, determined from (1) the change of remanence with tension, and from (2) the initial permeability, after plastically straining with a stress σ_0.

The reversible permeability μ_r, at the remanence point, is closely associated with initial permeability. When internal strains are high, the only difference is that the angle between I_s and H may vary from 0 to 180° when $I = 0$ and from 0 to 90° when $I = I_r$. Since $\overline{\sin^2 \theta_0}$ is the same in each case, $\mu_0 = \mu_r$. Dahl, Pfaffenberger, and Sprung [33D2] have confirmed this equality in severely cold worked material.

A further estimate of σ_i may be made from the area designated by shading in Fig. 13–41. At saturation I_s is parallel to H in each domain, and the strain energy is

$$E_{\sigma_i} = \tfrac{3}{2}\lambda_s \sigma_i \sin^2 \theta.$$

For a material with randomly directed strains $\overline{\sin^2 \theta} = \tfrac{2}{3}$, and

$$E_{\sigma_i} = \lambda_s \sigma_i.$$

When H is reduced to zero, θ and E_{σ_i} also go to zero, and for the material as a whole $I = I_r$. The energy decrease is then represented by the shaded area $I_r I_s S$ of Fig. 13–41 and is

$$E_{\sigma_i} = \int_{I_r}^{I_s} H dI.$$

Thus σ_i can be evaluated:

$$\sigma_i = E_{\sigma_i}/\lambda_s = (1/\lambda_s) \int_{I_r}^{I_s} H dI.$$

A comparison of the internal strain calculated by this method and also from measurements of μ_0 has been made by Kersten [32K11], who found 7×10^8 and 6×10^8 dynes/cm^2, respectively, in cold worked nickel. A check on the randomness of the strain orientations can be made by comparing I_r/I_s with the theoretical value $\overline{\cos \theta_0} = \frac{1}{2}$.

Thiessen [40T4] has related the change of resistivity with tension (see Chap. 16) with the magnitude of the internal strain in highly strained

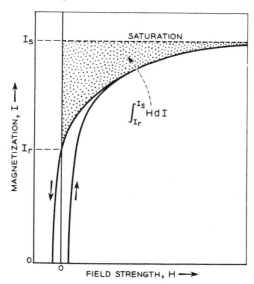

Fig. 13-41. Shaded area used for evaluation of internal stresses.

nickel and has compared the values of σ_i thus obtained with those derived from initial permeability and change of remanence with tension. The results of the various methods are in good agreement.

When internal strains are not as large as assumed above, and crystal anisotropy effects some directing force, calculation of the strain is more difficult and uncertain. This is discussed in some detail by Becker and Döring [39B5]. We will consider here only the *lower limit* for σ_i. Kersten [31K1] has pointed out that in a carefully annealed material there will be some residual strains resulting from the magnetostrictive change in shape that occurs as the material cools down from the Curie point. The magnitude of the stresses is taken to be $\sigma_i = \lambda_s E$, E being Young's modulus. In such a material μ_0 will be given by a relation similar to that mentioned previously, but with σ_i replaced by $\lambda_s E$:

$$\mu_0 - 1 = \frac{8\pi I_s^2}{9\lambda_s^2 E}.$$

This equation really gives a theoretical upper limit to μ_0. The theoretical values of μ_0 for the face-centered iron-nickel alloys, calculated from the known values of I_s, λ_s, and E, are shown by the broken line of Fig. 13-42 [35B1] and may be compared with the highest experimental values of these

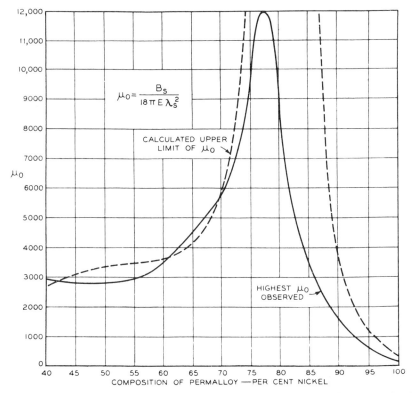

FIG. 13-42. Comparison of the highest observed initial permeabilities with the theoretical upper limit. (But see Fig. 5-25.)

alloys. This agreement gives a quantitative expression of the idea proposed earlier by McKeehan [26M3] that the high permeability of Permalloy is associated with its very small magnetostriction.

MAGNETOSTRICTION

Closely associated with the change of magnetization with stress is the Joule magnetostriction—the change in the length of a ferromagnetic body, of the order of a few parts per million—that occurs during magnetization. The term *magnetostriction* usually refers to this effect, but more generally it means any change of dimensions caused by magnetization. The more important kinds are the Joule magnetostriction, the transverse magnetostriction (the change in the dimension at right angles to the field), and the volume magnetostriction. The Wiedemann effect is the twist that occurs upon magnetization of a rod carrying a current; it can be interpreted in terms of the Joule effect by taking account of the field produced by the current in the wire. For experimental data the reader is referred to the summary by von Auwers [36A1].

Until recently magnetostriction has been largely of theoretical interest and of limited technological importance. There have been some practical applications to high-frequency oscillators [29P2] and to generators of supersound. In the echo depth-recorder [35W2] one magnetostrictively vibrating unit acts as a source of supersound and a similar unit as a microphone for the detection of the reflected sound waves. Much use was made of magnetostriction for underwater sound projectors and detectors during World War II.

The theoretical importance of magnetostriction has been emphasized particularly by McKeehan [26M2, 26M3], Akulov [28A3, 33A4], and Becker and his colleagues [39B5].

Experimental Methods.—Most of the older methods of measurement depended on the use of mechanical and optical levers, and a representative arrangement is shown in Fig. 13-43 [27W1]. The change in length of the specimen is communicated to a strip or thread that rests on and turns a carefully constructed roller on which is mounted a mirror. Rotation of the mirror is observed with telescope or light and scale. When the specimen rests directly on the roller, an expansion or contraction of about one part in a million can be detected.

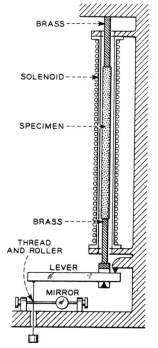

FIG. 13-43. Measurement of magnetostriction. Change in length of specimen moves lever arm to which is attached a wire that turns mirror that moves light beam on scale.

In 1920 Whiddington [20P1] constructed his "ultramicrometer" in which the displacement to be measured is communicated to one plate of a condenser that is part of a circuit oscillating at high frequency. The change in frequency resulting from the change in capacitance is measured by noting the change in the frequency of beating between the above-mentioned "primary" oscillator and a second constant-frequency "beating" oscillator. Displacements of 10^{-8} cm were so determined. In a somewhat different arrangement [35H1], radiation from the primary oscillator is picked up by a wave-meter which is tuned slightly off-resonance so that its response

EXPERIMENTAL METHODS IN MAGNETOSTRICTION

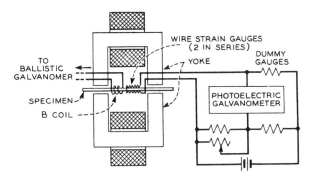

Fig. 13–44. Magnetostriction determined by measuring change in resistance in fine wires fixed to specimen. Bridge arrangement increases accuracy.

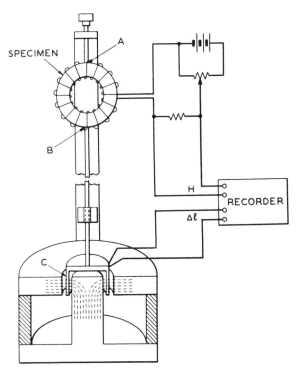

Fig. 13–45. Arrangement for automatic recording of magnetostriction ($\Delta l/l$ vs H). Change in length of specimen causes movement of coil in magnetic field, the integrated voltage of which is registered.

is sensitive to changes in frequency. The point of operation is near the steepest part of the resonance curve. Changes in the wave-meter output are read with a galvanometer. A further improvement has been made by using a quartz crystal as a piezoelectric oscillator in the wavemeter circuit and thereby sharpening the resonance curve and increasing the sensitivity.

A convenient method of determining change in length is to measure the change in resistance of a wire that is firmly cemented to the test specimen and expands and contracts with it. Such a strain gauge, composed of wire "folded" to a length of a few centimeters, has been used by Goldman [49G1], and in conjunction with a photoelectric galvanometer and Wheatstone bridge has a sensitivity of 2×10^{-8} cm/cm. The method of use is shown by the diagram of Fig. 13–44.

A method that permits recording of the magnetostriction vs field strength curve has been described by Nesbitt and Williams [48N2]. The specimen is connected rigidly to a coil that is in a strong radial magnetic field. Movement of the coil causes a voltage to appear at the ends of the coil, and this voltage is amplified and recorded. A sketch of the apparatus is shown in Fig. 13–45.

Measurements of changes in volume attending magnetization are almost invariably made by immersing the specimen in a liquid in a small, closed vessel and observing with a microscope the rise in level of the liquid in a glass capillary sealed to the top of the vessel. The arrangement used in 1900 by Nagaoka and Honda is sketched in Fig. 13–46 [28H2]. In apparatus of this sort more recently described by Masiyama [31M1], one scale division corresponds to a change of 2×10^{-8} cm^3 in the volume of the specimen, which was about 1 cm^3. Refinements in technique have been described by Döring [36D4], Snoek [37S6] and Vautier [50V1]. Becker [39B5] has shown that the shape of the specimen must be taken into account.

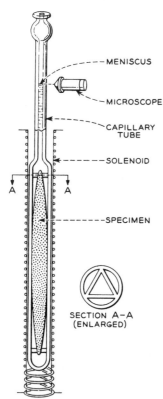

FIG. 13–46. Determination of volume magnetostriction by movement of meniscus of liquid.

Brief Survey.—The measurements of Joule, reported in 1842 and 1847 [1842J1, 1847J1], established the fact that iron, when magnetized, increased in length in the direction of magnetization and contracted at right angles thereto. As nearly as he could determine, its volume did not change. In 1882 Barret [82B1] showed that nickel contracts in all fields, and soon thereafter Bidwell [86B1] found that iron begins to contract in high fields and ultimately becomes shorter than the unmagnetized specimen. Barrett [82B1] observed for the first time a definite change in volume. In the meantime Wiedemann [1862W1] had observed the twist in a wire subjected

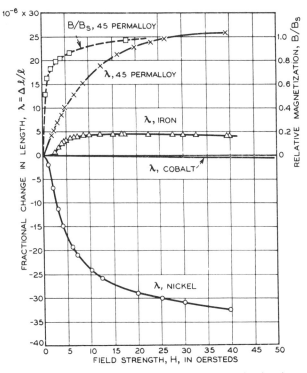

FIG. 13-47. Magnetostriction of some common materials showing expansion or contraction. Also, induction of 46 Permalloy for reference.

to an axial field and to a circular field due to the passage of a current, and thus opened a new field of measurements of this and related effects. Bidwell [90B1] showed that tension decreases the expansion of iron and that a sufficient force will prevent any expansion at all and will increase the contraction.

After the turn of the century other experimenters, including Nagaoka and Honda and their students, investigated the magnetostriction of iron, cobalt, and nickel at high and low temperatures and began to prepare and measure the properties of the various alloys of these elements. In 1925 Webster [25W1] reported the first systematic study of the magnetostriction of single crystals.

In Figs. 13-47 and 48 magnetostriction is plotted against the magnetic field strength for four materials of different kinds. The fractional change in length, $\Delta l/l$, represented by the symbol λ, is measured in the same direction as that in which H is applied. In weak fields (Fig. 13-47), iron and some of the iron-nickel alloys expand, and cobalt and nickel contract. In higher fields (Fig. 13-48) iron begins to contract, and at about $H = 200$ to 500 it becomes shorter than it was before magnetization. All curves begin tangent to the H-axis. In strong fields they approach a limiting value, superposed on which is a relatively small length change that is known to be a volume effect—the material expands equally in all directions. Except for such small changes the change in length measured at right angles to the field, the "transverse" magnetostriction, $(\Delta l/l)_t = \lambda_t$, is about half of the longitudinal magnetostriction and is opposite in sign: $\lambda_t = -\lambda/2$.

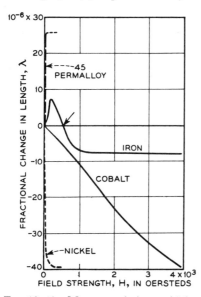

FIG. 13-48. Magnetostriction at high field strengths.

The magnetostriction of three materials is plotted against the intensity of magnetization in Fig. 13-49; no data of this kind for cobalt have been found.

Very little change in length occurs before the magnetization rises to the steepest part of the magnetization curve. This is shown by comparison with the B,H curve of Fig. 13-47 and is due to the fact that most of the magnetostriction takes place during the *rotation* of domains, which begins to occur at the knee of the B,H curve. For similar reasons the magnetostriction at remanence is quite low; this is illustrated in Fig. 13-50, which shows also the hysteresis in the λ vs B relation for iron. When the magnetization approaches saturation, the magnetostriction also approaches its limiting value, λ_s, the *saturation magnetostriction*.

Magnetostriction usually decreases in magnitude with increase in *temperature*. Döring's data for nickel (Fig. 13-51) show a particularly simple relation: $\lambda_s \propto I_s^2$. The relation for iron is more complicated.

Magnetostriction may be increased or decreased by stress. When $\lambda > 0$ tension usually decreases λ, and when $\lambda < 0$ tension increases the absolute magnitude of λ, making it more negative. This is easily explained in terms of domain theory, which will now be discussed.

BRIEF SURVEY OF MAGNETOSTRICTION

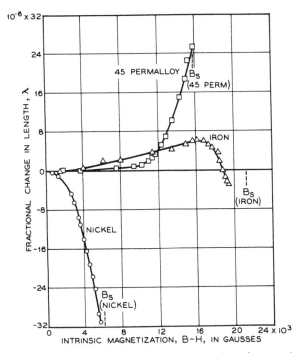

Fig. 13-49. Magnetostriction as dependent on intensity of magnetization. Saturation magnetizations noted. (Schulze [28S3], Cioffi).

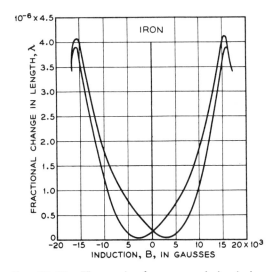

Fig. 13-50. Hysteresis of magnetostriction in iron.

Domain Theory.—Magnetostriction is associated with domain orientation—the change in dimensions of a *single* domain can be related in a simple quantitative way (see Chap. 18) to the change in direction of magnetization in the domain, as follows:

$$\lambda_1 = \tfrac{3}{2}\lambda_s (\cos^2 \theta - \tfrac{1}{3}).$$

Here θ is the angle between the direction of magnetization and the direction in which the change in length is measured. It is assumed now that magnetostriction is independent of the crystallographic direction of magnetization and that the change in volume is zero. The zero point of λ_1 is chosen so that it is equal to the longitudinal change in length, λ_s, when

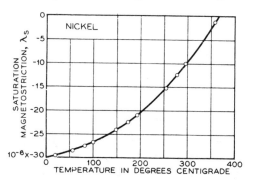

FIG. 13–51. Change of magnetostriction of nickel with temperature.

$\theta = 0$ (at saturation). When the length is measured at right angles to the direction of magnetization (transverse effect), $\theta = 90°$ and the change in length is $\lambda_t = -\lambda_s/2$.

In an *ensemble* of domains initially oriented at random, the same relation is applicable if one uses the average of $\cos^2 \theta$ over all the domains. When the material is unmagnetized, $\langle\cos^2 \theta\rangle_{av} = \tfrac{1}{3}$ and $\lambda = 0$; upon application of a strong field θ becomes zero and $\lambda = \lambda_s$.

If the domains are not initially random, one can use the relation

$$\lambda = \tfrac{3}{2}\lambda_s(\langle\cos^2 \theta\rangle_{av} - \tfrac{1}{3}) - \tfrac{3}{2}\lambda_s(\langle\cos^2 \theta\rangle_0 - \tfrac{1}{3})$$

or

$$\lambda = \tfrac{3}{2}\lambda_s(\langle\cos^2 \theta\rangle_{av} - \langle\cos^2 \theta\rangle_0),$$

$\langle\cos^2 \theta\rangle_0$ referring to the initial domain distribution and $\langle\cos^2 \theta\rangle_{av}$ to the distribution at any time. If the domains are oriented originally so that $\theta = 0$ for half of them and $\theta = 180°$ for the other half, $\langle\cos^2 \theta\rangle_0 = \langle\cos^2 \theta\rangle_{av} = 0$ and $\lambda = 0$ (the reference point); in a strong field $\theta = 0$ and there is no change in $\cos^2 \theta$ and again $\lambda = 0$. When used in this sense, λ depends decidedly on the initial domain distribution, while λ_s is a constant of the material. The constant λ_s can be determined in any specimen by measuring λ when a saturating field is applied first parallel and then at 90° to the direction of measurement of λ. The total change in length caused by the change in field is then $3\lambda_s/2$, independent of the initial domain distribution.

The magnetostriction of materials with various domain distribution is shown by the diagrams of Fig. 13-52. It is convenient to consider the

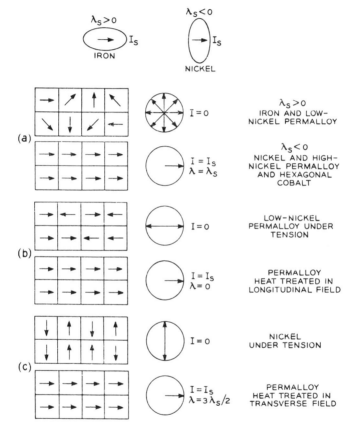

FIG. 13-52. Domain diagrams showing the changes in length of various materials, having positive or negative magnetostriction, when magnetized to saturation.

"magnetostriction figure" of a domain in terms of an ellipsoid, as shown. The figure of a material such as iron can be regarded as a prolate ellipsoid, with the magnetization parallel to the long axis, while in nickel it is an oblate ellipsoid with the magnetization parallel to the short axis. When the direction of magnetization in a domain is changed, the length of the domain changes (as measured in any given direction), and the total change in length of the material is the summation of all changes in length of the domains. The equation given previously is the mathematical expression of this process. In closer approximation the magnetostriction figure is more complicated than an ellipsoid, and it depends on the position of the crystal axes; this is discussed in a later section. The ellipsoid is a better approximation for nickel than for iron.

Preferred Domain Orientations.—As indicated in Fig. 13–52, the magnetostriction of a material should depend on the initial distribution of its domains. Experimental confirmation of this view will now be given for three cases, in which special distributions of orientations are produced by strain, by heat treatment in a magnetic field, and by special crystal orientation.

The effect of tension and compression on the magnetostriction of nickel is shown clearly by the measurements of Kirchner [36K1], reproduced in

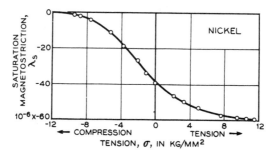

FIG. 13–53. Magnetostriction of nickel measured under tension or compression. Note limits of zero and 3/2 of magnetostriction at zero stress.

Fig. 13–53. Under tension the saturation magnetostriction increases until it is $\frac{3}{2}$ of the normal value (40×10^{-6}). The tension orients the domains at 90° to the axis (which is parallel to specimen, tension, and field); consequently, when the field is applied, the magnetization changes $\overline{\cos^2 \theta}$ from 0 to 1 so that λ changes from $-\lambda_s/2$ to λ_s and the magnetostriction, $\Delta l/l$, is $3\lambda_s/2$ [Fig. 13–52(c)]. Compression along the axis, on the contrary, orients all the domains parallel to the axis and $\theta = 0$ or 180° so that $\overline{\cos^2 \theta} = 1$. The field changes the magnetization by 180° reversals so that $\cos^2 \theta$ is not changed and the magnetostriction is zero (Fig. 13–52(b)].

The tensile or compressive stress necessary to align the domains must be large compared to the forces maintaining alignment of the domains in the unstrained material. These forces are due primarily to crystal anisotropy, and the crystal anisotropy constant K of nickel is about 5×10^4. The energy of the applied stress must then be greater than that of the crystal anisotropy:

$$\tfrac{3}{2}\lambda_s \sigma > K$$

or, $\sigma > 8 \times 10^8$ dynes/cm^2 or 8 kg/mm^2, for nickel. In agreement with theory the experiments show that a stress of this magnitude is able to change the saturation magnetostriction most of the way from its value without stress to its value with very high stress (Fig. 13–53).

When the applied tension is much greater than 8 kg/mm^2, and practically all domains are oriented at $\theta = 90°$, application of a field caused the magnetization to change by domain rotation alone. Then the magnetization is

$$I = I_s \cos \theta$$

and the magnetostriction is

$$\lambda = \tfrac{3}{2}\lambda_s \cos^2 \theta;$$

consequently

$$\lambda = \tfrac{3}{2}\lambda_s I^2/I_s^2.$$

Figure 13–54 shows that λ is proportional to I^2 for $\sigma = 10.4$ kg/mm^2, but not for tensions less than about 5 kg/mm^2.

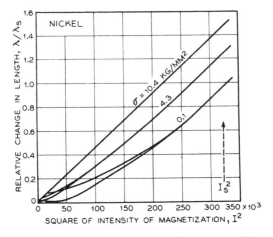

Fig. 13–54. Dependence of magnetostriction on the square of the magnetization, in nickel under various degrees of tension [36K1].

When 68 Permalloy has been heat-treated in a magnetic field, the direction of easy magnetization is that in which the field was applied during treatment. The domains have been stabilized parallel to the specimen axis, and as a consequence we should expect a lowering or disappearance of the magnetostriction measured in this direction [Fig. 3–52(b)]. Such is the case, as illustrated by the data of Fig. 13–55. The small magnetostriction observed is attributable to the lack of complete alignment of the domains in the unmagnetized specimen. Similarly the magnetostriction of the specimen heat-treated in a transverse field shows a considerable increase over that of the specimen annealed in the absence of a field [Fig. 13–52(c)]. In the transversely annealed specimen one finds also the expected linear relation between λ and $(B-H)^2$, as depicted in Fig. 13–56.

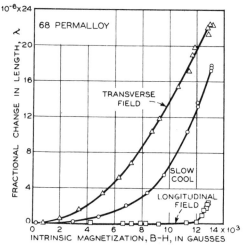

FIG. 13–55. Variation of the magnetostriction of 68 Permalloy with heat treatment [41W1].

The special crystal orientation usually present in rolled sheets often gives rise to a direction of easy magnetization in the sheet along which the

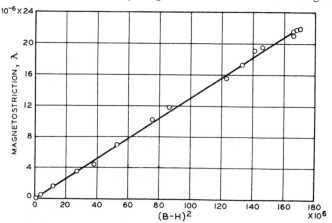

FIG. 13–56. Magnetostriction of 68 Permalloy heat-treated in a transverse field.

domains are aligned in the unmagnetized material. It is to be expected that such a non-random distribution of domain orientation will markedly affect the magnetostriction. As an example, consider the 30% iron-cobalt alloy, in the form of 0.002-in. tape that has been cold-rolled from a thickness of 0.090 in. A torque magnetometer shows readily that the easy direction is 90° to the direction of the tape. A possible explanation is that the crystals are aligned with a (011) plane in the rolling plane, [$\bar{2}$11] and [111] directions lying respectively parallel and perpendicular to the direction of rolling. Since [111] is an easy direction in this crystal ($K < 0$), domains in the unmagnetized material will lie in the transverse position, half in each direction, and the process of magnetization parallel to the tape will be one of domain rotation alone. The saturation magnetostriction will then have $\frac{3}{2}$ of the usual value for randomly oriented material. This is confirmed

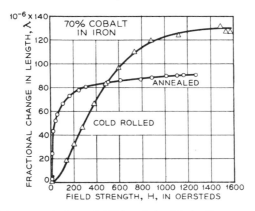

Fig. 13–57. Magnetostriction of 70% cobalt-nickel alloy as cold rolled and as annealed.

by Nesbitt's [50N1] curve (Fig. 13–57) which attains the value $\lambda = 130 \times 10^{-6}$ and is to be compared with 90×10^{-6}, the highest value previously reported (Masiyama [32M2]).

Reversible Magnetostriction.—This term is applied to the changes in length or other dimensions resulting from a cyclic change in field strength of very small amplitude, usually superposed upon a steady or "polarizing" field. Such changes are thermodynamically related to the corresponding changes of magnetization with stress and are of practical importance in the operation of the so-called magnetostriction microphone. Very few direct measurements of these changes in length have been reported, and data are available for only 45 and 78.5 Permalloys and for iron. Measurements of McKeehan and Cioffi [26M2] on the two latter materials are given in Fig. 13–58 and illustrate the general characteristics of the relation between reversible change in length, $(1/l)(\Delta l/\Delta H)_r$, and the steady polarizing field-strength H_0. From the origin ($H_0 = 0$) where the ordinate and slope are both zero, the curve rises to a maximum with increasing field strength and then decreases asymptotically to zero. The amplitude of the alternating field strength was less than 0.01 oersted.

640 STRESS AND MAGNETOSTRICTION

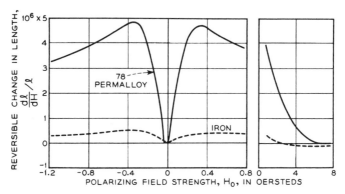

Fig. 13–58. Magnetostriction accompanying small cyclical changes in field (reversible magnetostriction) at various biasing fields. Iron and 78 Permalloy.

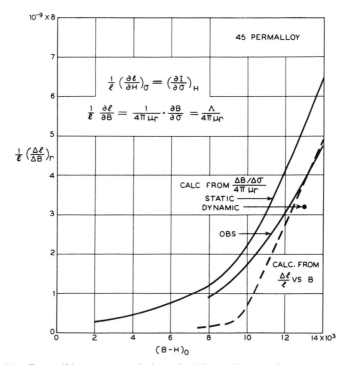

Fig. 13–59. Reversible magnetostriction of 45 Permalloy as observed, and as calculated from (a) measurements on change of magnetization with stress, and from (b) slope of normal magnetostriction curve.

The curve representing the reversible change in length per unit change in induction, $(1/l)(\Delta l/\Delta B)_r$, as a function of polarizing ferric induction $(B - H)_0$, differs from the previously mentioned curves in that it rises continuously from the origin to its highest value which occurs at saturation. Previously unpublished data of White and Sivian for 45 Permalloy are plotted in Fig. 13–59, curve "Obs."

At high inductions the quantities $(1/l)(\Delta l/\Delta H)_r$ and $(1/l)(\Delta l/\Delta B)_r$ approach as limits the slopes of the Joule magnetostriction curves obtained by plotting $\Delta l/l$ vs H and B, respectively. Figure 13–59 shows the line representing the slope of the $\Delta l/l$ vs B curve (see Fig. 13–49) for a specimen of similar composition and heat treatment, and this line coincides closely with that for the reversible magnetostriction above $\frac{3}{4}$ of saturation, while it departs by a large factor at lower inductions, as one would expect where hysteresis effects are important.

Another comparison can be made between the directly observed reversible magnetostriction and other quantities. The thermodynamic relation between magnetostriction and stress already mentioned can be put into the forms

$$\frac{1}{l}\frac{(\Delta l)}{(\Delta H)_r} = \frac{1}{4\pi}\frac{(\Delta B)}{(\Delta \sigma)_r}$$

and

$$\frac{1}{l}\frac{(\Delta l)}{(\Delta B)_r} = \frac{1}{4\pi\mu_r}\frac{(\Delta B)}{(\Delta \sigma)_r}$$

in which μ_r is the reversible permeability and $\Delta\sigma$ the small cyclic stress applied to produce the change in induction ΔB. The upper curve of Fig. 13–59, calculated from the data previously presented in Fig. 13–28, represents the right-hand side of the latter equation and shows fair agreement with the observed $\Delta l/\Delta B$. A comparison of the magnetostriction with the stress data can also be made for iron and 78.5 Permalloy as to order of magnitude. For one specimen of iron [26M2] the maximum value of $(1/l)(\Delta l/\Delta H)_r$ is 0.6×10^{-6}, whereas in another specimen the maximum of $(1/4\pi)(\Delta B/\Delta\sigma)_r$ is 0.1×10^{-6}. For 78.5 Permalloy $(1/l)(\Delta l/\Delta H)_r = 5 \times 10^{-6}$ at the maximum at $H_0 = 0.35$ and at the same field strength, for comparison, $(1/4\pi)(\Delta B/\Delta\sigma)_r = 1.4 \times 10^{-6}$. These discrepancies are not surprising in view of the wide variations in the composition and heat treatment of the materials used.

Volume Magnetostriction.—The change in volume caused by the application of a magnetic field is usually much smaller than the change in length. The fractional decrease in the length of nickel is about 30×10^{-6} in a field of 25 oersteds and is practically at its final value when a few hundred oersteds have been applied, while the fractional change in volume, designated ω, is only 0.1×10^{-6} in a field of 1000 oersteds.

The general course of the ω vs H curve of iron is shown in Fig. 13–60, in relation to the length magnetostriction, λ. Here $\omega/3$ is plotted instead of ω, to bring out the fact that the change in length measured in high fields

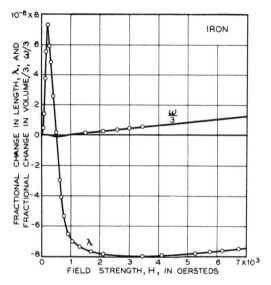

Fig. 13–60. Length and volume magnetostriction of iron.

(above about 4000) is due mainly to a change in volume. In low fields the transverse contraction is about half of the longitudinal expansion, whereas in the highest fields the transverse and longitudinal changes are both expansions and are equal.

In some materials, for example the 36% nickel-iron alloy, a fractional increase in volume of $\omega = 15 \times 10^{-6}$ is observed in a field of 1000 oersteds. This exceptionally large value is associated with the fact that the measurements are made near the Curie point of the material, which is just above room temperature. The length and volume magnetostrictions of this material are given in Fig. 13–61 according to the data of Nagaoka and Honda [04N1]. The value of ω at $H = 1050$ is shown in Fig. 13–62 for the iron-nickel alloys, according to Masiyama [31M1]. The composition at the maximum is nearly the same as that for which the material becomes non-magnetic at room temperature.

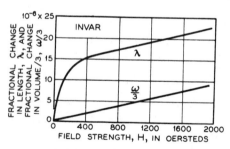

Fig. 13–61. Length and volume magnetostriction of 36 Permalloy, showing that in high fields the magnetostriction is all a volume expansion.

There is a close connection also between the maximum of the volume magnetostriction curve and the minimum of the curve showing thermal expansion as dependent on composition (Fig. 13–63). Döring [36D4] has

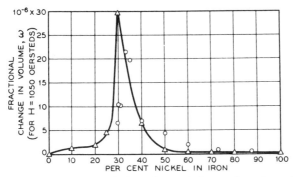

FIG. 13–62. Volume magnetostriction observed for various iron-nickel alloys at $H = 1050$ (triangles), and as calculated from the change of magnetization with hydrostatic pressure (circles).

pointed out that there is a thermodynamic relation between the thermal expansion coefficient α_I that a material would have if it were non-magnetic

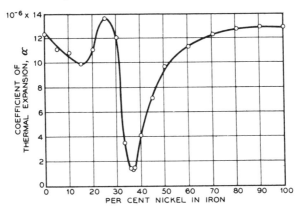

FIG. 13–63. Thermal expansion coefficient of iron-nickel alloys [31M2].

(or ferromagnetic with constant magnetization) and the coefficient α_H that it has in constant field:

$$\alpha_H = \alpha_I - \frac{1}{3}\left(\frac{\partial \omega}{\partial H}\right)_T \left(\frac{\partial H}{\partial T}\right)_I.$$

Here $\partial \omega/\partial H$ is the volume magnetostriction, and $(\partial H/\partial T)_I$ is a measure of the field necessary to maintain the saturation magnetization constant as it tends to decrease with increasing temperature (Fig. 14–7). Since $(\partial H/\partial T)_I$ is always positive, then $\alpha_H < \alpha_I$ when $(\partial \omega/\partial H)$ is positive.

In that case, α_H is less than the normal expansion coefficient, as it is in the technically important alloy Invar (36% Ni-Fe) and other alloys of this type. In some alloys α_H is actually negative, the material contracting when heated [34M4, 37K3].

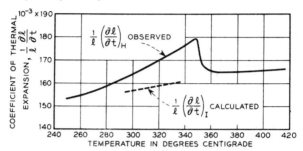

Fig. 13-64. Thermal expansion of nickel through the Curie point, and expansion calculated for the non-magnetic state (no change of magnetization with temperature). Solid line is from experiments by C. Williams [34W1].

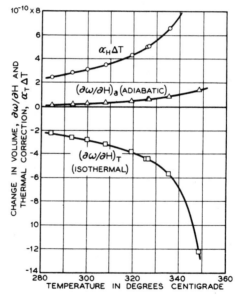

Fig. 13-65. Change of volume with field as determined adiabatically, $(\partial \omega/\partial H)_a$, and that at constant temperature, $(\partial \omega/\partial H)_T$, calculated from it using the temperature coefficient α_H in constant field and the increase in temperature ΔT resulting from the magnetocaloric effect.

In nickel at elevated temperatures $\partial \omega/\partial H$ is negative and α_H is therefore abnormally large, as shown by the data of Williams of Fig. 13-64 and the curves of Fig. 10-15. Döring has determined $\partial \omega/\partial H$ (Fig. 13-65) in the range of temperature in which α_H is abnormally large and has calculated the difference between α_H and α_I by means of the foregoing equation, using his data and the values of $\partial H/\partial T$ determined by Weiss and Forrer [26W4]. The broken line of Fig. 13-64 shows that the α_I, so calculated from α_H, has the course expected.

The measurements of Ebert and Kussmann [37E1] on the change of magnetization with hydrostatic pressure may be used to calculate the volume magnetostriction, according to the thermodynamic relation

$$\left(\frac{\partial \omega}{\partial H}\right)_p = -\frac{1}{V_0}\frac{\partial (VI)}{\partial p},$$

p now denoting hydrostatic pressure. This is analogous to the corresponding expression (see p. 610)

$$\frac{1}{l}\frac{\partial l}{\partial H} = \frac{\partial I}{\partial \sigma},$$

in which σ stands for linear stress (tension). The volume V is included in the first equation because this is affected by pressure. For some of the iron-nickel alloys the values of $\partial \omega/\partial H$ thus calculated are plotted in Fig. 13-62, where they compare favorably with the quantities directly observed.

These experiments on the effect of hydrostatic pressure indicate that at room temperature $\partial \omega/\partial H$ for nickel is negative. In his direct measurement of $\partial \omega/\partial H$ Döring has shown that it is important to correct the observed adiabatic magnitude $(\partial \omega/\partial H)_{ad}$, in order to obtain the isothermal $(\partial \omega/\partial H)_T$. Near the Curie point the amount of the heating that results from the application of the field is known for nickel from the work of Weiss and Forrer [26W4] (magnetocaloric effect), and the correction is shown in Fig. 13-65. It is concluded that the corrected (isothermal) $(\partial \omega/\partial H)_T$ is negative at elevated temperatures (Fig. 13-65) but is positive for nickel at room temperature, in contradiction to the pressure experiments. A negative $\partial \omega/\partial H$ is consistent with the position usually assigned to nickel on the Bethe interaction curve (Fig. 10-13). The contradiction at room temperature has so far not been resolved.

Magnetostriction of Single Crystals.—The magnetostriction of single crystals of *iron* was measured in 1925 and 1926 by Webster [25W4] and Honda and Masiyama [26H3], and later by Kaya and Takaki [36K7]. Data for the principal directions [100], [110], and [111] are shown in Figs. 13-66 and 67 as dependent on the intensity of magnetization. There is a rather large difference in the curves for the direction of easy magnetization, [100], obtained by different observers, and this reflects the difference in the domain structure of the demagnetized state, to be discussed later. The different experiments agree in that there is little or no magnetostriction in the [111] direction in iron until the magnetization is over half of saturation; this is to be expected, for the change of domain orientation from one [100] direction to another will not change the length in the [111] direction, because all [100] directions are equally inclined to [111]; con-

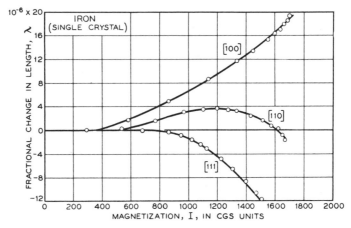

FIG. 13-66. Magnetostriction of single crystals of iron in various crystallographic directions according to Webster.

sequently magnetostriction will occur only when domain rotation sets in.

The magnetostriction of *nickel* crystals, as dependent on field, was studied by Masiyama [28M6]. Data for λ vs I are not available. Curves for

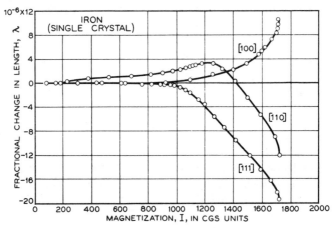

FIG. 13-67. Magnetostriction of single crystals of iron, according to Kaya and Takaki.

the three principal directions are given in Fig. 13-68. In contrast to the curves for iron, in nickel the change in length is always negative. A reasonably complete description of the magnetostriction at saturation, for any direction of the applied field, can be given in terms of four constants, to be discussed in the next section.

In *cobalt* the description of the results is more difficult because of the hexagonal symmetry. The longitudinal magnetostriction in the three principal directions [00·1], [10·0], and [11·0], as measured by Nishiyama [29N1], is given in Fig. 13-69 as dependent on field strength up to about

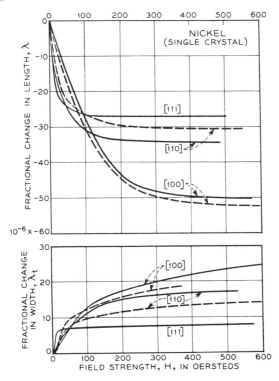

FIG. 13-68. Longitudinal and transverse magnetostriction in single crystals of nickel. The two sets of curves for [110] are from measurements of disks cut in different planes.

$H = 7000$, a field insufficient to saturate in the [10·0] and [11·0] directions. The greatest longitudinal change in length is a contraction of about 30×10^{-6} in the [11·0] direction. In contrast to this, there is an expansion of about 80×10^{-6} in the transverse effect: in a disk cut parallel to the (10·0) plane, with H applied 67.5° from the hexagonal axis, the disk expands in a direction 90° to H by about 80×10^{-6}. The transverse magnetostriction in some directions is given in Fig. 13-70. Here it will be noted that, when H is applied parallel to [10·0], the crystal contracts in this direction and at the same time it contracts in [11·0] and expands in [00·1]. The large contraction in polycrystalline cobalt (Fig. 13–69) is larger than the contraction in any simple crystallographic direction.

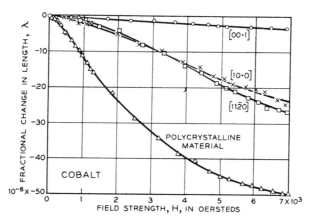

Fig. 13–69. Magnetostriction in single crystal of cobalt and polycrystalline material.

There is a rather large volume magnetostriction in cobalt, as deduced from measurements on polycrystalline material [29N1].

In the *iron-nickel* alloys containing 34–100% nickel the longitudinal magnetostriction at saturation has been measured in the three principal

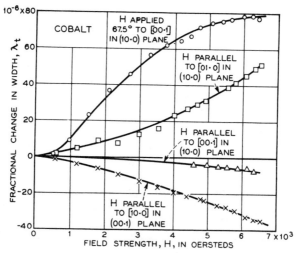

Fig. 13–70. Transverse magnetostriction in single crystal of cobalt.

directions by Lichtenberger [32L1] (see Fig. 13–71). The low-nickel alloys (30–45% nickel) contract in the [100] direction but expand in the [110] and [111] directions. In consequence, the magnetization of crystals oriented in a certain way will be increased by tension, whereas those oriented in other ways will be decreased. This has been observed indirectly by measurement of change of resistance of a polycrystalline specimen with tension and field [46B1]. Alloys of 60 and 85% nickel are practically isotropic. Lichtenberger's values for nickel confirm fairly well the measurements of Masiyama [28M6].

Some early measurements by Heaps [23H2] on *iron-silicon* alloys show that the magnetostriction depends markedly on crystal orientations. Shturkin [47S3] measured the magnetostriction of single crystals of silicon-iron containing 3.5% silicon. His values of the saturation magnetostriction in the two principal directions are: $\lambda_{100} = 24 \times 10^{-6}$, $\lambda_{111} = -2.3 \times 10^{-6}$, at 20°C. With increase in temperature λ_{100} increases to 480°C, then decreases, whereas the absolute value of λ_{111} decreases continually. Carr measured both λ_{100} and λ_{111} as dependent on silicon concentration up to 8% silicon, and his results are shown in Fig. 13-72. The author is indebted to Dr. Carr for furnishing the data before publication.

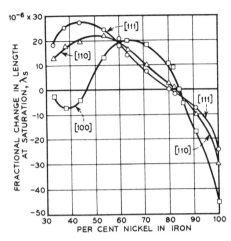

Fig. 13-71. Magnetostriction in single crystals of iron-nickel alloys.

Heaps [24H1] also measured the magnetostriction of a single crystal of *magnetite*. In a field of about 5000 oersteds (practical saturation) the longitudinal changes in length in the [100], [110], and [111] directions are respectively -4, $+30$, and $+12 \times 10^{-6}$; the transverse changes are respectively $+4$, -44, and -28×10^{-6}.

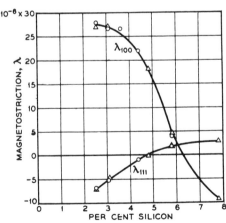

Fig. 13-72. Magnetostriction in single crystals of iron-silicon alloys.

Theory of Saturation in Single Crystals.—After it was established that the saturation magnetostriction depends on crystal direction, an empirical theory for cubic crystals, based on two constants, was developed by

Akulov [28A3, 30A3]. Later Gans and von Harlem [33G9] used an expression involving five constants and applied it to iron and nickel. We use here the equivalent equation given by Becker and Döring [39B5]:

$$\lambda_s = h_1(\alpha_1^2\beta_1^2 + \alpha_2^2\beta_2^2 + \alpha_3^2\beta_3^2 - \tfrac{1}{3})$$

$$+ 2h_2(\alpha_1\alpha_2\beta_1\beta_2 + \alpha_2\alpha_3\beta_2\beta_3 + \alpha_3\alpha_1\beta_3\beta_1)$$

$$+ h_4(\alpha_1^4\beta_1^2 + \alpha_2^4\beta_2^2 + \alpha_3^4\beta_3^2 + 2s/3 - 1/3)$$

$$+ 2h_5(\alpha_1\alpha_2\alpha_3^2\beta_1\beta_2 + \alpha_2\alpha_3\alpha_1^2\beta_2\beta_3 + \alpha_3\alpha_1\alpha_2^2\beta_3\beta_1$$

$$+ h_3(s - \tfrac{1}{3}) \text{ for nickel, } + h_3 s \text{ for iron.}$$

Here the α's are the direction cosines of the magnetization with respect to the crystal axes, and the β's the direction cosines of the measured change in length, and

$$s = \alpha_1^2\alpha_2^2 + \alpha_2^2\alpha_3^2 + \alpha_3^2\alpha_1^2.$$

The fractional change in volume, caused by domain rotation, is

$$\omega = 3h_3(s - \tfrac{1}{3}) \text{ for nickel,}$$

$$\omega = 3h_3 s \text{ for iron.}$$

The expressions for these two materials are different because the directions of easy magnetization, and therefore the distribution of domain directions for the demagnetized state, are different.

Sometimes the data can be satisfied sufficiently well by the use of a formula containing only two constants:

$$\lambda_s = \tfrac{3}{2}\lambda_{100}(\alpha_1^2\beta_1^2 + \alpha_2^2\beta_2^2 + \alpha_3^2\beta_3^2 - \tfrac{1}{3})$$

$$+ 3\lambda_{111}(\alpha_1\alpha_2\beta_1\beta_2 + \alpha_2\alpha_3\beta_2\beta_3 + \alpha_3\alpha_1\beta_3\beta_1).$$

Here λ_{100} and λ_{111} are the longitudinal magnetostriction for the [100] and [111] directions, and in relation to the constants in the 5-constant equation

$$\lambda_{100} = 2h_1/3,$$

$$\lambda_{111} = 2h_2/3,$$

provided $\qquad h_3 = h_4 = h_5 = 0.$

Becker and Döring have applied both equations to Masiyama's [28M6] data for *nickel*. They compared the difference between longitudinal and transverse magnetostriction at saturation ($\lambda_l - \lambda_t$) with the theory, after selecting the best values of the constants in each case. This difference is chosen because it is independent of the domain structure of the demagnetized state. The agreement between theory and experiment is illustrated in Fig. 13-73. Values of the constants are:

$h_1 = -24 \times 10^{-6}; h_2 = -47 \times 10^{-6}; h_4 = -51 \times 10^{-6}; h_5 = +52 \times 10^{-6}.$

SATURATION MAGNETOSTRICTION IN SINGLE CRYSTALS 651

The constants of the 2-constant equation that give the best agreement with experiment are:

$$\lambda_{100} = -46 \times 10^6; \quad \lambda_{111} = -25 \times 10^{-6}.$$

It was assumed that the volume change and, therefore, h_3 were zero. Naturally the variation of $\lambda_l - \lambda_t$ with azimuth in the (100) and (110) planes is lost when only one constant is used. Also, one notes that a

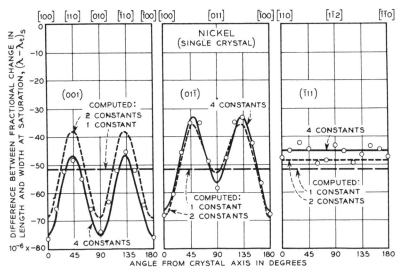

FIG. 13-73. Difference between longitudinal and transverse magnetostriction according to theory and experiment. Theoretical curves fitted by use of 1, 2 or 4 constants, as indicated.

periodic change in the (111) plane is observed which cannot be accounted for even by the 4-constant relation but requires a higher term in the expansion of the direction cosines.

By averaging, Becker and Döring have calculated from the values of the constants the magnetostriction of a polycrystalline material containing a random distribution of crystallites. They obtain

$$\overline{\lambda}_s = \frac{3\lambda_s}{2} (\cos^2 \theta - \tfrac{1}{3}),$$

θ being the angle between magnetization and measurement, with

$$\lambda_s = 4h_1/15 + 2h_2/5 + 8h_4/35 + 2h_5/35.$$

For nickel the calculated value of the longitudinal magnetostriction is

$$\overline{\lambda}_s = -34 \times 10^{-6},$$

in good agreement with -33×10^{-6}, the average of a number of experimental determinations.

Using the 2-constant relation (taking $h_4 = h_5 = 0$), the foregoing expressions for λ_s, h_1, and h_2 give the very useful equation:

$$\lambda_s = \frac{2\lambda_{100} + 3\lambda_{111}}{5}.$$

The data for iron do not warrant a comparison with the 5-constant formula. Kaya and Takaki [36K7] have measured the magnetostriction at saturation and at remanence in eight single crystal rods, and from the difference derived values of λ_{100} and λ_{111} of 25.5 and -18.8×10^{-6}, respectively. Becker and Döring prefer a value of $\lambda_{100} = 19 \times 10^{-6}$, more consistent with the larger value for polycrystalline iron observed, and calculated from the relation

$$\lambda_s = \frac{2\lambda_{100} + 3\lambda_{111}}{5} = -4 \times 10^{-6}.$$

Measurements of the difference in λ between saturation and remanence have been made on iron crystals at temperatures up to 750°C by Takaki [37T3]. At the highest temperatures the magnetostriction is positive in all directions.

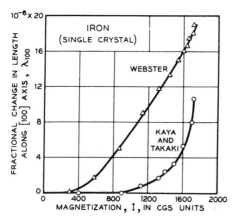

FIG. 13-74. Magnetostriction measured parallel to [100] in iron, according to different experimenters. Difference probably due to difference in initial domain distributions.

Magnetostriction in Unsaturated Crystals.—The magnetostriction measured on various specimens, by different experimenters, is far from being the same. Differences occur not only in the magnetostriction obtained at saturation but also in the form of the λ vs I curve. For example, λ_{100} in iron as observed by Webster [25W4] and by Kaya and Takaki [36K7] are, respectively, 20 and 11×10^{-6}, as given in Fig. 13-74. Such a difference is to be attributed to a difference in the domain structure of the demagnetized specimen, in which the domains may not be distributed equally among the six directions of easy magnetization.

The frequently observed difference in the form of the λ vs I curve is emphasized by the different assumptions made by Akulov [31A3] and

Heisenberg [31H6] regarding the change in distribution of domain directions with magnetization. On the one hand, Akulov assumes that when H is applied in the [100] direction the first change of magnetization takes place by 180° reversal of the domains oriented antiparallel to the field. Since these contribute nothing to the change in length, the predicted value of λ will be zero up to $I = I_s/3$. Onset of 90° changes in domain orientation (by boundary displacement) will then increase λ linearly with I to $I = I_s$. On the other hand, Heisenberg assumes that the distribution of domains among the various ⟨100⟩ directions will be governed by probability considerations, consistent with a given component of I in the direction of H. He assumes that there is no difference in energy between 90° and 180° changes of domain orientation. The data are intermediate between the

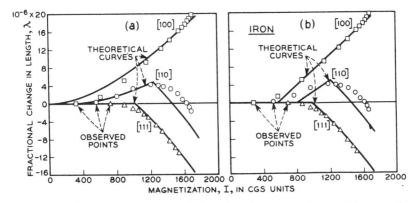

FIG. 13-75. Variation of magnetostriction of single crystals of iron with magnetization according to Heisenberg and to Akulov, as compared with observation.

two theoretical curves. The two assumptions are apparently extremes between which lies the actual situation. The presence of weak internal strains is likely to give some preference to initial 180° reversals but will not cause them all to be of this character.

Similar considerations apply to the initial part of the λ_{110} vs I curve. There is no disagreement as to λ_{111}, because here both the 90° and 180° changes from one ⟨100⟩ to another cause no change in length. After these processes are complete, the domain rotation will cause all of the magnetostrictive deformation. Of course, one expects some rounding of the curve, because some rotation will occur before all of the 90° and 180° changes are effected.

The theoretical curves of Heisenberg and Akulov are given in Fig. 13-75, with experimental points of Webster [25W1].

Calculation of the change of length during the rotation process, for [110] and [111], can be made with some certainty, and comparison of theory and experiment is made in Fig. 13–76, using the data of Kaya and Takaki [36K7] and the calculations of Becker and Döring [39B5].

Kaya and Takaki have calculated from their measurements the distribution of the domains among the various ⟨100⟩ directions, at $I = I_r$.

The Heisenberg theory has been extended and developed in three papers by Brown [37B7, 38B4, 38B8], for the calculation of magnetostriction and

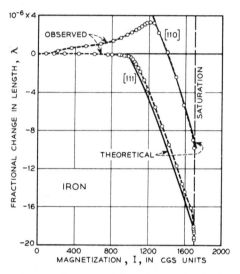

FIG. 13–76. Comparison of theory and experiment for single crystals of iron in [110] and [111] direction, at high magnetizations.

other effects. Various aspects of the theory have been reviewed by Takagi [39T3].

Origin of Magnetostriction.—The magnetic or quasi-magnetic forces between atoms give rise to an expansion (or contraction) of the lattice by opposing the purely elastic forces between atoms. The equilibrium distortion, or magnetostriction, occurs when the sum of the two corresponding energies is a minimum. Akulov [28A3], Becker [30B6], and Powell [31P6] have calculated the linear magnetostriction assuming that the magnetic forces between atoms can be simulated by magnetic dipole moments alone. The calculated magnetostriction is too small by an order of magnitude. Van Vleck [37V1] showed that spin-orbit coupling that accounts in order of magnitude for the crystal anisotropy (see preceding chapter) gives rise to quasi-magnetic interactions of the right order of magnitude. Calculations by Vonsovsky [40V4] also indicate that spin-orbit coupling can account for the observed magnetostriction.

It is thus apparent that a close connection exists between the magnetic crystal anisotropy constant and the linear magnetostriction; Kittel [49K1] has given the analytical expressions for the magnetostriction as a function of change of anisotropy with strain. If the anisotropy is independent of the state of strain, there will be no linear magnetostriction.

MAGNETOSTRICTION DATA

Iron.—Figure 13–47 has shown how the magnetostriction of iron increases with the field strength. On a different scale for H, the data of

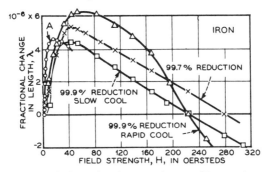

FIG. 13–77. Magnetostriction of various specimens of iron at low and intermediate magnetization.

Fig. 13–77 show how the length increases until it reaches a maximum, declines to its original length at between 200 and 300 oersteds, and con-

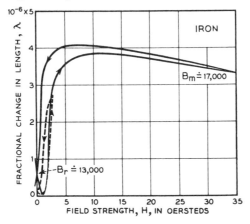

FIG. 13–78. Portions of hysteresis loops of magnetostriction in iron.

tinues to shorten in still higher fields. Purity, heat treatment, and dimensional ratio of specimen affect the shape of the curve, but its general character remains the same. It was noted in Fig. 13–60 that a minimum in λ occurs at about $H = 4000$ and that, thereafter, there is a slow lengthening associated with a change in volume.

The hysteresis loop of Fig. 13–78 in which λ is plotted against H should be compared with Fig. 13–50, which shows λ vs B, according to Masiyama. The residual magnetostriction is characteristically low.

The effect of tension within the elastic limit is shown in Fig. 13–79, plotted from measurements reported in 1902 [28H2]. The effect of stretch-

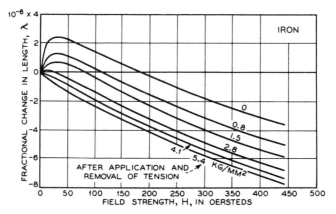

Fig. 13–79. Effect of tension on the magnetostriction of iron.

ing beyond the elastic limit or of cold rolling is seen, in Fig. 13–80, to cause an increase rather than a decrease in the magnetostriction in a given field

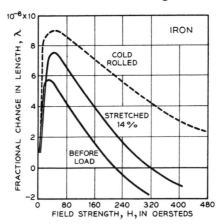

Fig. 13–80. Effect of plastic deformation on the magnetostriction of iron.

[31D1, 36B1]. Similar results have been reported also by Rankin [30R3, 31R2].

The effect of temperature has not been carefully investigated since 1903; data then reported by Honda and Shimizu [03H1] are given in Fig. 13–81.

Kornetzki's [34K10] measurements of the change of volume have already been described (Fig. 13–60). The fractional change in volume per unit of field strength, $\partial\omega/\partial H$, has been evaluated as 0.6×10^{-9} at room temperature and 1.0×10^{-9} at 85°C [35K2].

Direct measurements of the transverse effect have been made [31M1, 31D1, 31D2] and, taken with the longitudinal measurements, indicate that the volume change in low and intermediate fields is small, in agreement with the direct observations. Earlier measurements of the volume effect [31M1] agree qualitatively with those already mentioned.

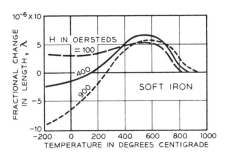

FIG. 13–81. Magnetostriction of iron at various temperatures.

Cobalt.—The variation of the magnetostriction of cast cobalt with the field strength is in one sense the reverse of that of iron—it shows [03H1] an initial shortening which changes to a lengthening at about $H = 150$, and above $H = 400$ the length is greater than the original length. After annealing there is a continual decrease in length up to the highest fields so far used.

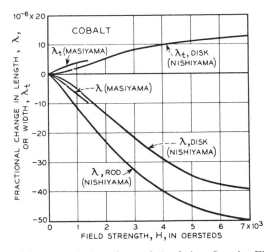

FIG. 13–82. Magnetostriction of annealed cobalt. See also Fig. 13–69.

In connection with a study of the magnetostriction of single crystals Nishiyama [29N1] has measured a polycrystalline specimen (rod of electrolytic cobalt) in fields as high as 7000 (Fig. 13–82) and here the contraction seems to approach a limit of $\lambda = 50$ to 60×10^{-6}. The same author measured both the longitudinal and the transverse magnetostriction in a disk of cobalt, and these results are also plotted in the figure, which in addition shows the data of Masiyama [32M2]. The latter author also measured the volume magnetostriction to field strengths of 1400 oersteds.

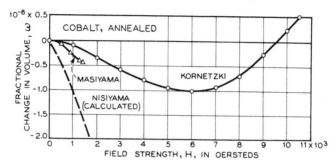

FIG. 13–83. Volume magnetostriction of annealed cobalt in high fields, according to Kornetzki, Masiyama and Nishiyama. See *Phys. Rev.* **96**, 311 (1954).

Kornetzki [34K10] has more recently measured the same quantity to $H = 10\,500$, and the results of both are plotted in Fig. 13–83. In the highest fields used, the fractional increase in volume is about 5×10^{-10} per oersted. The change in volume calculated from Nishiyama's measurements of longitudinal and transverse magnetostriction is shown in the figure

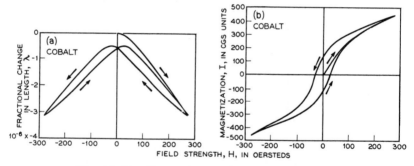

FIG. 13–84. Hysteresis of magnetostriction in cobalt.

by a broken line and is of the same sign but much greater in magnitude than the values directly observed. In fields of 6000 and 7000 oersteds the high calculated value of -14×10^{-6} is in disagreement with Kornetzki's measurements.

Fricke [33F1] has reported the transverse effect in specimens of Heraeus cobalt annealed and slowly cooled and having a coercive force of 10 oersteds.

A magnetostriction hysteresis loop [37M5] with λ plotted against H, and the corresponding magnetic hysteresis loop (B vs H), is shown in Fig. 13–84. Specimens were made from electrolytic cobalt containing only

0.01% carbon and traces of other elements and were annealed at 1050°C and slowly cooled.

There are no recent data on the effect of temperature on the magnetostriction of cobalt. The results of Honda and Shimizu [03H1] for cast cobalt (probably 93% Co, 5% Ni, 1% Fe, 1.4% C), and for annealed

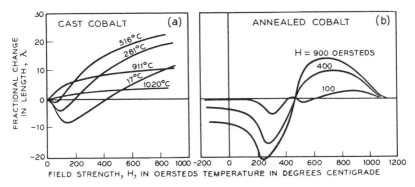

FIG. 13–85. Magnetostriction of cast and of annealed cobalt at various temperatures.

cobalt, are given in Fig. 13–85, which shows graphically the effect of the phase transformation from hexagonal to cubic structure at about 400°C. Such definite evidence of a transformation is absent in the data for cast cobalt, probably because the rapid cooling of the material following the casting tends to preserve the high temperature structure. The minima at 200–250°C in the annealed state may be connected with the change in the sign of the crystal anistropy in that range in temperature.

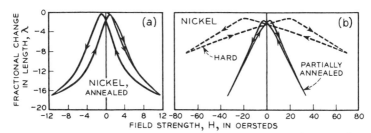

FIG. 13–86. Hysteresis of magnetostriction in nickel in various conditions.

Nickel.—The relatively large contraction of nickel has already been shown in Fig. 13–47. Other data indicate a contraction in the highest fields of $\lambda_s = 25$ to 47×10^{-6} [31S4, 27M1] depending on the composition, heat treatment, and methods of measurement and extrapolation; the best value is taken to be 34×10^{-6}. In weak fields the curve is characteristically flat, and when $B = 0.5B_s$, λ is only $0.2\lambda_s$. A hysteresis loop for a well-annealed specimen [37M5] and loops for hard-drawn ($H_c = 27$) and partially annealed specimens (690°C, $H_c = 4.7$) are shown in Fig. 13–86. The transverse expansion of nickel has been carefully measured by Fricke [33F1] in low as well as in high fields. His results are reproduced

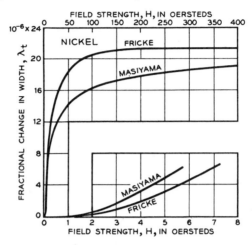

Fig. 13–87. Transverse magnetostriction of nickel, according to Fricke and Masiyama.

in Fig. 13–87 and those of Masiyama [31M1] are added. The latter found the magnitude of the transverse effect to be about half that of the longitudinal effect.

The effect of stress on the saturation magnetostriction has already been reported (Fig. 13–53). Kirchner's λ vs B curves for various tensions are

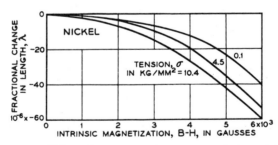

Fig. 13–88. Effect of tension on the magnetostriction of nickel.

reproduced in Fig. 13–88. With pressure there is a decrease in the magnitude of λ at all field strengths as well as at all inductions [36K1]. Kirchner's paper should be referred to for additional data including those associated with torsion, which decreases the maximum contraction to about $\frac{3}{4}$ of the value it has when torsion-free.

When nickel is strained beyond the elastic limit, the magnitude of the contraction in fields up to 300–400 oersteds is decreased, but that measured at a given induction (including saturation) is increased, according to Schulze [31S4]. Others have found that plastic deformation causes a reduction in the contraction in all fields used for measurement but have not extrapolated the magnetostriction to saturation; Dietsch [31D1] stretched wires by 10% and measured in fields up to 400 (Fig. 13–89); Williams [33W1] reduced over 90% in area by rolling and measured in fields up to $H = 1150$; but in neither of these investigations was the induction reported. Measurements have been reported also by Rankin [31R2].

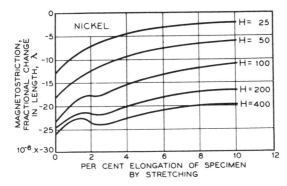

FIG. 13–89. Effect of plastic deformation (by stretching) on the magnetostriction of nickel.

In the highest fields nickel increases in volume. In applied fields below about 500 it may either increase or decrease depending on its purity, treatment, and shape. Kornetzki's [35K1] data of Fig. 13–90 illustrate this point. In fields higher than 3000 it may be expected that the fractional volume will continue to increase at the rate of about 2×10^{-10} per oersted. Masiyama had previously found a decrease in volume up to $H = 1440$, his highest field. All of these measurements were carried out adiabatically, without reference to the magnetocaloric heating.

In regard to the effect of temperature on the longitudinal magnetostriction of nickel, Döring [36D4] has found that λ_s is proportional to the square of the intensity of magnetization:

$$\lambda_s = aI_s^2$$

from 0°C to the Curie point. If this holds to 0°K, the data of Weiss [36W2] indicate that at saturation at this temperature there is a further contraction of 12% of the contraction at 0°C. Data similar to Döring's have been obtained by Kirkham [37K1], and his values for various field strengths from 0.8 to 740 are plotted in Fig. 13–91 along with Döring's.

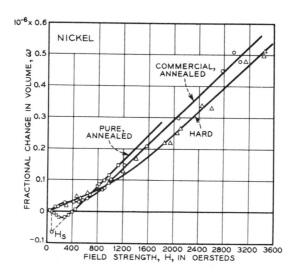

Fig. 13-90. Volume magnetostriction of various specimens of nickel: (1) pure nickel, annealed; (2) commercial nickel, annealed; (3) hard nickel.

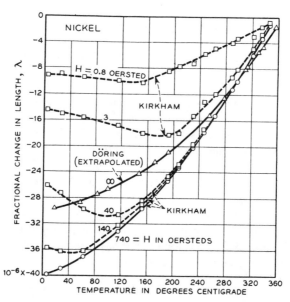

Fig. 13-91. Temperature-dependence of the magnetostriction of nickel, according to Kirkham and to Döring.

Kirkham's results show that higher fields are necessary to saturate at lower temperatures, and this is in agreement with the older measurements [03H1] which were made down to liquid air temperatures.

Iron-Cobalt Alloys.—The early experiments on the magnetostriction of these alloys were made by Honda and Kido [20H1] and were followed by those of Schulze [27S3]. More recently there have been two important investigations, both published in 1932. Williams [32W3] used specimens made by mixing the powdered elements obtained from the reduced oxides,

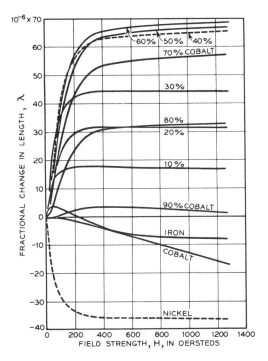

FIG. 13-92. Magnetostriction of iron-cobalt alloys plotted against field strength.

pressing, sintering, and hot swaging. The final treatment was annealing for 2 hours in hydrogen at 1000°C and cooling slowly. No analyses are given. Masiyama [32M2] used electrolytic iron and cobalt melted in vacuum with the addition of 0.5% manganese, cast, forged into rods and machined to ellipsoids. The chief impurities were reported as "traces." Annealing was at 1050°C in vacuum for 1.5 hours, and cooling was slow. The maximum field strength used in each investigation was about 1300 oersteds.

Williams' λ vs H curves for various compositions are reproduced in Fig. 13-92; Masiyama's curves of λ vs composition for various field strengths are shown in Fig. 13-93. Taken together the results show that in moderately high fields there is an unusually large expansion of 60 to 90 × 10^{-6} in alloys containing 40–70% cobalt. The highest magnetostriction of any material, reported at the time of this writing, has been observed by Nesbitt [50N1]. He observed a λ of 130 × 10^{-6} in specimens of hard rolled tape containing 70% cobalt (Fig. 13-94). Apparently the rolling orients the domains transversely to the direction of rolling, so that magnetization

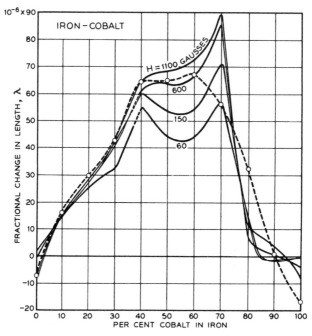

Fig. 13-93. Magnetostriction of iron-cobalt alloys vs composition according to Masiyama, and the saturation values according to Williams.

produces the maximum effect on the magnetostriction, which is $\frac{3}{2}$ of the normal value previously observed (see Fig. 13-57).

Masiyama [32M2] has measured the transverse and volume changes in addition to the longitudinal expansion, and his observed volume effect is compared in Fig. 13-95 with that calculated from the longitudinal expansion and the transverse contraction. There is a discrepancy between the two of a factor of about three.

Iron-Nickel Alloys.—These alloys have a special interest on account of the scientific and technical importance of the permalloys, and the related fact that the magnetostriction becomes very small as the nickel content of the alloy approaches 81%.

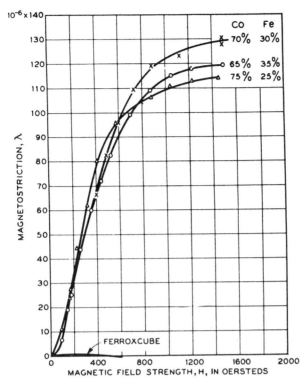

FIG. 13–94. High magnetostriction observed in some cold-rolled iron-cobalt alloys. Also, low magnetostriction in Ferroxcube III.

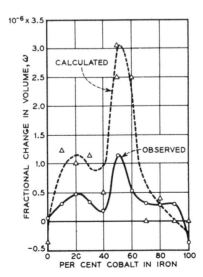

FIG. 13–95. Volume magnetostriction as observed and as calculated from the longitudinal and transverse effects.

666   STRESS AND MAGNETOSTRICTION

The magnetostriction and magnetization curves of the 46% nickel alloy have been given in Figs. 13–47 to 49. The behavior of three alloys in very low fields is shown in Fig. 13–96. These data were taken by McKeehan and Cioffi [26M4] with special care, proper corrections being made for the demagnetizing effects of the ends by simultaneous measurement of magneto-

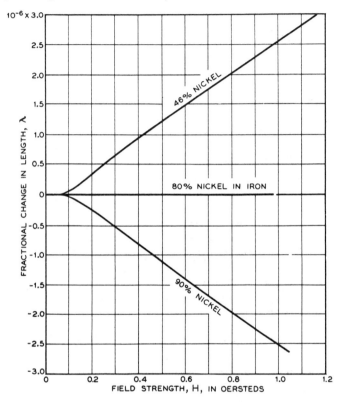

Fig. 13–96. Magnetostriction of 46, 80, and 90 Permalloys at low fields.

striction and induction. Their curves show accurately for the first time the small change of length with field in low fields and the horizontal tangent at the origin.

Figure 13–97 summarizes Schulze's data [28S3] for applied field strengths up to 300. His specimens were prepared from electrolytic iron and 99.2% pure nickel, with the addition of manganese that appeared to the extent of 0.4–1.7% in the final rods, of length 33 cm and diameter 6 mm. Similar results have been obtained by Masiyama [31M1].

The effect of composition on the magnetostriction can be seen easily in Fig. 13–98 where Schulze's data are replotted, the line for $I/I_s = 1.0$ being obtained by extrapolation. One observed here the characteristic change in sign near 81% nickel and the two maxima—one near 45% nickel, associated with the maximum in the saturation magnetization of γ-phase alloys, the

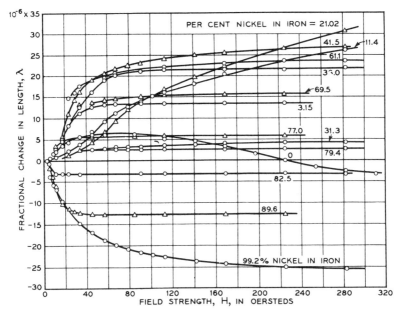

FIG. 13-97. Magnetostriction of various iron-nickel alloys.

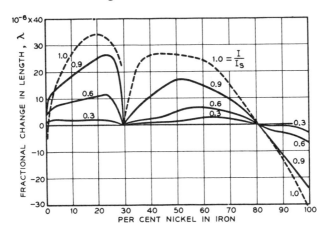

FIG. 13-98. Magnetostriction of iron-nickel alloys at various fractions of saturation.

other near 20% nickel in the α-phase region. In the latter iron-rich alloys the addition of nickel, which has negative magnetostriction, causes an unexpected increase in the length change.

The addition of only 1–3% of nickel is required to eliminate the high-

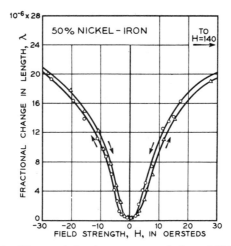

Fig. 13–99. Hysteresis in the magnetostriction of 50 Permalloy.

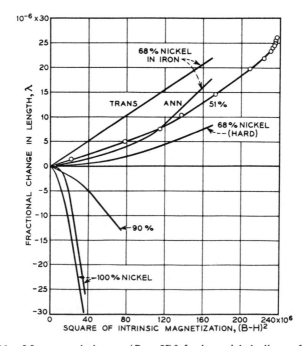

Fig. 13–100. Magnetostriction vs $(B - H)^2$ for iron-nickel alloys of various compositions and treatments: ANN, annealed at 1000° C; TRANS, annealed in a transverse field.

field contraction of iron. Rankin [29R2] has reported measurements on alloys containing 1–5% nickel. The minimum in the curves near 30% nickel is, of course, associated with the low or negligible saturation magnetization obtaining here at room temperature. The same general behavior has been found also in iron-nickel alloys prepared without the addition of manganese [28S3], some of the latter alloys showing slightly smaller changes in dimensions.

Magnetostriction-hysteresis loops are shown for the 50% nickel alloy in Fig. 13–99, selected from measurements by Masiyama [37M6].

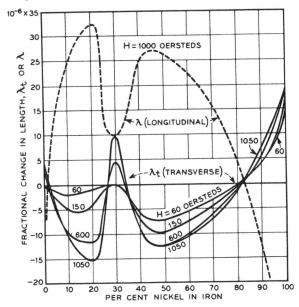

FIG. 13–101. Transverse magnetostriction of iron-nickel alloys. Also longitudinal magnetostriction for comparison.

Some of the alloys, particularly those between 50 and 80% nickel, are unusually sensitive to heat treatment. The most extensive experiments have been made on the 68% nickel alloy and the results have already been summarized in Fig. 13–55. Effects of a similar nature have been reported by Schulze [28S3] and by Kaya [38K2].

Finally, selected data on the longitudinal magnetostriction are plotted against the square of the ferric induction in Fig. 13–100. There is a tendency for the lines to approach linearity at both high and low inductions, but there is definite departure from this tendency at high inductions in some alloys, for example, that containing 51% nickel. As pointed out by Becker and Döring [39B5], this seems to be associated definitely with the isotropic volume expansion.

Masiyama has made the most complete report on the *transverse* and *volume* magnetostriction of these alloys (Figs. 13-101 and 102). Figure

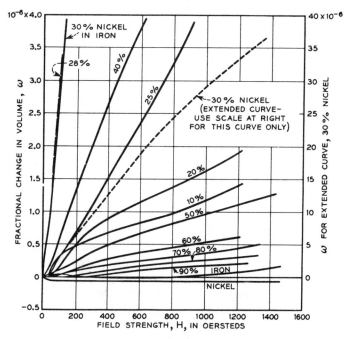

Fig. 13-102. Volume magnetostriction of some iron-nickel alloys.

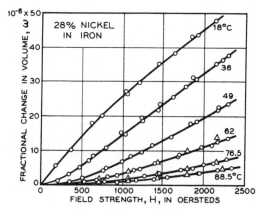

Fig. 13-103. Change of volume magnetostriction with temperature (28% Ni-Fe).

13-62 has already emphasized graphically the "enormous" changes in volume (20 to 30 × 10^{-9} per oersted) that take place in the alloys having compositions near 30% nickel, alloys that have also a Curie point near to the temperature at which the magnetostriction measurements were made. Kornetzki's data on the 28% nickel alloy have been added to Fig. 13-102 (broken line) and agree well with Masiyama's. Additional data of the former investigator, for temperatures from 18 to 88°C, are given in Fig. 13-103.

No important investigation of the effect of temperature on the longitudinal magnetostriction has been made since the paper of Honda and Shimizu published in 1905 [05H1]. Their measurements (Fig. 13-104) were carried to −186°C and showed, as one might expect, that the effect of cooling is to increase the expansion, the increase being small for the 70% nickel alloy and considerably larger (over 50%) for the 35% alloy. When

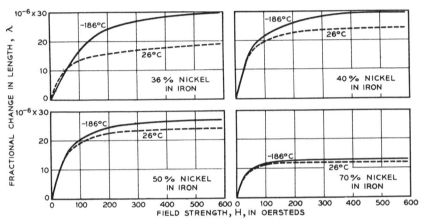

FIG. 13-104. Change of magnetostriction of some iron-nickel alloys on cooling to −186° C.

the nickel content was below 30%, cooling the cast specimen to −186°C increased the expansion by a large factor, and subsequent heating to room temperature caused a definite decrease in the magnetostrictive expansion which was, however, considerably more than the original lengthening at room temperature.

The effect of *tension*, compression, and torsion on the change in length of 50 Permalloy was investigated by Kirchner [36K1] whose work on nickel (p. 636) has already been discussed. As shown in Fig. 13-105 the magnetostrictive expansion of 50 Permalloy is decreased by tension and increased by pressure toward a limiting value 50% above its value without stress, as expected from domain theory. Tension and torsion both decrease the elongation in the 15% nickel alloy also.

Honda and Shimizu [02H2] have found that tension markedly decreases the magnetostriction of 45 Permalloy, and that under certain circumstances the magnetostriction becomes negative. In explanation of the small negative magnetostriction Braunewell and Vogt [49B10] suggest the two possibilities: (1) that there is a special orientation of crystals in the specimen such that [100] axes (for which λ_s is slightly negative, as shown in Fig.

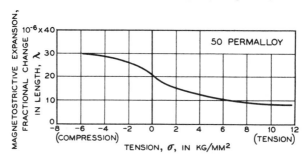

FIG. 13-105. Effect of stress on the saturation magnetostriction of 50 Permalloy.

13-71) are aligned parallel to the axis of the specimen; and (2) that the magnetostriction of a given domain configuration is actually somewhat changed when stress is applied. The first explanation is similar to that given by Bozorth [46B1] in a report on experiments on magnetoresistance.

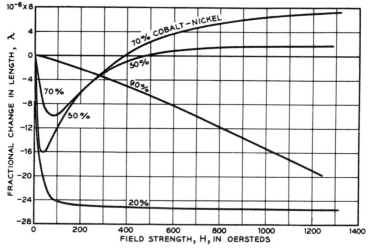

FIG. 13-106. Magnetostriction of cobalt-nickel alloys.

Cobalt-Nickel Alloys.—In the annealed state both cobalt and nickel show a longitudinal contraction at all field strengths, and this holds true for most of their alloys with each other. However, according to the most recent investigation, when the cobalt content lies between 50 and 75% there is an expansion when the field strength exceeds 400 oersteds. Masiyama's data, on which these statements are based, are shown in Figs. 13-106 and 107.

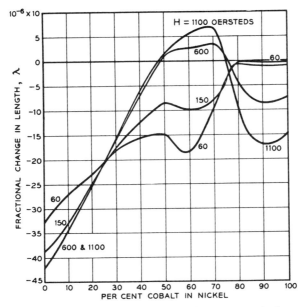

Fig. 13-107. Magnetostriction of cobalt-nickel alloys.

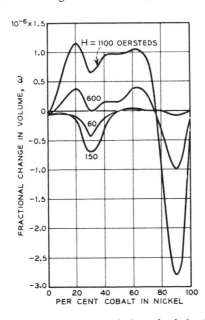

Fig. 13-108. Volume magnetostriction of cobalt-nickel alloys.

His specimens were prepared from electrolytic cobalt and nickel and were machined from forged rods and annealed for 1.5 hours in vacuum at 1050°C. The results are in general agreement with the earlier measurements of Masumoto [27M2]. On the contrary, Schulze [27S3] found no positive magnetostriction, although in the highest fields used by him ($H = 300$) there was some indication that several of the alloys would have positive magnetostriction in higher fields.

The transverse and volume magnetostriction have been determined by Masiyama [33M1], and some measurements are given in Fig: 13–108. The change in volume calculated from the longitudinal and transverse changes is in good agreement with the direct observations.

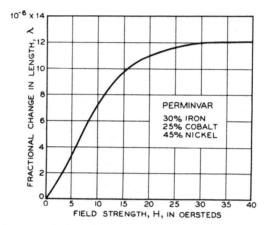

FIG. 13–109. Magnetostriction of standard Perminvar (30% Fe, 25% Co, 45% Ni).

Other Iron-Cobalt-Nickel Alloys.—The ternary iron-cobalt-nickel alloys have not been investigated with the same thoroughness as the binary alloys of these elements. Previously unpublished data (Cioffi, Fig. 13–109) of the longitudinal magnetostriction of Perminvar (30 Fe, 20 Co, 45 Ni) shows this to be positive and to have a saturation of about 12×10^{-6}.

Von Auwers [33A2] has measured the volume magnetostriction of 15 ternary alloys and combined his data with those of Masiyama [31M1, 32M2, 33M1] for the binary alloys to form the ternary diagram of Fig. 13–110. The lines indicate the expansion that has occurred up to $H = 1100$. Ordinary Perminvar is observed to lie in a "saddle" on the space-diagram.

Ide [34I1] used several alloys (36% Ni, 12–20% Co) as magnetostriction oscillators to stabilize the frequency of generators, and he found the temperature coefficient to pass through zero when the cobalt content was 12–13%.

A peculiar alternation of expansion and contraction has been observed

[34M4] in *Stainless Invar* (36.5% Fe, 54% Co, 9.5% Cr), an alloy having low thermal expansion and high corrosion resistance. Expansion occurs in low fields, a contraction in intermediate fields, and finally an expansion to $\lambda = 6.2 \times 10^{-6}$ at $H = 1415$.

According to unpublished data by H. J. Williams, *Molybdenum Permalloy* (4% Mo, 79% Ni, 17% Fe) has the low magnetostrictive expansion of

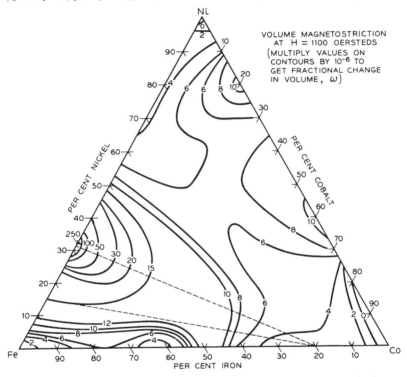

FIG. 13–110. Volume magnetostriction of Fe-Co-Ni alloys. Lines indicate expansion occurring at $H = 1100$ oersteds.

$\lambda = 2 \times 10^{-6}$ as might be expected on account of its high initial permeability.

Iron-nickel-chromium alloys have been investigated from several different points of view. Dean's work [30D1] indicates that the only alloys that have magnetostriction are the binary iron-nickel and iron-chromium alloys and the ternary alloys containing 5–20% of the third component. The limiting compositions for zero magnetostriction at $H = 100$, the maximum field used, include the well-known points on the iron-nickel side at 30 and 81% nickel and points on the iron-chromium side at 18 and 72% chromium. The magnetostriction of the nickel-chromium alloys vanishes when the chromium content is 10%, the composition for which the ferromagnetism is known to disappear at room temperature. Dietsch [31D1] measured a Krupp stainless steel (14.2% Cr, 0.17% C, 1.8% Ni, balance iron) and found a maximum expansion of 14 × 10^{-6}.

Modified *Invars* containing 25–40% nickel and up to 12% chromium and 20% cobalt were investigated by Ide [34I1] as to their temperature coefficients of magnetostrictive oscillation frequencies. When properly heat-treated, the alloys with 6–8% chromium and 36–38% nickel were found to be the best and to limit frequency variations to one cycle per million per degree centigrade.

Von Auwers and Neumann [35A1] investigated that portion of the *iron-nickel-copper* ternary system that is associated with alloys of high

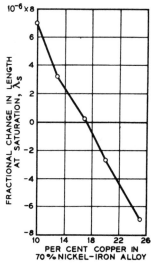

FIG. 13-111. Magnetostriction of some Fe-Ni-Cu alloys containing 70 per cent nickel.

initial permeability. Compositions were about 45–80% nickel and up to 50% copper. The line of zero magnetostriction runs along a somewhat irregular course from 81% Ni, 19% Fe to 45% Ni, 5% Fe, 50% Cu, and follows the same general trend as the line of highest initial permeability. Fig. 13–111 summarizes the results for alloys containing 70% nickel. Zero magnetostriction occurs at about 17% copper. The triangular diagram for the volume effect, given in the original article, shows a "saddle" similar to that previously referred to in discussing the Fe-Co-Ni alloys.

Other Iron-Alloys.—In general, anything that increases the hardness of an *iron-carbon* alloy decreases the magnetostrictive expansion or increases the contraction occurring in low or intermediate fields. This was noted first by Joule (see [10D1]) and is illustrated by Fig. 13–112 showing the data of Dorsey [10D1] for the hardening produced in annealed alloys by the

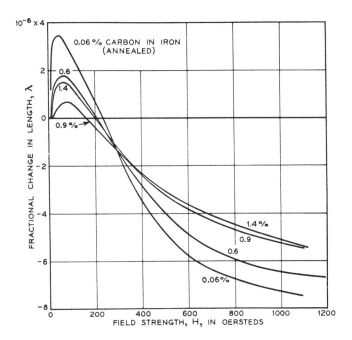

FIG. 13–112. Magnetostriction of annealed iron-carbon alloys of varying carbon content.

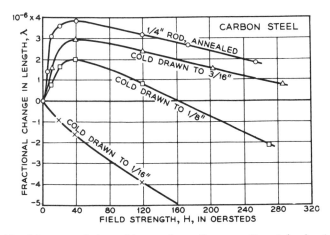

FIG. 13–113. Magnetostriction of iron-carbon alloys as affected by hardening by cold drawing.

addition of carbon up to about 1%. Similar data have been reported by Rankin [29R2, 30R3] and by Williams [33W1]. Hardening an annealed carbon-steel rod by cold drawing depresses the magnetostriction curve in

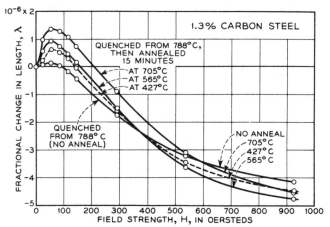

FIG. 13–114. Magnetostriction of steel rod, containing 1.3 per cent carbon, after quenching from 788° C (lowest maximum) and annealing at various temperatures.

all fields [30R3, 31R2, 32R2], as indicated in Fig. 13–113. Finally Fig. 13–114 [33W1] shows that the annealing of a quenched rod (1.3% C) at

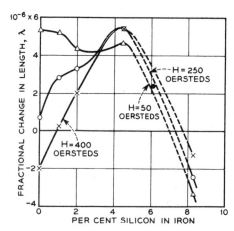

FIG. 13–115. Magnetostriction of iron-silicon alloys. The single point ● is derived from single crystal measurements of Carr (see Fig. 13–72) and is for saturation.

successively higher temperatures increases the elongation in fields up to 400–500 oersteds, and concurrent measurements of hardness showed that this treatment decreased the Scleroscope hardness from 70 to 35.

Schulze [28S3] studied the binary alloys of iron with *silicon* and *manganese* containing up to 10–12% of these elements, and his results for silicon are summarized in Fig. 13–115. The special interest in the silicon alloys lies in the disappearance of magnetostriction in the neighborhood of 7% silicon and the disappearance of the Villari reversal at lower silicon contents. The magnetostriction of cold-rolled, grain-oriented electrical sheet (annealed), containing 3% silicon, varies considerably with the direction of measurement in the sheet [42B3]. In the manganese alloys the Villari reversal disappears at about 5%, and ferromagnetism at about 12–14%.

Schulze [28S3] studied also the magnetostriction of the iron-aluminum alloys containing as much as 10% *aluminum*, and he found the magneto-

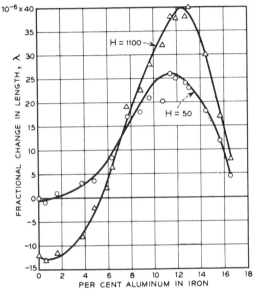

Fig. 13–116. Magnetostriction of annealed iron-aluminum alloys.

striction to increase rapidly with aluminum content to 35×10^{-6}. The Villari reversal disappeared at 4–5% aluminum. More recently Honda, Masumoto, Shirakawa, and Kobayashi [45H1] carried measurements to over 16% aluminum and observed a maximum in the curve at about 13% aluminum, when the magnetostriction was 40×10^{-6} (see Fig. 13–116). This material, called *Alfer*, was useful to the Japanese during World War II when nickel was difficult to obtain. The dependence of magnetostriction on field and on intrinsic induction are shown in Fig. 13–117. The magnetostriction is always positive when the aluminum content is 6% and increases continuously with field when the content is 9% or greater. Specimens were annealed at 1000°C and slowly cooled. Masumoto and Saito [45M1] observed lower magnetostriction in alloys quenched at 450–900°C. The alloys become non-magnetic at 17–18% aluminum (see Chap. 7).

Messkin, Somin, and Nekhampkin [41M1] have measured the magnetostriction of various alloys of iron as slowly cooled or as quenched in water.

Results on slowly cooled alloys at 1000 oersteds are given in Table 1. Their publication of the results on the highly magnetostrictively iron-aluminum alloys preceded that of the Japanese.

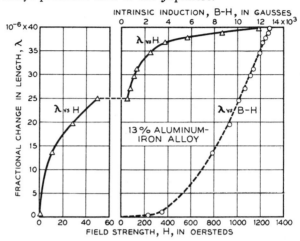

Fig. 13–117. Magnetostriction of Alfer (13% Al in Fe).

Table 1. Magnetostriction λ, of Various Alloys of Iron, as Annealed and Slowly Cooled and Measured in a Field of 1000 Oersteds

ALUMINUM		MOLYBDENUM		PHOSPHORUS	
Per Cent	$\lambda \cdot 10^6$	Per Cent	$\lambda \cdot 10^6$	Per Cent	$\lambda \cdot 10^6$
4	−0.3	2	−6.5	0.5	−8.0
6	1.7	4	−4.7	0.8	−5.5
8	12.2	8	1.2	1.0	−4.2
10	22.4	12	−3.2	1.2	−3.4
14	37.0	16	−1.7	1.5	−6.0
16	5.8	20	−3.0	1.7	−4.3
		24	0.7		

SILICON		TITANIUM		TUNGSTEN	
Per Cent	$\lambda \cdot 10^6$	Per Cent	$\lambda \cdot 10^6$	Per Cent	$\lambda \cdot 10^6$
2	−1.0	2	−6.2	8	−6.4
4	4.6	4	−1.8	12	−2.6
5	2.4	6	−1.0	20	−3.5
6	0.8	8	0.0	24	−3.0
7	−1.7				

Other Soft Materials.—Of the *nickel-copper* alloys *Monel metal* has been studied in both the annealed and hard rolled states. Schulze [28S4] reports the surprising fact that, although several specimens of annealed Monels (about 65% Ni, 30% Cu, 1.0% Mn, 1.5% Fe) exhibit a small contraction of 1 to 3.5×10^{-6}, the hard rolled alloy is completely inactive as nearly as can be observed. There is no statement as to whether or not

OTHER SOFT MATERIALS

the rolling causes the material to become non-magnetic. The Curie point is about 60°C for the annealed condition. Similar results have been reported by Williams [33W1]. Ide [34I1] has used Monel in magnetostriction oscillators.

Volume changes in a binary *nickel-copper* alloy containing 33% copper have been measured by Kornetzki [35K2] at temperatures ranging from 38 to 77°C. The largest change (measured adiabatically) is at the lowest temperature and is 1.4×10^{-9} per oersted, a value not inconsistent with that of von Auwers and Neumann, reported as part of their investigation of the ternary Fe-Ni-Cu system.

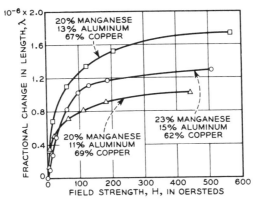

Fig. 13-118. Magnetostriction of some Heusler alloys.

The addition of 2% of *beryllium* to *nickel* reduces the contraction of the latter to $\frac{1}{3}$ or $\frac{1}{4}$ of its original amount in fields of over 100 oersteds [41M1, 31D1].

Schulze [33S1] has reported that a *nickel-tungsten* alloy containing 12% tungsten has a saturation magnetostriction of -16×10^{-6}. The curve obtained by plotting λ against $(B-H)^2$ is a straight line from $(B-H)^2/B_s^2 = 0.4$ to 1.0.

The magnetostrictive change in length of a single crystal sphere of *magnetite*, Fe_3O_4, was found by Heaps [24H1] to be as high as 30×10^{-6} in some directions in the crystal. Takei (private communication) measured $\lambda = 32 \times 10^{-6}$ in a specimen that had been pressed from powder and then annealed for several hours at 900°C. Kornetzki [35K1] measured the change in volume of polycrystalline Fe_3O_4 in fields up to $H = 10\,000$ oersteds and found that above $H = 5000$ there was a volume *decrease* of 0.07×10^{-9} per oersted. (For ferrites, see Vautier [50V1].)

Several reports, none of them recent, have been made on the longitudinal magnetostriction of the *Heusler alloys* [06G1, 04A1, 07M1]. Figure 13-118 gives representative data on three alloys having compositions near

20% manganese, 13% aluminum, and 67% copper, with the principal impurities 1.0% iron, 0.1% silicon, and 0.1-3% lead. The expansion is small in all specimens.

Permanent Magnet Materials.—Nesbitt [48N2, 50N1] has measured some of the standard permanent magnet materials, as well as a number of alloys near in composition to Alnico 5 and the Mishima alloys. Data for the standard materials are given in Fig. 13-119.

The effect of various heat treatments on the magnetostriction of alloys having the composition of Alnico 5 is shown in Fig. 13-120. The data are

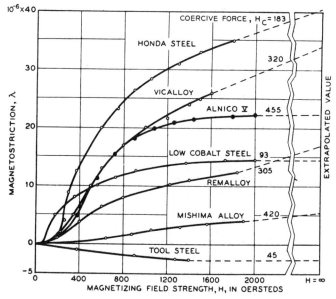

FIG. 13-119. Magnetostriction of some permanent magnet alloys.

similar to those obtained by Hoselitz and McCaig [49H1] and show the result, expected on the basis of domain theory, that the saturation magnetostriction measured parallel to the direction of the field present during heat treatment is very much smaller, and perpendicular thereto it is higher, than that in material heat-treated in zero field. Other data by Nesbitt [50N1] are given in Fig. 13-121. Hoselitz and McCaig believe that the data support the assumption that in the polycrystalline material the domains are initially magnetized parallel to that cubic axis which is most nearly parallel to the field present during heat treatment. This idea is also tested by McCaig [49M2] in a study of material having columnar crystals, all the crystals having a cubic axis parallel to a certain direction in the specimen.

Nesbitt [50N1] varied the composition of iron-nickel-aluminum alloys

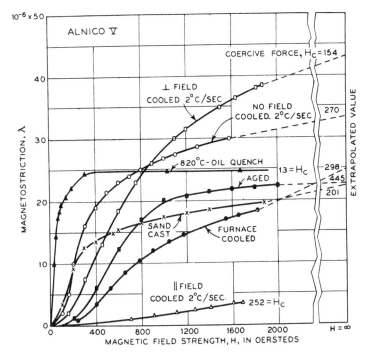

Fig. 13–120. Effect of heat treatment on the magnetostriction of Alnico 5.

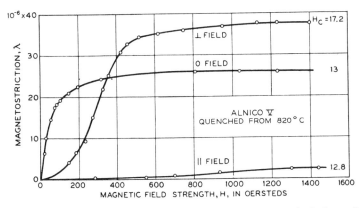

Fig. 13–121. Anisotropic magnetostriction of Alnico 5 quenched from 820° C. Coercive forces are indicated.

of the Mishima type, containing 29% nickel and various amounts of iron and aluminum. Near 12% aluminum the magnetostriction goes through zero, as shown in Fig. 13-122. This occurs even when the coercive force is rather high—over 400—and seems to prove that the strain theory of

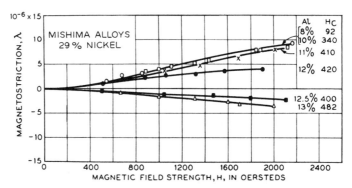

FIG. 13-122. Effect of small changes in composition on the magnetostriction of of some Fe-Ni-Al alloys.

coercive force, as discussed in Chap. 18, will not explain the high coercive force of this material, but that some mechanism operates to prevent the displacement of domain boundaries. See also pp. 393 and 832.

CHANGE OF ELASTIC MODULUS

Introduction.—When tension is applied to an unmagnetized ferromagnetic material, its length increases as a result of (1) a purely elastic expansion of the kind that occurs generally in solids and (2) an expansion resulting from the orientation of domains under stress. The increase in length of the second kind results from the magnetostriction associated with domain orientation, and it is responsible for the fact that in ferromagnetic materials Young's modulus, E, depends on amplitude of strain and on intensity of magnetization. The change of Young's modulus with magnetization is often called the ΔE *effect*.

The possible existence of a change in elasticity with magnetization was reported more than one hundred years ago [1846G1], but it was not definitely established until near the end of the last century [87B1, 02H1]. The older literature is reviewed by Auerbach [20A2]. It has been only since about 1930, however, that much progress has been made in understanding this and related phenomena. The first explanation in terms of domain theory was given by Akulov and Kondorsky [33A4], and the theory was developed in detail by Kersten [33K1, 34K8], Becker and Döring [39B5, 38D7], and others.

In addition to changes in the elastic moduli, observations have been made on the *damping* of mechanical oscillations in ferromagnetic materials. The damping is normally much larger than in non-magnetic materials, and the associated losses can now be understood in a general way and will be discussed in a later section.

The fractional increase in length, ϵ, with tension σ, predicted by domain theory, is shown schematically in Fig. 13-123. The line ON represents the extension in a non-magnetic material or in a ferromagnetic material in which the domain orientation is fixed (e.g., by a high field). In a soft magnetic material the curve OSM indicates that at low strain amplitudes the slight orientation of magnetization in the domains is responsible for some additional *elongation*, and that at high amplitudes the elongation approaches a limit which exceeds that of a nonmagnetic material by the absolute value of λ_s, the limiting *elongation or contraction* which occurs when the strain-free material is magnetized in a strong field. When the domains are less mobile, as they are in a hard magnetic material with high internal stress, greater stress is required to reorient the domains, and curve OHM applies. If the material is initially magnetized, so that some

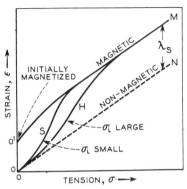

Fig. 13-123. Schematic stress-strain curves of non-magnetic (N) and magnetic (M) material, showing increased expansion resulting from magnetostriction λ_s.

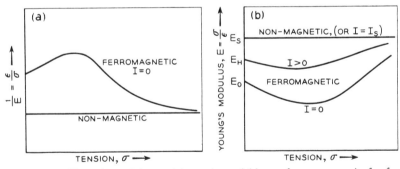

Fig. 13-124. Young's modulus and its reciprocal (slope of ϵ vs σ curve) of a ferromagnetic material depend on the amplitude of the stress.

magnetostrictive expansion has already occurred, the stress-strain curve follows the course $O'M$. Corresponding changes in $1/E$ and in E are illustrated in Fig. 13-124. We see here a change of E with amplitude of stress, and also an increase of E with field (H) when measurements are made at low stress amplitudes (i.e., $E_H > E_0$). The highest measured change of modulus with amplitude of stress is about 20% [48K1]. Changes of E with magnetization of over 35% have been recorded by Engler [38E1].

Experimental Methods.—Young's modulus, E, is readily determined from the resonance frequency of vibration of a specimen of given geometry. For rods vibrating in the longitudinal mode this frequency is given by the relation

$$f = \frac{n}{2l}\left(\frac{E}{\rho}\right)^{1/2}$$

in which n is any small integer, l the length of the rod, and ρ its density. For transverse vibrations in a circular rod having $l \gg d$,

$$f = \frac{0.8902md}{l^2}\left(\frac{E}{\rho}\right)^{1/2}$$

Here d is the diameter and m has the values 1, 2.757, 5.404, 8.933, 13.345, $\cdots$. In a rod of rectangular section, of thickness d, the same formula applies except that the numerical coefficient is multiplied by $2/\sqrt{3}$. Corrections applicable when $l \approx d$ have been tabulated [42M1]. Other relations have been derived for torsional vibrations.

Vibrations can be excited by application of a small mechanical force [37F4], by acting on the ends of the specimen with a force due to an electromagnet excited by a-c [35W5], by excitation with an alternating field using the magnetostriction of the specimen [29P2, 33A5], or by placing the specimen in contact with a piezoelectrically excited crystal [36C1, 41W1]. Torsional modes can be similarly excited. Resonance is found by varying the frequency of the applied force until the amplitude of vibration is a maximum. The direct measurement of stress and strain can also be used for determining moduli, especially in experiments involving torsion [34B5, 48K1].

Another method is to send a pulse of high-frequency waves through a specimen and to determine the time necessary for it to travel a measured distance. The velocity v, thus determined, is related to the elastic modulus C by the well-known relation

$$v = (C/\rho)^{1/2},$$

the particular modulus depending on the mode of vibration. The pulse is usually transmitted to the specimen by means of a piezoelectric quartz crystal cut in the appropriate manner, and it may be of 10-megacycle frequency and 1-millisecond duration. This method has been used for determining the various elastic constants of single crystals of nickel and investigating their change with magnetization [49B5].

The logarithmic decrement may be determined directly (1) from the decay with time of the amplitude of oscillations after the exciting source has been removed, (2) from the width of the resonance curve, following the analysis given in a later section, or (3) from the measurement of the Q of the system by electrical methods.

Some Experimental Results.—Dependence of E on *intensity of magnetization* is shown for various materials in Fig. 13-125, according to Siegel

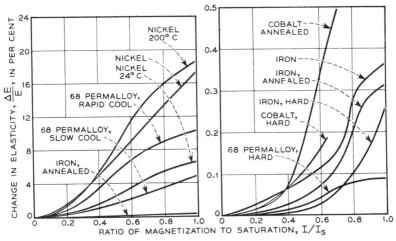

FIG. 13-125. Change of Young's modulus with magnetization for various materials.

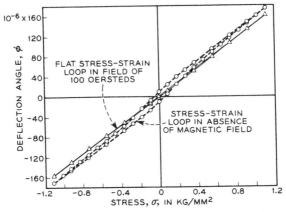

FIG. 13-126. Non-linearity of stress-strain curve (variation of Young's modulus) in carbonyl iron in torsion, with and without the presence of a field.

and Quimby [36S10], Nakamura [35N1], Williams *et al.* [41W1], Street [48S2], Yamamoto [38Y1, 41Y1], and Cook [36C1]. Generally $\Delta E/E$ is large for materials with large magnetostriction and small internal strain, such as annealed nickel; in iron and cobalt, and in work-hardened material, the effect is often less than 1%. Considerable variation from specimen to specimen is observed, as one would expect for a structure-sensitive property.

Variation of elastic constant with *strain amplitude* was also the subject of the older researches [00S1], but it was only in 1934 that Becker and Kornetzki [34B5] demonstrated plainly the nature of the effect in its relation to other magnetic phenomena. In their experiment (see Fig. 13–126) a carbonyl iron wire was twisted so that the maximum fiber stress in the surface was about $\sigma = 1$ kg/mm^2, and the torsion angle φ was measured as a function of the stress for a complete stress-strain cycle. When the specimen is unmagnetized, the stress-strain relation is non-linear, and the

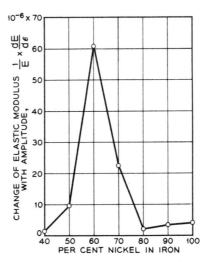

FIG. 13-127. Relative change of Young's modulus with stress amplitude, in iron-nickel alloys.

loop has a well-defined area. If a field is then applied, the lines become almost straight, and the loop becomes almost closed. (This shows, incidentally, that the area of the previous loop was not due to plastic deformation.)

Measurements of Förster and Köster [37F3] on a series of iron-nickel alloys have demonstrated that E depends linearly on amplitude of strain, ϵ, at very low amplitudes (10^{-8} to 10^{-7}). The slopes of the E vs ϵ lines for alloys of various compositions are plotted in Fig. 13–127.

Kornetzki [38K11] has noted the analogy between the change of E with σ (or ϵ) and change of μ with H in low fields. The relation

$$\frac{1}{E} = \frac{1}{E_0} + \alpha\sigma, \qquad \alpha = d(1/E)/d\sigma$$

is similar to Rayleigh's relation

$$\mu = \mu_0 + \nu H, \qquad \nu = d\mu/dH,$$

and the analogy extends also to remanent induction and remanent strain, and to the areas of the B,H and ϵ,σ loops. The relation between α and the magnetomechanical losses (damping) will be considered in a later section.

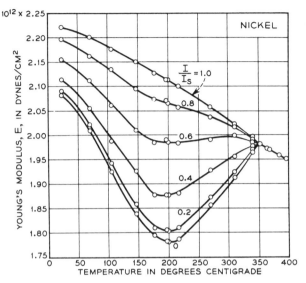

Fig. 13–128. Young's modulus of nickel as dependent on magnetization and temperature.

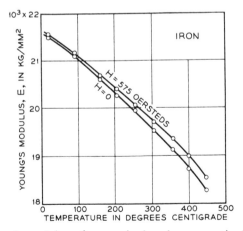

Fig. 13–129. Young's modulus of magnetized and unmagnetized iron at various temperatures.

The *effect of temperature* on ΔE is illustrated in Figs. 13–128 to 130, taken from data by Siegel and Quimby [36S10] for nickel and by Engler [38E1] for iron and some alloys of iron and nickel.

In nickel the magnitude of ΔE increases with temperature to 200°C, then decreases to zero at the Curie point. Values of E for saturation magnetization lie on a curve which merges into the curve for paramagnetic nickel without any break at the Curie point. This is to be expected from simple theory, according to which there is no change of magnetization with mechanical force in a fully magnetized material and hence no magnetostrictive expansion. The same result is found for iron-nickel alloys containing 93, 78, and 50% nickel [38E1]. In the 42% alloy, however, there is a marked change in the slope of the E vs T curve at the Curie point for all

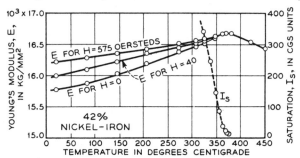

Fig. 13-130. Young's modulus of 42 Permalloy as dependent on magnetization and temperature. Note change in curve at Curie point.

values of the impressed field. This is attributed by Döring [38D7, 39B5] to the volume magnetostriction, which is especially large in this alloy; the theory will be given in more detail below (p. 697).

Theory for Large Internal Strain.—Kersten [33K1] first related the ΔE effect to internal strain and saturation magnetostriction. Consider a material in which the energy of internal strain is large compared to the energy of crystal anisotropy ($\lambda_s \sigma_i \gg K$). At one point in this material let the internal stress σ_i and the applied stress σ be inclined to each other at the angle θ_0 (see Fig. 13-39). If the magnetostriction λ_s is positive, the local magnetization I_s will be parallel to σ_i when $\sigma = 0$, and application of a given σ will rotate I_s by the angle θ. The problem is to calculate the extension ϵ_m in the direction of σ, as a result of the change in domain orientation and the concomitant magnetostriction. This is to be added to the ordinary lattice expansion ϵ_s, to give the total expansion $\epsilon = \epsilon_m + \epsilon_s$.

The magnetic energies associated with σ_i and σ, as evaluated in Chap. 18, are
$$E_\sigma = \tfrac{3}{2}\lambda_s \sigma \sin^2(\theta_0 - \theta)$$
and
$$E_{\sigma_i} = \tfrac{3}{2}\lambda_s \sigma_i \sin^2 \theta.$$

At equilibrium the value of θ is given by
$$(d/d\theta)(E_\sigma + E_{\sigma_i}) = 0.$$

When θ is small it is
$$\theta = (\sigma/\sigma_i) \sin \theta_0 \cos \theta_0.$$
The expansion connected with this rotation of a domain by the angle θ can be calculated from the relation for variation of magnetostriction with domain orientation (p. 634):
$$\lambda = \tfrac{3}{2}\lambda_s (\cos^2 \theta - \tfrac{1}{3}).$$
The expansion for a small angle θ is then
$$\epsilon_m = \theta \, d\lambda/d\theta = 3\lambda_s \theta \sin \theta_0 \cos \theta_0$$
$$= 3\lambda_s (\sigma/\sigma_i) \sin^2 \theta_0 \cos^2 \theta_0.$$
For randomly oriented internal strains
$$\sin^2 \theta_0 \cos^2 \theta_0 = \tfrac{2}{15}$$
and then
$$\epsilon_m = \frac{2\lambda_s \sigma}{5\sigma_i}.$$
In this and similar equations λ_s is always the absolute magnitude of the magnetostriction, and ϵ_m is always positive when σ is positive (tension), irrespective of the sign of λ_s. The total expansion is then
$$\epsilon = \epsilon_s + \epsilon_m = \frac{\sigma}{E_0} = \frac{\sigma}{E_s} + \frac{2\lambda_s \sigma}{5\sigma_i}$$
or
$$\frac{1}{E_0} - \frac{1}{E_s} = \Delta\left(\frac{1}{E}\right) = \frac{2\lambda_s}{5\sigma_i}$$
or
$$\frac{E_s - E_0}{E_0} = \frac{\Delta E}{E_0} = \frac{2\lambda_s E_s}{5\sigma_i}.$$

E_0 and E_s are the moduli at zero magnetization and at saturation. At saturation, domain orientation is not affected by small stresses acting parallel to the direction of magnetization, and then E has the value associated with the ordinary expansion of the lattice.

Kersten's formula for initial permeability is also related to internal strains and may be combined with the foregoing equation. When $\mu_0 \gg 1$,
$$\mu_0 = \frac{8\pi I_s^2}{9\lambda_s \sigma_i}$$
and therefore
$$\frac{\Delta E}{E_0} = \frac{9\mu_0 E_s \lambda_s^2}{20\pi I_s^2}.$$

A more extended analysis by Becker and Döring [39D4, 39B5], taking account of shear strains and Poisson's ratio, leads to this same result for a material having isotropic magnetostriction and isotropic elasticity. When λ_s is anisotropic, the situation is more complicated and no complete solution has been found. By averaging over all orientations, assuming the *stress* to be homogeneous, Döring [39D4] obtains

$$\frac{\Delta E}{E_0} = \frac{9\mu_0 E_s}{20\pi I_s^2} \cdot \frac{2\lambda_{100}^2 + 3\lambda_{111}^2}{5}.$$

When the *strain* is assumed homogeneous, he finds

$$\frac{\Delta E}{E_0} = \frac{9\mu_0 E_s}{4\pi I_s^2} \cdot \frac{2C_2^2 \lambda_{100}^2 + 3C_3^2 \lambda_{111}^2}{(2C_2 + 3C_3)^2},$$

C_2 and C_3 being shear moduli and having for iron the values $0.48 \cdot 10^{12}$ and $1.12 \cdot 10^{12}$ dynes/cm^2. Also for iron, $\lambda_{100} = 19 \cdot 10^{-6}$ and $\lambda_{111} \approx -19 \cdot 10^6$.

Small Internal Strains.—We have considered in the foregoing paragraphs the rotational process appropriate for a material having internal strains so high that the associated magnetic energy is higher than the crystal energy: $\lambda_s \sigma_i \gg K$. When the strains are so small that $\lambda_s \sigma_i \approx K$ or $\lambda_s \sigma_i < K$, the movements of domain boundaries play an important part in magnetization in weak fields, and the preceding calculation of ΔE does not apply. Becker and Döring [39B5, 39D4] have carried out a calculation for materials in which $\lambda \sigma_i \ll K$. Since the areas of the domain walls are not known and not subject to direct determination, the formulas derived must be used with care. The ΔE effect is influenced by 90° wall displacement only; the initial permeability is a measure of wall displacement in both 90° and 180° walls (in iron). The calculations of ΔE are made on the doubtful assumption that 90° wall displacements are predominant when the change in magnetization is small.

The analysis will not be given in detail here. It can be understood that in the direction of easy magnetization, [100], ΔE will depend on λ_{100} and E_{100}, and the relation derived (for $\mu_0 \gg 1$) is

$$\frac{\Delta E_{100}}{E_{100}} = \frac{9\mu_0 E_{100} \lambda_{100}^2}{16\pi I_s^2}.$$

In the [111] direction, however, $\Delta E_{111} = 0$ for the condition assumed; stress applied in this direction is equally inclined to all of the directions of easy magnetization and hence will not cause the magnetization to change from one easy direction to another. Averaging of ΔE over all directions of internal strain, on the assumption that the *stress* is homogeneous, gives

$$\frac{\Delta E}{E_0} = \frac{9\mu_0 E_s \lambda_{100}^2}{40\pi I_s^2}.$$

This is not much different from the formula for rotation against high internal strain.

When $\lambda\sigma_i$ and K are comparable in magnitude, ΔE may be affected by rotation of the local magnetization out of a direction of easy magnetization, [100]. Calculation of the effect [39B5] for a direction of difficult magnetization gives

$$\frac{\Delta E_{111}}{E_{111}} = \frac{E_{111}\lambda_{111}^2}{K}.$$

This is smaller than $\Delta E_{100}/E_{100}$ when $K > I_s^2/\mu_0$ (as in iron), provided $E_{111} = E_{100}$.

Summarizing, Becker and Döring point out that in an easy direction the formula for $\Delta E/E$ in terms of σ_i is about the same whether the internal strains are large or small. In the hard direction, however, ΔE depends on σ_i only when σ_i is large, and when σ_i is small an upper limit of ΔE is reached for which ΔE depends on K and is independent of σ_i. Thus for a material with crystal axes and internal strains distributed at random with respect to σ, for large σ_i the magnitude of ΔE is inversely proportional to σ_i, while for small σ_i the value of ΔE will be somewhat smaller than the expression for large internal strains would indicate.

Domain theory of the ΔE effect has also been discussed by Takagi [39T3].

Comparison with Data.—Kersten's theory requires that the change of Young's modulus with magnetization be directly proportional to λ_s and inversely proportional to σ_i. A convenient way to test the first requirement is to measure ΔE for the iron-nickel alloy series. This has been done by several authors, especially by Nakamura [35N1], Förster and Köster [37F3] and Köster [43K9]. The results of Köster are given in Fig. 13–131 and there compared with the values of λ_s determined by Schulze [38S3]. There is close correspondence between ΔE and λ_s, as would be expected if σ_i is nearly the same for all alloys. This is probably the case, for all specimens were annealed at 700°C, a temperature too low to erase entirely the effect of the previous cold working. Calculated values of σ_i are 1 to 2 kg/mm².

Another kind of comparison of theory and experiment has been made by Köster [43K7] by annealing cold worked nickel at various temperatures to produce different amounts of internal strain, and measuring E as dependent on temperature. Results are shown in Figs. 13–132 (E vs T) and 13–133 ($\Delta E/E$ vs T). It is obvious that the harder material has the

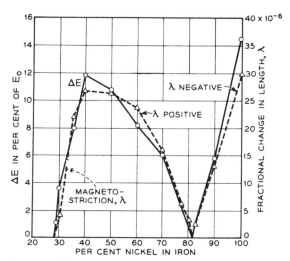

Fig. 13-131. Change of Young's modulus compared with the magnetostriction of iron-nickel alloys.

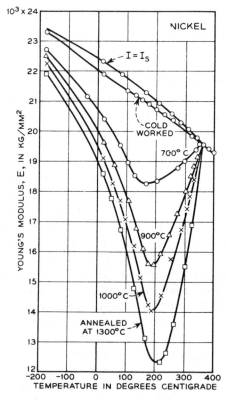

Fig. 13-132. Effect of annealing temperature on form of the E vs T curve of nickel. Measurements are for unmagnetized specimens, except in the case of the top curve.

smaller ΔE, as expected. Calculation of σ_i from ΔE, using Döring's [36D4] values of λ_s, shows that with decreasing temperature σ_i is constant down to about 200°C, the temperature of the maximum of ΔE as shown in Fig. 13-133. At lower temperatures the calculated σ_i apparently becomes larger. This increase is difficult to explain according to the theory given previously. Köster proposes that the apparent σ_i is of two kinds, one the real internal elastic stress σ_s, the other a kind of stress, σ_k, that has a similar power for directing the local magnetization and that can be identified with the magnetic crystal anisotropy. Accordingly he writes

$$\sigma_i = \sigma_s + \sigma_k.$$

When the anisotropy is small, the σ_i calculated from Kersten's formula is considered to be the true internal stress, and when K is comparable in

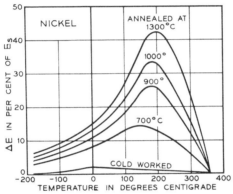

Fig. 13-133. Effect of annealing temperature and temperature of measurement on ΔE in nickel.

magnitude to $\lambda_s \sigma_i$ the calculated σ_i is believed to be larger than the true internal stress σ_s. In agreement with this idea it is known that K begins to increase rapidly with decreasing temperature below 200°C, the temperature at which the calculated σ_i also begins to increase. Using this hypothesis Köster has determined σ_s and σ_k from his measurements on nickel and iron-nickel alloys. The reader is referred to the original papers [43K7, 43K8, 43K9] for further discussion and for the values of σ_s and σ_k so obtained.

Köster [43K8] has also measured μ_0 and ΔE on the same specimens of nickel. Using the formulas of Kersten for large internal strain:

$$\mu_0 - 1 = \frac{I_s^2}{18\pi\lambda_s\sigma_i}$$

$$\frac{\Delta E}{E_0} = \frac{2\lambda_s E_s}{5\sigma_i},$$

Köster has calculated σ_i for material that has been work-hardened, or hardened and then annealed at temperatures of 700–1300°C. Table 2

TABLE 2. VALUES OF INTERNAL STRESS, σ_i, IN KG/MM^2, CALCULATED BY KÖSTER [43K8] FROM MEASUREMENTS OF μ_0 AND $\Delta E/E_0$

(Material originally cold reduced 80%.)

Temperature of Anneal (°C)	obs. μ_0	obs. $\Delta E/E_0$	σ_i calc. from:	
			μ_0	$\Delta E/E_0$
Hard.........	1.2	0.015	15.0	15.0
700..........	6.0	.085	3.03	2.82
900..........	8.2	.120	2.05	1.95
1000.........	11.2	.145	1.62	1.58
1300.........	15.6	.168	1.17	1.30

shows that the values obtained from the two relations are in good agreement, especially for the harder materials.

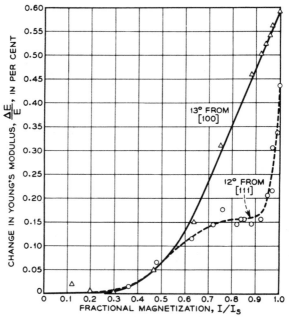

FIG. 13–134. Change of Young's modulus with magnetization in single crystals of iron: (a) crystal oriented 13° from [100]; (b) 12° from [111].

The change of Young's modulus with magnetization in iron-cobalt alloys has been measured by Yamamoto [41Y1], who found an increase in ΔE as large as 22% in the 50% alloy. This is higher than the change observed at room temperature in nickel, about 20% [43K7], and is consistent with the high magnetostriction of this alloy.

Kimura [39K7, 40K4] has measured the ΔE effect in 20 single crystals of iron of various orientations. A clear test of the prediction that E in the [111] direction is unaffected by weak magnetization is not possible, for none of his crystals is oriented very near to [111]. However, curves of ΔE vs I/I_s for two crystals, one oriented 13° from [100] and the other 12° from [111], are reproduced in Fig. 13–134; the latter is flatter at low intensities of magnetization and does not bend much until the domain-rotation process sets in at about two-thirds of saturation. Thus the prediction is supported in a qualitative way. A more definite test is desirable.

Change of E Near Curie Point.—When a specimen of nickel is subjected to a high field and heated, its Young's modulus shows no distinct change at the Curie point—the E vs T curve is flat in this region. This is illustrated in Fig. 13–128 and is in accordance with the theory already given; under these circumstances there is no magnetic expansion resulting from domain rotation, and therefore no change in E is to be expected. Most materials investigated show the same behavior. On the contrary, Engler [38E1] observed a rather large change in slope in a 42% nickel-iron alloy (see Fig. 13–130). Döring [38D7] has suggested that this effect is associated with the volume magnetostriction, which is especially large in this alloy. A change in length and a corresponding change in E are associated with volume magnetostriction in much the same way that the usual ΔE is associated with Joule magnetostriction. Döring's treatment is developed thermodynamically as follows:

Let l_0 be the length of a specimen in zero field under zero stress, and let $\epsilon = (l - l_0)/l_0$ be the fractional increase in length caused by field H, magnetization I, or stress σ. Young's modulus at constant magnetization, E_I, and at constant field, E_H, is given by

$$\frac{1}{E_I} = \left(\frac{\partial \epsilon}{\partial \sigma}\right)_I, \qquad \frac{1}{E_H} = \left(\frac{\partial \epsilon}{\partial \sigma}\right)_H.$$

By changing variables we have

$$\left(\frac{\partial \epsilon}{\partial \sigma}\right)_H = \left(\frac{\partial \epsilon}{\partial \sigma}\right)_I + \left(\frac{\partial \epsilon}{\partial I}\right)_\sigma \left(\frac{\partial I}{\partial \sigma}\right)_H.$$

The reciprocal relation between magnetostriction and stress may be written (see Chap. 15):

$$\left(\frac{\partial I}{\partial \sigma}\right)_H = \left(\frac{\partial \epsilon}{\partial H}\right)_\sigma.$$

Combining these relations, and using the equality

$$\left(\frac{\partial \epsilon}{\partial I}\right)_\sigma = \frac{(\partial \epsilon/\partial H)_\sigma}{(\partial I/\partial H)_\sigma},$$

we have

$$\frac{1}{E_H} - \frac{1}{E_I} = \frac{(\partial \epsilon/\partial H)_\sigma^2}{(\partial I/\partial H)_\sigma}.$$

If we are dealing with volume magnetostriction, and use ω for the fractional change of volume with field,

$$(\partial \epsilon/\partial H) = (\partial \omega/\partial H)/3.$$

The change in spontaneous magnetization with field, $(\partial I/\partial H)$, can be estimated from the Weiss theory. Numerical values for room temperature are approximately $(\partial \omega/\partial H) \approx 0.7 \times 10^{-8}$ and $(\partial I/\partial H) \approx 0.8 \times 10^{-4}$; therefore

$$\frac{1}{E_H} - \frac{1}{E_I} \approx 6 \times 10^{-14} \text{ cm}^2/\text{dyne},$$

and

$$\frac{E_I - E_H}{E_I} \approx 0.1,$$

since E_H is about 1.6×10^{12} dynes/cm^2.

If one extrapolates the E vs T curve from above the Curie point to room temperature, the value $E_I = 1.8 \times 10^{12}$ is obtained. The "experimental" value of $(E_I - E_H)/E_I$ is therefore $(1.8 - 1.6)/1.6 \approx 0.1$, the same as the calculated value. This agreement in order of magnitude indicates that volume magnetostriction is, in fact, responsible for the peculiar behavior of the modulus of this material near the Curie point. Most substances have a much smaller volume magnetostriction than this alloy, and since ΔE depends on $(\partial \omega/\partial H)^2$ it will normally be very much smaller than in the alloy containing 42% nickel. Köster [43K9] has called attention to the correlation between $E_I - E_H$ and volume magnetostriction in the iron-nickel alloys containing 30–100% nickel.

The change in E with temperature is closely associated with the unusually low temperature coefficient of elasticity of *Elinvar*, the iron-nickel alloy discovered by Guillaume [20G1]. This property is useful in maintaining constancy of frequency in mechanical systems vibrating at resonance. In an isotropic rod undergoing longitudinal vibration the natural period

$$f = \frac{1}{2l}\left(\frac{E}{\rho}\right)^{1/2}$$

has zero temperature coefficient when

$$\frac{1}{E}\frac{\partial E}{\partial T} = -\frac{1}{l}\frac{\partial l}{\partial T}$$

as pointed out by Pierce [29P2]. Normally $(1/E)\partial E/\partial T$ is negative and $(1/l)\partial l/\partial T$ positive, with the absolute value of $(1/E)\partial E/\partial T$ much the larger. In ferromagnetic materials $(1/E)\partial E/\partial T$ may be abnormally small or even positive. A negative value equal in magnitude to $(1/l)\partial l/\partial T$ is found in ferromagnetic material having the right amount of internal strain (Kersten [33K1]) or a suitably large volume magnetostriction (Döring [38E1]). $(1/E)\partial E/\partial T$ goes through zero in annealed iron-nickel alloys at about 30 and 45% nickel and has a suitable magnitude near these compositions but only over a small range. Addition of chromium and other elements makes the desired composition less critical [20G1, 32S5].

Other Data.—Variation of ΔE with temperature has also been studied in nickel by Möbius [32M4, 34M9], Kimura [39K7], Hibi [40H6], and Köster [43K7]; in iron and cobalt by Engler [38E1] and Köster [48K3]; and in iron-nickel alloys by Hibi [40H6]. Other papers on the ΔE effect include those by Ide [31I1], Giebe and Blechschmidt [31G2], Nakamura [35N3, 36N2], Williams, Bozorth and Christensen [41W1], and Yamamoto [42Y1, 43Y1].

MAGNETOMECHANICAL DAMPING

There is a loss in energy associated with elastic vibration of any material. In ferromagnetic substances the losses are generally much larger than in non-magnetic materials, and they can be interpreted in terms of other magnetic phenomena. As a result of the work of Becker and Kornetzki [34B5], Kersten [34K8], and Becker and Döring [39B5], the "magnetic" losses can be separated into the following parts:

1. Macro eddy currents.
2. Micro eddy currents.
3. Hysteresis.

The first kind is due to the eddy currents excited by the change of the induction of the specimen as a whole, resulting from the strain occurring during the vibration. This will naturally be zero if the magnetization is zero. The second kind results from the fact that each domain reacts to the strain of vibration in a characteristic manner, so that there are *local* changes in flux, and the associated eddy currents are different from one place to another in the specimen. Also, hysteresis is present when stress effects a change of local magnetization by irreversible displacement of the boundaries between domains.

A good illustration of the losses in magnetic material and of their variation with the state of magnetization comes from the simple experiment of Becker and Kornetzki [34B5] on the torsional oscillations of a carbonyl iron wire. Figure 13–135 shows the decay of amplitude with time after removal of the exciting force. The upper trace is with zero magnetization; the lower, with a field of 100 oersteds acting on the specimen. The field permits only a slight change in the orientation of the domains and therefore only a relatively small energy loss per cycle.

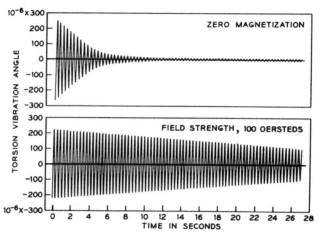

Fig. 13–135. Damping of torsion oscillations in a carbonyl iron wire as unmagnetized (upper) or magnetized in a field of 100 oersteds (below).

Damping Constants.—Measurements of energy losses are often carried out by noting the rate of decay of amplitude with successive vibrations, as illustrated in Fig. 13–135 (damped oscillations). Alternatively one can measure the amplitude of vibration as dependent on the frequency in the neighborhood of resonance, with constant force of excitation (forced oscillations).

The differential equation describing the damped oscillations of a vibrating system [36M15] is

$$\frac{d^2x}{dt^2} + 2f_0\delta \frac{dx}{dt} + 4\pi^2 f_0^2 x = 0,$$

and the amplitude of motion is

$$x = A_0 e^{-f_0 t \delta} \cos (2\pi f t - \varphi),$$

provided the energy dissipated is small. Here f is the frequency of vibration, f_0 the frequency of resonance, and δ the logarithmic decrement, equal to the natural logarithm of the ratio of the amplitudes of successive oscillations (one full period apart). Calculation of the energy stored in the system W and the energy dissipated per second P leads to the relation

$$\delta = \frac{P}{2f_0 W}.$$

Calculation shows that near resonance $[(f_0 - f)/f_0 \ll 1]$ the amplitude of *forced oscillations* is proportional to

$$[(1 - f^2/f_0^2)^2 + \delta^2/\pi^2]^{-\frac{1}{2}}.$$

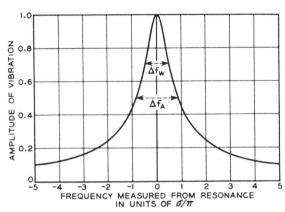

FIG. 13-136. Theoretical variation of amplitude of vibrating body near resonance, as affected by log decrement, showing frequency widths for half-maximum amplitude (Δf_a) and energy (Δf_w).

The resonance curve then has the form shown in Fig. 13-136. The full width of this curve at half-maximum amplitude is

$$\Delta f_A = \sqrt{3} f_0 \delta/\pi.$$

When energy (square of amplitude) is plotted against frequency, the width at half-maximum is

$$\Delta f_w = f_0 \delta/\pi.$$

One or the other of these quantities is determined by experiment. Results are often expressed in terms of Q rather than δ; then

$$\delta = \frac{\pi}{Q}.$$

Damping in Some Materials.—The change of damping with magnetization in iron is shown in Fig. 13-137. This illustrates the normal trend for ferromagnetic materials—a rise to a maximum that occurs at rather high values of I and then a drop to a low value at saturation. A similar maximum has been observed by von Auwers and others. Some specimens of nickel, for example those used by von Auwers [33A5] and Köster [43K10], show the same trend; the specimen used by Siegel and Quimby [36S10], however, shows a continually decreasing damping with increasing mag-

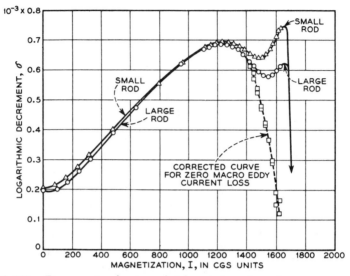

FIG. 13–137. Decrements of unannealed iron rod as dependent on magnetization. Circles, large rod; triangles, small rod; squares, after correction for macro-eddy-currents loss.

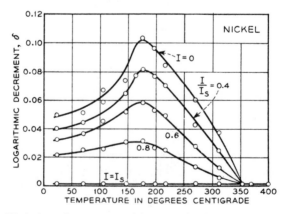

FIG. 13–138. Variation of decrement with magnetization and temperature in nickel.

netization. Other data of theirs for nickel (Fig. 13–138) illustrate the variation of δ with temperature—there is a maximum at 150–200°C and then a rapid fall to the Curie point, a course confirmed by Köster [43K10]. This maximum recalls the maximum in ΔE at about the same temperature and both may be related [39K7] to the disappearance of crystal anisotropy at this temperature. In iron-nickel alloys Köster [43K10] found a "normal" course for the δ vs I curve—a maximum at high values of I—while Williams, Bozorth, and Christensen [41W1] found (Fig. 13-139) that

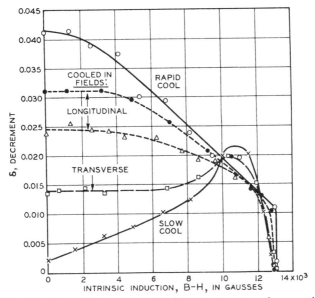

FIG. 13-139. Variation of decrement with heat treatment and magnetization in 68 Permalloy. For the unannealed material the decrement is 0.00053 at $B = 0$.

the curve depends on heat treatment and sometimes has no maximum at all but a continual decline with increase in magnetization, as previously observed in nickel.

Macro Eddy Currents.—Kersten [34K8] first suggested that an important part of the damping of ferromagnetic materials is due to the energy loss of the eddy currents caused by the (gross) alternating magnetization. His calculations have been extended by Brown [36B2] and Zener [38Z2] and discussed by Becker and Döring [39B5] and are followed here. Consider a rod magnetized to an induction B_0 and then subjected to a small stress σ. The total induction will then be

$$B = B_0 + \alpha\sigma, \quad \alpha = dB/d\sigma.$$

If now σ is changed periodically with time, there will be a periodic change in flux which will induce eddy currents opposing the change in flux. If the *frequency is low* enough so that the distribution of flux change is uniform over the cross section, it is not difficult to calculate the loss in energy per second per cm³ of specimen [39B5]:

$$P = \pi^2\alpha^2 ER^2 f^2 W/2\rho.$$

Here E is Young's modulus, R the radius of the rod, f the frequency of vibration, ρ the resistivity (all in cgs emu) and W the strain energy per cm^3 of material:

$$W = \sigma^2/2E,$$

σ being the maximum stress amplitude. The decrement is then

$$\delta_a = P/(2fW) = \pi^2 E\alpha^2 R^2 f/4\rho,$$

independent of σ.

If the *frequency is high*, the decrement is found to be

$$\delta_a = \frac{E\alpha^2}{8\pi R\mu}\sqrt{\frac{\rho}{\mu f}}.$$

Thus at low frequencies $\delta_a \propto R^2 f$ and at high frequencies $\delta_a \propto 1/\sqrt{R^2 f}$. In the latter case the change in flux is confined to a layer near the surface, and as f increases the layer becomes thinner and the losses correspondingly lower.

The decrements for rods or sheets at any frequency can be calculated from the following equations, derived in equivalent form by Zener [38Z2]: for rods,

$$\delta_a = \frac{\alpha^2 E}{2\mu x} \cdot \frac{\operatorname{ber} x\,\operatorname{ber}' x + \operatorname{bei} x\,\operatorname{bei}' x}{\operatorname{ber}^2 x + \operatorname{bei}^2 x}$$

where

$$x^2 = 8\pi^2 \mu f R^2/\rho;$$

and for sheets,

$$\delta_a = \frac{\alpha^2 E}{4\mu y} \cdot \frac{\sinh y - \sin y}{\cosh y + \cos y}$$

with

$$y^2 = 4\pi^2 d^2 \mu f/\rho,$$

d being the thickness of the sheet. Both of these functions are plotted in Fig. 13–140, in which d represents both the thickness of the sheet and the diameter ($2R$) of the rod. The frequency of maximum δ_a can easily be read from the plot, and for rods and sheets is respectively 0.33 and 0.14 times $\rho/(d^2\mu)$.

Cooke [36C1] and Brown [36B2] have measured the decrements on two similar specimens of unannealed iron of different diameters (see Fig. 13–137). At high inductions the curves differ, as one would expect if macro eddy currents contribute substantially to the vibrational losses. Under the conditions of the experiment ($f = 56\,000$, $\rho = 10\,000$) it is apparent that the skin effect is large and therefore $\delta \propto 1/d$. On this assumption Brown has accounted satisfactorily for the difference in the curves and has plotted the curve for material without macro eddy losses. When the frequency of vibration is high, the eddy-current shielding is large and almost all of the rod is at constant B, the changing induction being confined to a thin surface layer. Under these conditions the change of induction $\alpha\sigma$, caused by the alternating stress, is just neutralized by the induction $\mu_r H_e$, created by the eddy-current field H_e. This field gives rise to magnetostriction and hence changes the ratio of stress to strain; consequently, E_B, Young's modulus at constant B, is different from E_H,

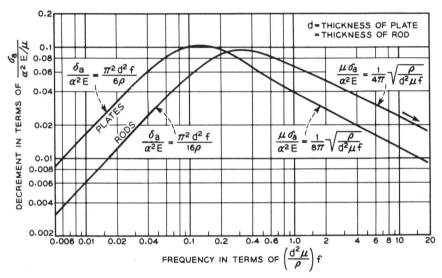

FIG. 13-140. Theoretical variation of macro-eddy-current decrement with frequency. Parameters are: d the thickness of plate or diameter of rod in cm, α the change in B for unit change in stress, E is Young's modulus, f the frequency of vibration, ρ the resistivity (c.g.s. units), μ the reversible permeability.

the modulus at constant H. The difference has been calculated by Brown [36B2, 39B5]. For high frequencies we have, for the reason just stated,

$$\alpha\sigma = -\mu_r H_e.$$

When H_e is small, the reciprocal relation between magnetostriction λ and stress may be written

$$\frac{\partial B}{\partial \sigma} = \alpha = \frac{4\pi\lambda}{H_e}.$$

Combining these gives

$$\lambda = \frac{\alpha H_e}{4\pi} = -\frac{\alpha^2 \sigma}{4\pi\mu_r}.$$

The total expansion resulting from the stress is then

$$\frac{\sigma}{E_B} = \frac{\sigma}{E_H} - \frac{\alpha^2 \sigma}{4\pi\mu_r}$$

or

$$\frac{E_B - E_H}{E_B} = \frac{\alpha^2 E_H}{4\pi\mu_r}.$$

This difference is large when the material is soft, for then H_e causes a relatively large change in induction and magnetostriction. It may be as large as 30% and therefore of the same order of magnitude as the change in E caused by magnetization. For the demagnetized and saturated states it is zero, for then $\alpha = 0$; consequently, $\Delta E = E_s - E_0$ does not depend on whether the skin effect is large or small and it is independent of the frequency of vibration.

Micro Eddy Currents.—Even though a vibrating specimen be in the demagnetized state, there will be local eddy losses caused by the local change of magnetization with stress. These changes in magnetization will add vectorially to zero, while the sum of the eddy-current losses will be finite.

Becker and Döring [39B5] have calculated the damping for several domain models. With high internal stresses, σ_i, the applied stress σ rotates the magnetization in accordance with Fig. 13–39. Internal strain is assumed to be distributed on a scale of wave length l; on the average, σ_i is reversed in sign in going this distance. A more rigorous (unpublished) analysis by L. A. MacColl gives the following expression for the energy loss per second per cm^3

$$P_i = \frac{42.6 l^2 I_s^2 f^2 \sigma^2}{\rho \sigma_i^2}$$

and, since $W = \sigma^2/(2E)$,

$$\delta_i = \frac{P_i}{2fW_i} = \frac{42.6 I_s^2 E l^2 f}{\rho \sigma_i^2},$$

proportional to frequency and independent of stress amplitude, σ.

A similar analysis, based on boundary displacement in a material having internal strains small compared to the crystal energy, has also been carried out with the result:

$$\delta_i = \frac{130 I_s^2 E l^2 f}{\rho \sigma_i^2},$$

not much different from the result for the rotation model.

Micro eddy damping is proportional to f at all frequencies, except when l^2 is comparable with $\rho/(\mu f)$. *Macro* eddy damping is proportional to f at low frequencies and to $f^{-\frac{1}{2}}$ at high frequencies. Both are independent of stress amplitude. An important difference is that δ_a, and not δ_i, is zero in a demagnetized specimen.

Micro eddy damping is inversely proportional to σ_i^2, according to Becker and Döring's analysis. One expects, then, that hard working of the material will reduce δ_i, and this expectation is borne out by the results of a number of experiments [33A5, 36C1, 41W1, 43K10].

Köster [43K10] has studied systematically the effect of temperature of anneal (700–1300°C) on δ for nickel, and his results are given in Fig. 13–141. Data have also been taken by Siegel and Quimby [36S10]. They refer to zero magnetization, so that macro eddy damping is excluded, but no separation of micro eddy and magnetomechanical damping has been made. If we assume that it is all micro eddy loss, we expect a relation between δ and $\Delta E/E$, and comparison of Figs. 13–133 and 141 indicates

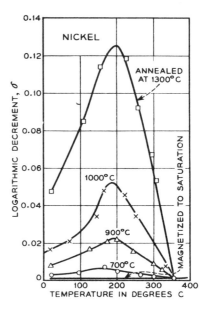

Fig. 13–141. The decrement in unmagnetized nickel as dependent on temperature of anneal and temperature of measurement.

that a relation exists. Combination of the equation relating $\Delta E/E$ to σ_i, namely,

$$\frac{\Delta E}{E} = \frac{2\lambda_s}{5\sigma_i},$$

with the equation given previously for δ_i, shows that

$$\delta_i \propto (\Delta E/E)^2.$$

Comparison of the maxima of Köster's curves showing ΔE and δ as functions of temperature (Figs. 13–133 and 141) shows that

$$\delta \propto (\Delta E/E)^{2.5}$$

approximately. In view of the assumptions made, the agreement with theory is all that can be expected.

Magnetomechanical Hysteresis.—It has already been mentioned (p. 688) that the relation between $1/E$ and σ in material subjected to small stresses is similar to the relation between μ and H (or κ and H) in materials subjected to weak fields. There is also an expression for magnetomechanical energy loss analogous to the formula for magnetic hysteresis loss

$$W = \frac{1}{3\pi}(d\mu/dH)H_m^3 = \frac{4}{3}(d\kappa/dH)H_m^3$$

(Rayleigh's law). The expressions for the magnetomechanical case are

$$\frac{1}{E} = \frac{1}{E_0} + b\sigma$$

$$P_h = \tfrac{4}{3} b\sigma^3 f$$

$$b = d(1/E)/d\sigma.$$

The magnetomechanical decrement δ_h is

$$\delta_h = \frac{P_h}{2f(\sigma^2/2E)} = \frac{4bE\sigma}{3}.$$

The linearity between δ_h and σ has been amply demonstrated by Förster and Köster [37F3] and by Snoek [41S8].

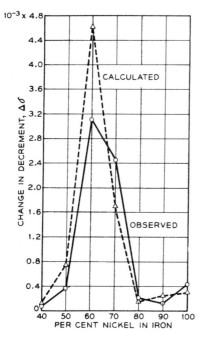

FIG. 13-142. Difference in decrement at different amplitudes, in iron-nickel alloys.

We now have a relation between δ_h and the change of E with σ. By eliminating b, we obtain

$$-\frac{1}{E}\frac{dE}{d\sigma} = \frac{3\delta_h}{4\sigma}.$$

This equation enabled Kornetzki [38K11] to use the data of Förster and Köster [37F3] to test the theory. They did not determine stress amplitude σ itself but the amplitude of transverse vibration A, which is proportional to σ. Then the relation

$$-\frac{1}{E}\frac{dE}{dA} = \frac{3}{4}\frac{d\delta_h}{dA}$$

should hold. Figure 13-142 shows the calculated and observed changes in δ for a change of A from 37 to 94 × 10^{-7} cm. The agreement gives strong support to the theory.

There is also a relation for residual strain ϵ_r, existing after the removal of stress σ, analogous to Rayleigh's relation for magnetism:

$$B_r = (d\mu/dH)H_m^2/2.$$

Richter [38R1] has tested the corresponding relation in torsion. If Φ_m and Φ_r are the maximum and residual twists, the analogy gives

$$\Phi_r \propto \Phi_m^2.$$

Richter's experiments on carbonyl iron show that Φ_r/Φ_m is proportional to Φ_m for small twists.

Separation of Losses.—The dependence of the three kinds of damping on frequency and on stress amplitude is summarized in Table 3. It has

TABLE 3. DEPENDENCE OF DAMPING ON FREQUENCY f AND AMPLITUDE OF STRESS σ

Kind of Damping	Symbol	Dependence of δ	
		on f	on σ
Macro eddy currents	δ_a	$\propto f$ (low f) to $\propto f^{-\frac{1}{2}}$ (high f)	Independent
Micro eddy currents	δ_i	$\propto f$	Independent
Magnetomechanical hysteresis	δ_h	Independent	$\propto \sigma$

already been hinted that proportionality between the magnetomechanical damping δ_h and stress amplitude makes it possible to separate δ_h from δ_a and δ_i. Such a separation has been made by Kornetzki [38K11] (his Fig. 7) using the data of Förster and Köster [37F3], and by Snoek [41S8], as indicated above.

Dependence on frequency can be used to separate micro eddy damping from magnetomechanical damping, provided macro eddy losses are zero, as they are in the demagnetized state. This separation has been carried out by Williams and Bozorth [41W3] (Fig. 13–143) for a specimen of annealed 68 Permalloy vibrated at various odd harmonics of the fundamental frequency (3000 cps) with constant stress amplitude. The slope of the δ vs f lines, multiplied by f, is equal to δ_i for that value of f, and the intercept on the δ-axis is δ_h for the stress used. The observed value of δ_i/f was 1.8×10^{-7}:

$$\frac{\delta_i}{f} = \frac{130 I_s^2 E l^2}{\rho \sigma_i^2} = 1.8 \cdot 10^{-7}$$

Using Kersten's relation

$$\mu_0 = \frac{B_s^2}{18\pi\lambda_s\sigma_i}$$

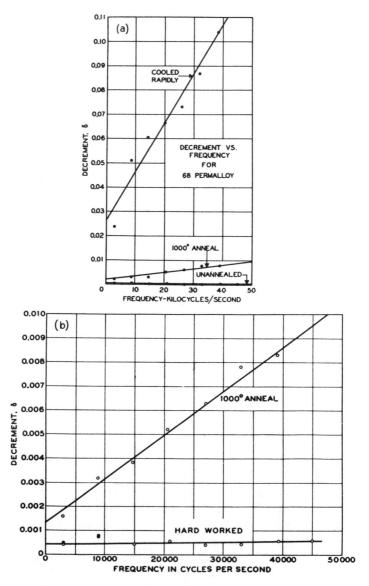

Fig. 13-143. Separation of losses in unmagnetized 68 Permalloy. At $f = 0$, loss is magnetomechanical. The slopes of the lines are proportional to the decrement due to the micro-eddy-current loss.

to calculate σ_i from μ_0, and knowing the other terms in the expression, they obtained

$$l \approx 10^{-3} \text{ cm}$$

and a volume l^3 of 2×10^{-9} cm³. This is about 10 times smaller than the volume of the Barkhausen discontinuities previously observed [30B1] in similar material. This and other evidence suggest [49B1, 49W2] that the movement of domain walls is closely connected with the scale of internal strains, but no definite relation has been established.

Köster [43K10] has related his measurements of damping in well-annealed, iron-nickel alloys to their magnetostriction. The peaks of his

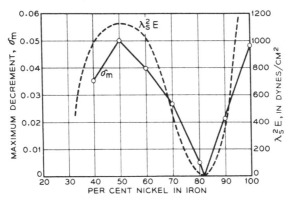

Fig. 13-144. Comparison of observed decrements with the theoretical $\lambda_s{}^2 E$, in iron-nickel alloys.

δ vs I/I_s curves are plotted against composition in Fig. 13-144. In comparing the data with theory we should consider both marco and micro eddy losses. In the former case

$$\delta_a \propto \alpha^2 E$$

as shown previously. But α is $dB/d\sigma$, and at low-stress amplitudes this is proportional to $d\lambda/dH$, the slope of the magnetostriction curve. At the peak of Köster's δ vs I/I_s curve, I/I_s is 0.8 to 0.9, and at such high induction $d\lambda/dH$ is roughly proportional to λ_s (see Fig. 13-97). Therefore one expects, approximately,

$$\delta_a \propto \lambda_s{}^2 E.$$

On the contrary, with micro eddy losses we have

$$\delta_i \propto \frac{E}{\sigma_i{}^2}.$$

In well-annealed material we assume with Kersten [31K1] that the residual internal strains are $\sigma_i = \lambda_s E$; therefore

$$\delta_i \propto \frac{1}{\lambda_s^2 E},$$

the inverse of the relation for δ_a. Köster data definitely support the idea that the decrement he measures here is due primarily to *macro* eddy currents, as one might expect. For comparison, $\lambda_s^2 E$ is plotted as the broken line in Fig. 13–144, using the data of Schulze [28S3] for λ_s and Marsh's collected data [38M2] for E. There is definite correlation between δ and $\lambda_s^2 E$.

Additional Data.—The older data on damping in magnetic materials have been reviewed by Auerbach [20A2]. Experiments in addition to those already cited have been reported by Ide [31I1], Parker [40P5], Snoek [41S5, 41S8], Thompson [44T1], Frommer and Murray [45F2], Scheil and Reinacher [44S2], Rotherham [46R2], Köster [48K3], and Kornetzki [43K11].

CHAPTER 14

TEMPERATURE AND THE CURIE POINT

Together with magnetic field and stress, temperature is one of the important factors in causing change in magnetization. The greatest influence of temperature is rarely near room temperature but rather just below the Curie point or near the temperature of a phase transformation. The way in which temperature modifies the magnetization curves of iron is shown in Fig. 14–1. At higher temperatures the curves rise more

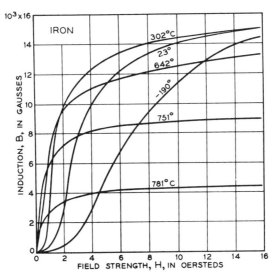

FIG. 14–1. Magnetization curves of iron measured at different temperatures, after annealing at 800° C.

quickly—at lower values of H—and then flatten out and saturate at lower inductions. The saturation continues to decrease and approaches zero at some temperature called the *Curie point*. Cobalt has the highest Curie point, 1120°C, of any known material.

When a magnetic material is subjected to a high constant field, an increase in temperature normally brings about a continuously accelerating decrease in induction; the induction comes down abruptly, almost to zero, at the Curie point (Fig. 14–2). The curve is retraced when the tempera-

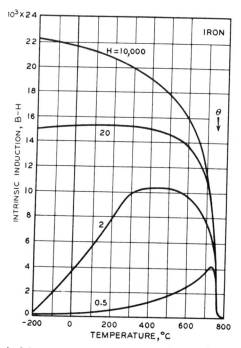

Fig. 14–2. In high fields, induction falls continually with increasing temperature, in low fields it first rises and then falls. Data are for iron.

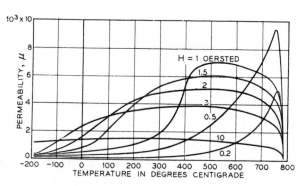

Fig. 14–3. Dependence of the permeability of iron on the temperature, the field strength being constant.

ture is lowered again. Conversely, when the iron is subjected to a weak field the induction will first increase with increase in temperature and, after passing through a maximum, will drop as before to a low value at the Curie point. Curves of permeability vs temperature, for various fixed values of field strength, derived from data like those of Fig. 14–2, have the characteristic shapes given in Fig. 14–3.

The way in which other magnetic quantities change with temperature has already been indicated in Fig. 3–8 for iron. In any material which may be called "normal" the curves are likely to have this same general character; e.g., the initial and maximum permeabilities first increase and then decrease with increasing temperature, and the coercive force and hysteresis loss continually decrease. The characteristic maximum in the

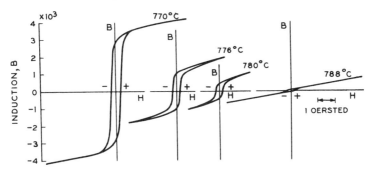

Fig. 14–4. Hysteresis loops of iron at temperatures near the Curie point.

initial and maximum permeabilities, just below the Curie point, is associated with the low magnetic anisotropy at this temperature. The way in which the size and shape of the hysteresis loop of iron change as the temperature approaches the Curie point is shown in Fig. 14–4 [32K2].

Effect of Phase Changes.—Iron and nickel are normal as to their magnetic behavior with change of temperature. No changes in their crystal structures occur between room temperature and the Curie point. On the contrary, such changes in structure do occur in many alloys, among which the so-called irreversible alloys of iron and nickel (5–30% nickel) investigated by Hopkinson, and some of the iron-cobalt alloys, may be considered as examples. In the 20–70% cobalt-iron alloys a change in phase, due to a rearrangement of the atoms in the crystal from a body-centered cubic (α) to a face-centered cubic (γ) form, occurs when the alloy is heated above 980°C (see Figs. 2–2 and 2–5). The α phase is ferromagnetic, the γ phase is non-magnetic (i.e., paramagnetic); consequently, the material loses its ferromagnetism when heated through the temperature of change of phase, 980°C, and the magnetization (measured in a high field) drops toward zero precipitously instead of in the normal fashion. When the temperature is lowered, this material recovers its magnetization without appreciable temperature hysteresis and the same B vs T curve is retraced. This is not always the case, however, when a phase change occurs. In the iron-nickel alloys just mentioned there is a considerable difference between the temperature at which magnetization disappears on heating and that at which it reappears on cooling (Fig. 14–5), sometimes as much as 500°C. Such a temperature lag in the B vs T curve, when observed in any material, indicates that some change has occurred in the metallurgical state of the material, such as a change in phase or in the state of order of the atoms (see Fig. 2–4), or a change in the solid

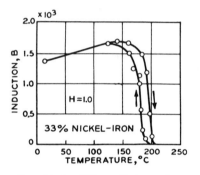

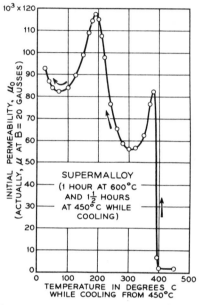

FIG. 14–5. Effect of phase change on the induction vs temperature curve.

FIG. 14–6. Dependence of the initial permeability of Supermalloy on the temperature.

solubility of some component. When no such lag occurs, the effect of temperature on the loss or recovery of magnetization is immediate.

A curve of unusual shape is obtained when the initial permeability of Supermalloy is plotted against the temperature (Fig. 14–6); this is attributed to a state of ordering that is intermediate between complete order and complete disorder. It will be noted that the temperature coefficient is abnormal in that it is negative at room temperature and changes twice at higher temperatures from negative to positive and back again. An unusual relation between permeability and temperature has also been observed for the Fe-Si-Al alloy Sendust [41Z1], in which atomic ordering also occurs (see Fig. 4–31).

Magnetization Near the Curie Point.—The change from the ferromagnetic to the paramagnetic state is not perfectly sharp, and it is difficult to define and determine the Curie point exactly. Careful work on this subject has been carried out by Weiss and his colleagues, and some of their results on nickel are shown in Fig. 14-7. The theoretical meaning of these data will not be considered.

According to the modified Weiss theory the curve relating the saturation magnetization I_s to the absolute temperature T is given by

$$\frac{I_s}{I_0} = \tanh \frac{I_s/I_0}{T/\theta}$$

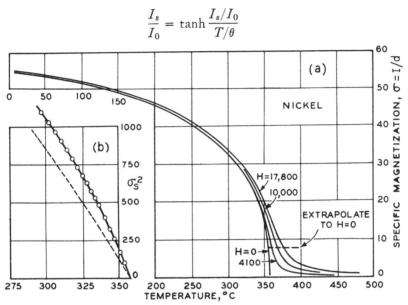

Fig. 14-7. Method of determining the saturation magnetization, σ_s, and the dependence of σ_s^2 on the temperature to determine the Curie point.

when the field strength is negligibly small. In high fields, especially when the temperature is near the Curie point, the more complete Weiss equation

$$\frac{I_s}{I_0} = \tanh \frac{I_s/I_0 + H/NI_0}{T/\theta}$$

must be used because then H/NI_0 is not negligibly small. In these equations, I_0 is the saturation magnetization at $T = 0$ and N is the molecular field constant.

At any temperature the local magnetization within a domain I_s is equal to the (true) magnetization I of the specimen as a whole when, and only when, the domains are oriented parallel. This state can be accomplished only by the application of a high field, which at the same time will produce a slight increase in the domain magnetization over its spontaneous value in weak fields. In order to make a proper estimate of the spontaneous magnetization, Weiss and Forrer [24W3] selected points on the I vs T curves [Fig. 14-7(a)] for which I has some chosen constant value and thus plotted a curve of T vs H for that value of I. Theory predicts that such a curve is linear; this is shown by rewriting the last equation in the form

$$\frac{T}{\theta} = \frac{I_s/I_0 + H/NI_0}{\operatorname{arc\,tanh}\,(I_s/I_0)}.$$

The theoretical relation between T/θ and H/NI_0 is drawn for several values of I_s/I_0 in Fig. 14–8, and Fig. 14–9 shows the data of Weiss and

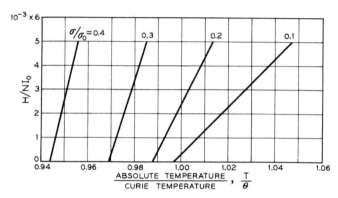

Fig. 14–8. Theoretical relation between spontaneous magnetization, temperature and field strength. Lines of constant magnetization.

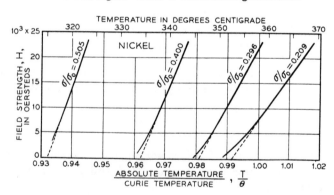

Fig. 14–9. Observed lines of constant magnetization for nickel.

Forrer [24W3] with T plotted against H for constant I. The curves follow the theoretical course except that they depart from linearity in the lower fields; this is to be expected, for a finite field is required to align the domains parallel to the field; thus a higher field will be necessary to produce a given magnetization at a given temperature than will be given by the Weiss equation. Linear extrapolation of the data for high fields, to $H = 0$ [see Fig. 14–7(a)] determines the temperature at which the chosen value of I is equal to the spontaneous magnetization, the magnetization of a domain in zero field. Except in high fields and at temperatures near the Curie point, it is practically indistinguishable from the saturation magnetization I_s, and this symbol is commonly used for both magnitudes.

The square of the spontaneous magnetization thus obtained is plotted against the temperature [Fig. 14–7(b)], and this curve is extrapolated to the point where it reaches the axis ($I_s = 0$), which is by Weiss' definition the Curie point: for nickel this point is at 358°C. The shape of the curve is in conformance with the theory, as indicated by the broken line drawn according to the theoretical equation given previously, but its actual position is at variance with theory. Extrapolation to the Curie point, $T = \theta$, is made easy by the fact that this line approaches linearity as T approaches θ.

The procedure just described is given here not only to show how to determine precisely the Curie point but to show also the success of the Weiss theory in predicting a variation of spontaneous magnetization with temperature in the range of temperatures just below the Curie point. The Curie point is not always defined in accordance with the Weiss theory but in other more empirical ways, e.g., as the temperature of the point of inflection of the I vs T curve.

The molecular field constant N can be determined from the slope of the H vs T curves of Fig. 14–9, using known values of I_0 and θ, and is found to be in general agreement with values obtained by other methods. It should be noted that the hyperbolic tangent law used previously is only one possible form of the relation between I_s, T, and H. Other forms give a linear H vs T relation, but the value of N deduced from the slope of the line depends on the particular kind of Langevin function used. A discussion of the actual values of N is given in Chap. 10.

Low Temperatures.—Some magnetic properties of iron at temperatures as low as -190°C are shown by the curves of Figs. 14–2 and 3, and Fig. 3–8 [10T1]. Careful measurements have been made of the intensity of magnetization in high fields at temperatures as low as -253°C (20°K) in order to determine the law of variation with temperature and to permit extrapolation to absolute zero so that the moment of the atom can be determined. The data for iron and for nickel (Fig. 14–10) [37W4] show that the Weiss theory in either its original or modified form is quite inadequate.

Bloch [30B3, 37O5] has derived from quantum theory a relation between magnetization and temperature that applies in the neighborhood of complete saturation, near absolute zero:

$$I_s = I_0(1 - AT^{3/2}).$$

For the derivation the reader is referred to the original papers. Data for iron and nickel show that the $\frac{3}{2}$ power is correct within experimental error, although a T^2 relation is almost as satisfactory. For further discussion see Chap. 10, p. 448.

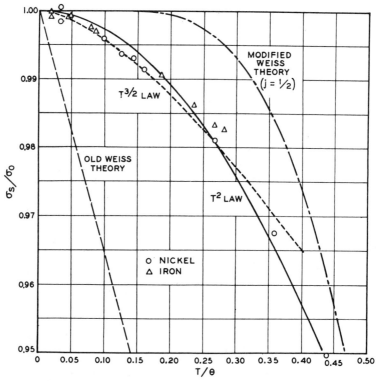

Fig. 14-10. Variation of spontaneous magnetization with temperature for nickel and iron, and comparison with theory.

Variation of Curie Point with Composition.—When one of the magnetic elements is alloyed with a non-magnetic element, e.g., nickel with copper, the Curie point is generally less than that of the pure element. As shown by Marian [37M2], many alloys of nickel show a systematic linear decrease of Curie point with atomic percentage of added element. Plots for nickel alloys and for iron alloys [36F1], of θ against atomic percentages, are shown in Figs. 14–11 and 12.

It is noted that for nickel the decrease in θ is similar to the decrease of average atomic moment (Fig. 10–11), and for most materials the decrease is linear and in proportion to the valence of the added element. For alloys of iron the curves are not as simple; this is caused partly by the formation of compounds and superstructures with some of the elements, and partly, perhaps, by the nature of the 3*d* shells of iron (see Chap. 10), neither of which is believed to be filled, as one shell is in nickel. The element manganese, which depresses but weakly the Curie point of nickel, strongly depresses that of iron. Conversely, vanadium actually raises θ for iron and depresses it strongly when alloyed with nickel.

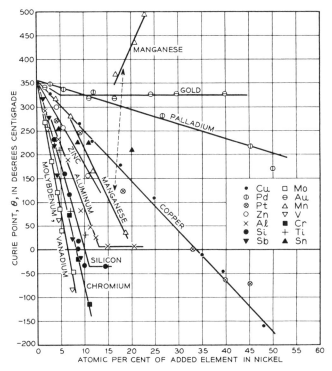

Fig. 14–11. Change of Curie point with the composition of nickel alloys (atomic per cent).

When the added element has limited solubility and forms a compound with the solvent metal, the Curie point usually decreases up to the point of limited solubility and, thereafter, remains constant as long as the compound (having a definite composition and Curie point) is present. This effect is different from that of composition on saturation; beyond the solubility limit the saturation magnetization drops linearly to the saturation of the pure compound, often zero, as the proportions of the phases change.

When a specimen is a mixture of phases, either an alloy or a gross mechanical mixture of two materials, the curve relating magnetization (in

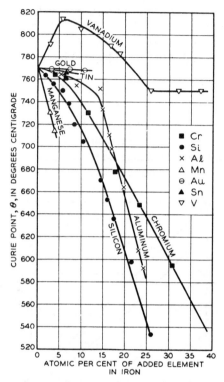

Fig. 14–12. Change of Curie point of iron alloys with composition (atomic per cent).

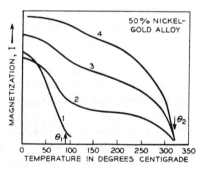

Fig. 14–13. Magnetization vs temperature curves of a nickel-gold alloy containing 50% gold. Curve 1, as quenched from 950° C; curves 2 to 4, after annealing for various lengths of time at 400° C.

constant field) to temperature has a characteristic break near the lower of the Curie points of the two phases. A curve of this kind for an alloy of nickel and gold [38G1] is reproduced in Fig. 14–13.

Curie points of a number of materials are given in Table 1.

TABLE 1. CURIE POINTS, IN °C, OF ELEMENTS AND COMPOUNDS (OR ORDERED STRUCTURES) AS REPORTED FROM VARIOUS SOURCES [33F2, 36F1, 37M2], INCLUDING A PRIVATE COMMUNICATION FROM L. W. MCKEEHAN

(Values are often approximate and should not be relied upon. Data for alloys are given in Chaps. 3 to 8.)

Elements		Compounds, Cont'd.	
Fe	770	MnBi	350
Co	1130	Mn_4N	470
Ni	358	MnP	25
Gd	16	MnSb	320
Dy	−168	Mn_2Sb	275
		Mn_3Sb_2	315
Compounds*		Mn_2Sn	0
Fe_3Al	500	Mn_4Sn	150
$FeBe_2$	520	AgF_2	−110
$FeBe_5$	<0	CrS_x	30; 100
Fe_2B	739	CrTe	100
Fe_3C	215	Fe_3O_3S	580
Fe_2Ce	116	$AlFe_2O_4$	339
Fe_4N	488	$BaFe_2O_4$	445
Fe_3P	420	$BeFe_2O_4$	190
FeS	320	$CdFe_2O_4$	250
FeS_2	>0	$CoFe_2O_4$	520
Fe_3O_4	575	$CuFe_2O_4$	490
Fe_2O_3	620	$MgFe_2O_4$	315
Fe_2P	420	$MnFe_2O_4$	295
Fe_4N	485	$NiFe_2O_4$	590
Fe_3Si_2	90	$PbFe_2O_4$	435
Co_2B	510	$SnFe_2O_4$	325
CoS_2	−180	$SrFe_2O_4$	450
CoZn	125	$La_2O_3 \cdot Fe_2O_3$	465
Co_4Zn	490	$Pr_2O_3 \cdot Fe_2O_3$	425
Ni_2Mg	235	$Nd_2O_3 \cdot Fe_2O_3$	300
Ni_3Mn	470	$Sm_2O_3 \cdot Fe_2O_3$	300
MnAs	45	$Er_2O_3 \cdot Fe_2O_3$	255
MnB	260	$Y_2O_3 \cdot Fe_2O_3$	275

* In addition to those listed, the following have been reported to be ferromagnetic at room temperature: $FeSn_2$, Co_5As_2, Co_2P, $NiHg_3$, Mn_3As_2, Mn_3C, Mn_3N_2, Mn_6N_2, Mn_5P_2, MnS, MnS_2, MnSe, MnTe, CrAs, Cr_2As. Also, $FeSn_2$ and CoPt are magnetic.

Curie Point and Pressure.—The variation of Curie point with pressure is of interest in connection with the magnetism of the earth. It has been argued that in the interior of the earth, where the temperature is above the Curie points of materials as we know them, the high pressures may permit these materials to be magnetic at considerably higher temperatures and hence contribute to the earth's magnetism.

In addition to the direct experimental approach to this problem, to be described later, there are two ways in which theory may be used. In the first method the Clapeyron equation is applied as if the Curie point were a phase-transformation temperature between the two phases of matter—the magnetic and the non-magnetic. The equation may be written

$$\frac{d\theta}{dp} = \frac{T\Delta V}{\Delta Q}$$

with ΔV the change in volume and ΔQ the heat absorbed at the transformation point. As discussed in Chap. 10, the change in volume at the Curie point is sometimes positive and sometimes negative, as shown by an upward or a downward cusp in the thermal expansion curve, and it follows that pressure causes the Curie point to be raised or lowered, respectively. From estimates of ΔV based on thermal expansion data, and ΔQ based on the entropy of a dipole having two possible orientations, Slater [40S3] calculates $d\theta/dp \approx 5 \times 10^{-5}$ degree/atmosphere for nickel and concludes that the interior of the earth cannot be ferromagnetic. A higher value is estimated by Leipunsky [38L1] and a lower value by DeBoer and Michels [38D6] (see Table 2).

The other theoretical approach, used by Slater and others, is more general in nature and depends on the quantum exchange interaction curve of Bethe, given in Fig. 10–13. This curve is, in effect, a plot of the Curie point (in degrees Kelvin) as dependent on the ratio D/d, of the lattice spacing to the diameter of the $3d$ electron shell responsible for ferromagnetism. The curve has a maximum, near cobalt, of about 1400–1500°K, and according to theory the Curie points of substances to the right (e.g., nickel) will increase with pressure and pass through the maximum and then decrease. Thus one would predict that even at very high pressures the Curie point of any known substance would not be substantially greater than that of cobalt at atmospheric pressure. This would indicate that the inside of the earth cannot be ferromagnetic if the temperatures there are several thousands of degrees centigrade, as estimated.

Kornetzki [35K2, 43K3] has pointed out that there is a close relation between volume magnetostriction and change of magnetization and of Curie point with hydrostatic pressure. Using certain thermodynamic relations, based on simple assumptions, he is able to calculate $d\theta/dp$ from measurements of volume magnetostriction, $d\omega/dH$. Thus the behavior of a substance can be determined without the difficulties involved in experimenting at high temperatures and high pressures. He concludes that $d\theta/dp$ is positive for iron and negative for nickel, in agreement with the theory of Bethe's interaction curve. His argument may be outlined as follows:

A change of volume with magnetizing field requires that there be a change of magnetization with pressure according to the thermodynamic relation

$$\left(\frac{\partial \omega}{\partial H}\right)_p = -\frac{1}{\sigma}\left(\frac{\partial \sigma}{\partial p}\right)_H,$$

in which $\partial \omega$ is the fractional change in volume and σ the magnetization per gram. The assumption is made that the magnetization at saturation is a function of T/θ:

$$\sigma_s = f(T/\theta)$$

and that θ may vary with the volume. Then one obtains by differentiation

$$T\frac{\partial \sigma_s}{\partial T} = -\theta\frac{\partial \sigma_s}{\partial \theta}$$

and

$$\frac{\partial \sigma_s}{\partial \omega} = \frac{\partial \sigma_s}{\partial \theta}\frac{\partial \theta}{\partial \omega} = -\frac{T}{\theta}\left(\frac{\partial \sigma_s}{\partial T}\right)_v \frac{\partial \theta}{\partial \omega}$$

or

$$\frac{1}{\theta}\frac{\partial \omega}{\partial \theta} = -\frac{\partial \sigma_s/\partial \omega}{T(\partial \sigma_s/\partial T)_v}.$$

Making the substitutions

$$\left(\frac{\partial \sigma_s}{\partial T}\right)_p = \left(\frac{\partial \sigma_s}{\partial T}\right)_v + 3\alpha \frac{\partial \sigma_s}{\partial \omega}$$

and

$$p = -\omega/\kappa$$

in which α is the linear expansion coefficient and κ the volume compressibility, we have

$$\frac{1}{\theta}\frac{\partial \theta}{\partial p} = \frac{\partial \omega/\partial H}{T(d\sigma_s/dT) - 3(\alpha T/\kappa)(\partial \omega/\partial H)}.$$

By measuring experimentally the quantities on the right side of this equation, at various temperatures, Kornetzki has calculated the shift of Curie point with pressure for iron, nickel, and the 30% nickel-iron and 33% copper-nickel alloys. Results are given in Table 2.

Some uncertainty exists in the values for nickel and the copper-nickel alloy, calculated from volume magnetostriction. This is due to the magnetocaloric effect, according to which an increase in spontaneous magnetization, caused by application of a strong field, is accompanied by an increase in temperature. This causes, in turn, an expansion of the

TABLE 2. SHIFT IN CURIE POINT IN DEGREES CENTIGRADE
PER ATMOSPHERE OF PRESSURE

Material	$\Delta\theta/\Delta P$, °C/atm	Reference
Fe	-5 to -10×10^{-3}	Kornetzki [43K3]
Ni	$+0.006 \times 10^{-3}$	DeBoer, Michels [38D6]
	$+0.05 \times 10^{-3}$	Slater [40S3]
30 Ni, 70 Fe	-4 to -5×10^{-3}	Kornetzki [43K3]
	$\ll 0$	Steinberger [33S2]
70 Ni, 30 Cu	$+0.065 \times 10^{-3}$	Michels et al. [37M7]
68 Ni, 29 Cu, 2 Fe	$+0.03 \times 10^{-3}$	Michels, Strijland [41M3]
Cd-Mg-Fe Spinel	$\pm 0.5 \times 10^{-3}$	Carnegie Institution [40C2]
67 Ni, 33 Cu	-0.4×10^{-3}	Kornetzki [43K3]

material. Consequently, in measuring the volume magnetostriction in high fields it becomes difficult to separate the expansion due to field at constant temperature $\partial\omega/\partial H$ from that due to magnetocaloric increase in temperature. Döring [36D6] and Snoek [37S6] have shown that, in the adiabatic measurement of the increase in volume with field strength in nickel, the magnetocaloric effect will change the temperature so much that the volume will increase with field strength, whereas when measured at constant temperature the volume will decrease. If $\partial\omega/\partial H$ is then negative, the thermodynamic relations show that $\partial\theta/\partial p$ is positive, and for these materials this is in accordance with the theory of the curve relating exchange interaction to atomic distance. However, the correction for thermal expansion is large, and the absolute value of $\partial\theta/\partial p$ deduced from the magnetostriction is therefore quite uncertain.

For iron the correction of magnetostriction for magnetocaloric expansion is much smaller, and $\partial\theta/\partial p$ therefore is more reliable. Kornetzki [35K2] has estimated that a decrease in the lattice constant of 2% causes a twofold decrease in exchange interaction and therefore in Curie point. This is consistent with the very steep rise of the interaction curve as it crosses the axis.

A further discussion of the relation between volume magnetostriction and change of Curie point (and molecular field constant) with pressure and volume is given by Smoluchowski [41S4].

Among the *direct determinations* of the change of Curie point with pressure, the first definite effect was observed by Steinberger [33S2]. He found that a 30% nickel-iron alloy was made paramagnetic ($\mu = 1.7$) under a pressure of 3000 atm. However, because of a possible influence of pressure on the α,γ-phase transformation, the effect on the true Curie point is uncertain, especially in view of the experiments of Ebert and Kussmann [38E5] who could not detect any certain change in Curie point of a similar alloy at 4000 atm. These workers, however, measured a change of saturation magnetization with pressure, and Kornetzki has shown that their data are consistent with a small decrease of Curie point.

Michels and co-workers [37M7] plotted the resistance-temperature curves of a 70% nickel-copper alloy at pressures up to 2600 atm, and they estimated $(d\theta/dp)$ to be 6×10^{-5} degree/atmosphere (see table).

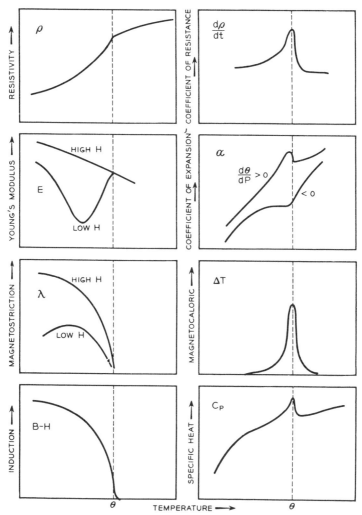

FIG. 14–14. Schematic representation of the change of various properties with temperature in the vicinity of the Curie point.

More recently it has been reported [40C2] that an increase of about 5°C in the Curie point of a Cd-Mg-Fe spinel occurs at 10 000 atmospheres. Measurements were made at low inductions using an a-c bridge. Details are not given in the brief note published.

Changes in Properties Near Curie Point.—When the saturation magnetization decreases rapidly and finally disappears at the Curie point, changes become evident in other physical properties, e.g., resistivity, specific heat, thermal expansivity, and elastic constants. These are discussed in some detail in other appropriate chapters of this book. However, in order to summarize such changes in properties here, Fig. 14–14 has been prepared. The curves are entirely schematic, and their forms have been discussed by Gerlach [39G4].

Applications.—It is often desirable to have a material with a definite positive or negative temperature coefficient of magnetization to use in a magnetic circuit to compensate for changes in flux that occur there as a result of changes in ambient temperature. The most common uses are in circuits containing permanent magnets and in loading coil cores. Alloys of iron-nickel, iron-nickel-molybdenum, nickel-copper, iron-nickel-chromium-silicon, and a composite of nickel and Invar have been used. They are discussed in the appropriate chapters on materials.

CHAPTER 15

ENERGY, SPECIFIC HEAT, AND MAGNETOCALORIC EFFECT

The change in energy associated with a given change in magnetization is of obvious importance in any discussion of specific heat and of magnetocaloric effect. It will be discussed here in setting up the thermodynamic expressions necessary in considering these phenomena and in deriving certain relations involving stress and magnetostriction.

Energy of Magnetization.—Three kinds of energy will be considered:

A_1: The work done in moving the material in a field.
A_2: The mutual energy between the material and the field.
A: The energy of magnetization of the material.

Expressions for these energies will now be given, and the relation between them will be pointed out.

Imagine an unmagnetized material in zero field and bring it into the field H of a permanent magnet (Fig. 15–1). The work necessary to do

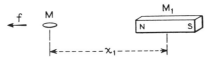

Fig. 15–1. Unmagnetized material M is brought into the field of a permanent magnet M_1. The force f at equilibrium is used in the calculation of the energy of magnetization.

this may be evaluated as follows: Let the moment of the permanent magnet be constant and equal to M_1, and let the variable moment of the magnetizable material be $M = Iv$. The field exerted by M_1 on M, when they are separated by x cm, is

$$H = \frac{2M_1}{x^3},$$

and the force between the bodies is

$$= f M \frac{\partial H}{\partial x} = \frac{6MM_1}{x^4}.$$

The work done by each cm³ of the material when it is displaced reversibly from $x = \infty$ to $x = x_1$ (into the field H) is then

$$A_1 = \frac{1}{v}\int_\infty^{x_1} f dx = \frac{1}{v}\int_\infty^{x_1} \frac{6MM_1 dx}{x^4} = \frac{1}{v}\int_\infty^{x_1} vI \frac{\partial H}{\partial x} dx = \int_0^H I dH.$$

This change in energy is shown schematically in Fig. 15-2.

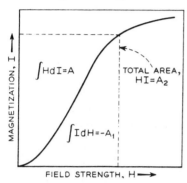

FIG. 15-2. Diagram of work and energy involved in magnetization.

Now consider the moment M to be fixed while the material is removed to zero field. The work done by an external force is

$$A_2 = \int_0^H I dH = HI.$$

This work, done with constant magnetization, is a measure of the mutual energy between the material and the field. A_1 is the energy associated with the change in both the magnetization of the material and mutual action between material and field, whereas A_2 is the energy associated only with the change in mutual action between material and field. The difference between A_1 and A_2 is then the energy caused by the change in magnetization alone and is called the *energy of magnetization A*:

$$A = A_2 - A_1 = HI - \int_0^H I dH = \int_0^I H dI.$$

The relation between A_1, A_2, and A is evident in Fig. 15-2.

The zero of energy is chosen for the unmagnetized material in zero field. When the material is drawn into a field, the total energy of the material and its mutual action with the field A_1 has decreased and is $-\int I dH$. The lowered energy is consistent with the principle that the more stable state has the lower energy. The change is illustrated in Fig. 15-3(a). When the fully magnetized material is removed from the field, the energy of the system *increases*, as shown in (b), because work has been done on it in removing the material. The total energy is now positive; and, since the only change from the original state of zero energy is the magnetization of the material, the energy of magnetization is positive (c). Its sign is easy to understand because energy has been absorbed during the process of magnetization, by rotating the local magnetization against the forces of crystal anisotropy, by moving domain boundaries, and so forth.

In determining the energetically stable state it is often desirable to minimize the quantity A_2, which contains both the energy of magnetization and the mutual energy between the material and the field. Since differences in energy, and not the position of zero energy, are important in determining the stable state, it is simplest to measure the energy from the state corresponding to the top line of Fig. 15–3. The energy A_2 is then $-HI$.

In the expressions used above, it has been assumed that H and I are parallel. If they are inclined to each other at the angle θ, the factor $\cos \theta$ must be added to the right side of all equations. The energy $-HI$ then becomes $-HI \cos \theta$. This quantity is often used in determining the stable

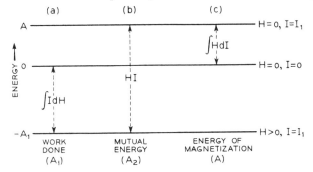

Fig. 15–3. Relation of energies to magnetization curve.

orientation of a domain when H and θ are subject to variation, I being constant and equal to I_s.

In an ensemble of magnetized particles, the field of each particle will act on the other particles. The mutual energy of the particles is then $(HI/2)$.

Thermodynamic Expressions.—According to the first law, the change in energy of unit volume of material is the sum of the heat absorbed by the body and the work done on it:

$$dU = dQ + dA.$$

The change in entropy is $dS = dQ/T$, and the energy of magnetization is $dA = HdI$; consequently

$$dU = TdS + HdI$$

when the changes are confined to heat and magnetization. When changes in volume occur, the term pdv/v_0, where p is the hydrostatic pressure, must be added to the expression for dU.

Other functions are especially suitable for deriving the relations useful in magnetism; two of these are the free energy, $F = U - TS$, and the thermodynamic potential, $\varphi = F - HI + pv/v_0$. In differential form these may be expressed as follows:

$$dF = -SdT + HdI - pdv/v_0$$

$$d\varphi = -SdT - IdH + vdp/v_0.$$

Here p is the hydrostatic pressure per unit area, and v_0 is the original volume of the material. These functions are perfect differentials and are treated as such in the following paragraphs.

Application to Magnetostriction.—Using the foregoing expression for $d\varphi$, one takes T and H as independent variables and performs the differentiations

$$\left(\frac{\partial \varphi}{\partial p}\right)_{T,H} = \frac{v}{v_0}, \qquad \left(\frac{\partial \varphi}{\partial H}\right)_{T,p} = -I.$$

Differentiation a second time gives

$$\frac{\partial^2 \varphi}{\partial p \partial H} = \frac{1}{v_0}\left(\frac{\partial v}{\partial H}\right)_{T,p} = -\left(\frac{\partial I}{\partial p}\right)_{T,H}$$

or, at constant temperature,

$$\frac{1}{v_0}\left(\frac{\partial v}{\partial H}\right)_p = -\left(\frac{\partial I}{\partial p}\right)_H = -\frac{1}{4\pi}\left(\frac{\partial B}{\partial p}\right)_H$$

An expression relating change of magnetization with linear pressure to change of length with field may be derived similarly. Denoting tension per unit area by σ (with positive sign), we have

$$\frac{1}{l_0}\left(\frac{\partial l}{\partial H}\right)_\sigma = \left(\frac{\partial I}{\partial \sigma}\right)_H = \frac{1}{4\pi}\left(\frac{\partial B}{\partial \sigma}\right)_H$$

at constant temperature.

These two equations are most useful in relating the phenomena of magnetostriction and stress and have already been applied in Chap. 13. They are based on the assumption that the changes expressed by the differentials are reversible, and application to finite ranges in H and σ must be made with care.

Heat Capacity.—When a ferromagnetic material is heated, there is generally a change in the energy of magnetization, and such a change affects the rate at which a given inflow of heat will raise the temperature of the material. The heat capacity (for *unit volume*) is defined as

$$c = \frac{\delta Q}{\delta T} = \frac{\delta U - \delta A}{\delta T}.$$

HEAT CAPACITY

The portion of the heat capacity that is due to magnetic changes is denoted by c_m:

$$c_m = \frac{\delta U_m - H\delta I}{\delta T}.$$

In order to make use of this expression we must have a physical theory of the change in magnetic energy, δU_m. According to the Weiss theory of an unmagnetized material, the local magnetization and field are equal to I_s and NI_s, respectively. Then, in zero applied field ($H = 0$),

$$c_m = \frac{\delta U_m}{\delta T} = -(NI_s)\frac{\delta I_s}{\delta T} = -\frac{N}{2}\cdot\frac{\delta I_s^2}{\delta T}.$$

This equation makes possible the evaluation of N from measurements of c and I_s at various temperatures, providing the magnetic heat capacity can be determined by subtracting the non-magnetic portion from the total measured heat capacity.

More specifically, if it is assumed that I_s follows the relation

$$\frac{I_s}{I_0} = \tanh\frac{I_s/I_0}{T/\theta},$$

then

$$\theta = \frac{NI_0\beta}{k},$$

β being the Bohr magneton and k Boltzmann's constant. If we represent by n_0 the number of Bohr magnetons per atom, θ may be expressed in terms of the moment of one atom, μ_A, as follows:

$$\theta = \frac{NI_0\mu_A}{kn_0} = \frac{NI_0^2 A}{n_0 R d}$$

where R is the gas constant per gram-atom, A the atomic weight, and d the density. Calculation of dI_s^2/dT and c_m then gives

$$c_m = \frac{n_0 R d f(x)}{A}$$

where

$$x = \frac{I_s/I_0}{T/\theta}$$

and

$$f(x) = \frac{x^2 \tanh x}{\tanh x \cosh^2 x - x}.$$

Then c_m should depend on T/θ as shown by the lowest curve of Fig. 15-4, drawn for $n_0 = 1$. The curve rises continually with increasing temperature until it reaches the Curie point, whereupon it drops suddenly to zero.

In a more general discussion, not limited to the tanh expression, we must use the Brillouin relation:

$$\frac{I_s}{I_0} = \frac{2j+1}{2j}\operatorname{ctnh}\frac{(2j+1)a}{2j} - \frac{1}{2j}\operatorname{ctnh}\frac{a}{2j}$$

in which $a = jg\beta NI_s/kT$. If we assume that the gyromagnetic ratio is $g = 2$, the atomic moment is $\mu_A = n_0\beta$ and $a = 2j\mu_A NI_s/n_0 kT$. Then $dI_s{}^2/dT$ and, therefore, c_m will depend on the value of the quantum number j, which determines the coupling between electron moments and the number of different orientations, $2j + 1$, which the moment may have with respect to the field. Curves are shown in Fig. 15–4 for c_m vs T/θ, with $n_0 = 1$ and $j = \tfrac{1}{2}$, 1, and ∞.

The magnitude of the discontinuity at the Curie point is

$$\Delta c_m = \frac{5j(j+1)}{j^2 + (j+1)^2} \cdot \frac{n_0 Rd}{A},$$

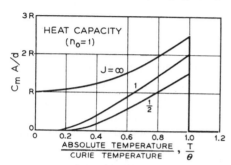

FIG. 15–4. Heat capacity calculated for various quantum states, according to simple theory.

A being the atomic weight and d the density. For $j = \tfrac{1}{2}$, 1, and ∞, the values of $\Delta c_m A/d$, the change in the heat capacity per gram-atom are respectively $3Rn_0/2$, $2Rn_0$ and $5Rn_0/2$, as may be seen in Fig. 15–4. In the more elaborate calculation of P. R. Weiss, who extended the Bethe-Peierls method to the evaluation of the molecular field, as described in Chap. 10, the discontinuity in the heat capacity per gram-atom in the body centered iron structure is $2.05R$ for $j = \tfrac{1}{2}$ and $3.40R$ for $j = 1$, values not much different from those derived from the Brillouin function.

Comparison with Experiment.—The experimental curves of heat capacity vs temperature are in agreement in their general features with the theory outlined above. In comparing with experiment it is convenient to use the *specific heat*, referred to one gram of material and designated by a capital C. Experiments are usually carried out at constant pressure, and the specific heat then measured is equal to the heat capacity per cm³ divided by the density:

$$C_p = c/d.$$

Fig. 15–5 shows data for C_p for nickel [29L1] and, for comparison, data for a non-magnetic element, copper [27K2].

The divergence of these curves at the higher temperatures makes it difficult to separate the magnetic specific heat from the remainder. Attempts to do this will be discussed later. A less difficult quantitative comparison with theory can be made of the discontinuity at the Curie point. Experiment shows that the decrease in specific heat takes place over a range in temperature, instead of precipitously. According to various authors, the range in temperature from the maximum to approxi-

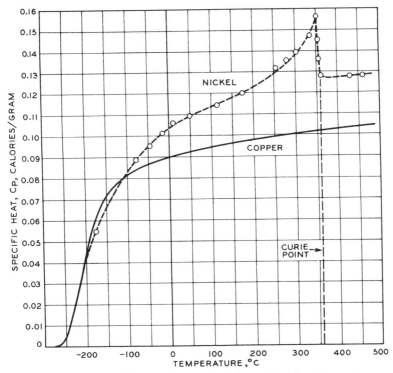

Fig. 15-5. Specific heat of magnetic material (nickel) is higher than a similar nonmagnetic material (copper).

mate minimum varies for nickel from 7 to 50°C, while according to simple theory the drop should be immediate. The drop in specific heat from maximum to minimum, as measured by the same authors is also subject to a considerable amount of variation. The data for iron and cobalt are less certain than those for nickel, but one can say that the changes at the Curie point are not known to be incompatible with the theory for any of the three elements.

Attempts have been made by Lapp [29L1], Stoner [36S4], and Sykes and Wilkinson [38S6] to separate out the magnetic part of the heat capacity of nickel. The observed heat capacity is assumed to have the following components:

Lattice vibration term, C_v.
Thermal expansion term, $C_p - C_v$.
Electronic term, C_e.
Magnetic term, C_m.

It is assumed that the specific heat at constant volume, caused by lattice vibrations, is given by the Debye theory. Its dependence on temperature is given for nickel by the curve C_v of Fig. 15–6, taken from Stoner; this is

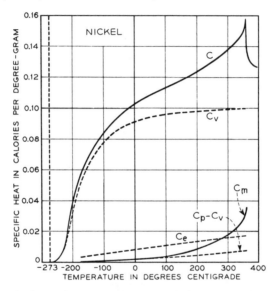

Fig. 15–6. Specific heat of nickel separated into its components: magnetic (c_m), electronic (c_e), lattice term (c_v) and term due to thermal expansion $(c_p - c_v)$.

calculated by well-known methods, using as parameter a characteristic temperature of $\theta_D = 400$. The thermal expansion term, $C_p - C_v$, is calculated from the linear coefficient of thermal expansion α, the volume compressibility β, and the density d, as a function of absolute temperature T:

$$C_p - C_v = (3\alpha)T/\beta d.$$

This term is small, as shown by the curve in the figure.

The electronic term, C_e, is exceptionally large in the ferromagnetic metals, a fact attributed by Mott and Jones [36M13] to the effect upon the heat capacity of the unfilled portions—the holes—of the d-shell. The calculation of its value has been the subject of several investigations, e.g., that of Stoner [39S10], to whom the reader is referred for further discussion. The magnetic term is obtained by subtraction of the three terms C_v, $C_p - C_v$, and C_e, from the observed C_p, and is also shown in the figure. The curve used for C_p is that derived by Sykes and Wilkinson from their own data and those of Klinkhardt [27K2], Lapp [29L1], Grew [34G3], Ahrens [34A1], Moser [36M4], and Ewert [36E2].

It will be noted at once that the magnetic specific heat curve, thus obtained, differs from the theoretical curves of Fig. 15–4 in two important respects: (1) the drop at the Curie point is not precipitous but extends over a range of temperature of about 40°C, as noted previously, and (2) the rising portion of the curve rises more rapidly than the theoretical curves. The explanation of the latter effect is associated with the difference between the experimental I_s, T curve and that calculated by the hyperbolic tangent expression, a difference that gives rise to different values of dI_s/dT for the two cases.

Since $C_m = -(N/2d)dI_s^2/dT$, the molecular field constant N may be calculated from the values of C_m and dI_s^2/dT derived from experiment. Sykes and Wilkinson found in this manner that N has a value of about 12 000 between 200 and 360°C. This compares favorably with values obtained by other methods, but in view of the uncertainties in the evaluation of C_m the results cannot be considered to have great accuracy.

As mentioned before, the decrease in C_m at the Curie point is in fair agreement with the theoretical value for $j = \frac{1}{2}$. However, since just below the Curie point dI_s/dT is greater than that predicted by the tanh law, one would expect that the drop in specific heat would be correspondingly greater than the 0.031 calculated with this law. This expected increase is probably compensated by the "unsharpness" of the maximum, which tends to reduce its height.

Measurements of specific heat have also been made in the vicinity of the Curie point for iron, cobalt, and several nickel-copper, iron-nickel, iron-cobalt-nickel, and nickel-manganese alloys. Of these, an attempt to separate the magnetic portion has been made only for iron, by Lapp [36L2]; she found that the descending portion of the curve covered a range of about 20°C in temperature and 0.079 in specific heat, but she showed that the discontinuity that would occur, if this interval of temperature were narrowed to zero, was close to the theoretical value of 0.118 derived from the Brillouin function using $j = \frac{1}{2}$, and the value 0.121 derived by P. R. Weiss for $j = 1$, but larger than the value 0.073 derived by Weiss for $j = \frac{1}{2}$.

The specific heat of iron is given in Fig. 15–7 for temperatures up to the melting point, according to Lapp, Zuithoff [38Z1], Jaeger, Rosenhohm, and Zuithoff [38J2], and Awberry and Griffiths [40A2]. This figure shows also the changes in heat content at the temperatures of the phase transformations α,γ and γ,δ. Recent data for temperatures between 50 and 200°K have been reported by Kelley [43K4].

Umino's curve [27U1] for *cobalt* shows the characteristic peak associated with magnetic change, at 1150°C, and another corresponding to the α,ϵ-phase transformation at 460°C. At the Curie point the drop in specific heat is estimated from Umino's curve to be about 0.003 cal/gram, a value considerably below that given by theory for $j = \frac{1}{2}$.

Grew's data [34G3] for the *nickel-copper* alloys are of special interest because they show that the electronic specific heat term, C_e, disappears

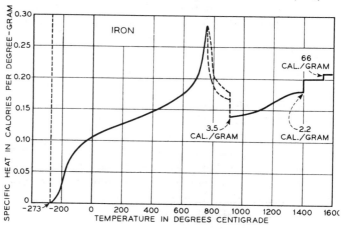

FIG. 15-7. Specific heat and heats of transition of iron as dependent on temperature.

when the copper content is great enough (60%) to fill the $3d$ shell and destroy ferromagnetism.

Kaya and Nakayama [39K3] have measured the specific heat of a number of *iron-nickel* and *iron-cobalt-nickel* alloys in the vicinity of their Curie points and have paid special attention to the changes that occur at the near-by order-disorder transformation temperatures. The energy associated with ordering is greatest at $FeNi_3$ and was not observable at $FeNi$, $FeCo_3$, $CoNi_3$ or Co_3Ni. Substitution of cobalt for nickel in $FeNi_3$ decreases rapidly the irregularity in the specific heat curve but does not influence noticeably the transformation temperature.

A change in the specific heat has been observed near the order-disorder transformation temperature of Ni_3Mn [40T1], 510°C. Some measurements have been made on the Heusler alloys [26S3].

Low Temperatures.—At low temperatures (0–20°K) the electronic specific heat, already referred to, may be separated from the total specific heat by subtracting the Debye heat, for here the magnetic heat is negligible, less than 1% of the total. Data are available for iron [39D2, 39K6, 43K4], cobalt [39D3], and nickel [35K9, 36C3], and the results of

Keesom and Kurrelmeyer for iron are given in Fig. 15–8. As there indicated, a good separation can be made of the Debye term C_p, proportional to T^3, using a characteristic temperature of 462°:

$$C_p = 4.70 \times 10^{-6} T^3,$$

and the electronic term, proportional to T:

$$C_e = 1.20 \times 10^{-3} T,$$

both referred to one gram-atom. The line through the experimental points is

$$C_A = 1.20 \times 10^{-3} T + 4.70 \times 10^{-6} T^3.$$

A calculation of the electronic term has been made for nickel by Stoner [39S10], assuming a simple form of the partially filled $3d$ band of electronic

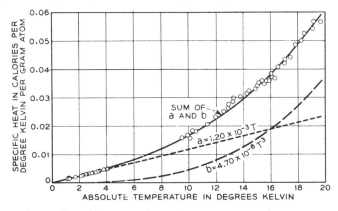

FIG. 15–8. Separation of heat capacity at low temperatures into terms proportional to T (electronic) and to T^3 (Debye).

energy levels. When his method is applied to iron, it is found [39K6] that the calculated ratio of C_e/T is four times too large; this probably means that the true form of the $3d$ band for iron is actually quite different from that assumed. Other evidence also points to such a difference for this element.

Data for cobalt can be described in similar fashion:

$$C_A = 1.20 \times 10^{-3} T + 5.34 \times 10^{-6} T^3$$

in calories per degree per gram-atom, and for nickel the corresponding relation is

$$C_A = 1.74 \times 10^{-3} T + 6.59 \times 10^{-6} T^3.$$

The theoretical value of the coefficient of T for nickel is 2.41×10^{-3}, according to Stoner's calculation [39S10]. The discrepancy between the calculated and observed coefficients for nickel and iron might well be used to obtain more information regarding the shape of the 3*d* band.

Magnetocaloric Effect.—Weiss and Piccard [18W1] showed that sudden application of a high field will increase the temperature of a ferromagnetic material, and that the amount of the increase is a measure of the molecular field. The thermodynamic relation, on which the measurements depend, is derived from the expression for the internal energy of a magnetic material (see p. 731):

$$dU = dQ + dA = TdS + HdI.$$

If we take S and H as independent variables, double differentiation gives

$$\frac{\partial^2 U}{\partial H \partial S} = \left(\frac{\partial T}{\partial H}\right)_S = -\left(\frac{\partial I}{\partial S}\right)_H = -\left(\frac{\partial T}{\partial S}\right)_H \left(\frac{\partial I}{\partial T}\right)_H.$$

By definition the heat capacity at constant field strength is

$$c_H = \left(\frac{\delta Q}{\delta T}\right)_H = T\left(\frac{\partial S}{\partial T}\right)_H.$$

Combining this with the preceding equation, we have

$$c_H = -T\left(\frac{\delta H}{\delta T}\right)_S \left(\frac{\delta I}{\delta T}\right)_H$$

or

$$(\Delta T)_S = -\frac{T}{c_H}\left(\frac{\partial I}{\partial T}\right)_H \Delta H.$$

Thus when H is applied adiabatically ($\Delta S = 0$), the resulting change in temperature depends on ΔH, T, c_H, and $(\partial I/\partial T)_H$. $(\partial I/\partial T)_H$ is well known to be negative and to be a maximum near the Curie point. The thermodynamic relation then leads us to expect an *increase* in temperature with application of a field, the increase to be greatest at temperatures near $T = \theta$. In nickel, Weiss and Forrer [26W4] found a maximum increase in temperature of about 1°C, upon application of a field of 10 000 to 20 000 oersteds, as depicted in Fig. 15-9. Potter's data for iron [34P2], given in the same figure, show an increase in temperature of the same order of magnitude.

If the foregoing expression for dU is differentiated with respect to S and I, the change in temperature may be expressed in terms of ΔI instead of ΔH:

$$(\Delta T)_S = \frac{T}{c_I}\left(\frac{\partial H}{\partial T}\right)_I \Delta I.$$

Here c_I is the heat capacity (per unit volume) at constant magnetization.

Measurements of the magnetocaloric effect are useful in determining the value of the molecular constant in the Weiss theory. This theory may be written in the following form, without stating explicitly the nature of the function L:

$$I_S/I_0 = L\left[\frac{\mu(H + NI)}{kT}\right].$$

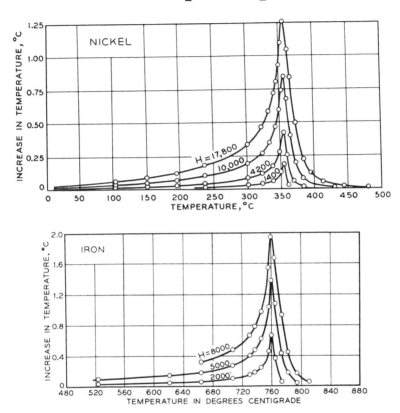

FIG. 15–9. Magnetocaloric effect in nickel and iron.

We may evaluate $(\partial H/\partial T)_I$, by differentiating with respect to H and T, holding I constant:

$$0 = \frac{\mu k T \partial H - \mu k(H + NI)\partial T}{k^2 T^2} \cdot L'.$$

Therefore

$$\left(\frac{\partial H}{\partial T}\right)_I = \frac{H + NI}{T}$$

and substitution in the thermodynamic expression gives:

$$\Delta T = \frac{H + NI}{c_I} \Delta I = \frac{(N + H/I)}{2c_I} \Delta I^2.$$

Weiss and Forrer, and Potter, have used this relation for the determination of N by measuring ΔT and ΔI at various temperatures. They found, as expected, that at any one temperature ΔT was proportional to the square

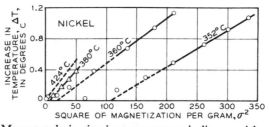

FIG. 15–10. Magnetocaloric rise in temperature is linear with square of specific magnetization, σ^2, at high magnetizations. Data for nickel.

of the magnetization I^2, as long as the field was high enough to effect saturation by orienting the domains parallel to the field. Data for several temperatures are shown for iron and for nickel in Figs. 15–10 and 11.

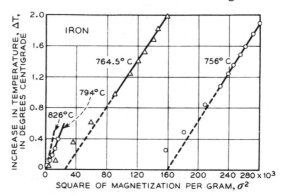

FIG. 15–11. Magnetocaloric rise in temperature as function of σ^2 in iron.

The linearity of the curves at high fields shows that, at any one temperature, N is independent of I_s. At various temperatures, however, Potter has shown that, for iron, N assumes different values when calculated from the relation:

$$N = \frac{2c_I \Delta T}{\Delta I^2},$$

valid when $NI \gg H$. Consideration of the data of Weiss and Forrer shows that the molecular field constant of nickel also varies with temperature. Potter's curves for both iron and nickel are shown in Fig. 15–12. In both elements the change is an increase that occurs most noticeably in a range of temperature 50–100°C immediately above the Curie point.

Kornetzki [35K2] has pointed out that a change in N is to be expected near the Curie point, where there is an anomaly in the thermal expansion vs temperature curve. The abnormal change in interatomic spacing with temperature will then cause a change in the energy of exchange interaction, and this will be shown by a change in the molecular field constant. The anomaly in the thermal expansion of nickel is a decrease in expansion at the Curie point. The exchange-interaction curve (Fig. 10–13) shows that a decrease in interatomic distance causes an increase in the energy of exchange. Consequently, an increase in the molecular field constant is to be expected when nickel is heated through the Curie point. For iron the anomaly in expansion is an increase, and the exchange curve indicates a

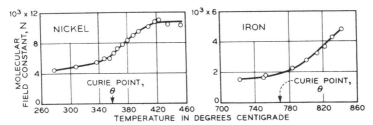

Fig. 15-12. Change in molecular field constant N (coefficient of I) in nickel and in iron.

corresponding increase in molecular field constant—the indicated change in the constant is thus an increase for iron as well as for nickel, in agreement with experiment.

Measurements of magnetocaloric increase in temperature may be used to evaluate the specific heat at constant field, c_H. Weiss and Forrer [24W5] have shown that the c_H vs T curve thus determined has a break at $T = \theta$, and the decrease in c_H at this temperature is about the same as that in the curve determined by direct measurement.

Temperature Change and Torque.—Akulov and Kirensky [40A1] have observed a new magnetocaloric effect related to the change of direction of magnetization with respect to the crystallographic axes in a single crystal. When a crystal is rotated adiabatically in a strong field, it will heat or cool, depending on the way in which the orientation is changed. If the single crystal is cut in the form of a disk, the torque L on the disk will depend on the angle θ by which it is turned in its own plane. Then the change in energy associated with the torque is

$$dA = L d\theta.$$

This quantity corresponds to the HdI of the classical magnetocaloric effect and may be used for a new expression for dU. This may then be differentiated with respect to S and θ, with the result:

$$(\Delta T)_S = \frac{T}{c}\left(\frac{\partial L}{\partial T}\right)_\theta \Delta\theta.$$

The dependence of L on temperature is determined from the following well-known relations:

$$A = K(\alpha_1^2\alpha_2^2 + \alpha_2^2\alpha_3^2 + \alpha_3^2\alpha_1^2)$$
$$K = K_0 e^{-aT^2},$$

in which K_0 and a are constants [37B5] independent of temperature. These combine with the relation $L = dA/d\theta$ to give:

$$\Delta T = -\frac{2aT^2K}{c}\frac{d}{d\theta}(\alpha_1^2\alpha_2^2 + \alpha_2^2\alpha_3^2 + \alpha_3^2\alpha_1^2).$$

Measurements on a nickel sphere show that the temperature changes in the expected way with orientation, except that the magnitude of the change is less than the theoretical value by a factor of 2.4. The greatest change in temperature, observed at the temperature of liquid nitrogen, was about 0.005°C.

CHAPTER 16

MAGNETISM AND ELECTRICAL PROPERTIES

It has been about 100 years since William Thomson [1857T1] (Lord Kelvin) showed that the electrical resistivity of iron changes when it is magnetized. In most magnetic materials this "magnetoresistance" is an increase of resistivity with magnetization when the current and magnetization are parallel and a decrease when they are at right angles to each other. At temperatures near the Curie point, however, magnetization in any direction with respect to the current is accompanied by a decrease in resistivity.

The magnitude of the change in resistivity caused by magnetization to saturation is usually a few per cent and rarely exceeds 5% at room temperature. At the temperature of liquid air 16% has been observed.

The change of resistivity is independent of the sense of field or induction and, in this respect, is like magnetostriction. Resistivity, as well as length, is therefore unaffected by the 180° reversal of a domain, and the greatest changes in these properties occur in the upper portion of the magnetization curve when changes in magnetization progress by domain rotations. Closely related to the change of resistivity with field strength (magnetoresistance) is its change with stress (elastoresistance).

Resistivity and Field Strength (Magnetoresistance).—As the field applied to a magnetic material increases in strength, the resistivity, measured with the current parallel or antiparallel to the field, rises and approaches a "saturation" limit. The rise is initially much slower than that of the magnetization curve, and the intensity of magnetization attains $\frac{3}{4}$ or $\frac{9}{10}$ of its saturation value before the resistivity has changed more than half of its ultimate amount. In Fig. 16–1 the data for 84 Permalloy [30M1] illustrate this point.

When the applied field is reduced to zero, the resistivity is still somewhat greater than its value when the specimen is unmagnetized. Let us designate the resistivity when unmagnetized as ρ, the increase in resistivity $\Delta\rho$, its value at saturation $\Delta\rho_h$, and at residual induction $(B = B_r)$ as $\Delta\rho_r$. It is observed that $\Delta\rho_r$ is a small fraction of $\Delta\rho_h$, whereas B_r is usually about $\frac{1}{2}$ of B_s. In the hysteresis loops for nickel [30S5] shown in Figs. 16–2 and 3, $\Delta\rho_r$ is only 10–15% of $\Delta\rho_h$, and in many of the iron-nickel alloys [30M1] it is small enough to escape observation. In all of these

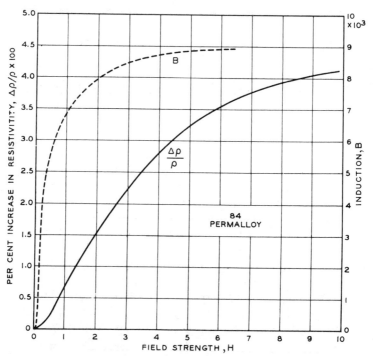

FIG. 16–1. Increase in resistivity with magnetization in 84 Permalloy. Most of increase occurs when magnetization is beyond the knee of the magnetization curve.

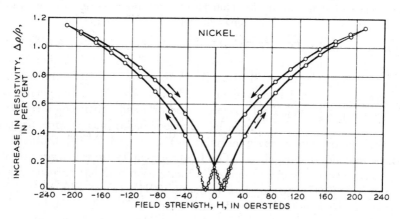

FIG. 16–2. Hysteresis of resistivity as plotted against field-strength.

materials, and in iron, $\Delta\rho$ is very close to zero when B is reduced to zero by applying a field in the reverse direction ($H = H_c$).

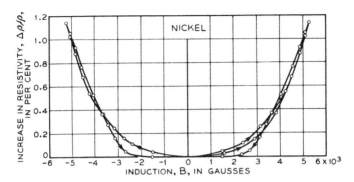

FIG. 16–3. Hysteresis of resistivity as plotted against induction.

The saturation value $\Delta\rho_h$ is best determined by plotting $\Delta\rho$ against $B-H$ or $(B-H)^2$, as in Fig. 16–4, and extrapolating to $B-H = B_s$.

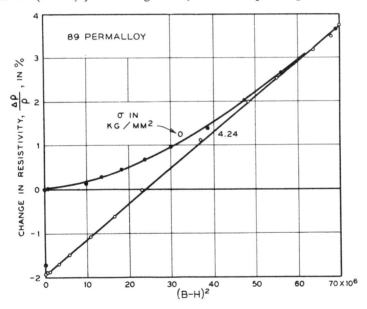

FIG. 16–4. In 89 Permalloy under tension, increase in resistivity increases linearly with square of magnetization (rotation process).

The upper part of the curve then approaches a straight line, and when $(B-H)^2$ is used as abscissa the curve is often straight over an extended region. Straight-line portions have been observed especially in materials with high internal strain [32G4], or under axial tension, as illustrated in Fig. 16–4.

The similarity in the phenomena of *magnetostriction* and change of resistivity has been known for many years [15H1]. Recent data by Kornetzki [43K2] bring out the linear relation between $\Delta l/l$ and $\Delta\rho/\rho$ for specimens of nickel and iron (Figs. 16–5 and 6).

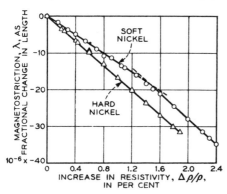

FIG. 16–5. Relation between magnetostriction and change of resistance in nickel.

When the current used to measure the resistivity is at *right angles* to the applied field, the resistivity usually *decreases* with increasing field. This transverse effect also has a limiting value of $\Delta\rho$, designated $\Delta\rho_t$. When using the high applied fields of over 15 000 oersteds, shown in Fig. 16–7, it is observed that both the longitudinal and the transverse resistivities begin to decrease with increasing field. Such an effect is attributed to the increase in spontaneous magnetization caused by these intense fields, and it is especially pronounced in its effect when the temperature of the specimen is near the Curie point; this will be discussed in some detail later. In determining $\Delta\rho_t$, account must be taken of the large demagnetizing field usually accompanying transverse magnetization. In the experiment of Fig. 16–7 the specimen was a wire; consequently at saturation the demagnetizing field is $B_s/2$, and the effect of change in spontaneous magnetization extrapolates to zero when the applied field H_a is equal to $B_s/2$. The value of $\Delta\rho_t/\rho$ thus corresponds to the point P for transverse magnetization, and $\Delta\rho_h/\rho$ to the point Q for longitudinal magnetization, for which the demagnetizing field is zero.

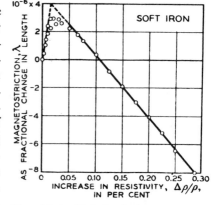

FIG. 16–6. Relation between magnetostriction and change of resistance in iron.

Effect of Tension.—The change in resistivity with magnetization is associated with domain orientation; such orientation is also effected by strain, and it is not surprising, therefore, to find that the resistivity of a magnetic material is changed by application of unidirectional tension or

EFFECT OF TENSION 749

compression. This change is often called "elastoresistance." Whereas longitudinal magnetization normally increases resistivity, longitudinal tension will either increase or decrease the resistivity, depending on

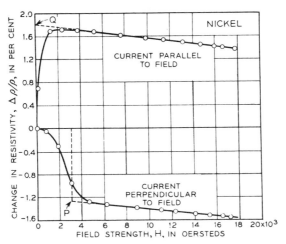

FIG. 16–7. Change of resistance of nickel in longitudinal and in transverse fields [32E1].

whether the magnetostriction is positive or negative. The effects of both field and tension on 89 Permalloy (negative magnetostriction) and

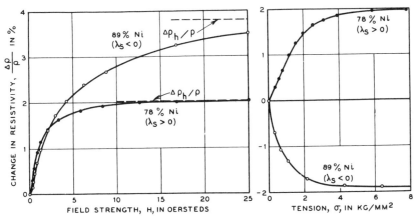

FIG. 16–8. Increase of resistance with magnetic field; and increase or decrease with tension depending on the sign of magnetostriction (positive for 78 Permalloy, negative for 89 Permalloy).

78 Permalloy (positive magnetostriction) are shown for comparison in Fig. 16–8. In 78 Permalloy the resistivity is increased by both field and tension to the same limiting value.

The effects of tension and field may be explained qualitatively by *domain theory* using the diagrams of Fig. 16–9. Domains having an initially haphazard distribution of orientations, shown schematically at (a), are aligned parallel by a strong field (b). Tension applied to an unmagnetized material with positive magnetostriction orients the domains along the axis of tension with half of the domains antiparallel to the others (c). Since the resistivity depends on magnetization and increases with increasing magnetization, each domain will contribute to the resis-

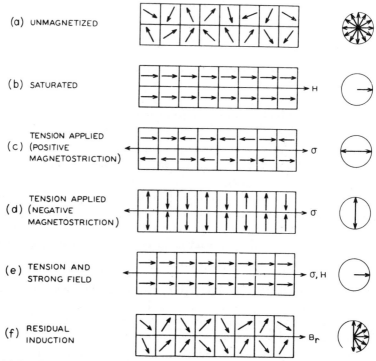

Fig. 16–9. Domain interpretation of effects of field and tension on resistance. At right, domain vectors have been moved to common origin.

tivity an amount depending on its orientation. This contribution will be the same for parallel and antiparallel orientations, and, for the specimen as a whole, $\Delta\rho$ will be the same for (b) and (c) and will be $\Delta\rho_s$. When magnetostriction is negative, tension orients the domains at right angles to the direction of tension and the resistivity decreases (d). A field applied subsequently causes the domains to rotate so as to increase the resistivity (d to e). When a material having positive magnetostriction is subjected to strong tension, subsequent application of a field will cause 180° rotations almost exclusively, and these will have no effect on the resistivity (c to e).

EFFECT OF TENSION

Experiments on materials with positive and negative magnetostriction (69 and 89 Permalloy), subjected to tension and magnetic field, are sum-

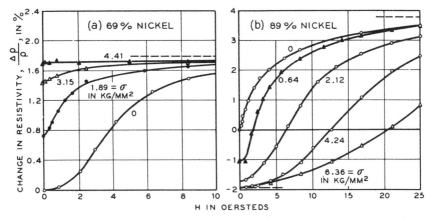

FIG. 16–10. Effect of magnetic field on resistivity of 69 and 89 Permalloys to which various tensions have been applied.

marized in Fig. 16–10(a) and (b) and confirm the theory. Data for the same specimens are plotted in Fig. 16–11 against ferric induction, $B-H$;

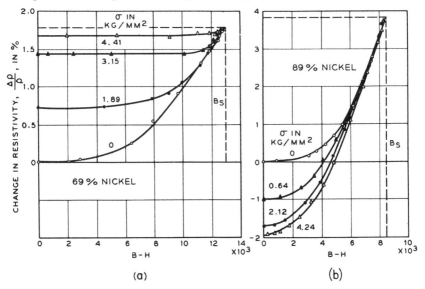

FIG. 16–11. Data like those of preceding figure, plotted against $B-H$ instead of H.

the curves for higher tension show that in 69 Permalloy there is a large change in magnetization without any appreciable change in resistivity, in accordance with Fig. 16–9(c) and (e). It is presumed that this material when heat-treated in a longitudinal field will also show little if any change of ρ with H. The change of ρ with $B-H$ for 89 Permalloy also corresponds to the theory.

Data for other alloys, however, show that $\Delta\rho_s$ and $\Delta\rho_\sigma$ are not equal (see Fig. 16–15). This may be attributed [46B1] to anisotropy in the magnetostriction. Lichtenberger [32L1] has shown that λ_s is sometimes positive, sometimes negative, depending on the crystallographic direction of the field used for measurement. In alloys having between 35 and 45% nickel the saturation magnetostriction (λ_s) is positive in the [111] direction (λ_{111}) and negative in the [100] direction (λ_{100}). In a polycrystalline specimen tension will therefore increase the resistivity of those crystals that have [111] directions lying near the axis of tension and decrease it in those having [100] near the axis. The resultant $\Delta\rho_\sigma$ will thus be less than $\Delta\rho_h$. Similarly Lichtenberger's curves show that λ_{111} and λ_{100} are opposite in sign between about 80 and 85% nickel, and this range may be expected to vary with heat treatment, which affects the amount of atomic order [40G1]. In other alloys, except those near 60 and 85% nickel, λ_{111} and λ_{100} are substantially different and prevent σ from causing complete alignment of all the domains.

Quantitative Aspects of Domain Theory. Polycrystalline Material.— The data of Fig. 16–10(b) are particularly adapted to treatment by domain theory. In Fig. 16–12 let A represent the orientation of a domain before application of field or stress, a stable position determined by the crystal structure and internal strain of the material. Since the magnetostriction is negative, a tension σ will cause rotation of the domain to some position B. A superposed field will then cause rotation in the opposite sense, and at some field strength H the domain will again take up orientation A. If the domain were oriented originally at A', it would change to B' with stress, and the field would cause it to invert and rotate to the same final position A. Since positions A and A' correspond to the same resistivity, the resistivity of the material will return to its original value when σ and H are exactly opposing. The relation between σ and H for this condition is calculable by domain theory as follows:

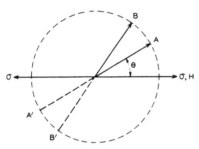

Fig. 16–12. Interpretation of preceding figures: stress σ rotates domain from A to B, and field H_0 rotates it back to A.

The energy of a domain due to stress (see Chap. 18) is

$$E_\sigma = \tfrac{3}{2}\lambda_s\sigma \sin^2 \theta$$

where λ_s is the saturation magnetostriction and θ the angle between the magnetization I_s and the stress σ (tension). The energy due to the field is

$$E_H = -HI_s \cos\theta.$$

The forces due to H and σ are equal when

$$(d/d\theta)(E_\sigma + E_H) = 0$$

or

$$\frac{H}{\sigma} = -\frac{3\lambda_\sigma \cos\theta}{I_s}.$$

The average value of $\cos\theta$ for domains oriented at random from $\theta = 0$ to $\theta = \pi/2$ is $\tfrac{1}{2}$. Thus

$$\frac{H}{\sigma} = -\frac{3\lambda_s}{2I_s}.$$

For 89 Permalloy, $\lambda_s = -13 \times 10^{-6}$ and $4\pi I_s = 8300$, and one calculates

$$\frac{H}{\sigma} = 3.0 \text{ oersted kg}^{-1} \text{ mm}^2.$$

In Fig. 16–13 the line is drawn with this theoretical slope, and the points represent the values of σ and H at which the curves of Fig. 16–10(b) cross the horizontal axis ($\Delta\rho = 0$). Domain theory is thus well supported by experiment. Other features of the curves of Fig. 16–10 can also be treated by domain theory, but first certain other elements of the theory must be considered.

The resistivity of a domain depends on the direction of its magnetization with respect to the electric current used for measurement and also on its orientation with respect to the crystal axes. The relation to the crystal axes will be discussed in a later section; here we are concerned only with variations in the angle between current and magnetization. The relation between magnetoresistance and orientation may then be expressed by a series of which we use only the first two terms:

$$\Delta\rho = A + B\cos^2\theta,$$

θ being the angle between the current i and the magnetization of the domain I_s. Odd powers of $\cos\theta$ are absent, to take account of the symmetry of the effect.

We may refer the zero of change of resistivity to the state in which the domains are distributed uniformly in all directions and for which $\overline{\cos^2\theta} = \tfrac{1}{3}$. The increase of resistivity from this particular state by magnetization will have its greatest value $\Delta\rho_h$ when $\theta = 0$. These two values determine the constants A and B and the foregoing relation becomes

$$\Delta\rho = \tfrac{3}{2}\Delta\rho_h(\cos^2\theta - \tfrac{1}{3}).$$

When $\theta = 90°$, $\Delta\rho = \Delta\rho_t = -\Delta\rho_h/2$; the change caused by transverse magnetization to saturation is thus half of $\Delta\rho_h$ and of opposite sign.

In many materials the distribution of domains in the unmagnetized state is *not at random*. Experimentally one can measure the change in resistivity in going from the demagnetized state to saturation with i and H

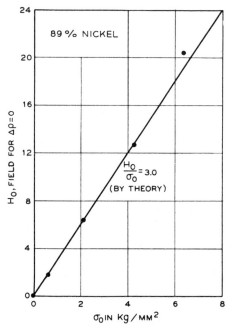

FIG. 16–13. Comparison of theory (line) and experiment (points) relating applied tension σ_0 and field H_0 necessary to bring resistivity back to value it had before application of tension. Material: 89 Permalloy.

parallel, and again with i and H at right angles. The change in resistivity may be expressed in terms of such longitudinal and transverse changes, $\Delta\rho_h$ and $\Delta\rho_t$:

$$\Delta\rho = \Delta\rho_t + (\Delta\rho_h - \Delta\rho_t) \cos^2 \theta.$$

The *important constant* of the material is then $\Delta\rho_h - \Delta\rho_t$, since this is independent of the initial configuration and measures the maximum amount that ρ will vary as the domain configuration changes.

The fact that no terms higher than that in $\cos^2 \theta$ are required in the series expansion given previously may be deduced from the data of Fig. 16–4 as follows: When a substantial tension is applied to a material with negative magnetostriction, the domains are aligned transversely to the axis of tension, and when a field is subsequently applied and the domains rotate, the induction is proportional to $\cos \theta$:

$$(B - H) = B_s \cos \theta.$$

If the accompanying change in resistivity is proportional to $\cos^2 \theta$, $\Delta \rho$ is proportional to $(B-H)^2$. That this is the case is confirmed by experiment (Fig. 16-4), with no apparent deviation of the kind that requires a $\cos^4 \theta$ term for the description of the behavior of this material, and as far as is known of any material.

The change in resistivity obtained in this same experiment may be plotted against H^2 instead of $(B-H)^2$ and compared with the predictions of the domain theory. Writing the expressions for the energies of strain and magnetization and putting $dE/d\theta = 0$, we again obtain the relation

$$\frac{H}{\sigma} = \frac{3\lambda_s \cos \theta}{I_s}.$$

The change in resistivity with orientation is given by the simple expression

$$\Delta \rho = \Delta \rho_h \cos^2 \theta,$$

provided $\Delta \rho$ is taken as zero when σ is already applied and $H = 0$. These expressions combine simply to give

$$\frac{\Delta \rho}{\Delta \rho_h} = \frac{I_s^2 H^2}{9 \lambda_s^2 \sigma^2}$$

which reduces to

$$\frac{\Delta \rho}{\rho} \times 100 = 4.3 \times 10^{-3} H^2$$

when known numerical values for 89 Permalloy are used. The solid line of Fig. 16-14 is drawn accordingly, and the observed points there shown correspond reasonably well to the theory. Deviations may be expected when H is large and $\Delta \rho$ approaches near to $\Delta \rho_s$, because the restraining forces due to crystal structure and internal strain have been neglected in the theory. When H is small, these forces will tend to increase ρ above the theoretical value given above, and when H is large they will cause the observed points to fall below the line, as they are observed to do.

The linearity between $\Delta \rho$ and $(B-H)^2$, shown in Fig. 16-4, is a result of domain rotation, and it is to be expected that this relation will hold generally in the region between $(B-H)/B_s = 0.5$ and 1.0 [32G4], even when no tension is applied. Nearer the beginning of the magnetization curve the changes in magnetization in a well-annealed material take place by boundary displacement between domains oriented (in iron) 90° or 180° from each other. When such changes occur, $\Delta \rho / \rho$ is not expected to be proportional to $(B-H)^2$ but will change less rapidly with $B-H$. Consequently, the relation

$$\frac{\Delta\rho}{\rho} = a + b(B-H)^2,$$

is to be expected and is confirmed by experiment [32G4].

Domain theory of single crystals and its relation to polycrystals are discussed in the second section below.

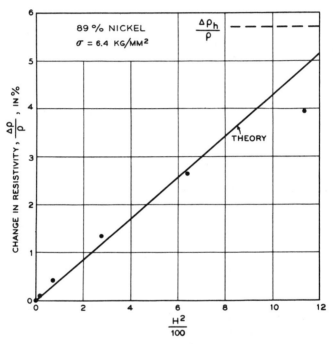

Fig. 16–14. Change of resistivity with field in 89 Permalloy subject to high elastic tension. Theoretical relation is followed except at the higher fields.

Results for Alloy Series.—Variation, with field and stress, of the resistivities of the iron-nickel alloys was studied in detail by McKeehan [30M1]. He established the general courses of the curves, and these showed $\Delta\rho_s$ (change of resistivity with applied field) to have its highest value when the nickel content is somewhat greater than 80%. He also pointed out the close connection between the change of resistivity with tension and the magnetostriction of the alloy and interpreted the changes in terms of the orientations of atomic magnets.

More recent data by Shirakawa [39S2], Yamanaka [40Y1], and Bozorth [46B1] are plotted in Fig. 16–15. The curves show the maximum fractional increase caused by a longitudinal magnetic field, $\Delta\rho_s/\rho$, and the maximum fractional change caused by tension $\Delta\rho_\sigma/\rho$.

The simple theory already discussed indicates that when the magnetostriction is positive (Ni < 81%), $\Delta\rho_h/\rho$ should equal $\Delta\rho_\sigma/\rho$. Deviations from this simple theory occur when the magnetostriction is anisotropic,

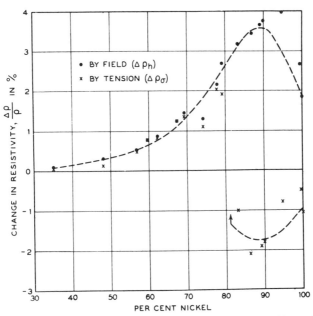

Fig. 16–15. Limiting effects of field and tension on resistivity of iron-nickel alloys. Broken line shows changes expected on simple theory.

and the inequality of $\Delta\rho_h$ and $\Delta\rho_\sigma$ is apparent in Fig. 16–15 for alloys which are known to have such anisotropy. When magnetostriction is negative (Ni > 81%), simple theory predicts that

$$\Delta\rho_\sigma/\rho = -\tfrac{1}{2}\Delta\rho_h/\rho,$$

provided that the domains in the unmagnetized specimens are distributed at random. The light broken lines indicate the curves to be expected on this basis.

Because it is independent of the initial domain distribution, a more fundamental quantity to evaluate is $\Delta\rho_h - \Delta\rho_t$, the difference between the changes due to longitudinal and to transverse fields. Measurements of this quantity are shown by the curve of Fig. 16–16. The broken lines of Fig. 16–15, already discussed, are derived from the data of Fig. 16–16 by putting $\Delta\rho_h = \tfrac{2}{3}(\Delta\rho_h - \Delta\rho_t)$, valid for initially random distribution, or by putting $\Delta\rho_\sigma = \Delta\rho_h$ or $\Delta\rho_t$, depending on whether the magnetostriction is positive or negative. It is noted that in contrast to Fig. 16–15, the points of Fig. 16–16 follow closely the curve drawn.

The points marked by crosses are for material annealed for 24 hours at 425°C, and they indicate that $\Delta\rho_h$ is changed by atomic ordering in the alloys near 70% nickel.

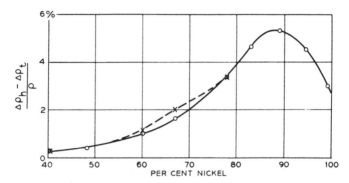

Fig. 16–16. Difference between changes in resistivity caused by longitudinal and transverse magnetic fields, plotted against per cent nickel.

Data [39S2] for $\Delta\rho_h$ measured by Shirakawa at temperatures above and below room temperature are shown in Fig. 16–17.

Observations on the change of resistivity with magnetization have been made on Fe-Si [39S9], Fe-Co [39S3], Co-Ni [36S18], and Ni-Cu [36M12]

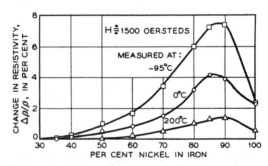

Fig. 16–17. Magnetoresistance of iron-nickel alloys at various temperatures.

alloys, and on Fe_3O_4 [41M2]. Data are reproduced in Figs. 16–18 to 21. In Fe-Si alloys and Fe_3O_4 a longitudinal field *decreases* the resistivity, at some temperatures far below the Curie point. Data for Fe_3O_4 refer to a single crystal cut parallel to a [111] direction. At 0°C the decrease was 0.3% when $H = 1000$, and at lower temperatures the decrease was less.

Some studies of the magnetoresistance of permanent magnets have been made by Bates [46B7], Drozhjina, Luzhinskaya, and Shur [48D4, 49D6],

RESULTS FOR ALLOY SERIES

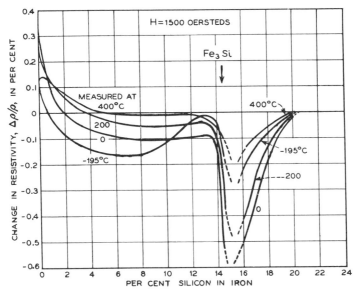

Fig. 16–18. Magnetoresistance of iron-silicon alloys.

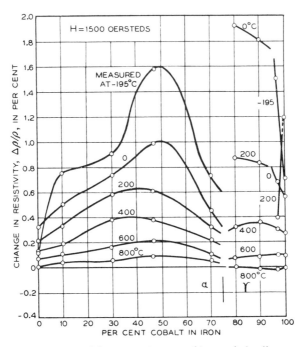

Fig. 16–19. Magnetoresistance of iron-cobalt alloys.

and Vonsovsky [48V1]. All of the measurements show a decrease of resistivity with magnetization when the measuring current is either parallel or perpendicular to the magnetization, in contrast to the usual behavior (Fig. 16–7).

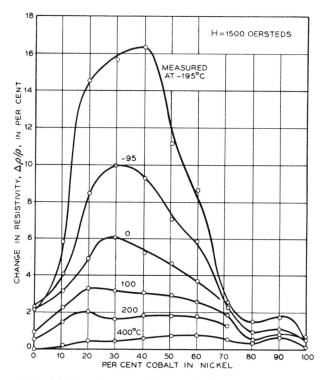

FIG. 16–20. Magnetoresistance of cobalt-nickel alloys.

Effect of Temperature.—Results of measurement on iron-nickel alloys (Fig. 16–17) illustrate the general rule that $\Delta\rho_s/\rho$ decreases as the temperature increases. At the temperature of liquid air the changes are relatively large, and Shirakawa [39S2] has observed an increase of resistivity of over 12% in a 90% nickel-iron alloy, and 16% in a cobalt-nickel alloy. In nickel Gerlach [33G4] has found that $\Delta\rho_h$ decreases with temperature at the same rate as B_s^2 decreases, but this relation does not appear to hold generally.

When the temperature is raised so that it approaches the Curie point, a new phenomenon appears, as shown for nickel in Fig. 16–22, taken from the work of Gerlach and Schneiderhan [30G5]. Similar data have been reported by Matuyama [34M11]. In sufficiently high fields the resistivity passes through a maximum and declines, and when the temperature is

EFFECT OF TEMPERATURE 761

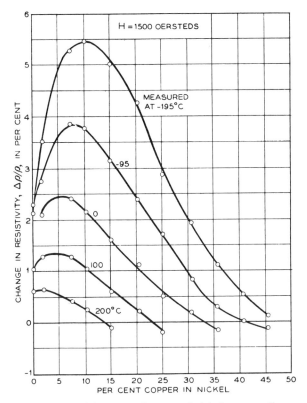

Fig. 16–21. Magnetoresistance of nickel-copper alloys.

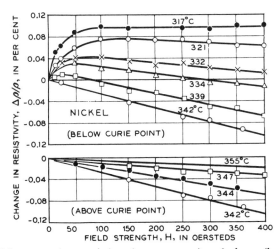

Fig. 16–22. Magnetoresistance below (upper curves) and above (lower curves) the Curie point.

within a few degrees of the Curie point a moderate field may be sufficient to cause the resistivity to decline to its original value. Just at the Curie point the maximum disappears and ρ decreases almost linearly with the field. Above the Curie point the decrease with field is less rapid.

Gerlach [30G6] has shown that this decrease of resistivity is associated with an increase in the intrinsic magnetization caused by the increasing field. The data of Gerlach and Schneiderhan [30G5] for nickel (Fig. 16-23), and similar data for iron and Heusler alloy by Potter [32P2], support this conclusion by showing that the slope of the $\Delta\rho$ vs H curve is a maximum at the Curie point; it is at this temperature that the intrinsic magnetization is most strongly affected by the field. Further evidence of the role of spontaneous magnetization is the linearity between $\Delta\rho$ and $(B-H)^2$; Englert's data [32E1,32E4] confirm this relation, and also Potter [32P2] found $\Delta\rho = aH^2$ immediately above the Curie point, when $(B-H)$ is proportional to H, in iron and Heusler alloy. In a transverse field, also, ρ decreases with increasing H, in accordance with theory [31P4, 32E5].

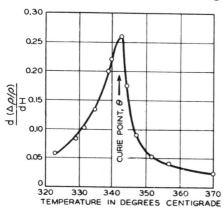

FIG. 16-23. Change of resistance per unit field in the neighborhood of the Curie point. Nickel.

As a result of the decrease of ρ with H near the Curie point, the $\Delta\rho$ vs T curve has a characteristic sharp dip below the temperature axis 50–100°C below the Curie point, before it rises finally to zero [05W1, 39S2].

The resistivity of a ferromagnetic material is abnormally low by virtue of the fact that it *is* ferromagnetic. This is well shown, as pointed out by Becker and Döring [39B5], by comparison of the resistivities of nickel and palladium, elements occupying similar positions in the periodic table (see Fig. 16-24). Data of Potter [37P1] and Conybeare [37C5] are replotted so that the curves for the two elements coincide at the Curie point of nickel. The separation of the curves at lower temperatures is attributed to the ferromagnetism of nickel, and theoretical calculations of Mott confirm this, as follows:

According to the band picture of the distribution of electrons in the atoms of the iron group, conduction takes place mainly by electrons in the s bands. These are disturbed by frequent transitions to the unfilled d bands, and as a result the resistivity of nickel is abnormally high as compared with the next element, copper, in which the d bands are filled.

However, when ferromagnetism occurs, one of the two d bands is filled; therefore transitions to it do not occur, and the scattering is thus cut in half. As a result, the resistivity of nickel below the Curie point is less

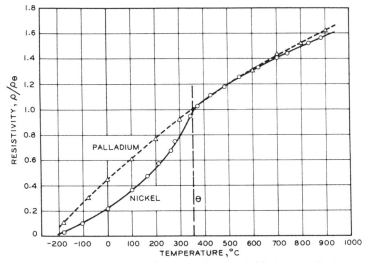

FIG. 16–24. The low resistivity of a magnetic material below the Curie point.

than that of an element, like palladium, which also has unfilled inner shells but is not ferromagnetic. When the temperature increases, the intrinsic

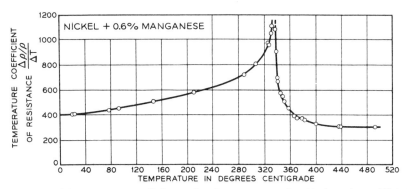

FIG. 16–25. Temperature coefficient of resistance near the Curie point. Nickel.

magnetization of nickel decreases, and its resistivity will approach more nearly the curve for a hypothetical non-magnetic nickel and will coincide with it at or above the Curie point.

By making certain simplifying assumptions regarding the nature of the bands, Mott [36M11] has calculated the effect of loss of intrinsic magnetization on resistivity, and his points are shown as circles in Fig. 16–24. Agreement with experiment is good.

The form of the ρ vs t curve may be useful in determining the Curie point. It is convenient then to plot the slope of this curve or the true temperature coefficient against the temperature. This has a sharp maximum at $t = \theta$, as shown in Fig. 16–25, taken from Gerlach [33G4]. He has pointed out that this curve is similar to that for specific heat; in both curves the ordinate drops to the "normal" value for a non-magnetic material only when the temperature is some tens of degrees about the Curie point proper.

Magnetoresistance in Single Crystals.—Variations of resistivity with fields in different crystallographic directions were first investigated in iron and nickel by Webster [26W3, 27W3, 30W3] and Kaya [28K1], and later by Döring [38D5] and Shirakawa [40S1, 40S2]. Domain theory was applied by Gans and von Harlem [33G5] and later extended by Döring [38D5] whose treatment is followed here. A less complete theory was previously proposed by Akulov [30A3].

Let α_1, α_2, and α_3 be the direction cosines determining the orientation of the magnetization I_s of a domain, and β_1, β_2, and β_3 those defining the direction of the current i used for measuring the resistance. The general expression for the magnetoresistance in any direction of a cubic crystal may then be written in series form, using no powers of α and β higher than the second. Döring has used the following form:

$$\frac{\Delta\rho}{\rho} = k_1(\alpha_1^2\beta_1^2 + \alpha_2^2\beta_2^2 + \alpha_3^2\beta_3^2 - \tfrac{1}{3})$$
$$+ 2k_2(\alpha_1\alpha_2\beta_1\beta_2 + \alpha_2\alpha_3\beta_2\beta_3 + \alpha_3\alpha_1\beta_3\beta_1)$$
$$+ k_3(s - \tfrac{1}{3})$$
$$+ k_4(\alpha_1^4\beta_1^2 + \alpha_2^4\beta_2^2 + \alpha_3^4\beta_3^2 + 2s/3 - \tfrac{1}{3})$$
$$+ 2k_5(\alpha_1\alpha_2\beta_1\beta_2\alpha_3^2 + \alpha_2\alpha_3\beta_2\beta_3\alpha_1^2 + \alpha_3\alpha_1\beta_3\beta_1\alpha_2^2)$$

in which $s = \alpha_1^2\alpha_2^2 + \alpha_2^2\alpha_3^2 + \alpha_3^2\alpha_1^2$. This expression is for a crystal, like nickel, in which in the initial state ($\Delta\rho = 0$) the domains are distributed equally among the eight [111] directions. Although the α's appear here in the fourth powers, these terms can be transformed, using the identity $\alpha_1^2 + \alpha_2^2 + \alpha_3^2 = 1$, to terms involving only second powers. For a crystal, like iron, having six [100] directions of easy magnetization, the expression is the same except that the term $k_3/3$ is absent.

For the longitudinal effect in the [100] direction in nickel ($\beta_1 = 1$), at saturation ($\alpha_1 = 1$) the expression reduces to

$$\frac{\Delta\rho_h}{\rho} = 2k_1/3 - k_3/3 + 2k_4/3,$$

and for transverse magnetization to saturation ($\alpha_1 = 0, \alpha_2 = \cos \varphi, \beta_1 = 1$) the corresponding expression is

$$\frac{\Delta \rho_t}{\rho} = \left(-\frac{k_1}{3} - \frac{5k_3}{24} - \frac{k_4}{4}\right) - \left(\frac{k_3}{8} + \frac{k_4}{12}\right) \cos 4\varphi,$$

dependent on the azimuth of the field.

The measurements of Kaya on the longitudinal and transverse magnetoresistance along the three principal crystallographic directions can be used for evaluation of the five constants of Döring's expression. Since the state of the specimen used for measurement may not correspond to the ideal state assumed, with domains distributed equally in the various directions of easy magnetization, sufficient data must be available to eliminate this uncertainty. Kaya's data on three crystals are just sufficient and from them Döring derived the values of the constants shown under "Kaya" below:

	Nickel (Kaya)	Nickel (Döring)	Iron (Hironi-Hori)
k_1	0.063	0.0654	0.00153
k_2	.029	.0266	.00593
k_3	−.036	−.032	.00194
k_4	−.051	−.054	.00053
k_5	.014	.020	.00269

These values are very close to those obtained from eight crystals measured by Döring, as shown in the next column.

In a similar way Hironi and Hori [42H3] have derived the constants for iron from the data of Webster [27W3, 28W3] and Shirakawa [40S1, 40S2], with the results shown in the last column.

Döring has related the crystal constants to the expression previously given for polycrystalline material,

$$\frac{\Delta \rho}{\rho} = A + B \cos^2 \theta.$$

If nickel crystals are oriented at random and the domains in the initial state are distributed uniformly *over all directions of easy magnetization*,

$$B = \frac{2k_1}{5} + \frac{3k_2}{5} + \frac{12k_4}{35} + \frac{3k_5}{35}$$

$$A = -\frac{2k_1}{15} - \frac{k_2}{5} - \frac{2k_3}{15} - \frac{4k_4}{35} - \frac{k_5}{35}.$$

The crystal constants of nickel give B the value 0.0254, whereas Englert observed 0.0305. This discrepancy is not large, and it may be the result of special crystal orientation in the specimen measured.

It will be noted that with such a uniform initial distribution of domains over all directions of easy magnetization,

$$A = -\frac{B}{3} - \frac{2k_3}{15}.$$

Using the values of the constants for nickel, one obtains $A = -0.00419$ and $B = 0.0254$, and the ratio of the longitudinal to transverse change in resistivity is

$$\frac{\Delta \rho_h}{\Delta \rho_t} = \frac{A+B}{A} = -5.1$$

and not -2 as assumed earlier in the discussion of polycrystalline material. However, Döring has pointed out that $A = -B/3$, and $\Delta \rho_s/\Delta \rho_t = -2$, if the domains are initially distributed over *all directions*, a situation that may occur when the crystal anisotropy is small.

Effect of Anisotropy of Magnetostriction.—Lichtenberger [32L1] has shown that in some of the iron-nickel alloys the magnetostriction is positive in some crystallographic directions and negative in others (Fig. 13–71). When a polycrystalline specimen of such an alloy is subjected to tension, it is to be expected that the resistivity of some domains will increase and others decrease. If also the specimen as a whole has positive magnetostriction (e.g., 35 Permalloy), the limiting elastoresistance ($\Delta \rho_\sigma$) will be less than the limiting magnetoresistance ($\Delta \rho_h$). The measurements [39S2, 46B1] are in agreement with this interpretation [46B1]; in 35 Permalloy $\Delta \rho_\sigma$ is less than one-half of $\Delta \rho_h$ [40Y1], and in other alloys the fraction is even less, while in the alloys having isotropic magnetostriction $\Delta \rho_\sigma$ and $\Delta \rho_h$ are equal.

In a quantitative evaluation of the effect of anisotropic magnetostriction [46B1] we start with the expression for the magnetic strain energy of a domain, as given by Becker and Döring [39B5]. Let α_1, α_2, α_3 be the direction cosines of the magnetization of the domain (I_s), and γ_1, γ_2, γ_3 the direction cosines of the stress (σ) with respect to the crystal axes. Then the energy is

$$F_\sigma = -\frac{3\sigma}{2}[\lambda_{100}(\alpha_1^2\gamma_1^2 + \alpha_2^2\gamma_2^2 + \alpha_3^2\gamma_3^2)$$
$$+ 2\lambda_{111}(\alpha_1\alpha_2\gamma_1\gamma_2 + \alpha_2\alpha_3\gamma_2\gamma_3 + \alpha_3\alpha_1\gamma_3\gamma_1)].$$

For given values of λ_{100}, λ_{111}, and the γ's, values of the α's can be found which make F_σ a minimum. The angle θ is then determined, and the problem is to evaluate $\overline{\cos^2 \theta}$ for a random distribution of crystals about the axis of tension. From this the difference between $\Delta \rho_\sigma$ and $\Delta \rho_h$ can readily be found.

In the principal crystallographic directions [100], [110], and [111],

$\theta = 0$ or $90°$ for all values of λ_{100} and λ_{111}, when stress is present. As an example of the effect expected in other directions θ is plotted in Fig. 16–26 for the [511] direction for various ratios, $r = \lambda_{111}/\lambda_{100}$, for positive and negative values of λ_{100}. Lichtenberger's data show that for some composition near 75–80% nickel, $\lambda_{100} > 0$, $r = 10$, and Fig. 16–26 indicates $\theta = 27°$. Averaging over all directions, $\overline{\cos^2 \theta} \approx 0.93$, and this means that one should observe $\Delta\rho_\sigma = (\tfrac{3}{2})(0.93 - 0.33)\Delta\rho_h = 0.89\Delta\rho_h$. The ratio 0.89 is not far from the ratio 0.85 observed by Yamanaka [40Y1] for 74%

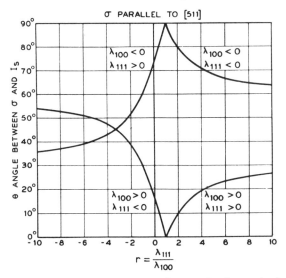

FIG. 16–26. The angle θ between directions of magnetization and of tension, when tension is applied parallel to a [511] direction, as dependent on magnetostriction constants λ_{111} and λ_{100}. Theoretical.

nickel (see Fig. 16–15). Values of λ_{100} and λ_{111} for any given specimen are probably dependent on heat treatment, and close comparison of calculated and observed values is not warranted.

Effect of Phase Change.—The normal course of the curve relating the resistivity of a ferromagnetic material to the temperature (Fig. 16–24) is altered when the material undergoes a change in phase, or when a change occurs in the ordering of atoms in an alloy. Examples of changes of each kind are given in Fig. 16–27. In the 20% nickel-iron alloy [39S2] the structure changes from body-centered cubic to face-centered cubic on heating and back to the body-centered form on cooling. The area between the curves depends markedly on the rate of heating and cooling and is generally larger when the temperature is changed rapidly. Similar phenomena are observed in connection with the phase changes occurring in iron and in cobalt.

The lower curve in this figure shows the effect of atomic ordering in 76 Permalloy. Special care must be exercised to observe the loop, and Kaya's [38K2] data here shown were taken by first cooling very slowly (during hundreds of hours) to 430°C and then heating at 2.5°C per minute. The two curves happen to coincide at the Curie point, where the usual

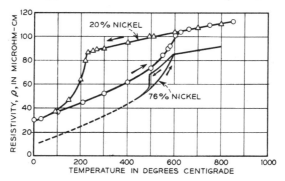

FIG. 16–27. Change of resistivity with change of phase (20% Ni) and change in ordering (76% Ni).

sudden change in slope is apparent. The rate of ordering at any temperature between about 500° and 550°C can be estimated by holding the temperature constant and noting the rate of rise of resistance.

Ordering and change of phase may also be studied by comparing the resistivities (at room temperature) of an alloy after quenching and after annealing for a long time at a relatively low temperature. Kaya and Kussmann [31K7] studied in this way the nickel-manganese alloys and observed resistivities of 39 and 76 microhm-cm for the same composition after cooling slowly and rapidly, respectively.

CHAPTER 17

CHANGE OF MAGNETIZATION WITH TIME

In many of their uses magnetic materials are subjected to an alternating field, produced by an alternating current in the wire associated with the material. The effects of the frequency of alternation on the magnetic properties of the material are primarily of three kinds: (1) the effective permeability of the material is reduced, (2) the energy loss in the material is increased, and (3) there is a time lag between the field strength and the corresponding induction. The time lag may be caused by eddy currents alone, or it may originate in other ways to be discussed later.

Effect of Eddy Currents.—When the magnetic flux in a conducting medium changes with time, an electromotive force is generated in the plane at right angles to the direction in which the flux is changing, and there is a resulting flow of currents within the material. These Foucault currents, or eddy currents, depend on the geometry of the material specimen, on its resistivity and permeability, and on the frequency of alternation of the flux. Their directions are always such as to counteract the change in field that produced them. Thus in Fig. 17–1 the flow of current in the eddies is opposite to that in the magnetizing coil. The net effect of the flow of eddy currents is to prevent the field from penetrating immediately to the interior of the material, and when the applied field is varying continually, as it is in much of the apparatus in which magnetic materials are used, the field strength in the interior may never be more than a small fraction of the field strength at the surface. The magnetic induction, therefore, decreases from the surface toward the interior. To avoid this reduction in field strength, and resultant decrease in alternating flux, the magnetic cores of transformers and other apparatus are laminated, each lamination being insulated electrically from its neighbor.

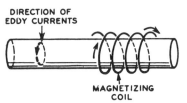

Fig. 17–1. Direction of eddy currents in solid bar is opposite to that of the applied magnetizing current giving rise to them.

In the first part of this chapter the calculable effects of eddy currents will be described and summarized. To a considerable extent this will involve a listing of the formulas relating field strength, induction, time and position in the specimen, for materials having the various geometrical forms commonly used. Formulas for the derived quantities—inductance, effective a-c resistance, and Q—are also included.

Eddy Currents in Sheets.—When the *permeability is constant*, independent of field strength, the effect of eddy currents in sheets or cylinders can be calculated accurately. In sheets the differential equation to be solved is

$$\frac{\partial H}{\partial t} = \frac{\rho}{4\pi\mu} \cdot \frac{\partial^2 H}{\partial x^2},$$

x being the distance measured perpendicular to the plane of the sheet of thickness δ, permeability μ, and resistivity ρ. This equation is the same as that used in the theory of diffusion. In this chapter the units are all cgs emu, unless otherwise stated.

Let a sinusoidally varying field, of amplitude H_0, be applied to a sheet parallel to its surface. The field amplitude inside the sheet diminishes with distance below the surface, according to the expression

$$\frac{H}{H_0} = \left[\frac{\cosh(2\theta x_0/\delta) + \cos(2\theta x_0/\delta)}{\cosh\theta + \cos\theta}\right]^{\frac{1}{2}}$$

in which x_0 is measured from the middle of the sheet, and

$$\theta = 2\pi\delta\sqrt{\mu f/\rho}.$$

For derivations of this and other formulas, the reader is referred to other sources [14R1, 25C2, 30S6].

It is of interest to know also the phase lag of the field (at any point inside the sheet) behind that at the surface, the average induction in the whole sheet, and the energy loss associated with the alternating flux. The peak of the flux wave progresses inward from the surface with the phase velocity

$$v = \frac{\sqrt{2f}}{\sqrt{\mu/\rho}},$$

proportional to the square root of the frequency, and for large values of θ it attenuates with distance x from the surface according to the law:

$$e^{-\theta x/\delta} = e^{-2\pi\sqrt{\mu f/\rho}\, x}.$$

The angle ϵ of lag behind the flux at the surface, at any point x, is given by

$$\tan\epsilon = 2\pi\sqrt{\mu f/\rho}\, x = \theta x/\delta.$$

The amplitude of induction at various points within the sheet is shown by the solid line in Fig. 17–2 for soft iron having the following characteristics: $\delta = 0.2$ cm, $\mu = 300$ (for low fields), $f = 1000$, and $\rho = 10\,000$

(10 microhm-cm); consequently for $\theta = 7$. The instantaneous value of B at any point in the sheet, at the instant when it has its maximum value at the surface, is shown by the broken lines in the same figure. In thick

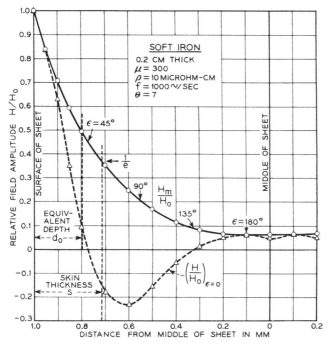

FIG. 17–2. Change in magnetic field strength (or induction) with position inside a sheet subjected to an alternating field of amplitude H_0; ε, phase angle of lag; s, skin thickness; d_0, equivalent depth. For θ, see p. 770.

sheets the induction at the middle may be opposite in sign to that at the surface.

As a measure of the penetration of the surface field into the interior, it is customary to use the "skin thickness," defined as

$$s = \frac{\delta}{\theta} = \frac{1}{2\pi\sqrt{\mu f/\rho}}.$$

This is usually expressed in the form

$$s = \frac{5030}{\sqrt{\mu f/(\rho \times 10^{-9})}},$$

where $\rho \times 10^{-9}$ is the resistivity in ohm-cm and μ and f are in the usual cgs units. At the distance s below the surface, the amplitude in a "thick" sheet is $1/e$th of that at the surface, and the phase angle is one radian.

For practical purposes it is sometimes desired to know the "effective" induction in the sheet, $\bar{B}_m$; this is defined as the maximum amplitude of the sinusoidally varying flux divided by the cross-sectional area:

$$\bar{B}_m = \frac{\varphi_m}{\delta} = \frac{\mu H_0 \sqrt{2}}{\theta} \left(\frac{\cosh\theta - \cos\theta}{\cosh\theta + \cos\theta}\right)^{1/2}$$

for unit width of sheet. When θ is large, this may be written

$$\frac{\bar{B}_m}{B_m} = \frac{\sqrt{2}}{\theta} = \frac{2d_0}{\delta}.$$

Here d_0 is called the "equivalent depth of uniform magnetization"; if two layers of depth d_0, one on each side of the sheet, were magnetized uniformly to an induction $B = \mu H_0$, the sheet would have the same amplitude as that observed.

Layers of depth d_0 and s are shown in Fig. 17-2 for specific conditions of magnetization ($\theta = 7$). The depth $s = d_0\sqrt{2}$.

If a coil, containing a core of laminated magnetic material, is measured on an inductance bridge (Fig. 19-9), balance is made with definite values of inductance L and resistance R in one arm of the bridge. We are now concerned with the relation of the L, so measured, to the flux in the laminations. If L_0 is the limiting value of L, approached at low frequencies,

$$\frac{L}{L_0} = \frac{1}{\theta} \cdot \frac{\sinh\theta + \sin\theta}{\cosh\theta + \cos\theta}.$$

The inductance L_0 of a toroidal core of mean diameter d and cross-sectional area A is $4N^2\mu A/d$, in emu. Since this is proportional, at low frequencies, to the permeability of the core material, an "apparent permeability" μ' may de defined by

$$\frac{L}{L_0} = \frac{\mu'}{\mu}.$$

At high frequencies, in the limit, one has

$$\frac{L}{L_0} = \frac{\mu'}{\mu} = \frac{1}{\theta} = \frac{s}{\delta}.$$

The ratio μ'/μ then differs from $\bar{B}_m/B_m$ by a factor of $\sqrt{2}$; the inductive component of the impedance, measured on the a-c bridge, is in quadrature with the voltage, and therefore only $1/\sqrt{2}$ of it is in phase with the current.

The contribution of eddy currents to the resistive component of the

EDDY CURRENTS IN SHEETS

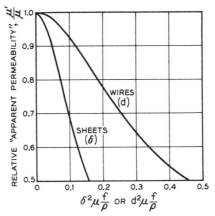

FIG. 17-3. Apparent permeability of wires of diameter d (in cm) or of sheets of thickness δ (in cm), as dependent on frequency f and resistivity ρ (c.g.s. units).

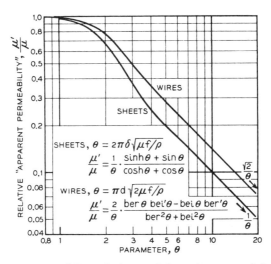

FIG. 17-4. Apparent permeability of wires and sheets in terms of the parameter θ.

impedance is determined from measurements on the bridge by deducting the d-c resistance R_0 from the measured resistance R. The difference ΔR caused by eddy currents (assuming no other losses are present) is then

$$\Delta R = 2\pi f L_0 \cdot \frac{1}{\theta} \cdot \frac{\sinh \theta - \sin \theta}{\cosh \theta + \cos \theta}.$$

At high frequencies this reduces to

$$\Delta R = 2\pi f L_0/\theta.$$

The quality factor Q, often used in engineering applications, is, considering eddy-current losses alone,

$$Q = \frac{2\pi f L}{\Delta R} = \frac{\sinh\theta + \sin\theta}{\sinh\theta - \sin\theta}$$

and approaches one at high frequencies. (Over a limited range in θ, Q is somewhat less than one.)

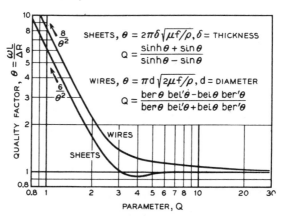

Fig. 17–5. Quality factor θ in terms of Q in wires and sheets.

In series form, expressions for L and R for $\theta \gg 1$ are:

$$\frac{L}{L_0} = \frac{\mu'}{\mu} = \frac{1}{\theta}[1 - 2e^{-\theta}(\cos\theta - \sin\theta)\cdots]$$

$$\Delta R = 2\pi f L_0 \cdot \frac{1}{\theta}[1 - 2e^{-\theta}(\cos\theta + \sin\theta)\cdots]$$

$$Q = 1 - 4e^{-\theta}\cos\theta\cdots$$

and for $\theta \ll 1$:

$$\frac{L}{L_0} = \frac{\mu'}{\mu} = 1 - \frac{\theta^4}{30} + \frac{31\theta^8}{22\,680}\cdots$$

$$\Delta R = 2\pi f L_0 \left(\frac{\theta^2}{6} - \frac{17\theta^6}{2520}\cdots\right)$$

$$Q = \frac{6}{\theta^2}\left(1 + \frac{\theta^2}{140}\cdots\right).$$

The expressions for μ'/μ and Q are shown graphically in Figs. 17–3, 4, and 5, and discrete values are given in Table 1.

Careful tests of the validity of these equations have been made by

TABLE 1. Effect of Eddy Currents on Apparent Permeability, μ', and on Quality Factor, Q, for Wires and Sheets, in Terms of the Parameter, θ

[d is the wire diameter, δ the sheet thickness (cm), f the frequency (cycles/sec), and ρ the resistivity (cgs).]

	Wires $(\theta = \pi d \sqrt{2\mu f/\rho})$			Sheets $(\theta = 2\pi\delta \sqrt{\mu f/\rho})$		
θ	μ'/μ	$\Delta R/\omega L_0$	$Q=\omega L/\Delta R$	μ'/μ	$\Delta R/\omega L_0$	$Q=\omega L/\Delta R$
0	1	0	∞	1	0	∞
0.1	1.000	.001	800.0	1.000	.002	600.0
0.2	1.000	.005	200.0	1.000	.007	150.0
0.4	.999	.020	50.0	.999	.027	37.5
0.6	.997	.045	22.2	.996	.060	16.7
1.0	.980	.122	8.06	.968	.160	6.04
1.4	.928	.221	4.20	.889	.283	3.14
1.8	.833	.312	2.67	.775	.380	1.99
2.2	.712	.368	1.94	.601	.417	1.44
2.6	.594	.377	1.58	.469	.402	1.17
3	.499	.360	1.39	.373	.363	1.03
3.4	.429	.333	1.29	.309	.319	0.97
4	.357	.292	1.22	.249	.263	0.95
5	.284	.242	1.18	.197	.202	0.97
7	.203	.181	1.12	.143	.142	1.00
10	.142	.131	1.08	.100	.100	1.00
20	.071	.068	1.04	.050	.050	1.00
40	.035	.035	1.02	.025	.025	1.00

Peterson and Wrathall [36P2]. They find good agreement with theory when the material is homogeneous, but observe that sheets often have a surface film differing from the rest of the material in conductivity or permeability, and that such films cause large apparent deviations from theory. Good agreement is again obtained by taking into account the properties of the film.

Eddy Currents in Cylinders.—When the specimen is in the form of a solid cylinder or wire, and sinusoidally varying magnetic field is applied parallel to the wire axis, the penetration of flux is governed by the differential equation:

$$\frac{1}{r}\frac{\partial}{\partial r}\left(r\frac{\partial H}{\partial r}\right) = \frac{4\pi\mu}{\rho}\cdot\frac{\partial H}{\partial t},$$

the solution of which is

$$H = H_0\left[\frac{\text{ber}^2(2\theta r/d) + \text{bei}^2(2\theta r/d)}{\text{ber}^2\theta + \text{bei}^2\theta}\right].$$

Here r is the distance from the cylinder axis to the point in question, d is the diameter of the cylinder, H_0 is the field strength at the surface, and

$$\theta = \pi d \sqrt{2\mu f/\rho}.$$

The ber and bei functions are combinations of Bessel functions of the first and second kinds, and have been tabulated [29D2].

Inductance and resistance, as measured on the a-c bridge, are:

$$\frac{L}{L_0} = \frac{\mu'}{\mu} = \frac{2}{\theta}\left(\frac{\text{ber}\,\theta\,\text{bei}'\,\theta - \text{bei}\,\theta\,\text{ber}'\,\theta}{\text{ber}^2\,\theta + \text{bei}^2\,\theta}\right)$$

$$\Delta R = 2\pi f L_0 \cdot \frac{2}{\theta}\left(\frac{\text{ber}\,\theta\,\text{ber}'\,\theta + \text{bei}\,\theta\,\text{bei}'\,\theta}{\text{ber}^2\,\theta + \text{bei}^2\,\theta}\right)$$

and

$$Q = \frac{2\pi f L}{\Delta R} = \frac{\text{ber}\,\theta\,\text{bei}'\,\theta - \text{bei}\,\theta\,\text{ber}'\,\theta}{\text{ber}\,\theta\,\text{bei}'\,\theta + \text{bei}\,\theta\,\text{ber}'\,\theta}.$$

These functions are shown in Figs. 17–3, 4, and 5, and in Table 1. At high frequencies ($\theta \gg 1$) they reduce to:

$$\frac{L}{L_0} = \frac{\mu'}{\mu} = \frac{\sqrt{2}}{\theta}\left(1 + \frac{1}{8\theta^2} + \frac{1}{4\sqrt{2}\,\theta}\cdots\right)$$

$$\frac{\Delta R}{2\pi f L_0} = \frac{\sqrt{2}}{\theta}\left(1 - \frac{1}{\sqrt{2}\,\theta} - \frac{1}{8\theta^2}\cdots\right)$$

$$Q = 1 + \frac{1}{\sqrt{2}\,\theta} + \frac{3}{4\theta^2}\cdots;$$

and when $\theta \ll 1$ to:

$$\frac{L}{L_0} = \frac{\mu'}{\mu} = 1 - \frac{\theta^4}{48} + \frac{19\theta^8}{30\,720}\cdots$$

$$\frac{\Delta R}{2\pi f L_0} = \frac{\theta^2}{8}\left(1 - \frac{11\theta^4}{384} + \frac{473\theta^8}{542\,960}\cdots\right)$$

$$Q = \frac{8}{\theta^2}\left(1 + \frac{\theta^4}{128}\cdots\right).$$

The behavior of eddy currents in wires of elliptical [27S6] and rectangular [07D1] sections, as well as tubes [29B3] of circular sections, has also been calculated. At high frequencies the effects are determined by the perimeter of the section, whatever its shape.

Wire Carrying Current.—The resistance and inductance of a wire carrying a high-frequency current have often been used for determination of permeability. As the frequency increases the resistance of a wire increases according to the equation

$$\frac{R}{R_0} = \frac{x}{2} \cdot \frac{\text{ber}\, x\, \text{bei}'\, x - \text{bei}\, x\, \text{ber}'\, x}{\text{ber}'^2\, x + \text{bei}'^2\, x}$$

in which

$$x = \pi d \sqrt{2\mu f/\rho}.$$

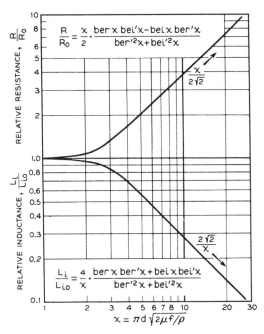

Fig. 17–6. Effective resistance and inductance of wire carrying current.

R/R_0 is plotted against x in Fig. 17–6 and may be expressed in series form for $x \ll 1$:

$$\frac{R}{R_0} = 1 + \frac{x^4}{192} \cdots ;$$

and for $x \gg 1$:

$$\frac{R_0}{R} = \frac{x}{2\sqrt{2}} + \frac{1}{4} + \frac{3\sqrt{2}}{32x} \cdots .$$

For the latter case the effective penetration of the current inward from the surface of the wire is confined to a "skin" thickness of $s = d/(x\sqrt{2}) = 5030/\sqrt{\mu f/\rho} \times 10^{-9}$.

As measured on an a-c bridge, the inductance of a straight wire at a given frequency, compared to that at zero frequency, is

$$\frac{L_i}{L_{i0}} = \frac{4}{x} \cdot \frac{\text{ber } x \text{ ber}' x + \text{bei } x \text{ bei}' x}{\text{ber}'^2 x + \text{bei}'^2 x},$$

a function [12R2] that decreases with increasing x, as shown in Fig. 17–6, and approaches the limiting value $2\sqrt{2}/x$. L_i is often called the inner inductance, and the μ determined from it is the inner permeability.

In actual measurements of R and L at high frequencies, a Lecher system is sometimes employed [32H4, 37P3], the material under investigation being in the form of two parallel wires placed l cm apart. In that case the measured inductance per unit length of the pair, L, will contain a geometrical term L_g, in addition to the inner inductance:

$$L = L_g + 2L_i$$

$$= 4 \ln \frac{2l}{d} + \mu(L_i/L_{i0}),$$

L_i/L_{i0} being the function given previously.

An alternative method of measurement of μ at very high frequencies makes use of a coaxial cable, the central wire of which is the material under investigation [39G5]. If d is the diameter of this wire and D that of the tube, and if $\mu\rho \times 10^9/D^2$ for the outer tube is much less than $\mu\rho \times 10^9/d^2$ for the wire, the inductance per unit length is

$$L = 2 \ln \frac{D}{d} + \frac{(\mu\rho/f)^{1/2}}{\pi d}.$$

L is determined by measuring the wave length of the electromagnetic wave traveling along the wires. The wave length depends on the inductance and capacitance of the cable, and the capacitance is calculated from the geometry.

Eddy-Current Losses.—Eddy currents give rise to an energy loss equal to the integral of ρI^2 over the volume of the material (I is the current density). In sheets of thickness δ subjected to sinusoidally varying field, analysis shows that when *flux penetration is complete* ($\theta < 1$), the power loss is

$$P = W_e f = \frac{\pi^2 \delta^2 B^2 f^2}{6\rho} \text{ (sheets)}$$

in ergs cm^{-3} sec^{-1}. For cylinders and spheres of diameter d the corresponding expressions are:

$$W_e f = \frac{\pi^2 d^2 B^2 f^2}{16\rho} \text{ (cylinders)}$$

and
$$W_e f = \frac{\pi^2 d^2 B^2 f^2}{20\rho} \text{ (spheres)}.$$

In considering eddy-current losses when the *penetration is incomplete* on account of shielding, it is convenient to use the relation between the losses and the ratio $\Delta R/L$ of effective a-c resistance to inductance as determined on a bridge. The inductance of material of effective permeability μ', cross-sectional area A, and length of magnetic path l, closely wound with N turns of wire connected to the bridge, is

$$L = 4\pi N^2 A \mu'/l.$$

The field strength at the surface of the material, caused by the sinusoidally varying current of rms value I_{rms} is

$$H = 4\pi N I_{\text{rms}} \sqrt{2}/l.$$

The power loss is
$$WfAl = \Delta R\, I_{\text{rms}}^2.$$
Therefore,
$$\frac{\Delta R}{L} = \frac{8\pi Wf}{\mu' H^2};$$

or, in terms of the effective induction, B', equal to $\mu' H$,

$$\frac{\Delta R}{L} = \frac{8\pi \mu' Wf}{B'^2}.$$

This is applicable to materials formed into sheets, wires, or any other form, and to losses due to any cause.

To determine the variation of W with dimensions, frequency, and resistivity (for any value of θ), use can now be made of the formulas for L/L_0 and $\Delta R/L_0$ given before, as long as the assumptions underlying these formulas are valid (constant μ, sinusoidal H). Thus the eddy-current power loss per cm³ of material is

$$P_e = W_e f = \frac{B'^2 f}{4\mu'} \cdot \frac{\Delta R}{2\pi f L_0} \cdot \frac{L_0}{L}.$$

When $\theta < 1$, use of these formulas confirms the expressions just given for sheets and cylinders. When $\theta \gg 1$, the formulas yield:

$$P_e = W_e f = \frac{B'^2 f}{4\mu'} = \frac{\pi B'^2 f^{3/2} \delta}{2\mu^{1/2} \rho^{1/2}} = \frac{B^2 \rho^{1/2} f^{1/2}}{8\pi \mu^{3/2} \delta} = \frac{H^2 (\mu \rho f)^{1/2}}{8\pi \delta} \text{ ergs/cm}^3$$

for sheets and

$$P_e = W_e f = \frac{B'^2 f}{4\mu'} = \frac{\pi B'^2 f^{3/2} d}{4\mu^{1/2} \rho^{1/2}} = \frac{B^2 \rho^{1/2} f^{1/2}}{4\pi \mu^{3/2} d} = \frac{H^2 (\mu \rho f)^{1/2}}{4\pi d}$$

for cylinders. Thus when a sinusoidal *field of constant amplitude* is applied to material having fixed values of μ and ρ, the power loss per cm^3 increases with the square root of the frequency. If the field is regulated so that there is *constant flux amplitude*, the loss varies as $f^{3/2}$. The latter case usually applies in practice, when a given voltage is required.

At high frequencies the eddy-current power loss depends only on the exposed surface of the material. For sinusoidally varying field strength of amplitude H_m, this loss is

$$P_e = H_m{}^2 (\mu \rho f)^{1/2} / 4\pi \text{ ergs/cm}^2/\text{sec.}$$

Losses in Low Fields.—With small eddy-current shielding, the eddy-current losses may be expressed as follows:

$$\frac{\Delta R}{\mu L f} = \frac{8\pi W}{B^2} = ef.$$

Here e is called the eddy-current resistance coefficient and is independent of f and B. It has different values for sheets, cylinders, and spheres, and these may be derived from equations given before.

A similar relation may be used for hysteresis loss, on the assumption that the losses of the two kinds are quite independent of each other. This will be a valid assumption if the fields are sufficiently low. Then Rayleigh's law may be used:

$$W_h f = \frac{(d\mu/dH) B^3 f}{3\pi\mu^3}$$

and for hysteresis and eddy currents

$$\frac{\Delta R}{\mu L f} = aB + ef,$$

a being the hysteresis resistance constant.

In practice it is found that the total observed losses may be accounted for only if a third term be added to this equation:

$$\frac{\Delta R}{\mu L f} = c + aB + ef.$$

This equation is used extensively to analyze the losses in materials used in communication, when they are subjected to low fields. The frequencies for which it is valid depends, of course, on the extent of the subdivision used in sheets, wires, or powder.

The constants are evaluated by measuring ΔR and L for fixed values of the current at various frequencies. Then $\Delta R/(\mu L f)$ is plotted against f for one value of the current (or B) and e determined from the slope of the line

(Fig. 17-7). The intercepts on the axis, obtained from these data and data for other values of B, are then plotted against B, and a and c evaluated from the slope and intercept of this line. The method is described in more detail by Legg [36L1] who refers to the earlier papers on the subject, among which those of Speed and Elmen [21S2] and of Jordan [24J1] should be

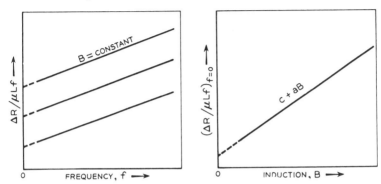

FIG. 17-7. Method of separation of losses occurring at low inductions.

especially mentioned. Detailed data have been published by Ellwood and Legg [37E2]. Results of measurements of various materials are summarized in Table 2, p. 797.

The loss corresponding to the constant c is called "residual loss," and has been the subject of some discussion. Its nature is discussed in a later section.

Some theoretical consideration has been given to the effect of hysteresis and eddy-current losses on each other. Cauer [25C2] has calculated the losses in a material having a hysteresis loop of the Rayleigh form. His results for sheets, expressed in series form in terms of θ and $\nu = d\mu/dH$, may be written as follows:

$$\frac{\mu'}{\mu} = \left(1 - \frac{\theta^4}{30} + \cdots\right) + \frac{\nu H}{\mu}\left(1 - \frac{4\theta^2}{9\pi} - \cdots\right)$$

$$\frac{\Delta R}{2\pi f L} = \frac{\theta^2}{6}\left(1 - \frac{17\theta^4}{420} + \cdots\right) + \frac{4\nu H}{3\pi\mu}\left(1 - \frac{\theta^4}{36}\cdots\right).$$

It will be noted that in addition to terms attributable separately to eddy currents and hysteresis, there are terms containing both ν and θ and therefore involving both phenomena. Use of the latter equation in separation and evaluation of losses has been discussed by Legg [36L1]. Calculation of the distortion caused by these two kinds of losses has been made by Kämmerer [49K7] and compared with experiment.

Losses at Intermediate and High Inductions.—The assumption of constant permeability, used in all discussions (except Cauer's) of eddy-current losses previously given in this chapter, can be applied in practice only when the inductions are very low. At the high inductions used in power transformers, when μ varies markedly with field, the assumption is commonly made that the eddy-current power loss is proportional to the square of the frequency, but no attempt is made to evaluate theoretically the proportionality factor. Since the hysteresis loss is proportional to the first power of the frequency, we have here the basis for separation of losses. Using the Steinmetz equation for hysteresis losses, usually applicable in the range $B = 2000$ to $15\,000$, we can express the total power losses as follows:

$$P = Wf = \eta B^{1.6} f + e B^2 f^2.$$

The procedure is then to measure P at various values of f, using sinusoidally varying flux, and plot W vs f. A straight line should then be obtained:

$$P/f = \eta B^{1.6} + e B^2 f.$$

The slope, multiplied by f^2, gives the eddy loss per second, eB^2f^2; and the intercept on the vertical axis, $\eta B^{1.6}$, is the hysteresis loss per cycle.

This method has been applied for many years to the separation of losses in commercial silicon-steel transformer sheet. It has been shown [12C2] to be reliable when the frequency and permeability are low ($f < 100$, $\mu < 5000$), by comparing the hysteresis loss thus obtained with that measured with a ballistic galvanometer. When the permeabilities or frequencies are higher, the separation is not feasible—the line ceases to be straight so that extrapolation to $f = 0$ is uncertain. Data for 45 Permalloy are given in Fig. 17-8 to illustrate this point. For these measurements (by J. A. Ashworth of the Bell Laboratories) $B_m = 10\,000$, $\mu = 20\,000$, and the thickness is 0.006 in. At $f = 0$ the cross shows the hysteresis loss measured ballistically. The expression given above may be considered valid as long as eddy-current shielding does not cause the flux near the middle of the specimen to lag too much behind the flux at the surface. At a given frequency, such a lag will cause the effective permeability (maximum flux amplitude divided by maximum field amplitude and area) to be substantially below that measured ballistically. At $f = 60$ a lowering of the effective permeability is marked when $\mu = 10\,000$ in sheet containing 3% silicon, or when $\mu = 5000$ in a 0.5% silicon alloy. The existence of the B^2f^2 term should not be affected by non-linearity between B and H because the effect of such non-linearity is equivalent to adding alternating fields of high frequencies (harmonics), and for these the losses are still proportional to B^2f^2, as long as the shielding is negligible. The effect of variation of permeability with induction has been discussed by Brailsford [48B1], Blake [49B6], and Butler and Mang [48B4].

In commercial silicon-iron sheet the total power loss can be represented approximately [43E1] by the empirical expression:

$$Wf = Cf^{1.36}B^{1.71}.$$

This may be used with caution over the range $B = 2000$ to 15 000, and $f = 30$ to 500. For a high-silicon alloy (4.2%) 0.014 in. thick, C has the value 3.3×10^{-10} when Wf is in watts per pound (7.3×10^{-10} when in watts/kg). Variation of total loss with thickness is approximately as the square of the thickness when the thickness is large (e.g., 0.060 in.), and as a smaller power when the thickness is smaller; in the latter case hys-

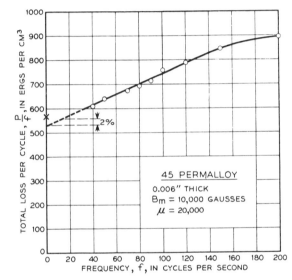

Fig. 17-8. Method of separation of hysteresis and eddy current losses occurring at high inductions. Intercept is to be compared with the hysteresis loss directly determined. Data are for 45 Permalloy.

teresis loss is a larger fraction of the total. In 0.014-in., hot rolled transformer sheet containing 4% silicon the hysteresis loss and eddy-current loss are about equal when $f = 60$, $B_m = 10\,000$.

Eddy Currents with Change in Field.—Here we have two cases: (1) the field changes suddenly from zero to a constant value, and (2) the field increases uniformly with time, or approximately so, so that a constant voltage is induced in a coil surrounding the magnetic material.

The permeability is assumed to be constant in both of these calculations. In the first case:

$$H = 0 \text{ for } t < 0$$
$$H = H_0 \text{ for } t > 0.$$

We designate by $\bar{B}$ the total flux per unit area. Then for *wires* [21W1] of radius r:

$$\frac{\bar{B}}{\mu H_0} = 1 - 4 \sum_{n=1}^{\infty} \frac{e^{-\lambda_n^2 \alpha t}}{\lambda_n^2}.$$

Here

$$\alpha = \rho/(4\pi\mu r^2),$$

and the λ's are the roots of Bessel functions of zero order, tabulated on p. 122 of Janke and Emde [23J1]. In series form we have

$$1 - \frac{\bar{B}}{\mu H_0} = 0.6917 e^{-5.783\alpha t} + 0.1312 e^{-30.47\alpha t}$$
$$+ 0.0535 e^{-74.8\alpha t} + \cdots.$$

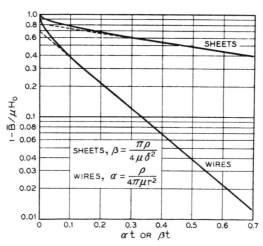

FIG. 17-9. Build-up of flux in wires and sheets subjected to sudden constant field.

When t is small, the rate of change in flux is very rapid and approaches infinity at $t = 0$. But soon ($\alpha t > 0.1$) only the first term in the series expansion is important and

$$d \ln \left(1 - \frac{\bar{B}}{\mu H_0}\right) \bigg/ dt = -5.78.$$

This is shown graphically in Fig. 17-9.

The problem of the decay of flux in *sheets* of thickness δ and of the associated energy loss has been solved, because of its importance in pulse transformers [46G1], by L. A. MacColl (unpublished). With sudden change of field,

$$\frac{\bar{B}}{\mu H_0} = 1 - \frac{8}{\pi^2} \sum_{n=1}^{\infty} \frac{\sin^2(n\pi/2)}{n^2} e^{-n^2\beta t}$$

or, in series form,
$$\frac{\bar{B}}{\mu H_0} = 1 - \frac{8}{\pi^2}(e^{-\beta t} + \tfrac{1}{9}e^{-9\beta t} + \cdots)$$
where
$$\beta = \pi\rho/(4\mu\delta^2).$$

The rate of build-up of flux is shown in Fig. 17–9, and after an initial period ($\beta t \approx 0.5$) it is represented by a single exponential, as may be seen from the series expansion.

If the field is applied for a specified time T and then removed, it is possible to calculate the flux remaining at time t after the first application. That is, when
$$H = 0 \text{ for } t < 0$$
$$H = H_0 \text{ from } t = 0 \text{ to } t = T$$
$$H = 0 \text{ for } t > T,$$
then, at time $t > T$,
$$\frac{\bar{B}}{\mu H_0} = \frac{8}{\pi^2}\sum_{n=1}^{\infty}\frac{\sin^2(n\pi/2)}{n^2}(1 - e^{-n^2\beta T})e^{-n^2\beta t}.$$

The total eddy-current loss that takes place when a field H is suddenly applied and held constant is
$$W = \frac{H_0^2\mu}{8\pi} \text{ ergs/cm}^3.$$

This is the familiar expression for the "energy of a field," and in this case it is equal to the energy dissipated by the eddy currents. When the field H_0 is applied at time $t = 0$, maintained constant until $t = T$ and then reduced to zero, the total eddy-current loss is
$$W = \frac{2H_0^2\mu}{\pi^3}\sum\frac{\sin^2(n\pi/2)}{n^2}(1 - e^{-n^2\beta T}) \text{ ergs/cm}^3.$$

When T is very large, the total loss resulting from applying and removing a field is
$$W = \frac{2H_0^2\mu}{\pi^3}\sum\frac{\sin^2(n\pi/2)}{n^2} = \frac{H_0^2\mu}{4\pi}$$

or twice the energy of the field H_0. Thus when the field is suddenly removed, the energy of the field is converted into eddy currents, and the loss is independent of the thickness and resistivity of the sheet.

We come now to the second case: the field is applied at a *constant rate*.
$$H = 0 \text{ for } t < 0$$
$$H = kt \text{ for } t > 0.$$

MacColl's solution for sheets is

$$\bar{B} = \mu H_0 \left[1 - \frac{8}{\pi^2 \beta} \sum_{n=1}^{\infty} \frac{\sin^2(n\pi/2)}{n^4} (1 - e^{-n^2\beta t}) \right].$$

When βt is larger than about 3, the average induction $\bar{B}$ is less than μH_0 by a constant amount, as shown in Fig. 17-10. Then

$$\bar{B} = \mu H_0 - \frac{\pi \delta^2 \mu k}{3\rho},$$

and the voltage induced in a coil surrounding the specimen is constant and proportional to $d\bar{B}/dt = d(\mu H_0)/dt = \mu k$.

If it is required that the *voltage* be constant from the time $t = 0$, then it is necessary to calculate H_0 as a function of time. Consider the more

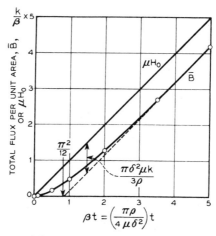

Fig. 17-10. Build-up of flux in sheets subjected to a field increasing uniformly with time.

general case for which the voltage is constant for a period extending from $t = 0$ to $t = T$, and otherwise is zero. For this constant-voltage pulse the field necessary to maintain this voltage is (MacColl)

$$H = E \left(\frac{\pi d}{3\rho} + \frac{t}{\mu d} - \frac{2d}{\pi \rho} \sum_{n=1}^{\infty} \frac{1}{n^2} e^{-4n^2\beta t} \right)$$

from $t = 0$ to $t = T$ and

$$H = E \left[\frac{T}{\mu \delta} - \frac{2}{\pi \rho} \sum_{n=1}^{\infty} \frac{1}{n^2} (1 - e^{-4n^2\beta t}) e^{-4n^2\beta t} \right]$$

when $t > T$. E is the constant electromotive force generated in a single turn of winding, in cgs emu.

The energy dissipated in a single pulse of this kind is

$$W = E^2 \rho T \left(\frac{1}{12} - \frac{\pi^2}{720 \beta T} + \frac{1}{8\pi^2 \beta T} \sum \frac{e^{-rn^2 \beta T}}{n^4} \right) \text{ergs/cm}^3.$$

Two other calculations of eddy-current decay and loss will be mentioned briefly. When a small volume of magnetic material, e.g., a Barkhausen domain, reverses its magnetization spontaneously, a definite energy loss will take place within the domain and in the surrounding magnetic material. As indicated previously, the total energy loss per cm^3 of material magnetized suddenly with a field H_0 is $H_0^2 \mu/8\pi$, or $B^2/8\pi\mu$. If the magnetization is changed spontaneously, from I_s to $-I_s$, the effective change in B is $2B_s = 8\pi I_s$, and the total loss per cm^3 is

$$W = 8\pi I_s^2/\mu.$$

This has been confirmed by a more detailed calculation by L. A. MacColl.

Finally, mention will be made of approximate calculations of time lag in materials having permeabilities that vary with induction. It has been shown previously that, when a field is suddenly applied to a cylinder of constant permeability and then held constant, the relation

$$\frac{d \ln (1 - B/B_0)}{dt} = \frac{5.76 \rho}{\pi d^2 \mu}$$

is valid except for an initial period of decay, during which, however, almost half of the ultimate change B_0 has taken place. In the approximate calculation μ is replaced by $dB/dH = f(B)$, a function known from experiment, and the equation becomes:

$$\frac{d \ln (1 - B/B_0)}{dt} = \frac{5.76 \rho}{\pi d^2 f(B)}.$$

The time corresponding to a definite value of B is then obtained by graphical integration. Putting $b = 1 - B/B_0$ we have

$$t = \frac{0.55 d^2}{\rho} \int_b^1 \frac{dB}{dH} \frac{db}{b}.$$

Experiments [28B1] on a material with large variations of dB/dH with B show agreement with theory as to order of magnitude and as to the general form of the b vs t curve.

Other methods of taking account of variable permeability have been used by Wolman and Kaden [32W5], Haberland [36H3], Kreielsheimer

[33K9], and others. Some progress in determining experimentally the distribution of flux in a lamination in which the permeability varies with induction has been reported by Brailsford [48B1]. The hope for exact solution of the differential equation involving variable permeability appears to lie in the use of mechanical or electronic computing machines.

Experimental determinations of apparent permeabilities and losses, including hysteresis, have been made on a variety of materials as a function of frequency by Dannatt [36D7] and others.

Propagation of Magnetic Waves.—Years ago there were many experimental studies of the spreading of magnetization from one small region in a specimen where a sudden change in field had been made. It was shown [06L1] that in a wire the amplitude of the change was exponential with distance, and the speed of propagation was approximately constant over a limited range and roughly inversely proportional to the area of cross section. Both of these properties depend on the material and on its magnetization. In soft iron the decay of amplitude F with distance x was given by

$$F = F_0 e^{-x/10},$$

and speeds of 20 000 cm/sec were observed in wire of 2-mm diameter; in other materials, different rates of decay and speeds were found.

More recently, Sixtus and Tonks have studied propagation in materials having square hysteresis loops. A description of their experiments is given in Chap. 11.

Magnetic Lag (not due to eddy currents).—As long ago as 1885 Ewing [89E1] observed that, when a magnetic field is applied to a magnetic material, an appreciable time is required for the magnetization to reach its final state. In many cases the time during which a change in B was observed was much too long to be accounted for by eddy currents; appreciable changes were observed in a 4-mm bar 10 minutes after the applied field became constant. In the period ending about 1910 many observations [87R1, 98M1, 08T1] were made of such lag, or "magnetic viscosity," or "after-effect" (Nachwirkung), using magnetometric methods, and of the analogous elastic lag observed during mechanical deformation [46Z1].

More recent experiments by Richter [38R1] and Snoek [39S11] have shown that one cause of magnetic lag is the presence of certain impurities in the iron investigated. This type of lag is strongly dependent on temperature and can be observed directly as a time lag in magnetization in low fields, or as an energy loss measured by a loss angle ϵ as mentioned previously. It is intimately connected with elastic lag.

Still another type of lag is associated with a loss angle that is only slightly dependent on temperature and frequency. The origin of this kind of lag is subject to speculation; it is sometimes referred to as "Jordan lag," because it was first reported by him [24J1].

The principal kinds of lag may then be listed under the following headings:

1. Eddy current lag.
2. Effect of impurities (temperature sensitive lag).
3. Jordan lag (constant phase angle).

The second item will now be considered.

Lag Dependent on Impurities.—For many years the existence of the lag observed by Ewing was not confirmed, and measurements made on various materials could be interpreted as due to eddy currents alone [21W1, 28B1]. In 1937 Richter [37R2, 38R1] and Wittke [38W3] confirmed the older findings by observing a well-defined lag in specimens of carbonyl iron annealed for 2 hours at about 1000°C and slowly cooled. It was shown later

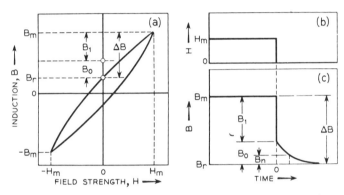

FIG. 17-11. Concepts used in the description of magnetic lag.

that their results could be attributed to the impurity of carbon in the material, and that the lag disappeared when the impurity was removed. Richter's measurements were made at low inductions by ballistic and magnetometric methods, and he observed the change in induction (to 200 gausses) after the removal of a rather low field (to 0.1 oersted). The initial permeabilities of the specimens were about 400–800. The results are described by the relations:

$$\Delta B = B_1 + B_0$$
$$B_n = B_0 \cdot \psi(t).$$

As indicated in Fig. 17-11, the rather weak field H_m is reduced suddenly to zero at time $t = 0$. The induction reduces immediately, except as delayed by eddy currents, by B_1 (from B_m to $B_r + B_0$), and then at a finite rate by the amount B_0 to its final value B_r. The lag is then $B_n = B_0 \cdot \psi(t)$, and it is desired to know both the total amount of lag B_0 and the function describing the change with time, $\psi(t)$. It is assumed that $\psi(0) = 1$ and $\psi(\infty) = 0$.

Results of measurements of B_0 are given in Fig. 17–12. The largest measured value is about 10 gausses when ΔB is 200 gausses. At low inductions the lag is 30% of the whole change in induction ($B_0/\Delta B = 0.30$), and at high inductions this percentage decreases to zero.

The form of the function $\psi(t)$ is shown in Fig. 17–13, where $B_0 \cdot \psi(t)$ is plotted against log t. The rate of decay is very sensitive to temperature

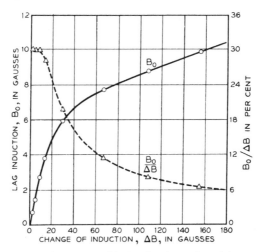

Fig. 17–12. Time lag in the decay of induction in carbonyl iron. Dependence of total lag B_0 on maximum change in induction ΔB.

(Fig. 17–14). The induction does not fall exponentially with time but decreases at first more slowly with time than a simple exponential law would require. In interpreting the data, Richter has found it necessary to replace such a simple exponential law,

$$\psi = e^{-t/\tau}$$

by a more general relation:

$$\psi = \int_0^\infty \frac{g(\tau)}{\tau} \cdot e^{-t/\tau} d\tau$$

in which $g(\tau)$ is a distribution function of time constants τ, normalized so that

$$\int_0^\infty \frac{g(\tau)\, d\tau}{\tau} = 1.$$

This corresponds physically to a material made of parts having different times of relaxation τ, the fraction of the material having a time constant between τ and $\tau + d\tau$ being $g(\tau)\, d\tau/\tau$. Richter has found that his curves may be fitted to a relatively simple form of the function $g(\tau)$, such that $g(\tau) = 0$ when $\tau < \tau_1$ and when $\tau > \tau_2$, and has a constant value equal to $1/\ln(\tau_2/\tau_1)$ when $\tau_1 < \tau < \tau_2$. Figure 17–15 shows $g(\tau)$ derived from one of the $\psi(t)$ curves, and in this case $\tau_2 = 0.14$, $\tau_1 = 0.0041$, $\tau_2/\tau_1 = 29$. The actual frequency distribution of τ's is presumably represented by a smooth curve having a single maximum lying between τ_1 and τ_2, and tapering toward zero at higher and lower τ's. Zener [46Z1] has derived a distribution function directly from the observed function $\psi(t)$, using the relation

$$g(\tau) = \frac{d\psi(t)}{d\ln t},$$

which he has shown to be valid, provided $g(\tau)$ is only a slowly varying function of τ.

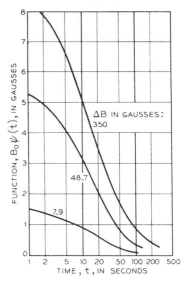

FIG. 17–13. Time rate of decay of induction in carbonyl iron for various lag amplitudes.

The effect of temperature is to shift the values of τ_2 (and τ_1) along the time axis so that τ_2/τ_1 remains constant. For the material investigated the values of τ_2 vary with absolute temperature T, according to the equation:

$$\tau_2 = 5 \times 10^{-15} e^{-10\,000/T},$$

so that the material has an "activation temperature" of 10 000°K, corresponding to an activation energy of about 1 electron volt.

Magnetic lag of the kind investigated by Richter has been shown to obey the superposition law. If, for example, a field H is applied from $t = -\infty$ to $t = -t_0$, reversed in sign at time $t = -t_0$, and brought to zero at $t = 0$, the following decay of induction will be governed by simple addition of the function ψ for the first application and that for the second application, the time for each being counted from its appropriate zero. Counting from $t = 0$, the total lag, $\psi_0(t)$, will then be given by

$$\psi_0(t) = \psi(t) - \psi(t + t_0).$$

Such a relation was found to hold experimentally, and fully accounted for the "anomalous" lag observed by Mitkevitch [36M14]; she observed that for certain values of t_0 the change in induction, taking place after $t = 0$, would change in direction and even in sign, before it sank to zero (Fig. 17–16).

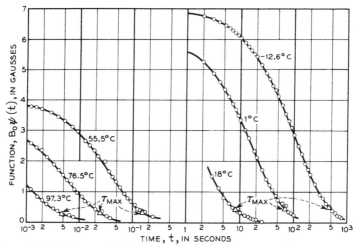

FIG. 17–14. Decay of induction in carbonyl iron at various temperatures.

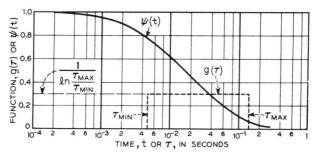

FIG. 17–15. Analysis of the function $\psi(t)$ describing the decay of induction (lag).

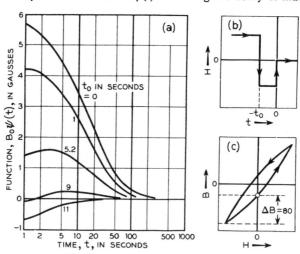

FIG. 17–16. Lag depending on previous magnetic history. "Anomalous" lag resulting from previous reversal of field.

There is a calculable relation between the measurements just described, and the power losses occurring in sinusoidally alternating fields. When the material obeys Rayleigh's law, the phase angle ϵ, between B and H, is related [37R2] to the distribution function $g(\tau)$ approximately as follows:

$$\tan \epsilon = \frac{2\pi f H}{\mu_0} \cdot \frac{d\mu}{dH} \int_0^\infty \frac{g(\tau)\, d\tau}{1 + (2\pi f \tau)^2}$$

and when $g(\tau)$ has the form assumed above,

$$\tan \epsilon = \frac{H\, d\mu/dH}{\mu \ln (\tau_2/\tau_1)} \arctan \frac{2\pi f(\tau_2 - \tau_1)}{4\pi^2 f^2 + \tau_1 \tau_2}.$$

The loss angle is related to the constant c of the loss equation:

$$\frac{R}{\mu f L} = c + aB + ef$$

by:

$$\tan \epsilon = \mu c / 2\pi.$$

Consequently, c can be calculated from $g(\tau)$ and, therefore, from $\psi(\tau)$ and the magnetic constants of the material (μ, $d\mu/dH$, etc.). Results of Richter's calculations for various temperatures are given in Fig. 17–17, and they agree with the values of c determined by experiment by Schulze [38S2] using the a-c bridge. (Following German custom, the ordinate used is n, equal to $1000\mu c$.)

Data showing the effects of both frequency and temperature have been reported by Schulze [38S1, 38S2], who used the same specimens of carbonyl iron as those investigated by Richter. When the measuring frequency is so low that the relaxation of the material is practically complete at all times ($f \ll 1/\tau_2$), the induction follows the field with negligible lag and c approaches zero. Also, when f is very high ($f \gg 1/\tau_1$), the time for relaxation is very small and c is again small. At intermediate frequencies c passes through a maximum, as already mentioned. When measurements are made at a given frequency and the temperature is varied, the relaxation times vary in their relation to frequency so that c again goes through a maximum.

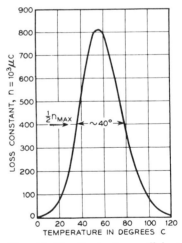

Fig. 17–17. Jordan lag coefficient as dependent on temperature in carbonyl iron.

It is to be expected that time lag in magnetization, of the kind under discussion, will cause a *dependence of apparent permeability on frequency;* when the frequency is low the induction for a given field strength will be relatively high. Such an effect was observed many years ago by Ewing [00E1] and has been considered again by Herrman [33H3] and by Schulze [38S1], some of whose data are given in Fig. 17-18.

Another common manifestation of this class of phenomena is the gradual *decrease of permeability with time* after demagnetization [26W5, 34W3, 34S7, 49F3], as shown [38S8] in Fig. 17-19. This effect is most marked at low induction and appears to be associated with the non-linearity of the

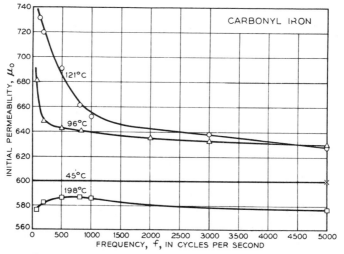

Fig. 17-18. Dependence of initial permeability of carbonyl iron on frequency, at different temperatures.

μ,H curve at low inductions (see Fig. 4-13). In materials that show these phenomena, the permeability and its change with time are especially sensitive to temperature and shock or, as Atorf [32A4] has shown, to a continually varying superposed field of very small amplitude. Fahlenbrach [48F4] observed that lag in different materials occurs at different temperatures, and that at these temperatures there are abnormalities in the permeability vs temperature curves.

A distinct contribution to the understanding of lag was made by Snoek [39S11, 39S12] when he showed its close connection with impurities of carbon and nitrogen in the iron. Before his work only one material— carbonyl iron—had been observed to show lag to a high degree. Snoek showed that the lag and related phenomena disappeared when the carbon was removed from the specimen by prolonged annealing in vacuum at a high temperature, and that the lag reappeared when less than 0.01% of carbon or nitrogen was reintroduced. The different permeability vs temperature relations of specimens respectively decarburized and carburized with 0.006% carbon are illustrated in Fig. 17-20. Similar specimens showed corresponding differences in phase angle of lag and in mechanical

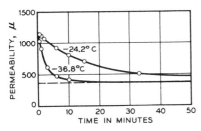

Fig. 17-19. Decrease of permeability of carbonyl iron with time after demagnetization, at two different temperatures.

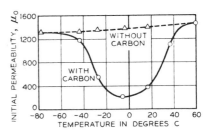

Fig. 17-20. Permeability vs temperature curve for carburized (0.006% C) and for decarburized iron.

losses. Also, purified specimens could be made to show various lag phenomena by the introduction in them of a few thousandths of a per cent of nitrogen.

The mechanism by which the presence of traces of carbon will cause elastic and magnetic lag has been discussed by Gorski [35G4], Snoek [41S5],

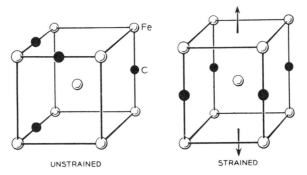

Fig. 17-21. Assumed positions of carbon atoms in iron before (random) and after (ordered) an elastic strain. Movement of carbon atoms is responsible for time lag.

and others [45P1, 46Z1], and the theory has received confirmation in a number of ways. It is assumed that carbon atoms normally take positions in the interstices of the iron structure, for example at the centers of the faces or edges of the unit cube of the lattice (Fig. 17-21). The carbon atoms, on account of their small size, have a mobility in the relatively loose body-centered structure of iron that permits them to go easily from one interstitial position to another, even at room temperature and below.

When the iron lattice is unstrained all interstitial sites of the kind mentioned are equally probable, but under strain the distances between some of the iron atoms will be relatively large and the sites between them will be more likely to be occupied by carbon atoms. Thus application of a strain will cause a redistribution of carbon atoms, and this will consume time and cause elastic lag.

In a quantitative consideration of this kind of elastic lag, Polder [45P1], following Gorski and Snoek, has shown how the relocation of mobile atoms with stress depends on the change of dimensions with concentration of carbon. This change of dimension is just that measured with X-rays for martensite, which is a solid solution of carbon in iron of the kind under consideration. For iron containing 0.01% carbon, Polder deduces from the X-ray measurements on martensite that when a stress is applied in a [100] direction the total extension resulting from a redistribution of carbon atoms is

$$\frac{\Delta l}{l} = 0.039$$

and that in a [111] direction the extension is zero. Experimentally Dijkstra [45P1] has observed respectively 0.043 and less than 0.002, in excellent agreement with theory. Physically, it is apparent that all carbon atoms contribute equally to an expansion of the lattice in the [111] direction; consequently there will be no redistribution under stress to cause a change in length and be responsible for elastic lag.

Polder has also derived the relation between the relaxation time τ and the diffusion constant of carbon in α-iron. The latter has not been determined directly, but values calculated from the τ's observed by Snoek [41S5] are of reasonable magnitude.

The foregoing considerations apply to elastic lag. Richter's experiments show that magnetic and elastic lag in carbonyl iron follow the same laws and must have the same origin, i.e., the diffusion of carbon or nitrogen or other impurities at the temperature of the experiment. The connection between mechanical lag and magnetic lag lies in the strain sensitivity of the magnetic properties, that is, in its magnetostriction.

Jordan Lag.—As mentioned above, this kind of lag is not sensitive to either temperature or frequency, and is not explained by the effect of impurities. The common occurrence of a loss of this sort was pointed out by Jordan [24J1] and had been recognized some years previously in the study of losses at low inductions in communication circuits.

Values of the lag coefficient c for some materials are given in Table 2 [37E2, 40L1] and are compared there with the hysteresis coefficient a. It is apparent that c and c/a vary considerably from one material to another, and that c is often larger in unannealed material.

The nature of the Jordan loss has been the subject of a number of investigations. Jordan has suggested that it is associated with lag which is not due to eddy currents but is analogous to viscosity. The work of Wittke [34W4], Preisach [35P2], and Ellwood and Legg [35E2, 37E2] showed that it could not be accounted for by deviations from Rayleigh's law of hysteresis at low inductions, that it was distinct from the hysteresis

TABLE 2. LAG AND HYSTERESIS LOSS COEFFICIENTS IN A VARIETY OF MATERIALS

Material	μ	$c \times 10^6$	$a \times 10^6$
Compressed powder cores			
2-81 Mo Permalloy................	125	30	1.6
	26	96	9.6
	14	143	11.4
81 Permalloy.....................	75	37	5.5
	26	108	11.5
Laminated cores			
38 Permalloy, hard................	100	118	9.6
38 Permalloy, annealed.............	2060	20	1.4
45 Permalloy, annealed.............	2550	14	0.43
78 Permalloy, annealed.............	3900	0	0.6
45-25 Perminvar, annealed..........	450	0	0.002
Isoperm (0.0017 in.)[34D2]..........	85	20	2.3

determined from B,H loops measured by the ballistic method, and that the loss was in fact due to a time lag of magnetization, as Jordan had suggested. Kindler [37K4] found that the loss was influenced strongly by either elastic or plastic deformation.

In lag of the Jordan type the angular lag of B behind H, and the erg loss per cycle, are independent of frequency. It was suggested by Ellwood [35E2] and by Preisach [35P2] that one mechanism that would account for this kind of loss is the eddy-current loss associated with sudden changes in magnetization. The magnitude of the observed loss is not inconsistent with what is known of the magnitude of such changes, but the connection between the two phenomena has not been established.

More extensive summaries of this phenomenon have been given by Kindler and Thoma [36K5] and by Becker and Döring [39B5].

Long Period Lag. Aging.—The effect of metallurgical changes on magnetic properties has already been considered in the discussion of permanent magnets (Chap. 9). In materials of high permeability, also, similar changes occur; in well-annealed Molybdenum Permalloy, for example, a change in maximum permeability of about 2% is observed after maintaining at 100°C for 150 hours. In commercial iron the effect is especially pronounced, and the coercive force may be doubled by this treatment. Changes of this kind are attributed to the slow precipitation of an impurity, such as carbide or nitride, and in iron there is good evidence [30K4] that the nitride is the active impurity at 100°C. Associated with the precipitation are internal strains, which develop during the precipitation and then are slowly relieved by diffusion. At room temperature these changes occur with various speeds; sometimes they take place with easily measurable speed and sometimes they are so slow as to be undetectable.

Relaxation periods of months and years have been reported by Kittel [46K1] for ships subjected to the earth's magnetic field. Since they are also subject to vibration, and this is known to change the magnetization of material subjected to constant field (see Chap. 13), it is not known whether the changes observed are caused by metallurgical changes (e.g., relief of strains) or by the alternating stresses attending the vibration.

Permeability at Very High Frequencies.—In considering the effect of frequency on permeability, it has been assumed that the true permeability μ is independent of the frequency of alternation of the field and that only the apparent permeability μ' diminishes in accordance with Maxwell's equations as the frequency increases (eddy-current shielding). At the frequency of visible or infrared light, however, it is known [03H2] that the permeability is one, so that at some frequency below this, and above that commonly used in radio, the permeability must fall off with increasing frequency. Experiments described later and theoretical considerations indicate that the greatest dispersion occurs in the neighborhood of 10^8–10^{10} cps, when the wave length is of the order of 10–100 cm.

The early experiments were carried out by Arkadiew [14A1] by reflecting short waves from wire gratings made of iron, nickel, or steel. In one experiment [19A1] the permeability decreased from $\mu = 93$ at a wave length $\lambda = 73$ cm to $\mu = 8$ at $\lambda = 1.3$ cm. More recently a number of other methods of measurement have been used, at frequencies above which the more conventional bridge methods begin to fail (10^6–10^7 cycles/sec.) These methods depend, directly or indirectly, on measurements of inductance or of energy loss, and the permeabilities derived from the measurements are then designated μ_L and μ_R, respectively. Usually the value of the ratio μ/ρ of the material determines the observed inductance, and the product $\mu\rho$ determines the energy loss or effective resistance.

A list of the methods most commonly used is as follows:

1. *Parallel wire system (Lecher wires).* Measure the ratio of a-c to d-c resistance or attenuation along wires (μ_R), or the wave length of standing waves (μ_L).
2. *Bridge arrangement of Lecher wire systems.* Measure resistance (μ_R) and inductance (μ_L).

3. *Coaxial cable.* Measure wave length of standing waves (μ_L) or transmission loss along central wire (μ_R).
4. *Resonance circuit.* Measure change of frequency related to inductance of specimen (μ_L).
5. *Thermal methods.* Measure rise in temperature caused by energy loss in material (μ_R).
6. *Microwave technique.* Measure attenuation along a wave guide, or Q of a resonating cavity (μ_R).

Measurements made by the various methods have been summarized by Allanson [45A2]. When μ_L and μ_R are determined on the same specimens, it is generally found that $\mu_R > \mu_L$. When eddy-current loss is predominant, the calculated phase lag between the resistive (R) and reactive (ωL) components of the impedance approaches 45° as a limit (see Fig. 17–22). If now there is an additional loss, caused by hysteresis or magnetic lag of unspecified kind, the phase angle will increase, and R will increase whereas ωL will decrease; as a result, the μ_R calculated from R will be greater than the μ_L.

In Figs. 17–23 and 24 are shown data derived from measurements made at frequencies of 10^8 cps or above on iron and nickel. There seems to be little doubt that there is here a decline of permeability toward one or zero. The frequencies at which the permeability is observed to drop to one-half of the normal value are estimated by Kittel [46K2] in Table 3. Also, data by Strutt [40S4] for iron at $-183°C$ indicate that

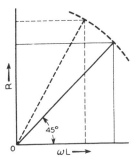

FIG. 17-22. Resistive and reactive components of impedance with high eddy-current loss, and the effect of additional loss on R and ωL.

μ drops to about half its normal value at $f = 1.5 \times 10^8$; since at this temperature the resistivity is about one-sixth of that at room temperature,

TABLE 3. FREQUENCIES AT WHICH PERMEABILITIES DERIVED FROM INDUCTANCE (μ_L) OR RESISTANCE (μ_R) DROP TO APPROXIMATELY HALF THEIR NORMAL VALUES [46K2]

Material	μ_R	μ_L
Iron	30×10^8	2×10^8
Nickel	15×10^8	1.7×10^8

the equivalent frequency for half-normal permeability at room temperature is about 9×10^8 (compare this with 2×10^8 in Table 3).

It should be emphasized that the apparent permeability at high frequencies depends very much on the character of the surface. Even at frequencies as low as 10^5 cps it has been shown by Peterson and Wrathall

800 CHANGE OF MAGNETIZATION WITH TIME

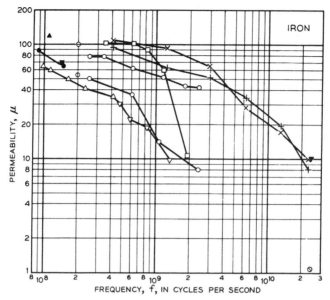

Fig. 17-23. Collection of data on the permeability of iron at high frequencies. Sources: O, Sänger [34S7]; + and ×, Arkadiew [19A1]; △, Hoag and Gottlieb [39H3]; ▽, Hoag and Jones [32H4]; □, Lindman; ●, Möhring [39M3]; ▲, Schwarz [32S8]; ⊘, Maxwell [46M1]; ■, Procopiu and d'Albon [37P2]; ⊖, Johnson, Rado and Maloof [47J1]; ⊙, Glathart [39G5].

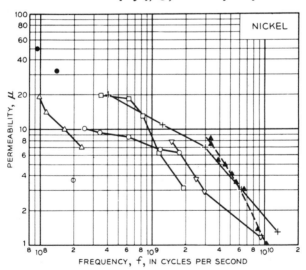

Fig. 17-24. Collection of data on the permeability of nickel at high frequencies. Sources: +, Arkadiew [19A1]; △, Hoag and Gottlieb [39H3]; □, Lindman [38L2]; ▽, Simon [46S1]; ●, Möhring [39M3]; ⊙, Glathart [39G5]; O, Potapenko and Sänger ([33P1] and private communication to C. Kittel); ▲, Hodsman et al. [49H2].

[36P2] that thin films of oxides, formed on the material during the heat treatment, may cause the apparent permeability to decrease by a factor of 10. At higher frequencies it is difficult or impossible to detect films of a thickness responsible for large changes in apparent permeability. Since the true permeability is calculated from the apparent permeability, the values reported may be subject to large errors.

At frequencies greater than 10^8 the penetration of the field is comparable to the thickness of the Barkhausen region in which the magnetization reverses as a unit, and this has been suggested as the cause of the decrease of permeability in this range. In iron of permeability 100 and resistivity 10 000, the skin thickness at various frequencies is as follows:

f in cps	10^8	10^{10}	10^{12}
s in cm	2×10^{-4}	2×10^{-5}	2×10^{-6}

Penetration of regions of thickness 10^{-4} cm will then occur when the frequency is about 10^8.

Several theoretical investigations have been made to estimate the effect of domain structure on permeability in this range of frequencies. Becker [38B6] has calculated the movement of a domain boundary between two antiparallel domains of cylindrical form. In weak fields, the (reversible) movement of the boundary, characteristic of initial permeability, is assumed to be restricted by eddy currents created by the change in magnetization resulting from the boundary displacement. At some critical frequency, f_0, the direct action of the field will be opposed by the eddy-current field so that movement of the boundary is limited. This occurs when

$$f_0 = \frac{3\rho}{\pi^2 \mu_0 l^2}.$$

In iron with an initial permeability of 100, resistivity 10 000 and domain thickness (Barkhausen region) $l = 10^{-4}$ cm, the critical frequency is

$$f_0 = 3 \times 10^9,$$

in excellent agreement with experiment.

A somewhat different basis for calculation has been used by Kittel [46K2]. Whereas Becker assumes that the surface domains are completely immersed in the applied alternating field, Kittel takes account of the reduction in field caused by the shielding within the domain itself, and he calculates the permeability of a layer one domain thick. Assuming a domain thickness (Barkhausen region) of 2.5×10^{-4}, $\mu_0 = 100$, and $\rho = 10\,000$ (for iron), μ_L and μ_R are found to have values consistent with the data. See also Polivanov, *J. Phys. USSR* 7, 18 (1943).

A still different point of view has been given by Döring [48D3], who has calculated the increased energy of the wall caused by its motion and found it large compared with the energy at rest. The wall behaves as if it had a mass inertia, because angular momentum is associated with the motion of the spins in the wall, and the apparent mass is evaluated in terms of the energies of exchange, crystal anisotropy, and field. The inertia effect is larger than the eddy-current effect, and as a consequence the resistivity does not enter into the expression for the critical frequency f_0 at which the permeability is substantially reduced. This is borne out by experiments of Birks [46B5] on iron oxide, in which the decrease in permeability is found to occur in the neighborhood of 10^9 cps, as in iron and nickel, even though the

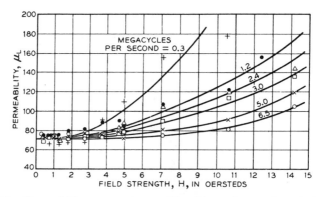

Fig. 17–25. Permeability curves (μ_L vs H) of iron wires at frequencies of 10^5 to 10^7 cps.

resistivity is higher than in these metals by many powers of ten. Döring's calculated value of f_0 for iron is 6×10^8 cps, in excellent agreement with experiment.

Recently Kittel [51K1] has shown that the damping term associated with spin precession in ferromagnetic resonance experiments (see below) is even more important than the inertia effect calculated by Döring. This kind of damping, of unknown origin, appears now to be the principal reason for the decrease of initial permeability with frequency. In effect the wall movement becomes highly damped at frequencies below those commonly used in microwave experiments.

At frequencies less than 10^8 cps, some investigators have found that μ varies greatly with frequency, whereas others found that there is practically no change. The early measurements, in which a number of resonant peaks were found at these frequencies in the μ vs f curves, have been shown to be in error [28W4]. In measurements carried out carefully at low fields, so that variation in μ with H does not affect the results, Michels [31M3] has found the permeability of iron to be independent of frequency from 0.2×10^8 to 0.8×10^8 cps and to be about equal to the permeability at low frequencies. When the field strength is as large as the coercive force, however, so that the changes in magnetization tend to occur discontinuously, there is evidence that the permeability changes with frequency when the latter is in the range 10^5–10^7. Kreielsheimer [33K9] derived from measurements of inductance the μ_L vs H curves for frequencies of 3×10^5 to 6×10^6, and his results, plotted in Fig. 17–25, show that, whereas μ_0 does not change noticeably with frequency, μ for $H = 10$ varies by more than a factor of 2. It must be pointed out, however, that Kreielsheimer was forced to calculate his values of permeability from the measured inductance by approximate methods, on account of the great variation of dB/dH with H over the hysteresis loop.

It may be expected that the permeability in fields of this magnitude will begin to decrease when the time necessary for the completion of a Barkhausen jump is comparable to the period of the applied alternating field. Becker has estimated theoretically the critical frequency f_i of the irreversible process in relation to the critical frequency f_0 of the reversible boundary displacement, and he derived the relation:

$$f_i = f_0 \frac{\mu_0 H}{3B_s}.$$

The expressions for f_0 and f_i may be written

$$\frac{\pi^2}{3} l^2 \mu_0 f_0 / \rho = 1$$

and

$$\pi^2 l^2 (B_s/H) f_i / \rho = 1$$

and compared with the similar expression for the skin thickness s:

$$4\pi^2 s^2 \mu f / \rho = 1.$$

Applying Becker's expression for f_i to iron with $H = 7$, $\mu_0 = 100$, and $B_s = 21\,600$, we find $f_i \approx f_0/100 \approx 2 \times 10^7$. That is, the irreversible process should fail at a frequency two orders of magnitude lower than that at which the reversible process declines, and should fail at lower frequencies when the alternating applied field is higher. This is in general agreement with Wien's [32W6] and Kreielsheimer's results.

An excellent review of ferromagnetic phenomena at microwave frequencies, including ferromagnetic resonance, has been published by Rado [50R1].

Ferromagnetic Resonance.—Investigation in the field of high frequencies has been stimulated by the development of the microwave technique of radar, and new phenomena have been uncovered.

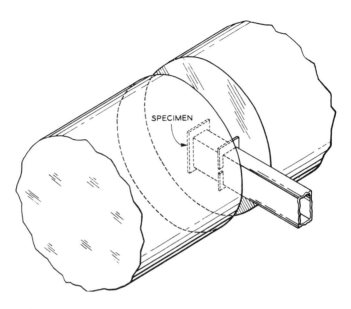

Fig. 17–26. Portion of experimental arrangement for observing ferromagnetic resonance.

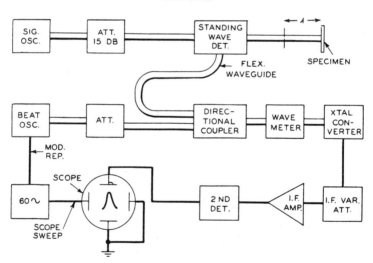

Fig. 17–27. Diagram of microwave circuit used in investigation of ferromagnetic resonance.

According to classical theory, if a magnetic field is applied to an electron revolving in its orbit, the orbit will precess around the field. If now an alternating field is applied in a direction at right angles to the original field, and its frequency of alternation is nearly equal to the frequency of precession, resonance will occur.

The first experiment on such an effect was reported by Griffiths [46G7] who used a wave guide containing a resonant cavity lined with magnetic material. A modified procedure, used by Yager and Bozorth [47Y1], is shown in Figs. 17–26 and 27. The magnetic material is in the form of a

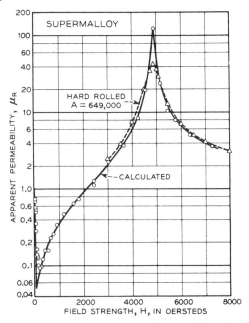

Fig. 17–28. Ferromagnetic resonance curve of annealed and of hard rolled Supermalloy, showing sharp maximum. Solid curve is calculated for single value of damping term λ. Wave length, 1.25 cm.

thin sheet and the constant applied magnetic field H_a is excited with an electromagnet, the poles of which are shown. The mode of excitation of the wave guide is such that the r-f field, H_{rf}, lies in the plane of the sheet at right angles to H_a. The circuit used is shown in the block diagram. The effective permeability can be calculated from the standing wave pattern, and in terms of the real and imaginary parts of the r-f permeability, μ_1 and μ_2, is

$$\mu_R = \sqrt{\mu_1^2 + \mu_2^2} + \mu_2.$$

Figures 17–28 and 29 show the relation between the effective permeability and the strength of the static magnetic field H observed for nickel and for Supermalloy by Yager and Bozorth [48Y3, 47Y1] at a frequency of 2.4 × 10^{10} (wave length, 1.25 cm). (The field H differs from the applied field H_a only by the demagnetizing field of the specimen.) At resonance the maximum effective permeability of nickel is about 40, at $H = 5300$.

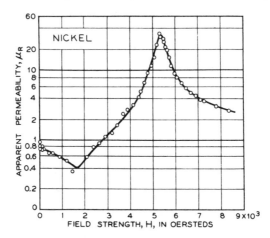

FIG. 17–29. Ferromagnetic resonance in nickel at a wave length of 1.25 cm.

In lower static fields the permeability is considerably less than one, and as the static field is reduced to zero the permeability increases to one.

In considering the theory, imagine an electron traveling in an orbit and suspended in space or "hung" by a thread from a fixed support as

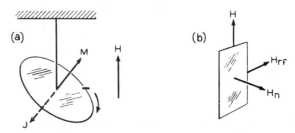

FIG. 17–30. Relation between (a) magnetic moment M and mechanical moment J of orbital electron, and (b) three components of field acting in the specimen.

indicated in Fig. 17–30(a). Associated with the orbital motion are a magnetic moment M, and an angular momentum J, indicated by vectors in the figure. The numerical relation between the moments is well known (Chap. 10) to be given by

$$M = \frac{geJ}{2mc} = \alpha J,$$

in which the symbols have their usual significance. For the orbital motion just described the Landé splitting factor g is one. If the magnetic and mechanical moments of electron *spin* alone are considered, then $g = 2$. When a field H is applied in the vertical direction, the magnetic and mechanical moments M and J will precess around the axis of H with the Larmor angular frequency

$$\omega_L = \frac{geH}{2mc} = \alpha H.$$

Now consider the thin specimen of magnetic material used in the wave guide. As indicated in Fig. 17–30(b) the static field H and the r-f field H_{rf} are mutually perpendicular in the plane of the specimen, and the normal component of the field is H_n. As the atomic moments in the material precess the magnetic vector M will have a fixed component parallel to H and time-variable components parallel to H_{rf} and to H_n. Kittel [48K9] has evaluated the components and solved the equations of motion using this classical model. The component H_n is evaluated by noting that in a thin sheet the magnetic induction perpendicular to the sheet is $B_n = 0$, since no field is applied in this direction. Then

$$B_n = H_n + 4\pi I_n = 0,$$

and the amplitude of the normal field is

$$H_n = -4\pi I_n.$$

The equation of motion is

$$dJ/dt = \mathbf{M} \times \mathbf{H} = \alpha \mathbf{J} \times \mathbf{H}$$

and the solution, assuming sinusoidal H_{rf}, is

$$\frac{I_{rf}}{H_{rf}} = \frac{I}{H} \cdot \frac{1}{1 - \omega^2/(\alpha^2 BH)}.$$

According to this the r-f permeability ($\mu_R = 1 + 4\pi I_{rf}/H_{rf}$) is a maximum when

$$\omega_0 = \alpha \sqrt{BH}.$$

This equation is similar to that for Larmor precession, except that H is now replaced by $\sqrt{BH}$.

The foregoing derivation is based on classical mechanics. A similar result is obtained by quantum mechanics, according to which the calculation has been carried through by Luttinger and Kittel [48L2] and others. A simple version may be appropriate here: In an effective internal magnetic field, H_{eff}, the vector representing the result of spin and orbit magnetization will have certain permissible orientations with respect to the field. The difference in energy between two adjacent orientations is

$$h\nu = g\beta H_{\text{eff}}.$$

Here h is Planck's constant, ν the frequency, and β the Bohr magneton, and for the specimen in the form of a sheet H_{eff} is $\sqrt{BH}$, as before. Using the expression for the Bohr magneton, $\beta = eh/(4\pi mc)$, and placing $\omega_0 = 2\pi\nu$ and $\alpha = ge/(2mc)$ as before, this reduces to the classical expression for ω_0.

Although the foregoing relations are derived only for thin sheets, in which the demagnetizing factor perpendicular to the sheet is 4π, similar relations can be derived for other shapes. Kittel has given the general expression, which reduces for spheres to

$$\omega_0 = \alpha H$$

(the Larmor value), and for an infinite cylinder to

$$\omega_0 = \alpha(H + 2\pi I).$$

The values of H are usually high enough to effect saturation, and then B can be replaced by $H + 4\pi I_s$.

According to the foregoing expression for $I_{\text{rf}}/H_{\text{rf}}$, this quantity approaches infinity at resonance. Naturally any energy loss associated with the resonance, apart from eddy-current loss which has already been taken into account, will reduce the maximum to a finite value. Yager [47Y1] has found that a damping term of the kind used by Frenkel [45F3] will fit the data rather well. This is of the form

$$-\lambda[(M/\chi_0) - H],$$

in which χ_0 stands for $I_{\text{rf}}/H_{\text{rf}}$, and must be added to the expression for dM/dt in the equation of motion. The degree to which a single value of the constant λ will describe the damping for all parts of the resonance curve may be seen in Fig. 17–28 by comparing the line with the observed points; some departure is noticeable at low values of applied field strength. For annealed Supermalloy the relaxation time was found to be 1.2×10^{-9} sec (29 cycles).

Experimental Values of g.—According to theory the values of the static field and of magnetic induction at resonance may be used in conjunction with the frequency to determine the gyromagnetic ratio or, more properly, the value of the Landé splitting factor g:

$$g = \frac{2mc\omega_0}{e\sqrt{BH}}.$$

This applies only to thin sheets of material magnetized parallel to the plane of the sheet, and modified relations are appropriate for other geometrical forms, as mentioned previously. Values of g thus determined may be compared with the similar quantity g', derived from the gyromagnetic experiments of Barnett and others (see p. 454). Comparison of the values determined by the two methods are made in Table 4. It is

TABLE 4. VALUES OF LANDÉ FACTORS g AND g' DETERMINED RESPECTIVELY FROM FERROMAGNETIC RESONANCE [47Y1, 48H3, 48Y3, 49B12, 49Y1] AND FROM GYROMAGNETIC EXPERIMENTS [49K5, 44B1]

	VALUES DERIVED FROM	
Material	Ferromagnetic Resonance: g	Gyromagnetic Ratio: g'
Iron	2.17	1.93
Cobalt	2.22	1.87
Nickel	2.19	1.91
Permalloy*	2.17, 2.08	1.91
Heusler alloy	2.01	2.00
Ferrite*	2.20, 2.12	2.04

* Specimens of rather different compositions were used for g and g'.

apparent here that usually $g > 2$, $g' < 2$. An explanation of this has been offered by Kittel [49K5]. It is assumed that ferromagnetism is associated primarily with electron spin ($g = 2$), but that some orbital motion ($g = 1$) enters through the medium of spin-orbit coupling. The effect of the latter is to modify the expressions for g and g' in different ways, so that in first approximation the expressions are

$$g = 2(1 + \epsilon),$$
$$g' = 2(1 - \epsilon).$$

Experimentally a given value of ϵ will explain the usual sign of the deviations from $g = 2$ but not their magnitudes, which are usually about twice as much for g as for g'. The average value of ϵ corresponds to the same order of magnitude of spin-orbit coupling as that derived from other sources.

Beljers and Polder [50B7] have measured g in the double ferrites of nickel and of manganese with zinc. Values progress from about 2.2 for the pure nickel or manganese ferrite to about 1.9 for pure zinc ferrite (which is paramagnetic).

Resonance in Single Crystals.—In the work on Supermalloy described previously, this material was chosen because the crystal anisotropy is very small and hence would not exert a disturbing effect on the precession of electron spins. A high magnetic anisotropy exerts a strong control on local magnetization and has much the same effect as a magnetic field in aligning the magnetization with respect to a crystal axis. In considering the energy of anisotropy in his analysis of ferromagnetic resonance, Kittel [48K9] has included a term containing the crystal anisotropy constant K_1 (see Chap. 12) and the saturation magnetization I_s. As expected, the term in K_1 adds to the terms in B and H. In the special case of a thin cubic crystal cut parallel to a (100) plane, the expression for resonance is

$$\omega_0 = \alpha \sqrt{[B + (K_1/2I_s)(3 + 4\cos 4\theta) + (K_2/2I_s)\sin^2 2\theta][H + (2K_1/I_s)\cos 4\theta]}$$

instead of

$$\omega_0 = \alpha \sqrt{BH}$$

as given before. Here θ is the angle between the strong constant field H and the crystallographic axis [001].

This relation has been tested by the experiments of Kip and Arnold [49K6] on a single crystal of iron containing some silicon. The variation

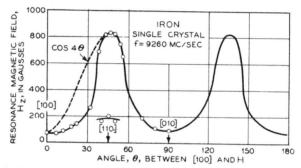

Fig. 17-31. Field at resonance in single crystal of iron, at various crystal orientations. θ is the angle between applied field and [100] direction, in (001) plane. Wave length, 3.2 cm.

of H with θ for a given ω is shown in Fig. 17-31. The theory is generally confirmed, and the deviations that occur can be explained by lack of saturation. The best value of K_1/I_s derived from the experiment agrees within experimental error with the value determined in the usual way.

Crystal anisotropy has also been observed by Yager [48Y3] in resonance experiments on cold rolled specimens of iron.

Bickford [49B4] has used the relation given above to evaluate the anisotropy constant of magnetite, Fe_3O_4, as dependent on temperature. By measuring the field necessary for resonance when a crystal was turned with H parallel to (a) a [100] direction and to (b) a [110] direction, the values of K_1 of Fig. 12-19 have been derived for temperatures between $-150°C$ and room temperature. K_1 passes through zero at 143°C.— Similar results were obtained in both natural and artificial crystals.

Reviews of ferromagnetic resonance have been published by Rado [50R1] and Standley [50S5].

CHAPTER 18

SPECIAL PROBLEMS IN DOMAIN THEORY

KINDS OF MAGNETIC ENERGY

Introduction.—In previous chapters, especially Chap. 11, the domain theory has been described and applied to a variety of problems. It will now be discussed in a more systematic and quantitative manner. First an evaluation of the various kinds of magnetic energy that are associated with domain structure and the change of intensity of magnetization is presented; then follows a description of the manner in which magnetization is changed by application of field, stress, and so forth; finally consideration is given to the nature of initial permeability and coercive force, and to the geometry of domain formation.

The energies having to do with domain structures are related to the following:

Crystal anisotropy
Strain
Magnetic poles (or energy of demagnetization)
Magnetic field (mutual energy between magnetization and external field)
Bloch wall

and expressions for them are given in succeeding paragraphs. Some of them have already been discussed briefly in Chap. 11.

Crystal Anisotropy.—The energy associated with magnetic crystal anisotropy has been evaluated in Chap. 12, and only the results will be described here. For a cubic crystal, such as iron or nickel, the energy per cm^3 is

$$E_k = K(\alpha_1{}^2\alpha_2{}^2 + \alpha_2{}^2\alpha_3{}^2 + \alpha_3{}^2\alpha_1{}^2).$$

Here the α's are the direction cosines of the local magnetization with respect to the cubic axes, and K is the anisotropy constant that measures the amount of anisotropy. The terms containing higher powers of the α's, seldom of importance, have been omitted.

For a hexagonal crystal,

$$E_k = K_1\alpha^2 + K_2\alpha^4,$$

α now being the direction cosine with respect to the hexagonal axis; or sines may be used instead of cosines. This expression gives a proper description of the anisotropy of cobalt, in which all directions at right angles to the hexagonal axis have equal energy, as nearly as have been determined by experiment. In crystals such as Mn_2Sb [43G2], having a tetragonal axis of easy magnetization, additional terms are required to take into account the variation of energy in directions lying in the plane perpendicular to the axis.

The crystal anisotropy is therefore associated with the turning of the magnetization of a domain out of the direction of easy magnetization. It is attributed to spin-orbit coupling.

Magnetic Strain Energy.—This energy is intimately related to the magnetostriction, and for its derivation requires knowledge of the relation between the magnetostrictive change in length and the angle between the magnetization and the direction in which the change in length is measured.

A derivation of this relation will now be made, following Becker and Döring [39B5], for a material having isotropic magnetostriction and no change of volume upon magnetization. Consider a small piece of this material cut in the form of a sphere of diameter l, while it is above the Curie point. Let it cool and become spontaneously magnetized into a single domain. Associated with the magnetization will be a change in shape that may be represented to a first approximation by the equation

$$l = A + B \cos^2 \theta,$$

l being the length of a diameter measured in a direction making the angle θ with the direction of magnetization.

Now consider an ensemble of such domains oriented at random. Then the average length of a domain, measured in any one direction in the ensemble, is

$$l_0 = A + B \overline{\cos^2 \theta} = A + B/3.$$

When the material is fully magnetized so that all domains are oriented at $\theta = 0$, the length of each domain in the direction of magnetization of the ensemble is $l_p = A + B$. Let

$$\lambda = \frac{l - l_0}{l_0}, \qquad \lambda_s = \frac{l_p - l_0}{l_0},$$

then

$$\frac{\lambda}{\lambda_s} = \tfrac{3}{2}(\cos^2 \theta - \tfrac{1}{3}).$$

In these relations λ represents the relative change in length (magnetostriction) of the material as the distribution of directions of the domains changes from an initially random one to one characterized by the angle θ, and λ_s the relative change when all domains are made parallel (saturation magnetostriction).

The strain energy for unit volume of material can now be evaluated by letting the change in length take place in the presence of an external constant tension, σ, as the domains are rotated from the original position to that for $\theta = 0$:

$$E_\sigma = -\sigma \int_0^{\lambda_s} d\lambda = \tfrac{3}{2}\lambda_s \sigma \sin^2 \theta,$$

the zero of energy being taken arbitrarily as that for saturation ($\theta = 0$). When magnetostriction is positive, as in the low-nickel Permalloys, and tension is applied ($\sigma > 0$), the energy is a minimum for $\theta = 0$; therefore tension causes the magnetization to be parallel to the axis of tension. When the magnetostriction is negative, as in nickel, the energy is a minimum for $\theta = 90°$ and therefore tension causes the magnetization to be at right angles to the axis.

When the magnetostriction is not isotropic, but has different values at saturation of λ_{100} and λ_{111} in the [100] and [111] directions of a cubic crystal, the expressions for the magnetostriction change in length and the strain energy have been shown [39B5] to be:

$$\lambda = \tfrac{3}{2}\lambda_{100}(\alpha_1{}^2\beta_1{}^2 + \alpha_2{}^2\beta_2{}^2 + \alpha_3{}^2\beta_3{}^2 - \tfrac{1}{3})$$
$$+ 3\lambda_{111}(\alpha_1\alpha_2\beta_1\beta_2 + \alpha_2\alpha_3\beta_2\beta_3 + \alpha_3\alpha_1\beta_3\beta_1)$$
$$E_\sigma = -\tfrac{3}{2}\sigma[\lambda_{100}(\alpha_1{}^2\gamma_1{}^2 + \alpha_2{}^2\gamma_2{}^2 + \alpha_3{}^2\gamma_3{}^2)$$
$$+ 2\lambda_{111}(\alpha_1\alpha_2\gamma_1\gamma_2 + \alpha_2\alpha_3\gamma_2\gamma_3 + \alpha_3\alpha_1\gamma_3\gamma_1)].$$

Here the direction cosines α refer to the magnetization, β to the magnetostrictive change in length, and γ to the axis of tension σ, all with respect to the crystallographic axes. Zero change in volume is assumed.

Mutual Energy Between Magnetization and External Field.—This energy has been discussed in Chap. 15 and is

$$E_H = -HI \cos \theta,$$

θ being the angle between H and I. Minimum of energy occurs when $\theta = 0$, when H and I are parallel, in accordance with the common experience that a magnet tends to align itself parallel to the field. Zero of energy is taken arbitrarily for $\theta = 90°$.

In domain theory one is concerned usually with regions that are saturated, and then

$$E_H = -HI_s \cos \theta.$$

Magnetostatic Energy.—This represents the work necessary to assemble magnetic poles in a given geometrical configuration. It may also be called the energy of surface charge or of demagnetization and can be calculated when the geometry has certain simple forms. It is fundamentally the integral over all space of the magnetic field,

$$E_p = \int \frac{H^2}{8\pi} dv$$

associated with the poles in unit volume of material.

For an ellipsoid the energy per unit volume of a saturated domain is

$$E_p = \tfrac{1}{2} N_d I_s^2$$

N_d being the demagnetizing factor. For an infinite sheet magnetized at right angles to the surface, $N_d = 4\pi$ and $E_p = 2\pi I_s^2$. When one ellipsoidal domain is completely enclosed by another magnetized in the opposite direction, the energy is quadrupled (because, effectively, I_s is doubled) and is therefore $2N_d I_s^2$. In the case of an enclosed spherical domain, the energy is $8\pi I_s^2/3$, and that of a spherical hole $2\pi I_s^2/3$ per cm^3 of hole. These results are all based on the use of the expression for mutual energy between magnetization and external field, given above.

Néel [44N3] and Kittel [46K3] have calculated the energy associated with coplanar strips of material of thickness d, magnetized to an intensity I alternately in opposite directions at right angles to the plane. Expressed as energy *per unit area* of surface they find

$$e_p = 0.85 d I^2.$$

Energy of Bloch Wall.—Bloch [32B2] showed that the boundary between domains is not sharp on an atomic scale but is spread over a region many atoms thick. Calculation indicates that less energy is required if the electron spins change direction gradually from atom to atom, as indicated in Fig. 18-1 for a wall between antiparallel domains. He pointed out that, whereas the exchange energy is a minimum when the spins are parallel to each other, and the energy of crystal anisotropy a minimum when the spins are parallel to a direction of easy magnetization, the sum of these two energies is least when the transition is as just described, so that the angle between neighboring spins is small. An approximate expression for the thickness and energy of a 180° wall can be made as follows, neglecting small numerical factors.

The exchange interaction (see Chap. 10) between two neighboring atoms is proportional to the thermal energy at the Curie point and to the (negative) cosine of the angle ϵ between their spins. For small angles the energy of two atoms is

$$w_{12} \approx k\theta \epsilon^2.$$

ENERGY OF BLOCH WALL 815

In a wall of thickness δ, with atoms separated by distance a, the angle between neighboring spins is approximately $2\pi a/\delta$ and the exchange energy per cm² of wall area is

$$e_e = w_{12}\delta/a^3 \approx k\theta/a\delta.$$

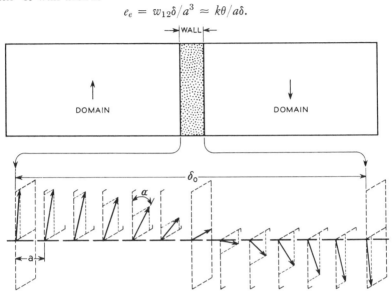

FIG. 18-1. Nature of the domain boundary (Bloch wall).

The energy of anisotropy is measured by the anisotropy constant K (which refers to 1 cm³ of material), and for 1 cm² of wall area is

$$e_k \approx K\delta.$$

The total energy per unit area of wall is then

$$e_w = e_e + e_k = \frac{k\theta}{a\delta} + K\delta$$

and is a minimum for a wall thickness of

$$\delta_0 = (k\theta/aK)^{1/2}.$$

The energies of interaction and crystal anisotropy in such a wall are equal:

$$e_e = e_k = (Kk\theta/a)^{1/2},$$

and the total energy per unit area of the wall is

$$e_w = 2e_e = 2K\delta_0 = (2K\theta)/(a\delta_0) = 2(Kk\theta/a)^{1/2}.$$

When the anisotropy is produced by strain, the magnetic strain energy $\frac{3}{2}\lambda_s\sigma$ must be added to the energy of crystal anisotropy, and the sum determines the wall energy. Thus the expression for wall energy is [38K1]

$$e_w = 2(k\theta/a)^{1/2}(K + 3\lambda_s\sigma/2)^{1/2},$$

with some uncertainty regarding small numerical factors.

Knowing the numerical values of the constants for iron one calculates the wall thickness to be about 1000 atom diameters thick.

More detailed calculations of the thickness and energy of walls of various forms have been made by Landau and Lifshitz [35L1], Becker and Döring [39B5], and Néel [44N1]. Néel considers as physically important only those walls in which the normal component of magnetization is constant across the wall, so that no poles occur at the wall and the magnetostatic energy is zero. His results are expressed in a somewhat different but equivalent form:

$$e_w = a(2EK/3)^{1/2},$$

E being the energy of the molecular field per cm^3, equal approximately to $6k\theta/a^3$. This analysis also shows that the exchange energy and the anisotropy energy are equal at every point in the wall.

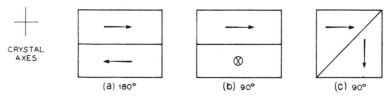

Fig. 18–2. Different kinds of domain boundaries: (a) 180° wall, with all spins parallel to the wall; (b) 90° wall with all spins parallel to the wall; (c) 90° wall in which spins are not parallel to wall.

Visual evidence of the variation of spin direction across the wall is afforded by the powder patterns of Williams, Bozorth, and Shockley [49W1]. The reader is referred to the original paper for a description of the experiment.

Evaluation of Wall Energy and Thickness.—In a cubic crystal having directions of easy magnetization parallel to the [100] crystallographic axes, domain walls are normally of the kinds designated 90° or 180° (see Fig. 18–2):

A *180° wall* can lie in any plane parallel to a [100] direction and still contain no poles. However, the lowest energy is associated with a wall parallel to a (001) plane (a), for in the wall the spins have all directions lying in the plane, and the crystal energy in the (001) plane is the lowest for any plane.

A *90° wall* can lie in any plane parallel to a [110] direction and can therefore be parallel to any one of the three principal planes (100), (110), or (111). The two important cases occur when the wall is parallel to either (b) a (001) or (c) a (110) plane. The wall of least energy is parallel to a (100) plane, and the energy is then just half of that of the 180° wall discussed previously. The energies of the three walls in terms of $a(2EK/3)^{1/2}$, and of some 180° walls, are given in Table 1.

TABLE 1. ENERGIES OF 90° AND 180° WALLS, ASSUMING ZERO MAGNETOSTRICTION

(E is molecular field energy, K the anistropy constant, and a the interatomic distance.)

Kind of Wall	Plane of Wall	Energy/cm², e_w	Values for iron in ergs/cm²
90°	(100)	$0.5a(2EK/3)^{1/2}$	0.8
90°	(110)	$0.86a(2EK/3)^{1/2}$	1.3
90°	(111)	$0.59a(2EK/3)^{1/2}$	0.9
180°	(100)	$1.0a(2EK/3)^{1/2}$	1.5
180°	(110)	$1.52a(2EK/3)^{1/2}$	2.3

Numerical values of wall energies will now be evaluated, assuming that no strains or magnetostrictive forces are present. For iron at room temperature, K is 4.2×10^5. There is some uncertainty about the value of E; Néel has chosen 0.86×10^{10} ergs/cm², but the value calculated from the Curie point (assuming $j = \frac{1}{2}$) is 1.3×10^{10} and that derived from Potter's [34P2] experiments on the magnetocaloric effect is 2.0×10^{10}. As a rough mean value we may choose 1.0×10^{10}. The distance between nearest neighbors is $a = 2.86 \times 10^{-8}$ cm. We then have

$$a(2EK/3)^{1/2} \approx 1.5 \text{ ergs/cm}^2.$$

Néel has also carried out a detailed calculation of the variation of spin direction with distance in going normally through the wall. For example let x be the distance measured from the center of the wall and θ the angle between the spin directions at point x and at the center of one of the domains ($x = -\infty$). For a 90° wall parallel to (100), he finds

$$x = a(E/6K)^{1/2} \ln \tan \theta.$$

This relation is plotted in Fig. 18–3 which shows that the edge of the wall is diffuse and has no exact limit. For convenience, the thickness δ_0 is designated as the distance between two planes at each of which θ has attained $\frac{3}{4}$ of its final value as measured from the middle. This distance is

$$\delta_0 = 3a(E/6K)^{1/2}$$

and is about 500×10^{-8} cm for iron.

The 180° wall parallel to the (100) plane may be regarded as composed of two 90° walls of the kind just described. Néel has pointed out that, if it be so regarded, the separation of the two halves is not definitely determined. If the magnetostriction energy of the wall be added to the

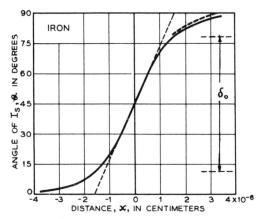

FIG. 18-3. Variation of spin direction with distance in penetrating a wall.

energy of crystal anisotropy, however, the two 90° portions are fixed together so that the total thickness is about

$$\delta_0 = 12a(E/6K)^{1/2}$$

or about 2000×10^{-8} cm for iron.

MAGNETIC PROCESSES

By a magnetic process we refer here to the manner in which the magnetization of a domain can be changed. As already discussed in Chap. 11 this change is commonly associated with *rotation* of the magnetization vector in a domain of fixed volume, or by a change in the volume of the domain by *movement of its boundaries*. The rotation process, characteristic of changes in magnetization in relatively high fields, will be considered first.

Rotational Processes.—Consider the action of a field on a single domain magnetized spontaneously in the direction of a crystal axis. The effect of the field is to rotate the direction of magnetization away from the axis toward the direction of the field. This interaction between *field* and *magnetization* is a fundamental process of magnetization and may be referred to as an $(H,K)_{\text{rot}}$ process because the energies involved are expressed in terms of the symbols H and K. This and other processes will now be discussed.

The rotation process (H,K) has already been considered in Chap. 11 but will be reviewed here briefly. As shown in Fig. 18-4(a), let θ_0 be the angle between the direction of easy magnetization and the field H, and θ the angle through which the magnetization is rotated by the field. In a cubic crystal, in the cube plane, the energies of anisotropy and magnetization (mutual energy between magnetization and field) are

$$E_k = K(1 - \cos 4\theta)/8$$
$$E_H = -HI_s \cos(\theta_0 - \theta).$$

Equilibrium will occur when

$$\frac{\partial}{\partial \theta}(E_k + E_H) = 0$$

or when

$$H = \frac{K \sin \theta}{2I_s \sin(\theta_0 - \theta)}.$$

The component of I_s parallel to H is then

$$I = I_s \cos(\theta_0 - \theta).$$

Similar expressions can be derived readily for other specific conditions from appropriate expressions for E_k obtained from the general expression.

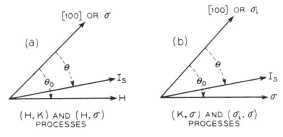

FIG. 18-4. Designation of vectors and angles for describing processes in which the field (H) and stress (σ) act on the crystal forces (anisotropy constant, K) and internal strain (σ_i).

The mutual effect of *field* and *stress*, (H,σ) process, can be similarly treated. Using Fig. 18-4(a) we have

$$E_\sigma = \tfrac{3}{2}\lambda_\sigma \sin^2 \theta$$
$$E_H = -HI_s \cos(\theta_0 - \theta)$$

and for minimum energy

$$H = \frac{3\lambda_s \sigma \sin 2\theta}{2I_s \sin(\theta_0 - \theta)}$$
$$I = I_s \cos(\theta_0 - \theta).$$

In the same way (K,σ) and (σ_i,σ) processes can be evaluated; these occur when an applied stress produces rotation against the forces of anisotropy or internal strain [Fig. 18–4(b)] and are described respectively by the relations:

$$\sigma = \frac{K \sin 4\theta}{3\lambda_s \sin 2(\theta_0 - \theta)}$$

and

$$\sigma = \frac{\sigma_i \sin 2\theta}{\sin 2(\theta_0 - \theta)}.$$

The change with field of many other quantities besides magnetization are calculable with the same general theory. These have been described in the chapters on stress, magnetostriction, crystals, resistivity, etc.

Boundary Displacement.—Becker and Döring [39B5] have calculated the movement of a 90° wall under the influence of a field applied at an angle of θ_1 with the direction of magnetization in one of the adjacent

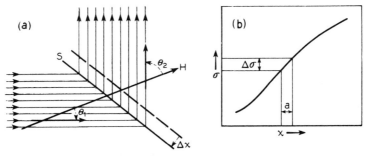

Fig. 18–5. Effect of magnetic field (H) on the displacement of a domain wall (S) restricted by internal stress (σ).

domains and θ_2 with the other, as indicated in Fig. 18–5. Let dx be the movement of the boundary caused by a field of strength H. The change in the mutual energy between magnetization and field, in a volume Δv of material, is

$$E_H \Delta v = -HI_s(\cos \theta_1 - \cos \theta_2) \Delta v.$$

By analogy with the expression for the energy associated with hydrostatic pressure, $p \Delta v$, the effect of the field may be said to exert a pressure on the wall of magnitude

$$p = HI_s(\cos \theta_1 - \cos \theta_2).$$

The force opposing this pressure is assumed to be a local stress, σ, having a gradient $d\sigma/dx$. When the wall is displaced by a distance Δx, the energy associated with the strain is

$$E_\sigma = \tfrac{3}{2}\lambda(d\sigma/dx)\,\Delta x$$

and this energy is supplied by the field (see E_H above); consequently

$$H = \frac{3\lambda(d\sigma/dx)\,\Delta x}{2I_s(\cos\theta_1 - \cos\theta_2)}.$$

Now let A be the total 90° wall area in 1 cm³ of material, so that A is the volume of material over which the wall has moved in each cm³ of the specimen. Then the change in magnetization is

$$I = I_s(\cos\theta_1 - \cos\theta_2)A\,\Delta x$$

and the susceptibility associated with this process of magnetization is

$$\kappa = \frac{I}{H} = \frac{2I_s^2(\cos\theta_1 - \cos\theta_2)^2 A}{3\lambda(d\sigma/dx)}.$$

Taking into account the values of $(\cos\theta_1 - \cos\theta_2)^2$ of all 90° boundaries in a material in which the local strains are oriented equally in all directions, Becker and Döring find a factor of 8 by which the foregoing value of κ should be multiplied. The magnetostriction λ is that for the [100] direction. The susceptibility is then

$$\kappa = \frac{16 I_s^2 A}{3\lambda_{100}(d\sigma/dx)}.$$

At the present time the usefulness of this equation is limited by our lack of knowledge of the magnitudes of A and $d\sigma/dx$. In order to be able to compare with experiment, Becker and Döring have made the simplifying assumptions described in the next section.

For discussion of the 180° boundary, see pp. 823-4.

APPLICATION TO SPECIFIC PROBLEMS

Initial Permeability.—This quantity will be calculated for the following conditions:

(a) Crystal energy large, internal strain small (e.g., well-annealed iron).
(b) Strain energy large, crystal energy small (e.g., unannealed Permalloy).

First (a) will be considered, assuming that the process is rotational. For this case we have the expressions (p. 577) for H and I in terms of the angle θ between magnetization and crystal axis, and from these it follows that when H and I are small

$$\kappa_0 = \left(\frac{dI}{dH}\right)_{\theta=0} = \left(\frac{dI/d\theta}{dH/d\theta}\right)_{\theta=0} = \frac{I_s^2 \sin^2\theta_0}{2K}.$$

When θ_0 is 45°

$$\kappa_0 = I_s^2/(4K).$$

For iron at room temperature, $I_s = 1720$, $K = 4.2 \times 10^5$ ergs/cm^3, and the foregoing relation gives $\mu_0 = 1 + 4\pi\kappa_0 = 20$. For nickel $I_s = 485$, $K = 35\,000$, and the calculated value of μ_0 is also about 20. Observed values of μ_0 for both iron and nickel in the annealed condition is about 100. As pointed out by Becker [33B3], this calculation shows that the mechanism responsible for the initial permeability of these metals is not rotation of the kind assumed. This conclusion led Becker [32B8] to adopt the mechanism of the moving boundary, also discussed by Bloch [32B2], Heisenberg [31H6], and Sixtus and Tonks [31S5] and described above.

The expression for μ_0 for a 90° *moving boundary* has been given in the preceding section. To make use of it Becker and Döring have assumed the internal strain to be sinusoidal in character, with a maximum amplitude of σ_i and a wave length equal to the dimension of the domains, which are taken quite arbitrarily to be cubical in form with an edge length of l cm. The total wall area is then $3/l$ per cm^3, the stress gradient is $\pi\sigma_i/l$, and the expression for κ_0 reduces to [32B8]

$$\kappa_0 = \frac{4}{3\pi} \frac{I_s^2}{\lambda_{100}\sigma_i}$$

independent of the size of the domains.

For condition (b), a rotation process, designated $(\sigma,H)_{\text{rot}}$, the appropriate equations have already been given (p. 819). From these it is easy to derive (dI/dH) for $\theta = 0$ and thus obtain [30B4]

$$\kappa_0 = \frac{2I_s^2}{9\lambda_s\sigma_i},$$

provided the internal stresses, $\sigma = \sigma_i$, are randomly oriented. Thus the two mechanisms just described, boundary displacement and rotation against internal stress, give nearly the same value of initial susceptibility. in each case it is the presence of internal strain that prevents κ_0 from becoming infinite.

In a pure, well-annealed material in which the domains are directed at random, the smallest internal strains to be expected are those associated with magnetostriction and brought into existence when the material becomes spontaneously magnetized as it cools through the Curie point. Kersten [31K1] has proposed that the stresses they have may then be as low as

$$\sigma_i = \lambda_s E,$$

E being Young's modulus. Use of this in the foregoing equation, in which μ_0 may replace $4\pi\kappa_0$, gives

$$\mu_0 = \frac{8\pi I_s^2}{9\lambda_s^2 E}.$$

This relation was tested by Kersten [31K1] by comparing the calculated values for the iron-nickel series of alloys, for which values of λ_s are known, with the highest observed initial permeabilities. Results [35B1] are shown in Fig. 13–42, and give good confirmation of the theory. It is to be expected that the theory will break down for $\lambda_s = 0$.

Initial permeability can also be calculated for a material subjected to a *uniform tension* and has been investigated by Becker and Kersten [30B4, 31K6]. Application of tension to nickel creates the energy

$$E_\sigma = \tfrac{3}{2}\lambda_s \sigma \sin^2 \theta$$

and since $\lambda_s < 0$ the minimum energy will occur when $\theta = 90°$, that is, when the domains are oriented at right angles to the axis of tension. The effect of the field, applied parallel to this axis, is to rotate the domains and give rise to an initial susceptibility. This is calculated in the same way as for case (b) above, except that $\sin \theta_0 = 1$, and is

$$\kappa_0 = \frac{I_s^2}{3\lambda_s \sigma}$$

or

$$\mu_0 - 1 = \frac{B_s^2}{12\pi\lambda_s} \cdot \frac{1}{\sigma}$$

when using the usual units, with σ in dynes/cm^2. For nickel, $B_s^2/(12\pi\lambda_s)$ is 2.9×10^{10}. Then when σ is in kg/mm^2,

$$\mu_0 - 1 = \frac{290}{\sigma}.$$

Good agreement between theory and experiment is obtained when the tension is large enough to overshadow residual internal stresses. Relevant data are plotted in Fig. 13–6.

Measurements of μ_0 under tension have also been made by Scharff [35S9] at various temperatures up to the Curie point. Just below the Curie point the theory fails to account quantitatively for the measurements by a factor of about 3. However, λ_s was not measured on the same specimen as μ_0, so that a deviation from theory has not been established.

Coercive Force.—The first important theory of magnetic hysteresis relates this phenomenon to strain gradients, $d\sigma/dx$, in the material, and results in an expression for H_c in terms of $d\sigma/dH$, λ_s, I_s, and δ_0. The calculation has been carried out by Kondorski [37K5, 38K1], essentially as follows: Consider a Bloch wall between two antiparallel domains, subject to a stress gradient $d\sigma/dx$ and a stress σ, both assumed to be practically constant across the narrow wall. When the wall of constant area A moves a distance dx under the influence of a field, there is a change in the energy of the wall that is balanced by the mutual energy of field and magnetization. Equating these energies we have:

$$A\, de_w = HI_s(\cos\theta_2 - \cos\theta_1) A\, dx.$$

For a 180° wall $\cos\theta_2 = 1$, $\cos\theta_1 = -1$, and $H = (de_w/dx)/(2I_s)$. Since the change of wall energy with distance is attributed to stress alone, we can use the expression given previously for wall energy in terms of stress and wall thickness, δ_0:

$$e_w = 3\lambda_s \sigma \delta_0.$$

Since the coercive force is the field required to move the wall, when the gradient has its maximum value, we can write

$$H_c \approx (\lambda_s \delta_0 / I_s)(d\sigma/dx)_m,$$

the desired expression for coercive force in terms of maximum strain gradient.

In reality the region over which σ and $d\sigma/dx$ are constant may not be large compared to the wall thickness. The effect of strain in homogeneities that occur in regions small compared to the wall thickness has been discussed by Kersten [38K1]. If a change in local stress $\Delta\sigma$ occurs in a distance $l \ll \delta_0$, the expression he derives is

$$H_c \approx (\lambda_s l / I_s)(\Delta\sigma/\delta_0).$$

When $l \approx \delta_0$, he finds

$$H_c \approx \lambda_s \Delta\sigma / I_s.$$

If the change in local stress is equal to the amplitude of the internal stress σ_i, the expression becomes

$$H_c \approx \lambda_s \sigma_i / I_s.$$

We shall see later that these formulas are based on a physical picture that is quite incomplete.

In a material having a large amount of internal stress we may represent the stress as a function of position as in Fig. 18–6; in the demagnetized state the wall will assume a position of zero gradient, as at (a), and with increasing field strength will move until the gradient becomes large enough to prevent further motion—until the "hill" is "too steep." The highest field necessary for wall displacement is required when the gradient is greatest, as at (b).

Another representation of wall movement is in Fig. 18–7, where the hills are shown by contour lines and the wall is shown in two positions, the right-hand position having been attained after an increased field has moved the wall over one of the hills. Figure 18–7 brings out one of the difficulties with the strain theory: if the wall is curved as indicated, and its position dependent on the shape of the hills of stress, the magnetization in the regions on either side of the wall must vary in direction from place to place. This necessitates the formation of magnetic poles associated with large amounts of energy, which are not taken into account in the theory (see

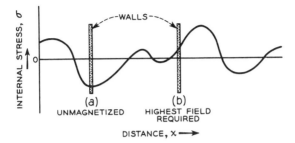

FIG. 18–6. Equilibrium positions of a domain wall in material in which stress varies with position.

below). Other difficulties are discussed later, and it is shown in Fig. 18–13 that there is no proportionality between H_c and λ_s in permanent magnet materials generally.

In 1943 Kersten [43K5, 43K6, 48K11] published his *foreign body theory*, based on a change of wall energy with change in *wall area* instead of

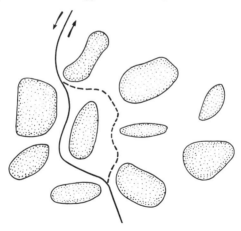

FIG. 18–7. Positions of minimum wall energy in material with internal strains shown by contours.

energy per unit area of wall. This is best described with the aid of Fig. 18–8, in which non-magnetic areas or particles of diameter d are represented by spheres arranged on a simple cubic space lattice having a repetition distance S_2. When the wall moves from position (a) to position (b), the wall area increases in the ratio $(A_2/A_1)^2$, and a calculable field is required to account for the corresponding increase in energy. The wall is like a soap film in the tendency to make its area a minimum and hence is stretched between the "foreign bodies" as at (a) or (c) unless moved by the force of a field. When the field is sufficient to move the wall so that it does not touch any particles, it moves freely until the next line of particles is reached.

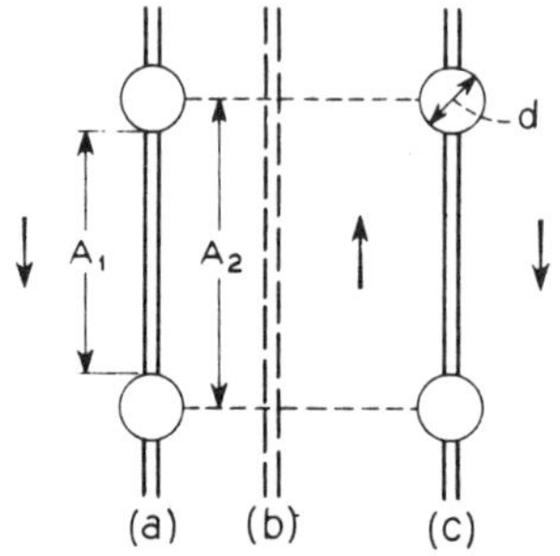

In deriving the expression for the coercive force, which in the model is the field necessary to move the wall from (a) to (b), the method is similar to that used in the strain theory except that the energy, instead of being equal to $A\,de_w$ is $d(Ae_w) = e_w\,dA$, because now e_w is constant while A is varied. The result is:

$$H_c \approx e_w(dA/dx)/(2I_sA).$$

FIG. 18-8. Kersten's foreign body theory. Wall area changed by cavities or inclusions.

When A is expressed in terms of A_2 and d (Fig. 18-8)

$$H_c \approx (e_w/I_s)(d/A_2^2).$$

Further, e_w may be expressed in terms of the wall thickness, which is now determined by the anisotropy constant (if K is high) and not by internal strain (assumed to be negligible). Then

$$H_c \approx (K\delta_0/I_s)(d/A_2^2).$$

If α represents the volume fraction of the material that is occupied by non-magnetic particles, Kersten finds

$$H_c \approx \frac{5K\delta_0\alpha^{2/3}}{I_sd},$$

provided $d \gg \delta_0$.

This equation has been compared with experiments on iron containing carbon. Assuming the carbon to be in the form of Fe_3C, which acts essentially as a non-ferromagnetic material, and that $d = 20\delta_0$, Kersten finds good agreement with the data of Köster [30K8] as shown in Fig. 18-9.

As stated by Néel [44N2], the same difficulties occur with this theory as with the strain theory—the energy associated with magnetic poles has been neglected.

In *Néel's theory* he has pointed out that there will be magnetic poles on the surface of the non-magnetic inclusions of Kersten's model, and he has calculated the magnitude of the corresponding energy. The difference in energy for the two positions of the wall (Fig. 18-8) is calculable and is found to be of the order of 100 times the energy of the wall when particles of 10^{-4} cm diameter are imbedded in iron [44N2].

In a later more comprehensive theory of coercive force Néel [46N2] has tried to calculate the fluctuations in the *direction* of magnetization caused by variations in elastic stress, and the fluctuations in *intensity* of magnetization caused by non-magnetic inclusions. As a consequence of fluctuations of either kind there will be magnetic poles (surface magnetic charge) distributed throughout the material, and associated with them will be an energy which will largely determine the positions of the domain walls and their movement with a field. In a material like iron most of the energy is associated with non-magnetic inclusions; in nickel, which has lower crystal anisotropy and higher magnetostriction, the energy is associated more with fluctuations in strain. For these two cases the expressions derived are as follows: for iron

$$H_c = 360\alpha + 2.1v,$$

and for nickel,

$$H_c = 97\alpha + 330v.$$

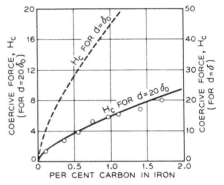

FIG. 18–9. Theoretical (line) and experimental increase of coercive force with cementite inclusions. Agreement occurs when particle diameter is 20 times the wall thickness.

Here α is the volume fraction of inclusions, as before, and v is the fraction of the material that is subject to a large disturbing stress, σ_i, which is chosen in these relations to be 3×10^9 dynes/cm^2, a value supposedly characteristic of severely strained material. The coefficients show the relative importance of strain in nickel and inclusions in iron. The fraction v is difficult to determine experimentally, but its coefficient is obviously of the right order of magnitude. Comparison of the relation for iron is more easily made and is shown in Fig. 18–10, using data collected by Kersten. The straight line is drawn according to theory, and agreement between theory and experiment is impressive. For more detailed consideration of Néel's theory the reader is referred to the original and subsequent [49N2, 49N3] articles. The general expressions, of which the foregoing are special cases, are:

$$H_c \approx (1.05v\lambda_s\sigma_i/I_s)[1.39 + \log(4.6I_s^2/\lambda_s\sigma_i)^{1/2}]$$

and

$$H_c \approx (2K\alpha/\pi I_s)[0.39 + \log(2\pi I_s^2/K)^{1/2}].$$

The symbols are those ordinarily used, K being the anisotropy constant.

At the present time we believe that coercive force is closely connected with some kind of inhomogeneity in the material, but the specific nature of the inhomogeneity and its exact relation to H_c are uncertain. A graphic illustration of the effect of a visible inhomogeneity on wall displacement is shown by the powder patterns and drawings of Fig. 18–11 taken from the work of Williams and Shockley [49W2]. When a principal domain wall moves past such a hole, superficial domain structure forms around it and unites with the wall as shown. As the principal wall continues to move as the result of an applied field, the walls extending to the hole become more and more extended and then suddenly break; a new structure is formed around the hole, and the principal wall becomes straight. The drag of the imperfection on the main wall obviously influences the field necessary to move the wall in its neighborhood, and similarly other imperfections, large or small, affect wall movements and may determine the coercive force.

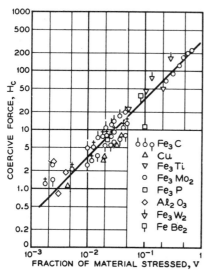

Fig. 18–10. Néel's theory (line) of the coercive force as dependent on the relative volume of inclusions.

The high coercive force of fine particles is discussed in the succeeding section.

Fine Particles.—During the last few years several independent investigations have shown that high coercive force can be attained in powders when the particle size is sufficiently small. In such particles there is no boundary formation, and the usual mechanism of magnetization by boundary displacement cannot occur; therefore, any change of magnetization must occur by domain rotation, a process that takes place to a considerable extent only in strong fields. Guillaud [43G2] first reported that the coercive force of MnBi increased as the particle size decreased and obtained a value of 12 000 (for $_IH_c$) when the particles were as small as 3×10^{-4} cm in diameter; he showed that the coercive force was higher the greater the crystal anisotropy of the material. Both Néel [47N1, 47N2] and Stoner and Wohlfarth [48S1] have shown theoretically that high coercive forces are to be expected when fine particles have either high crystal anisotropy or anisotropy of particle shape (non-spherical form).

All of this work was foreshadowed in a little-known paper by Antik and Kubyschkina [34A3], who showed that according to Akulov's theory fine particles should have coercive force and hysteresis loss dependent on crystal anisotropy. They found experimentally that iron dispersed in mercury ("amalgam") had hysteresis loss proportional to the crystal anisotropy constant, in accord with theory. Also, Frenkel and Dorfman [30F5] proposed in 1930 that small particles should be single domains. In 1935 Gottschalk [35G6] found that the coercive force of magnetite was increased when the particles were made finer. Soon thereafter Dean and Davis [41D1] made a permanent magnet by compressing fine powder of

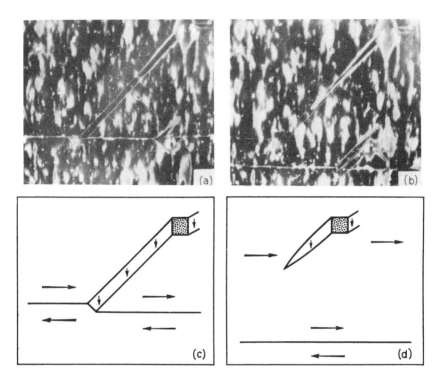

Fig. 18–11. Diagram of domain structure around cavity, before and after breaking connection with moving wall, and corresponding experimental patterns.

iron and other metals prepared by electrodeposition on a mercury electrode.

In 1946 Kittel [46K3], independently of previous work, showed by energy considerations that fine particles as well as thin films and wires should be composed of single domains and should have high coercive forces. Following methods similar to his, the energetics of domain structure is illustrated in Fig. 18–12. The cube (a) of edge l is a single domain and has poles on two opposite surfaces. Cube (b) is composed of four domains, magnetized parallel to the directions of easy magnetization in the (cubic) crystal and arranged so that there is complete flux closure and no poles. The magnetostatic energy in (a) is

$$E_p l^3 = N_d I_s^2 l^3 / 2 = 2\pi I_s^2 l^3 / 3,$$

the demagnetizing factor N_d being taken as $4\pi/3$. The energy of the Bloch walls in (b) is

$$2\sqrt{2}\, e_w l^2.$$

These two energies are equal when the cube edge is

$$l_0 = 3\sqrt{2}\, e_w / (\pi I_s^2).$$

When $l \ll l_0$ the single domain particle has less energy and is the stable one, and when $l \gg l_0$ the particle will have the domain structure of bulk

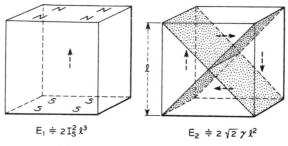

$$E_1 \doteq 2 I_s^2 \ell^3 \qquad E_2 \doteq 2\sqrt{2}\,\gamma \ell^2$$
$$E_1 = E_2 \text{ WHEN } \ell \doteq 10^{-6} \text{ CM IN IRON}$$

Fig. 18-12. Two possible domain configurations in a small particle: (a) is stable for very small particles; (b), for large particles.

material. In iron we may take $e_w \approx 2$ ergs/cm^2, $I_s \approx 1700$, and then find $l_0 \approx 10^{-6}$ cm. This estimate is only a rough one, and the diagrams of Fig. 18-12 should not be taken too literally because near $l = l_0$ the width of the Bloch wall may be of the same order of magnitude as the domain dimensions and, therefore, not as sharply defined as in the model. However, more extensive analyses [47N1, 48S1, 48K2] also contain approximations and give about the same result.

Stoner and Wohlfarth [48S1] consider anisotropies of three kinds: (1) crystal, (2) shape, and (3) strain; and they derive expressions for initial permeability, remanence, and coercive force of particles below critical size. One can infer immediately that in most materials, e.g., iron, the coercive force will be greatest when there is anisotropy of shape, for here there is the greatest difference in the energies of demagnetization in different directions. If the particles are considered to be ellipsoids of revolution, and N_a and N_b are the demagnetizing factors in the directions of the long and short axes, the difference in the energies in these two directions is

$$E_p = (N_b - N_a) I_s^2 / 2,$$

and in the extreme case of very long thin ellipsoids this is

$$E_p = \pi I_s^2.$$

The corresponding expression for crystal anisotropy—the greatest difference in energy in different directions in a cubic crystal—is

$$E_k = K/\sqrt{3},$$

and the relation for strain anisotropy is

$$E_\sigma = 3\lambda_s\sigma/2.$$

Consideration of the observed values of I_s, K, λ_s and σ shows that in iron $E_p > E_k > E_\sigma$ (taking $\sigma = 2 \times 10^8$ for a highly strained material).

The most detailed results of calculation are given by Stoner and Wohlfarth for anisotropy of shape. The coercive force (for $I = 0$) varies with the orientation of the ellipsoid from 0 to $(N_b - N_a)I_s$ and, for an ensemble of randomly oriented ellipsoids, has the value

$$H_c = 0.479(N_b - N_a)I_s.$$

Initial permeability and retentivity for such an ensemble are

$$\mu_0 = 8\pi/3(N_b - N_a) + 1$$

$$I_r = I_s/2.$$

(The remanence I_r and saturation I_s are understood to refer to the particles themselves and do not include the space between particles.) The relation

$$H_c = 4I_s/\mu_0$$

is approximately valid.

Similar relations apply to crystal and strain anisotropies, but then H_c is expressed in terms of the crystal anisotropy constant K or the product of saturation magnetostriction and stress, $\lambda_s\sigma$. When the crystal is uniaxial in symmetry, like cobalt,

$$H_c = 0.96K/I_s.$$

For a cubic crystal, like iron, having spherical particles and directions of easy magnetization along the cubic axes the highest coercive force is

$$H_c = 2K/I_s.$$

For an ensemble of such particles oriented at random Néel [47N1] obtains

$$H_c = 0.64K/I_s.$$

When stress is uniaxial and there is no anisotropy of form or crystal [48S1],

$$H_c = 1.44\lambda_s\sigma/I_s.$$

In all of the foregoing calculations the mutual magnetic forces between particles have been neglected.

In iron powders Weil [48W1] has observed coercive forces of more than 1000. Crystal anisotropy can account for only $0.62K/I_s \approx 150$ oersteds in a randomly oriented material, and $2K/I_s = 490$ in material having particles favorably oriented. The higher value actually observed is attributed to anisotropy of shape, which may theoretically push H_c as high

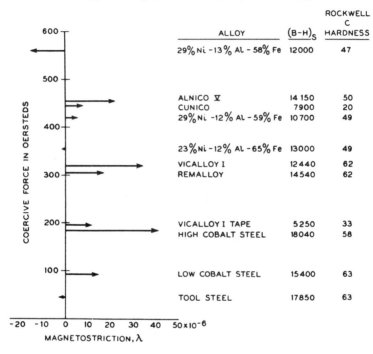

Fig. 18-13. Saturation magnetostriction of some permanent magnet materials, showing lack of proportionality between H_c and λ_s.

as $0.479 \times 2\pi I_s = 5200$ in long, thin particles randomly oriented. Weil has also reported coercive forces of 1200 in iron-cobalt alloys, 500 in cobalt, and 200 in nickel.

Stoner and Wohlfarth, and Néel [47N3] have suggested that the high coercive forces of modern permanent magnet materials such as the Alnicos may be attributed to the anisotropy of the fine particles or magnetic regions composing them. The strain theory, in which $H_c \approx \lambda_s \sigma_i/I_s$, cannot account for the coercive forces of these materials, for Nesbitt [50N1] has shown that H_c may be as high as 400 when λ_s is practically zero (Chap. 13). Figure 18-13 shows λ_s and H_c for some common materials used in permanent magnets.

When crystal anisotropy is high, as in MnBi, it may be the cause of the high coercive force in fine particles. In many materials anisotropy of shape seems the only possibility. It has recently been suggested [48H1] that this applies also to stainless steel of the 18-8 variety, which becomes magnetic only when work-hardening causes the precipitation of a new phase.

The variation of coercive force with temperature gives some indication of the type of anisotropy. If it is crystal anisotropy, one expects to find a decrease of H_c with temperature, when the crystal anisotropy constant K of the material decreases rapidly in this same interval of temperature. If the anisotropy is caused by shape, a less rapid change of H_c with temperature is to be expected. Experiments of this kind, by Weil and Marfoure [47W2], are illustrated in Fig. 18-14. It is concluded [47W2, 47N1] that

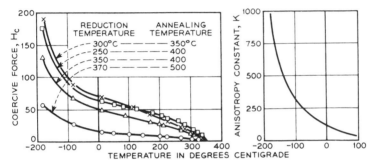

FIG. 18-14. Dependence of the coercive force of fine particles of nickel on the temperature. Powder prepared by reduction of nickel oxalate at temperatures of 370°, 350°, 250° and 200° C, and annealed at 500°, 400°, 400° and 350°, in samples I to IV, respectively. Also, for comparison, the change of anisotropy with temperature.

crystal anisotropy accounts for much, but not all, of the coercive force of the specimens of powdered nickel. A similar course of H_c vs T curves for precipitation-hardened alloys has been observed by Gerlach [47G1] and his results indicate that the origin of the coercive force in these materials may be the crystal anisotropy of the fine particles precipitated in the alloy.

Further data on the variation of coercive force with temperature have also been obtained by Galt [50G4] on fine particles of 68 Permalloy, material which in bulk form is known to have very small crystal anisotropy (Fig. 12-14). The curve for this material also shows the characteristic rise with decreasing temperature and so supports the conclusion that factors other than crystal anisotropy are of major importance.

Bertaut [49B11] has shown experimentally that particles of iron of various sizes, formed from iron formate by reducing at various temperatures, have maximum coercive force when their diameter is about 150A, and that the coercive force approaches zero for very much smaller particles. Néel [49N5] has shown theoretically that if particles are smaller than a certain critical size the coercive force will be less, because thermal agitation of small particles will prevent the existence of a stable magnetization. Calculation indicates this critical size to be about 160A for iron particles at room temperature, in good agreement with Bertaut's data. The critical size will be smaller at lower temperatures, therefore the coercive force of particles smaller than the critical size will increase rapidly as the temperature decreases. This is consistent with the data of Fig. 18–14, and with additional data obtained by Weil [49W3] using nickel prepared by the Raney process and estimated to have a particle size of about 60A. Similar considerations have been proposed independently by Ekstein and Gilbert [50E1].

König [46K5] has estimated that particles as small as 10–12A (4 atom diameters) in thickness are ferromagnetic. This is based on experiments

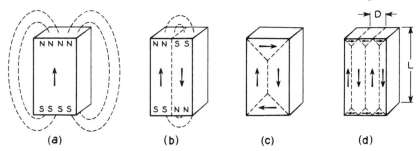

Fig. 18–15. Possible domain structures, showing large magnetostatic energy associated with isolated domain (a), and successively lower energies associated with (b), (c), and (d). The last represents the kind of domain structure actually observed.

on the Faraday optical rotation, which is as high as 7×10^5 degrees/cm in "thick" films and much less than this when the critical size is approached.

Domain Geometry.—Powder patterns of domain boundaries have been described in Chap. 11, and it was there stated that the energetics of the simpler patterns could be understood. Quantitative consideration of domain geometry of unmagnetized material has been given recently by Néel [44N3] and Lifshitz [44L1], and Williams, Bozorth and Shockley [49W1] have made progress by comparing their experiments with theory.

The kinds of energy that must be evaluated are illustrated in Fig. 18–15. Here (a) is a single domain block of material like iron, magnetized upward and having magnetic poles on its top and bottom. The magnetostatic energy associated with it can be reduced by inserting a Bloch wall as in (b). At the same time the energy associated with the wall has been added, but the total energy is lower and the domain arrangement more stable, provided the particle size is not too small. A further reduction in total energy can be made by inserting walls as shown in (c). This introduces some elastic or magnetostriction energy, because the top and bottom domains have been elongated in the direction of magnetization by magnetostriction and hence do not fit into the spaces allotted to them without causing some strain. The magnetostriction can be reduced by introducing more walls as in (d), and this process can be continued until the added wall energy just balances the reduced magnetostriction energy. Domain structures of this kind are observed in powder patterns, as shown in Fig. 11–52, when the surfaces are parallel to (100) planes.

The energies illustrated in Fig. 18–15, namely,

> Magnetostatic energy,
> Magnetostriction energy,
> Wall energy,

are believed to have a minimum sum for the stable domain arrangement. Calculations have been carried out [49W1] for the tree patterns such as those reproduced in Fig. 11–50(b). As illustrated in Fig. 18–16, the flux

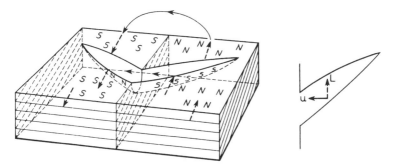

Fig. 18–16. Interpretation of the "tree" pattern. Lines on side surfaces are traces of (100) planes.

in one of the domains follows a direction of easy magnetization, shown by the dotted arrows on the right half of the figure, and where it cuts the surface there will be north (positive) poles. If it passes through the air as shown and enters the crystal in the adjacent domain, south (negative) poles are formed. The magnetostatic energy associated with the poles is then quite large; it can be reduced if the flux passes transversely, along a direction of easy magnetization, within the "branch" of the tree as indicated. The known energy of the walls of the branches, plus the magnetostatic energy, is found to be a minimum in the patterns observed, and the relative areas of "branches" and "sky" are observed to change in the calculated way as the angle between the surface and the crystallographic (100) plane is varied.

A further test of theory is afforded by the powder patterns around holes or inclusions in the material, as illustrated in Fig. 18–17 [49W1]. Néel

[44N2] considered these theoretically before any such observations had been made, and the form of the pattern is a striking confirmation of his proposals. Imagine a hole in a domain, as shown at (1) of Fig. 18-17(c). Then magnetic poles will exist on the surfaces normal to the direction of magnetization, and the magnetostatic energy (or demagnetizing energy) associated with them can be calculated. If domains are formed as at (2), poles will not be present at the edges of the cavity but will be distributed along the domain boundaries as indicated by the N's and S's. In this case there will be energy associated both with the demagnetization (a volume effect) and with the boundaries (an area effect). The minimum

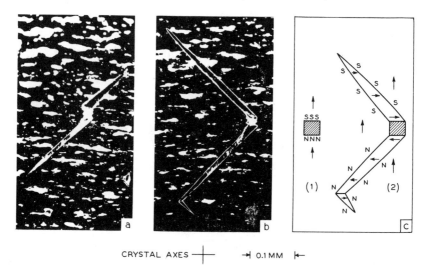

Fig. 18-17. Domain structure around cavities in a crystal; (a) and (b), patterns observed; (c1), diagram showing square hole; (c2), domain walls formed as postulated by Néel, with consequent distribution of poles over tapered domains.

of the total energy (demagnetization plus wall energy) determines the scale of the domains. Calculation for a hole 0.001 cm in diameter leads to a ratio of length to width of domain of about 100, whereas 40 is observed, a reasonably good agreement theory and experiment. Details of the calculation may be found in the original article [49W1].

Another test of the theory may be made by calculating the distance between domain boundaries in the configuration of Fig. 18-15(d). As stated previously the principal energies involved are the magnetostrictive energy and the wall energy. The former is

$$E_\sigma = E_{100}\lambda_{100}^2/2$$

ergs per cm^3 of volume of the domains of closure. For silicon-iron, Young's modulus is $E_{100} = 1.2 \times 10^{12}$ ergs/cm^2 [46B3]; the saturation magnetostriction in the [100] direction is $\lambda_{100} = 24 \times 10^{-6}$ [47S3], and the domain wall energy is about $e_w = 1$ erg per cm^2 of wall. For 1 cm^3 of a crystal slab composed of domains L cm long and D cm wide (Fig. 18–15), the volume of domains of closure is

$$V_c = D/(2L)$$

and the wall area is

$$A = [L + (2\sqrt{2} - 1)D]/(LD).$$

The sum of the energies is then

$$E = V_c E_\sigma + A e_w$$

per cm^3 of crystal, which is a minimum for

$$DL^{-\frac{1}{2}} = (2/\lambda_{100})(e_w/E_{100})^{\frac{1}{2}},$$
$$= 0.08$$

for silicon-iron. The calculation may be compared with observation for the crystal of Fig. 11–52. Here $L = 0.22$ cm, and the calculated and observed value of D are 0.04 and 0.05 cm, respectively. The agreement is better than could be expected considering the approximations made.

CHAPTER 19

MEASUREMENT OF MAGNETIC QUANTITIES

BASIC RELATIONS

As a prelude to a discussion of the production and measurement of magnetic fields and of intensity of magnetization, it is desirable to consider quantitatively the fields and forces excited by currents and magnets. These may be discussed according to the following list:

(a) The fields produced by magnets.
(b) The fields produced by currents.
(c) The force on a magnet in a field.
(d) The force on a current in a field.
(e) The electromotive forces produced by changing induction.

It is on these relations that most of the measurements of induction and field strength are based.

Field of a Magnet.—Consider a bar magnet NS (Fig. 19-1) of pole strength m and interpolar distance l and therefore magnetic moment $M = lm$. Using the fundamental relation according to which the field strength varies inversely as the square of the distance d from a pole,*

$$H = m/d^2, \qquad (1)$$

and adding vectorially the fields produced by both poles it may be shown that the field strength at distant r from the center of the magnet given by

$$H = (M/r^3)(1 + 3\cos^2\theta)^{1/2},$$

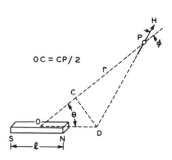

FIG. 19-1. Field produced at any point P by presence of bar magnet NS depends on distance r and angle θ.

provided r is large compared to l. Here θ is the angle between the axis SN of the magnet and the line OP drawn from its center to the point P, at which the field is observed. The direction of the field at this point is defined by φ, the angle it makes with

---
* This relation applies exactly when the magnet is surrounded by a vacuum; when the surrounding medium is air a very small modification may be necessary. In general, when surrounded by a medium of permeability μ, the law is $H = m/\mu d^2$.

OP prolonged, and is determined by

$$2 \tan \varphi = \tan \theta.$$

A graphical construction based on this relation may be made easily by trisecting OP so that $OC = CP/2$, drawing CD at right angles to OP to cut SN produced at D. Then DP is the direction of the field at P.

Important special cases are the end-on position (first Gaussian position) for $\theta = 0$, and the broadside position (second Gaussian) for $\theta = 90°$. For these the fields are, respectively,

$$H_1 = (2M/r^3)(1 + l^2/2r^2),$$

$$H_2 = (M/r^3)(1 - 3l^2/4r^2).$$

When two magnets are placed very closely together end-to-end, with opposite poles m_1 and m_2 separated by a distance, d, small compared to the extent of the surfaces, the two magnets are attracted with a force that may be very strong. For each portion of one of the near-by surfaces the corresponding part of the other has an attraction given by the fundamental relation

$$f = \frac{m_1 m_2}{d^2}.$$

Summing for the effect of all parts of one surface on all parts of the other, one obtains

$$f = AH^2/8\pi$$

proportional to the pole area A but independent of the distance between as long as this is small. Since under these circumstances the lines of flux will go through the surfaces without change, the field strength H in the space will be equal to the induction B in the material at the surface, and the force may then be expressed

$$f = AB^2/8\pi.$$

Here, as elsewhere in this chapter, B is expressed in gausses, H in oersteds, i in amperes, dimensions in cm, and forces in dynes.

Fields Produced by Currents.—The simplest way of producing a magnetic field of known strength and direction is by the use of a long coil or solenoid of wire. The field is parallel to the axis of the coil, and when a current of i amperes flows in the windings (having n turns per cm) the field strength within the coil is

$$H = 4\pi n i/10. \quad (2)$$

When many layers of wire are used in a solenoid that is not very long in comparison with its diameter, the field strength varies with the position and, for a point on the axis, x cm from the middle of the coil, may be calculated as follows:

$$H = \frac{4\pi ni}{10}\left[\frac{l+x}{2(a-b)}\log_e\frac{a+s_2}{b+s_1} + \frac{l-x}{2(a-b)}\log_e\frac{a+r_2}{b+r_1}\right].$$

Here $2l$ is the length of the solenoid, a the radius of the outer windings in cm, b the radius of the inside windings, and the other quantities are given by:

$$r_1^2 = b^2 + (l-x)^2 \qquad s_1^2 = b^2 + (l+x)^2$$
$$r_2^2 = a^2 + (l-x)^2 \qquad s_2^2 = a^2 + (l+x)^2.$$

When the solenoid has a single layer of wire ($a = b$), this formula reduces to

$$H = \frac{4\pi ni}{10}\left[\frac{l+x}{2\sqrt{a^2+(l+x)^2}} + \frac{l-x}{2\sqrt{a^2+(l-x)^2}}\right].$$

When the solenoid is in the form of a ring or toroid, without ends, it reduces again to

$$H = 4\pi ni/10.$$

In this case, n is the number of turns of wire per cm of circumference measured around the ring and, of course, varies somewhat from the inside to the outside of the ring in a way easily calculable.

When a field is produced by the current in a long straight wire of radius a, the direction of the field is everywhere at right angles to the wire axis, and at distance x from the axis the field strength is

$$H = 2i/10x$$

outside the wire, while inside it is

$$H = 2ix/10a^2.$$

To create a relatively uniform field in a large volume, Helmholtz proposed the arrangement of two thin, circular, coaxial coils of diameter $2a$ and axial separation a. At a point P, x cm on the axis from the center of symmetry, the field strength is

$$H = \frac{\pi Ni}{5a}\left\{\frac{1}{[1+(\frac{1}{2}+x/a)^2]^{3/2}} + \frac{1}{[1+(\frac{1}{2}-x/a)^2]^{3/2}}\right\}$$

and at the center ($x = 0$),

$$H = (16\pi/5^{3/2})Ni/a = 0.899Ni/a.$$

N is the number of turns of wire in each coil.

Other forms of coils may be used, and the strength and direction of the field can be calculated by summing the fields due to each current element.

The magnitude of the field dH caused by a current i flowing in an element of wire dl cm long, at a point r cm from the wire, is given by the equation

$$dH = (i/10r^2)\, dl \sin \theta,$$

θ being the angle between the direction of the current and that of the line r connecting it to the point. The direction of the field is at right angles to both the current and r.

Force on Magnet in Field.—The force on a single magnetic pole of strength m in a field of strength H is

$$f = Hm$$

acting parallel to the field. In a uniform field a magnet of moment M' is acted on by a torque tending to turn the magnet upon an axis at right angles to its length and to the field so that it will lie parallel to the field. The magnitude of this torque is

$$L = M'H \sin \varphi$$

where φ is the angle between the field and the length of the magnet. If the field is produced at P by a magnet of moment M placed at O (Fig. 19-1), the torque acting on M' is

$$L = MM'(1 + 3\cos^2 \theta)^{1/2} (\sin \alpha)/r^3.$$

If a pivoted magnet is placed in a field, it will have a natural period of oscillations equal to

$$T = 2\pi (K/MH)^{1/2} \text{ seconds,}$$

if K is the moment of inertia about its point of suspension. If an additional known moment of inertia, k, is added to that already present, without changing the magnet moment, and this increases the period to T_1, the expression

$$MH = 4\pi^2 k/(T_1^2 - T^2)$$

can be used to determine either H or M, provided the other is known.

In a non-uniform field a magnet will experience a translational force as well as a torque, because the force pulling one pole parallel to the field will be opposed by a larger force caused by a larger field acting on the other pole. If the gradient of the field is dH/dx, the force on a magnet of moment, M, is

$$f = M\, dH/dx. \tag{3}$$

When the moment is induced by the field and is therefore

$$M = Iv = \kappa Hv,$$

the force acting on the volume v is

$$f = \kappa v H dH/dx. \tag{4}$$

Force on Current in Field.—When a current i flows in an element of wire of length dl in a field of strength H, a force acts on the wire of magnitude

$$f = Hi\,dl(\sin\theta)/10. \tag{5}$$

Here θ is the angle between dl and H, and the direction of the force is at right angles to both. This principle is used in the Cotton balance [00C1] for measuring high field strengths.

In two long, parallel, straight wires carrying currents the field of each acts on the current in the other to produce an attractive force, if the currents are in the same direction, and a repulsive force if they are antiparallel, of magnitude

$$f = 2ii'/(100a)$$

in dynes/cm length of wire, a being the distance between wire axes.

Electromotive Force and Magnetomotive Force.—When an electric circuit encloses a changing magnetic flux φ, an electromotive force is generated in the circuit proportional to the time rate of change of flux $d\varphi/dt$, and to the number of times, N, the circuit threads the flux. Thus if N turns of wire are wrapped around a piece of iron of cross-sectional area A, and a change in the induction B is produced in the iron, a voltage is produced at the ends of the wire equal to $10^{-8}NA$ times the rate of change of induction in gausses per second:

$$E = 10^{-8}NA\,dB/dt. \tag{6}$$

The total change in B that has occurred in a specified time may be evaluated by integrating this equation with the result

$$B = 10^8 \int (E/NA)\,dt. \tag{7}$$

Along any closed path in a magnetic field the *magnetomotive force* or the *line integral of the magnetic field* is

$$F = \oint H \cos\theta\,dl,$$

θ being the angle that the field makes with the line element dl. F may be regarded as the agent which causes the magnetic flux φ to flow in a magnetic circuit having a reluctance $\mathcal{R}$, just as, in an electric circuit, the electromotive force may be said to cause a current I to flow through a resistance R. In a magnetic circuit or portion of a circuit of cross-sectional area A, length l, and permeability μ, the flux and reluctance are

$$\varphi = BA, \quad \mathcal{R} = \frac{l}{\mu A}$$

and the magnetomotive force, measured along a line parallel to H, is

$$F = \mathcal{R}\varphi = BA \cdot \frac{l}{\mu A} = Hl.$$

Magnetic circuits can then be analyzed in much the same way as electrical circuits are, with the important limitations that μ is normally dependent on H and that the ratio of permeabilities in iron and in air is not as great as the ratio of electrical resistivities in copper and air. The flux in a closed circuit is then

$$\varphi = \frac{F}{\mathcal{R}} = \frac{F}{\mathcal{R}_1 + \mathcal{R}_2 + \cdots}$$

$$= \frac{F}{\dfrac{l_1}{\mu_1 A_1} + \dfrac{l_2}{\mu_2 A_2} + \cdots},$$

the reluctances for the portions of the circuit being summed as indicated. The name *reluctivity* is often applied to $1/\mu$, and *permeance*, to $1/\mathcal{R} = \mu A/l$. When the air path is not very small, calculation of the flux in air is usually difficult and inexact; for such calculations the reader is referred to other sources [41R1, 43E1].

COMMON METHODS

Ballistic Method with Ring.—In determining magnetization curves and hysteresis loops the ballistic method, employing a ring specimen, is generally the most satisfactory. Rowland used this method in 1873, and he was the first to express the results of measurement in an absolute system. A ring is cut with a radial thickness small compared with its diameter, and two windings of wire are applied (Fig. 19-2); a secondary winding (S), usually consisting of many turns of fine wire, is wound closely to the specimen and connected to a ballistic galvanometer or fluxmeter (G), and a primary winding (P) of turns evenly spaced is applied and connected to a current source. In the primary circuit, means are provided for adjusting the current and for changing it rapidly from one value to another in the same or opposite direction. In the secondary circuit a mutual inductance (M) with an air core is provided for calibration and a switch for short-circuiting the galvanometer. The field strength H is calculated, according to Eq. (2), from the number of turns, the dimensions of the coil, and the current indicated by the ammeter A.

When H is changed suddenly from one value to another, the resulting change in B induces a voltage in the coil S and causes a deflection of the galvanometer that is proportional to the change in B, according to Eq. (7) (the ballistic galvanometer has a deflection proportional to $\int E\,dt$, provided the voltage impulse occurs in an interval of time short compared

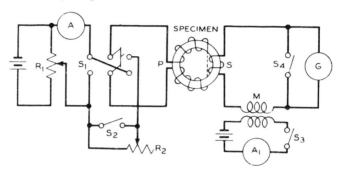

FIG. 19–2. Ballistic method of measurement of B in ring specimen (see text).

with the natural period of oscillation of the galvanometer coil and mirror). Since ballistic measurements always involve differences, one must start with a known value of B. The most common procedure for accomplishing this is first to reduce the magnetization to zero, or *demagnetize* the core, by subjecting it to an alternating field of high strength and gradually reducing the amplitude to zero. To determine a point such as M in Fig. 19–3, current corresponding to H_1 is applied and reversed several times with the galvanometer short-circuited, to establish a steady cyclic state, and then the galvanometer is connected in and the deflection for reversal of H_1 noted. This deflection then corresponds to a change in B of $2B_1$, and B_1 can easily be calculated from the constant obtained by calibration. The point M on the normal magnetization curve is thus established, and other points on the curve are determined similarly.

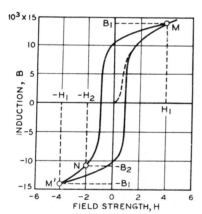

FIG. 19–3. Hysteresis loop illustrating ballistic method of measurement of B.

To determine a point, N, on a hysteresis loop, the field is alternated several times between the points of the loop (e.g., MM'); then H_1 is changed suddenly to $-H_2$, the deflection is noted, and the corresponding change in B is subtracted from B_1, already determined, to give the value of $-B_2$ at N. By varying H_2, a sufficient number of points on the whole loop can be obtained.

Calibration is made by interrupting a measured current through the mutual inductance M, using switch S_3, with *specimen removed* from the search coil S. If a galvanometer deflection δ_0 is obtained by interrupting a current of i amperes flowing in series through a mutual inductance of L henries and a search coil of N turns, the constant K may be evaluated:

$$K = \frac{Li \cdot 10^8}{N\delta_0}.$$

Then the deflections δ obtained during measurement will be related to change in flux through the search coil by

$$\Delta\varphi = K\delta.$$

The ring specimen is not adapted to rapid testing, for some time is consumed in machining the ring and in winding it with wire. However, several methods are available for the rapid application of turns. Stiff wires may be held in fixed positions and arranged so that the ring will slip over many of them at once, the wires being connected together by a plug-and-socket arrangement or with mercury contacts to make a single circuit. Also, toroidal winding machines are made that apply more than 25 turns a second to a closed ring.

Ordinarily the ballistic galvanometer and ring specimen are used for fields of intermediate strength. When the field strength is high, e.g., 1000 oersteds, the number of turns in the primary winding or the current through it must be large, and the consequent heat generated may interfere with the measurements. If the field strength is too low, many turns must be wound on the secondary in order to obtain a sufficiently large galvanometer deflection; this process is tedious and under some circumstances there may not be enough space available inside the ring. In very low fields a-c methods should be considered.

Rod Specimens. Demagnetizing Factors.—The ballistic method may be used with a straight specimen in the form of a rod, this being magnetized with the same procedure as that used for a ring. The specimen may easily be slipped in a long solenoid previously prepared, and the secondary winding or *search coil* should be placed around a small central portion of the specimen. Under these circumstances the most difficult quantity to determine is the field strength, because the field created by the solenoid will be disturbed by the magnetic poles of the specimen. The field created

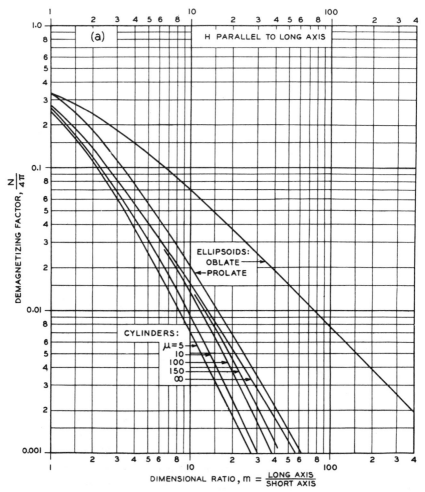

Fig. 19-4(a). Demagnetizing factors of ellipsoids and cylinders.

by the specimen itself is the *demagnetizing field* ΔH and is nearly proportional to the intensity of magnetization. It is usually specified by the *demagnetizing factor* N, which depends on the ratio length/diameter of the rod:

$$\Delta H = NI.$$

The field acting on the middle of the rod is then the resultant of the field in the solenoid H_0 and the demagnetizing field ΔH and, when I is replaced by $(B - H)/4\pi$, is

$$H = H_0 - \frac{N}{4\pi}(B - H).$$

ROD SPECIMENS. DEMAGNETIZING FACTORS

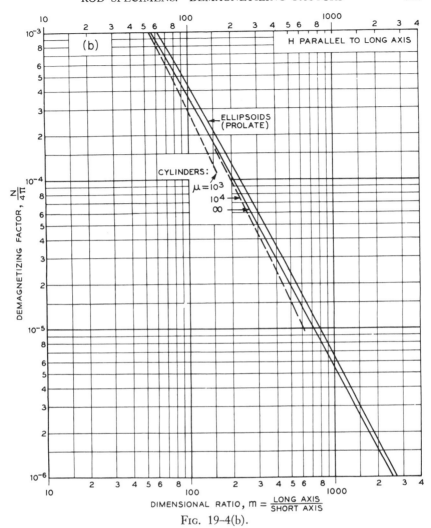

Fig. 19-4(b).

Values of $N/4\pi$ have been determined by a number of experimenters and selected values are given in Table 1 and Fig. 19-4.

The term *apparent permeability* (μ') is often applied to the ratio B/H_0, and its relation to the (true) permeability μ can be derived from the foregoing equation:

$$1/\mu = 1/\mu' - N/4\pi$$

when $1/\mu$ is negligibly small compared to one. The relation between μ and μ' for rods of various dimensional ratios $m = $ length/diameter is shown graphically [42B1] in Fig. 19-5. It is apparent that rods having dimensional ratios as large as 1000 must be used to determine accurately the value of the permeability when it is over 100 000.

In rods the induction varies from place to place in the bar, decreasing from the middle toward the ends. It is only when the specimen is in the form of an ellipsoid, and has been placed in a uniform field, that the induction is uniform throughout. In specimens of such form the demagnetiz-

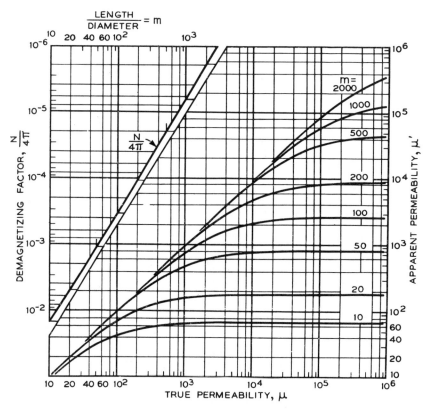

FIG. 19-5. Chart for converting apparent to true permeability, μ' to μ, of cylinders of given ratio m of length to diameter.

ing factor may be calculated accurately from the lengths of the three axes of the ellipsoid, and the direction of the induction may be determined. Tables and curves have been prepared by Stoner [45S3] and Osborn [45O1]. Equations are given below for the two most useful cases: (1) the prolate ellipsoid having the major axis m times the two equal minor axes, and the specimen magnetized parallel to its major axis; and (2) the oblate ellipsoid or ellipsoidal disk having two long axes each m times the short axis, the specimen being magnetized parallel to a long axis.

Prolate:

$$\frac{N}{4\pi} = \frac{1}{m^2 - 1}\left[\frac{m}{\sqrt{m^2 - 1}} \log_e (m + \sqrt{m^2 - 1}) - 1\right];$$

Oblate:

$$\frac{N}{4\pi} = \frac{1}{2}\left[\frac{m^2}{(m^2 - 1)^{3/2}} \arcsin \frac{\sqrt{m^2 - 1}}{m} - \frac{1}{m^2 - 1}\right].$$

Values of $N/4\pi$ for various values of m are given in Table 1.

TABLE 1. DEMAGNETIZING FACTORS, $N/4\pi$, FOR RODS AND ELLIPSOIDS MAGNETIZED PARALLEL TO LONG AXIS

Dimensional Ratio (length/diameter)	Rod	Prolate Ellipsoid	Oblate Ellipsoid
0	1.0	1.0	1.0
1	.27	.3333	.3333
2	.14	.1735	.2364
5	.040	.0558	.1248
10	.0172	.0203	.0696
20	.00617	.00675	.0369
50	.00129	.00144	.01532
100	.00036	.000430	.00776
200	.000090	.000125	.00390
500	.000014	.0000236	.001567
1000	.0000036	.0000066	.000784
2000	.0000009	.0000019	.000392

Yokes and Permeameters.—When the specimen is in the form of a straight rod or tube or tape, the demagnetizing field may be partially or completely annulled by connecting the two ends, outside the magnetizing solenoid, with magnetic material of high permeability and large cross-sectional area. Such a *yoke* may be used for specimens of high dimensional ratio, and measurements may be made with considerable accuracy.

More often, in the testing of permanent magnets or commercial materials in the form of bars or strips, it is desirable to use a yoke with a relatively short specimen, and in this case the demagnetizing field must be taken into account or annulled in some way. In the Babbitt permeameter [28B3] the field is measured with a ballistic galvanometer connected to a coil of many turns placed in the magnetizing coil near the middle of the specimen but not surrounding it (Fig. 19–6), and at the same time the yoke is magnetized with the same current that passes through the magnetizing coil. The turns on the yoke are adjusted to overcome the reluctance of the yoke and the air gaps that exist at each end of the specimen no matter how much care is taken to obtain good joints at these points. The search coil used in determining B is wound on the middle of the specimen as usual.

A further step toward increased accuracy was taken in the construction of the *Burrows permeameter* [09B4, 46S4]. This type requires two samples clamped between two connecting yokes of high permeability material that complete the square-shaped magnetic circuit. In addition to the magnetizing coils around the samples there are compensating coils around each end of the samples to give more adequate corrections for the effect of the air gaps at the joints. B coils are wound around the middle of each sample, and two search coils are placed on either side of each B coil. With proper adjustment the conditions of test approach closely those in the ring test, but the procedure is time-consuming because of the number of adjustments required. The *Illiovici* permeameter [27S4] is also used.

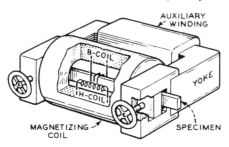

Fig. 19–6. Simple yoke method of compensating demagnetizing effects of ends of specimen.

Another scheme for adjusting the compensation in a yoke has been used by Niwa [24N1] and is based on the magnetic potentiometer described by Chattock [87C1] in 1887. A strip of non-magnetic material or material in which hysteresis is negligible is wound uniformly over its length by many turns of fine wire, and its ends are placed against two points of a specimen. If a ballistic galvanometer attached to this coil shows no net change in flux when the magnetizing current is reversed, this indicates that the magnetization in the bar is uniform and that H can be calculated from the current in the magnetizing coil surrounding the specimen. Niwa prepared his Chattock coil with a thin core of high permeability material and put magnetizing coils on the four sides of the permeameter.

The *Fahy permeameter* [18F1, 46S4] is commonly used for testing materials like iron and silicon-iron and some of the magnet steels of relatively low coercive force, and it is suitable for tests at magnetizing forces up to about 300 oersteds. This instrument, shown in Fig. 19–7, has one large magnetizing winding on a yoke of silicon-iron. Pole-pieces extending from either end of the yoke are arranged so that bar samples can be clamped to them. The magnetizing force is measured by an air-core solenoid (H coil) mounted across the ends of the pole pieces and above the sample. A winding enclosing the sample acts as a secondary (B coil) and measures the induction with the aid of a galvanometer as in Fig. 19–2.

The *saturation permeameter* is very similar to the Babbitt permeameter except that the magnetizing coil is larger and artificially cooled, and no compensating coils are used on the yoke. Magnetizing forces as high as 2500 oersteds are readily obtained with this instrument.

In determining the magnetic properties of materials in high fields, of the order of 300 to 10 000 or more, it is usually desirable to use some kind of an

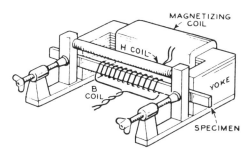

FIG. 19-7. Fahy permeameter.

electromagnet to create the field. A modification of the *isthmus method*, used originally by Ewing and Low [00E1], has been made by Sanford and Bennett [39S8] and is shown in Fig. 19-8. A similar instrument has been used by Neumann [34N2]. The heavy yokes are wound with many turns of heavy wire and when energized create a high field in the space between the pole faces, N and S. The specimen P is clamped between the pole-pieces as indicated, and it is surrounded by a search coil connected to a galvanometer for determining B. The field strength H is determined by measuring the change in the air flux in a coil placed near the specimen (H_1) and then in the same coil placed somewhat farther away (H_2). The field in this region varies approximately linearly with the distance from the surface of the specimen, and its value at the surface is obtained by extrapolation from the values for H_1

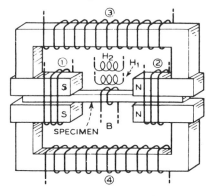

FIG. 19-8. Permeameter for measuring B and H in strong fields.

and H_2. By making the core of the electromagnet of laminated material the magnetizing field can be changed quickly enough to make satisfactory ballistic measurements, provided the period of the galvanometer is large. In investigations of properties in still higher fields, e.g., 10 000–50 000, special methods have been used (see below).

Another method of testing with a yoke was proposed by *Koepsel* [94K1, 98K1] and has been used extensively for routine testing of samples of standardized materials of moderate or low permeability. It has the advantage of a direct-reading instrument since it requires no galvanometer. To indicate the induction in the specimen S, a coil of wire is placed on a delicate bearing in an air gap in the yoke, and when a constant current is passed through this coil it deflects in accordance with Eq. (5) by an amount proportional to the field strength in the gap, which in turn depends on the flux in the specimen. The magnetizing coil is placed around the specimen and compensating coils wound on the yoke. The instrument is not generally of high accuracy and is used mainly for rapid comparison of permanent magnets with a standard as a control in manufacture. An arrangement for increasing its absolute accuracy has been used by Oertel [2904].

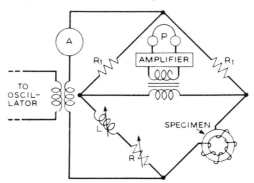

FIG. 19-9. Inductance bridge for measuring resistance (loss) and inductance of winding surrounding specimen. Alternatively, L may be omitted and a variable condenser C placed in parallel with R. Specimen may be connected through transformer having a single term secondary looped with the specimen.

Alternating-Current Methods.—The principal uses of a-c testing methods are for determining (1) permeability and energy losses of thin sheet at low inductions, usually not over a few hundred gausses, using an inductance bridge, and (2) energy losses at high inductions, usually $B = 10\,000$ or $15\,000$, using the Epstein method.

A diagram of the *inductance bridge* [36L1] is shown in Fig. 19-9. Current of the desired frequency f is supplied by an oscillator and fed to opposite corners of the bridge through a transformer and thermal ammeter, A. Two equal fixed resistances R_1 constitute two arms of the bridge, and in the others are the specimen S and variable inductance L and resistance R. When the bridge is balanced by adjusting L and R until no voltage is detected by the phones P or other instrument connected as shown through a transformer and amplifier, then L and R are equal to the inductance and a-c resistance of the specimen and its winding. The permeability can be calculated from L and the dimensions of the specimen (see Chap. 17), the field strength from the current through the core winding, and the energy loss in the specimen from the resistance R, correction being made for the losses in the copper wire.

To save the application of many turns of wire to the specimen, necessary when low inductions are measured, Kelsall [24K2] has used a device that may be used with the bridge of Fig. 19-9. Instead of connecting the windings of a specimen directly to one arm of the bridge, in the usual way, there is substituted a transformer having a many-turn primary and a single-secondary. This turn is also wound on the specimen, and the bridge is balanced as before. The core of the transformer is made of high-quality laminated material and has low losses which, however, must be taken into account and subtracted from the bridge measurement to obtain the loss in the specimen. This instrument has been adapted for use at high temperatures [24K1] as well as at ordinary temperatures.

In interpreting the permeability and energy losses measured by a-c methods one must consider the apparent reduction of permeability due to skin-effect or shielding caused by eddy currents, as well as the separable losses attributable to eddy currents and hysteresis (see Chap. 17). The data obtained with the a-c bridge are especially important in the study of materials used in communication engineering where small magnetic fields are important in the transmission of weak or undistorted signals.

The *Epstein* [00E2] or *Lloyd-Fisher* [09L1, 46A1] *apparatus* is used primarily for measuring magnetic losses in material for power and other transformers operating at high inductions. The material, in strip form, is built into a hollow square. An alternating field is applied, in one winding, to magnetize it to a high induction measured by the voltage induced in a second winding, and the loss is indicated by a wattmeter, W, connected to the circuits of both windings (Fig. 19-10). Both windings are placed evenly on forms with square sections, and four such forms are placed to form a hollow square. The strips are slipped into the forms and carefully clamped at the corners after being abutted or overlapped there. Standard specifications control the manner of cutting and stacking the specimens, of correcting for the air flux in the coils, and of taking into account the wave form of the exciting current. Measurements are commonly made at $B = 10\,000$ and $B = 15\,000$ and at a frequency of 60 cps in the United States and 50 cps in England.

In principle a simple method of testing a specimen by a-c is to determine the current-voltage characteristics of a specimen having two windings, the current being a measure of H and the voltage a measure of B. Several ways have been used for accomplishing this, involving potentiometric measurements or rectification [34T2, 46A1]. From the results one can determine not only B and H, but also by taking account of the phase can determine the total energy loss in the material. The circuit of one instrument of this kind, the *ferrometer*, is given in Fig. 19-11.

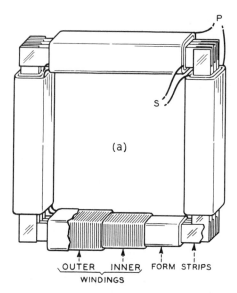
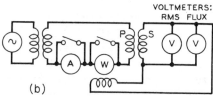

FIG. 19-10. Epstein, or Lloyd-Fisher, apparatus for measuring power dissipated in a specimen formed from strips.

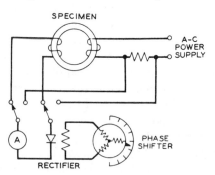

FIG. 19-11. Basic circuit of the ferrometer, for determining current-voltage characteristics of a material, using a phase-shifter and rectifier.

Cathode-ray oscilloscopes are sometimes used to give rapid indications of the a-c properties of materials. By means of a simple integrating circuit, hysteresis loops can be produced on the screen. A simplified circuit of this type is shown in Fig. 19-12. This test is not usually as precise as

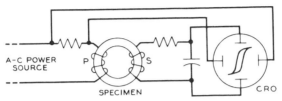

FIG. 19-12. Simple integrating circuit for showing B vs H loops on screen of cathode-ray oscilloscope (CRO).

those described previously, but because of its rapidity it finds frequent application in certain types of production testing. It is also used for the measurement of small specimens (see below).

SPECIAL METHODS

Methods other than those already described may be used advantageously for special investigations. Measurements of very small quantities of materials, of single crystals in definite directions, of paramagnetic or diamagnetic solids, liquids, or gases, all require specific arrangement for obtaining results of high accuracy with a minimum of effort.

Production of High Fields.—For many laboratory experiments *electromagnets* are useful when fields of 1000–50 000 are required. A simple design of the Weiss type is shown in Fig. 19-13. Often iron-cobalt alloy is used in the core where the induction is highest—at the pole tips when these are tapered, and at the centers of the coils when the air gap is large. The rest of the circuit is usually of soft iron, the section being such that the iron operates at a point just above the knee of the magnetization curve where μ begins to decrease rapidly with increasing B. In a representative magnet having cores 3 in. in diameter and 12 in. long, a field of 13 000 was obtained in a gap of 1 in., and 17 500 in a gap of $\frac{1}{2}$ in. Using conical pole tips of 1-in. diameter, made of a 50% iron-cobalt alloy and attached to the same core, the field was increased to about 25 000 in a $\frac{1}{2}$-in. gap.

Higher fields in larger volumes are obtained in the Paris (Bellevue), [28C1], and Dreyfus [35D8] magnets. In the former the cores weigh about 8 tons, and a field of 70 000 is produced in a volume of about 15 mm^2; a field of 46 000, in a volume of about 25 cm^3.

Air core coils have been used successfully by Bitter [39B12] for the creation of fields of 40 000–100 000. The windings were constructed so as to permit easy flow of water for cooling and were made either of copper strip in concentric cylinders or of disks stacked with the plane of the disk at right angles to the coil axis and cut so that each disk was a single turn of the coil. Water was forced through holes formed by perforations in the disks. In one design 50 000 oersteds were maintained in a 4-in. diameter space about 4 in. long with a current of 10 000 amperes at 115 volts. In another design 100 000 oersteds ($\pm 1\%$) was produced in a space $1\frac{1}{8}$ in. in diameter and 4 in. long with a power of 1700 kw.

Higher fields, up to 300 000, have been obtained by Kapitza [27K4] for a period of 0.01 second at a time. He stored electrical energy slowly and released it rapidly, by charging and discharging storage batteries or condensers, and, most successfully, by suddenly applying the load to a special

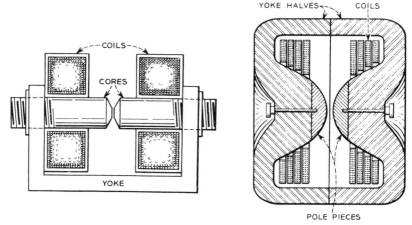

Fig. 19-13. Two designs of electromagnets; (a) Weiss design, (b) Arthur D. Little design.

a-c generator after it had attained its maximum speed. Measurements were made during the first half cycle of the generated current. The original paper must be consulted for details.

Measurements of the strengths of high fields are often made with the conventional search coil, noting the deflection of a ballistic galvanometer when the coil is removed from or reversed in the field. Another method is to measure the change in resistance of a bismuth spiral placed at right angles to the field [09B5, 23S1]; an increase of over 50% occurs in a field of 10 000 oersteds. Still another, used with the Paris magnet, depends on the mechanical force acting on a wire carrying current, when the wire is placed in the field to be measured (Cotton balance [00C1]). Direct-reading instruments based on crystal anisotropy (Dupouy [27D2]) and Hall effect (Pearson [48P1]) have been used where high accuracy is not required. Rather high accuracy can be obtained by rotating a small coil in the field with a known frequency and measuring the voltage produced at its ends, after rectifying through slip rings (Kohaut [37K6], Cole [38C2]). In a similar arrangement a wire is vibrated with known frequency, and the alternating voltage thus generated is amplified and measured (Thuras [37T2]).

Magnetometers.—Years ago the intensity of magnetization of a specimen was usually measured by the torque it exerted on a small magnetic needle of known moment suspended near by. The deflection of a light beam by a mirror attached to the delicate suspended system is still a very sensitive indicator of the state of magnetization of a specimen under investigation, but for most purposes the uncertainty in the positions of the magnetic poles of the specimen makes the magnetometer an instrument of low precision. For some purposes, however, it has decided advantages.

In principle the magnetometer measures the field strength at a known distance from the poles of the specimen, in accordance with Eq. (1). It has been used for measuring, with not too great accuracy, the magnetic properties of very small specimens in the form of fine wires or thin films deposited by evaporation or electrodeposition. In the arrangement proposed by Tobusch [08T1] and modified by others [25B3] (Fig. 19-14), two oppositely directed needles (N, S) of equal moment are placed one above the other on the same suspension, and the specimen is made with a length about equal to the distance between the needles and is placed parallel to the suspension. In this arrangement one pole of the specimen produces a twist on the nearest needle and the other pole a twist in the same sense on the other needle. The resultant twist is neutralized by a small solenoid (B-coil) placed opposite the specimen and energized with a current that is proportional to the magnetization of the specimen at balance. The magnetizing coils M_1, M_2 are also balanced so as to produce no effect on the needles. In this way [29H2] films as thin as 10^{-5} cm, weighing considerably less than a milligram, have been investigated in fields of various strengths and at different temperatures.

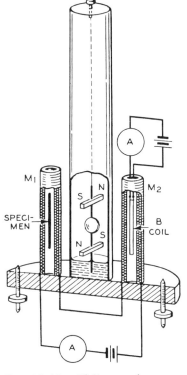

Fig. 19-14. Null astatic magnetometer for small specimens.

Para- and Diamagnetic Materials.—Measurements of weakly magnetic substances usually depend on the translational force exerted on a magnetized body in a non-uniform magnetic field. Faraday suspended the material to be measured between the poles of an electromagnet, displaced from the central position so that the field gradient was large, and used a torsion balance to measure the force pulling the material toward the stronger part of the field or repelling it therefrom. This method was developed by P. Curie [95C1] and used in his classical researches on the effect of temperature on paramagnetic and diamagnetic materials.

When the field gradient is dH/dx, the force on material of magnetic moment M is $M\, dH/dx$ (Eq. 3), and when expressed in terms of susceptibility and volume (Eq. 4) it is

$$f = \kappa v H\, dH/dx.$$

The force is zero at the central position between the poles of the magnet because here dH/dx is zero, and the force is a maximum at some point indicated by P_1 in Fig. 19–15 and falls off again toward zero at points distant from the magnet.

To avoid laborious mapping of the field in order to determine the product $H\, dH/dx$, and the careful setting of the sample at a given position, Curie and Chéveneau [03C1, 10C1, 32G5] proposed that the *magnet* be moved until the sample deviates a maximum

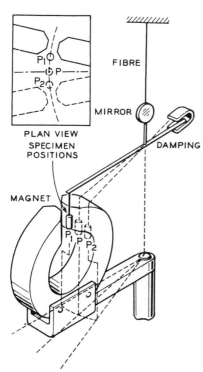

FIG. 19–15. Curie-Chéveneau balance for paramagnetic and diamagnetic measurements.

amount from its zero position P, against the restoring force of the torsion balance shown in the upper part of the figure. The deflection is then proportional to the susceptibility and to the maximum value of $H\, dH/dx$, and the proportionality factor is determined by calibration with a substance of known susceptibility.

Various refinements of these methods have been suggested. In one [31G3] the torsion is measured electrically by the current in a solenoid that reacts with the field of a permanent magnet to produce a torque on the suspending fiber.

Théodoridès [22T1] has used a translational instead of a torsional balance, following the method of Weiss and Foëx described later.

An ingenious quick-reading device has been used by Sucksmith [38S10,

39S16] and is described in principle in Fig. 19–16. A thin phosphor-bronze strip is formed into a ring and supported at point P_1. At P_2 a light rod leads to the specimen S placed in the non-uniform field of an electromagnet E. Two mirrors, M_1 and M_2, attached to the ring, lie in the path of a light beam L_1L_2 which registers on a scale. When the field is applied to the specimen, the ring is deformed, and the resulting deflection of the light beam is proportional to the force. The movement of the specimen is so small that $H\,dH/dx$ can be made substantially independent of displacement. The deflection is then proportional to the susceptibility. Although the sensitivity is not high, the method is convenient and accurate. It has been adapted to the measurement of the intensity of magnetization of small specimens of ferromagnetic materials in high fields and at high temperatures.

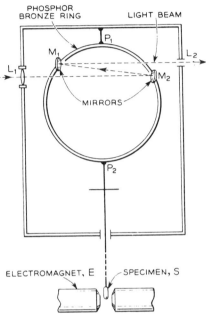

Fig. 19–16. Sucksmith balance. Pull on specimen in inhomogeneous field produces deflection of light beam L_1L_2.

Liquids and Gases.—The force exerted upon a magnetic material in an inhomogeneous magnetic field is also the basis of the methods of Gouy [89G1] and Quincke [85Q1] for measuring the susceptibilities of liquids. There are numerous modifications of the original methods, and adaption has been made to the measurement of gases. The force on the liquid in the field is measured by its rise in a capillary tube, against the force of gravity. In a simple modification, an electromagnet produces a high field at the surface of the liquid, the movement of which is measured by a microscope when the field is applied or removed. In using the method for gases [24W6] a standard liquid of known susceptibility is employed and the gas fills the remainder of the tube. The force on the gas, caused by the field, adds algebraically to that on the liquid, and the displacement of the meniscus is a measure of the resultant.

The force on a column of liquid (or gas) extending from a place where the field is H_1 to where it is H_0 may be obtained by integration of Eq. (3) to give

$$F = \kappa A(H_1^2 - H_0^2)/2,$$

A being the cross-sectional area of the column. If a column of liquid (susceptibility κ_1) is opposed by a column of gas (susceptibility κ_2), the relation becomes

$$F = (\kappa_1 - \kappa_2)A(H_1^2 - H_0^2)/2.$$

Usually H_1 is the field strength directly between the poles of the magnet at the boundary between liquid and gas, and H_0 at the other end of the column is negligibly small. The susceptibility of gases has also been measured with the aid of a sensitive torsion balance [30B5].

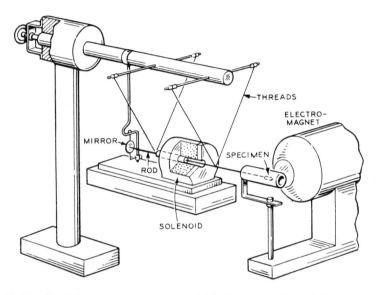

FIG. 19-17. Pendulum magnetometer, in which force on specimen in inhomogeneous field is balanced by current through solenoid carrying magnetic plunger. Only half of electromagnet is shown.

Other Special Measurements.—The measurement of the saturation induction of ferromagnetic materials has been made with great accuracy by Weiss and his students in fields up to 20 000 oersteds, at temperatures ranging from about −250 to +800°C. For this purpose they [11W2] have used some modification of the scheme shown in Fig. 19-17, in which the force is produced by a field gradient as in the Faraday methods, but is measured as a translation instead of a torsion. Such a *pendulum magnetometer* has a light horizontal beam with the specimen mounted at one end in a region of high field gradient. The force on the specimen in the field is annulled by adjusting a current in a solenoid in which a plunger is attached to the other end of the beam, and the point of zero deflection is indicated by a tilting mirror. The current is then a measure of the force being determined. In a refinement of this method used by McKeehan

[34M10], the specimen (e.g., of a single crystal) is mounted in a strong uniform field, easily measurable, and a readily calculable gradient is applied by means of two thin coils mounted near each other with fields opposing. In this way one avoids the mapping of the field, as well as the uncertainty in the direction of the field and the rotational hysteresis in the specimen, difficulties likely to be encountered when a torsion balance is used.

Other methods of measurement in strong fields deserve brief mention. DuBois [90D1] measured the intensity of magnetization by observing the rotation of the polarization of light reflecting from a magnetized surface (Kerr effect). The strength of a field can be measured by noting the change in the electrical resistance of a bismuth wire [34M11, 35K14] immersed in the field, or by observing the force (opposing gravity) produced on a conductor carrying a known current (Eq. 5). Special methods have been devised for measuring the magnetic properties of materials in which these vary with direction, as in single crystals [37W3].

In *weak fields*, such as that of the earth, the methods of geomagnetism can be used. Aschenbrenner and Goubau [36A4] constructed an apparatus for measuring and recording rapid changes in fields as small as 1 gamma (10^{-5} oersted), and during World War II this method was developed to even higher sensitivity [46F1]. A piece of high permeability material is subjected to an alternating field of moderate amplitude, and if an additional steady field (that to be measured) is also present an alternating voltage of twice the exciting frequency will be produced and can be amplified and recorded. Its amplitude is proportional to the field strength to be determined.

A number of *curve tracers* have been made for taking magnetization curves and hysteresis loops so that a permanent accurate record may be obtained in a short time. Among the more accurate and convenient of these are the devices of Haworth [31H5], Thal [34T2], and Cioffi [45C1, 37E4]. Alternating-current methods have been used [46C2, 47W3, 49C5] to measure the magnetization curve and hysteresis loops of specimens of the order of 10^{-5} gram. A magnetizing coil is excited by alternating current and contains two search coils, in which are produced opposing emf's that are balanced when no specimen is present. Introduction of the specimen in one of the coils causes a large, unbalanced emf which is integrated and amplified and then applied to one pair of plates of a cathode-ray oscillograph, the other pair of which registers the magnetizing field. Refinements and details of the circuit may be found in the original papers.

References to other methods of measurement will be found in summaries by Spooner [27S4], McKeehan [29M4], Neumann [38N2], Steinhaus [43S3], Sanford [46S4], and others.

APPENDIX 1

SYMBOLS USED IN THE TEXT

A Atomic weight. Area.
B Magnetic induction (B_m, maximum induction; B_r, residual induction; B_s, saturation induction; B_b, biasing induction; B_Δ, incremental induction).
C Heat capacity per gram (C_A, per gram-atom). Curie constant per cm³ (C_A, per gram-atom).
E Young's modulus. Energy per unit volume (E_k, anisotropy energy; E_σ, stress energy; E_p, magnetostatic energy, E_H, energy in a field).
F Gibbs' free energy.
H Magnetic field strength (H_a, applied field; H_c, coercive force; H_m, maximum field; H_Δ, incremental field).
I Intensity of magnetization (I_s, saturation intensity of magnetization; I_0, saturation at 0°K).
J Quantum number. Angular momentum.
K Crystal anisotropy constant.
L Torque. Self-inductance.
M Magnetic moment. Molecular weight.
N Weiss molecular field constant (multiplies I). Demagnetizing factor.
N_0 Avogadro number.
P Power loss per unit volume of material.
Q Heat absorbed per unit volume. Quality factor.
R Gas constant per gram-atom.
S Entropy.
T Temperature (Kelvin).
U Energy content per unit volume.
a Hysteresis loss coefficient.
c Heat capacity per unit volume. Residual loss constant.
d Density. Diameter.
e Electronic charge. Energy per unit area (e_w, wall energy; e_p, magnetostatic energy). Eddy current loss constant.
f Frequency, cps.
g Landé factor.

h	Planck constant.
i	Electric current.
k	Boltzmann constant.
l	Length.
m	Mass of one electron.
n_0	Bohr magneton number.
s	Skin thickness (eddy-current penetration).
t	Time.
α	Direction cosine. Coefficient of thermal expansion.
β	Bohr magneton ($he/4\pi m$). Temperature coefficient of resistance. Volume compressibility.
δ	Logarithmic decrement (δ_i, micro eddies; δ_a, macro eddies; δ_h, hysteresis). Thickness of sheet.
δ_0	Thickness of Bloch wall.
ϵ	Increase in length per unit length. Lag angle.
η	Steinmetz coefficient.
θ	Curie point. Eddy-current parameter. Angle.
κ	Susceptibility referred to 1 cm^3.
λ	Joule magnetostriction expansion, increase in length per unit length (λ_s, at saturation; λ_t, transverse).
μ	Permeability (μ_0, initial; μ_m, maximum; μ_r, reversible; μ_Δ, incremental; μ_L, μ derived from inductance; μ_R, μ derived from loss measurements).
ν	Change of permeability with field strength ($d\mu/dH$).
ρ	Electrical resistivity. Gyromagnetic ratio.
σ	Magnetic moment per gram (σ_s, saturation, σ_0, saturation at 0°K). Stress (tension).
τ	Time constant.
φ	Magnetic flux. Angle. Thermodynamic potential.
χ	Susceptibility referred to 1 gram. (χ_A, susceptibility referred to 1 gram-atom).
ω	Magnetostriction (volume expansion per unit volume).

APPENDIX 2
SOME PHYSICAL PROPERTIES OF THE ELEMENTS

Element	Symbol	Atomic Number	Atomic Weight (1947)	Density at 20°C (g/cm³)	Melting Point (°C)	Boiling Point (°C)	Electrical Resistivity (microhm-cm)	Modulus of Elasticity (10^{12} dynes per cm²)	Crystal Structure	Lattice Constants (kx) and Axial Angles at 20°C	Closest Approach of Atoms (kx)	Element
Actinium	Ac	89	227.05		1600	2060			Face-centered cubic	4.0408	2.856	Actinium
Aluminum	Al	13	26.97	2.699	660.2		2.655 (20C)	10				Aluminum
Americium	Am	95	241									Americium
Antimony	Sb	51	121.76	6.62	630.5	1440	39.0 (0C)	11.3	Rhombohedral	4.4974, 57°6.5'	2.898	Antimony
Argon	A	18	39.944	0.00166	−189.4	−185.8			Face-centered cubic	5.42 (−233C)	3.83	Argon
Arsenic	As	33	74.91	5.73	814	610	35 (0C)		Rhombohedral*	4.151, 53°49'	2.50	Arsenic
Astatine	At	85	211									Astatine
Barium	Ba	56	137.36	3.5	704	1640			Body-centered cubic*	5.015	4.34	Barium
Beryllium	Be	4	9.02	1.82	1289	2770	5.9 (0C)	37	Close-packed hexagonal*	2.2810, 3.5771	2.221	Beryllium
Bismuth	Bi	83	209.00	9.80	271.3	1420	106.8 (0C)	4.6	Rhombohedral	4.7361, 57°14.2'	3.105	Bismuth
Boron	B	5	10.82	2.3	2300		1.8×10¹² (0C)		Orthorhombic	17.86, 8.93, 10.13		Boron
Bromine	Br	35	79.916	3.12	−7.2	58			Orthorhombic	4.48, 6.67, 8.72 (−150C)	2.27	Bromine
Cadmium	Cd	48	112.41	8.65	320.9	765	6.83 (0C)	8	Close-packed hexagonal	2.9727, 5.606	2.972	Cadmium
Calcium	Ca	20	40.08	1.55	850	1440	3.43 (0C)	3	Face-centered cubic*	5.56	3.93	Calcium
Carbon (graphite)	C	6	12.010	2.22	3700	4830	1375	0.7	Hexagonal*	2.4564, 6.6906	1.42	Carbon (graphite)
Cerium	Ce	58	140.13	6.9	630		78 (20C)		Face-centered cubic*	5.143	3.63	Cerium
Cesium	Cs	55	132.91	1.9	28	690	18.83 (0C)		Body-centered cubic	6.05 (−173C)	5.24	Cesium
Chlorine	Cl	17	35.457		−101	−34.7			Tetragonal	8.56, 6.12 (−185C)	1.81	Chlorine
Chromium	Cr	24	52.01	7.19	1590	2500	13 (28C)	36	Body-centered cubic*	2.8787	2.493	Chromium
Cobalt	Co	27	58.94	8.9	1495	2900	6.24 (20C)	30	Close-packed hexagonal*	2.502, 4.061	2.501	Cobalt
Copper	Cu	29	63.54	8.96	1083.0	2600	1.673 (20C)	16	Face-centered cubic	3.6080	2.551	Copper
Curium	Cm	96	242									Curium
Dysprosium	Dy	66	162.46	8.56					Close-packed hexagonal	3.578, 5.648	3.499	Dysprosium
Erbium	Er	68	167.2	9.16					Close-packed hexagonal*	3.532, 5.589	3.459	Erbium
Europium	Eu	63	152.0	5.24					Body-centered cubic	4.573	3.960	Europium
Fluorine	F	9	19.00		−223	−188.2						Fluorine
Francium	Fr	87	223									Francium
Gadolinium	Gd	64	156.9	7.98					Close-packed hexagonal	3.622, 5.748	3.554	Gadolinium
Gallium	Ga	31	69.72	5.91	29.78	2070	53.4 (0C)		One-face-centered orthorhombic	4.517, 4.511, 7.645	2.437	Gallium
Germanium	Ge	32	72.60	5.36	958		89,000 (0C)		Diamond cubic	5.647	2.445	Germanium
Gold	Au	79	197.2	19.32	1063.0	2970	2.19 (0C)	12	Face-centered cubic	4.0701	2.878	Gold
Hafnium	Hf	72	178.6	11.4	1700				Close-packed hexagonal*	3.200, 5.077	3.14	Hafnium
Helium	He	2	4.003	0.000166	−271.4	−268.9			Close-packed hexagonal	3.57, 5.83 (−271.5C)	3.57	Helium
Holmium	Ho	67	164.94	8.76					Close-packed hexagonal	3.557, 5.620	3.480	Holmium
Hydrogen	H	1	1.0080	0.000084	−259.4	−252.7			Hexagonal	3.75, 6.12 (−271C)		Hydrogen
Illinium†	Il	61	147									Illinium†
Indium	In	49	114.76	7.31	156.4		8.37 (0C)		Face-centered tetragonal	4.585, 4.941	3.24	Indium
Iodine	I	53	126.92	4.93	114	183	1.3×10¹⁵ (20C)		Orthorhombic	4.777, 7.251, 9.773	2.70	Iodine
Iridium	Ir	77	193.1	22.5	2454	5300	5.3 (20C)	75	Face-centered cubic	3.8312	2.709	Iridium
Iron	Fe	26	55.85	7.87	1539	2740	9.71 (20C)	28.5	Body-centered cubic*	2.8606	2.476	Iron
Krypton	Kr	36	83.7	0.00349	−157	−152			Close-packed hexagonal*	5.68 (−191C)	4.02	Krypton
Lanthanum	La	57	138.92	6.15	826		59 (18C)		Close-packed hexagonal*	3.754, 6.063	3.73	Lanthanum
Lead	Pb	82	207.21	11.34	327.4	1740	20.65 (20C)	2.6	Face-centered cubic	4.9395	3.493	Lead

APPENDIX 2

Element	Sym.	At. No.	At. Wt.	Density	m.p.	b.p.			Crystal Structure	Lattice Parameters		Element
Lithium	Li	3	6.940	0.53	186	1370	8.55	(0C)	Body-centered cubic	3.5019	3.033	Lithium
Lutecium	Lu	71	174.99	9.74	1652				Close-packed hexagonal	3.509, 5.559	3.439	Lutecium
Magnesium	Mg	12	24.32	1.74	650	1110	4.46	(20C)	Close-packed hexagonal	3.2028, 5.1998	3.190	Magnesium
Manganese	Mn	25	54.93	7.43	1245	2150	185.1	(20C)	Cubic* (complex)	8.894	3.24	Manganese
Mercury	Hg	80	200.61	13.55	−38.87	357	94.1	(0C)	Rhombohedral	2.999, 70°31.7′(−46C)	2.999	Mercury
Molybdenum	Mo	42	95.95	10.2	2625	4800	5.17	(0C)	Body-centered cubic	3.140	2.720	Molybdenum
Neodymium	Nd	60	144.27	7.05	840		79	(18C)	Close-packed hexagonal*	3.650, 5.890	3.62	Neodymium
Neon	Ne	10	20.183	0.00084	−248.6	−246.0			Face-centered cubic	4.52 (−268C)	3.20	Neon
Neptunium	Np	93	237									Neptunium
Nickel	Ni	28	58.69	8.90	1455	2730	6.84	(20C)	Face-centered cubic	3.5167	2.486	Nickel
Niobium	Nb	41	92.91	8.57	2415		13.1	(18C)	Body-centered cubic	3.2941	2.853	Niobium
Nitrogen	N	7	14.008	0.00116	−210.0	−195.8			Hexagonal*	4.03, 6.59 (−234C)		Nitrogen
Osmium	Os	76	190.2	22.5	2700	5530	9.5	(20C)	Close-packed hexagonal	2.7298, 4.3104	2.670	Osmium
Oxygen	O	8	16.0000	0.00133	−218.8	−183.0	10.8	(20C)	Cubic*	6.83 (−225C)		Oxygen
Palladium	Pd	46	106.7	12.0	1554	4000			Face-centered cubic	3.8824	2.745	Palladium
Phosphorus Yellow	P	15	30.98	1.82	44.1	280	10¹·⁷	(11C)	Cubic*	7.17 (−35C)		Phosphorus Yellow
Platinum	Pt	78	195.23	21.45	1773.5	4410	9.83	(0C)	Face-centered cubic	3.9158	2.769	Platinum
Plutonium	Pu	94	239						Monoclinic			Plutonium
Polonium	Po	84	210						Body-centered cubic	7.42, 4.29, 14.10, β = 92°	3.4	Polonium
Potassium	K	19	39.096	0.86	63	770	6.15	(0C)	Body-centered cubic*	5.333	4.618	Potassium
Praseodymium	Pr	59	140.92	6.63	940		88	(18C)		3.662, 5.908	3.633	Praseodymium
Protactinium	Pa	91	231		3000							Protactinium
Radium	Ra	88	226.05	5.0	700							Radium
Radon	Rn	86	222		−71	−61.8						Radon
Rhenium	Re	75	186.31	4.40	3170	4500	4.5	(20C)	Close-packed hexagonal	2.7553, 4.4493	2.734	Rhenium
Rhodium	Rh	45	102.91	20.44	1966	2300	12.5	(20C)	Face-centered cubic*	3.7957	2.684	Rhodium
Rubidium	Rb	37	85.48	1.53	39	680	7.6	(0C)	Body-centered cubic	5.62 (−173C)	4.87	Rubidium
Ruthenium	Ru	44	101.7	12.2	2500	4900			Close-packed hexagonal*	2.6984, 4.2730	2.644	Ruthenium
Samarium	Sm	62	150.43	7.7	>1300							Samarium
Scandium	Sc	21	45.10	2.5	1200	2300			Face-centered cubic*	4.532	3.205	Scandium
Selenium	Se	34	78.96	4.81	220	680	10⁵	(0C)	Hexagonal*	4.3552, 4.9494	2.32	Selenium
Silicon	Si	14	28.06	2.33	1430	2210	1.59	(20C)	Diamond cubic	5.4173	2.346	Silicon
Silver	Ag	47	107.880	10.49	960.5	1950	4.2	(20C)	Face-centered cubic	4.0774	2.882	Silver
Sodium	Na	11	22.997	0.97	97.7	892	23	(20C)	Body-centered cubic	4.2820	3.708	Sodium
Strontium	Sr	38	87.63	2.6	770	1380	2×10²³	(18C)	Face-centered cubic*	6.075	4.30	Strontium
Yellow Sulfur	S	16	32.066	2.07	119.0	444.6	12.4	(20C)	Face-centered orthorhombic*	10.48, 12.92, 24.55	2.12	Yellow Sulfur
Tantalum	Ta	73	180.88	16.6	2996			(18C)	Body-centered cubic	3.2959	2.854	Tantalum
Technetium	Tc	43	99		2700		2×10⁶	(19.6C)				Technetium
Tellurium	Te	52	127.61	6.24	450	1390	6		Hexagonal	4.4469, 5.9149	2.86	Tellurium
Terbium	Tb	65	159.2	8.33	327	1460			Close-packed hexagonal*	3.585, 5.664	3.508	Terbium
Thallium	Tl	81	204.39	11.5	300	1800	18	(0C)	Close-packed hexagonal*	3.450, 5.514	3.401	Thallium
Thorium	Th	90	232.12	11.5			19	(20C)	Face-centered cubic	5.077	3.59	Thorium
Thulium	Tm	69	169.4	9.35		2270			Close-packed tetragonal	3.523, 5.564	3.446	Thulium
Tin	Sn	50	118.70	7.298	231.9		11.5	(20C)	Body-centered tetragonal*	5.8194, 3.1753	3.016	Tin
Titanium	Ti	47	47.90	4.54	1820	5930	80	(0C)	Close-packed hexagonal*	2.953, 4.729	2.91	Titanium
Tungsten	W	74	183.92	19.3	3410		5.5	(20C)	Body-centered cubic	3.1585	2.734	Tungsten
Uranium	U	92	238.07	18.7	1130	3400	60	(18C)	Orthorhombic*	2.852, 5.865, 4.945	2.76	Uranium
Vanadium	V	23	50.95	6.0	1735		26	(20C)	Body-centered cubic	3.033	2.627	Vanadium
Xenon	Xe	54	131.3	0.00550	−112	−108.0			Face-centered cubic	6.24 (−185C)	4.41	Xenon
Ytterbium	Yb	70	173.04	7.01	1490				Face-centered cubic	5.468	3.866	Ytterbium
Yttrium	Y	39	88.92	5.51	1490	906	5.916	(20C)	Close-packed hexagonal	3.663, 5.814	3.59	Yttrium
Zinc	Zn	30	65.38	7.133	419.46		41.0	(0C)	Close-packed hexagonal	2.659, 4.935	2.659	Zinc
Zirconium	Zr	40	91.22	6.5	1750				Close-packed hexagonal*	3.223, 5.123	3.16	Zirconium

* Other forms known or suspected.
† Promethium, Pm (61).

APPENDIX 3
VALUES OF SOME CONSTANTS*

Quantity	Symbol	Value
Electronic charge	e	1.6020×10^{-20} emu, 4.802×10^{-10} esu
Electronic mass	m	9.107×10^{-28} g
Ratio charge/mass	e/m	1.7589×10^{7} emu g^{-1}
Velocity of light	c	2.9979×10^{10} cm sec^{-1}
Planck's constant	h	6.624×10^{-27} erg sec
$h/2\pi$	$\hbar$	1.0542×10^{-27} erg sec
Avogadro number	N_0	6.025×10^{23} mole^{-1}
Loschmidt number	L	2.6874×10^{19} cm^{-3}
Boltzmann constant	k	1.3803×10^{-16} erg deg^{-1}
Radius of first Bohr orbit	a_0	0.5292×10^{-8} cm
Bohr magneton	β	0.9271×10^{-20} erg gauss^{-1}
Bohr magneton/mole	M_B	5586 erg gauss^{-1} mole^{-1}
0°C on Kelvin scale	T_0	273.16°K
Gyromagnetic ratio for $g=2$	ρ_s	5685×10^{-8} g emu^{-1}
Energy of 1 electron volt ($=11\,606k$)	ev	1.6019×10^{-12} erg
Energy kT at 0°C	kT_0	3.771×10^{-14} erg
Curie constant/mole for 1 Bohr magneton/molecule	C_M	0.3752 deg mole^{-1}

Quantity	Symbol	Iron	Cobalt	Nickel
Saturation moment per gram, 20°C	σ_s	218.0	161	54.39
Saturation moment per cm^3, 20°C	I_s	1714	1422	484.1
$4\pi I_s$, 20°C	B_s	21580	17900	6084
σ_s at 0°K	σ_0	221.9	162.3	57.50
Bohr magnetons/atom	n_0	2.218	1.715	0.604
Curie Point, °K	θ	1043	1388	631

* References: [41B6, 45B2, 48D2, 49D3] and *Phys. Rev.* **82**, 555–6 (1951) (DuMond and Cohen).

APPENDIX 4

MAGNETIC PROPERTIES OF VARIOUS MATERIALS

TABLE 1. SOME STANDARD ALLOYS OF HIGH PERMEABILITY
Hot Rolled Iron-Silicon Alloys Are Omitted (see Table 4-3, p. 78).

Name	Composition (%)*	Company
Allegheny Electric Metal (4750)	48 Ni	Allegheny Ludlum Steel Corp.
1040 Alloy	3 Mo, 14 Cu, 72 Ni	Vacuumschmelze, Hanau
Alfer	13 Al	Japanese
Alperm	16 Al	Japanese
Anhyster, B	36 Ni	Aciéries d'Imphy
Anhyster, D	48-49 Ni	Aciéries d'Imphy
Armco Electric Alloy No. 48	48 Ni	Armco Steel Corp.
Audiolloy No. 1	48 Ni	Crucible Steel Co.
Calmalloy No 1	66 Ni, 30 Cu, 2 Fe	General Electric Co.
Calmalloy No. 2	88 Ni, 10 Cu, 2 Fe	General Electric Co.
Carpenter 49	49 Ni	Carpenter Steel Co.
Cerromag	Ferrite	Stackpole Carbon Co.
Chrome Permalloy	3.8 Cr, 78 Ni	Western Electric Co.
Compensator Alloy	30 Ni	Carpenter Steel Co.
Conpernik	50 Ni	Westinghouse Electric Corp.
Crystalloy	3 Si	Transformer Steels, Ltd.
Deltamax	50 Ni	Allegheny Ludlum Steel Corp.
Ferramic	Ferrite	General Ceramics Co.
Ferroxcube 1.	Cu-Zn ferrite	Philips Gloeilampenfabrieken
Ferroxcube 2.	Mg-Zn ferrite	Philips Gloeilampenfabrieken
Ferroxcube 3.	Mn-Zn ferrite	Philips Gloeilampenfabrieken
Ferroxcube 4.	Ni-Zn ferrite	Philips Gloeilampenfabrieken
H. C. R.	50 Ni	Telegraph, Construction and Maintenance Co., Ltd.
Hiperco	0.5 Cr, 35 Co	Westinghouse Electric Corp.
Hipernik	50 Ni	Westinghouse Electric Corp.
Hipernik V	50 Ni	Westinghouse Electric Corp.
Hipersil	3 Si	Westinghouse Electric Corp.
Hymu 80	4 Mo, 79 Ni	Carpenter Steel Co.
Hyperm	50 Ni	Friedrich Krupp A. G.
Isoperm (precipitation)	36 Ni, 9 Cu	{ Allgemeine Elektricitäts
Isoperm (texture)	50 Ni	{ Gesellschaft
4-79 Molybdenum Permalloy	4 Mo, 79 Ni	Western Electric Co.
2-81 Molybdenum Permalloy	2 Mo, 81 Ni	Western Electric Co.
Monimax	3 Mo, 47 Ni	General Electric Co.

* Remainder iron and minor constituents.

Table 1. (Con't.)

Name	Composition (%)*	Company
Mumetal	5 Cu, 2 Cr, 77 Ni	Allegheny Ludlum Steel Co.
Mumetal	4 Mo, 78 Ni	Telephone Construction and Maintenance Co., Ltd.
Mumetal	5 Cu, 4 Mo, 78 Ni	Aciéries d'Imphy
Nicaloi	49 Ni	General Electric Co.
Orthonik	50 Ni	Armco Steel Corp.
Orthonol	50 Ni	Naval Ordnance Laboratory
78 Permalloy	78.5 Ni	Western Electric Co.
45 Permalloy	45 Ni	Western Electric Co.
65 Permalloy	65 Ni	Bell Telephone Laboratories
68 Permalloy	68 Ni	Bell Telephone Laboratories
81 Permalloy	81 Ni	Bell Telephone Laboratories
Permalloy A	78 Ni	Standard Telephone & Cable Co.
Permalloy B	45 Ni	Standard Telephone & Cable Co.
Permalloy C	4 Mo, 78 Ni	Standard Telephone & Cable Co.
Permanite	50 Ni	I.T.E. Circuit Breaker Co.
Permendur	50 Co	Western Electric Co.
Permenorm 4801	48 Ni	Vacuumschmelze, Hanau
Permenorm 5000-Z	50 Ni	Vacuumschmelze, Hanau
Permenorm 4001	40 Ni	Vacuumschmelze, Hanau
Permenorm 3601	36 Ni	Vacuumschmelze, Hanau
45-25 Perminvar	25 Co, 45 Ni	Bell Telephone Laboratories
7-70 Perminvar	7 Co, 70 Ni	Bell Telephone Laboratories
7.5-45-25 Perminvar	7.5 Mo, 25 Co, 45 Ni	Bell Telephone Laboratories
3-34-29 Perminvar	3 Mo, 29 Co, 34 Ni	Bell Telephone Laboratories
Radiometal	45 Ni, 5 Cu†	Telegraph Construction and Maintenance Co., Ltd.
Rhometal	40-45 Ni, 5 Cr, 3 Si†	Telegraph Construction and Maintenance Co., Ltd.
Sendust	5 Al, 10 Si	Japanese; Siemens & Halske A.G.
Sinimax	3 Si, 43 Ni	General Electric Co.
Supermalloy	5 Mo, 79 Ni	Western Electric Co.
Superpermalloy	1 Si, 77 Ni	Japanese
Super-perminvar	3 Cu, 23 Co, 9 Ni	Japanese
Thermalloy	66 Ni, 30 Cu, 2 Fe	General Electric Co.
Thermoperm	30 Ni	Friederich Krupp A. G.
Trafoperm 25N1	3 Si	Vacuumschmelze, Hanau.
Trancor X, 3X	3 Si	Armco Steel Corp.
Vanadium Permendur	1.8 V, 49 Co	Western Electric Co.

* Remainder iron and minor constituents.
† Other compositions have been reported.

TABLE 2. SOME PROPERTIES OF

Considerable variation from one specimen to another is to be expected.

Name	Composition[1]	Heat Treatment[2] (°C)	Initial Permeability	Maximum Permeability
Mild steel	0.2 C	950	120	2 000
Iron	0.2 (impurity)	950	150	5 000
Purified Iron	0.05 (impurity)	1480 (H_2), 880	10 000	200 000
Silicon-iron	4 Si	800	500[4]	7 000
Grain-oriented Fe-Si	3 Si	800	1 500[4]	40 000
Thermoperm	30 Ni	1000		
45 Permalloy	45 Ni	1050	2 500	25 000
Hipernik	50 Ni	1200 (H_2)	4 000	70 000
Permenorm 5000Z	50 Ni	(CR)[5]	500	40 000
Monimax	3 Mo, 47 Ni	1125 (H_2)	2 000	35 000
Sinimax	3 Si, 43 Ni	1125 (H_2)	3 000	35 000
Radio Metal	5 Cu, 45 Ni	1050	2 000	20 000
Megaperm 6510	65 Ni, 10 Mn		4 800	25 000
68 Permalloy	68 Ni	1000 (F)	1 200	250 000
78 Permalloy	78.5 Ni	1050, 600Q	8 000	100 000
4-79 Permalloy	4 Mo, 79 Ni	1100 (C)	20 000	100 000
Supermalloy	5 Mo, 79 Ni	1300 (H_2, C)	100 000	1 000 000
Mumetal	5 Cu, 2 Cr, 77 Ni	1175 (H_2)	20 000	100 000
1040 Alloy	3 Mo, 14 Cu, 72 Ni	1100 (H_2)	40 000	100 000
Permendur	50 Co	800	800	5 000
Vanadium Permendur	1.8 V, 49 Co	800	800	4 500
Hiperco	0.5 Cr, 35 Co	850	650	10 000
45-25 Perminvar	25 Co, 45 Ni	1000, 400	400	2 000
7-70 Perminvar	7 Co, 70 Ni	1000, 425	850	4 000
Thermalloy	67 Ni, 30 Cu, 2 Fe			
Alperm	16 Al	600 Q	3 000	55 000
Cobalt	99 Co	1000	70	250
Nickel	99 Ni	1000	110	600
Sendust	5 Al, 10 Si	Cast	30 000	120 000
50 Isoperm	50 Ni	1100, CR	90	100
36 Isoperm	9 Cu, 36 Ni		60	65
Ferroxcube 3	Mn-Zn-Ferrite	1100–1200	1 000	1 500
Ferroxcube 4	Ni-Zn-Ferrite	1100–1200	100	
Heusler Alloy	10 Al, 15 Mn, 75 Cu	600Q, 200		100–1000
2-81 Permalloy (powder)	2 Mo, 81 Ni	Pressed; 650	125	130
Pressed carbonyl iron powder	40% air and filler	Pressed	20	

[1] Remainder iron and impurities. Presence of 0.3–0.5% Mn not noted.

[2] (H_2), annealed in atmosphere of pure hydrogen; (CR), severely cold rolled; (C), controlled cooling rate; (F), cooled in presence of magnetic field; (Q), quenched from indicated temperature.

[3] From saturation.

APPENDIX 4

HIGH-PERMEABILITY MATERIALS

Some iron-silicon alloys have been omitted (see Table 4–3, p. 78).

Coercive Force[3] (oersteds)	Saturation Induction (gausses)	Saturation Hysteresis (ergs/cm³)	Curie Point	Electrical Resistivity (microhm-cm)	Density (g/cm³)
1.8	21 200		770	10	7.8
1.0	21 500	5 000	770	10	7.88
0.05	21 500	300	770	10	7.88
0.5	19 700	3 500	690	60	7.65
0.1	20 000	700	740	47	7.67
	2 000[6]				
0.3	16 000	1 200		45	8.17
0.05	16 000	220	500	45	8.25
0.3	15 000	350	500	40	8.22
0.1	14 500	800		80	8.27
0.1	11 000	400		85	7.95
0.4	15 600	1 100		55	8.3
0.08	8 600			58	
0.03	13 000	120	600	20	8.45
0.05	10 800	580	600	16	8.60
0.05	8 700	200	460	55	8.72
0.002	7 900	8	400	60	8.77
0.05	6 500			62	8.58
0.02	6 000	200	290	56	8.76
2.0	24 500	12 000	980	7	8.3
2.0	24 000	6 000	980	26	8.2
1.0	24 200	3 300	970	20	8.0
1.2	15 500	2 500[7]	715	19	
0.6	12 500	[7]	650	16	8.6
2	2 000[6]				
0.04	8 000	1 500	400	140	6.5
10	17 900	2 000	1120	9	8.9
0.7	6 100	2 000	358	7	8.90
0.05	10 000	100	500	60	7.0
6	16 000		500	40	8.25
6			300	70	8.2
0.1	2 500	130		10⁸	
			130	10¹²	5.3
1–200	6 000	300–450			
	7 000		480	10⁶	7.7
15	15 000		770	10¹⁰	6

[4] Permeability at $B = 20$ instead of at $B = 0$.
[5] Square hysteresis loop.
[6] Varies markedly with temperature, and saturation is not definite. Curie point is just above room temperature.
[7] Very low hysteresis loss at low inductions.

TABLE 3. PROPERTIES OF SOME COMMERCIAL TYPES OF PERMANENT MAGNET MATERIALS

References to original reports have been given by Chegwidden [48C1], Oliver and Hadfield [48O1], Scott [48S5], Ruder [46R3], and others.

Name	Composition*	H_c	B_r	$(BH)_m \times 10^{-6}$	Preparation	Heat Treatment‡	Mechanical Properties**	Density (g/cm^3)
Carbon Steel........	0.9 C, 1 Mn	50	10 000	0.2		Q800		7.8
Tungsten Steel......	0.7 C, 0.3 Mn, 5 W	70	10 300	.3		Q850		8.1
Tungsten Steel†.....	0.4 C, 6 W	65	10 500	.3		AQ750, Q800		8.1
Chrome Steel.......	0.6 C, 1 Cr	45	9 000	.2		Q800		7.8
Chrome Steel.......	0.9 C, 0.3 Mn, 3.5 Cr	65	9 700	.3		Q830		7.7
Chrome Steel†......	1.05 C, 6 Cr	70	9 800	.3		AQ750, Q840		7.75
3 Co Steel..........	1.0 C, 3 Co, 4 Cr, 0.4 Mo	80	10 000	.4		Q860		7.8
3 Co Steel†.........	1.05 C, 3 Co, 9 Cr, 1.5 Mo	130	7 200	.35	Hot roll, machine, punch	FC780, AQ1000	Hard, strong	7.7
6 Co Steel†.........	1.05 C, 6 Co, 9 Cr, 1.5 Mo	145	7 500	.45		FC780, AQ1000		7.75
9 Co Steel†.........	1.05 C, 9 Co, 9 Cr, 1.5 Mo	160	7 800	.5		AQ1150, FC780, AQ1000		7.8
15 Co Steel†........	1.05 C, 15 Co, 9 Cr, 1.5 Mo	180	8 200	.6		AQ1150, FC780, AQ1000		7.9
17 Co Steel.........	0.75 C, 17 Co, 2.5 Cr, 8 W	150	9 500	.65		Q930		8.35
35 Co Steel†........	0.85 C, 35 Co, 6 Cr, 4 W	250	9 500	1.0		AQ1150, FC780, Q950		8.15
36 Co Steel.........	0.7 C, 36 Co, 4 Cr, 5 W	240	9 500	1.0		Q930		8.2
Alni† (MK).........	25 Ni, 13 Al, 4 Cu	500	6 000	1.3				7.0
Oerstit 120.........	27 Ni, 12.5 Al	500	5 700	1.1				
Alnico†.............	13 Co, 18 Ni, 10 Al, 6 Cu	510	7 000	1.6		Q1200, B600		7.45
Alnico 1............	5 Co, 20 Ni, 12 Al	440	7 200	1.4		Q1200, B700		6.9
Alnico 2............	12.5 Co, 17 Ni, 10 Al, 6 Cu	540	7 200	1.6		Q1200, B600		7.1
Alnico 4............	5 Co, 28 Ni, 12 Al	700	5 500	1.3		Q1200, B650		7.0
Alnico 5 (Ticonal)....	24 Co, 14 Ni, 8 Al, 3 Cu	575	12 500	5.0	Cast and ground	AF1300, B600	Hard, brittle	7.3
Alnico 5 (DG).......	24 Co, 14 Ni, 8 Al, 3 Cu	640	13 100	5.5		AF1300, B600		7.3
Alcomax 1†.........	25 Co, 11 Ni, 7 Al, 3 Cu, 1.5 Ti	475	12 000	3.5		AF1300, B600		7.3
Alcomax 2†.........	25 Co, 11 Ni, 8 Al, 6 Cu	570	12 400	4.3		AF1300, B600		7.55
Alcomax 3†.........	25 Co, 13 Ni, 8 Al, 3 Cu, 0.7 Nb	650	13 200	5.0		AF1300, B600		7.35
Alcomax 4†.........	25 Co, 13 Ni, 8 Al, 3 Cu, 2.7 Nb	760	11 800	4.7		AF1300, B600		7.35
Alnico 6............	24 Co, 15 Ni, 8 Al, 3 Cu, 1.25 Ti	730	10 700	3.8		AF1300, B600		7.4

* Weight per cent, remainder iron.
† English manufacture.
‡ Q quenched from indicated temperature (°C) in oil or water, AQ quenched in air, FC furnace cooled, AF cooled in magnetic field, B baked, D drawn (wire).
** Before final heat treatment.

APPENDIX 4

TABLE 3 (Con't.)

Name	Composition*	H_c	B_r	$(BH)_m \times 10^{-6}$	Preparation	Heat Treatment‡	Mechanical Properties**	Density (g/cm³)
Alnico 12	35 Co, 18 Ni, 6 Al, 8 Ti	950	5 800	1.5	Cast and ground	AQ1200, Q650	Hard, brittle	7.2
New KS	27 Co, 18 Ni, 4 Al, 7 Ti	900	6 000	2.0				7.4
Nipermag	32 Ni, 12 Al, 0.4 Ti	675	5 500	1.3				6.9
Hycomax†	20 Co, 21 Ni, 9 Al, 2 Cu	830	9 000	3.2				7.25
Remalloy (Comol)	12 Co, 17 Mo	250	10 500	1.1	Hot roll, machine, punch	Q1200, B700	Hard, malleable	8.15
Remalloy 2	12 Co, 20 Mo	360	9 200	1.5		Q1250, B700		
Vicalloy 1	52 Co, 10 V	300	8 800	1.0	Cold roll, machine, punch	B600		8.2
Vicalloy 2	52 Co, 14 V	510	10 000	3.5	Cold roll, draw, machine, punch	D, B600	Ductile	8.1
Cunife 1 (Magnetoflex)	20 Ni, 60 Cu	590	5 800	1.9		Q1070, B700, D, B600		8.6
Cunife 2	2.5 Co, 20 Ni, 50 Cu	260	7 300	0.8				8.6
Cunico 1	29 Co, 21 Ni, 50 Cu	700	3 400	0.9	Cold roll, machine, punch	Q1080, B625		8.3
Cunico 2	41 Co, 24 Ni, 35 Cu	450	5 300	1.0				8.3
Platinum-iron	78 Pt	1570	5 800	3.0	Cold roll, machine	Q1300, B	Malleable Ductile	10
Platinum-cobalt	23 Co, 77 Pt	2600	4 500	8.0		Q1200, B650		11
Silmanal	9 Mn, 4 Al, 87 Ag	6000 ($_IH_C$)	550	0.08	Cold roll, machine, punch	B250		
Alnico 2, sintered	2.5 Co, 17 Ni, 10 Al, 6 Cu	540	6 900	1.4		AQ1300		7
Alnico 5, sintered	24 Co, 15 Ni, 8 Al, 3 Cu, 1 Ti	575	10 000	3.5				6.6
Oerstit 1000	19 Co, 17.5 Ni, 7.5 Al, 3 Cu, 7 Ti	975	5 200	1.1	Sintered, ground		Hard, strong	
Alcomax, sintered†	21 Co, 11 Ni, 8 Al, 4 Cu	560	11 200	3.3				7
Permet†	30 Co, 25 Ni, 45 Cu	800	2 500	0.5		AQ1100		8
Indalloy	12 Co, 17 Mo	240	9 000	0.9				3
Vectolite, OP	30 Fe₂O₃, 44 Fe₃O₄, 26 Co₂O₃	900	1 600	0.5		AF1000	Brittle	
Caslox	17 Co, 27 O₂	700	1 100	0.2		AF1000		3.2
Tromolit	11 Co, 24 Ni, 11 Al, 3.5 Cu	615	3 700	0.7	Bonded 200°	None		5
Hyflux		390	6 600	1.0	Powder	Pressed	Weak	
Powdered iron (PF)	4 O₂	600	5 000	1.0	Powder	Pressed		4.3
Powdered iron cobalt (PF)	26 Co, 6O₂	500	7 500	1.7	Powder	Pressed		

* Weight per cent, remainder iron.

† English manufacture. Oerstit 200 and Oerstit 1000 are German manufacture.

‡ Q quenched from indicated temperature (°C) in oil or water, AQ quenched in air, FC furnace cooled, AF cooled in magnetic field, B baked, D drawn (wire).

** Before final heat treatment. Cunife is ductile after treatment.

The following references are to some of the more important advances made since the first printing of "Ferromagnetism."

Ferrites.—Many new compositions have been studied, as summarized by Gorter,[1] Smit and Wijn,[2] and others.[3,4] Compounds of new types have interesting ferromagnetic and antiferromagnetic properties: especially the rare-earth orthoferrites[5] (*e.g.*, $GdFeO_3$), which are orthorhombic, and the rare-earth compounds of the garnet type[6] (*e.g.*, $Gd_3Fe_5O_{12}$).

High Frequencies.—Ferromagnetic and paramagnetic resonance, the microwave Faraday effect, and domain wall motion and damping at high frequencies, are in this category. A number of review papers on the behavior of ferrites at high frequencies are given by N. Bloembergen, C. L. Hogan, and others.[7]

Antiferromagnetism.—Many new compounds, and the elements Mn and Cr, have been shown to be antiferromagnetic. Data and theories, including a discussion of the transition from the antiferromagnetic to the paramagnetic state at high fields and low temperatures, have been summarized by Nagamuya *et al.*[8]

Fine Particle Magnets.—It has been established that the high coercive force of Alnico 5 is due to the shape anisotropy of the fine particles which compose it and which are directed by the magnetic anneal.[9] Magnets of compressed powder have been produced with particles of high dimensional ratio and correspondingly high energy product, according to theory, by T. O. Paine, F. E. Luborsky, and L. I. Mendelsohn.[10]

Process of Magnetization.—Domains have been observed in thin evaporated metal films 50 to 10,000 A thick; dense concentrations of walls and new details of domain growth were noted.[11] Experiments on domain walls have enabled Rado and Weertman[12] and C. P. Bean to evaluate exchange interaction and domain wall energy.

A summary of magnetic data will soon appear in the American Institute of Physics Handbook. A general review article on Ferromagnetism is published as a chapter in Recent Advances in Science, edited by M. H. Shamos and G. M. Murphy (1956).

[1] E. W. Gorter, *Philips Res. Repts.* 9, 295–365, 403–43 (1954).
[2] J. Smit and H. P. J. Wijn, *Advances in Electronics and Electron Physics* 6, 69–136 (1954).
[3] R. M. Bozorth, E. F. Tilden and A. J. Williams, *Phys. Rev.* 99, 1788–98 (1955).
[4] G. H. Jonker, H. P. J. Wijn, and P. B. Braun, ref. 10.
[5] R. M. Bozorth, H. J. Williams and D. E. Walsh, *Phys. Rev.* 103, 572–8 (1956).
[6] M. A. Gilleo, *J. Chem. Phys.* 24, 1239–43 (1956).
[7] *Proc. I.R.E.*, October (1956).
[8] T. Nagamuya, K. Yosida, and R. Kubo, *Advances in Physics* 4, 1–112 (1955).
[9] E. A. Nesbitt and R. D. Heidenreich, *J. App. Phys.* 23, 352–71 (1952).
[10] Proc. Boston Conf. on Magnetism and Magnetic Material, A.I.E.E. (1956).
[11] H. J. Williams and R. C. Sherwood, *J. App. Phys.*, in press.
[12] G. T. Rado and J. R. Weertman, *Phys. Rev.* 94, 1386 (1954).

BIBLIOGRAPHY*

(Pages in the text on which the author's name or his work is mentioned are given at the end of each entry.)

1842–1889

1842J1 JOULE, J. P. *Ann. Electr. Magn. Chem.* **8**, 219–24. On a new class of magnetic forces. 630
1846G1 GUILLEMIN, A. *Compt. rend.* **22**, 264–5, 432–3. Change of elasticity of soft iron. 684
1847J1 JOULE, J. P. *Phil. Mag.* [3] **30**, 76–87, 225–41. On the effects of magnetism upon the dimensions of iron and steel bars. 630
1857T1 THOMSON, W. *Proc. Roy. Soc.* (London) **8**, 546–50. Effects of magnetization on electrical conductivity of Ni and Fe. 745
1859W1 WÖHLER, F. *Ann. chim. phys.* **56**, 501–6. Observations on chromium. 342
1862W1 WIEDEMANN, G. *Pogg. Ann.* **117**, 193–217. Magnetic investigations. 631
1865V1 VILLARI, E. *Ann. Phys. Chem.* **126**, 87–122. Change of magnetization by tension and by electric current. 602
1867L1 LAMONT, J. Voss, Leipzig, 1–468. Handbuch des Magnetismus. 484
81F1 FRÖLICH, O. *Electrotech. Z.* **2**, 134–41. Investigations of dynamoelectric machines and electric power transmission and theoretical conclusions therefrom. 484
81W1 WARBURG, E. *Ann. Physik* [3] **13**, 141–64. Magnetic investigations. 508
82B1 BARRETT, W. F. *Nature* **26**, 585–6. On the alterations in the dimensions of the magnetic metals by the act of magnetization. 631
85H1 HOPKINSON, J. *Trans. Roy. Soc.* (London) A **176**, 455–69. Magnetization of iron. 67, 83, 347, 371, 374
85Q1 QUINCKE, G. *Ann. Physik* [3] **24**, 347–416. Electrical investigations. 859
86B1 BIDWELL, S. *Proc. Roy. Soc.* (London) **40**, 109–34. Change in length of Fe, Ni and steel by magnetization. 631
87B1 BARUS, C. *Am. J. Sci.* [3] **34**, 175–86. Effect of magnetization on viscosity and rigidity in iron and steel. 684
87C1 CHATTOCK, A. P. *Phil. Mag.* [5] **24**, 94–6. On a magnetic potentiometer. 850
87R1 RAYLEIGH, LORD. *Phil. Mag.* [5] **23**, 225–45. The behavior of iron and steel under the operation of feeble magnetic forces. 489, 788
89E1 EWING, J. A. *Proc. Roy. Soc.* (London) **46**, 269–86. Time lag in the magnetization of iron. 788
89G1 GOUY, L. G. *Compt. rend.* **109**, 935–7. On magnetic potential energy and the measurement of the coefficients of magnetization. 859

* Titles are not exactly as published.

89H1 HOPKINSON, J. *Proc. Roy. Soc.* (London) **47**, 23–4. Magnetic properties of alloys of nickel and iron. 107

89H2 HADFIELD, R. A. *J. Iron Steel Inst.* (London), 222–55. On alloys of iron and silicon. 67

1890–1899

90B1 BIDWELL, S. *Proc. Roy. Soc.* (London) **47**, 469–80. Effect of tension on magnetostriction of Fe, Co and Ni. 631

90D1 DuBois, H. E. J. G. *Phil. Mag.* [5] **29**, 293–306. Magnetization in strong fields at different temperatures. 861

90H1 HOPKINSON, J. *Proc. Roy. Soc.* (London) **48**, 1–13. Magnetic properties of alloys of nickel and iron. 107

91F1 FINZI, G. *Electrician* **26**, 672–3. On hysteresis in the presence of alternating currents. 552

91K1 KENNELLY, A. E. *Trans. Am. Inst. Elec. Engrs.* **8**, 485–517. Magnetic reluctance. 484

91S1 STEINMETZ, C. P. *Electrician* **26**, 261–2. Note on the law of hysteresis. 509

92S1 STEINMETZ, C. P. *Trans. Am. Inst. Elec. Engrs.* **9**, 3–51. On the law of hysteresis. 509

92H1 HADFIELD, R. A. *J. Iron Steel Inst.* (London) **42**, 49–175. Alloys of Fe and Cr. 347

93E1 EWING, J. A., KLAASSEN, H. G. *Trans. Roy. Soc.* (London) A **184A**, 985–1039. Magnetic qualities of iron. 423

94K1 KOEPSEL, A. *Elektrotech. Z.* **15**, 214–6. Apparatus for determination of magnetic properties of Fe in absolute measure with direct reading. 851

95C1 CURIE, P. *Ann. chim. phys.* [7] **5**, 289–405. Magnetic properties of bodies at various temperatures. 457, 858

96B1 BAILY, F. G. *Trans. Roy. Soc.* (London) A **187A**, 715–46. Hysteresis of iron and steel in a rotating magnetic field. 515

96N1 NAGAOKA, H. *Ann. Physik* [3] **59**, 66–83. On dilute ferromagnetic amalgams. 236, 292

96W1 WEISS, P. *Éclairage élec.* **8**, 248–54, 306–14. Magnetization of alloys of Fe and Sb. 216

96W2 WEISS, P. *J. Phys.* [3] **5**, 435–53. Magnetization of crystalline magnetite. 555, 574

97A1 ASHWORTH, J. R. *Proc. Roy. Soc.* (London) **62**, 210–23. On methods of making magnets independent of temperature. 357

97C1 CURIE, M. *Compt. rend.* **125**, 1166–9. Magnetic properties of tempered steel. 372, 379

98C1 CURIE, M. *Bull. Soc. d'Encour. l'Ind. Nat.* **97**, 36–76. Magnetic properties of quenched steels. 372, 374, 382

98K1 KATH, H. *Elektrotech. Z.* **19**, 411–5. A new apparatus for magnetizing. 851

98M1 MAURAIN, C. *J. physique* [3] **7**, 461–6. A true lag exists, independent of eddy-currents. 788

1900–1904

00B1 BARRETT, W. F., BROWN, W., HADFIELD, R. A. *Sci. Trans. Roy. Dublin Soc.* **7**, 67–126. Electrical conductivity and magnetic permeability of various alloys of Fe. 67, 210

00C1 COTTON, A. *Éclairage Élec.* 24, 257–66. Magnetometer. 842, 856
00E1 EWING, J. A. The Electrician, London, 3d ed., 1–393. Magnetic induction in iron and other metals. 266, 356, 423, 604, 794, 851
00E2 EPSTEIN, J. *Elektrotech, Z.*, 21, 303–7. Magnetic testing of iron sheet. 853
00F1 FRANK, H. *Ann. Physik* [4] 2, 338–58. Influence of hardening and temperature cycling on permanent magnet steels. 357
00S1 STEVENS, J. S. *Proc. Roy. Soc.* (London) 11, 95–100. Effect of magnetization upon the modulus of elasticity. 688
01B1 BEATTIE, R. *Phil. Mag.* [6] 1, 642–7. Hysteresis of Ni and Co in a rotating magnetic field. 518
02B1 BARRETT, W. F., BROWN, W., HADFIELD, R. A. *J. Inst. Elec. Engrs.* 31, 674–729. Researches on the electrical conductivity and magnetic properties of upwards of one hundred alloys of iron. 67, 107, 217
02H1 HONDA, K., SHIMIZU, S., KUSAKABE, S. *Phil. Mag.* [6] 4, 459–68, 537–46. Change of modulus of rigidity by magnetization. 684
02H2 HONDA, K., SHIMIZU, S. *Phil. Mag.* [6] 4, 338–46. Change in length of ferromagnetic wires under tension by magnetization. 671
02N1 NAGAOKA, H., HONDA, K. *Phil. Mag.* [6] 4, 45–72. On the magnetostriction of steel, nickel, cobalt and nickel-steels. 612
03C1 CURIE, P., CHÉVENEAU, C. *J. physique* [4] 2, 796–802. Apparatus for determination of magnetic constants. 858
03H1 HONDA, K., SHIMIZU, S. *Phil. Mag.* [6] 6, 392–400. Change in length of ferromagnetic substances under high and low temperatures and magnetization. 656, 657, 658, 663
03H2 HAGEN, E., RUBENS, H. *Ann. Physik* [4] 11, 873–901. Relation of reflection and emissive power of metals to the electrical conductivity. 798
03H3 HADFIELD, R. A. U.S.P. 745 829 (Appl. 6/12/03). Magnetic composition and method of making same. 67, 83
03H4 HEUSLER, F. *Verhandl. deut. physik. Ges.* 5, 219. Magnetic Mn alloys. 328
03H5 HEUSLER, F., STARCK, W., HAUPT, E. *Verhandl. deut. physik. Ges.* 5, 220–32. Magnetochemical studies. 328
04A1 AUSTIN, L. W. *Verhandl. deut. physik. Ges.* 6, 211–6. Magnetic expansion of Heusler's alloys. 681
04H1 HEUSLER, F. *Z. angew. Chem.* 17, 260–4. Manganese-bronzes, and magnetic alloy of non-magnetic metals. 328, 334, 337, 339
04N1 NAGAOKA, H., HONDA, K. *J. phys.* [4] 3, 613–20. Magnetization and magnetostriction of nickel steels. 642

1905–1909

05G1 GUMLICH, E. *Ann. Physik* [4] 16, 535–50. Investigation of Heusler Cu-Mn-Al alloys. 333
05H1 HONDA, K., SHIMIZU, S. *Phil. Mag.* [6] 10, 548–74, 642–61. On the magnetization and the magnetic change of length in ferromagnetic metals and alloys at temperatures ranging from $-186°C$ to $+1200°C$. 265, 671
05L1 LANGEVIN, P. *Ann. chim. phys.* [8] 5, 70–127. Magnetism and electron theory. 458
05W1 WILLIAMS, W. E. *Phil. Mag.* [6] 9, 77–85. Magnetic change of resistance of Fe, Ni and nickel steel at various temperatures. 762
05W2 WEISS, P. *J. phys.* [4] 4, 469–508. Magnetic properties of pyrrhotite. 555, 574

06B1 BINET DU JASSONEIX, A. *Compt. rend.* **142**, 1336-8. Magnetic properties of Mn-B alloys. 337

06G1 GUTHE, K. E., AUSTIN, L. W. *Natl. Bur. Standards* (U.S.) *Bull.* **2**, 297-316. Experiments on Heusler Magnetic alloys. 681

06L1 LYLE, T. R., BALDWIN, J. M. *Phil. Mag.* [6] **12**, 433-68. Experiments on propagation of longitudinal waves of magnetic flux along Fe wires and rods. 788

06L2 LOSSEW, K. *Z. anorg. allgem. Chem.* **49**, 58-71. Alloys of Ni with Sb. 300

07D1 DEBYE, P. *Z. Math. u. Physik*, **54**, 418-37. Eddies in rods of rectangular section. 776

07M1 MCLENNAN, J. C. *Phys. Rev.* **24**, 449-473. On the magnetic properties of Heusler's alloys. 681

07W1 WEISS, P. *J. phys.* [4] **6**, 661-90. Hypothesis of the molecular field and ferromagnetic properties. 427, 477

07W3 WILLIAMS, R. S. *Z. anorg. allgem. Chem.* **55**, 1-33. Alloys of Sb with Mn, Cr, Si and Sn; of Bi with Cr & Si; of Mn with Sn & Pb. 334, 339

07W4 WEDEKIND, E. *Ber. deut. chem. Ges.* **40**, 1259-69. Ferromagnetic compounds of Mn with B, Sb and P. 337, 339

08A1 ASTEROTH, P. *Verhandl. deut. physik. Ges.* **10**, 21-32. Effect of thermal and mechanical history on the magnetic properties of Heusler alloys. 328

08F1 FRIEDRICH, K. *Metallurgie* **5**, 212-5. Phase diagram of Co-S alloys. 295

08F2 FRIEDRICH, K. *Metallurgie* **5**, 150-7. Equilibrium diagram of Co-As alloys. 285

08G1 GWYER, A. G. C. *Z. anorg. allgem. Chem.* **57**, 113-53. Alloys of Al with Cu, Fe, Ni, Co, Pb and Ca. 299

08K1 KONSTANTINOV, N. *Z. anorg. allgem. Chem.* **60**, 405-15. Compounds of Ni and P. 322

08R1 RICHARDSON, O. W. *Phys. Rev.* **26**, 248-253. A mechanical effect accompanying magnetization. 452

08T1 TOBUSCH, H. *Ann. Physik* [4] **26**, 439-82. On elastic and magnetic lag. 788, 857

08V1 VOSS, G. *Z. anorg. allgem. Chem.* **57**, 34-71. Alloys of Ni with Sn, Pb, Tl, Bi, Cr, Mg, Zn and Cd. 315, 323

08W1 WEISS, P., PLANER, V. *J. physique* [4] **7**, 5-27. Hysteresis in rotating fields. 515

08W2 WEDEKIND, E., VEIT, T. *Ber. deut. chem. Ges.* **41**, 3769-73. Ferromagnetic nitrogen compounds of Mn. 338

09B2 BURGESS, C. F., ASTON, J. *Trans. Am. Electrochem. Soc.* **15**, 369-89. Observations on alloys of electrolytic Fe with As and Bi. 220

09B3 BURGESS, C. F., ASTON, J. *Chem. and Met. Eng.* **7**, 403-5. Influence of As and Sn on the magnetic properties of Fe. 255

09B4 BURROWS, C. W. *Natl. Bur. Standards* (U. S.) *Bull.* **6**, 31-88. Determination of magnetic induction in straight bars. 850

09B5 BLAKE, F. C. *Ann. Physik* [4] **28**, 449-75. Influence of temperature and transverse magnetization on the d.c. resistance of Bi and Ni. 856

09G1 GUMLICH, E. *Z. Electrochem.* **15**, 599-600. Measurements of high inductions. Some physical properties of Fe-Si alloys. 69

09H1 HEUSLER, F., RICHARZ, F. *Z. anorg. allgem. Chem.* **61**, 265-79. Magnetizable manganese alloys. 330

09H2 HILPERT, S. *Ber. deut. chem. Ges.* **42**, 2248-61. Magnetic properties of ferrites. 245

09L1 LLOYD, M. G., FISHER, J. V. S. *Natl. Bur. Standards* (U. S.) *Bull.* **5**, 453-82. Testing of transformer steel. 853

09M1 MARS, G. *Stahl u. Eisen* **29**, 1673–78, 1769–81. Magnet steels and permanent magnetism. 374, 378, 379, 382
09Q1 QUITTNER, V. *Ann. Physik* [4] **30**, 289–325. Magnetic properties of magnetite crystals. 242
09W1 WEDEKIND, E. *Z. physik. Chem.* **66**, 614–32. Magnetic compounds of non-magnetic elements. 334
09Z1 ZEMCZUZNY, S., SCHEPELEW, J. *Z. anorg. allgem. Chem.* **64**, 245–57. Phosphorus compounds of cobalt. 293

1910–1914

10B1 BURGESS, C. F., ASTON, J. *Met. Chem. Eng.* **8**, 23–26. The magnetic and electrical properties of the iron-nickel alloys. 107, 222
10B2 BROWN, W. *Proc. Roy. Dublin Soc.* **12**, 349–53. Chrome steel permanent magnets. 346
10C1 CHEVENEAU, C. *Phil. Mag.* [6] **20**, 357–66. Magnetic balance of P. Curie and C. Chéveneau. 858
10D1 DORSEY, H. G. *Phys. Rev.* **30**, 698–719. Magnetostriction in Fe-C alloys. 676
10G1 GANS, R. *Physik. Z.* **11**, 988–91. Magnetic corresponding states. 543, 545
10H1 HEGG, F. *Arch. sci. phys. nat.* [4] **29**, 592–617; **30**, 15–45. Thermomagnetic study of iron-nickel alloys. 57, 108, 111, 501
10H2 HONDA, K. *Ann. Physik* [4] **32**, 1027–63. Thermomagnetic properties of the elements. 457
10H3 HONDA, K. *Ann. Physik* [4] **32**, 1003–26. Magnetization of some alloys as a function of composition and temperature. 339
10M1 MCWILLIAMS, A., BARNES, E. J. *J. Iron Steel Inst.* (London) **81**, 246–67, 276–86. Some physical properties of 2% Cr steels. 378
10P1 PANEBIANCO, G. *Rend. accad. sci.* (Napoli) [3a] **16**, 216–221. Magnetic susceptibility of ferromagnetic materials in weak fields. 109
10P2 PASCAL, P. *Ann. chim. phys.* [8] **19**, 5–70. Magnetochemical researches. 460
10R1 RICHTER, R. *Elektrotech. Z.* **31**, 1241–6. Proposal for description of hysteresis loss. 511
10R2 ROSS, A. D., GRAY, R. C. *Proc. Roy. Soc. Edinburgh* **31**, 85–99. Magnetism of Cu-Mn-Sn alloys under varying treatment. 328
10S1 SIEVERTS, A., KRUMBHAAR, W. *Ber. deut. chem. Ges.* **43**, 893–900. Solubility of gases in metals and alloys. 321
10T1 TERRY, E. M. *Phys. Rev.* **30**, 133–60. Effect of temperature on magnetic properties of electrolytic iron. 57, 501, 719
10W1 WEISS, P. *J. phys.* [4] **9**, 373–93. Absolute value of intensity of magnetization at saturation. 484
10W2 WAHL, W. *Z. anorg. allgem. Chem.* **66**, 60–72. Co-Au alloys. 290
10W3 WEISS, P., ONNES, H. K. *Compt. rend.* **150**, 686–9. Magnetic properties of Mn, V and Cr. 337, 338, 341
11G1 GANS, R. *Physik. Z.* **12**, 1053–4. Equation of the curve of reversible susceptibility. 545
11H2 HILPERT, S., DIECKMANN, T. *Ber. deut. chem. Ges.* **44**, 2831–5. Ferromagnetic compounds of Mn with P, As, Sb and Bi. 335, 337
11T1 TAKE, E. *Abh. kön. Ges. Wiss. Gött.* **8**, No. 2, 1–127. Aging and transformation studies in Heusler Al-Mn bronzes. 333
11W1 WOODRIDGE, W. J. *Trans. Am. Inst. Elec. Engrs.* **30**, 215–7. Hysteresis and eddy current exponents for Si steel. 510

11W2 WEISS, P., FOËX, G. *J. phys.* [5] **1**, 274–87. Study of the magnetization of ferrous bodies above Curie point. 860
11W3 WEDEKIND, E. Bornträger, Berlin, 1–114. Magnetochemie. 337, 339
12B1 BLOCH, O. *Arch. sci. phys. nat.* **33**, 293–308. Magnetization of alloys of Ni and Co. 264, 271, 276, 278, 280
12B2 BINET DU JASSONEIX, A. *8th Int. Cong. App. Chem.* **2**, 165–70. Magnetic properties of alloys of Fe, Co, Ni and Mn with B. 224
12B3 BOECKER, G. *Metallurgie* **9**, 296–303. System Co-C. 287
12C2 CAMPBELL, A., BOOTH, H. C., DYE, D. W. *J. Inst. Elec. Engrs.* **48**, 269–80. Report on 5 samples of magnetic sheet material tested for total loss and hysteresis at the Phys.-Tech. Reichsanstalt and Nat. Phys. Lab.
12F1 FRIEDRICH, K. *Metall u. Erz.* **10**, 659–71. On the knowledge of the freezing points of Co-Ni-As. 281, 418
12G1 GUMLICH, E. *Ferrum* **10**, 33–44. Magnetic properties of Fe-C and Fe-Si alloys. 73
12G2 GUMLICH, E., GOERENS, P. *Trans. Faraday Soc.* **8**, 98–114. Magnetic properties of Fe-C and Fe-Si alloys. 67, 68
12G3 GRAY, A. *Phil. Mag.* **24**, 1–14. Magnetic properties of graded series of Ni-Mn alloys. 315, 319
12H1 HEUSLER, F., TAKE, E. *Trans. Faraday Soc.* **8**, 169–84. Nature of the Heusler alloys. 328
12L1 LIEDGENS, J. *Stahl u. Eisen* **32**, 2109–15. Influence of As on properties of iron. 221, 222
12O1 OWEN, M. *Ann. Physik* [4] **37**, 657–99. Magnetochemical investigations II. 457
12P1 PREUSS, A. Dissertation, Zurich (See 29W2). Magnetic properties of Fe-Co alloys at different temperatures. 190, 193, 195
12P2 PRING, J. N., FAIRLIE, D. M. *J. Chem. Soc.* **101**, 91–103. The methane equilibrium.
12R1 RUER, R., KANEKO, K. *Metallurgie* **9**, 419–22. Nickel-cobalt system. 276
12R2 ROSA, E. B., GROVER, F. W. *Natl. Bur. Standards* (U. S.) *Bull.* **8**, 1–237. Formulas and tables for calculation of mutual and self inductance. 778
12R3 ROSS, A. D. *Trans. Faraday Soc.* **8**, 92–101, 185–94. Magnetic properties and microstructure of the Heusler alloys. 328
12S1 SMITH, S. W. J., GUILD, J. *Proc. Phys. Soc.* (London) **24**, 344–9. Self-demagnetization of annealed steel rods. 370
12W1 WEISS, P., PREUSS, A. *Trans. Faraday Soc.* **8**, 154–6. Magnetic properties of Fe-Co alloys. 190
12W2 WEDEKIND, E. *Trans. Faraday Soc.* **8**, 160–8. Magnetic properties of compounds in relation to stoichiometric composition. 245, 336, 337
13H1 HILPERT, S. *Jahrb. Radioakt.* **10**, 91–120. Ferromagnetic properties and chemical structure. 334
13R1 RUER, R., FICK, K. *Ferrum* **11**, 39–51. System Fe-Cu. 231
13S1 SWINDEN, T. J. *Iron Steel Inst.* (London), *Carneg. Schol. Mem.* **5**, 100–68. Study of the constitution of C-Mo steels. 378, 379
14A1 ARKADIEW, W. *Ann. Physik* [4] **45**, 133–46. Reflection of electromagnetic waves by wires. 798
14F1 FREUDENREICH, J. de. *Arch. sci. phys. nat.* [4] **38**, 36–45. Preparation of alloys for magnetic research. 555
14L1 LAMORT, J. *Ferrum* **11**, 225–37. Titanium-iron alloys. 256
14R1 RUSSELL, A. Univ. Press, Cambridge, 1–534. Treatise on the Theory of Alternating Currents. I. 2d ed. 770

14R2 RUDER, W. E. U.S.P. 1 110 010 (Appl. 6/22/12). Reduction of impurities by annealing at high temperature. 86
14S1 SCHWARZ, M. V. *Ferrum* 11, 80–90, 112–7. Ferrosilicon. 74
14Y1 YENSEN, T. D. *Trans. Am. Inst. Elec. Engrs.* 33, 451–75. Magnetic and other properties of electrolytic iron melted in vacuo. 83

1915–1919

15A1 APPLEGATE, K. P. *Rensselaer Polytech. Inst. Bull., Eng. Sci. Ser.*, No. 5, 1–19. Effect of Ti on the magnetic properties of Fe. 256
15B1 BARNETT, S. J. *Phys. Rev.* 6, 239–70. Magnetization by rotation. 450, 451, 454
15B2 BALL, J. D. *Trans. Am. Inst. Elec. Engrs.* 34, 2693–715. The unsymmetrical hysteresis loop. 549
15C1 CHUBB, L. W., SPOONER, T. *Trans. Am. Inst. Elec. Engrs.* 34, 2671–92. Effect of displaced magnetic pulsations on hysteresis loss of sheet steel. 549
15E1 EINSTEIN, A., HAAS, W. J. de. *Verhandl. deut. physik. Ges.* 17, 152–70. Experimental proof of Ampères molecular currents. 450, 452, 454
15G1 GANS, R., LOYARTE, R. G. *Arch. Elektrotech.* 3, 139–50. On rotational hysteresis. 515
15G2 GIEBELHAUSEN, H. v. *Z. anorg. allgem. Chem.* 91, 251–62. Behavior of V with Si, Ni, Cu and Ag, and B with Ni. 305, 326
15H1 HEAPS, C. W. *Phys. Rev.* 6, 34–42. Magnetostriction and resistance of Fe and Ni. 748
15S1 STEINHAUS, W., GUMLICH, E. *Verhandl. deut. physik. Ges.* 17, 369–84. Ideal or hysteresis-free magnetization. 549
15W1 WILLIAMS, E. H. *Phys. Rev.* 6, 404–9. Saturation value of intensity of magnetization and the theory of the hysteresis loop. 194
15Y1 YENSEN, T. D. *Trans. Am. Inst. Elec. Engrs.* 34, 2601–41. Magnetic and other properties of some iron alloys melted in vacuo. 23, 68, 69, 75, 76, 194, 198
15Y2 YENSEN, T. D. *U. of Ill. Bull.* 12, No. 29, 1–17. Effect of boron on magnetic and other properties of electrolytic Fe melted in vacuum. 225
16A1 ALDER, M. Thesis, Zurich (see 29W2). Magnetic properties of Ni-Cu alloys. 271, 310
16B1 BALL, J. D. *Gen. Elec. Rev.* 19, 369–90. Investigation of magnetic laws for steel and other materials. 485, 510
16G1 GUMLICH, E. *Elektrotech. Z.* 37, 592. Report of some investigations on magnetic properties and stability of Cr steel magnets. 346, 347, 375
16I1 ISHIWARA, T. *Sci. Repts. Tôhoku Imp. Univ.* 5, 53–61. Magnetic susceptibility of nitrided manganese. 338
16W1 WEISS, P., FREUDENREICH, J. de. *Arch. sci. phys. nat.* 49, 5–13, 449–70. Initial magnetization as a function of temperature. 196
17B1 BARNETT, S. J. *Phys. Rev.* 10, 7–21. Magnetization of Fe, Co and Ni by rotation and the nature of the elementary magnet. 454
17C1 CHEVENARD, P. *Rev. de Mét., Mém.*, 14, 610–40. The differential dilatometer. 447
17E1 ELMEN, G. W. Can. Patent 180 539 (Appl. 10/4/16). Magnetic material. 110
17H1 HADFIELD, R. A., CHÉVENEAU, C., GÉNEAU, C. *Proc. Roy. Soc.* (London) 94, 65–87. Magnetic properties of Mn and some Mn-steels. 337
17V1 VOGEL, R. *Z. anorg. allgem. Chem.* 99, 25–49. Cerium-iron alloys. 225
17Y1 YENSEN, T. D., GATWARD, W. A. *Univ. of Ill. Bull.* 14, No. 22, 1–50. Magnetic and other properties of Fe-Al alloys melted in vacuo. 51, 215, 216, 217

18B1 BECK, K. *Zürich naturforsch. Ges.* **63**, 116–86. The magnetic properties of Fe crystals at ordinary temperature. 567, 573
18F1 FAHY, F. P. *Chem. and Met. Eng.* **19**, 339–42. Permeameter for general magnetic analysis. 850
18G2 GUMLICH, E. *Wiss. Abhandl. physik-tech. Reichsanstalt.* **4**, 267–410. Dependence of magnetic properties, specific resistance and density of Fe alloys on chemical composition and heat treatment. 68, 76, 79, 83, 217, 234, 235, 367, 369
18S1 STEWART, J. Q. *Phys. Rev.* **11**, 100–20. Moment of momentum accompanying magnetic moment in Fe and Ni. 452, 454
18W1 WEISS, P., PICCARD, A. *Compt. rend.* **166**, 352–4. On a new magnetocaloric phenomenon. 740
19A1 ARKADIEW, W. *Ann. Physik* [4] **58**, 105–38. Absorption of electromagnetic waves in two parallel wires. 798, 800
19B1 BECK, E. *Ann. Physik* **60**, 109–48. On an experimental test of Ampères molecular currents. 454
19B2 BARKHAUSEN, H. *Physik. Z.* **20**, 401–3. Two phenomena uncovered with help of the new amplifiers. 524
19G1 GUMLICH, E. *Ann. Physik* [4] **59**, 668–88. Temperature coefficients of permanent magnets and shape of magnet. 357
19H1 HONDA, K. *Sci. Repts. Tôhoku Imp. Univ.* **8**, 51–8. Some physical constants of Fe-Co alloys. 191, 192, 194
19V1 VOURNASOS, A. C. *Compt. rend.* **168**, 889–91. Nitrides of Ni and Co. 293, 321

1920–1925

20A1 ARVIDSSON, G. *Physik. Z.* **21**, 88–91. An investigation of Ampère molecular currents by method of Einstein and de Haas. 454
20A2 AUERBACH, F. *Graetz, Hdb. Elect. & Magn.* **4**, 712–937. Magnetism of various materials. 684, 712
20A3 AUWERS, O. v. *Jahrb. Radioakt.* **17**, 181–229. Magnetism and atomic structure. 334
20E1 EVERSHED, E. *J. Inst. Elec. Engrs.* **58**, 780–837. Permanent magnets in theory and practice. 348
20E2 EDWARDS, C. A., NORBURY, A. L. *J. Iron Steel Inst.* (London) **101**, 447–82. Chromium steels II, effect of heat treatment on electrical resistivity. 377
20G1 GUILLAUME, C. E. *Proc. Phys. Soc.* (London) **32**, 374–404. The anomaly of the nickel steels. 147, 698, 699
20G2 GANS, R. *Ann. Physik* [4] **61**, 379–95. Reversible permeability on the ideal magnetization curve. 549
20H1 HONDA, K., KIDO, K. *Sci. Repts. Tôhoku Imp. Univ.* **9**, 221–31. Change of length by magnetization in Fe-Ni and Fe-Co. 663
20H2 HONDA, K., SAITO, S. *Sci. Repts. Tôhoku Imp. Univ.* **9**, 417–22. On K-S magnet steel. 346, 347, 379, 380, 381
20M1 MONYPENNY, J. H. G. *J. Iron Steel Inst.* (London) **101**, 493–525. Structure of some chromium steels. 376
20P1 WHIDDINGTON, R. *Phil. Mag.* **40**, 634–9. The ultramicrometer. 628
20S1 SANFORD, R. L., CHENEY, W. L. *Natl. Bur. of Standards* (U. S.) *Sci. Papers* **16**, 291–8. Variation of residual induction and coercive force with magnetizing force. 501
20Y1 YENSEN, T. D. *J. Am. Inst. Elec. Engrs.* **39**, 396–405. Magnetic and electrical properties of iron-nickel alloys. 110, 113, 189

21B2	Burrows, C. W. *Elec. World* **78**, 115-6. Monel metal has definite magnetic properties. 311	
21F1	Fondiller, W., Martin, W. H. *J. Inst. Elec. Engrs.* **40**, 553-79. Hysteresis effects with varying superposed magnetizing forces. 540	
21H1	Holborn, L. *Z. Metallkunde* **8**, 58-62. Dependence of resistance of pure metals on the temperature. 262	
21H2	Honda, K., Matumura, T. *Sci. Repts. Tôhoku Imp. Univ.* **10**, 417-21. On dependency of the temperature coefficient of a permanent magnet on its dimensions. 357	
21M1	Murakami, T. *Sci. Repts. Tôhoku Imp. Univ.* **10**, 79-92. Equilibrium diagram of Fe-Si system. 72	
21S2	Speed, B., Elmen, G. W. *Jl. Am. Inst. Elec. Engrs.* **40**, 596-609. Magnetic properties of compressed powdered iron. 781	
21V1	Vogel, R. *Z. anorg. allgem. Chem.* **116**, 231-42. Ni-W alloys. 326	
21W1	Wwedensky, B. *Ann. Physik* [4] **64**, 609-20. Eddy-currents for spontaneous change of magnetization. 784, 789	
22A1	Anderson, N. L., Lance, T. M. C. *Engineering* **114**, 351-2. Relation between magnetic hysteresis loss and coercivity. 511	
22C2	Cheney, W. L. *Natl. Bur. Standards* (U. S.) *Sci. Papers* **18**, 609-35. Magnetic properties of Fe-C alloys as affected by heat treatment and carbon content. 369	
22C3	Claasen, H. Diss., Hamburg, 1-8 (see *Phys. Ber.* **4**, 359). Einstein- de Haas effect for detection of Ampèrian molecular currents. 454	
22E1	Elmen, G. W. Canadian Pat. 221 525 (Appl. 5/30/21). Loading of telephone conductors. 110	
22E2	Ewing, J. A. *Phil. Mag.* [6] **43**, 493-503. New model of ferromagnetic induction. 426	
22G1	Gumlich, E. *Stahl u. Eisen* **42**, 41-6; 97-103. Investigation of Cr-C steels for permanent magnets. 376	
22M1	Matsushita, T. *Sci. Repts. Tôhoku Imp. Univ.* **11**, 471-85. On the magnetic hardness of quenched steels. 369	
22T1	Théodoridès, P. *J. phys. radium* [6] **3**, 1-19. Anhydrous paramagnetic compounds. 858	
23A2	Arnold, H. D., Elmen, G. W. *J. Franklin Inst.* **195**, 621-32. Permalloy, an alloy of remarkable magnetic properties. 110, 115	
23C1	Chattock, A. P., Bates, L. F. *Trans. Roy. Soc.* (London) **223A**, 257-88. On the Richardson gyromagnetic effect. 452, 454	
23F1	Fry, A. *Stahl u. Eisen* **43**, 1039-44. Diffusion in solid iron of elements present in commercial iron. 64	
23G1	Gumlich, E. *Elektrotech. Z.* **44**, 147-151. A new material for permanent magnets. 346, 375, 380	
23H1	Hunter, M. A., Sebast, E. M., Jones, A. *Trans. Am. Inst. Mining Met. Engrs.* **68**, 750-6. Some electrical properties of Ni and monel wires. 269	
23H2	Heaps, C. W. *Phys. Rev.* **22**, 486-501. Effect of crystal structure on magnetostriction. 649	
23J1	Jahnke, E., Emde, F. Teubner, Leipzig, 1-176. Funktionentafeln mit formeln und kurven. 424, 510, 784	
23K1	Kaiser, J. F. *Engineering* **135**, 57, 83-4. Cobalt steels for permanent magnets. 381	
23P1	Pilling, N. B. *Trans. Am. Inst. Mining Met. Engrs.* **69**, 780-90. Low temperature brittleness in silicon steels. 76	

23R1 RICHARDSON, O. W. *Inst. intern. de. phys. Solvay* (1921 meeting) 216–21. Discussion of report by W. J. de Haas. 450
23S1 SMITHSONIAN INSTITUTION, Washington. Smithsonian Physical Tables. 856
23S2 SUCKSMITH, W., BATES, L. F. *Proc. Roy. Soc.* (London) **104A**, 499–511. On a null method of measuring gyromagnetic ratio. 452, 454
23S3 SPOONER, T. *Trans. Am. Inst. Elec. Engrs.* **42**, 340–6. Permeability. 547
23W1 WATSON, E. A. *J. Inst. Elec. Engrs.* **61**, 641–60. Permanent magnets, and the relation of their properties to the constitution of magnet steels. 350
23W2 WHITELEY, J. H., BRAITHWAITE, A. *J. Iron Steel Inst.* (London) **107**, 161–9. Some observations on the effect of small quantities of Sn in steel. 255
23W3 WOLTJER, H. R., ONNES, H. K. *Proc. Acad. Sci.* Amsterdam **26**, 626–34. Magnetization of gadolinium sulfate at liquid He temperatures. 466
24D1 DREIBHOLZ. *Z. physik. Chem.* **108**, 1–50. Investigation of binary and ternary alloys of Mo. 313, 321, 418
24E2 EDWARDS, C. A., PFEIL, L. B. *J. Iron Steel Inst.* (London) **109**, 129–47. Production of large crystals by annealing strained iron. 555
24G1 GOERENS, P. *Stahl u. Eisen* **44**, 1645–9. Properties of high quality steels. 75
24G2 GERLACH, W., STERN, O. *Ann. Physik* [4] **74**, 673–99. Directional quantization in a magnetic field. 468
24H1 HEAPS, C. W. *Phys. Rev.* **24**, 60–7. Magnetostriction of a magnetite crystal. 649, 681
24H2 HANNACK, G. *Stahl u. Eisen* **44**, 1237–43. On magnetic steels with specific reference to relation between carbon and magnetic properties. 346, 347, 371, 376
24J1 JORDAN, H. *Elek. Nachr. Tech.* **1**, 7–29. Ferromagnetic constants for weak fields. 489, 781, 788, 796
24K1 KELSALL, G. A. *J. Opt. Soc. Am.* **8**, 699–74. Furnace permeameter for alternating current measurements at small magnetizing forces. 168, 197, 853
24K2 KELSALL, G. A. *J. Opt. Soc. Am.* **8**, 329–38. Permeameter for small a.c. losses at small magnetizing forces. 853
24N1 NIWA, Y. *Researches Electrotech. Lab.* (Tokyo) No. 142, 1–36. Null method for testing magnetic properties of material. 850
24O1 OBERHOFFER, P., DAEVES, K., RAPATZ, F. *Stahl u. Eisen* **44**, 432–5. Consideration of solubility line for carbon in Cr and W steels. 371
24P1 PARKIN, A. M. *Iron Steel Inst.* (London), *Carnegie Schol. Mem.* **13**, 1–46. Effect of heat treatment and of variable C content on a W magnet steel of fixed W content. 378
24S1 SMITH, W. S., GARNETT, H. J. Br. Pat. 224 972 (Appl. 8/25/23). Nickel alloys of high magnetic permeability. 110, 153, 159
24S2 SMITH, S. W. J., DEE, A. A., MAYNEORD, W. V. *Proc. Phys. Soc.* (London) **37**, 1–14. Magnetism of annealed carbon steels. 370
24T1 TYNDALL, E. P. T. *Phys. Rev.* **24**, 439–51. The Barkhausen effect. 524
24W1 WATSON, E. A. *J. Inst. Elec. Engrs.* **63**, 822–38. Economic aspect of the utilization of permanent magnets in electrical apparatus.
24W2 WATSON, E. A. *Engineering* **118**, 274–6, 302–4. Cobalt magnet steels 380
24W3 WEISS, P., FORRER, R. *Compt. rend.* **178**, 1670–3. Spontaneous magnetization of nickel. 717, 718

24W5 WEISS, P., FORRER, R. *Compt. rend.* **178**, 1347–51. Magnetocaloric phenomena and specific heat of Ni. 743
24W6 WILLS, A. P., HECTOR, L. G. *Phys. Rev.* **23**, 209–20. Magnetic susceptibility of O, H and He. 859
24W7 WEDEKIND, E. *Z. angew. Chem.* **37**, 87–9. Magnetochemical researches. 334
24Y1 YENSEN, T. D. *Trans. Am. Inst. Elec. Engrs.* **43**, 145–75. Magnetic properties of ternary Fe-Si-C alloys. 68, 75, 83, 84, 87, 367

1925–1927

25A1 ANDREW, J. H., FISHER, M. S., ROBERTSON, J. M. *J. Roy. Tech. Coll.* (Glasgow) **2**, 70–8. Specific volume determination of C and Cr steels. 377
25B1 BARNETT, S. J., BARNETT, L. J. H. *Proc. Am. Acad. Arts Sci.* **60**, 127–216. Magnetization by rotation. 454
25B2 BUCKLEY, O. E., MCKEEHAN, L. W. *Phys. Rev.* **26**, 261–73. Effect of tension upon magnetization and magnetic hysteresis in permalloy. 595
25B3 BOZORTH, R. M. *J. Opt. Soc. Am.* **10**, 591–8. Null reading astatic magnetometer. 857
25B4 BRIDGMAN, P. W. *Proc. Am. Acad. Arts Sci.* **60**, 305–83. Physical properties of single crystals of W, Sb, Bi, Te, Cd, Zn and Sn. 555
25C2 CAUER, W. *Arch. Elektrotech.* **15**, 308–19. Effective permeability and iron loss in sheets and wires with weak magnetic fields. 770, 781
25E1 EVERSHED, S. *J. Inst. Elec. Engrs.* **63**, 725–821. Permanent magnets in theory and practise. 372, 373, 374
25F1 FORESTIER, H., CHAUDRON, G. *Compt. rend.* **181**, 509–11. Magnetic transformation points in system Fe_2O_3–MgO. 245
25G1 GERLACH, W. *Ann. Physik* [4] **76**, 163–97. Directional quantization in a magnetic field. 468
25H1 HUND, F. *Z. Physik* **33**, 855–9. Theoretical meaning of the magnetism of the rare earths. 464
25I1 ISING, E. *Z. Physik* **31**, 253–8. Contribution to the theory of ferromagnetism. 446
25M1 MAURER, E. *Stahl u. Eisen* **45**, 1629–32. On vanadium steels. 259
25M2 MERICA, P. D., WALTENBURG, R. G. *Trans. Am. Inst. Mining Met. Engrs.* **71**, 709–16. Malleability and metallography of Ni. 323
25O1 OBERHOFFER, P., EMICKE, O. *Stahl u. Eisen* **45**, 537–40. On Cr-Steel for permanent magnets. 377
25P1 PESCHARD, M. *Compt. rend.* **180**, 1475–78; **180**, 1836–38; **181**, 99–101; **181**, 854–5. Magnetism of the ferronickels. 441
25R1 RUDER, W. E. *Trans. Am. Soc. Steel Treating* **8**, 23–9. Magnetization and crystal orientation. 571
25S1 SUCKSMITH, W. *Proc. Roy. Soc.* (London) **108A**, 638–42. Gyromagnetic ratio for magnetite and cobalt. 269, 452, 454
25S3 SCHÖNERT, K., HANNACK, G. *Ber. der Werkstoffausschusses des Vereins deut. Eisenhüttenleute*, No. 73, 1–3. Carbon and manganese in tungsten magnet steel. 372
25S4 SCHLUMBERGER, E. *Chem. Ztg.* **49**, 913–5. Physico-chemical methods of rapid control of melting in electric furnaces. 74
25W1 WEBSTER, W. L. *Proc. Roy. Soc.* (London) **109A**, 570–84. Magnetostriction in iron crystals. 631, 653
25W3 WEVER, F., REINECKEN, W. *Mitt. Kaiser-Wilhelm Inst. Eisenforsch. Düsseldorf* **7**, 69–79. System Fe-Sn. 255

25W4	WEBSTER, W. L. *Proc. Roy. Soc.* (London) **107A**, 496–509. Magnetic properties of Fe crystals. 567, 645, 650	
25Y1	YENSEN, T. D. *J. Franklin Inst.* **199**, 333–42. Magnetic properties of the fifty per cent iron-nickel alloys. 110, 121, 130	
26A1	ADAMS, J. R., GOECKLER, F. E. *Trans. Am. Soc. Steel Treating* **10**, 173–94. Some factors affecting H_c and B_r of some magnet steels. 373	
26C1	CAMPBELL, E. D. *J. Iron and Steel Inst.* (London) **113**, 375–92. Specific resistance and thermoelectric potential of some steels differing only in carbon content. 367	
26E1	ELMEN, G. W. U.S.P. 1 586 884 (Appl. 7/24/16 and 5/31/21). Magnetic material.	
26F1	FORRER, R. *J. phys. radium* [6] **7**, 109–24. Artificial magnetic anisotropy of Ni. 496, 504, 532, 610	
26G1	GOUDSMIT, S., UHLENBECK, G. E. *Nature* **117**, 264–265. Spinning electrons and the structure of spectra. 450	
26G2	GERLACH, W. *Z. Physik* **38**, 828–40. Iron single crystals. 567	
26H1	HADFIELD, R. A. Van Nostrand, N. Y., 1–388. Metallurgy and its influence on modern progress. 67, 68	
26H2	HONDA, K., KAYA, S. *Sci. Repts. Tôhoku Imp. Univ.* **15**, 721–53. On the magnetization of single crystals of iron. 567, 582	
26H3	HONDA, K., MASIYAMA, Y. *Sci. Repts. Tôhoku Imp. Univ.* **15**, 755–76. Magnetostriction of iron crystals. 645	
26M1	MASUMOTO, H. *Sci. Repts. Tôhoku Imp. Univ., Sendai*, 449–77. On a new transformation of cobalt and the equilibrium diagrams of nickel-cobalt and iron-cobalt. 262, 264, 276	
26M2	McKEEHAN, L. W. *J. Franklin Inst.* **202**, 737–73. Magnetostriction. 628, 639, 641	
26M3	McKEEHAN, L. W. *Phys. Rev.* **28**, 158–66. The significance of magnetostriction in Permalloy. 627, 628	
26M4	McKEEHAN, L. W., CIOFFI, P. P. *Phys. Rev.* **28**, 146–57. Magnetostriction in Permalloy. 666	
26M5	MAHAJANI, G. S. *Proc. Cambridge Phil. Soc.* **23**, 136–43. Contribution to the theory of ferromagnetism. 592	
26P1	PHRAGMÉN, G. *J. Iron Steel Inst.* (London) **114**, 397–404. Constitution of Fe-Si alloys. 73, 74	
26P2	PAULI, W. *Z. Physik* **41**, 81–102. Paramagnetism of a degenerate gas. 467	
26R1	RIBBECK, F. *Z. Physik* **38**, 772–87, 887–907; **39**, 787–812. Dependence of electrical resistivity of nickel steels on composition, temperature and heat treatment. 55	
26S1	SCHULZ, E. H., JENGE, W. *Stahl u. Eisen* **46**, 11–3. On the question of the heat treatment and testing of Cr magnet steels. 377	
26S2	SCHULZ, E. H., JENGE, W., BAUERFELD, F. *Z. Metallkunde* **18**, 155–6. New advances in the field of high alloy materials. 380, 381	
26S3	SUCKSMITH, W., POTTER, H. H. *Proc. Roy. Soc.* (London) **112**, 157–76. On the specific heat of ferromagnetic substances. 738	
26W1	WEISS, P., FOËX, G. Colin, Paris, 1–215. Le Magnetisme. 493	
26W2	WEBB, C. E. *J. Inst. Elec. Engrs.* **64**, 409–27. Power losses in magnetic sheet material at high flux densities. 510	
26W3	WEBSTER, W. L. *Proc. Roy. Soc.* (London) **113**, 196–207. Longitudinal magnetoresistance effect in single crystals of Fe. 764	

26W4 WEISS, P., FORRER, R. *Ann. physique* [10] **5**, 153–213. Magnetization of Ni and the magnetocaloric effect. 645, 740

26W5 WILD, G., PERRIER, A. *Arch. sci. phys. nat.* **7**, 209–12. Law of magnetic aging and recovery of iron used in telephony. 794

27A1 ADELSBERGER, U. *Ann. Physik* **83**, 184–212. On hysteresis heat and magnetic energy in ferromagnetic bodies. 518

27D1 DICKIE, H. A. *J. Iron Steel Inst.* (London) **116**, 223–43. Magnetic and other changes concerned in the temper brittleness of Ni-Cr steels. 147

27D2 DUPOUY, G. *Compt. rend.* **184**, 375–7. Demonstration apparatus for direct measurement of magnetic field. 856

27E2 ELLIS, W. C. *Rensselaer Polytech. Inst. Bull., Eng. Sci. Ser.* **16**, 1–57. Study of physical properties of electrolytic Co and its alloys with Fe. 163, 190, 191, 192

27E3 EILENDER, W., OERTEL, W. *Stahl u. Eisen* **47**, 1558–61. Influence of oxygen on the properties of steel. 85, 372

27G1 GIAUQUE, W. F. *J. Am. Chem. Soc.* **49**, 1870–7. Low temperature magnetic susceptibility of gadolinium sulfate. 466

27G2 GOLDSCHMIDT, V. M. *Ber. deut. chem. Ges.* **60**, 1263–96. Crystal structure and chemical constitution. 342

27G3 GUMLICH, E. *Handb. d. Physik.* **15**, 222–70. Ferromagnetic materials. 328, 330

27H1 HADFIELD, R. A. *J. Iron Steel Inst.* (London) **115**, 297–363. Alloys of iron and manganese containing low carbon. 236

27H2 HAUGHTON, J. L. *J. Iron Steel Inst.* (London) **115**, 417–33. Constitution of the alloys of Fe and P. 251

27I1 ISHIGAKI, T. *Sci. Repts. Tôhoku Imp. Univ.* **16**, 295–302. Determination of density of cementite. 367

27J1 JONG, W. F. DE, WILLEMS, H. W. V. *Physica* **7**, 74–9. Compounds with the lattice structure of pyrrhotite. 284, 293

27K1 KASÉ, T. *Sci. Repts. Tôhoku Imp. Univ.* **16**, 491–513. On the equilibrium diagram of the Fe-Co-Ni system. 161, 276

27K2 KLINKHARDT, H. *Ann. Physik* [4] **84**, 167–200. Measurement of true specific heat at high temperatures by heating with thermal electrons. 734, 737

27K4 KAPITZA, P. *Proc. Roy. Soc.* (London) **115A**, 658–83. Method of obtaining strong magnetic field. 856

27L1 LEU, A. *Z. Physik* **41**, 551–62 (1927). Deflection of molecular rays in a magnetic field. 468

27M1 MASUMOTO, H., NARA, S. *Sci. Repts. Tôhoku Imp. Univ.* **16**, 333–41. On the coefficient of thermal expansion in nickel-cobalt and iron-cobalt alloys, and the magnetostriction of iron-nickel alloys. 659

27M2 MASUMOTO, H. *Sci. Repts. Tôhoku Imp. Univ.* **16**, 321–32. On the magnetic, electric and thermal properties of Ni-Co alloys. 278, 674

27O1 OFTEDAL, I. *Z. physik. Chem.* **128**, 135–53. Some crystal structures of the NiAs type. 295, 342

27S1 SCHULZE, A. *Physik. Z.* **28**, 669–73. Thermal expansion of cobalt-nickel, cobalt-iron, and iron-nickel alloys. 278

27S2 SCHULZE, A. *Z. tech. Physik* **8**, 365–70. Some physical properties of cobalt. 262

27S3 SCHULZE, A. *Z. tech. Physik* **8**, 495–502. Measurements of magnetostriction in some alloy series. 663, 674, 712

27S4 SPOONER, T. McGraw-Hill, N. Y., 1–385. Properties and testing of magnetic materials. 861
27S5 SANFORD, R. L. *Natl. Bur. Standards* (U. S.) *Sci. Papers* **22**, 557–67. Some principles governing the choice and utilization of permanent magnet steels. 352
27S6 STRUTT, M. J. O. *Ann. Physik* [4] **84**, 485–506. Eddy-currents in elliptical cylinders. 776
27T1 TAMMANN, G., KOLLMANN, K. *Z. anorg. allgem. Chem.* **160**, 242–8. Solubility of metals of the Fe group and Cu in Hg. 291
27U1 UMINO, S. *Sci. Repts. Tôhoku Imp. Univ.* **16**, 593–611. On the heat of transformation of nickel and cobalt. 738
27W1 WILLIAMS, S. R. *J. Opt. Soc. Am.* **14**, 383–408. Some experimental methods in magnetostriction. 628
27W2 WILLIAMS, S. R. *Trans. Am. Soc. Steel Treating* **11**, 885–98. Correlation of magnetic properties with mechanical hardness in cold-worked metals. 275
27W3 WEBSTER, W. L. *Proc. Roy. Soc.* (London) **114A**, 611–9. Transverse magneto-resistance effect in single crystal of Fe. 764, 765

1928

28A1 AUWERS, O. v. *Z. tech. Physik* **9**, 475–8 (1928). Influence of grain size on magnetic properties. 87
28A2 ARNFELT, H. *Iron Steel Inst.* (London) *Carnegie Schol. Mem.* **17**, 221 pp. (1928). On the constitution of the Fe-W & Fe-Mo alloys. 237
28A3 AKULOV, N. S. *Z. Physik* **52**, 389–405. Magnetostriction of iron crystals. 628, 650, 654
28A4 Anonymous. *Elektrotech. Z.* **49**, 828–9. Properties of various magnetic alloys. 153, 311
28A5 ALLIBONE, T. E., SYKES, C. *J. Inst. Metals* **39**, 173–86. Alloys of zirconium. 328
28B1 BOZORTH, R. M. *Phys. Rev.* **32**, 124–32. Time-lag in magnetization. 787, 789
28B2 BLOCH, F. *Z. Physik* **52**, 555–600. Quantum mechanics of electrons in crystal lattices. 437
28B3 BABBITT, B. J. *J. Opt. Soc. Am.* **17**, 47–58. Improved permeameter for testing magnet steel. 849
28B4 BATES, L. F., BROWN, R. C. *Nature* **122**, 240. Laboratory uses of monel metal. 311
28B5 BATES, L. F. *Phil. Mag.* [7] **6**, 593–7. Experiments on a ferromagnetic compound of Mn and As. 336
28C1 COTTON, A. *Compt. rend.* **177**, 77–89. The large electromagnet of the Academy of Sciences. 855
28C2 CHEVENARD, P. *Rev. de mét.* **25**, 14–34. Alloys having high Ni and Cr contents. 149, 150, 278
28C3 CONSTANT, F. W. *Phys. Rev.* **32**, 486–93. Distribution of heat emission in magnetic hysteresis cycle. 522
28C4 CROSS, H. C., HILL, E. E. *Natl. Bur. Standards* (U. S.) *Sci. Papers* **22**, 451–66. Density of hot rolled and heat treated carbon steels. 367
28C5 CORSON, M. G. *Trans. Am. Inst. Mining Met. Engrs.* **80**, 249–300. Constitution of Fe-Si alloys. 75, 76
28D1 DUSSLER, E. *Z. Physik* **50**, 195–214. Magnetization of Fe Single Crystal as function of temperature. 567

BIBLIOGRAPHY

28E1 ELMEN, G. W. *J. Franklin Inst.* **206**, 317–38. Magnetic properties of perminvar. 110, 160, 163, 177, 498

28G1 GUMLICH, E., STEINHAUS, W., KUSSMANN, A., SCHARNOW, B. *Elek. Nachr.-Tech.* **5**, 83–100. Materials of high initial permeability. 180

28G2 GEWECKE, H. *Z. tech. Physik* **9**, 57–60. Temperature dependence of remanent magnetism. 356, 357

28G3 GERLACH, W. *Atti congresso intern. fisici.*, Como 1927, **1**, 77–94. Magnetic properties of gases and vapors. 466

28H1 HEISENBERG, W. *Z. Physik* **49**, 619–36. On the theory of ferromagnetism. 443, 446, 447

28H2 HONDA, K. Syokwabo and Co., Tokyo, 1–256. Magnetic properties of matter. 630, 656

28H3 HONDA, K., MASUMOTO, H., KAYA, S. *Sci. Repts. Tôhoku Imp. Univ.* **17**, 111–30. Magnetization of single crystal of Fe at high temperatures. 567

28H4 HEUSLER, O. *Z. anorg. allgem. Chem.* **171**, 126–42. On Heusler alloys. 328

28K1 KAYA, S. *Sci. Repts. Tôhoku Imp. Univ.* **17**, 1027–37. Magneto-resistance effect in single crystal of Ni. 764

28K2 KAYA, S. *Sci. Repts. Tôhoku Imp. Univ.* **17**, 639–63. On the magnetization of single crystal of Ni. 567, 568

28K3 KAYA, S. *Sci. Repts. Tôhoku Imp. Univ.* **17**, 1157–77. On the magnetization of single crystals of Co. 568

28K4 KUSSMANN, A. *Z. Metallkunde* **20**, 406–7. Magnetization curves of Monel metal. 311

28M1 MITTASCH, A. *Z. anorg. allgem. Chem.* **41**, 827–33. Iron carbonyl and carbonyl iron. 52

28M4 MAMES, E., NIENHAUS, H. *Stahl u. Eisen* **48**, 996–1005. The inner structure of chromium steels. 376

28M5 MASING, G. *Z. Metallkunde* **20**, 19–21. Alloys of Be with Cu, Ni, Co and Fe. 224, 286

28M6 MASIYAMA, Y. *Sci. Repts. Tôhoku Imp. Univ.* **17**, 945–61. Magnetostriction of a single crystal of nickel. 646, 649, 650

28M7 MILLAR, R. W. *J. Am. Chem. Soc.* **50**, 1875–83. Specific heats at low temperatures of MnO, MnO_2 and Mn_3O_4. 470

28P1 PÖLZGUTER, F. *Stahl u. Eisen* **48**, 1100–2. Influence of Si on W magnet steel. 373

28P2 PARTRIDGE, J. H. *Iron Steel Inst.* (London) *Carnegie Schol. Mem.* **17**, 157–90. The magnetic and electrical properties of cast iron. 95

28P3 PERSSON, E. *Naturwissenschaften* **16**, 613. X-ray analysis of Heusler alloys and theory based thereon. 328

28S1 SHACKELTON, W. J., BARBER, I. G. *Trans. Am. Inst. Elec. Engrs.* **47**, 429–37. Compressed powdered permalloy. 125, 133

28S2 SAMUEL, M. *Ann. Physik* [4] **86**, 798–824. Magnetic properties of cobalt. 266, 267

28S3 SCHULZE, A. *Z. Physik* **50**, 448–505. Magnetostriction I. 633, 666, 669, 678, 679, 693

28S4 SCHULZE, A. *Z. Metallkunde* **20**, 403–6. On Monel metal. I. Electrical & thermal properties and magnetostriction. 680

28S5 STÄBLEIN, F., SCHROETER, K. *Z. anorg. allgem. Chem.* **174**, 193–215. Determination of magnetic saturation of iron carbide. 367

28S6 SWAN, J. *J. Iron Steel Inst.* (London) **117**, 369–82. Effect of Si on W magnet steel. 372

28S7 STOGOFF, A. F., MESSKIN, W. S. *Arch. Elektrotech.* **2**, 321–31. Copper steel with high carbide content. 382
28S8 SCHULZE, A. *Z. tech. Physik* **9**, 338–43. Thermal expansion of alloys of iron. 75, 212, 215, 257
28S9 SLATER, J. C. *Phys. Rev.* **32**, 349–60. Normal state of helium. 459
28S10 SUCKSMITH, W., POTTER, H. H., BROADWAY, L. *Proc. Roy. Soc.* (London) **117A**, 471–85. Magnetic properties of a single crystal of Ni. 567, 568
28T1 TAKEI, T. *Kinzoku no Kenkyu* **5**, 364, 79. Diagram of Co-Mo system. 292
28W1 WESTGREN, A., PHRAGMÉN, G. *Trans. Am. Soc. Steel Treating* **13**, 539–54. On the double carbide of high speed steel. 372
28W2 WESTGREN, A., PHRAGMÉN, G., NEGRESCO, T. *J. Iron Steel Inst.* (London) **117**, 383–400. The structure of the Fe-Cr-C system. 376
28W3 WEBSTER, W. L. *Proc. Roy. Soc.* (London) **114**, 611–9. Transverse magnetoresistance in single crystals of iron. 765
28W4 WAIT, G. R., BRICKWEDDE, F. G., HALL, E. L. *Phys. Rev.* **32**, 967–73. Electrical resistance and magnetic permeability of Fe wire at radio frequency. 802

1929

29A2 AUWERS, O. V. *Wiss. Veröffentl. Siemens Konzern* **8**, 236–47. Magnetic measurements of Fe-Be alloys. 224
29A3 AKULOV, N. S. *Z. Physik* **54**, 582–7. Atomic theory of ferromagnetism. 563
29B1 BOZORTH, R. M. *Phys. Rev.* **34**, 772–84. Barkhausen effect in Fe, Ni and Permalloy. 524, 525
29B2 BRACE, P. H. *Elec. J.* **26**, 111–21. Cobalt magnet steel. 381, 524
29B3 BUCHHOLZ, H. *Arch. Elektrotech.* **22**, 360–74. Screening action and eddy-current loss of a hollow cylindrical conductor in alternating magnetic fields. 776
29B4 BATES, L. F. *Phil. Mag.* [7] **8**, 714–32. Magnetic properties of some compounds of Mn. 334, 336, 337, 339
29D1 DEUTSCHMANN, W. *Z. tech. Physik* **10**, 511–5. The flutter effect in Pupin coils. 540
29D2 DWIGHT, H. B. *Trans. Am. Inst. Elec. Engrs.* **48**, 812–20. Bessel functions for a.c. problems. 776
29D3 DAEVES, K. *Z. tech. Physik* **10**, 67–8. Influence of grain size on magnetic properties. 87
29E1 ELMEN, G. W. *J. Franklin Inst.* **207**, 583–617. Magnetic alloys of iron, nickel and cobalt. 112, 115, 117, 118, 163
29E4 ELMEN, G. W. U.S.P. 1 739 752 (Appl. 12/28/27). Magnetic material and appliance. 190, 197
29F1 FOSTER, D. D. *Phys. Rev.* **33**, 1071. Magnetic properties of iron crystals. 567
29G1 GOULD, J. E. *Proc. World Eng. Cong.*, Tokyo, **34**, 273–92. Aging of permanent magnet steels. 373, 382
29G2 GRIES, H., ESSER, H. *Arch. Eisenhüttenw.* **2**, 749–61. Single crystals of Fe. 567
29H1 HÄGG, G. *Z. Krist.* **71**, 134–6. X-ray studies of the system Fe-As. 221
29H2 HOWEY, J. H. *Phys. Rev.* **34**, 1440–7. Magnetic behavior of nickel and iron films. 857

29H3	HÄGG, G. *Z. physik. Chem.* (B) **4**, 346–70. X-ray study of the nitrides of Mn. 338	
29H4	HONDA, K., OKUBO, J., HIRONI, T. *Sci. Repts. Tôhoku Imp. Univ.* **18**, 409–17. Heat evolution during the magnetization of steels. 523	
29I2	INGLIS, D. R. *Instruments* **2**, 129–32. Some magnetic properties of Monel metal. 311	
29K1	KUSSMANN, A., SCHARNOW, B. *Z. Physik* **54**, 1–15. Coercive force and mechanical hardness. 157, 231, 232, 397	
29K2	KROLL, W. *Wiss. Veröffentl. Siemens-Konzern* **8** (1), 220–35. Alloys of Be with Fe. 189, 224, 382, 417	
29K3	KÖSTER, W. *Z. anorg. allgem. Chem.* **179**, 297–308. Influence of finely divided precipitate on coercive force. 369	
29K5	KRUPKOWSKI, A. *Rev. de. mét.* **26**, 131–53, 193–208. Nickel-copper alloys. 308, 574	
29L1	LAPP, E. *Ann. physique* [10] **12**, 442–521. Determination of the true specific heats of nickel by a direct electrical method. 734, 735, 737	
29M1	MASUMOTO, H. *Sci. Repts. Tôhoku Imp. Univ.* **18**, 195–229. On the intensity of magnetization in iron-nickel-cobalt alloys. 113, 163, 194, 195, 279, 280	
29M2	MAHAJANI, G. *Trans. Roy. Soc.* (London) **228A**, 63–114. Theory of ferromagnetic crystals. 423, 426, 563, 592	
29M3	MESSKIN, W. S. *Arch. Eisenhüttenw.* **3**, 417–25. Influence of cold working on the magnetic properties of a carbon steel. 370	
29M4	MCKEEHAN, L. W. *J. Opt. Soc. Am.* **19**, 213–42. Measurement of magnetic quantities. 861	
29M5	MASING, G., DAHL, O. *Wiss. Veröffentl. Siemens-Konzern.* **8** (1), 211–9. Be-Ni alloys. 302	
29M6	MILLAR, R. W. *J. Am. Chem. Soc.* **51**, 215–22. Heat capacities at low temperatures of FeO, Fe$_3$O$_4$, Cu$_2$O and CuO. 470	
29N1	NISHIYAMA, Z. *Sci. Repts. Tôhoku Imp. Univ.* **18**, 341–57. On the magnetostriction of single crystals of Co. 647, 648, 657	
29N2	NISHIYAMA, Z. *Sci. Repts. Tôhoku Imp. Univ.* **18**, 359–400. Measurement of elastic constant, lattice constant and density of binary alloys in the range of solid solution. 256, 258	
29O1	ONNES, H. K., TUYN, W. *Intl. Crit. Tables* **6**, 124–35. Electrical resistance of elementary substances at temperatures below −80°C. 55	
29O2	OBERHOFFER, P., KREUTZER, C. *Arch. Eisenhüttenw.* **2**, 449–56. Contribution to the systems Fe-Si, Fe-Cr & Fe-P. 72	
29O3	OSAWA, A., OYA, S. *Sci. Repts. Tôhoku Imp. Univ.* **18**, 727–31. X-ray analysis of Fe-V alloy. 258	
29O4	OERTEL, W. *Stahl u. Eisen* **49**, 1449–54. Testing of permanent magnet steel. 852	
29P1	PREISACH, F. *Ann. Physik* [5] **3**, 737–99. Investigations of the Barkhausen effect. 496, 524, 610	
29P2	PIERCE, G. W. *Proc. Inst. Radio Engrs.* **17**, 42–88. Magnetostriction oscillators. 628, 686, 699	
29P3	POTTER, H. H. *Proc. Phys. Soc.* (London) **41**, 135–42. X-ray structure & magnetic properties of single crystals of Heusler alloy. 328, 574	
29P4	PERSSON, E. *Z. Physik* **57**, 115–33. On the structure of Heusler alloys. 328, 330	
29R1	ROGERS, B. A. *Trans. Am. Electrochem. Soc.* **56**, 225. Discussion of Paper by T. D. Yensen (29Y1). 60	

29R2 RANKIN, J. S. *J. Roy. Tech. Coll. Glasgow* **2**, 12–9. Magnetostriction of various steels. 669, 676
29S1 SIZOO, J. G. *Ann. Physik* [5] **3**, 270–6. On effective and reversible permeability. 543
29S2 STOGOFF, A. F., MESSKIN, W. S. Moscow, N.T.U. (see ref. 32M3, p. 237). On the temperature-dependence of remanent magnetism. 373
29S3 STOGOFF, A. F., MESSKIN, W. S. *Arch. Eisenhüttenw.* **2**, 595–600. Mo steels and their use in permanent magnets. 378, 379
29S4 STÄBLEIN, F. *Arch. Elektrotech.* **3**, 301–5. Physical properties of pure Cr and W steels. 227, 228, 256, 257
29S5 SCHMIDT, W. *Arch. Eisenhüttenw.* **3**, 293–300. X-ray investigation of the system Fe-Mn. 235
29S6 STONER, E. C. *Proc. Leeds Phil. Lit. Soc., Sci. Sect.* **1**, 484–90. Diamagnetism and space charge distribution of atoms and ions. 459
29S7 SIZOO, J. G. *Z. Physik* **56**, 649–70. Magnetization diagram of single Fe crystals. 567
29S8 SIEVERTS, A. *Z. Metallkunde* **21**, 37–46. Absorption of gases by metals. 24, 314
29S9 SYKES, C. *J. Inst. Metals* **41**, 179–89. Alloys of zirconium. 328
29T1 TAKEI, T., MURAKAMI, T. *Sci. Repts. Tôhoku Imp. Univ.* **18**, 135–53. Equilibrium diagram of the Fe-Mo system. 33, 237
29V1 VAN VLECK, J. H. FRANK, A. *Proc. Natl. Acad. Sci.* (U. S.) **15**, 539–44. Diamagnetism of the normal H_2 molecule. 461
29W1 WEISS, P., FORRER, R. *Ann. physique* [10] **12**, 279–374. Absolute saturation of ferromagnetics and law of approach as a function of H and T. 190, 193, 194, 224, 242, 243
29W2 WEISS, P., FOËX, G. *Intl. Crit. Tables* **6**, 366–44. Ferromagnetism. 195, 242, 253, 271, 280, 337, 574
29W3 WEISS, P., FORRER, M., BIRCH, F. *Compt. rend.* **189**, 789–91. Saturation magnetization of Co-Ni and atomic moments of Co and of Ni. 264, 279, 280, 441
29W4 WEVER, F. *Naturwissenschaften* **17**, 304–9. Relation between the influence of elements on the polymorphism of iron and their position in the periodic system. 233
29W6 WOLTJER, H. R., COPPOOLSE, C. W., WIERSMA, E. C. *Proc. Acad. Sci. Amsterdam* **32**, 1329–33. Magnetic susceptibility of O_2 as function of temperature and density. 467
29W7 WEVER, F., HASHIMOTO, U. *Mitt. Kaiser Wilhelm Inst. Eisenforsch. Düsseldorf* **11**, 293–330. System Co-Cr. 288
29W8 WISE, E. M. *Trans. Am. Inst. Mining Met. Engrs.* 384–403. High strength gold alloys. 313
29Y1 YENSEN, T. D. *Trans. Am. Electrochem. Soc.* **56**, 215–23. On the road to pure iron and some of its indicated properties. 60
29Y2 YENSEN, T. D. *J. Iron Steel Inst.* (London) **120**, 187–206. Fe-Si-C alloys: Constitutional diagrams and magnetic properties. 73

1930

30A3 AKULOV, N. S. *Z. Physik* **59**, 254–64. On a law which relates different properties of ferromagnetic crystals with each other. 650, 764
30A4 AGEEW, N. W., VHER, O. I. *J. Inst. Metals* **44**, 83–96. The diffusion of Al into Fe. 212

30B1	BOZORTH, R. M., DILLINGER, J. F. *Phys. Rev.* **35**, 733–52. Barkhausen effect II. Determination of the average size of the discontinuities in magnetization. 525, 527, 711	
30B2	BRITZKE, E. V., KAPUSTINSKY, A. F. *Z. anorg. allgem. Chem.* **194**, 323–50. Affinity of metals for sulfur. 66, 253	
30B3	BLOCH, F. *Z. Physik* **61**, 206–19. On the theory of ferromagnetism. 445, 448, 719	
30B4	BECKER, R., KERSTEN, M. *Z. Physik* **64**, 660–81. Magnetization of Ni wire under large stress. 612, 822, 823	
30B5	BITTER, F. *Phys. Rev.* **33**, 389–97. Magnetic susceptibility of several organic gases. 860	
30B6	BECKER, R. *Z. Physik* **62**, 253–69. Theory of the magnetization curve. 654	
30B7	BURGERS, W. G., BASART, J. M. *Z. Krist.* **75**, 155–7. Lattice constants of mixed crystal series Cu-Ni. 310	
30B8	BATES, L. F. *Proc. Phys. Soc.* (London) **42**, 441–8. Specific heats of ferromagnetic substances. 336	
30C1	CIOFFI, P. P. *Nature* **126**, 200–1. Hydrogenized iron. 60	
30C2	CONSTANT, F. W. *Phys. Rev.* **36**, 1654–60. Magnetic properties of Pt-Co and Pd-Co alloys. 414	
30D1	DEAN, W. A. *Rensselaer Polytechnic Inst. Bull., Eng. Sci. Ser.* **26**, 31–55. An investigation of some of the physical properties of the iron-nickel-chromium system. 150, 151, 675, 676	
30E1	EUCKEN, A., WERTH, H. *Z. anorg. allgem. Chem.* **188**, 152–72. The specific heat of some metals and alloys at low temperature.	
30E2	ELMEN, G. W. U.S.P. 1 768 443 (Appl. 8/5/25). Magnetic material and appliance. 110	
30E4	EBINGER, A. *Z. tech. Physik* **11**, 221–7. Investigation of permeability of Fe with alternating magnetization. 547	
30E5	ELLWOOD, W. B. *Phys. Rev.* **36**, 1066–82. Change in temperature accompanying change of magnetization of iron. 523	
30F1	FISCHER, F. K. *Rensselaer Polytechnic Inst. Bull., Eng. Sci. Ser.* **28**, 1–32. An investigation of the magnetic and electrical properties of some iron-chromium alloys. 228, 229	
30F2	FORRER, R. *J. phys. radium* [7] **1**, 49–64. The problem of two Curie points. 190, 195, 448	
30F3	FOSTER, D., BOZORTH, R. M. *Nature* **125**, 525. Nature of the magnetization curve of single Fe crystals. 567	
30F4	FORRER, R. *J. phys. radium* [7] **1**, 325–39. Atomic moments in ferromagnetic alloys. 441	
30F5	FRENKEL, J., DORFMAN, J. *Nature* **126**, 274–5. Spontaneous and induced magnetization in ferromagnetic crystals. 829	
30G2	GUMLICH, E., STEINHAUS, W., KUSSMANN, A., SCHARNOW, B. *Elek. Nachr. Tech.* **7**, 231–35. Materials of high initial permeability. 183, 189, 417	
30G3	GOLDSCHMIDT, R. *Z. tech. Physik* **11**, 8–12. Superposition of strong and weak fields in magnetic materials. 543	
30G4	GIER, J. R. *Elec. J.* **27**, 236–37. Cobalt-steel magnets for compass needles. 381	
30G5	GERLACH, W., SCHNEIDERHAN, K. *Ann. Physik* [5] **6**, 772–84. Resistance, magnetic change of resistance and true magnetism at Curie point. 760, 762	

30G6 GERLACH, W. *Z. Physik* **59**, 847-9. Magnetic change of resistance and spontaneous magnetization. 762
30G7 GERLACH, W. *Z. Physik* **64**, 502-6. Magnetic characteristics of iron crystals. 567
30H2 HEISENBERG, W. *Metallwirtschaft* **9**, 843-4. Progress in the theory of ferromagnetism. 592
30I1 ISHIWARA, T. *Sci. Repts. Tôhoku Imp. Univ.* **19**, 499-519. Equilibrium diagram of Al-Mn and Fe-Mn systems. 235
30J1 JORDAN, L., SWANGER, W. H. *J. Research Natl. Bur. Standards* (U.S.) **5**, 1291-1307, 1309. Properties of pure nickel. 269, 270
30K1 KOSTING, P. R. *Rensselaer Polytechnic Inst. Bull., Eng. Sci. Ser.* **26**, 1-27. The nickel-iron-copper system. 154
30K3 KÜHLEWEIN, H. *Physik. Z.* **31**, 626-40. The ternary iron-nickel-cobalt alloys. 278
30K4 KÖSTER, W. *Arch. Eisenhüttenw.* **3**, 637-58. On the question of nitrogen in technical iron. 60, 798
30K5 KUSSMANN, A., SCHARNOW, B., MESSKIN, W. S. *Stahl u. Eisen* **50**, 1194-7. Cuprous steel for dynamo and transformer sheet. 87
30K6 KÖSTER, W. *Stahl u. Eisen* **50**, 687-95. On the hardening of copper-alloyed steels. 231
30K7 KÖSTER, W. *Z. Metallkunde* **22**, 289-96. Question of recovery based on experiments on Fe alloys. 233
30K8 KÖSTER, W. *Arch. Eisenhüttenw.* **4**, 289-94. Effect of cold forming and nitrogen precipitation on magnetic properties. 240, 826
30K9 KINNARD, I. F., FAUS, H. T. *Trans. Am. Inst. Elec. Engrs.* **49**, 949-51. Self-compensating temperature indicator. 311
30M1 McKEEHAN, L. W. *Phys. Rev.* **36**, 948-77. Electrical resistance of nickel and permalloy wires as affected by longitudinal magnetization and tension. 745, 756
30O1 OSAWA, A. *Sci. Repts. Tôhoku Imp. Univ.* **19**, 109-21. X-ray investigation of alloys of Ni-Co and Fe-Co systems. 276
30O2 OYA, M. *Sci. Repts. Tôhoku Imp. Univ.* **19**, 235-45. Equilibrium diagram of the Fe-V system. 258
30P2 PROCOPIU, S. *J. phys. radium* [7] **1**, 365-72. Magnetization of a ferromagnetic body under influence of an alternating field. 549
30P3 POWELL, F. C. *Proc. Roy. Soc.* (London) **130A**, 167-81. Direction of magnetization of single ferromagnetic crystals. 592
30R3 RANKIN, J. S. *J. Roy. Tech. Coll.* (Glasgow) **2**, 173-87. Effect of tensile overstrain on the magnetostriction of steel. 656, 676
30S1 SCOTT, H. *Trans. Am. Inst. Mining. Met. Engrs.* **89**, 506-37. Expansion properties of low-expansion Fe-Co-Ni alloys. 160
30S2 STONER, E. C. *Phil. Mag.* [7] **10**, 27-48. Magnetic and magneto-thermal properties of ferromagnetics. 447
30S3 SLATER, J. C. *Phys. Rev.* **36**, 57-64. Atomic shielding constants. 331, 444
30S5 STIERSTADT, O. *Physik. Z.* **31**, 561-74. Change of electrical conductance of ferromagnetic materials in longitudinal fields. 745
30S6 SCOTT, K. L. *Proc. Inst. Radio Engrs.* **18**, 1750-64. Variation of inductance of coils, due to magnetic shielding effect of eddy currents in the cores. 770
30S8 SADRON, C. *Compt. rend.* **190**, 1339-40. On the ferromagnetism of alloys of Ni and Cr. 307

BIBLIOGRAPHY

30T1 TAMMANN, G., OELSEN, W. *Z. anorg. allgem. Chem.* **186**, 257–88. Dependence of concentration of saturated mixed crystals on temperature. 232, 286, 289, 290, 292, 294, 323

30W1 WEVER, F., JELLINGHAUS, W. *Mitt. Kaiser-Wilhelm Inst. Eisenforsch.* (Düsseldorf) **12**, 317–22. The binary system iron-vanadium. 187, 258

30W2 WOOD, W. A. *Phil. Mag.* [7] **10**, 659–67. X-ray study of some W magnet steel residues. 372, 373

30W3 WEBSTER, W. L. *Proc. Phys. Soc.* (London) **42**, 431–40. Magnetostriction and change of resistance in single crystals of Fe & Ni. 576, 764

30W4 WEVER, F., MÜLLER, A. *Z. anorg. allgem. Chem.* **192**, 337–45. Structure of mixed crystals of Fe with Be & Al. 222

30W5 WEVER, F., MÜLLER, A. *Z. anorg. allgem. Chem.* **192**, 317–36. Two component system Fe-B and structure of Fe boride Fe_4B_2. 224

30W6 WEVER, F., LANGE, H. *Mitt. Kaiser-Wilhelm Inst. Eisenforsch. Düsseldorf* **12**, 353–63. Dependence of magnetic properties of Co-Cr mixed crystals on temperature. 288, 289

30W7 WESTGREN, A., EKMAN, W. *Arkiv. Kemi, Miner. Geol.* (B) **10** (11), 1–6. Structure analogies of intermetallic phases. 313

1931

31A3 AKULOV, N. *Z. Physik* **69**, 78–99. Theory of magnetization curve of single crystals. 577

31A4 AKULOV, N. S. *Z. Physik* **69**, 822–31. On the course of the magnetization curve in strong fields. 486, 581

31A5 ADCOCK, F. *J. Iron Steel Inst.* (London) **124**, 99–139. The Cr-Fe constitutional diagram. 227, 228

31B1 BREDIG, G., SCHWARTZ VON BERGKAMPF, E. *Z. Physik. Chem.* (*Bodenstein*) 172–176. On hexagonal nickel. 315, 321

31B2 BARNETT, S. J. *Proc. Am. Acad. Arts Sci.* **66**, 274–348. The rotation of permalloy and soft iron by magnetization. 454

31B3 BRUNAUER, S., JEFFERSON, M. E., EMMETT, P. H., HENDRICKS, S. B. *J. Am. Chem. Soc.* **53**, 1778–86. Equilibria in the Fe-N system. 66, 240

31B4 BOZORTH, R. M., DILLINGER, J. F. *Nature* **127**, 777. Propagation of magnetic disturbances along wires. 496, 497, 531, 532

31B5 BITTER, F. *Phys. Rev.* **38**, 1903–5. On inhomogeneities in the magnetization of ferromagnetic materials. 533

31B6 BLOCH, F., GENTILE, G. *Z. Physik* **70**, 395–408. Anisotropy of magnetization of ferromagnetic single crystals. 592

31D1 DIETSCH, G. *Z. tech. Physik* **12**, 380–9. Magnetostriction of ferromagnetic substances. 656, 657, 661, 681

31D2 DIETSCH, G., FRICKE, W. *Physik. Z.* **32**, 640. On the transverse effect of magnetostriction. 657

31D3 DAHL, O., PFAFFENBERGER, J. *Z. Physik* **71**, 93–105. Anisotropy in magnetic materials. 586

31E2 ELMEN, G. W. *Bell Lab. Record* **10**, 2–5. New permalloys. 136

31F1 FORRER, R. *J. phys. radium* [7] **2**, 312–20. Some experimental verifications of the problem of two Curie points. 196, 245

31F2 FOWLER, R. H., POWELL, F. C. *Proc. Cambridge Phil. Soc.* **27**, 280–9. Note on ferromagnetism. 592

31G1 GEROLD, E. *Stahl u. Eisen* **51**, 613–5. Dependence of magnetic induction of structural steel on chemical composition. 75

31G2 GIEBE, E., BLECHSCHMIDT, E. *Ann. Physik* [5] **11**, 905–36. Influence of magnetization on elastic modulus with rotation-oscillation of ferromagnetic materials. 699
31G3 GUTHRIE, A. N., BOURLAND, L. T. *Phys. Rev.* **37**, 303–8. Magnetic susceptibilities in the Pd & Pt groups. 858
31H1 HERAEUS-VACUUMSCHMELZE, A. G. *Arch. tech. Messen* **1**, Z 913–2. Highly magnetic alloys of nickel-iron for transformers and meters. 160
31H2 HONDA, K., MASUMOTO, H. *Sci. Repts. Tôhoku Imp. Univ.* **20**, 323–41. On the magnetization of single crystals of Co at high temperatures. 264, 265, 568
31H3 HÁMOS, L. v., THIESSEN, P. A. *Z. Physik* **71**, 442–4. Making visible the regions of different magnetic states of solid bodies. 533
31H4 HOUGARDY, H. *Arch. Eisenhüttenw.* **4**, 497–505. Contribution to the system Fe-C-V. 259
31H5 HAWORTH, F. E. *Bell System Tech. J.* **10**, 20–32. A magnetic curve tracer. 861
31H6 HEISENBERG, W. *Z. Physik* **69**, 287–97. Theory of magnetostriction and the magnetization curve. 576, 582, 583, 653, 822
31I1 IDE, J. M. *Proc. Inst. Radio Engrs.* **19**, 1216–32. Measurements on magnetostriction vibrators. 699, 712
31K1 KERSTEN, M. *Z. tech. Physik* **12**, 665–9. The effect of elastic tension on the magnitude of initial permeability. 621, 626, 711, 822, 823
31K4 KÜHLEWEIN, H. *Wiss. Veröffentl. Siemens-Konzern* **10**, No. 2, 72–88. The magnetic properties of perminvars. 169, 171
31K5 KEINATH, G. *Arch. tech. Messen* Z913–1. High permeability alloys of nickel-iron. 183
31K6 KERSTEN, M. *Z. Physik* **71**, 553–92. On the dependence of magnetic properties of nickel on elastic tension. 504, 598, 607, 612, 823
31K7 KAYA, S., KUSSMANN, A. *Z. Physik* **72**, 293–309. Ferromagnetism and phase-structure in the 2 component system Ni-Mn. 313, 768
31K8 KÖSTER, W. *Arch. Eisenhüttenw.* **4**, 609–11. Precipitation hardening of Fe-P alloys. 253
31M1 MASIYAMA, Y. *Sci. Repts. Tôhoku Imp. Univ.* **20**, 574–93. On the magnetostriction of iron-nickel alloys. 630, 642, 657, 660, 666, 674
31M2 MASUMOTO, H. *Sci. Repts. Tôhoku Imp. Univ.* **20**, 101–23. On the thermal expansion of the alloys of iron, nickel, and cobalt. 161, 162, 163, 278, 643
31M3 MICHELS, R. *Ann. Physik* [5] **8**, 877–98. The suppression of initial permeability with short electric waves. 802
31M4 MATSUNAGA, Y. *Kinzoku no Kenkyu* **8**, 549–64. Equilibrium diagram of the Co-Cr system. 288, 289
31M5 MISHIMA, T. *Proc. World Eng. Cong. Tokyo* **36**, 215–6. Influence of C on "annealing-brittleness" of Ni and its alloys. 313
31P3 POTTER, H. H. *Phil. Mag.* [7] **12**, 255–64. Some magnetic alloys and their properties. 333, 340, 415, 418
31P4 POTTER, H. H. *Proc. Roy. Soc.* (London) **132A**, 560–9. On the change of resistance of Ni in a magnetic field. 762
31P5 PRESTON, G. D. *J. Iron Steel Inst.* (London) **124**, 139–41. X-ray examination of Cr-Fe alloys. 227
31P6 POWELL, F. C. *Proc. Cambridge Phil. Soc.* **27**, 561–9. Magnetostriction in single crystals of Fe and Ni. 654

31R2 RANKIN, J. S. *J. Roy. Tech. Coll.* (Glasgow) **2**, 385–95. Further experiments on the magnetostriction of overstrained materials. 656, 661, 676

31S2 SÖHNCHEN, E., PIWOWARSKY, E. *Arch. Eisenhüttenw.* **5**, 111–21. The effect of the alloying elements nickel, silicon, aluminum, and phosphorus on the solubility of carbon in iron.

31S3 SUCKSMITH, W. *Proc. Roy. Soc.* (London) **133A**, 179–88. Gyromagnetic effect for paramagnetic substances. II. 455

31S4 SCHULZE, A. *Ann. Physik* [5] **11**, 937–48. Hysteresis phenomena with magnetostriction. 659, 661

31S5 SIXTUS, K. J., TONKS, L. *Phys. Rev.* **37**, 930–58. Propagation of large Barkhausen discontinuities. 494, 822

31S6 SUGIURA, J. *Researches Electrotech. Lab.* (Tokyo), No. 300, 1–2. On the magnetic properties of iron polarized in any direction. 553

31S7 SATO, T. *Tech. Repts. Tôhoku Imp. Univ.* **9**, 515–65. A metallographic investigation of the Fe-Si-C alloys. 72

31S8 STÖSSEL, R. *Ann. Physik* [5] **10**, 393–436. Temperature-dependent magnetic moment of NO. 467

31T1 TRAUTMANN, B. *Z. Metallkunde* **23**, 86. A new metal for high temperatures. 161

31T2 TAKEDA, S. *Tech. Repts. Tôhoku Imp. Univ.* (Tokyo) **9**, 483–514, 627–64; **10**, 42–92. Metallographic investigations of the ternary alloys of the Fe-W-C system. 371

31W1 WEVER, F., HEINZEL, A. *Mitt. Kaiser-Wilhelm Inst. Eisenforsch. Düsseldorf* **13**, 193–7. Two examples of 3-component systems of Fe with closed gamma loop. 418

31W2 WEVER, F., HINDRICKS, J. *Mitt. Kaiser-Wilhelm Inst. Eisenforsch. Düsseldorf* **13**, 273–89. Production of Si-Al steel for dynamo and transformer sheet in high frequency induction furnace. 87

31W3 WEBB, C. E. *J. Iron Steel Inst.* (London) **124**, 141–5. Magnetic tests on Cr-Fe alloys. 228, 229

31W4 WARTENBURG, H. v., GURR, W. *Z. anorg. allgem. Chem.* **196**, 374–83. Melting diagrams of oxides stable at high temperatures. 293

31Y1 YENSEN, T. D. *Elec. J.* **28**, 386–8. Permeability of Hipernik reaches 167 000. 110, 123, 130

31Y2 YENSEN, T. D. U.S.P. 1 807 021 (Appl. 3/29/24). Magnetic product and method of making the same. 86, 123

1932

32A3 AKULOV, N. S., DEGTIAR, M. *Ann. Physik* **15**, 750–6. On the complicated structure of ferromagnetic single crystals. 533

32A4 ATORF, H. *Z. Physik* **76**, 513–26. Disaccommodation of small symmetrical and unsymmetrical hysteresis loops. 794

32A5 AKULOV, N. S., BRÜCHATOV, N. *Ann. Physik* **15**, 741–9. Method for quantitative investigation of rolling texture. 587

32B2 BLOCH, F. *Z. Physik* **74**, 295–335. Theory of the exchange problem and of residual ferromagnetism. 479, 814, 822

32B3 BITTER, F. *Phys. Rev.* **41**, 507–15. Experiments on the nature of ferromagnetism. 538

32B4 BOZORTH, R. M., DILLINGER, J. *Phys. Rev.* **41**, 345–55. Barkhausen effect. III. 530

32B5   BECK, F. J., MCKEEHAN, L. W. *Phys. Rev.* **2**, 42, 714–20. Mono-crystal Barkhausen effects in rotating fields. 531
32B6   BARTH, T. F. W., POSNJAK, E. *Z. Krist.* **82**, 325–41. Spinel structures. 244
32B7   BRADLEY, A. J., JAY, A. H. *J. Iron Steel Inst.* (London) **125**, 3390; *Proc. Roy. Soc.* (London) **136A**, 210–32. The formation of superstructures in alloys. 211, 212
32B8   BECKER, R. *Physik. Z.* **33**, 905–13. Elastic strains and magnetic properties. 479, 822
32B10 BOZORTH, R. M. *Phys. Rev.* **42**, 882–92. Theory of ferromagnetic anisotropy of single crystals. 582
32B11 BRILL, R., HAAG, W. *Z. Electrochem.* **38**, 211–2. On Fe-Hg and Ni-Hg. 320
32C2   CIOFFI, P. P. *Phys. Rev.* **39**, 363–7. Hydrogenized iron. 36, 60, 123
32C3   COETERIER, F., SCHERRER, P. *Helv. Phys. Acta* **5**, 217–23. Measurement of gyromagnetic effect in pyrrhotite. 454
32C4   CZERLINSKY, E. *Ann. Physik* [5] **13**, 80–100. On magnetic saturation. 487, 567, 569
32C5   CIOFFI, P. P. U.S.P. 1 866 925 (Appl. 7/31/30). Magnetic material. 69
32D1   DAHL, O. *Z. Metallkunde* **24**, 107–11. Supercooling and transformation of iron-nickel alloys. 163, 182, 185, 417
32E1   ENGLERT, E. *Ann. Physik* [5] **14**, 589–612. Change of electric resistivity in longitudinal and transverse magnetic fields. 762
32E2   ELENBAAS, W. *Physica* **12**, 125–32. Relation between hysteresis curve and virgin curve of ferromagnetic substances. 511
32E3   ELENBAAS, W. *Elec. Engg.* **51**, 668. Magnet steels and permanent magnets. 353
32E4   ENGLERT, E. *Z. Physik* **74**, 748–56. Change of resistance and magnetization at the Curie point. 762
32F1   FORRER, R., MARTAK, J. *J. phys. radium* [7] **3**, 408–36. Magnetic multiplets, their mechanism and experimental determination. 532
32G2   GREGG, J. L. McGraw-Hill, N. Y., 1–507. The alloys of iron and molybdenum. 236
32G3   GANS, R. *Ann. Physik* [5] **15**, 28–44. On the magnetic properties of isotropic magnetic materials. 486, 563, 567, 577, 581, 582
32G4   GERLACH, W. *Ann. Physik* [5] **12**, 849–64. Ferromagnetism & electrical properties IV. 748, 755, 756, 764
32G5   GRAY, F. W., FARQUHARSON, J. *J. Sci. Instruments* **9**, 1–5. Improvements in the Curie-Chéveneau balance. 858
32H1   HASHIMOTO, U. *Kinzoku no Kenkyu* **9**, 57–73. Co-Ni alloys. 264, 276
32H2   HARLEM, J. VON. *Ann. Physik* [5] **14**, 667–74. On rotating hysteresis. 515
32H3   HÁMOS, L. V., THIESSEN, P. A. *Z. Physik* **75**, 562. Making visible the regions of different magnetic states of solid bodies. 533
32H4   HOAG, J. B., JONES, H. *Phys. Rev.* **42**, 571–6. Permeability of iron at ultra-radio frequencies. 778, 800
32K1   KROLL, W. *Metallwirtschaft* **11**, 435–7. Gallium alloys. 313, 417
32K2   KÜHLEWEIN, H. *Wiss. Veröffentl. Siemens-Konzern* **11**, 124–40. The magnetic properties of iron, nickel, cobalt, and some alloys at elevated temperatures. 58, 169, 170, 197, 266, 274, 501, 715
32K3   KUSSMANN, A., SCHARNOW, B., SCHULZE, A. *Z. tech. Physik* **13**, 449–60. Physical properties and structure of binary system Fe-Co. 191, 194

BIBLIOGRAPHY 899

32K4 KÖSTER, W., TONN, W. *Arch. Eisenhüttenw.* **5**, 627–30. System Fe-Co-Mo. 205, 382, 416

32K5 KÖSTER, W., TONN, W. *Arch. Eisenhüttenw.* **5**, 431–40. The system Fe-Co-W. 205, 208, 382, 416

32K6 KÖSTER, W. *Arch. Eisenhüttenw.* **6**, 113–6. System Fe-Co-Cr. 205, 206, 416

32K7 KÜHLEWEIN, H. *Physik. Z.* **33**, 348–51. Relation between virgin and remanent magnetization. 511

32K8 KÖSTER, W. *Arch. Eisenhüttenw.* **6**, 17–23. Mechanical and magnetic precipitation hardening of Fe-Co-W and Fe-Co-Mo alloys. 205, 208, 347, 382, 419

32K9 KÖSTER, W. *Z. Electrochem.* **38**, 549–53. Relation of magnetic properties to the structure of alloys. 418

32K10 KŘÍŽ, A., POBOŘIL, F. *J. Iron Steel Inst.* (London) **126**, 323–49. Further contributions on constitution of Fe-C-Si system. 72

32K11 KERSTEN, M. *Z. Physik* **76**, 505–12. Magnetic analysis of internal stresses. 621, 626

32K12 KÖSTER, W., TONN, W. *Z. Metallkunde* **24**, 296–9. Binary systems Co-W and Co-Mo. 292, 297

32L1 LICHTENBERGER, F. *Ann. Physik* [5] **15**, 45–71. Magnetostriction and magnetization of single crystals of iron-nickel alloys. 569, 648, 752, 766

32M1 MISHIMA, T. *Ohm* **19**, 353. Nickel-aluminum steel for permanent magnets. 183, 346, 347, 385

32M2 MASIYAMA, Y. *Sci. Repts. Tôhoku Imp. Univ.* **21**, 394–410. On magnetostriction of Fe-Co alloys. 639, 658, 663, 664, 674

32M3 MESSKIN, W. S., KUSSMANN, A. Springer, Berlin, 1–418. Die Ferromagnetischen Legierungen. 95, 253, 322, 334, 371, 373, 378, 382

32M4 MÖBIUS, W. *Physik. Z.* **33**, 411–7. Torsion modulus of Ni at high temperatures with simultaneous magnetization. 699

32N1 NEMILOV, V. A. *Z. anorg. allgem. Chem.* **204**, 49–59. Hardness, microstructure and temperature coefficient of resistance of Fe-Pt. alloys. 410

32N2 NÉEL, L. *Ann. physique* [10] **17**, 5–105. Influence of fluctuations of molecular field on magnetic properties of bodies. 472

32O1 OCHSENFELD, R. *Ann. Physik* [5] **12**, 353–84. Occurrence of ferromagnetism in system Mn-N. 338

32P1 PREISACH, F. *Physik. Z.* **33**, 913–23. Permeability and hysteresis in the direction of easiest magnetization. 496, 610

32P2 POTTER, H. H. *Phil. Mag.* [7] **13**, 233–48. Magneto-resistance and magneto-caloric effects in Fe and Heusler alloys. 762

32R2 RANKIN, J. S. *J. Roy. Tech. Coll.* (Glasgow) **2**, 587–9. Further experiments on the magnetostriction of cold-drawn wire. 676

32S3 SADRON, C. *Ann. physique* [10] **17**, 371–452. Ferromagnetic moments of the elements and the periodic system. 235, 291, 299, 300, 313, 328, 441

32S4 SUCKSMITH, W. *Proc. Roy. Soc.* (London) **135A**, 276–81. The gyromagnetic ratio for paramagnetic substances. III. 455

32S5 SHUBROOKS, G. E. *Metal Progress* **21**, No. 2, 58–63. Elinvar hairsprings for watches. 147, 699

32S6 SELJESATER, K. S., ROGERS, B. A. *Trans. Am. Soc. Steel Treating* **19**, 553–76. Magnetic and mechanical hardness of dispersion hardened iron alloys. 34, 224, 238, 256, 346, 347, 382, 385, 416, 417, 418, 419

32S7 SCOTT, K. L. *Elec. Engg.* **51**, 320–3. Relationships among the magnetic properties of magnet steels and permanent magnets. 352

32S8 SCHWARZ, H. *Hochfrequ. u. Elektroakust.* **39**, 160–71. Current measurements with very high frequencies. 800
32T1 TAMARU, K. *Sci. Repts. Tôhoku Imp. Univ.* **21**, 344–63. Equilibrium diagram of Ni-Zn system. 327
32V1 VOGEL, R., SUNDERMANN, W. *Arch. Eisenhüttenw.* **6**, 35–8. System Fe-Co-C. 379
32V2 VOGEL, R., TONN, W. *Arch. Eisenhüttenw.* **5**, 387–9. Constitutional diagram of Fe-Zr alloys. 260
32V3 VAN VLECK, J. H. Clarendon Press, Oxford, 1–384. Theory of electric and magnetic susceptibilities. 431, 459, 464, 466, 467
32W1 WINKLER, O., VOGEL, R. *Arch. Eisenhüttenw.* **6**, 165–172. The iron-nickel-tungsten diagram. 186, 417
32W2 WHITE, J. H., WAHL, C. V. U.S.P. 1 862 559 (Appl. 8/14/31). Workable magnetic compositions containing principally iron and cobalt. 190, 202, 208
32W3 WILLIAMS, S. R. *Rev. Sci. Instruments* **3**, 675–83. Joule magnetostriction effect in a group of Fe-Co alloys. 663
32W4 WOOD, W. A., WAINRIGHT, C. *Phil. Mag.* [7] **14**, 191–8. Lattice distortion and carbide formation in W magnet steels. 373
32W5 WOLMAN, W., KADEN, H. *Z. tech. Physik* **13**, 330–5. On eddy-current delay accompanying sudden magnetic changes. 787
32W6 WIEN, M. *Physik. Z.* **33**, 173–5. Dependence of the permeability of iron wires on the field at high frequencies. 803
32Z1 ZIEGLER, N. A. *Trans. Am. Inst. Mining Met. Engrs.* **100**, 267–71. Resistance of Fe-Al alloys to oxidation at high temperatures. 211
32Z2 ZIEGLER, N. A. *Trans. Am. Soc. Steel Treating* **20**, 73–84. Solubility of oxygen in iron. 241

1933

33A2 AUWERS, O. V. *Physik. Z.* **34**, 824–7. Volume magnetostriction in poly- and monocrystals. 674
33A3 AUWERS, O. V., KÜHLEWEIN, H. *Ann. Physik* [5] **17**, 121–45. Contribution to stereomagnetism II. The perminvar problem. 169
33A4 AKULOV, N., KONDORSKY, E. *Z. Physik* **78**, 801–7. Mechanostriction and ΔE effect. 628, 684
33A5 AUWERS, O. V. *Ann. Physik* [5] **17**, 83–106. Dependence of E and δ of ferromagnetic material on magnetization. 686, 702
33A6 AUER, H. *Ann. Physik* [5] **18**, 593–612. Absolute magnetic susceptibility of H_2O and its temperature dependence. 461
33B2 BETHE, H. *Handb. d. Physik* **24**, pt. 2, 595–8. Ferromagnetism. 443
33B3 BECKER, R. Magnetismus (Debye, Leipzig), 82–90. The technical magnetization curve. 479, 822
33B4 BJURSTRÖM, T. *Arkiv Kemi, Miner. Geol.* A, **11**, No. 5, 1–12. X-ray analysis of systems Fe-B, Co-B, and Ni-B. 287, 305
33B5 BÜSSEM, W., GROSS, F. *Z. Physik* **86**, 135–6. Structure and gas content of sputtered Ni. 315
33D2 DAHL, O., PFAFFENBERGER, J., SPRUNG, H. *Elek. Nachr. Tech.* **10**, 317–32. New magnetic materials for Pupin coils. 159, 183, 184, 186, 625
33D3 DEAN, R. S. U.S.P. 1 904 859 (Appl. 3/24/30). Ferrous alloy. 253, 347, 382
33F1 FRICKE, W. *Z. Physik* **80**, 324–41. On the transverse effect of magnetostriction. 658, 660

33F2	FORRER, R. *J. phys. radium* [7] **4**, 427–39, 501–12. Ferromagnetic Curie points. 723	
33G4	GERLACH, W. *Physik. Z.* **33**, 953–7. Change of electrical resistance with magnetization. 760, 764	
33G5	GANS, R., HARLEM, J. v. *Ann. Physik* **15**, 516–26. Resistance change in ferromagnetic crystals. 764	
33G7	GORTER, C. J. *Nature* **132**, 517–8. Remanence in single crystals of iron. 562	
33G8	GANS, R., CZERLINSKY, E. *Ann. Physik* [5] **16**, 625–35. Theory of magnetization curve of ferromagnetic single crystals. 567, 577	
33G9	GANS, R., HARLEM, J. v. *Ann. Physik* [5] **16**, 162–73. Magnetostriction of ferromagnetic crystals. 650	
33H1	HAGEMANN, H., HIEMENZ, H. Heraeus Vacuumschmelze, Albertis, Hanau, 181–200. The effect of thickness of strip, of annealing atmosphere, etc., on initial and maximum permeabilities of iron-nickel alloys. 117	
33H2	HOUDREMONT, E., SCHRÄDER, H. *Arch. Eisenhüttenw.* **7**, 49–59. On temper brittleness. 147	
33H3	HERMANN, P. C. *Z. Physik* **84**, 565–70. On magnetic lag. 794	
33H4	HAVENS, C. G. *Phys. Rev.* **43**, 992–1000. Magnetic susceptibilities of some common gases. 459, 461	
33H5	HEUSLER, O. *Z. Metallkunde* **25**, 274–7. Crystal structure and ferromagnetism of Mn-Al-Cu alloys. 328	
33J2	JETTE, E. R., GREINER, E. S. *Trans. Am. Inst. Mining Met. Engrs.* **105** 259–73. X-ray study of Fe-Si alloys containing 0 to 15% Si. 73, 74	
33K1	KERSTEN, M. *Z. Physik* **85**, 708–16. The temperature coefficient of the modulus of elasticity of ferromagnetic substances. 684, 690, 699	
33K2	KÖSTER, W. *Stahl u. Eisen* **53**, 849–56. Permanent magnets on the basis of precipitation hardening. 386, 418	
33K3	KÖSTER, W. *Arch. Eisenhüttenw.* **7**, 257–62. The iron-nickel-aluminum system. 385	
33K4	KUSSMANN, A., SCHARNOW, B., STEINHAUS, W. Heraeus Vacuumschmelze, Albertis, Hanau, 310–338. The permalloy problem. 182, 183, 417	
33K5	KÖSTER, W. *Arch. Eisenhüttenw.* **7**, 263–4. System Fe-Co-Al. 205, 207, 389, 416	
33K6	KÖSTER, W., SCHMIDT, W. *Arch. Eisenhüttenw.* **7**, 121–6. The system Fe-Co-Mn. 205, 208	
33K7	KATO, Y., TAKEI, T. *J. Inst. Elec. Engrs.* (Japan) **53**, 408–12. Permanent oxide magnet and its characteristics. 421	
33K8	KÖSTER, W., TONN, W. *Arch. Eisenhüttenw.* **7**, 365–6. The Fe corner of the system Fe-Mn-Al. 418	
33K9	KREIELSHEIMER, K. *Ann. Physik* [5] **17**, 293–333. Magnetic permeability of Fe wires at wavelengths 46 to 1000 m. 788, 803	
33K10	KAYA, S. *Z. Physik* **84**, 705–16. On the remanence of iron single crystals. 561, 562	
33K11	KERSTEN, M. *Z. Physik* **82**, 723–8. Magnetic analysis of internal strains. 621, 625	
33M1	MASIYAMA, Y. *Sci. Rept. Tôhoku Imp. Univ.* **22**, 338–53. On the magnetostriction of Ni-Co alloys. 674	
33M2	MOELLER, C. *Z. Physik* **82**, 559–67. Theory of exchange problem & ferromagnetization at low temperatures. 449	
33M3	MISHIMA, T. Br. P. 392 661 (Appl. 8/27/31). Alloy steel for permanent magnets. 347	

BIBLIOGRAPHY

33P1 POTAPENKO, G., SÄNGER, R. *Naturwissenschaften* **21**, 818–9. Magnetic permeability of ferromagnetic metals at very high frequencies. 800

33P2 PFAFFENBERGER, J. *Arch. Eisenhüttenw.* **7**, 117-20. Magnetic properties of small sheet specimens. 185

33R3 ROGERS, B. A. *Metals and Alloys* **4**, 69–73. Magnetic properties of Fe-Co-W alloys. 209, 382, 385

33R4 ROHN, W. Heraeus Vacuumschmelze, Albertis, Hanau, 381–7. Rolling material with thin work-rolls. 25

33R5 ROHN, W. Heraeus Vacuumschmelze, Albertis, Hanau, 356–80. Induction furnace for a.c. of power frequency. 23.

33S1 SCHULZE, A. *Z. Physik* **82**, 674–83. Magnetostriction—III. 681

33S2 STEINBERGER, R. L. *Physics* **4**, 153–61. Magnetic properties of iron-nickel alloys under hydrostatic pressure. 445, 726

33S3 STONER, E. C. *Phil. Mag.* [7] **15**, 1018–34. Atomic moments in ferromagnetic metals and alloys with non-ferromagnetic elements. 440, 441, 442

33S4 SYKES, W. P. *Trans. Am. Soc. Steel Treating* **21**, 385–423. The Co-W system. 297

33S5 SCHMIDT, W. *Ergeb. tech. Röntgenkunde* III, 194–201. Thermal extension measurements with X-rays. 53

33S6 SIXTUS, K. J. *Phys. Rev.* **44**, 46–51. On irregularities in magnetization. 533

33S7 SIDHU, S. S. *Elec. Engg.* **52**, 625–30. Formulae for magnetic hysteresis losses. 510, 549

33S8 SMITH, W. S., GARNETT, H. J., RANDALL, W. F. U.S.P. 1 915 766 (Appl. 10/31/30). Manufacture of magnetic alloys. 89

33S9 SCHENCK, R., KORTENGRÄBER, A. *Z. anorg. allgem. Chem.* **210**, 273–85. The system Mn-N. 338

33T2 TAMMANN, G., ROCHA, H. J. *Ann. Physik* [5] **16**, 861-4. Change of magnetic induction at constant field strength by cold working and recovery upon heating. 60

33T3 THERKELSEN, E. *Metals and Alloys* **4**, 105-8. Properties of the alloys of Ni and Ta. 323

33V1 VALENTINER, S., BECKER, G. *Z. Physik* **83**, 371–403. Investigation of Heusler alloys. 328

33W1 WILLIAMS, S. R. *Trans. Am. Soc. Steel Treating* **21**, 741–68. Mechanical hardness, influenced by magnetism and measured by magnetostrictive effects. 661, 676, 681

1934

34A1 AHRENS, E. *Ann. Physik* **21**, 169–81. The effect of temperature on the true specific heat of nickel. 737

34A2 AUWERS, O. v. *Gmelin's Handbook anorg. Chem.* 8th, Fe, Pt.A, No. 7, 1421–1634. Magnetic and electrical properties of pure iron and iron containing carbon. 489

34A3 ANTIK, I., KUBYSCHKINA, T. *Wiss. Ber. Univ. Mosk.* **11**, 143–50. On the hysteresis loss in liquid ferromagnetics. 236, 828

34B3 BOZORTH, R. M., DILLINGER, J. F., KELSALL, G. A. *Phys. Rev.* **45**, 742–3. Theory of the heat treatment of magnetic materials. 117, 348, 496

34B4 BARNETT, S. J. *Proc. Am. Acad. Arts Sci.* **69**, 119–35. Researches on the rotation of permalloy and soft iron by magnetization and the nature of the elementary magnet. 454

34B5 BECKER, R., KORNETZKI, M. *Z. Physik* **88**, 634–46. Some magnetoelastic torsion investigations. 686, 688, 699, 700

34B6 BORÉN, B. *Arkiv Kemi, Miner. Geol. A*, **11**, No. 10, 1–28. X-ray investigation of alloys of Si with Cr, Mn, Co and Ni. 294

34B7 BATES, L. F. *Phil. Mag.* [7] **17**, 783–93. Resistance of manganese arsenide. 336

34B8 BRADLEY, A. J., RODGERS, J. W. *Proc. Roy. Soc.* (London) **144A**, 340–59. Structure of the Heusler alloys. 328, 333, 334

34C1 CIOFFI, P. P. *Phys. Rev.* **45**, 742. New high permeabilities in hydrogen treated iron. 60, 62, 66, 86

34D1 DAHL, O., PFAFFENBERGER, J. *Metallwirtschaft* **13**, 527–30, 543–9, 559–63. Special magnetic behavior of cold-rolled iron-nickel alloys. 57, 159, 187, 189

34D2 DAHL, O., PFAFFENBERGER, J. *Z. tech. Physik* **15**, 99–106. Low-hysteresis materials of great magnetic stability. 797

34D3 DOWDELL, R. L. *Trans. Am. Soc. Metals.* **22**, 19–30. Investigation of treatment of steel for permanent magnets. 373

34E1 ELLIS, W. C., SCHUMACHER, E. E. *Metals and Alloys* **5**, 269–76. Magnetic materials, a survey in relation to structure. 144

34E2 ELLWOOD, W. B. *Rev. Sci. Instruments* **5**, 300–5. A new ballistic galvanometer operating in high vacuum. 493

34E3 EILENDER, W., OERTEL, W. *Stahl u. Eisen* **54**, 409–14. Status of the production of dynamo and transformer sheet. 83

34E4 ELLEFSON, B. S., TAYLOR, N. W. *J. Chem. Phys.* **2**, 58–64. Crystal structures and expansion anomalies of MnO, MnS, FeO, Fe_3O_4 between 100 and 200°K. 470

34F1 FALLOT, M. *Compt. rend.* **199**, 128. The Fe-Pt alloys; Curie points and magnetic moments. 410

34G1 GREGG, J. L. McGraw-Hill, N. Y., 1–511. The alloys of iron and tungsten. 372

34G3 GREW, K. E. *Proc. Roy. Soc.* (London) **145A**, 509–22. The specific heat of nickel and of some nickel-copper alloys. 737, 738

34G4 GOSS, N. P. U.S.P. 1 965 559 (Appl. 8/7/33). Electrical sheet and method and apparatus for its manufacture and test. 68, 69, 70, 88

34H1 HONDA, K. H., MASUMOTO, H., SHIRAKAWA, Y. *Sci. Repts. Tôhoku Imp. Univ.* **23**, 365–73. On new KS permanent magnet. 395, 421

34H2 HULL, A. W., BURGER, E. E. *Physics* **5**, 384–405. Glass to metal seals. 160

34H3 HAUGHTON, J. L., PAYNE, R. J. M. *J. Inst. Metals* **54**, 275–83. Constitution of Mg-rich alloys of Mg and Ni. 315

34H4 HARALDSEN, H., KLEMM, W. *Z. anorg. allgem. Chem.* **220**, 183–92. Manganese sulfide and related compounds. 339

34H5 HEUSLER, O. *Ann. Physik* [5] **19**, 155–201 (1934). Crystal structure and ferromagnetism of Mn-Al-Cu alloys. 328, 330

34I1 IDE, J. M. *Proc. Inst. Radio Engrs.* **22**, 177–90. Magnetostrictive alloys with low temperature coefficients of frequency. 674, 676, 681

34I2 INGLIS, D. R. *Phys. Rev.* **45**, 118–9. Magnetic and gyromagnetic properties of pyrrhotite. 454

34J1 JETTE, E. R., BRUNER, W. L., FOOTE, F. *Trans. Am. Inst. Elec. Engrs.* **111**, 354–9. X-ray study of Au-Fe alloys. 412

34J2 JETTE, E. R., NORDSTROM, V. H., QUENAU, B., FOOTE, F. *Trans. Am.*

Inst. Mining Met. Engrs. 111, 361–71. X-ray studies of the Ni-Cr system. 307

34K1 KELSALL, G. A. *Physics* 5, 169–72. Permeability changes in ferromagnetic materials heat treated in magnetic fields. 117, 119, 171, 348

34K4 KERSTEN, M. *Z. tech. Physik* 15, 249–57. Physical investigation of new magnetic materials. 504

34K5 KÖSTER, W. *Arch. Eisenhüttenw.* 8, 169–171. The iron-nickel-molybdenum system. 134, 417

34K7 KÜHLEWEIN, H. *Z. anorg. allgem. Chem.* 218, 65–8. The properties of the ferromagnetic alloys of the ternary iron-nickel-vanadium system. 187, 188, 189, 417

34K8 KERSTEN, M. *Z. tech. Physik* 15, 463–7. Significance of mechanical damping of ferromagnetic material by magnetization. 684, 699, 703

34K9 KERSTEN, M. *Wiss. Veröffentl. Siemens-Konzern* 13 (3), 1–9. Some anomalous properties of new magnetic materials. 157

34K10 KORNETZKI, M. *Z. Physik* 87, 560–79. On the magnetostriction of ferromagnetic ellipsoids. 657, 658

34K11 KAMURA, H. Br. Pat. 420 543 (Appl. 9/5/32). Improvements relating to magnetic iron alloys. 252

34K12 KAYA, S. *Z. Physik* 90, 551–8. Powder figures on magnetized iron crystals. 533

34K14 KLEMM, W., HASS, K. *Z. anorg. allgem. Chem.* 219, 82–6. On NiO. 321

34K15 KRAMERS, H. A. *Physica* 1, 182–92. Interaction between magnetic atoms in a paramagnetic crystal. 475

34M1 McKEEHAN, L. W. *Trans. Am. Inst. Mining Met. Engrs.* 111, 11–52. Ferromagnetism in metallic crystals. 567

34M3 McKEEHAN, L. W., ELMORE, W. C. *Phys. Rev.* 46, 226–8. Surface magnetization in ferromagnetic crystals. 533

34M4 MASUMOTO, H. *Sci. Repts. Tôhoku Imp. Univ.* 23, 265–80. Thermal expansion of alloys of Co, Fe and Cr, and a new alloy "stainless invar." 207, 644, 675

34M5 McKEEHAN, L. W., CLASH, R. F. *Phys. Rev.* 45, 839–40. Directions of discontinuous changes of magnetization in rotating mono-crystals of Si-Fe. 531

34M6 MÜLLER, N., STEINBERG, D. *Tech. Phys.* (USSR) 1, 205–11. On the layer-like magnetization in magnetic crystals. 533

34M7 McKEEHAN, L. W., ELMORE, W. C. *Phys. Rev.* 46, 529–31. Surface magnetization in ferromagnetic crystals. 533

34M8 MORRAL, F. R. *J. Iron Steel Inst.* (London) 130, 419–28. The constitution of Fe-rich Fe-Al-C alloys. 212

34M9 MÖBIUS, W. *Physik. Z.* 35, 806–11. Torsion modulus of Ni at high temperatures and simultaneous magnetization. 699

34M10 McKEEHAN, L. W. *Rev. Sci. Instruments* 5, 265–8. Pendulum magnetometer for crystal ferromagnets. 861

34M11 MATUYAMA, Y. *Sci. Repts. Tôhoku Imp. Univ.* 23, 537–88. Magnetoresistance of Bi, Ni, Fe and Heusler alloy. 760, 861

34N1 NEUMANN, H. *Arch. tech. Messen* 4, Z913–5, T168. The new magnetic alloy "1040" with high initial permeability. 160

34N2 NEUMANN, H. *Arch. tech. Messen* J66–2, T151. Testing apparatus for permanent magnet steels and finished magnets. 851

34O1 OROWAN, E. *Z. Physik* 89, 605–59. Plasticity of crystals. 621

34O2	OWEN, E. A., PICKUP, L. *Z. Krist.* **88**, 116–21. Parameter values of Cu-Ni alloys. 310	
34P1	POLANYI, M. *Z. Physik* **89**, 660–2. Lattice distortion originating plastic flow. 621	
34P2	POTTER, H. H. *Proc. Roy. Soc.* (London) **146**, 362–87. Magnetocaloric effect and other magnetic phenomena in Fe. 433, 740, 817	
34R1	RUDER, W. E. U.S.P. 1 968 569 (Appl. 6/3/33). Permanent magnet and method of making it. 347, 388	
34R2	RUDER, W. E. U.S.P. 1 947 274 (Appl. 2/1/33). Permanent magnet and method of making it. 385	
34R3	RUDER, W. E. *Trans. Am. Soc. Metals* **22**, 1120–32. Influence of grain size on magnetic properties. 69, 87	
34S3	STÄBLEIN, F. *Tech. Mitt. Krupp* **2**, 127–8. Thermoperm, a material with magnetization changing with temperature. 111, 133	
34S5	SIX, W., SNOEK, J. L., BURGERS, W. G. *De Ingenieur* **49**, E195–200. New magnetic material for Pupin coils. 588	
34S6	SCHEIL, E., BISCHOFF, K., SCHULZ, E. H. *Arch. Eisenhüttenw.* **7**, 637–40. Precipitation hardening of Fe-Cr-Mo and Fe-Cr-W alloys. 418, 419	
34S7	SÄNGER, R. *Helv. Phys. Acta* **7**, 478–80. Frequency-dependence of permeability of Fe, Co and Ni. 800	
34S8	SANFORD, R. L. *J. Research, Natl. Bur. Standards* **13**, 371–6. Drift of permeability at low inductions after demagnetization. 794	
34S9	SYKES, C., EVANS, H. *Proc. Roy. Soc.* (London) **145A**, 529–39. Some peculiarities in the physical properties of Fe-Al alloys. 212, 215	
34S10	SYKES, C., BAMPFYLDE, J. W. *J. Iron Steel Inst.* **130**, 389–418. Physical properties of Fe-Al alloys. 212, 215.	
34S12	SHIH, J. W. *Phys. Rev.* **46**, 139–42. Magnetic properties of single crystals of Fe-Co. 570, 571	
34S13	SIEVERTS, A., HAGEN, H. *Z. Physik. Chem.* (A) **169**, 237–40. Absorption of H and N by Co. 290	
34T1	TAYLOR, G. I. *Proc. Roy. Soc.* (London) **145A**, 362–404. The mechanism of plastic deformation of crystals. 621	
34T2	THAL, W. *Z. tech. Physik* **15**, 469–73. A new magnetic apparatus for measurement of soft iron. 853, 861	
34W1	WILLIAMS, C. *Phys. Rev.* **46**, 1011–4. Thermal expansion and the ferromagnetic change in volume of nickel. 644	
34W2	WEISS, P. *Compt. rend.* **198**, 1893–5. Variation of magnetic saturation at low temperatures. 111	
34W3	WEBB, C. E., FORD, L. H. *J. Inst. Elec. Engrs.* **75**, 787–97. Time-decrease of permeability at low magnetizing forces. 794	
34W4	WITTKE, H. *Ann. Physik* [5] **20**, 106–12. Quasi-static magnetic cycles in weak fields.	

1935

35A1	AUWERS, O. v., NEUMANN, H. *Wiss. Veröffentl. Siemens-Werken* **14**, 93–108. Iron-nickel-copper alloys of high initial permeability. 52, 155	
35B1	BOZORTH, R. M. *Elec. Engg.* **54**, 1251–61. Present status of ferromagnetic theory. 115, 627, 823	
35B2	BOZORTH, R. M., DILLINGER, J. F. *Physics* **6**, 285–91. Heat treatment of magnetic materials in a magnetic field.' 119, 172	
35B4	BURGERS, W. G., SNOEK, J. L. *Z. Metallkunde* **27**, 158–60. Structures of rolled and recrystallized Fe-Ni alloys. 570	

35B5 BARNETT, S. J. *Rev. Mod. Phys.* **7**, 129–66. Gyromagnetic and electron-inertia effects. 451, 452

35B6 BOZORTH, R. M., DILLINGER, J. F. U.S.P. 2 002 689 (Appl. 3/2/34). Magnetic material and method of treating magnetic materials. 118

35B7 BRONIEWSKI, W., PIETREK, W. *Compt. rend.* **201**, 206–8. On the structure of Ni-Co alloys. 276, 278

35B8 BRAMLEY, A., HAYWOOD, F. W., COOPER, A. T., WATTS, J. T. *Trans. Faraday Soc.* **31**, 707–34. Diffusion of non-metallic elements in iron and steel. 64, 65

35B9 BOZORTH, R. M. *Trans. Am. Soc. Metals* **23**, 1107–11. Orientation of crystals in silicon-iron. 88, 90, 587

35C1 COETERIER, F. *Helv. Phys. Acta* **8**, 522–64. Measurement of gyromagnetic effect in pyrrhotite. 454

35C2 CRAWFORD, C. H., THOMAS, E. J. *Elec. Engg.* **54**, 1348–53. Silicon steel in communication equipment. 68

35C3 CLEAVES, H. E., THOMPSON, J. G. McGraw-Hill, N. Y., 1–574. The Metal—Iron. 51, 52

35C4 CLASH, R. F., BECK, F. J. *Phys. Rev.* **47**, 158–65. Directions of discontinuous changes in magnetization in mono-crystal bars and disks of Si-Fe. 531

35C5 CONE, E. F. *Steel* **96**, 48–50. Production, properties, analysis, and uses of cobalt magnet steel. 380

35D1 DAHL, O., PAWLEK, F. *Z. Physik* **94**, 504–22. Effect of fiber structure and of cooling in a magnetic field on magnetization. 118

35D2 DAHL, O., PAWLEK, F., PFAFFENBERGER, J. *Arch. Eisenhüttenw.* **9**, 103–12. The magnetic properties of electrolytic iron sheets. 220, 221, 255, 260

35D3 DAHL, O., PFAFFENBERGER, J., SCHWARTZ, N. *Metallwirtschaft* **14**, 665–70. On iron nickel alloys. 156, 398

35D4 DILLINGER, J. F., BOZORTH, R. M. *Physics* **6**, 279–84. Heat treatment of magnetic materials in a magnetic field. 118, 171, 172, 175

35D5 DROZZINA, V., JANUS, R. *Nature* **135**, 36–7. New magnetic alloy with high coercive force. 238

35D6 DAHL, O., PFAFFENBERGER, J. *Metallwirtschaft* **14**, 25–8. Special magnetic properties of cold-rolled Fe-Ni alloys (Isoperm). 133

35D7 DAVENPORT, E. S., BAIN, E. C. *Trans. Am. Soc. Metals* **23**, 1047–96. The aging of steel. 86

35D8 DREYFUS, L. *ASEA J.* **12**, 8–16. Electromagnet of Upsala University. 855

35E1 ELMEN, G. W. *Elec. Engg.* **54**, 1292–9. Magnetic alloys of iron, nickel and cobalt. 110, 180, 198, 498

35E2 ELLWOOD, W. B. *Physics* **6**, 215–26. Magnetic hysteresis at low flux densities. 493, 797

35E3 EDGAR, R. F. *Gen. Elec. Rev.* **38**, 466–9. Permanent magnets. 364

35F1 FRIEDERICK, E., KUSSMANN, A. *Physik. Z.* **36**, 185–92. On the ferromagnetism of Pt-Cr alloys. 415

35F2 FETZ, E., JETTE, E. R. *J. Chem. Phys.* **4**, 537. Note on phase relations in Ni-Sn system. 324

35G2 GRAF, L., KUSSMANN, A. *Physik. Z.* **36**, 544–51. Equilibrium diagram and magnetic properties of Pt-Fe alloys. 410

35G3 GRUBE, G., WINKLER, O. *Z. Electrochem.* **41**, 52–60. Magnetic investigations in system Co-Pd. 414

35G4	GORSKI, W. S. *Physik. Z. Sowjetunion* **8**, 457–71. Theory of elastic lag in unordered mixed crystals. 795	
35G5	GOSS, N. P. *Trans. Am. Soc. Metals* **23**, 515–31. New development in electrical strip steels characterized by fine grain structure approaching properties of a single crystal. 88, 587	
35G6	GOTTSCHALK, V. H. *Physics* **6**, 127–32. Coercive force of magnetite powders. 243, 348, 829	
35H1	HAYNES, J. R. *Bell Lab. Record* **13**, 337–42. Measuring displacements of microphone contacts. 628	
35H2	HORSBURGH, C. D. L., TETLEY, F. W. Br. Pat. 431 660 (Appl. 5/23/34). Improvements in alloys for permanent magnets. 346, 347, 388	
35H4	HONDA, K., MASUMOTO, H., SHIRAKAWA, Y. *Sci. Repts. Tôhoku Imp. Univ.* **24**, 391–410. On the magnetization of single crystal of Ni at various temperatures. 567, 568	
35H5	HARALDSEN, H., KOWALSKI, E. *Z. anorg. allgem. Chem.* **224**, 329–36. Magnetic properties of CrS, CrSe and CrTe. 342	
35H6	HARALDSEN, H., KLEMM, W. *Z. anorg. allgem. Chem.* **223**, 409–16. Magnetic properties of some sulfides of the FeS_2 type. 339	
35J1	JETTE, E. T., FOOTE, F. *J. Chem. Phys.* **3**, 605–16. Precision determination of lattice constants. 52	
35J2	JELLINGHAUS, W. *Z. anorg. allgem. Chem.* **223**, 362–4. On the knowledge of the two-component system Fe-Ta. 253, 254	
35K1	KORNETZKI, M. *Z. Physik* **97**, 662–6. The volume magnetostrictive effect of nickel and of magnetite. 661, 681	
35K2	KORNETZKI, M. *Z. Physik* **98**, 289–313. The dependence of volume magnetostriction and of the Weiss factor on temperature and lattice constant. 448, 657, 681, 724, 726, 743	
35K3	KÖSTER, W., DANNÖHL, W. *Z. Metallkunde* **27**, 220–6. The copper-nickel-iron system. 154, 397	
35K4	KRUTTER, H. *Phys. Rev.* **48**, 664–71. Energy bands in copper. 439	
35K5	KÖSTER, W. *Arch. Eisenhüttenw.* **8**, 491–8. Changes in properties of irreversible ternary iron alloys upon heat-treatment. 205, 206, 208, 416	
35K6	KINZOKU ZAIRYO KENKYUSHO. Br. Pat. 428 288 (Appl. 5/1/33). Improvements in alloys for permanent magnets. 395	
35K7	KÖSTER, W., GELLER, W. *Arch. Eisenhüttenw.* **8**, 471–2. The system Fe-Co-Ti. 394, 416	
35K8	KÖSTER, W., GELLER, W. *Arch. Eisenhüttenw.* **8**, 557–60. System Fe-Co-Sn. 416	
35K9	KEESOM, W. H., CLARK, C. W. *Physica* **2**, 513–20. Atomic heat of Ni from 1.1 to 19.0°K. 738	
35K10	KAMURA, H. *Metals and Alloys* **6**, M-A 502. Effect of P on magnetic properties of Fe. 251	
35K11	KAMURA, H. Br. Pat. 431 975 (Appl. 9/24/33). Improvements relating to magnetic iron alloys. 252	
35K13	KLEMM, W., SODOMANN, H. *Z. anorg. allgem. Chem.* **255**, 273–80. Magnetic properties of potassium polyoxide and sulfide. 342	
35K14	KOHLRAUSCH, F. 17th ed., Teubner, Leipzig. Praktische Physik. 861	
35L1	LANDAU, L., LIFSHITZ, E. *Physik. Z. Sowjetunion* **8**, 153–69. Theory of dispersion of magnetic permeability in ferromagnetic bodies. 816	
35L2	LAVES, F., WITTE, H. *Metallwirtschaft* **14**, 645–9, 1002. Crystal structure of $MgNi_2$. 315	

35M1 MOTT, N. F. *Proc. Phys. Soc.* (London) **47**, 571–88. Discussion of the transition metals on the basis of quantum mechanics. 442
35M2 MISCH, LORE. *Z. physik. Chem. (B)* **29**, 42–58. Structure of intermetallic compounds of Be with Cu, Ni and Fe. 224, 576
35N1 NAKAMURA, K. *Sci. Repts. Tôhoku Imp. Univ., Sendai* **24**, 303–31. The effect of magnetization on Young's modulus of elasticity of some ferromagnetic substances. 687, 693
35N2 NORTON, J. T. *Am. Inst. Mining Met. Engrs. Tech. Pub.* **586**, 1–9. Solubility of Cu in Fe and lattice changes during aging. 230
35N3 NAKAMURA, K. *Z. Physik* **94**, 707–16. Variation of E of Fe-Ni alloys by magnetization. 699
35P2 PREISACH, F. *Z. Physik* **94**, 277–302. On magnetic lag. 494, 797
35P3 PAWLEK, F. *Z. Metallkunde* **27**, 160–5. Structures of rolled and of recrystallized iron-nickel alloys and their relation to magnetic properties. 587
35P4 PREISACH, F. *Z. Physik* **93**, 245–69. Magnetic investigations of precipitation-hardening Fe-Ni alloys. 189
35R1 RAY-CHAUDHURI, D. P. *Indian J. Phys.* **9**, 383–414. Researches on the gyromagnetic effect of some ferromagnetic compounds. 454
35S2 SNOEK, J. L. *Physica* **2**, 403–12. Magnetic investigation of Pupin coils of modern construction. 504
35S3 SUCKSMITH, W. *Helv. Phys. Acta* **8**, 205–10. Gyromagnetic effect of a ferromagnetic substance above its Curie point. 454, 455
35S4 SEYBOLT, A. U., MATHEWSON, C. H. *Trans. Am. Inst. Mining Met. Engrs.* **117**, 156–72. Solubility of oxygen in solid Co and the upper transformation point of the metal. 262, 293
35S5 STEINWEHR, H. E. V., SCHULZE, A. *Physik. Z.* **36**, 307–11. Heat exchange in metallic transformations. Cobalt. 262, 263
35S6 SIXTUS, K. J. *Phys. Rev.* **48**, 425–30. Propagation of large Barkhausen discontinuities. 496
35S7 SEQUENZ, H. *Arch. Elektrotech.* **29**, 387–94. Contribution to equations of hysteresis loops. 511
35S8 SYKES, C., EVANS, H. *J. Iron Steel Inst.* (London) **131**, 225–47. Transformations in Fe-Al alloys. 212, 215
35S9 SCHARFF, G. *Z. Physik* **97**, 73–82. Becker equation for μ_0 of strongly stretched Ni wire. 612, 823
35S10 SYKES, W. P., GRAFF, H. F. *Am. Soc. Metals* **23**, 249–83. The Co-Mo system. 292
35S11 SIXTUS, K. J. *Physics* **6**, 105–11. Magnetic anistropy in silicon steel. 587
35S12 SNOEK, J. L. *Physica* **2**, 403–12. Magnetic investigation of new kind of Pupin coil. 587
35T2 TOWNSEND, A. *Phys. Rev.* **47**, 306–10. Change in thermal energy which accompanies a change in magnetization in Ni. 520
35U1 URBAIN, G., WEISS, P., TROMBE, F. *Compt. rend.* **200**, 2132–2134. Gadolinium, a new ferromagnetic metal. 342
35V1 VOGEL, R., ROSENTHAL, K. *Arch. Eisenhüttenw.* **9**, 293–9. System iron-cobalt, cobalt silicide, iron silicide. 416
35V2 VALENTINER, S., BECKER, G. *Z. Physik.* **93**, 795–803. System Ni-Mn. 320
35V3 VERWEY, E. J. W. *J. Chem. Phys.* **3**, 592–3. Incomplete atomic arrangement in crystals. 242
35W2 WOOD, A. B., SMITH, F. D., MCGEACHY, J. A. *J. Inst. Elec. Engrs.* **76**, 550–66. A magnetostriction depth recorder. 628

35W4　WALTERS, F. M., WELLS, C.　*Trans. Am. Soc. Metals* **23**, 727–50.　Alloys of Fe & Mn—Part 13.　235
35W5　WEGEL, R. L., WALTHER, H.　*Physics* **6**, 141–57.　Internal dissipation in solids.　686
35Z1　ZUMBUSCH, W.　*Stahl u. Eisen* **55**, 860.　Discussion of paper by F. Pölzgutter.　412

1936

36A1　AUWERS, O. v.　Verlag Chemie, G.m.b.H., Berlin, 8 ed., 1–466.　Gmelin's Hdb. Anorg. Chem. Iron. Part D.　Magnetic and electric properties of alloys.　182, 210, 628
36A3　ANONYMOUS.　*Electronics* **9** (May), 30–2, 35.　The new permanent magnet alloys.　396
36A4　ASCHENBRENNER, H., GOUBAU, G.　*Hochfrequ. u. Elektroakust.* **47**, 177–81.　Arrangement for recording rapid magnetic disturbances.　861
36B1　BOZORTH, R. M.　*Phys. Rev.* **50**, 1076–81.　Determination of ferromagnetic anisotropy in single crystals and in polycrystalline sheets.　569, 579, 581, 587, 656
36B2　BROWN, W. F., JR.　*Phys. Rev.* **50**, 1165–72.　Variation of internal friction and elastic constants with magnetization in Fe.　703, 704, 705
36C1　COOKE, W. T.　*Phys. Rev.* **50**, 1158–64.　Variation of internal friction and elastic constants with magnetization in Fe.　686, 687, 704, 707
36C2　CATHERALL, A. C.　U.S.P. 2 096 670 (Appl. 10/17/35).　Permanent magnet.　395, 396
36C3　CLUSIUS, K., GOLDMANN, J.　*Z. Physik. Chem.* (B) **31**, 256–62.　Atomic heat of Ni at low temperatures.　738
36D1　DAHL, O.　*Z. Metallkunde* **28**, 133–8.　Cold deformation and recovery of alloys with ordered atomic distribution.　103, 189, 319
36D2　DAHL, O., PAWLEK, F.　*Z. Metallkunde* **28**, 230–3.　Structures of rolled and of recrystallized iron-nickel alloys and their relation to magnetic properties.　43
36D5　DAHL, O., PFAFFENBERGER, J.　*Jahrb. Forsch-Inst.* A.E.G. **4**, 1–10.　Development of Isoperm.　125, 126, 133
36D6　DÖRING, W.　*Z. Physik* **103**, 560–82.　On the temperature dependence of magnetostriction of nickel.　125, 612, 630, 643, 661, 695, 726
36D7　DANNATT, C.　*J. Inst. Elec. Engrs.* **79**, 667–80.　Variation of magnetic properties of ferromagnetic laminae with frequency.　788
36D8　DEHLINGER, U.　*Z. Metallkunde* **28**, 194–6.　Change of volume with magnetization and the Invar alloys.　448
36E2　EWERT, M.　*Proc. Royal Acad. Sci. Amsterdam* **39**, 833–8.　Specific heat and allotropy of nickel between 0° and 1000°C.　737
36E3　ELMORE, W. C., MCKEEHAN, L. W.　*Trans. Am. Inst. Mining Met. Engrs.* **120**, 236–52.　Surface magnetization and block structure of ferrite.　533
36F1　FALLOT, M.　*Ann. physique* [11] **6**, 305–87.　Ferromagnetism of alloys of iron.　72, 79, 210, 215, 228, 258, 412, 441, 720, 723
36G1　GERLACH, W.　*Z. Metallkunde* **28**, 80–3.　Magnetic investigation of hardening of Ni-Be alloys.　304
36G2　GRUBE, G., KÄSTNER, H.　*Z. Electrochem.* **42**, 156–60.　Electrical conductivity and phase diagram of binary alloys.　414
36G3　GENDERS, R., HARRISON, R.　*J. Iron Steel Inst.* (London) **134**, 173–212.　Ta-Fe alloys and Ta steels.　253, 417

36H1 HANSEN, M. Julius Springer, Berlin, 1–1100. Constitution of binary alloys. 103, 180, 187, 220, 234, 240, 253, 264, 284, 290, 291, 293, 294, 295, 300, 306, 308, 320, 323, 324, 334, 335, 337, 339, 410, 411

36H3 HABERLAND, G., HABERLAND, F. *Arch. Elektrotech.* **30**, 126–33. Eddy currents in massive iron. 787

36H4 HÜLSMANN, O., WEIBKE, F. *Z. anorg. allgem. Chem.* **227**, 113–23. Constitution diagram of the system Co-CoS. 294

36H5 HALLA, F., NOWOTNY, H. *Z. Physik. Chem.* **34B**, 141–4. X-ray study of the system Mn-Sb. 334

36H6 HONDA, K., NISHINA, T. *Z. Physik* **103**, 728–37. Temperature variation of magnetization in weak fields. 559

36I1 IWASÉ, K., OKAMOTO, M. *Sci. Repts. Tôhuku Imp. Univ.* (Honda) 777–92. Diagram of the Ni-Si system. 322

36J1 JETTE, E., FOOTE, F. *Trans. Am. Inst. Mining Met. Engrs.* **120**, 259–76. X-ray study of iron-nickel alloys. 103

36J2 JELLINGHAUS, W. *Z. tech. Physik* **17**, 33–6. New alloys with high coercive force. 346, 410, 412

36J3 JELLINGHAUS, W. *Hochfrequ. u. Elektroakust.* **48**, 58–9. The oxide magnet of Kato and Takei. 421

36J4 JELLINGHAUS, W. *Arch. Eisenhüttenw.* **10**, 115–8. System Fe-Co-Cu. 416

36K1 KIRCHNER, H. *Ann. Physik* [5] **27**, 49–69. The effect of tension, compression, and torsion on longitudinal magnetostriction. 636, 637, 660, 671

36K2 KLEIS, J. D. *Phys. Rev.* **50**, 1178–81. Ferromagnetic anisotropy of nickel-iron crystals at various temperatures. 569, 570, 579

36K3 KINZOKU ZAIRYO KENKYUSHO. Br. Pat. 450 619 (Appl. 8/12/35). Improvements in alloys for permanent magnets. 395

36K4 KLEMM, W. Akad. Verl., Leipzig, 1–262. Magnetochemie. 334

36K5 KINDLER, H., THOMA, A. *Arch. Elektrotech.* **30**, 514–27. On magnetic lag. 797

36K6 KROLL, W. *Metals and Alloys* **7**, 24–7. Be-Fe alloys. 224

36K7 KAYA, S., TAKAKI, H. *Sci. Repts. Tôhoku Imp. Univ.* (Honda) 314–28. Hysteresis loops and magnetostriction of Fe crystals 561, 645, 652, 654

36K8 KÖSTER, W., DANNÖHL, W. *Z. Metallkunde* **28**, 248–53. Hardening of Ni-Au alloys. 314, 415

36L1 LEGG, V. E. *Bell System Tech. J.* **16**, 39–62. Magnetic measurements at low flux densities using the alternating current bridge. 126, 781, 852

36L2 LAPP, E. *Ann. physique* **6**, 826–55. Specific heat of iron. 737

36L3 LORIG, C. H., KRAUSE, D. E. *Metals and Alloys* **7**, 9–13, 51–6, 69–73. Phosphorus as an alloying element in low-carbon alloy steels. 251, 252

36M4 MOSER, H. *Physik. Z.* **37**, 737–53. An improved method for measuring the true specific heat of silver, nickel, β-brass, crystalline and fused silica. 737

36M5 MARICK, L. *Phys. Rev.* **49**, 831–7. Variation of resistance and structure of Co with temperature and discussion of its photoelectric emission. 262, 276

36M6 MISHIMA, T. U.S.P. 2 027 996 (Appl. 8/27/31). Strong permanent magnet with cobalt. 347, 388

36M7 MISHIMA, T. U.S.P. 2 027 994 (Appl. 3/9/31). Magnet steel containing Ni and Al. 347, 385

36M8 MASUMOTO, H. *Sci. Repts. Tôhoku Imp. Univ.* (Honda) 388–402. On a new alloy "Sendust" and its magnetic and electric properties. 95, 418

36M9 MADDOCKS, W. R., CLAUSSEN, G. E. *Iron Steel Inst.* (London) *Rep. No. 14*, 97–124. Alloys of Fe-Cu-Co. 229, 416

36M10 MESSKIN, W. S., SOMIN, B. E. *Z. Physik* **98**, 610–23. Experimental test of Akulov's theory of H_c. 238, 256, 418, 574

36M11 MOTT, N. F. *Proc. Roy. Soc.* (London) **156A**, 368–82. Resistance and thermoelectric properties of the transition metals. 763

36M12 MASUMOTO, H., SHIRAKAWA, Y. *Sci. Repts. Tôhoku Imp. Univ.* **25**, 104–27. On the longitudinal magneto-resistance effect at various temperatures in Ni-Cu alloys. 758

36M13 MOTT, N. F., JONES, H. Clarendon Press, Oxford, 1–326. Theory of the Properties of Metals and Alloys. 433, 437, 442, 736

36M14 MITKEVITCH, A. *J. phys. radium* [7] **7**, 133–7. Anomalous magnetic viscosity.

36M15 MORSE, P. M. McGraw-Hill, N. Y., 1–468. Vibration and sound. 700

36N1 NISHINA, T. *Sci. Repts. Tôhoku Imp. Univ.* (Honda) 344–361. An investigation on some magnetic alloys. 151, 171, 185

36N2 NAKAMURA, K. *Sci. Repts. Tôhoku Imp. Univ.* **25**, 415–25. Effect of temperature on Young's modulus of elasticity in Cu-Ni alloys. 699

36N3 NÉEL, L. *Ann. physique* [11] **5**, 232–79. Magnetic properties of metallic state and energy of interaction between ions. 472

36O1 OWEN, E. A., YATES, E. L. *Phil. Mag.* [7] **21**, 809–18. X-Ray measurements of the thermal expansion of pure nickel. 269, 276

36O2 OKAMURA, T. *Sci. Repts. Tôhoku Imp. Univ.* **24**, 745–807. Change of thermal energy due to magnetization. 518, 521

36P2 PETERSON, E., WRATHALL, L. R. *Proc. Inst. Radio Engrs.* **24**, 275–86. Eddy currents in composite laminations. 775, 801

36P3 PIETY, R. J. *Phys. Rev.* **50**, 1173–7. Ferromagnetic anisotropy of Fe crystals, at various temperatures. 567

36S4 STONER, E. C. *Phil. Mag.* [7] **22**, 81–106. The specific heat of nickel. 735

36S5 SLATER, J. C. *Phys. Rev.* **49**, 537–45. The ferromagnetism of nickel. 439, 442

36S6 SNOEK, J. L. *Physica* **3**, 463–83. Magnetic and electrical properties of ferrites. 243, 421, 422

36S7 SIMPSON, K. M., BANNISTER, R. J. *Metals and Alloys* **7**, 88–94. Alloys of copper and iron. 231

36S8 SCHUMACHER, E. E., SOUDEN, A. G. *Metals and Alloys* **7**, 95–101. Some alloys of copper and iron. 231

36S9 SCHRAMM, J. *Z. Metallkunde* **28**, 203–7. System Fe-Zn. 259

36S10 SIEGEL, S., QUIMBY, S. L. *Phys. Rev.* **49**, 663–70. Variation of Young's modulus with magnetization and temperature in Ni. 687, 689, 702, 707

36S12 SLATER, J. C. *Phys. Rev.* **49**, 931–7. Ferromagnetism of Nickel II. 442

36S13 SHOENBERG, D., UDDIN, M. Z. *Proc. Roy. Soc.* (London) **156A**, 701–20. Magnetic properties of Bi. 456

36S14 SHIH, J. W. *Phys. Rev.* **50**, 376–9. Magnetic anisotropy of Ni-Co single crystals. 570, 571

36S15 SCHLECHTWEG, H. *Ann. Physik* [5] **27**, 573–96. Magnetic anisotropy of single crystals of Fe and Ni. 566, 573, 579

36S16 SVENSSON, B. *Ann. Physik* [5], **25**, 263–71. Ferromagnetic change of resistance of Ni-Cu alloys. 310

36S17 SEYBOLT, A. U. Dissertation, Yale. System Ni-O. 321

36S18 SHIRAKAWA, Y. *Sci. Repts. Tôhoku Imp. Univ.* (Honda vol.) 1–26. Magnetoresistance in Co-Ni alloys. 758

36V1 VAN ARKEL, A. E., VERWEY, E. J. W., VAN BRUGGEN, M. G. *Rec. trav. chim. pays-bas* **55**, 331–9. Ferrites I. 245

36V2 VERWEY, E. J. W., VAN ARKEL, A. E., VAN BRUGGEN, M. G. *Rec. trav. chim. pays-bas* **55**, 340–7. Ferrites II. 245

36V3 VERWEY, E. J. W., DE BOER, J. *Rec. trav. chim. pays-bas* **55**, 531–40. Cation arrangements in some oxides of the spinel type. 242

36Y1 YENSEN, T. D. *Stahl. u. Eisen* **56**, 1545–50. Development of transformer steel in North American. 67, 69, 83

36Y2 YENSEN, T. D. U.S.P. 2 050 408 (Appl. 10/23/35). Process of treating magnetic material. 86

36Y3 YENSEN, T. D., ZIEGLER, N. A. *Trans. Am. Soc. Metals* **24**, 337–58. Effect of C, O and grain size on magnetic properties of Si-Fe. 36

1937

37A1 ALEXANDER, W. O., VAUGHAN, N. B. *J. Inst. Metals* **61**, 247–60. Constitution of the Ni-Al system. 299

37B1 BOZORTH, R. M. *J. Applied Phys.* **8**, 575–88. Directional ferromagnetic properties of metals. 60, 559, 567, 568, 586

37B2 BRADLEY, A. J., JAY, A. H., TAYLOR, A. *Phil. Mag.* [7], **23**, 545–57. The lattice spacing of iron-nickel alloys. 103

37B4 BITTER, F. McGraw-Hill, N. Y., 1–314. Introduction to ferromagnetism. 579

37B5 BRUKHATOV, N. L., KIRENSKY, L. V. *Physik. Z. Sowjet Union* **12**, 602–9. Anisotropy of magnetic energy in single crystals of Ni as a function of temperature. 567, 568, 744

37B6 BITTER, F. *Phys. Rev.* **54**, 79–86. Generalization of theory of ferromagnetism. 472

37B7 BROWN, W. F. *Phys. Rev.* **52**, 325–34. Domain theory of ferromagnetics under stress. 654

37B8 BOZORTH, R. M., MCKEEHAN, L. W. *Phys. Rev.* **51**, 216. Explanation of directions of easy magnetization in ferromagnetic cubic cyrstals. 592

37B9 BRADLEY, A. J., TAYLOR, A. *Proc. Roy. Soc.* (London) **159A**, 56–72. X-ray analysis of Ni-Al system. 299

37B10 BATES, L. F., ILLSLEY, P. F. *Proc. Phys. Soc.* (London) **49**, 611–8. Magnetic properties of iron amalgams. 236

37C2 CIOFFI, P. P., WILLIAMS, H. J., BOZORTH, R. M. *Phys. Rev.* **51**, 1009. Single crystals with exceptionally high magnetic permeabilies. 60, 555, 556

37C3 CHIPMAN, J., LI, T. *Trans. Am. Soc. Metals* **25**, 435–65. Equilibrium in reaction of H with FeS in liquid Fe, and thermodynamics of desulfurization. 66

37C4 CHEGWIDDEN, R. A. Thesis, Polytech. Inst. Brooklyn. Aging of magnets. 358

37C5 CONYBEARE, J. G. G. *Proc. Phys. Soc.* (London) **49**, 29–37. Resistance of Pd and Pd-Au alloys. 762

37E1 EBERT, H., KUSSMANN, A. *Physik. Z.* **38**, 437–45. Change of saturation magnetization owing to unilateral compression. 645

37E2 ELLWOOD, W. B., LEGG, V. E. *J. Applied Phys.* **8**, 351–8. Study of magnetic losses at low flux densities in 35 permalloy sheet. 781, 796

37E3 ELMORE, W. C. *Phys. Rev.* **51**, 982–8. Properties of the surface magnetization in ferromagnetic crystals. 533

37E4 EDGAR, R. F. *Elec. Engg.* **56**, 805–9. New photoelectric hysteresigraph. 861

37F1	FALLOT, M. *Ann. physique* [11] 7, 420–8. Magnetic properties of alloys of Fe and Zn. 259	
37F3	FÖRSTER, F., KÖSTER, W. *Naturwissenschaften.* 25, 436–9. Dependence of Young's modulus and damping of transversely oscillating metal reeds on the amplitude. 688, 693, 708, 709	
37F4	FÖRSTER, F. *Z. Metallkunde* 29, 109–15. New method for determination of elasticity modulus and damping. 686	
37F5	FARCAS, T. *Ann. physique* [11] 8, 146–52. Ferromagnetic moments of some alloys of Co with Cr, Al, W, Mo. 282, 284, 289, 292, 297, 441	
37G2	GOSS, N. P. U.S.P. 2 084 336–7 (Appl. 1/30/34 and 12/1/34). Magnetic material and method of manufacture. 88, 89	
37G3	GERLACH, W. *Z. Metallkunde* 29, 124–31. Hardening of Ni-Be alloys. 302, 304	
37G4	GREINER, E. S., JETTE, E. R. *Trans. Am. Inst. Mining Met. Engrs.* 125, 473–81. X-ray study of constitution of Fe-Si alloys. 72	
37G5	GRUNER, E., KLEMM, W. *Naturwissenschaften* 25, 59–60. Magnetic properties of AgF_2. 342	
37H1	HAWORTH, F. E. *Phys. Rev.* 52, 613–20. Energy of lattice distortion in cold-worked permalloy. 621, 624	
37H3	HASHIMOTO, U. *Nippon Kinzoku Gakkai-Si* 1, 135–43. The Co-Si equilibrium diagram. 282, 294	
37H4	HASHIMOTO, U. *Nippon Kinzoku Gakkai-Si* 1, 19–26. The Co-Cu equilibrium diagram. 282–289	
37H5	HASHIMOTO, U. *Nippon Kinzoku Gakkai-Si* 1, 177–90. Effect of additions to Co on its allotropic transformation. 282, 283, 284, 285, 286, 289, 290, 294, 295, 297, 299	
37J1	JANUS, R., SHUR, J. S. *Physik. Z. Sowjetunion* 12, 383–8. On magnetic hysteresis in single crystals. 562	
37J2	JENKINS, C. H. M., BUCKNALL, E. H., AUSTIN, C. R., MELLOR, G. A. *J. Iron Steel Inst.* (London) 136, 187–220. Constitution of the alloys of Ni, Cr and Fe. 307	
37K1	KIRKHAM, D. *Phys. Rev.* 52, 1162–7. Variation of susceptibility and magnetostriction with temperature and magnetization in nickel. 273, 661	
37K2	KINZEL, A. B., CRAFTS, W. McGraw-Hill, N. Y., 1–535. Alloys of Fe and Cr. Vol. I—Low Cr Alloys. 376	
37K3	KUSSMANN, A. *Physik. Z.* 38, 41–2. An extension-anomaly in Fe-Pt alloys. 410, 644	
37K4	KINDLER, H. *Ann. Physik* [5] 28, 375–84. Dependence of magnetic lag on inner tension. 797	
37K5	KONDORSKI, E. *Physik. Z. Sowjetunion* 11, 597–620. H_c and irreversible changes in magnetization. 823	
37K6	KOHAUT, A. *Z. tech. Physik* 18, 198–9. Measuring apparatus for magnetic fields. 856	
37K7	KÖSTER, W., WAGNER, E. *Z. Metallkunde* 29, 230–2. Influence of Al, Cu, Sb, Sn, Ti, V, Zn on transformation of Co. 281, 283, 284, 289, 295, 296, 298, 299	
37K8	KÖSTER, W., SCHMID, E. *Z. Metallkunde* 29, 232–3. Influence of Be, C and Si on transformation of Co. 281, 285, 287, 294	
37K9	KELLEY, K. K. *U. S. Bur. of Mines, Bull. No.* 407, 1–66. Thermodynamic properties of metal carbides and nitrides. 240, 287	
37L2	LEGAT, H. *Metallwirtschaft* 16, 743–9. Magnetic properties of Ni-Cu-Fe alloys. 400	

37M1 MCKEEHAN, L. W. *Phys. Rev.* **51**, 136–9. Ferromagnetic anisotropy in nickel-cobalt-iron crystals at various temperatures. 163, 567, 570, 571

37M2 MARIAN, V. *Ann. physique* [11] **7**, 459–527. Ferromagnetic Curie points and the absolute saturation of some nickel alloys. 271, 299, 300, 307, 313, 321, 322, 325, 326, 327, 414, 441, 720, 723

37M3 MINER, D. F., SEASTONE, J. B. *Metal Progress* **31**, 611–7. Melting and annealing of electrical alloys. 130

37M4 MEYER, W. F. *Z. Kryst.* **97**, 145–69. Cobalt and the Co-C system. 276, 287

37M5 MASIYAMA, Y. *Sci. Repts. Tôhoku Imp. Univ.* **26**, 1–39. On hysteresis of magnetostriction of Fe, Ni, Co and single crystals of Fe. 658, 660

37M6 MASIYAMA, Y. *Sci. Repts. Tôhoku Imp. Univ.* **26**, 65–85. On hysteresis of magnetostriction for alloys of systems Fe-Ni, Ni-Co, Fe-Co. 669

37M7 MICHELS, A., JASPERS, A., DE BOER, J., STRIJLAND, J. *Physica* **4**, 1007–16. Influence of pressure on Curie point of nickel-copper alloy. 726, 727

37M8 MCKEEHAN, L. W. *Phys. Rev.* **52**, 18–30. Magnetic interaction and anisotropy in ferromagnetic crystals. 592

37M9 MIKULAS, W., THOMASSEN, L., UPTHEGROVE, C. *Trans. Am. Inst. Mining Met. Engrs.* **124**, 111–33. Equilibrium relations in the Ni-Sn system. 324

37N1 NEUMANN, H. *Arch. tech. Messen* Z912-1, T38–43. Permanent-magnet materials. 395, 410, 413

37N2 NEUMANN, H., BÜCHNER, A., REINBOTH, H. *Z. Metallkunde* **29**, 173–85. Mechanically soft iron-nickel-copper permanent-magnet alloys. 346, 348, 398, 400

37N3 NÉEL, L. *Ann. physique* [11] **8**, 237–308. Moment and molecular field of ferromagnetic substances. 434, 444

37N4 NOWOTNY, H., ÅRSTAD, F. *Z. physik. Chem.* (B) **38**, 461–5. X-ray investigation of system Cr-CrAs. 341

37O3 OWEN, E. A., YATES, E. L., SULLY, A. H. *Proc. Phys. Soc.* (London) **49**, 315–22. An X-ray investigation of pure iron-nickel alloys—IV. The variation of lattice parameter with composition. 55, 103

37O5 OPECHOWSKI, W. *Physica* **4**, 715–22. Temperature dependence of magnetization of ferromagnetics at low temperatures. 448, 719

37P1 POTTER, H. H. *Proc. Phys. Soc.* (London) **49**, 671–8. The electrical resistance of ferromagnetics. 279, 762

37P2 PROCOPIU, S., D'ALBON, G. *Compt. rend.* **205**, 1373–5. Magnetic μ, at high frequencies, of thin films of electrolytically deposited Fe. 800

37P3 POTAPENKO, G., SÄNGER, R. *Z. Physik* **104**, 779–803. New method of measurement of ferromagnetic properties of metals in region of very high frequencies. 778

37R1 RANDALL, W. F. *J. Inst. Elec. Engrs.* **80**, 647–58. Nickel-iron alloys of high permeability, with special reference to mumetal. 159, 160

37R2 RICHTER, G. *Ann. Physik* [5] **29**, 605–35. Magnetic after-effects in carbonyl iron. 789, 793

37S4 STEINHAUS, W., KUSSMANN, A., SCHOEN, E. *Physik. Z.* **38**, 777–85. Saturation magnetization and the law of approach of iron. 56, 252, 485

37S5 SOLLER, T. *Z. Physik* **106**, 485–98. The properties of Bitter stripes under elastic strain. 533

37S6 SNOEK, J. L. *Physica* **4**, 853–62. Volume magnetostriction of iron and nickel. 630, 726

37S7 SVECHNIKOV, V. N., GRIDNEV, V. N. *Metallurg* **12**, 35–9. Allotropic transformation of Fe in Fe-Zn alloys (in Russian). 259

37S8 SLATER, J. C. *J. Applied Phys.* **8**, 385–90. Electronic structures in alloys. 441
37S9 SIXTUS, K. J. *Phys. Rev.* **52**, 347–52. Coercive force in single crystals. 562
37T1 TARASOV, L. P., BITTER, F. *Phys. Rev.* **52**, 353–60. Precise magnetic torque measurements on single crystals of iron. 573, 579
37T2 THURAS, A. L. *J. Acoust. Soc. Am.* **9**, 74. A sensitive method of measuring magnetic flux in small areas. 857
37T3 TAKAKI, H. *Z. Physik* **105**, 92–103. Magnetostriction of Fe crystals at high temperatures. 652
37T4 TROMBE, F. *Ann. physique* [11] **7**, 385–419. Magnetic properties of rare metals. 342
37V1 VAN VLECK, J. H. *Phys. Rev.* **52**, 1178–98. On the anisotropy of cubic ferromagnetic crystals. 574, 593, 654
37W1 WILLIAMS, H. J. *Phys. Rev.* **52**, 1004–5. Variation of initial permeability with direction in single crystals of Si-Fe. 94
37W2 WISE, E. M. *Proc. Inst. Radio Engrs.* **25**, 714–52. Nickel in the radio industry. 269
37W3 WILLIAMS, H. J. *Rev. Sci. Instruments* **8**, 56–60. Some uses of the torque magnetometer. 515, 558, 861
37W4 WEISS, P. *Extr. Actes VII Cong. Int. Froid* **1**, 508–14. Law of approach of magnetization to absolute saturation and determination of atomic moment. 57, 269, 271, 661, 719
37W5 WILLIAMS, H. J. *Phys. Rev.* **52**, 747–51. Magnetic properties of single crystals of Si-Fe. 20, 69, 90, 556, 558, 562, 573, 579
37Y1 YENSEN, T. D. Bitter's Introduction to Ferromagnetism, p. 67–125. Magnetic materials and their preparation. 52

1938

38A1 ADAMS, J. Q., G. E. *Rev.* **41**, 518–22. Alnico. 359
38A2 ASHWORTH, J. R. Taylor & Francis, Ltd., London, 1–97. Ferromagnetism. 356, 549
38B1 BRADLEY, A. J., TAYLOR, A. *Proc. Roy. Soc.* (London) **166A**, 353–75. X-ray examination of iron-nickel-aluminum ternary equilibrium system. 35, 184, 212, 385
38B2 BITTEL, H. *Ann. Physik* [5] **32**, 608–24. Effect of cold-working and heat-treatment on the electrical and magnetic properties of pure Ni. 275
38B3 BRAILSFORD, F. *J. Inst. Elec. Engrs.* **83**, 566–75. Rotational hysteresis loss in electrical sheet steel. 515, 518
38B4 BROWN, W. F. *Phys. Rev.* **54**, 279–87. Domain theory of ferromagnetics under stress III. Reversible susceptibility. 545, 654
38B5 BUMM, H., MÜLLER, H. C. *Wiss. Veröffentl. Siemens-Werke* **17** (2), 63–73. Relation of precipitation processes to magnetic hardness of permanent magnet alloys of systems Fe-Ni-Al, Fe-Ni-Cu. 387
38B6 BECKER, R. *Z. tech. Physik* **19**, 542–6. Ferromagnetism in high-frequency alternating fields. 801
38B7 BRUCHATOW, N., KIRENSKY, L. *Tech. Phys.* (USSR) **5**, 171–83. Influence of temperature on magnetic anisotropy energy of ferromagnetic crystals. 568
38B8 BROWN, W. F. *Phys. Rev.* **53**, 482–9. Domain theory of ferromagnetics under stress II. 65
38B9 BILTZ, W., HEIMBRECHT, M. *Z. anorg. allgem. Chem.* **237**, 132–44. Phosphides of nickel. 322

38C1	CIOFFI, P. P. U.S.P. 2 110 569 (Appl. 8/19/32). Magnetic material. 86, 92, 138	
38C2	COLE, R. H. *Rev. Sci. Instruments* **9**, 215–7. Magnetic field meter. 856	
38D1	DANNÖHL, W., NEUMANN, H. *Z. Metallkunde* **30**, 217–31. On permanent magnet alloys of Co, Cu and Ni. 280, 402, 405	
38D2	DÖRING, W. *Probleme der Technischen Magnetisierungskurve*, Springer, Berlin, 26–41. The growth of reverse-magnetization nuclei during large Barkhausen jumps. 496	
38D3	DANNÖHL, W. *Z. Metallkunde* **30**, 95–9. Magnetic properties of hardening ferritic Fe-Ni-Cu alloys. 400	
38D4	DANNÖHL, W. *Wiss. Veröffentl. Siemens-Werke* **17** (2), 1–13. Alloys of Fe, Cu and Mo. 418	
38D5	DÖRING, W. *Ann. Physik* **32**, 259–76. Dependence of resistance of Ni crystals on direction of spontaneous magnetization. 764	
38D6	DE BOER, J., MICHELS, A. *Physica* **5**, 775–6. Ferromagnetic Curie point as a phase transition of 2d kind. 724, 726	
38D7	DÖRING, W. *Ann. Physik* **32**, 465–70. Temperature-dependence of Young's modulus of ferromagnetic substances. 684, 690, 697	
38E1	ENGLER, O. *Ann. Physik* [5] **31**, 145–63. Effects of temperature and of magnetic field on the modulus of elasticity of ferromagnetic materials. 686, 688, 690, 697, 699	
38E2	ESSER, H., EILENDER, W., BUNGARDT, K. *Arch. Eisenhüttenw.* **12**, 157–61. X-ray investigation of metals at high temperatures. 55, 269	
38E3	ELMORE, W. C. *Phys. Rev.* **53**, 757–64. Magnetic structure of cobalt. 538	
38E4	ELMORE, W. C. *Phys. Rev.* **54**, 309–10. Ferromagnetic colloid for studying magnetic structures. 533	
38E5	EVERT, F., KUSSMANN, A. *Physik. Z.* **39**, 598–605. Influence of hydrostatic pressure on the Curie point. 726	
38E6	EGGERS, H., PETER, W. *Mitt. Kaiser-Wilhelm Inst. Eisenforsch. Düsseldorf.* **20**, 199–203. Binary system Fe-Nb. 239	
38E7	ELMORE, W. C. *Phys. Rev.* **53**, 933–4. Motion photomicrographs of magnetic colloid patterns on cobalt crystals. 538	
38F1	FALLOT, M. *Ann. physique* [11] **10**, 291–332. Alloys of Fe with metals of Pt group. 249, 250, 410, 412, 441	
38F2	FREY, A. A., BITTER, F. U.S.P. 2 112 084 (Appl. 11/1/34). Magnetic material and method of producing the same. 89	
38G1	GERLACH, W. *Probleme der Technischen Magnetisierungskurve*, Springer, Berlin, 141–56. Analysis of precipitation hardening with ferromagnetic measurements. 304, 314, 723	
38G3	GRUBE, G., SCHLECHT, H. *Z. Electrochem.* **44**, 413–22. System Ni-Mo. 320, 321	
38G4	GRUBE, G., WINKLER, O. *Z. Electrochem.* **44**, 423–8. Magnetic properties of Ni-Mo alloys. 321	
38H1	HARDY, T. C., QUIMBY, S. L. *Phys. Rev.* **54**, 217–23. Change in thermal energy with adiabatic change of magnetization in Fe, Ni and carbon steel. 522, 524	
38H2	HEIMENZ, H. U.S.P. 2 113 537 (Appl. 7/2/36). Method of rolling and treating silicon steel. 89	
38H3	HASHIMOTO, U. *Nippon Kinzoku Gakkai-Si* **2**, 67–77. Effect of various elements on the α-β transformation point of Co. 282, 293, 294, 296, 299	

38H4	HASHIMOTO, U., KAWAI, N. *Nippon Kinzoku Gakkai-Si* **2**, 26–8. *Metals and Alloys* **5**, 475. Equilibrium diagram of the Co-C system. 287
38H5	HILPERT, R. S., MAIER, K. H., HOFFMANN, A. *Ber. deut. Chem. Ges.* **71**, 2682–5. Sulfo-magnetite. 253
38I2	INGERSON, W. E., BECK, F. J. *Rev. Sci. Instruments* **9**, 31–5. Magnetic anisotropy in sheet steel. 515, 591
38I3	IWASÉ, K., OKAMOTO, M., AMEMIYA, T. *Sci. Repts. Tôhoku Imp. Univ.* **26**, 618–40. Formation of 2 liquid layers in Cu-Fe alloys. 229
38J1	JACKSON, L. R., RUSSELL, H. W. *Instruments* **11**, 280–2. Temperature-sensitive magnetic alloys and their uses. 149, 151, 153
38J2	JAEGER, F. M., ROSENBOHM, E., ZUITHOFF, A. J. *Rec. trav. chim. pays-bas* **57**, 1313–40. Specific heat, electrical resistivity, thermoelectric behavior and thermal expansion of electrical Fe. 737
38K1	KERSTEN, M. *Probleme der Technischen Magnetisierungskurve*, Springer, Berlin, 42–72. On the significance of coercive force. 816, 823, 824
38K2	KAYA, S. *J. Faculty Sci. Hokkaido Imp. Univ.* **2**, 29–53. Superstructure formation in Ni-Fe alloys and the permalloy problem. 103, 106, 669, 768
38K3	KÖSTER, W., LANG, K. *Metallkunde* **30**, 350–2. The cobalt corner of the system Fe-Co-V. 201, 416
38K4	KERSTEN, M. *Physik. Z.* **39**, 860–4. Investigation of reversible and irreversible wall displacements between antiparallel domains. 503
38K6	KUSSMANN, A., NITKA, H. *Physik. Z.* **39**, 373–5. Pt-Ni alloys. 415
38K7	KUSSMANN, A., NITKA, H. *Physik. Z.* **39**, 208–12. Ferromagnetism and structure of manganese ferrite. 421, 422
38K8	KÖSTER, W., GEBHARDT, E. *Z. Metallkunde* **30**, 281–6. System Co-Mn-Al. 291, 418
38K9	KÖSTER, W., WAGNER, E. *Z. Metallkunde* **30**, 352–3. Ternary system Co-Mn-Cu. 291, 418
38K10	KÖSTER, W., GEBHARDT, E. *Z. Metallkunde* **30**, 291–3. System Ni-Mn-Al. 418
38K11	KORNETSKI, M. *Wiss. Veröffentl. Siemens-Werken* **17** (4), 48–62. Connection between Young's modulus and damping of ferromagnetic materials. 688, 708, 709
38K12	KÖSTER, W., GEBHARDT, E. *Z. Metallkunde* **30**, 286–90. Magnetic properties of Co-Mn-Al alloys. 291, 418
38K13	KONDORSKI, E. *Phys. Rev.* **53**, 319–20. Magnetic anisotropy in ferromagnetic crystals in weak fields. 559
38K14	KÖSTER, W., MULFINGER, W. *Z. Metallkunde* **30**, 348–50. Systems of Co with As, B, Cb, Ta, Zr. 281, 285, 286, 293, 295, 299
38K15	KAHAN, T. *Ann. physique* [11] **9**, 105–76. Researches on initial magnetization. 493
38L1	LEIPUNSKY, O. I. *J. Exp. Theoret. Phys.* (U.S.S.R.) **8**, 1026–30. Displacement of Curie point by pressure. 724
38L2	LINDMAN, K. F. *Z. tech. Physik* **19**, 323–4. On the magnetic permeability of Ni for Herzian Oscillations. 800
38L3	LÖHBERG, K., SCHMIDT, W. *Arch. Eisenhüttenw.* **11**, 607–14. The Fe corner of the Fe-Al-C system. 212
38L4	LONSDALE, K. *Science Progress* **32**, 677–93. Magnetic anisotropy of crystals. 594
38L5	LANGE, H. *Probleme der Technischen Magnetisierungskurve*, Springer, Berlin, 157–70. Magnetic measuring methods for metallurgical investigations. 306

38L6 LUNDQUIST, D., WESTGREN, A. *Z. anorg. allgem. Chem.* **239**, 85–8. X-ray study of Co-S system. 294
38L7 LEGAT, H. *Metallwirtschaft* **17**, 277–8. Magnetic investigation of Fe-Ni-Sn alloys. 417
38M2 MARSH, J. S. McGraw-Hill, N. Y., 1–593. Alloys of Iron and Nickel, 1. 103, 106, 111, 187, 712
38M3 MANNING, M. F., GOLDBERG, L. *Phys. Rev.* **53**, 662–7. Self-consistant field for iron. 444
38N2 NEUMANN, H. *Arch. Eisenhüttenw.* **11**, 483–96. Testing methods for permanent magnet materials. 861
38N4 NEMILOV, V. A., VORONOV, N. M. *Bull. acad. sci. U.R.S.S., Classe Sci. Math. Nat., Sér. Chim.* **21**, 905–12. Tantalum-iron alloys. 253
38N5 NOWOTNY, H., HEGLEIN, E. *Z. physik. Chem.* **40**(B), 281–4. X-ray investigation in system Ni-P. 322
38N6 NIAL, O. *Z. anorg. allgem. Chem.* **238**, 287–96. X-ray investigation of Co-Sn alloys and comparison with Fe-Sn and Ni-Sn. 296
38O1 OLIVER, D. A. *Magnetism*, Inst. of Phys., London, 69–88. Permanent magnets. 380
38O2 OLIVER, D. A., SHEDDEN, J. W. *Nature* **142**, 209. Cooling of permanent magnet alloys in a constant magnetic field. 348, 389
38P1 PAULING, L. *Phys. Rev.* **54**, 899–904. The nature of the interatomic forces in metals. 441
38P2 VAN PEYPE, W. F. *Physica* **5**, 465–82. On the theory of magnetic anisotropy of cubic crystals at absolute zero. 593
38P3 PRICE, G. H. S., SMITHELLS, C. J., WILLIAMS, S. V. *J. Inst. Metals* **62**, 239–54. Sintered Ni-Cu-W alloys. 313
38R1 RICHTER, G. *Probleme der Technischen Magnetisierungskurve*, Springer' Berlin, 93–113. On magnetic and mechanical lag. 709, 788, 789
38S1 SCHULZE, H. *Wiss. Veröffentl. Siemans-Werke* **17** (2), 39–73. Magnetic lag in weak fields. 793, 794
38S2 SCHULZE, H. *Probleme der Technischen Magnetisierungskurve*, Springer, Berlin, 114–128 (38). Investigation of magnetic lag with a.c. 793
38S4 SIXTUS, K. J. *Probleme der Technischen Magnetisierungskurve*, Springer, Berlin, 9–25. Large Barkhausen jumps. 496
38S5 SNOEK, J. L. *Probleme der Technischen Magnetisierungskurve*, Springer, Berlin, 73–92. Mechanism of the increase of coercive force in 2 ternary alloys. 385
38S6 SYKES, C., WILKINSON, H. *Proc. Phys. Soc.* (London) **50**, 834–851. Specific heat of Ni from 100 to 600° C. 735
38S7 SNOEK, J. L. *Physica* **5**, 663–88. Time effects in magnetization. 794
38S8 STONER, E. C. *Proc. Roy. Soc.* (London) **165A**, 372–414. Collective electron ferromagnetism. 442
38S9 SQUIRE, C. F., BIZETTE, H., TSAI, B. *Compt. rend.* **207**, 449–50. Transition (Λ) point of magnetic susceptibility of MnO. 470
38S10 SUCKSMITH, W., PEARCE, R. R. *Proc. Roy. Soc.* (London) **167A**, 189–204. Paramagnetism of the ferromagnetic elements. 858
38T2 TARASOV, L. P. *J. Applied Phys.* **9**, 192–6. Quantitative measurements of texture by magnetic torque method. 587
38T3 TENGER, S. *Z. anorg. allgem. Chem.* **239**, 126–32. Diselenides and ditellurides of Fe, Co and Ni. 322, 323

38V1 VOGEL, R., WALLBAUM, H. J. *Arch. Eisenhüttenw.* 12, 299–304. The system Fe-Ni, Ni-titanide. 394, 417

38V2 VOLK, K. E., DANNÖHL, W., MASING, G. *Z. Metallkunde* 30, 113–22. Segregation process in Co-Cu-Ni alloys in solid state. 402

38V3 VOGEL, R., ERGANG, R. *Arch. Eisenhüttenw.* 12, 149–53. System Fe, Fe_3W_2, Fe_2Ti. 256

38W2 WEBB, J. S. *Nature* 142, 795. Variation in longitudinal incremental permeability due to superimposed circular field. 554

38W3 WITTKE, H. *Ann. Physik* [5] 31, 97–115. Change of magnetization with time, in weak fields. 789

38Y1 YAMAMOTO, M. *Sci. Repts. Tôhoku Imp. Univ.* 27, 115–36. Change of E by magnetization in Fe and Fe-C. 687

38Z1 ZUITHOFF, A. J. *Proc. Acad. Sci. Amsterdam* 41, 264–274. Specific heat of pure Fe between 25 and 1500°C. 737

38Z2 ZENER, C. *Phys. Rev.* 53, 1010–3. General theory of macroscopic eddy currents. 703

1939

39A1 ALEXANDER, W. O. *J. Inst. Metals* 64, 93–109. Solubility of Cu in alloys containing Cr. 313

39A2 AMERICAN SOCIETY FOR METALS, Cleveland, Ohio. *Metals Handbook.* 147, 269

39A3 ALLEGHENY LUDLUM STEEL CORP. *Tech. Bull. EM-1*, Pittsburgh, 1–32. Allegheny Ludlum magnetic core materials. 131, 159

39A4 ALAVERDOV, G., SHAVLO, S. *J. Tech. Phys.* (USSR) 9, 211–4. X-ray study of transformation in Fe-Pd alloys in range of 45–75 at. %. 411

39A5 AUER, H. *Z. Electrochem.* 45, 609–15. Kinetics of precipitation on basis of magnetic measurements. 304

39B2 BRISTOW, C.A. *Iron & Steel Inst. Spec. Rep.* 24, 1–8. Constitutional diagram of alloys of Fe and Ni. I—The delta-region. 103

39B5 BECKER, R., DÖRING, W. Springer, Berlin, 1–440. Ferromagnetismus. 448, 486, 494, 559, 581, 582, 626, 628, 630, 650, 654, 669, 684, 690, 692, 699, 703, 705, 706, 762, 766, 797, 812, 813, 816, 820

39B6 BRAILSFORD, F. *J. Inst. Elec. Engrs.* 84, 399–407. Alternating hysteresis loss in electrolytic sheet steel. 518

39B7 BULLENS, D. K., Engg. and Spec. Purpose Alloys, Batelle, Columbus, Ohio, 460–9. Permanent magnet steels and their heat treatment. 378

39B8 BETTERIDGE, W. *J. Iron Steel Inst.* (London) 139, 187–208. Nickel-iron-aluminum permanent magnet alloys. 387, 388

39B9 BRADLEY, A. J., GOLDSCHMIDT, H. J. *J. Inst. Metals* 65, 389–401. X-ray study of slowly cooled Fe-Cu-Al alloys. 418

39B10 BITTER, F., KAUFMANN, A. R. *Phys. Rev.* 56, 1044–51. Magnetic studies of solid solutions. 232

39B12 BITTER, F. *Rev. Sci. Instruments* 10, 373–81. Design of powerful electromagnets. 855

39B15 BRADLEY, A. J., SEAGER, G. C. *J. Inst. Metals* 64, 81–8. X-ray investigation of Co-Al alloys. 283

39B16 BATES, L. F., TAYLOR, G. G. *Proc. Phys. Soc.* (London) 51, 33–6. Ferromagnetic compounds of Cr. 342

39B17 BABICH, M. M., KISLJAKOVA, E. N., UMANSKIJ, J. S. *J. Tech. Phys.*

	(USSR) **9**, 533–6. Intermetallic compounds in Co-W and Co-Mo systems (III). 297
39B19	BUMM, H. *Z. Metallkunde* **31**, 318–21. Formation of binary superstructures. 103
39C1	CIOFFI, P. P., BOOTHBY, O. L. *Phys. Rev.* **55**, 673. Preparation of single crystals of Fe, Co, Ni and their alloys. 556
39C2	COLE, G. H., DAVIDSON, R. L. U.S.P. 2 158 065 (Appl. 1/9/35). Art of producing magnetic materials. 86, 89
39D2	DUYCKAERTS, G. *Physica* **6**, 401–7. Specific heat of iron from 1.5 to 20°K. 739
39D3	DUYCKAERTS, G. *Physica* **6**, 817–22. Specific heat of Co, 2 to 18°K. 738
39D4	DÖRING, W. *Z. Physik* **114**, 579–601. Reversible processes in magnetic materials with small inner strains. 692
39F1	FORESTIER, H., VETTER, M. *Compt. rend.* **209**, 164–7. Study of ferrite systems. 245
39F2	FOËX, G., GRAFF, M. *Compt. rend.* **209**, 160–2. Experimental study of some cases of antiferromagnetism. 470
39F3	FETZ, E. *Trans. Am. Soc. Metals* **27**, 106–24. New 70/30 Ni-Cu alloy subject to precipitation hardening. 313
39F4	FALLOT, M. Private communication. 103
39G2	GALAVICS, F. *Helv. Phys. Act.* **12**, 581–608. Measurement of gyromagnetic effect in Mn-Sb and Fe-Se alloys. 454
39G3	GELLER, W. *Arch. Eisenhüttenw.* **13**, 263–6. System Fe-Co-Sb. 207, 416
39G4	GERLACH, W. *Z. Electrochem.* **45**, 151–66. Ferromagnetic transformations. The problem of the Curie-temperature. 728
39G5	GLATHART, L. *Phys. Rev.* **55**, 833–8. The inner, initial magnetic permeability of Fe and Ni. 778, 800
39G6	GENDERS, R., HARRISON, R. *J. Iron Steel Inst.* (London) **140**, 29–37. Niobium-iron alloys. 239
39H1	HULTGREN, R., ZAPFFE, C. A. *Trans. Am. Inst. Mining Met. Engrs.* **133**, 58–68. X-ray study of Fe-Pd and Ni-Pd systems. 411, 415
39H2	HAWORTH, F. E. *Phys. Rev.* **56**, 289. Superstructure in $FeNi_3$. 103
39H3	HOAG, J. B., GOTTLIEB, N. *Phys. Rev.* **55**, 410. Inner, initial permeability of Fe & Ni from 98 to 410 × 10^6 cps. 800
39H4	HOCART, R., GUILLAUD, C. *Compt. rend.* **209**, 443. The alloy MnBi. 337
39H5	HARALDSEN, H., NYGAARD, E. *Z. Electrochem.* **45**, 686–8. Magnetic investigation of the system Cr-As. 341
39H6	HALPERN, O., JOHNSON, M. H. *Phys. Rev.* **55**, 898–923. Magnetic scattering of neutrons. 473
39K1	KAYA, S., MIYAHARA, S. *Sci. Repts. Tôhoku Imp. Univ.* **27**, 450–8. Magnetization of a pyrrhotite crystal. 574
39K3	KAYA, S., NAKAYAMA, M. *Z. Physik* **112**, 420–9. Superstructure in Fe-Ni-Co alloys and the perminvar problem. 162, 163, 738
39K4	KÖSTER, W., BECKER, G. *Arch. Eisenhüttenw.* **13**, 93–4. The system Fe-Co-Ta. 205, 209, 416
39K5	KÖSTER, W. *Arch. Eisenhüttenw.* **13**, 227–30. The system Fe-Co-Be. 205, 208, 416
39K6	KEESOM, W. H., KURRELMEYER, B. *Physica* **6**, 633–647. Atomic heat of Fe from 1.1 to 20.4°K. 738, 739
39K7	KIMURA, R. *Proc. Math-Phys. Soc. Japan* **21**, 686–706, 786–99. Elastic moduli of ferromagnetic materials. 697, 699, 702

39K8	KELLEY, K. K. *J. Am. Chem. Soc.* **61**, 203–7. Specific heats at low temperatures of Mn, MnSe, MnTe. 472	
39K9	KANZLER, M. *Ann. Physik* [5] **36**, 38–46. The validity of Curie's law for O_2 at high densities. 467	
39L1	LEECH, P., SYKES, C. *Phil. Mag.* [7] **27**, 742–53. Evidence for superlattice in the Ni-Fe alloy Ni_3Fe. 103, 162, 184	
39L2	LEGG, V. E. *Bell System Tech. J.* **18**, 438–64. Survey of magnetic materials and applications in the telephone system. 416	
39L3	LAVES, F., WALLBAUM, H. J. *Naturwissenschaften* **27**, 674–5. Crystal chemistry of Ti alloys. 296	
39M1	McKEEHAN, L. W., GRABBE, E. M. *Phys. Rev.* **55**, 505. Ferromagnetic anisotropy in Ni-Fe crystals. 103, 111, 570	
39M2	MORIWAKI, K. *Tetsu to Hagane* **25**, 396–403. Constitution of the ternary system Fe-Cr-Cu. 418	
39M3	MÖHRING, D. *Hochfrequ. u. Elektroakust.* **53**, 196–9. Permeability of magnetic metals in region of high frequency. 800	
39M4	MESSKIN, W. E., MARGOLIN, J. M. *Metals and Alloys* **10**, 26–31. Effect of very small additions to transformer steel. 88	
39O1	OWEN, E. A., SULLY, A. H. *Phil. Mag.* [7] **27**, 614–36. Equilibrium diagram of Fe-Ni alloys. 103	
39O3	OKAMURA, T., HIRONE, T. *Phys. Rev.* **55**, 102. On the change of magnetic properties of the single crystal of Ni due to temperature. 561	
39O4	OKAMOTO, M. *Nippon Kinzoku Gakkai-Si* **3**, 365–402. Equilibrium diagram of the system Ni-Cu-Si. 313	
39O5	OSAWA, A., OKAMOTO, M. *Sci. Repts. Tôhoku Imp. Univ.* **27**, 326–47. X-ray analysis of alloys of the Ni-Si system. 322	
39P1	PAN, S. T. *Phys. Rev.* **56**, 933–6. Magnetic test for superstructure in Permalloy. 103	
39P3	POWELL, R. W. *Proc. Phys. Soc.* (London) **51**, 407–18. Further measurements of the thermal and electrical conductivity of iron at high temperatures. 55	
39P4	POLLEY, H. *Ann. Physik* [5] **36**, 625–50. Approach of magnetization to saturation in nickel. 486, 487, 567, 569, 582	
39R1	RODGERS, J. W., MADDOCKS, W. R. *Iron and Steel Inst. Rep. No. 24*, 167–77. Influence of the alloying elements on the A_3 point in iron-cobalt and other alloys. 162, 190	
39R3	RABI, I. I., MILLMAN, S., KUSCH, P. *Phys. Rev.* **55**, 526–35. Molecular beam resonance method for measuring nuclear magnetic moments. 469	
39S1	SHOCKLEY, W. *Bell System Tech. J.* **18**, 645–723. Quantum theory of solids. 441, 448	
39S2	SHIRAKAWA, Y. *Sci. Repts. Tôhoku Imp. Univ.* **27**, 485–531. Longitudinal magneto-resistance effect at various temperatures in Fe-Ni. 104, 756, 758, 760, 762, 766, 767	
39S3	SHIRAKAWA, Y. *Sci. Repts. Tôhoku Imp. Univ.* **27**, 532–60. Longitudinal magneto-resistance effect at various temperatures in Fe-Co. 192, 758	
39S4	SCHAFMEISTER, P., ERGANG, R. *Arch. Eisenhüttenw.* **12**, 459–64. Equilibrium diagram of Fe-Ni-Cr alloys. 146, 417	
39S5	SCOTT, K. L. *Metals Handbook*, Am. Soc. Metals 498–503. Permanent magnet alloys. 346, 371, 380	
39S6	SNOEK, J. L. *Physica* **6**, 321–31. Magnetic studies in ternary Fe-Ni-Al alloys. 387	

39S7 SCHAFMEISTER, P., ERGANG, R. *Arch. Eisenhüttenw.* **13**, 95–103. Phase diagram of Fe-Ni-Sn. 417
39S8 SANFORD, R. L., BENNETT, E. G. *J. Research Natl. Bur. Standards* **23**, 415–25. Apparatus for magnetic testing up to 5000 oersteds. 851
39S9 SHIRAKAWA, Y. *Sci. Repts. Tôhoku Imp. Univ.* **27**, 255–77. On the longitudinal magneto-resistance effect at various temperatures in Fe-Si alloys. 758
39S10 STONER, E. C. *Proc. Roy. Soc.* (London) **169A**, 339–71. Collective electron ferromagnetism. II. Energy and specific heat. 737, 739
39S11 SNOEK, J. L. *Physica* **6**, 161–70. Magnetic after-effect and chemical constitution. 788, 794
39S12 SNOEK, J. L. *Physica* **6**, 797–805. Magnetic after-effects at higher inductions. 794
39S13 SUCKSMITH, W. *Proc. Roy. Soc.* (London) **171A**, 525–40. Magnetic study of Fe-Ni-Al system. 216
39S14 SYKES, W. P. *Metals Handbook*, Am. Soc. Metals, Cleveland, 401–2 (1939). Constitution of Fe-W alloys. 256
39S15 SQUIRE, C. F. *Phys. Rev.* **56**, 922–5. Antiferromagnetism of some Mn compounds. 470
39S16 SUCKSMITH, W. *Proc. Roy. Soc.* (London) **170A**, 551–60. Measurement of magnetic saturation intensities at different temperatures. 859
39S17 SCHAARWÄCHTER, C., RUPPELT, A. USP 2 167 188 (Appl. 2/26/37). Sound recording and reproducing element. 401
39T1 THOMPSON, J. G., CLEAVES, H. E. *J. Research, Natl. Bur. Standards* **23**, 163–74. Preparation of High Purity Iron. 52
39T2 TARASOV, L. P. *Phys. Rev.* **56**, 1231–40. Ferromagnetic anisotropy of iron and iron-rich silicon alloys. 506, 567, 571
39T3 TAKAGI, M. *Sci. Repts. Tôhoku Imp. Univ.* **28**, 20–127. Statistical domain theory of ferromagnetic crystals. 654, 693
39T4 TARASOV, L. P. *Phys. Rev.* **56**, 1224–30. Dependence of ferromagnetic anisotropy on field-strength. 556, 566
39T5 TARASOV, L. P. *Phys. Rev.* **56**, 1245–6. Ferromagnetic anisotropy of low Ni alloys of Fe. 571
39T6 TARASOV, L. P. *Trans. Am. Inst. Mining Met. Engrs.* **135**, 353–71. Magnetic torque studies of iron-silicon alloys. 587
39V1 VERWEY, E. J. W. *Nature* **144**, 327–8. Electronic conductivity and transition point of magnetite. 243
39W1 WELLS, C. *Metals Handbook*, Am. Soc. Metals, Cleveland, 409–415. Constitution of iron-carbon-manganese alloys. 371
39W2 WILLIAMS, H. J., BOZORTH, R. M. *Phys. Rev.* **55**, 673. Magnetic anisotropy of Fe-Ni and Cu-Ni alloys. 556, 567, 568, 570, 573, 593
39W3 WEISS, P. R., VAN VLECK, J. H. *Phys. Rev.* **55**, 673–4. A theory of ferromagnetism. 445
39W4 WILLIAMS, H. J., BOZORTH, R. M. *Phys. Rev.* **56**, 837. Magnetic anisotropy of Ni at 20°K. 568
39W5 WEIBKE, F. *Z. Metallkunde* **31**, 228–30. Alloys of Ga and In. 315
39Y1 YENSEN, T. D. *Trans. Am. Soc. Metals* **27**, 797–820. Magnetically soft materials. 110, 123, 130

1940

40A1 AKULOV, N. S., KIRENSKY, L. W. *J. Phys.* (USSR) **3**, 31–4. On a new magnetocaloric effect. 743

40A2 AWBERRY, J. H., GRIFFITHS, E. *Proc. Roy. Soc.* (London) **174A**, 1–15. Thermal capacity of pure Fe. 737

40B1 BATELLE MEMORIAL INSTITUTE, Cleveland, Ohio. Project Comm. #1 on Metallurgical Applications of Electricity, Utilities Coordinated Research, Inc. 153

40B2 BECKER, R., POLLEY, H. *Ann. Physik* [5] **37**, 534–40. Influence of internal strain on the law of approach to saturation for nickel. 488

40B3 BOZORTH, R. M. *Bell System Tech. J.* **19**, 1–39. The physical basis of ferromagnetism. 444

40B4 BOZORTH, R. M., WILLIAMS, H. J., MORRIS, R. J. *Phys. Rev.* **58**, 203. Magnetic properties of Fe-Al alloys. 217

40B5 BROOKS, H. *Phys. Rev.* **58**, 909–18. Ferromagnetic anisotropy & the itinerant electron model. 453–592

40B6 BARNETT, S. J. *Proc. Am. Acad. Arts Sci.* **73**, 401–55. Gyromagnetic ratios for ferromagnetic substances. 454

40B7 BATES, L. F., BAKER, C. J. W. *Proc. Phys. Soc.* (London) **52**, 436–42. Magnetic properties of Ni amalgams. 320

40C1 CAMP, J. M., FRANCIS, C. B. Carneg.-Ill. Steel Corp., Pittsburgh, Penna., 1–1440. Making, Shaping and Treating of Steel (5th ed.). 51

40C2 CARNEGIE INST. WASH., *Year Book*, 1939. Annual Report of Director of Department of Terrestrial Magnetization. 726, 727

40C3 CONRADT, H. W., DAHL, O., SIXTUS, K. J. *Z. Metallkunde* **32**, 231–8. Magnetic anisotropy in rolled Fe-Ni alloys. 586, 588, 590

40E1 ELLINGER, F. H., SYKES, W. P. *Trans. Am. Soc. Metals* **28**, 619–45. The Ni-W system. 325

40F1 FORSYTH, A. C., DOWDELL, R. L. *Trans. Am. Inst. Mining Met. Engrs.* **137**, 373–87. Co-Ni-Si system between 0 and 20% Si. 281, 418

40G1 GRABBE, E. M. *Phys. Rev.* **57**, 728–34. Ferromagnetic anisotropy, magnetization at saturation, and superstructure in Ni_3Fe and related compositions. 103, 570, 752

40G3 GEBHARDT, E., KÖSTER, W. *Z. Metallkunde* **32**, 253–61. The system Pt-Co, with particular reference to phase CoPt. 413, 414, 415

40G4 GORDEN, R. B., COHEN, M. Symposium on Age-Hardening, Am. Soc. Metals, Cleveland, 161–84. Age hardening of a Co-Cu and a Cu-Fe alloy. 231

40G5 GORTER, C. J., KAHN, B. *Physica* **7**, 753–64. On the theory of the gyromagnetic effects. 454

40H1 HOUGARDY, H. *Metals Progress* **37**, 64–65, 68. The brittle constituent in Cr-Ni-Fe alloys. 147

40H2 HORNFECK, A. J., EDGAR, R. F. *Trans. Am. Inst. Elec. Engrs.* **59**, 1017–24. Output and optimum design of permanent magnets subjected to demagnetizing forces. 363, 364

40H3 HOWE, G. H. U.S.P. 2 192 741 (Appl. 9/17/37). Method of making a sintered alloy. 417

40H4 HOWE, G. H. *Iron Age* **145** (2), 27–31. Sintering of Alnico. 417

40H5 HENDUS, H., SCHEUFELE, E. *Z. Metallkunde* **32**, 275–7. Velocity of transition from partial to complete ordering in Fe_2NiAl. 387

40H6 HIBI, T. *Sci. Repts. Tôhoku Imp. Univ.* **28**, 435–49. Change of torsional modulus of some ferromagnetic substances with temperature and magnetization. 699

40H7 HOLSTEIN, T., PRIMAKOFF, H. *Phys. Rev.* **58**, 1098–1113. Field dependence of the intrinsic domain magnetization of a ferromagnet. 486

40J1 JOHNSON, H. L., MARSHALL, A. L. *J. Am. Chem. Soc.* **62**, 1382–90. Vapor pressures of Ni and NiO. 322

40K2 KINZEL, A. B., FRANKS, R. McGraw-Hill, New York, v. 2. The alloys of iron and chromium. 149

40K3 KÖSTER, W., MULFINGER, W. *Z. Electrochem.* **46**, 135–41. Systems Cu-Ni-S and Cu-Ni-As. 300, 302, 313, 323, 418

40K4 KIMURA, R. *Proc. Phys.-Math. Soc., Japan* **22**, 45–60, 219–32, 233–50. Elastic moduli of ferromagnetic materials. 697

40K5 KAYA, S., NAKAYAMA, M. *Proc. Phys.-Math. Soc., Japan* **22**, 126–41. Ordering in Ni-Mn alloys. 313, 319

40K6 KIKOIN, I. K., GOOBAR, S. W. *J. Phys.* (USSR) **3**, 333–54. Gyromagnetic effect in superconductors. 454

40K7 KRAMERS, H. A. Le Magnetisme (Strasbourg Conference, 1939) **3**, 45–64. Interaction between magnetic ions in a crystal. 475

40L1 LEGG, V. E., GIVEN, F. J. *Bell System Tech. J.* **19**, 385–406. Compressed powdered molybdenum permalloy for high quality inductance coils. 125, 144, 249, 796

40M1 MUSSMANN, H., SCHLECHTWEG, H. *Ann. Physik* [5] **38**, 215–31. Rotation moment of cubic recrystallization textures in magnetic field. 587

40N1 NESBITT, E. A., KELSALL, G. A. *Phys. Rev.* **58**, 203. (See also U.S.P. 2 190 667.) Vicalloy. 346, 348, 405

40O1 OSAWA, A., MURATA, T. *Nippon Kinzoku Gakkai-Si* **4**, 228–42. Equilibrium diagram of Fe-Si system. 72

40O2 OKAMOTO, M. *Japan Nickel Rev.* **8**, 125–31. Ni-Be constitution. 303

40O3 OSAWA, A., SHIBATA, N. *Nippon Kinzoku Gakkai-Si* **4**, 362–8. X-ray investigation of Ni-Sb alloys. 300

40P1 PICKLES, A. T., SUCKSMITH, W. *Proc. Roy. Soc.* (London) **175**, 331–44. Magnetic study of 2-phase Fe-Ni alloys. 103

40P2 PHILIPS GLOEILAMPENFABRIEKEN. Br. Pat. 522 731 (Appl. 12/7/38). Permanent magnet material. 348, 389, 390, 395, 396

40P3 PILLING, N. B., TALBOT, A. M. Age-hardening of Metals. Am. Soc. Metals 231–61. Dispersion-hardening alloys of Ni and Fe-Ni-Ti. 394

40P4 PARKER, E. R. *Trans. Am. Soc. Metals* **28**, 797–807. Development of alloys for use at temperatures above 1000°F. 239

40P5 PARKER, E. R. *Trans. Am. Soc. Metals* **28**, 661–8. The influence of magnetic fields on damping capacity. 712

40R1 RUSSELL, H. W., JACKSON, L. R. U.S.P. 2 207 685 (Appl. 7/17/39). Magnetic material and method of producing same. 388

40R2 RUTTEWIT, K., MASING, G. *Z. Metallkunde* **32**, 52–6. The alloys of Ge with Bi, Sb, Fe and Ni. 233, 313

40S1 SHIRAKAWA, Y. *Sci. Repts. Tôhoku Imp. Univ.* **29**, 152–161. On the longitudinal magneto-resistance effect at low temperatures of single crystals of Fe. 764, 765

40S2 SHIRAKAWA, Y. *Sci. Repts. Tôhoku Imp. Univ.* **29**, 132–151. On the longitudinal magneto-resistance effect of single crystal of Fe. 764, 765

40S3 SLATER, J. C. *Phys. Rev.* **58**, 54–56. Note on effect of pressure on Curie point of iron-nickel alloys. 724, 726

40S4 STRUTT, M. J. O., KNOL, K. S. *Physica* **7**, 145–54, 635–54. Resistance measurements of Fe wires in frequency region 10^7 to 3×10^8 herz. 799

40S5 SAMPSON, J. B., SEITZ, F. *Phys. Rev.* **58**, 633–9. Theoretical magnetic susceptibility of metallic Li and Na. 467

40S6 SACHS, G., VAN HORN, K. R. Am. Soc. Metals, Cleveland, 1–567. Practical Metallurgy. 23.

BIBLIOGRAPHY 925

40S8 STARR, C., BITTER, F., KAUFMANN, A. R. *Phys. Rev.* **58**, 977–83. Magnetic properties of Fe group chlorides at low temperatures, I. 470
40T1 THOMPSON, N. *Proc. Phys. Soc.* (London) **52**, 217–28. Order-disorder transformation in alloy Ni_3Mn. 181, 319
40T2 TAKEI, T., YASUDA, T., ISHIHARA, S. *Electrotech. J.* (Japan) **4**, 75–8. High-temperature magnetization of the ferrites. 421
40T3 TAKEDA, S., MUTUZAKI, K. *J. Iron Steel Inst. Japan* **26**, 335–61. Equilibrium diagram of Fe-Al-Si system. 418
40T4 THIESSEN, G. *Ann. Physik* [5] **38**, 153–76. Elementary processes of magnetization in the region of initial permeability. 626
40T5 THIELMANN, K. *Ann. Physik* [5] **37**, 41–62. Ferromagnetism and its carrier in system Mn-Bi. 337
40V1 VAN URK, A. T. *Philips Tech. Rev.* **5**, 29–35. Use of modern steels for permanent magnets. 346, 348, 364
40V2 VONSOVSKY, S. V. *J. Phys.* (USSR) **2**, 11–18. On theory of technical magnetization curve in ferromagnetic single crystals. 562
40V3 VONSOVSKY, S. V. *J. Phys.* (USSR) **3**, 83–93. Temperature dependence of magnetic anisotropy of Co single crystals. 594
40V4 VONSOVSKY, S. V. *J. Phys.* (USSR) **3**, 181–90. On quantum theory of magnetostriction of ferromagnetic single crystals. 654
40W1 WELLS, C., MEHL, R. F. *Trans. Am. Inst. Mining Met. Engrs.* **140**, 279–306. Rate of diffusion of C in plain carbon, in Ni and in Mn steels. 64, 65
40Y1 YAMANAKA, N. *Sci. Repts. Tôhoku Imp. Univ.* **29**, 36–68. On the elasto-resistance change of Ni, Fe and some Ni-Fe alloys. 756, 766, 767

1941

41A1 AVERBACH, B. L. *Metals and Alloys* **13**, 730–3. Electrolytic Mn alloyed with Cu and Ni. 313
41A3 AUSTIN, C. R., SAMANS, C. H. *Trans. Am. Inst. Mining Met. Engrs.* **143**, 216–27. Study of metallographic and certain physical properties of some alloys of Co, Fe and Ti. 394
41B1 BARRER, R. M. Univ. Press, Cambridge, 1–464. Diffusion in and through solids. 64
41B2 BROWN, W. F. *Phys. Rev.* **60**, 139–47. Effect of dislocations on magnetization near saturation. 621
41B3 BATES, L. F., WESTON, J. C. *Proc. Phys. Soc.* (London) **53**, 5–34 (1941). Energy changes accompanying magnetization. 522
41B4 BRADLEY, A. J., COX, W. F., GOLDSCHMIDT, H. J. *J. Inst. Metals* **67**, 189–201. X-ray study of Fe-Cu-Ni equilibrium diagram at various temperatures. 153, 397
41B5 BOZORTH, R. M., WILLIAMS, H. J. *Phys. Rev.* **59**, 827–33. Calculations of torque on a ferromagnetic single crystal in a magnetic field. 566, 579, 584
41B6 BIRGE, R. T. *Rev. Mod. Phys* **13**, 233–9. New table of values of general physical constants. 867
41B7 BATES, L. F. *Proc. Phys. Soc.* (London) **53**, 113–5. Note on concentrated ferromagnetic amalgams. 320
41B8 BOHLEN-HALBACH, C. V., LEITGEBEL, W. *Tech. Mitt. Krupp* **4**, 37–44. Action of oxygen on Fe-Ni and Cu-Ni melts. 313
41C1 CARPENTER, V. W. U.S.P. 2 236 519 (Appl. 1/22/36). Method of producing Si steel sheet or strip. 86, 89

41D1 DEAN, R. S., DAVIS, C. W. U.S.P. 2 239 144 (Appl. 7/11/38). Permanent magnet. 236, 348, 419, 829
41E1 ELLIS, W. C., GREINER, E. S. *Trans. Am. Soc. Metals* 29, 415–32. Equilibrium relations in the solid state of the iron-cobalt system. 162, 190
41E2 EDWARDS, O. S. *J. Inst. Metals* 67, 67–77. X-ray investigation of Al-Co-Fe systems. 190, 207, 389, 416
41E3 ESSER, H., EUSTERBROCK, H. *Arch. Eisenhüttenw.* 14, 341–55. Investigation of thermal expansion of some metals and alloys with an improved dilatometer. 53
41F1 FÖRSTER, F., WETZEL, H. *Z. Metallkunde* 33, 115–23. On question of magnetic reversals in Fe and Ni. 528, 531
41F2 FÖRSTER, F., STAMBKE, K. *Z. Metallkunde* 33, 97–114. Magnetic investigations of inner strain. 625
41G1 GORTER, C. J. *Phys. Rev.* 60, 836. On the gyromagnetic effects in ferromagnetic substances. 455
41G2 GOTTSCHALK, V. H. *U.S. Bur. Mines, Bull. No.* 425, 88–95. Coercive force of magnetite powders. 243
41H1 HULL, A. W., BURGESS, E. E., NAVIAS, L. *J. Applied Physics* 12, 698–707. Glass-to-metal seals. 160
41H2 HOLSTEIN, T., PRIMAKOFF, H. *Phys. Rev.* 59, 388–94. Magnetization near saturation in polycrystalline ferromagnets. 486
41H3 HOWE, G. H. U.S.P. 2 264 038 (Appl. 6/14/40). Permanent Magnets containing Ti. 395, 396
41H4 HULTGREN, R., JAFFEE, R. I. *J. Applied Phys.* 12, 501–2. Preliminary X-ray study of binary alloys of Pt with Co, Mo and W. 413
41H6 HARALDSEN, H. *Z. anorg. allgem Chem.* 246, 169–194, 195–226. FeS mixed crystals, and their high temperature transformation. 253
41I1 IWASÉ, K., OKAMOTO, M. *Nippon Kinzoku Gakkai-Si* 5, 82–91. Equilibrium diagrams of Cu-Be and Ni-Cu-Be systems. 313
41J1 JONAS, B., EMDEN, H. J. M. v. *Philips Tech. Rev.* 6, 8–11. New kinds of steel of high magnetic power. 346, 348, 389, 393
41K1 KÖSTER, J., DAVIS, H. W. Report for Nat. Acad. Sci., pp. 1–33. Uses and possible substitutes for cobalt. 261
41K2 KIUTI, S. *Japan Nickel Rev.* 9, 78–104. X-ray study of a new α' phase in magnetic Fe-Ni-Al system. 385
41K3 KRAMERS, H. A., WANNIER, G. H. *Phys. Rev.* 60, 252–76. Statistics of 2-dimensional ferromagnet. 446
41L1 LIVSHITS, B. G. *Stal* 1, 40–50. Effect of composition of Ni-Al steels on their magnetic properties. 388
41L2 LIPSON, H., SHOENBERG, D., STUPART, G. V. *J. Inst. Metals* 67, 333–40. Relation between atomic arrangement and H_c in an alloy of Fe and Pt. 410
41M1 MESSKIN, V. S., SOMIN, B. E., NEKHAMKIN, A. S. *J. Tech. Phys.* (USSR) 11, 918–35. Magnetostriction of alloys. 679, 681
41M2 MASUMOTO, H., SHIRAKAWA, Y. *Phys. Rev.* 60, 835. Change in electrical resistance of a single crystal of Fe_3O_4 by a magnetic field at low temperatures. 758
41M3 MICHELS, A., STRIJLAND, J. *Physica* 8, 53–7 (1941). Effect of pressure on the Curie point of a Monel alloy. 726
41N1 NIX, F. C., MACNAIR, D. *Phys. Rev.* 60, 597–605. Thermal expansion of pure metals: Cu, Au, Al, Ni and Fe. 269

41O1 OWEN, E. A., SULLY, A. H. *Phil. Mag.* [7] **31**, 314–38. Migration of atoms in Fe-Ni alloys. 103
41R1 ROTERS, H. C. Wiley, N. Y., 1–561. Electromagnetic devices. 349, 364, 843
41R2 RATHENAU, G. W., SNOEK, J. L. *Physica* **8**, 555–75. Magnetic anisotropy phenomena in cold-rolled Ni-Fe. 588
41R3 RADO, G. T., KAUFMANN, A. R. *Phys Rev.* **60**, 336–9. Saturation magnetization of Ni-Sb and Ni-Ta alloys. 300, 323
41S2 SCHRAMM, J. *Z. Metallkunde* **33**, 403–12. Ternary system Co-Ni-Al. 281, 418
41S3 SHIBATA, N. *Nippon Kinzoku Gakkai-Si* **5** (2), 46–9. Magnetic investigation of ternary alloys of Cu-Sb-Ni system. 418
41S4 SMOLUCHOWSKI, R. *Phys. Rev.* **59**, 309–317. On the theory of volume magnetostriction. 434, 726
41S5 SNOEK, J. L. *Physica* **8**, 711–33. Effect of small quantities of C and N on elastic and plastic properties of Fe. 712, 795, 796
41S6 SELISSKY, I. P. *J. Phys.* (USSR) **4**, 567–8. High permeability and superstructure in Fe-Si-Al alloys of sendust type. 97
41S7 SMITH, CHARLES S. *J. Applied Phys.* **12**, 817–22. Precipitation hardening in the Fe-W system. 256
41S8 SNOEK, J. L. *Physica* **8**, 745–7. A mechanical counterpart to the Rayleigh law of ferromagnetic hysteresis. 708, 709, 712
41S9 SHIMIZU, Y. *Proc. Imp. Acad.* (Tokyo) **17**, 100–1. Magnetization of Fe single crystals, polycrystals of Fe, Ni and their alloys in very weak fields. 559
41S10 SHUR, J. S. *J. Phys.* (USSR) **4**, 439–47. Effect of thermal treatment in magnetic field upon character of the anisotropy of H_c in ferromagnetic single crystals. 562
41S11 SCHRAMM, J. *Z. Metallkunde* **33**, 381–7. Part-system Co-CoAl. 283
41S12 SCHRAMM, J. *Z. Metallkunde* **33**, 46–8. The Co-Zn system. 298
41S13 SCHRAMM, J. *Z. Metallkunde* **33**, 347–55. The partial system Ni-NiAl. 299
41S14 SHIBATA, N. *Tr. Inst. Met.* (Japan) **5**, 46–9. Magnetic investigation of ternary alloys of Cu-Sb-Ni system. 313
41S15 SHIBATA, N. *Sci. Repts. Tôhoku Imp. Univ.* **29**, 697–727. Equilibrium diagram of the Ni-Sb system. 300
41T1 TREW, V. C. G. *Trans. Faraday Soc.* **37**, 476–92. Ionic diamagnetic susceptibility & diamagnetic correcting constants. 459
41V1 VAN VLECK, J. H. *J. Chem. Phys.* **9**, 85–90. Theory of antiferromagnetism. 472
41V2 VOLKENSTEIN, N., KOMAR, A. *J. exp. theoret. Phys.* (USSR) **11**, 723–4. Coercive force and saturation of Ni_3Mn in relation to atomic order. 317, 319
41V3 VERWEY, E. J. W., HAAYMAN, P. W. *Physica* **8**, 979–87. Electronic conductivity and transition point of magnetite. 243
41W1 WILLIAMS, H. J., BOZORTH, R. M., CHRISTENSEN, H. *Phys. Rev.* **59**, 1005–12. Magnetostriction Young's modulus and damping of 68 permalloy as dependent on magnetization and heat treatment. 624, 638, 686, 687, 699, 703, 707
41W2 WALLBAUM, H. J. *Arch. Eisenhüttenw.* **14**, 521–6. Systems of Fe metals with Ti, Zr, Nb, and Ta. 253, 255, 260, 325, 394

41W3 WILLIAMS, H. J., BOZORTH, R. M. *Phys. Rev.* **59**, 939. Magnetic domain size from measurements of damping in 68 permalloy. 709

41Y1 YAMAMOTO, M. *Phys. Rev.* **59**, 768. Young's modulus of elasticity and its change with magnetism in Fe-Co alloys. 687, 696

41Z1 ZAIMOVSKY, A. S. *J. Phys.* (USSR) **4**, 569–572. On the temperature dependence of magnetic permeability in weak fields. 100, 716

41Z2 ZAIMOVSKY, A. S., SELISSKY, I. P. *J. Phys.* (USSR) **4**, 563–5. On the cause of high permeability of alloys Fe-Si-Al. 101, 574

1942

42A1 AUWERS, O. v. *Elektrotech. Z.* **63**, 341–8. Magnetically soft and hard materials for testing and communication engineering.

42A2 AMERICAN INSTITUTE OF ELECTRICAL ENGINEERS (AIEE) N.Y., 1–311. American standard definitions of electrical terms. 1

42A3 AUSTIN, J. B., DAY, M. J. Controlled Atmospheres, Am. Soc. Metals, 20–49. Chemical equilibrium and control of furnace atmospheres. 66

42B1 BOZORTH, R. M., CHAPIN, D. M. *J. Applied Phys.* **13**, 320–6. Demagnetizing factors of rods. 352, 847

42B2 BOZORTH, R. M., WILLIAMS, H. J. U.S.P. 2 300 336 (Appl. 8/7/40). Magnetic alloy of iron and aluminum. 217

42B3 BRAILSFORD, F., MARTINDALE, R. G. *J. Inst. Elec. Engrs.* **89**, 225–31. Magnetostriction in electrical sheet steel. 679

42C1 CLEAVES, H. E., HIEGEL, J. M. *J. Research Natl. Bur. Standards* **28**, 643–67. Properties of high purity iron. 53

42C2 CARPENTER, V. W., JACKSON, J. M. U.S.P. 2 287 467 (Appl. 1/3/40). Process of producing silicon steel. 70, 86

42C3 CARPENTER, V. W. U.S.P. 2 287 466 (Appl. 12/5/39). Process of producing high permeability silicon steel. 89

42C4 CONRADT, H. W., SIXTUS, K. *Z. tech. Physik* **23**, 39–49. Magnetic anisotropy in rolled Fe-Ni alloys. 590

42C5 CARAPELLA, L. A., HULTGREN, R. *Trans. Am. Inst. Mining Met. Engrs.* **147**, 232–42. Ferromagnetic nature of β phase in Cu-Mn-Sn system. 328, 330

42D1 DEHLER, H. *Stahl u. Eisen* **62**, 983–6. Pressed magnets with artificial binding materials. 418

42E1 EDWARDS, O. S., LIPSON, H. *Proc. Roy. Soc.* (London) **180A**, 268–77. Imperfections in the structure of cobalt. 264

42E2 ELMORE, W. C. *Phys. Rev.* **62**, 486–93. Magnetic structure of iron crystals. 533

42E3 ELLINGER, F. H. *Trans. Am. Soc. Metals* **30**, 607–38. The Ni-Mo system. 320

42G1 GERMER, L. H. *Phys. Rev.* **62**, 295. Stray magnetic fields from cobalt. 538

42H1 HOWE, G. H. *Powder Metallurgy*, Am. Soc. Metals, 530–6. Sintered Alnico. 417

42H2 HAUGHTON, J. L. *Institute of Metals Monograph and Report Series*, London, No. 2, 1–163. Bibliography of literature relating to constitutional diagrams of alloys. 210

42H3 HIRONE, T., HORI, N. *Sci. Repts. Tôhoku Imp. Univ.* **30**, 125–36. Changes of resistivity with magnetization in single crystals of iron and nickel. 765

42M1 MASON, W. P. Van Nostrand, N.Y., 1–333. Electromechanical transducers and wave filters. 686

42N1 NESBITT, E. A. U.S.P. 2 298 225 (Appl. 7/30/39). Permanent magnet, material and production thereof. 405
42N2 NÉEL, L. *Cahiers phys.* **12**, 1–20. Theory of Rayleigh's law of magnetization. 494
42P1 PAN, S. T., KAUFMANN, A. R., BITTER, F. *J. Chem. Phys.* **10**, 318–21. Ferromagnetic Au-Fe alloys. 412
42R1 RUDER, W. E. *Proc. Inst. Radio Engrs.* 437–40. New magnetic materials. 79
42S1 SMITH, B. M. *Gen. Elec. Rev.* **45**, 210–3. Alnico, properties and equipment for magnetization and test. 346, 387, 419
42T1 THOMPSON, M. DEK., Electroch. Soc., N. Y., 1–89. Total and free energies of formation of oxides of 32 metals. 293
42W2 WISE, E. M., SHAEFER, R. H. *Metals and Alloys* **16**, 424-8, 891–3, 1067–71. Effect of alloying elements on nickel. 40
42W3 WILSON, A. J. C. *Proc. Roy. Soc.* (London) **180**, 177–85. Imperfections in the structure of cobalt. Mathematical treatment of proposed structure. 264
42W4 WEYGANDT, C. N., CHARP, S. *Elec. Engg.* **61**, 387–8. An analytic approach to the hysteresis loop. 511
42W5 WEVER, F., PETER, W. *Arch. Eisenhüttenw.* **15**, 357–61. Precipitation hardening and creep limit of Fe-Nb alloys and Nb-alloyed steels. 239
42W6 WALTERS, F. M., KRAMER, I. R., LORING, B. M. *Trans. Am. Soc. Mining Met. Engrs.* **150**, 401–3. Mechanical properties of Fe-Mn alloys. 234, 235
42Y1 YAMAMOTO, M. *J. Inst. Metals* (Japan) **6**, 249–57. Young's modulus and its variation with magnetization in Ni-Cu alloys. 699

1943

43A1 ANONYMOUS. *Metal Progress* **43**, 560. Properties of important wrought Cr-Fe alloys. 228
43A2 ANDERSEN, A. G. H., KINGSBURY, A. W. *Trans. Am. Soc. Mining Met. Engrs.* **152**, 38–47. Phase diagram of the Cu-Fe-Si system from 90 to 100% Cu. 230
43A3 ARMCO STEEL CORP. *Bulletin* 1–12. Magnetic ingot iron. 59
43A4 ARMBRUSTER, M. H. *J. Am. Chem. Soc.* **65**, 1043–54. Solubility of H in Fe, Ni and steel at 400 to 600°C. 314
43B1 BATES, L. F., HEALEY, D. R. *Proc. Phys. Soc.* (London) **55**, 188–202. Adiabatic temperature change accompanying magnetization in Fe in low and intermediate fields. 522
43B2 BRAILSFORD, F. *J. Inst. Elec. Engrs.* **90** (II), 307–22. Survey of electrical sheet steels for power plants and the factors affecting their magnetic properties. 79
43B3 BOWMAN, F. E., PARKE, R. M., HERZIG, A. J. *Trans. Am. Soc. Metals* **31**, 487–500. Alpha-Fe lattice parameter as affected by Mo. 237
43B5 BIZETTE, H., TSAI, B. *Compt. rend.* **217**, 390-2. Antiferromagnetism of ferrous iron. Susceptibility at low temperatures of FeO. 470
43B6 BARRETT, C. S. McGraw-Hill, N. Y., 1–567. Structure of metals. 42
43C1 CLARK, J. R., PAN, S. T., KAUFMANN, A. R. *Phys. Rev.* **63**, 139. Magnetic properties of Ce-Fe alloys. 225
43C2 COOK, A. J., JONES, F. W. *J. Iron Steel Inst.* (London) **148**, 217–23. Brittle constituent of the Fe-Cr system—I. 227
43C3 CONSTANT, F. W., FAIRES, R. E., LENANDER, H. *Phys. Rev.* **63**, 441–4. Investigation of ferromagnetic impurities. 232

930 BIBLIOGRAPHY

43D1   DANIEL, V., LIPSON, H. *Proc. Roy. Soc.* (London) **181**, 368–78. X-ray study of the dissociation of an alloy of Fe, Ni and Cu. 154, 398

43E1   ELECTRICAL ENGINEERING STAFF, Mass. Inst. of Tech., Wiley, N. Y., 1–718. Magnetic circuits and transformers. 349, 362, 364, 783, 843

43E2   EHRET, W. F., GURINSKY, D. C. *J. Am. Chem. Soc.* **65**, 1226–30. Thermal diagram of system Fe-Sn. 255

43E3   EDWARDS, O. S., LIPSON, H. *J. Inst. Metals* **69**, 177–88. X-ray study of transformation of Co. 264

43G1   GREINER, E. S., JETTE, E. *Trans. Am. Soc. Mining Met. Engrs.* **152**, 48–64. Constitution of the Fe-rich Fe-Ni-Si alloys at 600°C. 184, 417

43G2   GUILLAUD, C. Thesis, Strasbourg, 1–129. Ferromagnetics of binary alloys of Mn. 45, 334, 335, 336, 337, 340, 341, 348, 420, 564, 575, 812, 828

43G3   GERLACH, W., RENNENKAMPFF, J. v. *Z. Electrochem.* **49**, 200–3. Magnetic investigation of some Ni-C alloys. 306

43H1   HOSELITZ, K., SUCKSMITH, W. *Proc. Roy. Soc.* (London) **171A**, 303–13. Magnetic study of two-phase iron and nickel alloys. 103

43H2   HERROUN, E. F. *Proc. Phys. Soc.* (London) **55**, 338–43. Magnetic Properties of solid and powdered magnetites. 243

43H3   HEUMANN, T. *Z. Metallkunde* **35**, 206–11. Equilibria in system Ni-Sn. 323

43H4   HARALDSEN, H., ROSENQUIST, T. *Tidsskr. Kjemi, Bergvesen Met.* **3**, 81–2. Magnetic relations and binding relations in Cr-Sb. 341

43H5   HOWE, G. H. Private Communication. 395

43J1   JELLINGHAUS, W. *Arch. Eisenhüttenw.* **16**, 247–51. Fe-Ni-Al-Co-Cu magnet alloys with preferred orientation. 390

43J2   JUZA, R., SACHSZE, W. *Z. anorg. allgem. Chem.* **251**, 201–12. On the Ni/N system. 321

43K1   KRAINER, H., RAIDL, F. *Arch. Eisenhüttenw.* **16**, 253–60. German magnet steels. 378

43K2   KORNETZKI, M. *Ann. Physik* [5] **43**, 203–219. On the remanence of magnetostriction of polycrystal Fe and Ni. 748.

43K3   KORNETZKI, M. *Physik. Z.* **44**, 296–302. Displacement of Curie temperature by hydrostatic pressure. 724, 726

43K4   KELLEY, K. K. *J. Chem. Phys.* **11**, 16–8. Specific heat of pure Fe at low temperatures. 738

43K5   KERSTEN, M. S. Hirzel, Leipzig, 1–88. Grundlagen einer Theorie der Ferromagnetischen Hysterese und Koerzitivkraft. 825

43K6   KERSTEN, M. *Physik. Z.* **44**, 63–77. Theory of ferromagnetic hysteresis and initial permeability. 825

43K7   KÖSTER, W. *Z. Metallkunde* **35**, 57–67. Contribution to knowledge of magnitude of σ_i on basis of measurements of ΔE-effect in Ni. 693, 695, 697, 699

43K8   KÖSTER, W. *Z. Metallkunde* **35**, 68–72. Significance of σ_i in relation to μ_0 and $dI_R/d\sigma$. 695

43K9   KÖSTER, W. *Z. Metallkunde* **35**, 194–9. Elastic modulus and ΔE effect in Fe-Ni alloys. 693, 695, 698

43K10   KÖSTER, W. *Z. Metallkunde* **35**, 246–9. Damping of Ni and Ni-Fe Alloys. 702, 703, 707, 711

43K11   KORNETZKI, M. *Z. Physik* **121**, 560–73. Damping of mechanical oscillations by magnetic hysteresis. 712

43K12   KLEMM, W., FRATONI, N. *Z. anorg. allgem. Chem.* **251**, 222–32. System NiTe—NiTe$_2$. 323

43K13 KAYA, S., NAKAYAMA, M., SAITO, H., *Proc. Phys.-Math. Soc. Japan* **25**, 179–97. Superstructure in Fe-Ni-Mn alloys. 181
43N1 NESBITT, E. A., U.S.P. 2 317 294 (Appl. 4/20/43). Magnetic materials. 416
43N2 NESBITT, E. A. U.S.P. 2,317,295 (Appl. 4/20/43). Magnetic materials. 419
43N3 NOWOTNY, H. *Z. Electrochemie* **49**, 254–60. Structure of metalloid alloys and ferromagnetism. 341
43P1 PAWLEK, F. *Arch. Eisenhüttenw.* **16**, 363–6. Fe-Si alloys of high μ_0 by special metallurgical treatment. 92
43P2 POGODIN, S. A., ZELIKMAN, A. N. *Ann. Secteur Anal. Phys.-Chim. Inst. Chim. Gen.* (USSR) **16**, 158–66. Phases of variable composition in the Ni-Niobium system. 321
43S1 STANLEY, J. K. *Iron Age* **152**, 42–4. Mechanical properties of Fe-P alloys. 251
43S2 SELWOOD, P. W. Interscience, N. Y., 1–287. Magnetochemistry. 293, 457, 463
43S3 STEINHAUS, W. F. Kohlrausch, *Praktische Physik* **2**, 72–123. Magnetic measurements. 861
43T1 TROIANO, A. R., MCGUIRE, F. T. *Trans. Am. Soc. Metals* **31**, 340–59. A study of the Fe-rich Fe-Mn alloys. 234
43W1 WEILL, A. R. *Nature* **152**, 413. Structure of η phase of Fe-Si system. 72
43Y1 YAMAMOTO, M. *Sci. Repts. Tôhoku Imp. Univ.* **31**, 101–16. The ΔE effect in iron, nickel and cobalt. 699

1944

44B1 BARNETT, S. J. *Proc. Am. Acad. Arts Sci.* **75**, 109–29. New Researches on magnetization by rotation. 454, 809
44C1 COLOMBANI, A. *Ann. physique* **19**, 272–326. Properties of ionoplastic Ni. 315
44D1 DANIEL, V., LIPSON, H. *Proc. Roy. Soc.* (London) **182A**, 378–87. Dissociation of an alloy of Cu, Fe and Ni. 154, 398
44D2 DEAN, R. S. *Trans. Am. Soc. Mining Met. Engrs.* **156**, 301–17. Present status of electrolytic Mn and its alloys. 313
44E1 EDWARDS, A., HOSELITZ, K. *Elec. Rev.* **135**, 165–9. Permanent magnet design. 363, 364
44E2 ENDTER, F., KLEMM, W. *Z. anorg. allgem. Chem.* **252**, 377–9. Crystal structure of Fe_2Gd and Mn_2Gd. 233
44E3 ESCH, U., SCHNEIDER, A. *Z. Electrochem.* **50**, 268–74. System Ni-Pt. 415
44F1 FINCH, O. J., WHITE, J. H. U.S.P. 2 347 817 (Appl. 9/17/41). Permanent magnet material. 389
44G1 GUILLAUD, C. *Compt. rend.* **219**, 614–6. Ferromagnetic properties of $MnNi_3$ and FeC_3. 319, 367, 441
44G2 GUILLAUD, C., WYART, J. *Compt. rend.* **219**, 203–5. Magnetic moments, Curie points and interatomic distances in ferromagnetic Mn-N alloys. 338
44G3 GUILLAUD, C., MICHEL, A., BÉNARD, J., FALLOT, M. *Compt. rend.* **219**, 58–60. Ferromagnetic properties of CrO_2. 342
44G4 GUILLAUD, C., WYART, J. *Compt. rend.* **219**, 393–4. Polymorphism of MnAs at the magnetic transformation temperature. 336
44H1 HOSELITZ, K. *J. Iron Steel Inst.* (London) **149**, 193–211. Iron-nickel phase diagram by magnetic analysis. 103

44H2 HESSENBRUCH, W., SCHICHTEL, K. *Z. Metallkunde* **36**, 127–30. Development of technical vacuum melting. 23
44H3 HARKER, D. *J. Chem. Phys.* **12**, 315–7. Crystal structure of Ni_4Mo. 320
44H4 HAUGHTON, J. L. *Supplement to Institute of Metals Monograph and Report Series*, No. 2, London, 1–14. Bibliography of literature relating to constitutional diagrams of alloys. 210
44K1 KROLL, W. J. *Metals and Alloys* **20**, 1604–6. Heat treatable Ni alloys. 315
44L1 LIFSHITZ, E. *J. Phys.* (USSR) **8**, 337–46. On the magnetic structure of iron. 834
44N1 NÉEL, L. *Cahiers phys.* **25**, 1–20. Some properties of boundaries between ferromagnetic domains. 814
44N2 NÉEL, L. *Cahiers phys.* **25**, 21–44. Effect of cavities and inclusions on the coercive force. 826, 836
44N3 NÉEL, L. *J. phys. radium* **5**, 241–76. Laws of magnetization and subdivision of elementary domains of iron. 814, 834
44O1 ONSAGER, L. *Phys. Rev.* **65**, 117–49. Two-dimensional model with order-disorder transition. 446
44S1 SANFORD, R. L. *Natl. Bur. Standards* (U. S.) *Circ.* **448**, 1–39. Permanent magnets. 380
44S2 SCHEIL, E., REINACHER, G. *Z. Metallkunde* **36**, 63–9. Elasticity modulus and damping of irreversible Fe-Ni alloys. 712
44T1 THOMPSON, F. C. *Res. Repts. of British Non-ferrous Metals Research Assn.*, 1–37. Damping capacity. 712
44T2 TROMBE, F. *Compt. rend.* **219**, 182–3. Magnetic properties of hydride, deuteride and carbide of Gd. 343
44U1 UNDERHILL, E. M. *Electronics* **17**, Jan., Feb., Apr. Permanent magnet design. 364

1945

45A1 ANONYMOUS. *Gen. Elec. Rev.* **48**, 61. Permanent magnets. 415
45A2 ALLANSON, J. T. *J. Inst. Elec. Engrs.* **92** (III), 247–55. Permeability of ferromagnetic materials at frequencies between 10^5 and 10^{10} cps. 799
45B1 BOZORTH, R. M., WILLIAMS, H. J. *Rev. Mod. Phys.* **17**, 72–80. Effect of small stresses on magnetic properties. 601, 615
45B2 BIRGE, R. T. *Am. J. Phys.* **13**, 63–73. 1944 values of certain atomic constants. 341, 867
45C1 CIOFFI, P. P. *Phys. Rev.* **67**, 200. A recording fluxmeter of high accuracy and sensitivity. 861
45C2 CARPENTER, V. W., BELL, S. A., HECK, J. E. U.S.P. 2 383 332 (Appl. 4/23/41). Production of silicon sheet steel having insulative surfaces. 71
45D1 DUNN, C. G. *Trans. Am. Soc. Mining Met. Engrs.* **161**, 98–113. Orientation changes during recrystallization in silicon ferrite. 92
45F1 FARQUAR, M. C. M., LIPSON, H., WEILL, A. R. *J. Iron Steel Inst.* (London) **152**, 457–62. X-ray study of Fe-rich Fe-Si alloys. 72, 73, 74, 75
45F2 FROMMER, L., MURRAY, A. *J. Iron Steel Inst.* (London) **151**, 45–53. Influence of heat treatment of steel on damping capacity at low stresses. 712
45F3 FRENKEL, J. *J. Phys.* (USSR) **9**, 299–311. Relaxation losses with magnetic resonance in solid bodies. 808
45H1 HONDA, K., MASUMOTO, H., SHIRAKAWA, Y., KOBAYASHI, T. Unpublished

	report, 19 November 1945. Magnetostriction of Fe-Al alloys and a new magnetostrictive alloy, Alfer. 679
45L1	LaBlanchetais, C. H. *Compt. rend.* **220**, 392–4. Magnetic properties of Fe-free Ce. 226
45M1	Masumoto, H., Saito, H. Unpublished report. Effect of heat treatment on the magnetic properties of Fe-Al alloys. 215, 220, 679
45N2	Nesbitt, E. A. U.S.P. 2 382 650 (Appl. 10/23/41). Magnetic materials. 417
45N3	Nesbitt, E. A. U.S.P. 2 382 651 (Appl. 10/23/41). Magnetic materials. 418
45N4	Nesbitt, E. A. U.S.P. 2 382 652 (Appl. 10/23/41). Magnetic materials. 419
45N5	Nesbitt, E. A. U.S.P. 2 382 653 (Appl. 10/23/41). Magnetic materials. 417
45N6	Nesbitt, E. A. U.S.P. 2 382 654 (Appl. 10/23/41). Magnetic materials. 417
45O1	Osborn, J. A. *Phys. Rev.* **67**, 351–7. Demagnetizing factors of the general ellipsoid. 848
45P1	Polder, D. *Philips Res. Repts.* **1**, 5–12. Theory of the elastic after effect and diffusion of C in Fe. 795, 796
45P2	Permanent Magnet Association. *J. Sci. Instruments* **22**, 56. Improved permanent magnet material. 393
45S1	Stanley, J. K. *Trans. Am. Inst. Mining Met. Engrs.* **162**, 116–40. Effect of variables on the recrystallization of silicon ferrite in terms of rates of nucleation and growth. 92
45S2	Sucksmith, W., Hoselitz, K., Heitler, H., Guggenheimer, K. *British & Allied Ind. Res. Assn. Report* N/T31, 1–16. Magnetic study of phase-change processes, Fe-Si. 72
45S3	Stoner, E. C. *Phil. Mag.* [7] **36**, 803–21. Demagnetizing factors for ellipsoids. 848
45T1	Trombe, F. *Compt. rend.* **221**, 19–21. Ferromagnetism and paramagnetism of metallic dysprosium. 341
45V1	VanVleck, J. H. *Rev. Mod. Phys.* **17**, 27–47. Survey of the theory of ferromagnetism. 443, 446

1946

46A1	Anonymous. *Am. Soc. Testing Materials, Standards* **1A**, 647–81. Standard methods of testing magnetic materials. 853
46B1	Bozorth, R. M. *Phys. Rev.* **70**, 923–32. Magnetoresistance and domain theory of iron-nickel alloys. 649, 672, 752, 756, 766
46B3	Benford, F. G. *Metals Progress* **46**, 94–5. Elastic anisotropy in electrical strip steel. 92, 836
46B4	Bizette, H. *Ann. physique* [12] **1**, 233–334. Orientation of some molecules and crystals by the magnetic field. 293, 322, 470
46B5	Birks, J. B. *Nature* **158**, 671–2. Magnetic dispersion of iron oxides at centimeter wave lengths. 802
46B6	Block, F., Hansen, W. W., Packard, M. *Phys. Rev.* **70**, 474–85. Nuclear induction experiment. 470
46B7	Bates, L. F. *Proc. Phys. Soc.* (London) **58**, 153–64. Magnetoresistance of high-coercivity alloys.
46C1	Cole, G. H., Burns, R. S. *Materials & Methods* **24**, 1457–60. A new radar transformer steel. 70

46C2 CRITTENDEN, E. C., SMITH, C. S., OLSEN, L. O. *Rev. Sci. Instruments* **17**, 372–4. B, H meter for samples of small cross-sectional area. 861

46D1 DEAN, R. S., LONG, J. R., GRAHAM, T. R., FEUSTEL, R. G. *Tr. Am. Soc. Metals* **36**, 116–36. Properties of Fe-Mn alloys containing 1 to 7% Mn. 234

46F1 FELCH, E. P., et al. *Trans. Am. Inst. Elec. Engrs.* **66**, 641–51. Air-borne magnetometers for search and survey. 861

46G1 GANZ, A. G. *Elec. Engg.* **65**, 177–83. Application of thin permalloy tape in wide band telephone and pulse transfers. 784

46G2 GUILLAUD, C. *Compt. rend.* **223**, 1110–2. Thermal variation of spontaneous magnetization. 240, 336, 339

46G3 GUILLAUD, C., WYART, J. *Compt. rend.* **222**, 71–3. Magnetic properties of nitrides of manganese. 338

46G4 GUILLAUD, C., BARBEZAT, S. *Compt. rend.* **222**, 386–90. Ferromagnetic properties of the definite compound CrTe. 342

46G5 GUILLAUD, C. *Compt. rend.* **222**, 1224–6. Magnetic isotherms of CrTe and the variation of its spontaneous magnetization with temperature. 342

46G6 GUILLAUD, C., CREVEAUX, H. *Compt. rend.* **222**, 1170–2. Preparation and properties of Fe_4N. 240

46G7 GRIFFITHS, J. H. E. *Nature* **158**, 670–1. Anomalous high frequency resistance of ferromagnetic metals. 805

46H1 HAAS, W. J. DE, WESTERDIJK, J. B. *Nature* **158**, 271–2. Strong magnetic fields.

46J1 JAFFEE, R. I., MCMULLEN, E. W., GONSER, B. W. *Trans. Am. Electrochem. Soc.* **89**, 277–89. Technology of Ge. 233

46K1 KITTEL, C. *Phys. Rev.* **69**, 640–4. Theory of long period magnetic relaxation. 798

46K2 KITTEL, C. *Phys. Rev.* **70**, 281–90. Theory of dispersion of permeability in ferromagnetic materials at microwave frequencies. 799, 801

46K3 KITTEL, C. *Phys. Rev.* **70**, 965–71. Theory of structure of ferromagnetic domains. 814, 829

46K4 KELLOGG, J. B. M., MILLMAN, S. *Rev. Mod. Phys.* **18**, 323–51. Molecular beam magnetic resonance method. 469

46K5 KÖNIG, H. *Naturwissenschaften* **33**, 71–5. Smallest elementary particle of iron. 834

46M1 MAXWELL, E. *MIT Rad. Lab. Rpt. 854*. No title available. 800

46N1 NESBITT, E. A. *Metals Tech.* **13**, No. 1973, 1–11. Vicalloy, a workable permanent magnet alloy. 405, 406, 416

46N2 NÉEL, L. *Ann. Univ. Grenoble* **22**, 299–343. Principles of a new general theory of the coercive field. 827

46N3 NOWOTNY, H., SCHUBERT, K. *Metallforschung* **1**, 17–23. System Mn (Fe)-Sn. 340

46P1 PURCELL, E. M., TORREY, H. C., POUND, R. V. *Phys. Rev.* **69**, 37–8. Resonance absorption by nuclear moments in a solid. 470

46R1 RICKETT, R. L., FICK, N. C. *Trans. Am. Inst. Mining Met. Engrs.* **167**, 346–54. Constitution of commercial low-carbon Fe-Si alloys. 72, 73

46R2 ROTHERHAM, L. *Metal Treatment* **12**, 215–22, 232. Damping capacity. 712

46R3 RUDER, W. E. *Iron Age* **157**, 65–9. Permanent magnet steels. 395, 396, 872

46S1 SIMON, I. *Nature* **157**, 735. Magnetic permeability of Ni in region of cm waves. 800

46S2 SELISSKY, V. P. *J. Phys. Chem.* (U.S.S.R.) **20**, 597–604. Crystal lattice constants of Fe-rich solutions of Fe, Si and Al. 97

46S3 SENDZIMIR, T. *Iron Steel Engr.* **23**, 53–9. The Sendzimir precision cold strip mill. 25, 70

46S4 SANFORD, R. L. *Natl. Bur. Standards* (U.S.) Circ. C456, 1–40. Magnetic testing. 850, 861

46S5 SHUR, J. S. *J. Phys.* (USSR) **10**, 299. Temperature dependence of coercive force in single crystals of transformer steel. 562

46S6 SHOENBERG, D., WILSON, A. J. C. *Nature* **157**, 548. Orientation of ferromagnetic domains near crystal surface. 559

46S7 STEINITZ, R. *Powder Met. Bull.* **1**, 45–7. Permet, a non-ferrous permanent magnet made from powders. 418

46Z1 ZENER, C. *Trans. Am. Soc. Mining Met. Engrs.* **167**, 155–89. Anelasticity of metals. 788, 791, 795.

1947

47B1 BOOTHBY, O. L., BOZORTH, R. M. *J. Applied Phys.* **18**, 173–6. New magnetic material of high permeability. 138, 142

47B3 BATES, L. F., EDMONDSON, A. S. *Proc. Phys. Soc.* (London) **59**, 329–43; **60**, 308. Changes accompanying magnetization in cobalt. 522

47B4 BARTON, J. P. *Elec. Mfg.* **39**, 105–7, 1946, 1948. New developments in electrical steel sheets and strips. 81

47B6 BOOTHBY, O. L., BOZORTH, R. M., WENNY, D. H. Belgian patent 474 121 (Appl. 8/23/46). Magnetic alloy and method of producing it. 138

47B7 BOTH, E. *Signal Corps. Eng. Lab., Tech. Memo M-1091.* Plant scale process for production of rectangular hysteresis loop material. 128

47D1 DÖRING, W. *Z. Physik* **124**, 501–13. Temperature dependence of μ_0 of Ni under tension. 613

47G1 GERLACH, W. *Metallforschung* **2**, 275–80. Dependence of ferromagnetic properties on temperature. 833

47G2 GUILLAUD, C., CREVEAUX, H. *Comp. Rend.* **224**, 268–70. Preparation and properties of manganese phosphide.

47H1 HARKER, D. *J. Sci. Instruments* **24**, 11. Deformation of metal crystals. 92

47H2 HORSTMAN, C. C., BARTLETT, C. H. *Steel Processing* **33**, 603–5, 644. Developments in magnetic steels for transformers. 68

47H3 HOFER, L. J. E., PEEBLES, W. C. *J. Am. Chem. Soc.* **69**, 893–9. Preparation and study of new cobalt carbide. 288

47H4 HOSELITZ, K., McCAIG, M. *Nature* **159**, 710. Spoiling of tungsten steel. 373

47J1 JOHNSON, M. H., RADO, G. T., MALOOF, M. *Phys. Rev.* **71**, 322–3 (47). Ferromagnetism at very high frequencies. 800

47K1 KÄLLBÄCH, O. *Ark. Matematik Astr. Fysik* **34B** (17), 1–6. Order-disorder point in FeNi₃. 103

47N1 NÉEL, L. *Compt. rend.* **224**, 1488–90. Properties of a cubic ferromagnetic having small grains. 828, 830, 831, 833

47N2 NÉEL, L. *Compt. rend.* **224**, 1550–1. H_c of ferromagnetic powders with anisotropic particles. 828

47N3 NÉEL, L. *Compt. rend.* **225**, 109–11. Anisotropy of magnetic steels heat treated in a magnetic field. 394, 832

47R1 RAUB, E., ENGEL, A. *Metallforschung* **2**, 147–58. Ordering of Au-Cu alloys in ternary system Au-Cu-Ni. 313

47S2 SNOEK, J. L. Elsevier, N. Y., 1–136. New developments in ferromagnetic materials. 101, 245, 249, 574

47S3 SHTURKIN, D. A. *Compt. Rend. Acad. Sci.* (USSR) **58**, 581–2. Magnetostriction of single crystals of Fe-Si. 649, 837

47S4 SLOTTMAN, G. V., LOUNSBERRY, F. B. *Iron Age* **159**, No. 8, 42–5. Use of oxygen in the openhearth bath. 68

47S5 STANLEY, J. K., YENSEN, T. D. *Trans. Am. Inst. Elec. Engrs.* **66**, 714–8. Hiperco—a magnetic alloy. 207

47S6 SOCIÉTÉ D'ÉLECTRO-CHIMIE, D'ÉLECTRO-MÉTALLURGIE ET DES ACIÉRIES ÉLECTRIQUE D'UGINE. Br. Pat. 590 392 (Appl. 4/7/42). Improvements in the manufacture of magnets. 421

47V1 VAN VLECK, J. H. *Ann. Inst. Henri Poincaré* **10**, 57–190. Some aspects of the theory of magnetism. 574, 594

47V2 VOGEL, R. *Metallforschung* **2**, 97–103. Systems Ce-Ni, La-Ni, Pr-Ni and Ce-Co. 288, 307

47V3 VALENTINER, S. *Naturwissenschaften* **34**, 123–4. Substitution of Al and Sn by In in Heusler alloys. 333

47V4 VERWEY, E. J. W., HEILMANN, E. L. *J. Chem. Phys.* **15**, 174–80. Cation arrangement in spinels. 245

47V5 VERWEY, E. J. W., HAAYMAN, P. W., ROMEIJN, F. C. *J. Chem. Phys.* **15**, 181–7. Electronic conductivity and cation arrangement in spinels. 245

47W2 WEIL, L., MARFOURE, S. *J. phys. radium* [8] **8**, 358–60. Thermal variation of coercive field of Ni agglomerates. 833

47W3 WIEGAND, D. W., HANSEN, W. W. *Tr. Am. Inst. Elec. Engrs.* **66**, 119–31. A 60-cycle hysteresis loop tracer. 861

47Y1 YAGER, W. A., BOZORTH, R. M. *Phys. Rev.* **72**, 80–1. Ferromagnetic resonance at microwave frequencies. 805, 806, 808, 809

47Y2 YAMAMOTO, T. *J. Inst. Elec. Engrs. Japan* **67**, 12–17, 33–8. Magnetic properties of Fe-Ni-Si alloys. 186

1948

48A1 ASTBURY, N. F., DEBARR, A. E. *Sheet Metal Ind.* **25**, 911–26, 921. Aeolotropy in steel sheet and strip. 92

48B1 BRAILSFORD, F. *J. Inst. Elec. Engrs.* **95** (II), 38–48. Investigation of the eddy-current anomaly in electrical sheet steels. 782, 788

48B2 BOZORTH, R. M. *Z. Physik* **124**, 519–27. On magnetic remanence. 504, 505

48B4 BUTLER, O. I., MANG, C. Y. *J. Inst. Elec. Engrs.* **95** (II), 25–37. Magnetic properties of ferromagnetic laminae at power and audio frequencies. 782

48B5 BATES, L. F., HARRISON, E. G. *Proc. Phys. Soc.* (London) **60**, 225–36. Temperature changes accompanying magnetization in iron. 522

48B6 BATES, L. F., DAVIS, J. H. *Proc. Phys. Soc.* (London) **60**, 307. Effect of temperature on the heat changes accompanying magnetization in a Ni-Si alloy. 522

48B7 BATES, L. F., HARRISON, E. G. *Proc. Phys. Soc.* (London) **60**, 213–25. Temperature changes accompanying magnetization of some ferromagnetic alloys. 522

48C1 CHEGWIDDEN, R. A. *Metal Progress* **54**, 705–14. Review of magnetic materials, especially for communication systems. 872

48C2 CIOFFI, P. P. *Trans. Am. Inst. Elec. Engrs.* **67**, 1540–3. Stabilized permanent magnets. 364

48D1 DARKEN, L. S., GURRY, R. W. *Metals Handbook*, Am. Soc. Metals, 1212–3. Iron-oxygen. 240
48D2 DUMOND, J. W. M., COHEN, E. R. *Rev. Mod. Phys.* **20**, 82–108. Atomic constants in 1947. 867
48D3 DÖRING, W. *Z. Naturforsch.* 3a, 373–9. Inertia of walls between Weiss domains. 802
48D4 DROZZINA, V. I., SHUR, J. S. *J. tech. Phys.* (USSR) **18**, 149–52. Change of resistivity in field of alloys of high coercive force. 758
48E1 ELLIS, W. C., GREINER, E. S. *Metals Handbook*, Am. Soc. Metals, 1136–7. Cobalt. 276
48E2 ELSEA, A. R., WESTERMAN, A. B. *Metals Tech.* **15**, T.P. 2393, 1–24. Co-Cr binary system. 288
48E3 EASH, J. T. *Metals Handbook*, Am. Soc. Metals, 1262. System Ni-Cu-Sn. 313
48F1 FINE, M. E., ELLIS, W. C. *Trans. Am. Inst. Mining Met. Engrs.* Thermal expansion properties of Fe-Co alloys. 191
48F2 FINK, W. L., WILLEY, L. A. *Metals Handbook*, Am. Soc. Metals, 1164. System Ni-Al. 299
48F3 FOËX, M. *Compt. rend.* **227**, 193–4. Transformation of bivalent oxides of Mn, Fe, Co, Ni. 293, 321, 470
48F4 FAHLENBRACH, H. *Ann. Physik* [6] **2**, 355–69. Temperature dependence of permeability and lag of ferromagnetic materials. 794
48G1 GUILLAUD, C., BERTRAND, R. *Compt. rend.* **227**, 47–8. Initial magnetization and coercive force of cobalt along the axis of easy and difficult magnetization. 561, 583
48G2 GEISLER, A. H., NEWKIRK, J. B. *Metals Tech.* **15**, No. 2444, 1–20. Mechanism of precipitation in a permanent magnet alloy. 402, 404
48G3 GUILLAUD, C. *Rev. de mét.* **45**, 271–6. Ferromagnetic binary alloy of Mn. 334
48H1 HOBSON, P. T., OSMOND, W. P. *Nature* **161**, 562–3. Ferromagnetic structure of cold worked austenitic stainless steels. 833
48H2 HAMES, F. A., EPPELSHEIMER, D. S. *Nature* **162**, 968. Some new ferromagnetic manganese alloys. 333, 337
48H3 HEWITT, W. H. *Phys. Rev.* **73**, 1118–9. Microwave resonance absorption in ferromagnetic semi-conductors. 809
48J1 JAFFEE, R. I. *J. Applied Phys.* **19**, 867–70. Magnetic properties of ordered Ni-Mn alloys. 319
48J2 JACK, K. H. *Proc. Roy Soc.* (London) **195A**, 34–40. Structures of Fe_4N and Fe_2N. 239
48J3 JACK, K. H. *Proc. Roy. Soc.* (London) **195A**, 41–55. Iron-carbon-nitrogen system. 240
48J4 JELLINGHAUS, W. *Z. Metallkunde* **39**, 52–6. Magnetization process of an Fe-Ni-Al-Co-Cn magnet alloy having preferred orientation. 390
48K1 KORNETZKI, M. *Ann. Physik* [6] **2**, 265–9. Stress-strain curve of ferromagnetic material. 684, 686
48K2 KITTEL, C. *Phys. Rev.* **73**, 810–1. Domain theory and dependence of coercive force on particle size. 830
48K3 KÖSTER, W. *Z. Metallkunde* **39**, 1–9. Temperature dependence of elasticity modulus of pure metals. 699, 712
48K4 KENYON, R. L. *Metals Handbook*, Am. Soc. Metals, 423–34. Physical and mechanical properties of iron. 43, 53
48K5 KÖSTER, W., ZWICKER, U., MOELLER, K. *Z. Metallkunde* **39**, 225–31. Microscopic & X-ray examinations of system Cu-Ni-Al. 313

48K6 KIHLGREN, T. E. *Metals Handbook*. Am. Soc. Metals, 1263. System Ni-Cu-Zn. 313

48K7 KIHLGREN, T. E., EASH, J. T. *Metals Handbook*, Am. Soc. Metals, 1183. System Ni-C. 306

48K8 KÖSTER, W., RAUSCHER, W. *Z. Metallkunde* **39**, 178–84. System Ni-Mn. 315

48K9 KITTEL, C. *Phys. Rev.* **73**, 155–61. Theory of ferromagnetic resonance absorption. 807, 810

48K10 KIRENSKY, L. V. *Izves. Akad. Nauk, Fiz. Ser.* (USSR) **12**, 121–5. Relation between magnetic anisotropy and magnetic field strength. 566

48K11 KERSTEN, M. *Z. Physik* **124**, 714–41. On the theory of coercive force.

48L1 LACY, C. E. *Metals Handbook*, Am. Soc. Metals 1236. System Ni-Zn. 327

48L2 LUTTINGER, J. M., KITTEL, C. *Helv. Phys. Acta* **21**, 480–2. Quantum theory of ferromagnetic resonance. 807

48M1 MORRILL, W. *Metals Progress* **54**, 675–8. Improved silicon iron for electrical equipment. 67, 86

48M2 *Metals Handbook*, Am. Soc. Metals, Cleveland, 1–1444. 106, 311, 334

48N1 NÉEL, L. *Ann. physique* [12] **3**, 137–98. Magnetic properties of ferrites: ferromagnetism and antiferromagnetism. 242, 246, 472

48N2 NESBITT, E. A., WILLIAMS, H. J. *Phys. Rev.* **73**, 1246. Measurement of magnetostriction with automatic recording. 630, 682

48N3 NORTON, J. T. *Metals Handbook*, Am. Soc. Metals, 1211. System Fe-N.

48N4 NÉEL, L. *J. phys. radium* [8] **9**, 184–99. Law of approach to saturation and a new theory of magnetic hardness. 487, 488

48O1 OLIVER, D. A., HADFIELD, D. *J. Inst. Elec. Engrs.* **95** (I), 531–9. Permanent magnet materials. 393, 872

48P1 PEARSON, G. L. *Rev. Sci. Instruments* **19**, 263–5. Magnetic field strength meter employing Hall effect in germanium. 856

48P2 POUILLARD, E. *Compt. rend.* **227**, 194–6. Preparation of new magnetites by replacement of ferric ions. 245

48S1 STONER, E. C., WOHLFARTH, E. P. *Trans. Roy. Soc.* (London) A **240**, 599–644. Mechanism of magnetic hysteresis in heterogeneous alloys. 394, 828, 830, 831

48S2 STREET, R. *Proc. Phys. Soc.* (London) **60**, 236–43. Variation with magnetization of Young's modulus for Co. 687

48S3 STONER, E. C. *Reports on progress in physics*, Physical Society, London, 43–112. Ferromagnetism. 443

48S4 SACHS, G. *Metals Handbook*, Am. Soc. Metals, 1211–2. Phase diagram of Fe-Ni alloys. 103

48S5 SCOTT, K. L. *Metals Handbook*, Am. Soc. Metals, 595–9. Permanent magnet materials. 872

48S6 SCHRAMM, J. *Z. Metallkunde* **39**, 71–8. Hardening of Fe-Zn. 259

48S7 SMITH, C. S. *Metals Handbook*, Am. Soc. Metals, 1191. Constitution of cobalt-copper. 289

48S9 SCHRAMM, J., MOHRNHEIM, A. *Z. Metallkunde* **39**, 81–8. Hardening of Fe-Zn and Co-Zn alloys. 299

48S10 SHAW, J. L. *Product Eng.* **19**, No. 6, 158–63. Curie temperature alloys. 311

48S11 SINIZER, D. I. *Metals Handbook*, Am. Soc. Metals, 1243. System Ni-Cu-Al. 313

48S12	SKINNER, E. N. *Metals Handbook*, Am. Soc. Metals, 1233. System Ni-Si. 322
48S13	SOCIÉTÉ D'ÉLECTRO-CHIMIE, D'ÉLECTRO-MÉTALLURGIE ET DES ACIÉRIES ÉLECTRIQUE D'UGINE. Br. Pat. 596 875 (Appl. 8/21/43). Improvements in the manufacture of magnets. 421
48S14	STANLEY, J. K. U.S.P. 2 442 219 (Appl. 10/30/46). Magnetic alloy. 207
48S15	SOCIÉTÉ D'ÉLECTRO-CHIMIE, D'ÉLECTRO-MÉTALLURGIE ET DES ACIÉRIES ÉLECTRIQUE D'UGINE. Br. Pat. 594 681 (Appl. 7/26/44). Improvements in preparation of powders for permanent magnets. 421
48S16	STEINITZ, R. *Powder Metallurgy Bull.* **3**, 124–7. Permanent magnets from pure iron powder. 421
48S17	SNOEK, J. L. U.S.P. 2 452 529 (Appl. 9/17/45). Magnet core. 245
48S18	SNOEK, J. L. U.S.P. 2 452 530 (Appl. 9/17/45). Magnet core. 245
48S19	SNOEK, J. L. U.S.P. 2 452 531 (Appl. 9/17/45). Magnetic material and core. See also *Physica* **14**, 207, 245
48T1	TROIANO, A. R., LOKICH, J. L. *Metals Tech.*, T. P. 2348, 1–2. Transformation of Co. 264
48V1	VONSOVSKY, S. V. *J. Tech. Phys.* (USSR) **18**, 145–8. Change of electrical resistance in a magnetic field for some ferromagnetic alloys. 759
48W1	WEIL, L. International Powder Metallurgy Day, Graz, July, 1948, No. 17. Special properties of finely divided ferromagnetic materials. 832
48W2	WEISS, P. R. *Phys. Rev.* **74**, 1493–1504. Application of Bethe-Peierls method to ferromagnetism. 445, 447
48W3	WISE, E. M. *Metals Handbook*, Am. Soc. Metals, 1046–7 Properties of nickel. 269
48Y2	YENSEN, T. D. *Metals Handbook*, Am. Soc. Metals, 587–95. Magnetically soft materials. 312
48Y3	YAGER, W. A. *Phys. Rev.* **73**, 1247. Ferromagnetic resonance absorption at microwave frequencies. 806, 809, 810

1949

49A1	AKULOV, N. S., ALIZADE, Z. I., BELOV, K. P. *Dokl. Akad. Nauk* (USSR) **65**, 815–8. Magnetostriction of Fe-Pt alloys. 410
49A2	AMERICAN SOCIETY for TESTING MATERIALS. Special Technical Publication No. 85, 191–3. Standard definitions of terms relating to magnetic testing. 1
49B1	BOZORTH, R. M. *Physica* **15**, 207–18. Ferromagnetic domains. 533, 711
49B2	BROWN, W. F. *Phys. Rev.* **75**, 147–54. Irreversible magnetic effects of stress. 619
49B3	BOZORTH, R. M. *Electr. Engg.* **68**, 471–7. Advances in the theory of ferromagnetism. 533
49B4	BICKFORD, L. R. *Phys. Rev.* **76**, 137–8. Ferromagnetic resonance absorption in magnetite. 810
49B5	BOZORTH, R. M., MASON, W. P., MCSKIMIN, H. J., WALKER, J. G. *Phys. Rev.* **12**, 1954–5. Elastic constants and loss of single nickel crystals. 686
49B6	BLAKE, L. R. *Jl. Inst. Elec. Engrs.* **96**, II, 705–18. Eddy current anomaly in ferromagnetic laminae. 782
49B7	BRADLEY, A. J. *J. Iron Steel Inst.* (London) **163**, 19–30. Microscopical studies of the Fe-Ni-Al system. 385, 394
49B8	BARNETT, S. J., GIAMBONI, L. A. *Phys. Rev.* **76**, 1542. New gyromagnetic effect. 454

49B9 BUHL, O. *Z. Physik* **126**, 84–97. Approach to saturation of nickel under tension. 488
49B10 BRAUNEWELL, W., VOGT, E. *Z. Naturforsch.* **4a**, 491–5. Reversal of sign of magnetostriction by tension. 672
49B11 BERTAUT, F. *Comp. rend.* **229**, 417–9. Coercive force and crystalline dimensions. 833
49B12 BELJERS, H. G. *Physica* **14**, 629–41. Gyromagnetic resonance of a ferrite using cavity resonators. 809
49B13 BATES, L. F. *J. phys. radium* [8] **10**, 353–63. Thermal effects of magnetization in weak fields. 522
49C1 CARNEGIE-ILLINOIS STEEL CORP. *Catalog*, 1–180. U.S.S. Electrical Sheet Steels. 83
49C2 COLES, B. R., HUME-ROTHERY, W., MEYERS, H. P. *Proc. Roy. Soc.* (London) **196A**, 125–33. Structure and properties of Cu_2MnIn. 333
49C3 CREDE, J. H., MARTIN, J. P. *Jl. App. Phys.* **20**, 966–71. Magnetic characteristics of oriented 50 per cent Ni-Fe alloy, Deltamax. 133
49C4 CARPENTER, V. W., BELL, S. A. U.S.P. 2 492 682 (Appl. 7/23/45). Process of producing glass coated silicon steel. 71
49C5 CRITTENDEN, E. C., STROUGH, R. I. *Phys. Rev.* **75**, 1630. Limiting sensitivity of instruments for displaying magnetic hysteresis loops. 861
49D1 DUSHMAN, S. Wiley, New York, 1–882. Vacuum Technique. 24, 234, 241, 293, 314
49D2 DOMENICALI, C. A. *Phys. Rev.* **76**, 460. Magnetic and electric properties of magnetite single crystals. 243
49D3 DUMOND, J. W. M., COHEN, E. R. *Rev. Mod. Phys.* **21**, 651–2. Erratum: Our knowledge of the atomic constants in 1947. 867
49D4 DRIGO, A., PIZZO, M. *Nuovo Cimento* **6**, 297–9. Ferromagnetic impurities and the magnetization of small particles. 232
49D5 DIJKSTRA, L. J., SNOEK, J. L. *Philips Res. Repts.* **4**, 334–56. Propagation of large Barkhausen discontinuities in Fe-Ni alloys. 496
49D6 DROZZINA, V. I., LUZHINSKAYA, M. G., SHUR, J. S. *J. tech. Phys.* (USSR) **19**, 95–9. Effect of magnetic anneal on the resistivity of Alnico alloys. 393, 758
49E1 EPREMIAN, E., HARKER, D. *Jl. Metals* **1**, 267–73. Crystal structure of Ni_4W. 326
49F1 FAHLENBRACH, H. *Arch. Eisenhüttenw.* **20**, 301–4. Magnetic properties and usefulness of permanent magnets produced by powder metallurgy. 418
49F2 FOËX, M., LA BLANCHATAIS, C. H. *Comp. rend.* **228**, 1579–80. Antiferromagnetism of NiO. 470
49F3 FAHLENBRACH, H., SIXTUS, K. J. *Z. Metallkunde* **40**, 187–93. Investigation of different unmagnetized states. 794
49G1 GOLDMAN, J. E., SMOLUCHOWSKI, R. *Phys. Rev.* **75**, 140–7. Magnetostriction and order-disorder. 630
49G2 GRINSTEAD, R. R., YOST, D. M. *Phys. Rev.* **75**, 984–5. Ferromagnetic alloys in the system Cu-Mn-In. 333
49G4 GERLACH, W. *Z. Metallkunde* **40**, 281–9. Precipitation in the system Au-Ni. 314
49G5 GALPERIN, F. M., PEREKALINA, T. M. *Jl. exp. theoret. Physics* (USSR) **19**, 470–2. Magnetic properties of the tellurides of chromium. 342

49G6	GUILLAUD, C. *Comp. rend* **229**, 818–9. Coercive force of ferromagnetic powder.	
49G7	GUILLAUD, C., BERTRAND, R., VAUTIER, R. *Comp. rend.* **228**, 1403–5. Initial magnetization of Mn_2Sb single crystal. 561, 583	
49G8	GUILLAUD, C., ROUX, M. *Comp. rend* **229**, 1133–5. Some magnetic properties of mixed ferrites of nickel and zinc. 247	
49G9	GUILLAUD, C. *Ann. physique* [12] **4**, 671–703. Magnetic properties of Mn-Sb and Mn-As alloys. 335, 336	
49H1	HOSELITZ, K., MCCAIG, M. *Proc. Phys. Soc.* (London) **B62**, 163–70; *Physica* **15**, 241–3. Cause of anisotropy in permanent magnet alloys. 393, 682	
49H2	HODSMAN, G. F., EICHHOLZ, G., MILLERSHIP, R. *Proc. Phys. Soc.* (London) **B62**, 377–90. Magnetic dispersion at microwave frequencies. 800	
49H3	HAMES, F. A., EPPELSHEIMER, D. S. *Jl. Metals* **1**, 495–9. Ferromagnetic alloys in systems Cu-Mn-In and Cu-Mn-Ga. 337	
49J1	JELLINGHAUS, W. *Arch. Eisenhüttenw.* **20**, 249–54. Heat treatment and magnetic properties of chrome magnet steel. 377	
49J2	JELLINGHAUS, W. *Z. Metallkunde* **40**, 339–44. Hysteresis of incompletely magnetized permanent magnet materials. 511	
49K1	KITTEL, C. *Rev. Mod. Phys.* **21**, 541–83. Physical theory of ferromagnetic domains. 449, 506, 654	
49K2	KUBASCHEWSKI, O., SCHNEIDER, A. *J. Inst. Metals*, **75**, 403–16. Oxidation-resistance of high melting point alloys. 321	
49K3	KUBASCHEWSKI, O., SPEIDEL, H. *J. Inst. Metals* **75**, 417–30. Phase relations in the Ni-Cr-Ta system. 323	
49K4	KUSSMANN, A., STEINWEHR, H. E. V. *Z. Metallkunde* **40**, 263–6. Superstructure in the system Pt-Ni. 415	
49K5	KITTEL, C. *Phys. Rev.* **76**, 743–8. Gyromagnetic ratio and spectroscopic splitting factor of ferromagnetic substances. 809	
49K6	KIP, A. F., ARNOLD, R. D. *Phys. Rev.* **75**, 15566–60. Ferromagnetic resonance at microwave frequencies in single crystals of Fe. 810	
49K7	KAMMERER, H. *Arch. elek. Übertr.* **3**, 249–56. Frequency dependence of the distortion factor of coils with iron cores. 78	
49K8	KIRENSKY, L. V. *Dokl. Ak. Nauk* (USSR) **64**, 53–6. Temperature relation of the magnetic anisotropy in nickel.	
49M1	MAXWELL, L. R., SMART, J. S., BRUNAUER, S. *Phys. Rev.* **76**, 459–60. Dependence of intensity of magnetization and Curie point of iron oxides on the ratio Fe^{++}/Fe^{+++}. 242	
49M2	MCCAIG, M. *Proc. Phys. Soc.* (London) **B62**, 652–6. Magnetostriction of anisotropic permanent magnet alloys. 393, 682	
49M3	MEYER, A. J. P. *Comp. rend.* **228**, 1934–5. Gyromagnetic constant of Fe and Ni. 454	
49M4	MEYER, A. J. P. *Comp. rend.* **229**, 707–8. Gyromagnetic effect in isoelectronic ferromagnetic alloys. 454	
49N1	NÉEL, L. *Ann. physique* [12] **4**, 249–69 (1949). Tentative explanation of magnetic properties of rhombohedral Fe_2O_3. 241	
49N2	NÉEL, L. *Physica* **15**, 225–34. New theory of coercive force. 827	
49N3	NÉEL, L. *Bull. soc. franc. électr.* [6] **9**, 308–15. Some practical aspects of ferromagnetic theory. 827	
49N5	NÉEL, L. *Comp. rend.* **228**, 664–6. Effect of thermal fluctuations on the magnetization of small particles. 833	

49O1   Owen, E. A., Liu, Y. H.   *J. Iron Steel Inst.* (London) **163**, 132-7. Further X-ray study of the equilibrium of the Fe-Ni system. 103

49P1   Pfisterer, H.   *Z. Metallkunde* **40**, 378-83. Phase structure of the system Co-Ge. 289

49P2   Puzei, I. M.   *J. Tech. Phys.* (USSR) **19**, 653-60. Magnetic anisotropy of binary alloys of nickel. 573

49R1   Rees, W. P., Burns, B. D., Cook, A. J.   *J. Iron Steel Inst.* (London) **162**, 325-36. Constitution of Fe-Ni-Cr alloys at 650° to 800°C. 146

49S2   Shull, C. G., Siegel, S.   *Phys. Rev.* **75**, 1008-10. Neutron diffraction studies of order-disorder in alloys. 315

49S3   Shull, C. G., Smart, J. S.   *Phys. Rev.* **76**, 1256-7. Detection of antiferromagnetism by neutron diffraction. 473, 475

49S4   Shoenberg, D.   *Nature* **164**, 225-6. Magnetic properties of metallic single crystals at low temperatures. 456

49S5   Schneider, A., Wunderlich, W.   *Z. Metallkunde* **40**, 260-3. The γ-mixed crystals in the Co-Mn system. 291

49S6   Stout, W., Griffel, M.   *Phys. Rev.* **76**, 144-5. Paramagnetic anisotropy of MnF_2. 472

49S7   Shubina, L. A., Shur, J. S.   *J. techn. Physics* (USSR) **19**, 88-93. Magnetic structures of high coercive force. 393, 394

49S8   Stoner, E. C., Rhodes, P.   *Phil. Mag.* [7] **40**, 481-522. Magnetothermal effects in ferromagnetics. 523

49S10   Scholefield, H. H.   *J. Sci. Instruments* **26**, 207-9. A recent development in soft magnetic materials. 133

49V1   Vogel, R., Au, R.   *Z. Metallkunde* **40**, 290-5. Relation of FeS to Co_6S_5 and Co_4S_3. 294

49W1   Williams, H. J., Bozorth, R. M., Shockley, W.   *Phys. Rev.* **75**, 155-78. Magnetic domains in single crystals of silicon ion. 533, 538, 816, 834, 835, 836

49W2   Williams, H. J., Shockley, W.   *Phys. Rev.* **75**, 178-83. Simple domain structure in an iron crystal showing direct correlation with magnetization. 533, 559, 711, 828

49W3   Weil, L.   *Comp. rend.* **229**, 584-5. Coercive force and granulometry of Raney nickel. 834

49W4   Walker, J. G., Williams, H. J., Bozorth, R. M.   *Rev. Sci. Instruments* **20**, 947-50. Growing and processing of single crystals of magnetic metals 556

49W5   Wohlfarth, E. P.   *Phil. Mag.* [7] **40**, 703-17. Collective electron ferromagnetism. 443

49W6   Wohlfarth, E. P.   *Proc. Roy. Soc.* (London) **195A**, 434-63. Collective electron ferromagnetism III. Ni and Ni-Cu alloys. 443

49Y1   Yager, W. A., Merritt, F. R.   *Phys. Rev.* **75**, 318. Ferromagnetic resonance in Heusler alloy. 809

1950

50A1   Astbury, N. F., Emmerson, T., McFarlane, J.   *Jl. Inst. Elec. Engrs.* **97** (II), 221-8. Routine testing of transformer steels.

50A2   Anderson, P. W.   *Phys. Rev.* **79**, 105-10. Molecular field theory of antiferromagnetism. 474

50A3   Auwarter, M., Kussmann, A.   *Ann. Physik* **7**, 169-72. Ferromagnetic Pt-Mn alloys.

50B1 BROCKMAN, F. G., DOWLING, P. H., STENECK, W. G. *Phys. Rev.* 77, 85–93. Dimensional effects resulting from a high dielectric constant in ferromagnetic ferrite. 247
50B2 BROCKMAN, F. G. *Phys. Rev.* 77, 841–2. Cation distribution in ferrites with spinel structure. 245
50B3 BUHL, O. *Z. Naturforschung* 4a, 601–5. The magnetization curve of nickel under tension.
50B4 BATES, L. F., NEALE, F. E. *Pr. Phys. Soc.* (London) A63, 374–88. Domain structure of single crystals of silicon iron by powder pattern technique.
50B5 BERAK, J., HEUMANN, T. *Z. Metallkunde* 41, 19–23. System manganese-phosphorus. 339
50B6 BARBIER, J. C. *Comp. rend.* 230, 1040–1. Magnetic lag in the Rayleigh region.
50B7 BELJERS, H. G., POLDER, D. *Nature* 165, 800. g-factors in ferrites. 809
50B8 BULGAKOV, N. V. *Dok. Ak. Nauk* 70, 205–6. Internal demagnetization factor of high H_c alloys.
50B9 BUTLER, O. I. *Jl. Inst. Elec. Engrs.* 97 (II), 215–20. Miniature Lloyd-Fisher square for power losses at audio frequencies.
50B10 BICKFORD, L. R. *Phys. Rev.* 78, 449–57. Ferromagnetic resonance absorption in magnetic crystals. 574
50B11 BLOEMBERGEN, N. *Phys. Rev.* 78, 572–80. Ferromagnetic resonance in nickel and Supermalloy.
50B12 BIRKS, J. B. *Pr. Phys. Soc.* (London) B63, 65–74. Properties of ferromagnetic compounds at centimeter wavelengths.
50B13 BERTAUT, F. *Compt. rend.* 230, 213–5. Nature of spinel ferrites.
50B14 BOZORTH, R. M. *Phys. Rev.* 79, 887. Atomic moments of ferromagnetic alloys.
50B15 BOZORTH, R. M., WALKER, J. G. *Phys. Rev.* 79, 888. Domain structure of a Co-Ni crystal.
50C1 CARTER, R. O., RICHARDS, D. L. *Jl. Inst. Elec. Engrs.* 97 (II), 199–214. Incremental magnetic properties of Fe-Si alloys.
50D1 DÖRING, W. *Z. Naturforschung* 4a, 605–10. The inner magnetic field in ferromagnetic materials.
50D2 DOMENICALI, C. A. *Phys. Rev.* 78, 458–67. Magnetic and electric properties of magnetite crystals.
50D3 DE KLERK, D., STEENLAND, M. J., GORTER, C. J. *Phys. Rev.* 78, 476. Thermodynamic temperature of 0.0015°K. 466
50E1 EKSTEIN, H., GILBERT, T. *Phys. Rev.* 79, 214–5. Mechanism of remagnetization in an initially saturated ferromagnet. 834
50F1 FORESTIER, H., GUIOT-GUILLAIN, G. *Comp. rend.* 230, 1844–5. New series of magnetic materials; ferrites of the rare earths. 245
50F2 FELDTKELLER, R. *Fernmelde tech. Z.* 3, 112–7. Magnetic lag.
50F3 FINE, M. E., ELLIS, W. E. *Tr. Am. Inst. Min. Met. Engrs.* 188, 1120–5. Young's modulus in 36 to 52% Ni alloys.
50G1 GEISLER, A. H. *Elec. Eng.* 69, 37–44. Structure and properties of the permanent magnet alloys. 393, 404
50G2 GORDON, P., KAUFMANN, A. R. *Jl. Metals* 188, 182–94. Uranium-aluminum and uranium-iron. 257
50G3 GORTER, E. W. *Comp. rend.* 230, 192–4. Spontaneous magnetization of ferrites having the spinel structure. 246, 247
50G4 GALT, J. K. *Phys. Rev.* 77, 845–6. Coercive force *vs* temperature in an alloy with zero crystal magnetic anisotropy. 833

50G5 GUILLAUD, C., CREVEAUX, H. *Comp. rend.* **230**, 1458–60. Magnetic properties of Co-Zn and Mn-Zn ferrites.
50G6 GORTER, E. W. *Nature* **165**, 798–800. Saturation magnetization in ferrites.
50G7 GUILLAUD, C., YAGER, W. A., MERRITT, F. R., KITTEL, C. *Phys. Rev.* **78**, 181. Ferromagnetic resonance in manganese ferrite.
50J1 JONKER, G. H., VAN SANTEN, J. H. *Physica* **16**, 337–49. Ferromagnetic compounds of manganese with perovskite structure. 339
50K1 KITTEL, C., NESBITT, E. A., SHOCKLEY, W. *Phys. Rev.* **77**, 739–40. Theory of magnetic properties and nucleation in Alnico 5. 394
50K2 KUSSMANN, A., RITTBERG, G. V. *Ann. Physik* **7**, 173–81. Magnetic properties of Fe-Pt alloys II. 410
50L1 LIBSCH, J. F., BOTH, E., BECKMAN, G. W., WARREN, D., FRANKLIN, R. J. *Trans. Am. Inst. Mining Met. Engrs.* **188**, 287–96. Effect of annealing in a magnetic field on Fe-Co and Fe-Co-Ni alloys. 199
50L2 LATIMER, K. E., MACDONALD, H. B. *Jl. Inst. Elec. Engrs.* **97** (II), 257–67. Survey of possible applications of ferrites.
50L3 LLIBOUTRY, L. *Comp. rend.* **230**, 1042–4. Some laws of magnetic lag.
50L4 LLIBOUTRY, L. *Compt. rend.* **230**, 1586.7. Detection of magnetic lag with the aid of shock.
50L5 LILLEY, B. A. *Phil. Mag.* [7] **41**, 792–813. Energies and widths of domain boundaries in ferromagnetics.
50M1 MCCAIG, M. *Nature* **165**, 969. Preferred domain orientation in permanent magnet alloys.
50M2 MICHELS, A., DEGROOT, S. R. *Physica* **16**, 249–52. Influence of pressure on the Curie point.
50M3 MELVILLE, W. S. *Jl. Inst. Elec. Engrs.* **97** (II), 165–98. Pulse magnetization of nickel-irons from 0.1 to 5 microseconds.
50N1 NESBITT, E. A. *Jl. Applied Phys.* **21**, 879–89. Magnetostriction of permanent magnet alloys. 386, 393, 639, 664, 832
50N2 NESBITT, E. A. *Phys. Rev.* **80**, 112–3. Mechanism of magnetization in Alnico 5. 393, 682
50N3 NÉEL, L. *Ann. Inst. Fourier* **1**, 163–83. Experimental proof of ferrimagnetism and antiferromagnetism. 246
50N4 NÉEL, L. *J. phys. radium* **11**, 49–61. Magnetic lag in solid ferromagnetics in the Rayleigh region.
50N5 NÉEL, L. *Physica* **16**, 350–2. Ferromagnetism of ferrites, or ferrimagnetism.
50N6 NÉEL, L. *Compt. rend.* **230**, 375–7. Saturation magnetization of nickel zinc ferrites.
50N7 NÉEL, L. *Comp. rend.* **230**, 190–2. Saturation magnetization in ferrites.
50P1 PARANJPE, V. G., COHEN, M., BEVER, M. B., FLOE, C. F. *Jl. Metals* **188**, 261–7. Iron-nitrogen system. 240
50P2 PAUTHENET, R. *Comp. rend.* **230**, 1842–3. Thermal variation of spontaneous magnetization of the ferrites of Mn, Fe, Co and Ni.
50P3 POLDER, D. *Jl. Inst. Elec. Engrs.* **97** (II), 246–56. Ferrite materials.
50R1 RADO, G. T. Advances in Electronics II (Marton, ed.), Academic Press, N.Y., 251–98. Ferromagnetic phenomena at microwave frequencies. 803, 810
50R2 RICHARDS, C. E., BUCKLEY, S. E., BARDELL, P. R., LYNCH, A. C. *Jl. Inst. Elec. Engrs.* **97** (II) 236–45. Properties and tests of magnetic powder cores.

50R3 RANDALL, W. F., SCHOLEFIELD, H. H. *Jl. Inst. Elec. Engrs.* **97** (II), 133–40. Some factors affecting the properties of soft magnetic materials.
50S1 SAGE, M., GUILLAUD, C. *Comp. rend.* **230**, 1749–51. Crystal structure and magnetic properties of mixed ferrites of Ni and of Zn.
50S2 STEWART, K. H. *Jl. Inst. Elec. Engrs.* **97** (II), 121–5. Losses in electrical sheet steel.
50S3 SCHMID, E., THOMAS, H. *Z. Metallkunde* **41**, 45–9. Texture in sheets of 50% Ni-Fe.
50S4 SNOEK, J. S. *Physica* **16**, 333–5. Weak ferromagnetism believed to be present in α-Fe_2O_3 and other antiferromagnetic compounds. 241
50S5 STANDLEY, K. J. *Sci. Progress* **38**, 231–45. Ferromagnetic resonance at microwave frequencies. 810
50S6 STEWART, K. H. *Pr. Phys. Soc.* (London) **A63**, 761–5. Domain wall movement in a single crystal.
50S8 SHOCKLEY, W., WILLIAMS, H. J., KITTEL, C. *Phys. Rev.* **78**, 341–2. Dynamic experiments with a simple domain boundary.
50S9 SUSSMANN, H., EHRLICH, S. L. *J. Acoustic Soc. Amer.* **22**, 499–506. Magnetostrictive properties of Hiperco.
50S10 STONER, E. C. *Phys. Soc. Rep. Prog. Physics* **13**, 83–183.
50T1 TEBBLE, R. S., SKIDMORE, I. C., CORNER, W. D. *Proc. Phys. Soc.* (London) **A63**, 739–61. Barkhausen effect. 528
50V1 VAUTIER, R. *J. Res. C.N.R.S.* No. 10, 23–7. Measurement of linear and volume magnetostriction. 630, 681
50W1 WEIL, L., BERTAUT, F., BOCHIROL, L. *J. phys. radium* **11**, 208–12. Magnetic properties and structure of tetragonal copper ferrite.
50W2 WELCH, A. J. E., NICKS, P. F., FAIRWEATHER, A., ROBERTS, F. F. *Phys. Rev.* **77**, 403. Natural ferromagnetic resonance.
50W3 WILLIAMS, H. J. *Elec. Eng.* **69**, 817–22. Ferromagnetic domains.
50W4 WILLIAMS, H. J. SHOCKLEY, W. *Phys. Rev.* **78**, 341. Memory in a simple domain crystal. See also *Phys. Rev.* **80**, 1090, 496

1951

The following papers on ferromagnetism were presented at the Grenoble Conference on Ferromagnetism and Antiferromagnetism on July 3–7, 1950. They are intended for publication during 1951; they were received too late for proper review in this book.

51B1 BOZORTH, R. M. Magnetic domain patterns. 538
51B2 BECKER, R. Dynamics of the Bloch wall and the permeability at high frequencies.
51B3 BATES, L. F. Thermal effects associated with ferromagnetism.
51B4 BARBIER, J. C. Irreversible lag in weak fields.
51B5 BERG, T. G. O. Ferromagnetism, paramagnetism and cohesive energy.
51C1 CHEVALLIER, R. Magnetic properties of rhombohedral iron oxide. 241
51E1 EPELBOIN, I. Recent theories and experiments on the Rayleigh region.
51F1 FORRER, R. New electronic conceptions of ferromagnetics.
51G1 GOLDMAN, J. E. New techniques and results in the measurement of magnetostriction.
51G2 GUILLAUD, C. Magnetic properties of ferrites.
51G3 GUILLAUD, C. Rotation of magnetization.
51G4 GOLDSCHMIDT, R. Ferromagnetism in weak fields.
51H1 HOSELITZ, K. Recent progress in the field of permanent magnets.

51J1 Josso, E. Structure and magnetic properties of alloys near FeNi$_3$. 103
51K1 Kittel, C. Ferromagnetic resonance.
51L1 Lliboutry, L. Magnetic effects of tension in weak fields.
51L2 Langevin, A. Permeability of steel under tension.
51M1 Meyer, A. J. P. Gyromagnetic ratios of some alloys of the iron group.
51M2 Michel, A., Chaudron, G., Bénard, J. Properties of non-metallic ferromagnetic compounds.
51N1 Néel, L. Magnetic lag.
51P1 Pauthenet, R., Bochirol, L. Spontaneous magnetization of ferrites.
51S1 Snoek, J. L. Magnetic properties of ferrites.
51S2 Shockley, W. Dynamic experiments with a simple domain boundary.
51S3 Stewart, K. H. Experiment on a specimen with large domains.
51S4 Stoner, E. C. Collective electron ferromagnetism.
51S5 Smoluchowski, R. Influence of order on magnetic properties.
51S6 Sucksmith, W. Magnetic saturation and some related phenomena. 264, 265
51T1 Taglang, P. Atomic moments and Curie points of isoelectronic alloys.
51T2 Trombe, F. Ferromagnetism of metallic dysprosium.
51V1 van Santen, J. H., Jonker, G. H. Ferromagnetic compounds of manganese with the perovskite structure.
51W1 Weil, L. Coercive force.
51W2 Weil, L. Expansion anomalies of ferrites.

NAME INDEX

Numbers refer to the preceding Bibliography, in which the pages of the text are given.

Adams, J. Q., 38A1
Adams, J. R., 26A1
Adcock, F., 31A5
Adelsberger, U., 27A1
Ageew, N. W., 30A4
Ahrens, E., 34A1
Akulov, N. S., 28A3, 29A3, 30A3, 31A3, 31A4, 32A3, 32A5, 33A4, 40A1, 40A2, 49A1
Alaverdov, G., 39A4
Alder, M., 16A1
Alexander, W. O., 37A1, 39A1
Alizade, Z. I., 49A1
Allanson, J. T., 45A2
Allegheny-Ludlum, **39A3**
Allibone, T. E., 28A5
Amemiya, T., 38I3
American Institute of Electrical Engineers, 42A2
American Society for Metals, 39A2
American Society for Testing Materials, 49A2
Andersen, A. G. H., 43A2
Anderson, N. L., 22A1
Anderson, P. W., 50A2
Andrew, J. H., 25A1
Antik, I., 34A3
Applegate, K. P., 15A1
Arkadiew, W., 14A1, 19A1
Armbruster, M. H., 43A4
Armco Bulletin, 43A3
Arnfelt, H., 28A2
Arnold, R. D., 49K6
Arnold, H. D., 23A2
Årstad, F., 37N4
Arvidsson, G., 20A1
Aschenbrenner, H., 36A4
Ashworth, J. A., p. 92, 137, 782
Ashworth, J. R., 38A2, 97A1
Astbury, N. F., 48A1, 50A1
Asteroth, P., 08A1
Aston, J., 09B2, 09B3, 10B1
Atorf, H., 32A4
Au, R., 49V1
Auer, H., 33A6, 39A5
Auerbach, F., 20A2
Austin, C. R., 37J2, 41A3
Austin, J. B., 42A3
Austin, L. W., 04A1, 06G1
Auwärter, M., 50A3
Auwers, O. v., 20A3, 28A1, 29A2, 33A2, 33A3, 33A5, 34A2, 35A1, 36A1, 42A1
Averbach, B. L., 41A1
Awberry, J. H., 40A2

Babbitt, B. J., 28B3
Babich, M. M., 39B17
Baily, F. G., 96B1
Bain, E. C., 35D7
Baker, C. J. W., 40B7
Baker, W. O., p. 533
Baldwin, J. M., 06L1
Ball, J. D., 15B2, 16B1
Bampfylde, J. W., 34S10
Bannister, R. J., 36S7
Barber, I. G., 28S1
Barbezat, S., 46G4
Barbier, J. C., 50B6, 51B4
Bardell, P. R., 50R2
Barkhausen, H., 19B2
Barnes, E. J., 10M1
Barnett, L. J. H., 25B1
Barnett, S. J., 15B1, 17B1, 25B1, 31B2, 34B4, 35B5, 40B6, 44B1, 49B8
Barrer, R. M., 41B1
Barrett, C. S., 43B6
Barrett, W. F., 82B1, 00B1, 02B1
Barth, T. F. W., 32B6
Bartlett, C. H., 47H2
Barton, J. P., 47B4
Barus, C., 87B1
Basart, J. M., 30B7
Batelle Mem. Inst., 40B1
Bates, L. F., 22C1, 23S2, 28B4, 28B5, 29B4, 30B8, 34B7, 37B10, 39B16, 40B7, 41B3, 41B7, 43B1, 46B7, 47B3, 48B5, 48B6, 48B7, 49B13, 50B4, 51B3
Beattie, R., 01B1
Beck, E., 19B1
Beck, F. J., 32B5, 35C4, 38I2
Beck, K., 18B1
Becker, G., 33V1, 35V2, 39K4
Becker, R., 30B4, 30B6, 32B8, 33B3, 34B5, 38B6, 39B5, 40B2, 51B2
Beckman, G. W., 50L1
Beljers, H. G., 49B12, 50B7
Bell, S. A., 49C4
Belov, K. P., 49A1
Bénard, J., 44G3, 51M2
Benford, F. G., 46B3
Bennett, E. G., 39S8
Berak, J., 50B5
Berg, T. G. O., 51B5
Bertaut, F., 49B11, 50W1, 50B13
Bertrand, R., 48G1, 49G7
Bethe, H., 33B2
Betteridge, W., 39B8

947

948   NAME INDEX

Bever, M. B., 50P1
Bickford, L. R., 49B4, 50B10
Bidwell, S., 86B1, 90B1
Biltz, W., 38B9
Binet du Jassoneix, A., 06B1, 12B2
Birch, F., 29W3
Birge, R. T., 41B6, 45B2
Birks, J. B., 46B5, 50B12
Bischoff, K., 34S6
Bittel, H., 38B2
Bitter, F., 30B5, 31B5, 32B3, 37B4, 37B6, 37T1, 38F2, 39B10, 39B12, 40S8, 42P1
Bittrich, G., p. 533
Bizette, H., 38S9, 43B5, 46B4
Bjurström, T., 33B4
Blake, F. C., 09B5
Blake, L. R., 49B6
Blechschmidt, E., 31G2
Bloch, F., 28B2, 30B3, 31B6, 32B2, 46B6
Bloch, O., 12B1
Bloembergen, N., 50B11
Bochirol, L., 50W1, 51P1
Boecker, G., 12B3
Bohlen-Halbach, C. v., 41B8
Booth, H. C., 12C2
Boothby, O. L., 39C1, 47B1, 47B6, pp. 69, 506
Borèn, B., 34B6
Both, E., 47B7, 50L1
Bourland, L. T., 31G3
Bowman, F. E., 43B3
Brace, P. H., 29B2, 32B9
Bradley, A. J., 32B7, 34B8, 37B2, 37B9, 38B1, 39B9, 39B15, 41B4, 49B7
Brailsford, E. F., 38B3, 39B6, 42B3, 43B2, 47B3, 48B1
Braithwaite, A., 23W2
Bramley, A., 35B8
Braunewell, W., 49B10
Bredig, G., 31B1
Brickwedde, F. G., 28W4
Bridgman, P. W., 25B4
Brill, R., 32B11
Bristow, C. A., 35A2, 39B2
Britzke, E. V., 30B2
Broadway, L., 28S10
Brockman, F. G., 50B1, 50B2
Broniewski, W., 35B7
Brooks, H., 40B5
Brown, R. C., 28B4
Brown, W., 00B1, 02B1, 10B2
Brown, W. F., 36B2, 37B7, 38B4, 38B8, 41B2, 49B2
Brüchatov, N. L., 32A5, 37B5, 38B7
Brunauer, S., 31B3, 49M1
Bruner, W. L., 34J1
Buchholz, H., 29B3
Buckley, O. E., 25B2
Buckley, S. E., 50R2
Bucknall, E. H., 37J2
Büchner, A., 37N2
Büssem, W., 33B5
Buhl, O., 49B9, 50B3
Bulgakov, N. V., 50B8
Bullens, D. K., 39B7

Bumm, H., 38B5, 39B19
Bungardt, K., 38E2
Burger, E. E., 34H2
Burgers, W. G., 30B7, 35B10
Burgers, W. J., 34S5
Burgess, C. F., 09B2, 09B3, 10B1
Burgess, E. E., 41H1
Burns, B. D., 49R1
Burns, R. S., 46C1
Burrows, C. W., 09B4, 21B2
Butler, O. I., 48B4, 50B9

Camp, J. M., 40C1
Campbell, A., 12C2
Campbell, E. D., 26C1
Carapella, L. A., 42C5
Carnegie-Illinois Steel Corp., 49C1
Carnegie Institution of Washington, 40C2
Carpenter, V. W., 41C1, 42C2, 42C3, 45C2, 49C4
Carr, W. J., p. 649, 678
Carter, R. O., 50C1
Catherall, A. C., 36C2
Cauer, W., 25C2
Chapin, D. M., 42B1
Charp, S., 42W4
Chattock, A. P., 22C1, 87C1
Chaudron, G., 25F1, 51M2
Chegwidden, R. A., 37C4, 41A1, 48C1, p. 92, 137
Cheney, W. L., 20S1, 22C2
Chevallier, R., 51C1
Chevenard, P., 17C1, 28C2
Chéveneau, C., 03C1, 10C1, 17H1
Chipman, J., 37C3
Christensen, H., 41W1
Chubb, L. W., 15C1
Cioffi, P. P., 26M4, 30C1, 32C2, 32C5, 34C1, 37C2, 38C1, 39C1, 45C1, 47C2, 48C2, p. 42, 361, 674
Classen, H., 22C3
Clark, C. W., 35K9
Clark, J. R., 43C1
Clash, R. F., 34M5, 35C4
Claussen, G. E., 36M9
Cleaves, H. E., 35C3, 39T1, 42C1
Clusius, K., 36C3
Coeterier, F., 32C3, 35C1
Cohen, E. R., 48D2, 49D3
Cohen, M., 40G4, 49T1, 50P1
Cole, G. H., 39C2, 46C1
Cole, R. H., 38C2
Coles, B. R., 49C2
Colombani, A., 44C1
Cone, E. F., 35C5
Conradt, H. W., 40C3, 42C4
Constant, F. W., 28C3, 30C2, 43C3
Conybeare, J. G. G., 37C5
Cook, A. J., 43C2, 49R1
Cooke, W. T., 36C1
Cooper, A. T., 35B8
Coppoolse, C. W., 29W6
Corner, W. D., 50S7
Corson, M. G., 28C5

Cotton, A., 00C1, 28C1
Cox, W. F., 41B4
Crafts, W., 37K2
Crawford, C. H., 35C2
Crede, J. H., 49C3
Creveaux, H., 46G6, 47G2, 50G5
Crittenden, E. C., 46C2, 49C5
Cross, H. C., 28C4
Curie, M., 97C1, 98C1
Curie, P., 03C1, 95C1
Czerlinsky, E., 32C4, 33G8

Daeves, K., 2401, 29D3
Dahl, O., 29M5, 31D3, 32D1, 33D2, 34D1, 34D2, 35D1, 35D2, 35D3, 35D6, 36D1, 36D2, 36D5, 40C3
D'Albon, G., 37P2
Daniel, V., 43D1, 44D1
Dannatt, C., 36D7
Dannöhl, W., 35K3, 36K8, 38D1, 38D3, 38D4, 38V2
Darken, L. S., 48D1
Davenport, E. S., 35D7
Davidson, R. L., 39C2
Davies, C. W., 41D1
Davis, H. W., 41K1
Davis, J. H., 48B6
Day, M. J., 42A3
Dean, R. S., 33D3, 41D1, 44D2, 46D1
Dean, W. A., 30D1
DeBarr, A. E., 48A1
DeBoer, J., 36V3, 37M7, 38D6
Debye, P., 07D1
Dee, A. A., 24S2
de Klerk, D., 50D3
DeGroot, S. R., 50M2
Degtiar, M., 32A3
Dehler, H., 42D1
Dehlinger, U., 36D8
Deutschmann, W., 29D1
Dickie, H. A., 27D1
Dieckmann, T., 11H2
Dietsch, G., 31D1, 31D2
Dijkstra, L. J., 49D5
Dillinger, J. F., 31B4, 32B4, 34B3, 35B2, 35B6, 35D4
Dirac, P. A. M., p. 450
Döring, W., 36D4, 36D6, 38D2, 38D5, 38D7, 39B5, 39D4, 47D1, 48D3, 50D1
Domenicali, C. A., 49D2, 50D2
Dorfman, J., 30F5
Dorsey, H. G., 10D1
Dowdell, R. L., 34D3, 40F1
Dowling, P. H., 50B1
Dreibholz, 24D1
Dreyfus, L., 35D8
Drigo, A., 49D4
Drozzina, V. I., 35D5, 48D4, 49D6
Druyckaerts, G., 39D2, 39D3
duBois, H. E. J. G., 90D1
DuMond, J. W. M., 48D2, 49D3
Dunn, C. G., 45D1
Dupouy, G., 27D2
Dushman, S., 49D1

Dussler, E., 28D1
Duyckaerts, G., 39D2, 39D3
Dwight, H. B., 29D2
Dye, D. W., 12C2

Eash, J. T., 48E3, 48K7
Ebert, F., 37E1, 38E5
Ebinger, A., 30E4
Edgar, R. F., 35E3, 37E4, 40H2
Edmondson, A. S., 47B3
Edwards, A., 44E1
Edwards, C. A., 20E2, 24E2
Edwards, O. S., 41E2, 42E1, 43E3
Eggers, H., 38E6
Ehret, W. F., 43E2
Ehrlich, S. L., 50S9
Eichholz, G., 49H2
Eilender, W., 27E3, 34E3, 38E2
Einstein, A., 15E1
Ekman, W., 30W7
Ekstein, H., 50E1
Electrical Engineering Staff, M.I.T., 43E1
Elenbaas, W., 32E2, 32E3
Ellefson, B. S., 34E4
Ellinger, F. H., 40E1, 42E3
Ellis, W. C., 27E2, 34E1, 41E1, 48E1, 48F1, 50F3, p. 201
Ellwood, W. B., 30E5, 34E2, 35E2, 37E2
Elmen, G. W., 17E1, 21S2, 22E1, 23A2, 26E1, 27E1, 28E1, 29E1, 29E4, 30E2, 31E2, 35E1
Elmore, W. C., 34M3, 34M7, 36E3, 37E3, 38E3, 38E4, 38E7, 42E2
Elsea, A. R., 48E2
Emde, F., 23J1
Emden, H. J. M. v., 41J1
Emicke, O., 2501
Emmerson, T., 50A1
Emmett, P. H., 31B3
Endter, F., 44E2
Engel, A., 47R1
Engler, O., 38E1
Englert, E., 32E1, 32E4
Epelboin, I., 51E1
Eppelsheimer, D. S., 48H1, 49H3
Epremian, E., 49E1
Epstein, J., 00E2
Ergang, R., 38V3, 39S4, 39S7
Esch, U., 44E3
Esser, H., 29G2, 38E2, 41E3
Eucken, A., 30E1
Eusterbrock, H., 41E3
Evans, H., 34S9, 35S8
Evershed, S., 20E1, 25E1
Evert, T., 38E5
Ewert, M., 36E2
Ewing, J. A., 89E1, 00E1, 22E2, 93E1

Fahlenbrach, H., 48F4, 49F1, 49F3
Fahy, F. P., 18F1
Faires, R. E., 43C3
Fairlie, D. M., 12P2
Fairweather, A., 50W2

Fallot, M., 34F1, 36F1, 37F1, 38F1, 39F1, 44G3
Farcas, T., 37F5
Farquar, M. C. M., 45F1
Farquharson, J., 32G5
Faus, H. T., 30K9
Felch, E. P., 46F1
Feldtkeller, R., 50F2
Fetz, E., 35F2, 39F3
Feustel, R. G., 46D1
Fick, K., 13R1
Fick, N. C., 46R1
Finch, O. J., 44F1
Fine, M. E., 48F1, 50F3
Fink, W. L., 48F2
Finzi, 91F1
Fischer, F. K., 30F1
Fisher, J. V. S., 09L1
Fisher, M. S., 25A1
Floe, C. F., 50P1
Förster, F., 37F3, 37F4, 41F1, 41F2
Foëx, G., 11W2, 26W1, 29W2, 39F2
Foëx, M., 34J2, 49F2
Fondiller, W., 21F1
Foote, F., 34J1, 35J1, 36J1
Ford, L. H., 34W3
Forestier, H., 25F1, 39F1, 50F1
Forrer, R., 24W3, 24W5, 26F1, 26W4, 29W1, 29W3, 30F2, 30F4, 31F1, 32F1, 33F2, 51F1
Forsyth, A. C., 40F1
Foster, D. D., 29F1, 30F3
Fowler, R. H., 31F2
Fráncis, C. B., 40C1
Frank, A., 29V1
Frank, H., 00F1
Franklin, R. J., 50L1
Franks, R., 40K2
Fratoni, N., 43K12
Frenkel, J., 30F5
Freudenreich, J. de, 14F1, 16W1
Frey, A. A., 38F2
Fricke, W., 31D2, 33F1
Friederich, E., 35F1
Friedrich, K., 08F1, 08F2, 12F1
Frölich, O., 81F1
Frommer, L., 45F2
Fry, A., 23F1

Galavics, F., 39G2
Galperin, F. M., 49G5
Galt, J. K., 50G4
Gans, R., 10G1, 11G1, 15G1, 20G2, 32G3, 33G5, 33G8, 33G9
Ganz, A. G., 46G1
Garnett, H. J., 24S1, 33S8
Gebhardt, E., 38K8, 38K10, 38K12, 40G3
Geisler, A. H., 48G2, 50G1, p. 413
Geller, W., 35K7, 35K8, 39G3
Géneau, C., 17H1
Genders, R., 36G3, 39G6
Gentile, G., 31B6
Gerlach, W., 24G2, 25G1, 26G2, 28G3, 30G5, 30G6, 30G7, 32G4, 33G4, 36G1, 37G3, 38G1, 39G4, 43G3, 47G1, 49G4
Germer, L. H., 42G1
Gerold, E., 31G1
Gewecke, H., 28G2
Giamboni, L. A., 49B8
Giauque, W. F., 27G1
Giebe, E., 31G2
Giebelhausen, H. v., 15G2
Gier, J. R., 30G4
Gilbert, T., 50E1
Given, F. J., 40L1
Glathart, J. L., 39G5
Goeckler, F. E., 26A1
Goerens, P., 12G2, 24G1
Goertz, Miss M., p. 175
Goldberg, L., 38M3
Goldman, J. E., 49G1, 51G1
Goldmann, J., 36C3
Goldschmidt, H. J., 39B9, 41B4
Goldschmidt, R., 30G3, 51G4
Goldschmidt, V. M., 27G2
Gonser, B. W., 46J1
Goobar, S. W., 38K12, 40K6
Gorden, R. B., 40G4
Gordon, P., 50G2
Gorski, W. S., 35G4
Gorter, C. J., 33G7, 40G5, 41G1, 50D3
Gorter, E. W., 50G3, 50G6
Goss, N. P., 34G4, 35G5, 37G2
Gottlieb, N., 39H3
Gottschalk, V. H., 35G6, 41G2
Goubau, G., 36A4
Goudsmit, S., 26G1
Gould, H. L. B., p. 178
Gould, J. E., 29G1
Gouy, L. G., 89G1
Grabbe, E. M., 39M1, 40G1
Graf, L., 35G2
Graff, H. F., 35S10
Graff, M., 39F2
Graham, T. R., 46D1
Gray, A. 12G3
Gray, F. W., 32G5
Gray, R. C., 10R2
Gregg, J. L., 32G2, 34G1
Greiner, E. S., 33J2, 37G4, 41E1, 41E2, 43G1, 48E1, p. 201, 383
Grew, K. E., 34G3
Gridnev, V. N., 37S7
Gries, H., 29G2
Griffel, M., 49S6
Griffiths, E., 40A2, 46G7
Griffiths, J. H. E., 46G7
Grinstead, R. R., 49G2
Gross, F., 33B5
Grover, F. W., 12R2
Grube, G., 35G3, 36G2, 38G3, 38G4
Gruner, E., 37G5
Guggenheimer, K., 45S2
Guild J., 12S1
Guillaud, C., 39H4, 43G2, 44G1, 44G2, 44G3, 44G4, 46G2, 46G3, 46G4, 46G5, 46G6, 48G1, 48G3, 47G2, 49G6, 49G7,

49G8, 49G9, 50G5, 50S1, 50G7, 51G2, 51G3
Guillaume, C. E., 20G1
Guiot-Guillain, G., 50F1
Gullemin, A., 1846G1
Gumlich, E., 05G1, 09G1, 12G1, 12G2, 15S1, 16G1, 18G2, 19G1, 22G1, 23G1, 27G3, 28G1, 30G2
Gurinsky, D. C., 43E2
Gurr, W., 31W4
Gurry, R. W., 48D1
Guthe, K. E., 06G1
Guthrie, A. N., 31G3
Gwyer, A. G. C., 08G1

Haag, W., 32B11
Haas, W. J. de, 15E1, 46H1
Haayman, P. W., 41V3, 47V5
Haberland, F., 36H3
Haberland, G., 36H3
Hadfield, D., 48O1
Hadfield, R. A., 89H2, 92H1, 00B1, 02B1, 03H3, 17H1, 26H1, 27H1
Hägg, G., 29H1, 29H3
Hagemann, H., 33H1
Hagen, E., 03H2
Hagen, H., 34S13
Hall, E. L., 28W4
Halla, F., 36H5
Halpern, O., 39H6
Hames, F. A., 48H2, 49H3
Hámos, L. v., 31H3, 32H3
Hannack, G., 24H2, 25S3
Hansen, M., 36H1
Hansen, W. W., 46B6, 47W3
Haraldsen, H., 34H4, 35H5, 35H6, 39H5, 41H6, 43H4
Hardy, T. C., 38H1
Harker, D., 44H3, 47H1, 49E1
Harlem, J. v., 32H2, 33G5, 33G9
Harrison, E. G., 48B5, 48B7
Harrison, R., 36G3, 39G6
Hashimoto, U., 29W7, 32H1, 37H3, 37H4, 37H5, 38H3, 38H4
Hass, K., 34K14
Haughton, J. L., 27H2, 34H3, 42H2, 44H4
Haupt, E., 03H5
Havens, C. G., 33H4
Haworth, F. E., 31H5, 37H1, 39H2
Haynes, J. R., 35H1
Haywood, F. W., 35B8
Healey, D. R., 43B1
Heaps, C. W., 15H1, 23H2, 24H1
Heck, J. E., 45C2
Hector, L. G., 24W6
Hegg, F., 10H1
Heglein, E., 38N5
Heidenreich, R. D., p. 533
Heilmann, E. L., 47V4
Heimbrecht, M., 38B9
Heinzel, A., 31W1
Heisenberg, W., 28H1, 30H2, 31H6
Heitler, H., 45S2
Hendricks, S. B., 31B3

Hendus, H., 40H5
Heimenz, H., 38H2
Heraeus-Vacuumschmelze, A. G., 31H1
Hermann, P. C., 33H3
Herroun, E. F., 43H2
Herzig, A. J., 43B3
Hessenbruch, W., 44H2
Heumann, T., 43H3, 50B5
Heusler, F., 03H4, 03H5, 04H1, 09H1, 12H1
Heusler, O., 28H4, 33H5, 34H5
Hewitt, W. H., 48H3
Hibi, T., 40H6
Hiegel, J. M., 42C1
Hiemenz, H., 33H1, 38H2
Hill, E. E., 28C4
Hilpert, R. S., 09H2, 11H2, 13H1, 38H5
Hindricks, J., 31W2
Hirone, T., 29H4, 39O3, 42H3
Hoag, J. B., 32H4, 39H3
Hobson, P. T., 48H1
Hocart, R., 39H4
Hodsman, G. F., 49H2
Hofer, L. J. E., 47H3
Hoffmann, A., 38H5
Holborn, L., 21H1
Holstein, T., 40H7, 41H2
Honda, K., 02H1, 02H2, 02N1, 03H1, 04N1, 05H1, 10H2, 10H3, 19H1, 20H1, 20H2, 21H2, 26H2, 26H3, 28H2, 28H3, 29H4, 31H2, 34H1, 35H4, 36H6, 45H1
Hopkinson, J., 85H1, 89H1, 90H1
Hori, N., 42H3
Hornfeck, A. J., 40H2
Horsburgh, C. D. L., 35H2
Horstman, C. C., 47H2
Hoselitz, K., 43H1, 44H1, 44E1, 45S2, 47H4, 49H1, 51H1
Houdremont, E., 33H2
Hougardy, H., 31H4, 40H1
Howe, G. H., 40H3, 40H4, 41H3, 42H1, 43H5, 44H1, p. 372
Howey, J. H., 29H2
Hülsmann, O., 36H2
Hull, A. W., 34H2, 41H1
Hultgren, R., 39H1, 41H4, 42C5
Hume-Rothery, W., 49C2
Hund, F., 25H1
Hunter, M. A., 23H1

Ide, J. M., 31I1, 34I1
Illsley, P. F., 37B10
Ingerson, W. E., 38I2
Inglis, D. R., 29I2, 34I2
International Critical Tables, 29I1
Ishigaki, T., 27I1
Ishihara, S., 40T2
Ishiwara, T., 16I1, 30I1
Ising, E., 25I1
Iwasé, K., 36I1, 38I3, 41I1

Jack, K. H., 48J2, 48J3
Jackson, J. M., 42C2
Jackson, L. R., 38J1, 40R1

952 NAME INDEX

Jaeger, F. M., 38J2
Jaffee, R. I., 41H4, 46J1, 48J1
Jahnke, E., 23J1
Janus, R., 35D5, 37J1
Jaspers, A., 37M7
Jay, A. H., 32B7, 37B2
Jefferson, M. E., 31B3
Jellinghaus, W., 30W1, 35J2, 36J2, 36J3, 36J4, 43J1, 48J4, 49J1, 49J2
Jenge, W., 26S1, 26S2
Jenkins, C. H. M., 37J2
Jette, E., 33J2, 34J1, 34J2, 35F2, 35J1, 36J1, 37G4, 43G1
Johnson, H. L., 40J1
Johnson, M. H., 39H6, 47J1
Jonas, B., 41J1
Jones, A., 23H1
Jones, F. W., 43C2
Jones, Hayden, 32H4
Jones, H., 36M13
Jong, W. F. de, 27J1
Jonker, G. H., 50J1, 51V1
Jordan, H., 24J1
Jordan, L., 30J1
Josso, E., 51J1
Joule, J. P., 1842J1, 1847J1
Juza, R., 43J2

Kaden, H., 32W5
Källbäck, O., 47K1
Kämmerer, H., 49K7
Kästner, H., 36G2
Kahan, T., 38K15
Kahn, B., 40G5
Kaiser, J. F., 23K1
Kamura, H., 34K11, 35K10, 35K11
Kaneko, K., 12R1
Kanzler, M., 39K9
Kapitza, P., 27K4
Kapustinsky, A. F., 30B2
Kasé, T., 27K1
Kath, H., 98K1
Kato, Y., 33K7
Kaufmann, A. R., 39B10, 40S8, 41R3, 42P1, 43C1, 50G2
Kawai, N., 38H4
Kaya, S., 26H2, 28H3, 28K1, 28K2, 31K7, 33K10, 34K12, 36K7, 38K2, 38K3, 39K1, 39K3, 40K5, 43K13
Keesom, W. H., 35K9, 39K6
Keinath, G., 31K5
Kelley, K. K., 37K9, 39K8, 43K4
Kellogg, J. B. M., 46K4
Kelsall, G. A., 24K1, 24K2, 34K1, 34B3, 35K4, 40N1, p. 183
Kennelly, A. E., 91K1
Kenyon, R. L., 48K4
Kersten, M., 30B4, 31K1, 31K6, 32K11, 33K1, 33K11, 34K3, 34K4, 34K8, 34K9, 38K1, 38K4, 43K5, 43K6, 48K11
Kido, K., 20H1
Kihlgren, T. E., 48K6, 48K7
Kikoin, I. K., 38K12, 40K6
Kimura, R., 39K7, 40K4

Kindler, H., 36K5, 37K4
Kingsbury, A. W., 43A2
Kinnard, I. F., 30K9
Kinzel, A. B., 37K2, 40K2
Kinzoku Zairyo Kenkyusho, 35K6, 36K3
Kip, A. F., 49K6
Kirchner, H., 36K1
Kirensky, L. W., 37B5, 38B7, 40A1, 48K10, 49K8
Kirkham, D., 37K1
Kisljakova, E. M., 39B17
Kittel, C., 46K1, 46K2, 46K3, 48K2, 48K9, 48L2, 49K1, 49K5, 50K1, 50G7, 50S8, 51K1
Kiuti, S., 41K2
Klaasen, H. G., 93E1
Kleis, J. D., 36K2
Klemm, W., 34H4, 34K14, 35H6, 35K13, 36K4, 37G5, 43K12, 44E2
Klinkhardt, H., 27K2
Knol, K. S., 40S4
Khbayashi, T., 45H1
König, H., 46K5
Koepsel, A., 94K1
Köster, W., 29K3, 30K4, 30K6, 30K7, 30K8, 31K8, 32K4, 32K5, 32K6, 32K8, 32K9, 32K12, 33K2, 33K3, 33K5, 33K6, 33K8, 34K5, 35K3, 35K5, 35K7, 35K8, 36K8, 37F3, 37K7, 37K8, 38K3, 38K8, 38K9, 38K10, 38K12, 38K14, 39K4, 39K5, 40G3, 40K3, 43K7, 43K8, 43K9, 43K10, 48K3, 48K5, 48K8
Kohaut, A., 37K6
Kohlrausch, F., 35K14
Kollmann, K., 27T1
Komar, A., 41V2, 33A4
Kondorsky, E., 37K5, 38K13
Konstantinov, N., 08K1
Kornetzki, M., 34B5, 34K10, 35K1, 35K2, 38K11, 43K2, 43K3, 43K11, 48K1
Kortengräber, A., 33S9
Koster, J., 41K1
Kosting, P. R., 30K1
Kowalski, E., 35H5
Krainer, H., 43K1
Kramer, I. R., 42W6
Kramers, H. A., 34K15, 40K7, 41K3
Krause, D. E., 36L3
Kreielsheimer, K., 33K9
Kreutzer, C., 2902
Kříž, A., 32K10
Kroll, W., 29K2, 32K1, 36K6, 44K1
Krumbhaar, W., 10S1
Krupkowski, A., 29K5
Krutter, H., 35K4
Kubaschewski, O., 49K2, 49K3
Kubyschkina, T., 34A3
Kuhlewein, H., 30K3, 31K4, 32K2, 32K7, 33A3, 34K7
Kurrelmeyer, B., 39K6
Kusakabe, S., 02H1
Kusch, P., 39R3
Kussmann, A., 28G1, 28K4, 29K1, 30G2,

30K5, 31K7, 32K3, 32M3, 33K4, 35F1, 35G2, 37E1, 37K3, 37S4, 38E5, 38K6, 38K7, 49K4, 50A3, 50K2

LaBlanchetais, C. H., 45L1, 49F2
Lacy, C. E., 48L1
Lamont, J., 1867L1
Lamort, J., 14L1
Lance, T. M. C., 22A1
Landau, L., 35L1
Lang, K., 38K3
Lange, H., 30W6, 38L5
Langevin, A., 51L2
Langevin, P., 05L1
Langmuir, I., p. 494
Lapp, E., 29L1, 36L2
Latimer, K. E., 50L2
Laves, F., 35L2, 39L3
Leech, P., 39L1
Legat, H., 37L2, 38L7
Legg, V. E., 36L1, 39L2, 40L1
Leipunsky, O. I., 38L1
Leitgebel, W., 41B8
Lenander, H., 43C3
Leu, A., 27L1
Li, T., 37C3
Libsch, J. F., 50L1
Lichtenberger, F., 32L1
Liedgens, J., 12L1
Lifshitz, E., 35L1, 44L1
Lilley, B. A., 50L5
Lindman, K. F., 38L2
Lipson, H., 41L2, 42E1, 43D1, 44D1, 45F1
Liu, Y. H., 49O1
Livshits, B. G., 41L1
Lliboutry, L., 50L3, 50L4, 51L1
Lloyd, M. G., 09L1
Löhberg, K., 38L3
Lokich, J. L., 48T1
Long, J. R., 46D1
Lonsdale, K., 38L4
Lorig, C. H., 36L3
Loring, B. M., 42W6
Lossew, K., 06L2
Lounsberry, F. B., 47S4
Loyarte, R. G., 15G1
Lundquist, D., 38L6
Luttinger, J. M., 48L2
Luzhinskaya, M. G., 49D6
Lyle, T. R., 06L1
Lynch, A. C., 50R2

MacColl, L. A., p. 706, 784, 786, 787
MacDonald, H. B., 50L2
MacNair, D., 41N1
Maddocks, W. R., 36M9, 39R1
Mahajani, G. S., 26M5, 29M2
Maier, K. H., 38H5
Maloof, M., 47J1
Mames, E., 28M4
Mang, C. Y., 48B4
Manning, M. F., 38M3
Marfoure, S., 47W2

Margolin, J. M., 39M4
Marian, V., 37M2
Marick, L., 36M5
Mars, G., 09M1
Marsh, J. S., 38M2
Marshall, A. L., 40J1
Martak, J., 32F1
Martin, D. L., p. 413
Martin, J. P., 49C3
Martin, W. H., 21F1
Martindale, R. G., 42B3
Masing, G., 28M5, 29M5, 38V2, 40R2
Masiyama, Y., 26H3, 28M6, 31M1, 33M1, 32M2, 37M5, 37M6
Mason, W. P., 42M1, 49B5
Masumoto, H., 26M1, 27M1, 27M2, 28M3, 29M1, 31H2, 31M2, 34H1, 34M4, 35H4, 36M8, 36M12, 41M2
Mathewson, C. H., 35S4
Matsunaga, Y., 31M4
Matsushita, T., 22M1
Matumura, T., 21H2
Matuyama, Y., 34M11
Maurain, C., 98M1
Maurer, E., 25M1
Maxwell, E., 46M1
Maxwell, L. R., 49M1
Mayneord, W. V., 24S2
McCaig, M., 47H4, 49H1, 49M2, 50M1
McFarlane, J., 50A1
McGeachy, J, A., 35W2
McGuire, F. T., 43T1
McKeehan, L. W., 25B2, 26M2, 26M3, 26M4, 29M4, 30M1, 32B5, 34M5, 34M7, 34M1, 34M3, 34M10, 36E3, 37B8, 37M1, 37M8, 39M1
McLennan, J. C., 07M1
McMullen, E. W., 46J1
McSkimin, H. J., 49B5
McWilliams, A., 10M1
Mehl, R. F., 40W1, p. 42
Mellor, G. A., 37J2
Melville, W. S., 50M3
Merica, P. D., 25M2
Merritt, F. R., 49Y1, 50G7
Messkin, W. S., 28S7, 29M3, 29S2, 29S3, 32M3, 34M2, 36M10, 30K5, 39M4, 41M1
Meyer, A. J. P., 49M3, 49M4, 51M1
Meyer, W. F., 37M4
Meyers, H. P., 49C2
Michel, A., 44G3, 51M2
Michels, A., 37M7, 38D6, 41M3, 50M2
Michels, R., 31M3
Mikulas, W., 37M9
Millar, R. W., 28M7
Millership, R., 49H2
Millman, S., 39R3, 46K4
Miner, D. F., 37M3
Misch, L., 35M2
Mishima, T., 31M5, 32M1, 33M3, 36M6, 36M7
Mitkevitch, A., 36M14
Mittasch, A., 28M1

NAME INDEX

Miyahara, S., 39K1, 39K10
Möbius, W., 32M4, 34M9
Möhring, D., 39M3
Moeller, C., 33M2
Moeller, K., 48K5
Mohrnheim, A., 48S9
Monypenny, J. H. G., 20M1
Moriwaki, 39M2
Morral, F. R., 34M8
Morrill, W., 48M1
Morris, R. J., 40B4
Morse, P. M., 36M15
Moser, H., 36M4
Mott, N. F., 35M1, 36M11, 36M13, 38M1
Müller, A., 30W4, 30W5
Müller, H. C., 38B5
Müller, N., 34M6
Mulfinger, W., 38K14, 40K3
Murakami, T., 21M1
Murata, T., 40O1
Murray, A., 45F2
Mussmann, H., 40M1
Mutuzaki, K., 40T3
Myers, H. P., 49C2

Nagaoka, H., 96N1, 02N1, 04N1
Nakamura, K., 35N1, 35N3, 36N2
Nakayama, M., 39K3, 40K5, 43K13
Nara, S., 27M1
Navias, L., 41H1
Neale, F. E., 50B4
Néel, L., 32N2, 36N3, 37N3, 42N2, 44N1, 44N2, 44N3, 46N2, 47N1, 47N2, 47N3, 48N1, 48N4, 49N1, 49N2, 49N3, 49N5, 50N2, 50N3, 50N4, 50N5, 50N6, 51N1, 51N2, 50N7, 51N1, 51N2
Negresco, T., 28W2
Nekhamkin, A. S., 41M1
Nemilov, V. A., 32N1, 38N4
Nesbitt, E. A., 40N1, 42N1, 43N1, 43N2, 46N1, 48N2, 49N1, 50K1, 50N1, 50N2, p. 383, 394, 401, 418
Neumann, H., 34N1, 34N2, 35A1, 37N1, 37N2, 38D1, 38N2, 38N5
Newkirk, J. B., 48G2, p. 413
Nial, O., 38N6
Nicks, P. F., 50W2
Nienhaus, H., 28M4
Nishina, T., 36H6, 36N1
Nishiyama, Z., 29N1, 29N2
Nitka, H., 38K6, 38K7
Niwa, Y., 24N1
Nix, F. C., 41N1
Norbury, A. L., 20E2
Nordstrom, V. H., 34J2
Norton, J. T., 35N2, 48N3
Nowotny, H., 36H5, 37N4, 38N5, 43N3, 46N3
Nygaard, E., 39H5

Oberhoffer, P., 24O1, 25O1, 29O2
Ochsenfeld, R., 32O1
Oelsen, W., 30T1

Oertel, W., 27E3, 29O4, 34E3
Oftedal, I., 27O1
Okamoto, M., 36I1, 38I3, 39O4, 39O5, 40O2, 41I1
Okamura, T., 36O2, 39O3
Okubo, J., 29H4
Oliver, D. A., 38O1, 38O2, 48O1
Olsen, L. O., 46C2
Onnes, H. K., 10W3, 23W3, 29O1
Onsager, L., 44O1
Opechowski, W., 37O5
Orowan, E., 34O1
Osawa, A., 29O3, 30O1, 39O3, 39O5, 40O1, 40O3
Osborn, J. A., 45O1
Osmond, W. P., 48H1
Owen, E. A., 34O2, 36O1, 37O3, 39O1, 41O1, 49O1
Owen, M., 12O1
Oya, M., 30O2
Oya, S., 29O3

Packard, M., 46B6
Pan, S. T., 39P1, 42P1, 43C1
Panebianco, G., 10P1
Paranjpe, V. G., 50P1
Parke, R. M., 43B3
Parker, E. R., 40P4, 40P5
Parkin, A. M., 24P1
Partridge, J. H., 28P2
Pascal, P., 10P2
Pauli, W., 26P2
Pauling, L., 38P1
Pauthenet, R., 50P2, 51P1
Pawlek, F., 35D1, 35D2, 35P1, 36D2, 43P1
Payne, R. J. M., 34H3
Pearce, R. R., 38S10
Pearson, G. L., 48P1
Peebles, W. C., 47H3
Perekalina, T. M., 49G5
Permanent Magnet Association, 45P2
Perrier, A., 26W5
Persson, E., 28P3, 29P4
Peschard, M., 25P1, 25P2
Peter, W., 38E6, 42W5
Peterson, E., 36P2
Peype, W. F. van., 38P2
Pfaffenberger, J., 31D3, 33D2, 33P2, 34D1, 34D2, 35D2, 35D3, 35D6, 36D5
Pfeil, L. B., 24E2
Pfisterer, H., 49P1
Philips Gloeilampen Fabrieken, 40P2
Phragmén, G., 26P1, 28W1, 28W2
Piccard, A., 18W1
Pickles, A. T., 40P1
Pickup, L., 34O2
Pierce, G. W., 29P2
Pietrek, W., 35B7
Piety, R. J., 36P3
Pilling, N. B., 23P1, 40P3
Piwowarsky, E., 31S2
Pizzo, M., 49D4
Planer, V., 08W1
Poboril, F., 32K10

Pölzguter, F., 28P1
Pogodin, S. A., 43P2
Polanyi, M., 34P1
Polder, D., 45P1, 50B7, 50P3
Polley, H., 39P4, 40B2
Posnjak, E., 32B6
Potapenko, G., 33P1, 37P3, 46P1
Potter, H. H., 26S3, 28S10, 29P3, 31P3, 31P4, 32P2, 34P2, 37P1
Pouillard, E., 48P2
Pound, R. V., 46P1
Powell, F. C., 30P3, 31F2, 31P6
Powell, R. W., 39P3
Preisach, F., 29P1, 32P1, 35P2, 35P4
Preston, G. D., 31P5
Preuss, A., 12P1, 12W1
Price, G. H. S., 38P3
Primakoff, H., 40H7, 41H2
Pring, J. N., 12P2
Procopiu, S., 30P2, 37P2
Purcell, E. M., 46P1
Puzei, I. M., 49P2

Quenau, B., 34J2
Quimby, S. L., 36S10, 38H1
Quincke, G., 85Q1
Quittner, V., 09Q1

Rabi, I. I., 39R3
Rado, G. T., 41R3, 47J1, 50R1
Raidle, F., 43K1
Randall, W. F., 33S8, 37R1, 50R3
Rankin, J. S., 29R2, 30R3, 31R2, 32R2
Rapatz, F., 24O1
Rathenau, G. W., 41R2
Raub, E., 47R1
Rauscher, W., 48K8
Ray-Chauduri, D. P., 35R1
Rayleigh, (Lord), 87R1
Rees, W. P., 49R1
Reinacher, G., 44S2
Reinboth, H., 37N2
Reinecken, W., 25W3
Rennenkampff, J. v., 43G3
Rhodes, P., 49S8
Ribbeck, F., 26R1
Richards, C. E., 50R2
Richards, D. L., 50C1
Richardson, O. W., 08R1, 23R1
Richarz, F., 09H1
Richter, G., 37R2, 38R1
Richter, R., 10R1
Rickett, R. L., 46R1
Rittberg, G. V., 50K2
Roberts, F. F., 50W2
Robertson, J. M., 25A1
Rocha, H. J., 33T2
Rodgers, J. W., 34B8, 39R1
Rogers, B. A., 29R1, 32S6, 33R3
Rohn, W., 33R4, 33R5
Romeijn, F. C., 47V5
Rosa, E. B., 12R2
Rosenbohm, E., 38J2
Rosenquist, T., 43H4

Rosenthal, K., 35V1
Rosin, S., 36S2
Ross, A. D., 10R2, 12R3
Roters, H. C., 41R1
Rotherham, L., 46R2
Roux, M., 49G8
Rubens, H., 03H2
Ruder, W. E., 14R2, 25R1, 34R1, 34R2, 34R3, 42R1, 46R3
Ruer, R., 12R1, 13R1
Ruppelt, A., 39S17
Russell, A., 14R1
Russell, H. W., 38J1, 40R1
Ruttewit, K., 40R2

Sachs, G., 40S6, 48S4
Sachsze, W., 43J2
Sadron, C., 30S8, 32S3
Sänger, R., 33P1, 34S7, 37P3
Sage, M., 50S1
Saimovski, A. S., 35S10
Saito, H., 43K13, 45M1
Saito, S., 20H2
Samans, C. H., 41A3
Sampson, J. B., 40S5
Samuel, M., 28S2
Sanford, R. L., 20S1, 27S5, 34S8, 39S8, 44S1, 46S4
Sato, T., 31S7
Scaff, J. H., p. 21, 86
Schaarwächter, 39S17
Schafmeister, P., 39S4, 39S7
Scharff, G., 35S9
Scharnow, B., 28G1, 29K1, 30G2, 30K5, 32K3, 33K4
Scheil, E., 34S6, 44S2
Schenck, R., 33S9
Schepelew, J., 09Z1
Scherrer, P., 32C3
Scheufele, E., 40H5
Schichtel, K., 44H2
Schlecht, H., 38G3
Schlechtweg, H., 36S15, 40M1
Schlumberger, E., 25S4
Schmid, E., 37K8, 50S3
Schmidt, W., 29S5, 33K6, 33S5, 38L3
Schneider, A., 44E3, 49K2, 49S5
Schneiderhan, K., 30G5
Schoen, E., 37S4
Schoenberg, D., 41L2
Scholefield, H. H., 49S10, 50R3
Schonert, K., 25S3
Schräder, H., 33H2
Schramm, J., 36S9, 41S2, 41S11, 41S12, 41S13, 48S6, 48S9
Schroeter, K., 28S5
Schubert, K., 46N3, 49P1
Schulz, E. H., 26S1, 26S2, 34S6
Schulze, A., 27S1, 27S2, 27S3, 28S3, 28S4, 28S8, 31S4, 32K3, 33S1, 35S5, 38S3
Schulze, H., 38S1, 38S2, 38S3
Schumacher, E. E., 34E1, 36S8
Schwartz, H., 32S8

NAME INDEX

Schwartz, N., 35D3
Schwartz von Bergkampf, E., 31B1
Schwarz, M. v., 14S1
Scott, H., 30S1
Scott, K. L., 30S6, 32S7, 39S5, 48S5
Seager, G. C., 39B15
Seastone, J. B., 37M3
Sebast, F. M., 23H1
Seitz, F., 40S5
Selissky, I. P., 41S6, 41Z2, 46S2
Seljesater, K. S., 32S6
Selwood, P. W., 43S2
Sendzimir, T., 46S3
Sequenz, H., 35S7
Seybolt, A. U., 35S4, 36S17
Shackelton, W. J., 28S1
Shaefer, R. H., 42W2
Shavlo, S., 39A4
Shaw, J. L., 48S10
Shedden, J. W., 3802
Shibata, N., 41S3, 41S14, 41S15, 40O3
Shih, J. W., 34S12, 36S14
Shimizu, S., 02H1, 02H2, 03H1, 05H1
Shimizu, Y., 41S9
Shirakawa, Y., 34H1, 35H4, 36M12, 39S2, 39S3, 39S8, 39S9, 41M2, 40S1, 40S2
Shockley, W., 39S1, 49W1, 49W2, 50K1, 50S8, 59W4, 51S2
Shoenberg, D., 36S13, 46S6, 49S4
Shturkin, D. A., 47S3
Shubina, L. A., 49S7
Shubrooks, G. E., 32S5
Shull, C. G., 49S2, 49S3
Shur, J. S., 37J1, 41S10, 46S5, 48D4, 49D6, 49S7
Sidhu, S. S., 33S7
Siegel, S., 36S10, 49S2
Sieverts, A., 10S1, 29S8
Simon, ı., 46S1
Simpson, K. M., 36S7
Sinizer, D. I., 48S11
Six, W., 34S5
Sixtus, K. J., 31S5, 32S9, 33S6, 35S6, 35S11, 37S9, 38S4, 40C3, 42C4, 49F3
Sizoo, J. G., 29S1
Skidmore, I. C., 50S7
Skinner, E. N., 48S12
Slater, J. C., 28S9, 30S3, 36S5, 36S12, 37S8, 40S3
Slottman, G. V., 47S4
Smart, J. S., 49M1, 49S3
Smith, B. M., 42S1
Smith, Charles S., 41S7, 46C2
Smith, Cyril S., 48S7
Smith, F. D., 35W2
Smith, S. W. J., 12S1, 24S2
Smith, W. S., 24S1, 33S8
Smithells, C. J., 38P3
Smithsonian Institution (Washington), 23S1
Smoluchowski, R., 41S4, 49G1, 51S5
Snoek, J. L., 34S5, 35B10, 35S2, 35S12, 36S6, 37S6, 38S5, 38S7, 39S6, 39S11, 39S12, 41R2, 41S5, 41S8, 47S2, 48S17, 48S18, 48S19, 49D5, 50S4, 51S1

Société d'Électro-Chimie, d'Electro-Métallurgie et des Aciéries Électrique d'Ugine, 47S6, 48S13, 48S15
Sodomann, H., 35K13
Söhnchen, E., 31S2
Soller, T., 37S5
Somin, B. E., 34M2, 36M10, 41M1
Souden, A. G., 36S8
Speed, B., 21S2
Speidel, H., 49K3
Spooner, T., 15C1, 23S3, 27S4
Sprung, H., 33D2
Squire, C. F., 38S9, 39S15
Stäblein, F., 28S5, 29S4, 34S3, 34S4
Stambke, K., 41F2
Standley, K. J., 50S5
Stanley, J. K., 43S1, 45S1, 47S5, 48S14
Starck, W., 03H5
Starr, C., 40S8
Steenland, M. J., 50D3
Steinberg, D., 34M6
Steinberger, R. L., 33S2
Steinhaus, W., 15S1, 28G1, 30G2, 33K4, 37S4, 43S3
Steinitz, R., 46S7, 48S16
Steinmetz, C. P., 91S1, 92S1
Steinwehr, H. E. v., 35S5, 49K4
Steneck, W. G., 50B1
Stern, O., 24G2
Stevens, J. S., 00S1
Stewart, J. Q., 18S1
Stewart, K. H., 50S2, 50S6, 51S3
Stierstadt, O., 30S5
Stössel, R., 31S8
Stogoff, A. F., 28S7, 29S2, 29S3
Stoner, E. C., 29S6, 30S2, 33S3, 36S4, 38S8, 39S10, 45S3, 48S1, 48S3, 49S8, 50S10, 51S4
Stout, W., 49S6
Strauss, B., 34S4
Street, R., 48S2, 49S9
Strijland, J., 37M7, 41M3
Strough, R. I., 49C5
Strutt, M. J. O., 27S6, 40S4
Stupart, G. V., 41L2
Sucksmith, W., 23S2, 25S1, 26S3, 28S10, 31S3, 32S4, 35S3, 38S10, 39S13, 39S16, 40P1, 43H1, 45S2, 51S6
Sugiura, J., 31S6
Sully, A. H., 37O3, 39O1, 41O1
Sumitomo, K., 379
Sundermann, W., 32V1
Sussmann, H., 50S9
Svensson, B., 36S16
Svetchnikov, V. N., 37S7
Swan, J. C., 28S6
Swanger, W. H., 30J1
Swinden, T., 13S1
Sykes, C., 28A5, 29S9, 34S9, 34S10, 35S8, 38S6, 39L1
Sykes, W. P., 33S4, 35S10, 39S14, 40E1

Taglang, P., 51T1
Takagi, M., 39T3

Takaki, H., 36K7, 37T3
Take, E., 11T1, 12H1
Takeda, S., 31T2, 40T3
Takei, T., 28T1, 29T1, 33K7, 40T2
Talbot, A. M., 40P3
Tamara, K., 32T1
Tammann, G., 27T1, 30T1, 33T2
Tarasov, L. P., 28T2, 37T1, 38T2, 39T2, 39T4, 39T5, 39T6
Taylor, A., 37B2, 37B9, 38B1
Taylor, G. G., 39B16
Taylor, G. I., 34T1
Taylor, N. W., 34E4
Tebble, R. S., 50S7
Tenger, S., 38T3
Terry, E. M., 10T1
Tetley, F. W., 35H2
Thal, W., 34T2
Théodorides, P., 22T1
Therkelsen, E, 33T3
Thielmann, K., 40T5
Thiessen, G., 40T4
Thiessen, P. A., 31H3, 32H3
Thoma, A., 36K5
Thomas, E. E., p. 16, 392
Thomas, E. J., 35C2
Thomas, H., 50S3
Thomassen, L., 37M9
Thompson, F. C., 44T1
Thompson, J. G., 35C3, 39T1
Thompson, M. de K., 42T1
Thompson, N., 40T1
Thomson, W., 1857T1
Thuras, A. L., 37T2
Tobusch, H., 08T1
Tolman, E. M., p. 383
Tonks, L., 31S5, 32S9, 35S6
Tonn, W., 32K4, 32K5, 32K12, 32V2, 33K8
Torrey, H. C., 46P1
Townsend, A., 35T2
Trautmann, B., 31T1
Trew, V. C. G., 41T1
Troiano, A. R., 43T1, 48T1
Trombe, F., 35U1, 37T4, 44T2, 45T1, 51T2
Tsai, B., 38S9, 43B5
Tuyn, W., 29O1
Tyndall, E. P. T., 24T1

Uddin, M. Z., 36S13
Uhlenbeck, G. E., 26G1
Umanskij, J. S., 39B17
Umino, S., 27U1
Underhill, E. M., 44U1
Upthegrove, C., 37M9
Urbain, G., 35U1

Valentiner, S., 33V1, 35V2, 47V3
Van Arkel, A. E., 36V1, 36V2
Van Bruggen, M. G., 36V1, 36V2
Van Horn, 40S6
van Peype, W. F., 38P2
van Santen, J. H., 50J1, 51V1
van Urk, A. T., 40V1

Van Vleck, J. H., 29V1, 32V3, 37V1, 39W3, 41V1, 45V1, 47V1
Vaughan, N. B., 37A1
Vautier, R., 49G7, 50V1
Veit, T., 08W2
Verwey, E. J. W., 35V3, 36V1, 36V2, 36V3, 39V1, 41V3, 47V4, 47V5
Vetter, M. 39F1
Vher, O. I., 30A4
Villari, E., 1865V1
Vogel, R., 17V1, 21V1, 32V1, 32V2, 32W1, 35V1, 38V3, 38V1, 47V2, 49V1
Vogt, E., 49B10
Volk, K. E., 38V2
Volkenstein, N., 41V2
Vonsovsky, S. V., 40V2, 40V3, 40V4, 48V1
Voronov, N. M., 38N4
Voss, G., 08V1, 08V2
Vournasos, A. C., 19V1

Wagner, E., 37K7, 38K9
Wahl, C. V., 32W2
Wahl, W., 10W2
Wainwright, C., 32W4
Wait, G. R., 28W4
Walker, J. G., 49B5, 49W4, 50W15
Wallbaum, H. J., 38V1, 39L3, 41W2
Waltenburg, R. G., 25M2
Walters, F. M., 35W4, 42W6
Walther, H., 35W5
Wannier, G. H., 41K3
Warburg, E., 81W1
Warren, D., 50L1
Wartenburg, H. v., 31W4
Watson, E. A., 23W1, 24W1, 24W2
Watts, J. T., 35B8
Webb, C. E., 26W2, 31W3, 34W3
Webb, J. S., 38W2
Weber, W. E., p. 423
Webster, W. L., 25W1, 25W4, 26W3, 27W3, 28W3, 30W3
Wedekind, E., 08W2, 07W4, 09W1, 11W3, 12W2, 24W7
Wegel, R. L., 35W5
Weibke, F., 36H4, 39W5
Weil, L., 47W2, 48W1, 49W3, 50W1, 51W1, 51W2
Weill, A. R., 43W1, 45F1
Weis, A., 35W1
Weiss, P., 96W1, 96W2, 05W2, 07W1, 08W1, 10W1, 10W3, 11W2, 12W1, 16W1, 18W1, 24W3, 24W5, 26W1, 26W4, 29W1, 29W2, 29W3, 34W2, 35U1, 36W2, 37W4
Weiss, P. R., 39W3, 48W2
Welch, A. J. E., 50W2
Wells, C., 35W4, 39W1, 40W1
Wenny, D. H., 47B6
Werth, H., 30E1
Westerdijk, J. B., 46H1
Westerman, A. B., 48E2
Westgren, A., 28W1, 28W2, 30W7, 38L6
Weston, J. C., 41B3
Wetzel, H., 41F1

Wever, F., 25W3, 26W4, 29W4, 29W7, 30W1, 30W4, 30W5, 30W6, 31W1, 31W2, 42W5
Weygandt, C. N., 42W4
Whiddington, R., 20P1
White, J. H., 32W2, 44F1
White, S. D., 37W1
Whiteley, J. H., 23W2
Wiedemann, G., 1862W1
Wiegand, D. W., 47W3
Wien, M., 32W6
Wiersma, E. C., 29W6
Wild, G., 26W5
Wilkinson, H., 38S6
Willems, H. W. V., 27J1
Willey, L. A., 48F2
Williams, C., 34W1
Williams, E. H., 15W1
Williams, H. J., 37C2, 37W1, 37W3, 37W5, 39W4, 39W2, 40B4, 41B5, 41W1, 41W3, 42B2, 42W2, 45B1, 48N2, 49W1, 49W2, 49W4, 50S8, 50W3, 50W4, p. 123, 538, 567
Williams, R. S., 07W3
Williams, S. R., 27W1, 27W2, 32W3, 33W1
Williams, S. V., 38P3
Williams, W. E., 05W1
Wills, A. P., 24W6
Wilson, A. J. C., 42W3, 46S6
Winkler, O., 32W1, 35G3, 38G4
Winslow, F. H., p. 533
Wise, E. M., 29E8, 37W2, 42W2, 48W3
Witte, H., 35L2
Wittke, H., 34W4, 38W3

Wöhler, F., 1859W1
Wohlfarth, E. P., 48S1, 49W5, 49W6
Wolman, W., 32W5
Woltjer, H. R., 23W3, 29W6
Wood, A. B., 35W2
Wood, W. A., 30W2, 32W4
Woodridge, W. J., 11W1
Wooley, J. C., 49S9
Wrathall, L. R., 36P2
Wunderlich, W., 49S5
Wwedensky, B., 21W1
Wyart, J., 44G2, 44G4, 46G3

Yager, W. A., 47Y1, 48Y3, 49Y1, 50G7
Yamamoto, M., 38Y1, 41Y1, 43Y1, 42Y1
Yamamoto, T., 38Y1, 47Y2
Yamanaka, N., 40Y1
Yasuda, T., 40T2
Yates, E. L., 36O1, 37O3
Yensen, T. D., 14Y1, 15Y1, 15Y2, 17Y1, 20Y1, 24Y1, 25Y1, 29Y1, 29Y2, 31Y1, 31Y2, 36Y1, 36Y2, 36Y3, 37Y1, 39Y1, 45Y1, 47S5, 48Y2
Yost, D. M., 49G2

Zaimovsky, A. S., 41Z1, 41Z2
Zapffe, C. A., 39H1
Zelikwan, A. N., 43P2
Zemczuzny, S., 09Z1
Zener, C., 38Z2, 46Z1
Ziegler, N. A., 32Z1, 32Z3, 36Y3
Zuithoff, A. J., 38J2, 38Z1
Zumbusch, W., 35Z1
Zwicker, U., 48K5

SUBJECT INDEX

(Alloy-compositions are listed in the order: Iron, Cobalt, Nickel, and then other elements in alphabetical order.)

Accommodation, 540
Advance, 153
Age hardening, see Hardening
Aging, 38, 59, 81, 358, 377, 797
Alcomax, 39, 346, 348, 393, 872
Alfer, 220, 679, 868
Allegheny Electric Metal, 130, 868
1040 Alloy, 160, 869, 870
Alni, 388, 872
Alnic, 387
Alnico, 261, 268, 346, 348, 385 ff, 872, 873
  Alnico 2, 39, 388, 417, 419
  Alnico 3, 385
  Alnico 4, 388
  Alnico 5, 39, 389, 682
  Alnico 5DG, 393
  Alnico 6, 396
  Alnico 12, 39, 394
Alperm, 220, 868, 870
Alternating-current methods, 852
Alternating fields, 549, 769 ff
Alternating hysteresis, see Hysteresis
Amalgams, cobalt, 291
  iron, 236, 829
  nickel, 320
Angular momentum, 449, 806
Anhyster, 868
Anhysteretic curve, 8, 549, 604
Anisotropy, elastic, 92, 692
Anisotropy, see Magnetic anisotropy
Annealing, 14, 28, 58, 60, 112, 115
  see also Magnetic anneal
Antiferromagnetism, 6, 470 ff
Apparent permeability, 847
Armco electric alloy, 868
Atomic constants, 864
Atomic diameters, 41, 864
Atomic moments, 54, 264, 270, 441, 867
Atomic numbers, 456, 864
Atomic ordering (see under materials)
Atomic structure, 434
Atomic weights, 864
Audiolloy, 868
Avogadro's number, 867

Babbitt permeameter, 849
Bainite, 366
Ballistic galvanometer, 843
Barkhausen effect, 524 ff, 538

Barnett effect, 450, 454
Bethe curve, 443, 447, 724, 726
Biasing field and induction, 6, 539
Bismuth, 456, 856
Bitter patterns, see Powder patterns
Bloch wall, see Domain walls
Bohr magneton, 8, 450, 463, 867
Bohr magneton number, 8
  (see also under materials)
Boundary displacement, 479, 820
  see also Domain walls
Brillouin function, 431
Burrows permeameter, 850
Butterfly loops, 541
  (see also under materials)

Calmalloy 311, 868
Carbon as impurity, 36, 62, 83, 392
  see also Iron-carbon alloys, etc.
Carbon steel, 39, 364 ff, 676
Carbonyl iron, 51
  powder, 146, 870
Carpenter 49, 130, 868
Caslon, 873
Casting, 20
Cementite, 366
Cerium, 341
Cerromag, 868
Chattock potentiometer, 850
Chrome Permalloy, 140, 146 ff, 868
Chrome steel, 39, 346, 347, 374 ff, 872
Chrome-tungsten steel, 378
Chromium, 341
Chromium alloys with antimony, 341
  arsenic, 341
  oxygen, 342
  platinum, 342, 415
  selenium, 342
  sulfur, 342
  tellurium, 342
Cioffi fluxmeter, 861
Cobalt, atomic moment, 441, 867
  crystal properties, 568
  elastic properties, 262, 687
  ferromagnetic resonance, 809
  magnetic properties, 264 ff
  magnetostriction, 647, 657 ff
  metallurgy, 261
  physical properties, 262

959

Cobalt, specific heat, 738
  structure, 263
  volume magnetostriction, 658
Cobalt alloys (binary) with aluminum, 283
  antimony, 284
  arsenic, 282, 285
  beryllium, 285
  bismuth, 286
  boron, 282, 287
  carbon, 287
  cerium, 288
  chromium, 288, 441
  copper, 289
  germanium, 289
  gold, 290
  hydrogen, 290
  iron, see Iron-cobalt alloys
  lead, 290
  manganese, 291, 441
  mercury, 291
  molybdenum, 292
  nickel, see Cobalt-nickel alloys
  niobium, 293
  nitrogen, 293
  oxygen, 293
  palladium, 414
  phosphorus, 282, 293
  platinum, 39, 282, 346, 412, 873
  selenium, 293
  silicon, 294
  silver, 294
  sulfur, 282, 294
  tantalum, 295
  tellurium, 295
  thallium, 295
  tin, 295
  titanium, 296
  tungsten, 297
  vanadium, 298
  zinc, 282, 298
  zirconium, 299
Cobalt-aluminum-manganese alloys, 418
Cobalt-copper-manganese alloys, 418
Cobalt-copper-molybdenum alloys, 418
Cobalt-nickel alloys, crystal anisotropy, 571
  magnetic properties, 278 ff
  magnetoresistance, 760
  magnetostriction, 672 ff
  physical properties, 276
  volume magnetostriction, 674
Cobalt-nickel-aluminum alloys, 281, 418
Cobalt-nickel-arsenic alloys, 281, 418
Cobalt-nickel-copper alloys, 280, 282, 402 ff, 418
Cobalt-nickel-silicon alloys, 281, 418
Cobalt-nickel-titanium alloys, 394
Cobalt steel, 39, 261, 346, 347, 379 ff, 872
Coercive force, 5, 499 ff, 823 ff
  (see also under materials)
Coercive force, fine particles, 348, 828 ff
Coercivity, 5, 499
Coinage alloy, 299, 308
Cold-rolled silicon-iron, see Grain-oriented silicon-iron

Cold working, 25, 619
Collective electron theory, 442
Commutation curve, 6
Comol, 39, 346, 383 ff
Compensator Alloy, 868
Composition, effect of, 29
Compressed Molybdenum Permalloy powder, 144
Compressed Permalloy powder, 131, 146, 870
Compression, see Stress
Conpernik, 130, 868
Constantan, 153, 299, 308
Constants, physical, 864
  (see also under materials)
Constitutional diagram, see Phase diagram
Cotton balance, 842, 856
Crystals, anisotropy, 11, 32, 478, 555 ff, 828
  anisotropy constants, 482, 567 ff
  anisotropy energy, 563 ff, 811
  coercive force, 561
  ferromagnetic resonance, 809
  low fields, 559
  magnetization, 478
  magnetoresistance, 764
  magnetostriction, 645 ff, 813
  permeability, 558 ff
  preferred orientations, 42, 88, 124
  remanence, 561
  torque curves, 565 ff
Cunico, 346, 402, 873
Cunife, 39, 346, 396 ff, 402, 873
Curie-Chéveneau balance, 858
Curie law, 428
Curie point, 6, 431, 713 ff
  effect of composition, 30, 720
  effect of pressure, 445, 723 ff
  magnetization near, 716
  table of values, 723
  virtual, 190
Curie-Weiss law, 429
Cyclic state, 6

Damping, of oscillations, 685, 699 ff
  macro-eddy-currents, 701
  magnetomechanical hysteresis, 707
  micro-eddy-currents, 706
  separation of losses, 709
  various materials, 701
ΔE effect, 684 ff
Deltamax, 133, 868
Demagnetization curve, 8, 344, 349
Demagnetizing, 6, 547
Demagnetizing factors, 10, 845 ff
Demagnetizing field, 9, 846
Diamagnetism, 5, 455 ff
  elements, 456
  gases, 459
  measurement, 857
  organic compounds, 461
  salts, 459
  theory, 458
  water, 461
Differential permeability, 7
Diffusion, 64

Dilatation, see Thermal expansion
Dipoles, assemblage of, 423, 592
Discontinuities in magnetization, see Barkhausen effect
Dislocations, 621
Dispersion hardening, see Hardening
Displacement factor, 549
Domain theory, 13, 477 ff, 811 ff
  antiferromagnetism, 470
  Barkhausen effect, 524
  coercive force, 823
  domain geometry, 834
  domain rotation, 479, 818
  fine particles, 828
  micro-eddy-currents, 706
  powder patterns, 532 ff, 834
  preferred domain orientations, 481, 586 ff
  resistivity, 750
  retentivity, 501
  stress, 609
  systematic treatment, 811
  Weiss, theory, 429, 477
  see also Domain walls
Domain walls, see also Powder patterns
  displacement, 479, 820
  energy, 814 ff
  structure, 815
  thickness, 815
Dreyfus magnet, 855
Dynamo steel, 78
Dysprosium, 341, 342

Easy directions of magnetization, see Magnetic anisotropy
Eddy currents, 769
  cylinders, 775
  high fields, 781
  high frequencies, 780
  lag, 769
  losses, 146, 768, 778, 781
  low fields, 127, 780
  sheets, 770
Effective permeability, 769 ff, 805
Effective resistance, 777
Einstein-de Haas effect, 450 ff, 454
Elastic loss, 685, 699 ff, 709
Elastic modulus, 684 ff
  near Curie point, 697
  temperature effect, 689, 697
  theory, large internal strain, 690
  theory, small internal strain, 692
  various materials, 687 ff
Elastoresistance, 749 ff
Electrical properties, 745 ff
Electromagnets, 855
Electron, angular momentum, 449, 806
Electron moment, 430, 449, 458, 867
Electron spin, 430, 435, 450, 462, 467
Elements, physical properties, 864
Elinvar, 147, 698
Energy, 729, 811
  of Bloch wall, 814
  of crystal anisotropy, 482
  of demagnetization, 814
  in field, 482, 729

Energy, free energy, 731
  of magnetization, 729, 813
  of stress, 482, 812
  thermodynamics, 731
Energy bands, 437, 442
Energy product, 8, 349
Entropy, 731
Epstein apparatus, 853
Equilibrium diagram, see Phase diagram
Ewing model, 424
Ewing theory, 423 ff
Exchange forces, 427, 443
Excitation curve, 83
Expansion, see Thermal expansion

Fabrication, 14, 24
Fahy permeameter, 850
Faraday effect, 834
Feralsi, 100
Fermi energy, 442
Fernico, 160
Ferramic, 868
Ferrichrome, 160
Ferric induction, see Intrinsic induction
Ferrites, 244 ff, 421, 681, 809
Ferromagnetic resonance, 803 ff
Ferromagnetism, 5
Ferrometer, 853
Ferroxcube, 18, 247 ff, 868, 870
Field, see Molecular field, Field-strength
Field-strength, 1, 838 ff
Fine particles, 45, 419, 828
Flutter effect, see Superposed fields
Force on magnetized body, 455, 841, 858
Foreign body theory, 825
Forrer loops, 610
Foucault currents, see Eddy currents
Free energy, 731
Frölich-Kennelly relation, 350, 476, 484, 542
Fullness factor, 351

Gadolinium, 341, 444
Gadolinium sulfate, 466
Gases, measurement of, 859
Gauss, 3
Gaussian positions, 839
Gouy method, 859
Grain orientation, 11, 586
Grain-oriented silicon-iron, 70, 88 ff, 870
Grain size, 86
Gyromagnetic effect, 449 ff
  values of g, 454, 808

Hadfield steel, 234
Hardening, 33, 347, 365
Hardness, coefficient of magnetic, 484
Haworth fluxmeter, 861
H. C. R., 133, 868
Heat capacity, 732 ff
Heat treatment, 14, 28, 58, 60, 112, 115
  see also Magnetic anneal
Heusler alloys, 328 ff, 870
  composition, 333
  crystal properties, 574
  ferromagnetic resonance, 809

962 SUBJECT INDEX

Heusler alloys, magnetostriction, 681
  permeability, 333
  saturation, 330
  structure, 328
High fields, production, see Magnetic fields
High frequency, eddy currents, 770 ff
  losses, 778, 802
  permeability, 798 ff
  see also Ferromagnetic resonance
Hiperco, 207, 868, 870
Hipernik, 129, 133, 868, 870
Hipersil, 88
History, gyromagnetic effect, 450, 451
  iron-silicon alloys, 67
  iron-nickel alloys, 106
  magnetostriction, 630
  permanent magnets, 345
H-Monel, 312
Holes (in electron shells), 437, 439
Honda steel, 347, 379 ff, 872
Hooke's law, variations from, 684
Hycomax, 873
Hydrogen treatment, 28, 60, 92, 121, 138, 142, 512
Hyflux, 873
Hymu 80, 868
Hyperm, 868
Hysteresis, 5, 507 ff
  alternating, 507
  heat, 508, 518
  loops, constricted, 498
  loops, rectangular, 494
  loss coefficients, 146, 797
  magnetomechanical, 604
  magnetostriction, 655, 658, 659, 668
  resistivity, 747
  rotational, 514, 590
  (see also under materials)

Ideal magnetization, 8, 549, 604
Illiovici permeameter, 850
Impurities, 35, 61, 83, 390
Incremental permeability, 6, 539 ff
Inductance, 772 ff
Inductance bridge, 776, 852
Induction, intrinsic (ferric), 7
  magnetic, 2
  residual, 5, 370, 499 ff, 623
  see also Magnetization
Initial permeability, see Permeability, initial
Instability, 126, 177
Intensity of magnetization, concept, 2
  measurement, 843 ff
  saturation, 7
  see also Magnetization
Internal strain, see Stress
Intrinsic induction, 7
Invar, 102, 448, 642, 644
Iron, 48 ff
  analyses, 51
  approach to saturation, 487
  atomic moment, 441, 867
  crystals, 555, 567
  Curie points of alloys, 722

Iron, damping, 701
  diffusion in, 64
  elastic modulus, 687
  fine powders, 832
  heat capacity, 737
  high fields, 487
  high frequency permeability, 799
  hydrogen treatment, 60
  impurities, 52, 61
  lag, 792
  low fields, 489
  magnetic properties, 55, 870, 873
  magnetocaloric effect, 741
  magnetoresistance, 748
  magnetostriction, 632, 639, 645, 652, 655 ff
  manufacture, 48
  physical properties, 53
  pressure and Curie point, 726
  saturation, 54, 867
  stress, 602
  structure, 434, 478
  superposed permeability, 539
  temperature, 59, 719
  volume magnetostriction, 642, 657
Iron alloys (binary) with aluminum, 210 ff, 574, 678
  antimony, 220
  arsenic, 220
  beryllium, 210, 222, 416
  boron, 210, 224
  carbon (see also Steel), 35, 37, 65, 676, 364 ff
  cerium, 210, 225
  chromium, 226, 441
  cobalt, see Iron-cobalt alloys
  copper, 229
  gadolinium, 233
  germanium, 233
  gold, 412
  hydrogen, 233
  iridium, 249
  manganese, 234, 441, 678
  mercury, 236
  molybdenum, 35, 236, 416, 680
  neodymium, 238
  nickel, see Iron-nickel alloys
  niobium, 238
  nitrogen, 35, 38, 64, 66, 239
  osmium, 249
  oxygen, 35, 64, 240, 473, 802
  palladium, 249, 411
  phosphorus, 64, 210, 251, 680
  platinum, 249, 410, 873
  rhodium, 249
  ruthenium, 249
  selenium, 253
  silver, 412
  sulfur, 66, 210
  tantalum, 253
  tin, 210, 254
  titanium, 255, 416, 680
  tungsten, 256, 416, 680
  uranium, 257
  vanadium, 258, 441

Iron alloys (binary), zinc, 259
  zirconium, 260
Iron-aluminum-cobalt alloys, see Iron-cobalt alloys
Iron-aluminum-copper alloys, 418
Iron-aluminum-manganese alloys, 418
Iron-aluminum-nickel alloys, see Iron-nickel alloys
Iron carbonyl, 51
Iron-chromium-copper alloys, 418
Iron-chromium-manganese alloys, 418
Iron-chromium-molybdenum alloys, 418
Iron-chromium-molybdenum-tungsten alloys, 418
Iron-chromium-titanium alloys, 418
Iron-chromium-titanium-tungsten alloys, 418
Iron-chromium-tungsten, 418
Iron-cobalt alloys (binary), 33, 190
  atomic moments, 441
  crystal properties, 571, 574
  fabrication, heat treatment, 192
  magnetic anneal, 199
  magnetic properties, 193
  magnetoresistance, 759
  magnetostriction, 663
  physical properties, 190
  structure, 15, 190
  temperature effect, 196
Iron-cobalt alloys containing aluminum, 205, 207, 389, 416
  antimony, 205, 207, 416
  beryllium, 205, 208, 416
  carbon, 200
  chromium, 200, 205, 416
  copper, 416
  manganese, 200, 205, 208, 416
  molybdenum, 200, 205, 346, 382 ff, 416
  nickel, see Iron-cobalt-nickel alloys
  niobium, 200
  silicon, 416
  tantalum, 205, 208, 416
  titanium, 200, 394, 416
  tungsten, 200, 205, 208, 382 ff, 416
  vanadium, 200, 346, 348, 405 ff, 416
Iron-cobalt-nickel alloys, crystal properties, 571
  magnetic properties, 160 ff, 200, 416
  magnetostriction, 674
  specific heat, 738
Iron-cobalt-nickel alloys containing aluminum, 389
  chromium, 160
  copper, 389
  manganese, 180
  titanium, 161
Iron-copper-molybdenum alloys, 418
Iron-manganese-titanium alloys, 418
Iron-manganese-vanadium alloys, 418
Iron-molybdenum-vanadium alloys, 418
Iron-nickel alloys (binary), atomic moments, 441
  commercial alloys, 128
  crystal properties, 569
  damping, 708

Iron-nickel alloys (binary), elastic properties, 106, 693 ff
  history, 106
  hydrogen anneal, 121
  magnetic anneal, 117
  magnetic properties, 111 ff
  magnetostriction, 632, 637, 648, 664 ff
  physical properties, 103
  pressure and Curie point, 726
  resistivity and magnetoresistance, 104, 757
  specific heat, 738
  structure, 102
  thermal expansion, 643
  volume magnetostriction, 639, 642, 670
Iron-nickel alloys containing aluminum, 183, 346, 385 ff, 417, 682
  beryllium, 189, 417
  chromium, 140, 146, 417
  cobalt, see Iron-cobalt-nickel alloys
  copper, 153 ff, 346, 397 ff, 417
  manganese, 180 ff, 417
  molybdenum, 134 ff, 417
  silicon, 184, 417
  silver, 184, 417
  tantalum, 186 ff, 417
  tin, 417
  titanium, 186 ff, 394 ff, 417
  tungsten, 186 ff, 417
  vanadium, 186 ff, 417
Iron-silicon alloys (binary), 30, 67 ff, 870
  crystals, 88, 571, 574
  history, 67
  impurities, 83, 87
  losses, 79, 513–4
  low fields, 92
  magnetic properties, 76 ff
  magnetoresistance, 758
  magnetostriction, 649, 678
  metallic additions, 87
  permeability, 79
  physical properties, 74
  production, 68
  saturation, 76
  structure, 71
Iron-silicon alloys containing aluminum, 87, 95, 418, 574
  beryllium, 88
  carbon, 84
  copper, 87
  manganese, 87
  nitrogen, 86
  oxygen, 86
  phosphorus, 86
  sulfur, 86
  titanium, 88
  vanadium, 88
Iron-titanium-tungsten alloys, 418
Iron-tungsten-vanadium alloys, 418
Irreversible alloys, 715
Isoperm, 125, 133 ff, 146, 153, 157, 183, 504, 797, 868, 870
Isthmus method, 851

Joule magnetostriction, 627

SUBJECT INDEX

Kerr effect, 861
K-Monel, 312
Koepsel permeameter, 851
Koerzit, 346
Konel, 161, 394
Kovar, 160, 163
KS steel, 261, 346, 379 ff

Lag, 769 ff, 788 ff
  coefficients, 146, 780, 797
  eddy currents, 769
  impurities, 789
  Jordan, 796
  long period, 797
  temperature sensitive, 789
Lamont's law, 484
Landé factor, 451, 808
Langevin theory
  diamagnetism, 458
  paramagnetism, 427, 462
Lanthanum, 341
Larmor precession, 807
Lattice constants, 41, 864
  (see also under materials)
Leakage flux, 349, 360 ff
Ledeburite, 366
Lines of induction, 3
Lloyd-Fisher apparatus, 853
Load line, 363
Loss angle, 770
Loss constants, *see* Isoperm, Molybdenum Permalloy powder, Ferroxcube, etc.
Loss equation, 127, 780
Low fields, 489 ff, 780–1
Low temperatures, 465–6

Macro eddy currents, 703
Magnetic anisotropy, 11, 32, 478, 482, 555 ff, 828
  crystals, 482, 555 ff, 809, 811
  magnetic anneal, 117 ff, 171, 389
  origin, 592
  polycrystalline material, 586
  strain, 482
  torque curves, 584
Magnetic anneal, 117, 171, 348, 389
Magnetic field, 1, 838 ff
Magnetic field, production, currents, 839 ff, 855
  electromagnets, 855
  high fields, 855
  magnets, 838, 855
Magnetic flux, 3
Magnetic hysteresis, *see* Hysteresis
Magnetic induction, 3
  see also Magnetization
Magnetic moment, 2
  see also Atomic moments, Electron moment
Magnetic viscosity, *see* Lag
Magnetic waves, 788
Magnetite, 245, 574, 649, 681, 758, 810
Magnetization, crystals, 478
  curve, 476 ff, 576
  elasticity, 684

Magnetization, energy, 729 ff, 811 ff
  ideal, 8, 549, 604
  measurement, 843
  processes, 476, 818
  by rotation, 451
  spontaneous, 12, 429, 477
  strong fields, 484
  superposed fields, 539
  temperature effect, 713
  tension effect, 595
Magnetizing coils, 839 ff
Magnetocaloric effect, 520, 645, 725, 740 ff
Magnetoflex, 346, 348, 396 ff, 401, 873
Magnetomechanical hysteresis, 604, 707
Magnetometer, 857
Magnetomotive force, 842
Magneton, *see* Bohr magneton
Magnetoresistance, 745 ff
Magnetostriction, 11, 595, 627 ff
  crystals, 645, 652, 766, 813
  domain theory, 634
  experimental methods, 628
  figure, 635
  history, 630
  hysteresis, 656 ff, 668
  Joule effect, 627, 630
  permanent magnets, 682, 832
  reversible, 639
  saturation, 632
  theory, atomic, 649, 654
  thermodynamics, 732
  volume, 630, 641 ff
  Wiedemann effect, 628
  (see also under materials)
Magnets, *see* Permanent magnets
Manganese, 341
Manganese alloys (binary) with antimony, 334, 341, 575
  arsenic, 334, 336, 341
  bismuth, 334, 337, 341, 420, 575, 833, 46
  boron, 334, 337
  carbon, 337
  cobalt, 291, 441
  fluoride, 472
  gallium, 337
  hydrogen, 337
  indium, 337
  iron, 234, 441, 678
  nickel, 315 ff, 441, 768
  nitrogen, 338
  oxygen, 338, 473
  phosphorus, 339
  selenium, 339, 470
  silicon, 339
  sulfur, 339
  tellurium, 339, 470
  tin, 339
Manganese-aluminum-copper alloys, 328 ff
Manganese-aluminum-silver alloys, 328, 333, 410, 415, 418
Manganese-copper-gallium alloys, 328, 333
Manganese-copper-indium alloys, 328, 333
Manganese-copper-tin alloys, 328 ff
Manganites of La, Ca, Sr, Ba, 339

Martensite, 366, 796
Maximum permeability, 3, 6
  (see also under materials)
Measurements, a-c methods, 852
  curve tracers, 861
  elastic constants, 686
  elastic losses, 686
  liquids and gases, 859
  magnetic quantities
    basic relations, 838
    common methods, 843
    special methods, 855
  magnetostriction, 628
  para- and diamagnetism, 857
  small specimens, 855, 857, 861
Megaperm, 183, 870
Melting, 20
Micro eddy currents, 706
Microphones, magnetic, 615
Microwave frequencies, 798, 803
Mishima alloys, 346, 347, 385 ff, 684, 832
MK magnet alloy, 385 ff
Molecular beams, 468
Molecular field, 427, 433, 443
Molybdenum Permalloy, 136 ff, 139 ff, 144 ff, 539, 542, 603, 607, 868, 870
Molybdenum Permalloy powder, 144
Molybdenum steel, 378
Moment
  angular (spin), 450
  atomic, alloys, 441 (see also under composition)
  atomic, elements, 54, 264, 269, 867
  magnetic, 2, 435, 450
  nuclear, 468
Monel metal, 153, 268, 299, 308, 311, 312, 680
Monimax, 144, 868, 870
Moniseal, 144
Moving boundary, see Domain walls
Mumetal, 153 ff, 159, 868, 870

Nachwirkung, see Lag
Neodymium, 341
New Honda steel, 261, 394 ff
New KS, 394 ff, 873
Nicaloi, 130, 869
Nickel, 267 ff, 870
  atomic moment, 441, 867
  crystal properties, 568
  Curie points of alloys, 721
  damping in, 702 ff
  elastic modulus, 687, 693 ff
  elastic properties, 689 ff
  ferromagnetic resonance, 806
  fine particles, 833
  high frequency permeability, 799 ff
  magnetocaloric effect, 741
  magnetic properties, 269 ff
  magnetostriction, 632, 636, 650, 659 ff
  magnetoresistance, 748
  metallurgy, 268
  physical properties, 269
  pressure and Curie point, 726

Nickel, specific heat, 734 ff
  temperature, 718
  volume magnetostriction, 641, 661
Nickel alloys (binary) with aluminum, 299, 441
  antimony, 300, 441
  arsenic, 300
  beryllium, 302, 681
  bismuth, 305
  boron, 305
  carbon, 305
  cerium, 307
  chromium, 307, 441
  cobalt, see Cobalt nickel alloys
  copper, 308 ff, 440, 574, 680, 726, 761
  gallium, 313
  germanium, 313
  gold, 313, 722
  hydrogen, 314
  indium, 315
  iron, see Iron nickel alloys
  magnesium, 315
  manganese, 315 ff, 441, 768
  mercury, 320
  molybdenum, 320
  niobium, 321
  nitrogen, 321
  oxygen, 321
  palladium, 415, 441
  phosphorus, 322
  platinum, 414, 441
  selenium, 322
  silicon, 322, 441
  silver, 323
  sulfur, 323
  tantalum, 323
  tellurium, 323
  thallium, 323
  tin, 323
  titanium, 325
  tungsten, 325, 681
  vanadium, 326, 441
  zinc, 327, 441
  zirconium, 328
Nickel-aluminum-manganese alloys, 418
Nickel-copper alloys, complex, 312, 313, 323, 418, 681
Nickel-silicon-, 299, 308
Nickel under tension, 612
Nipermag, 394, 396, 873
Niwa permeameter, 850
Normal permeability, 6
Nuclear moments, 468

Oersted, 2
Oerstit, 872
O.P. Magnet, 873
Order-disorder (see under materials)
Orientations, preferred, 42, 88, 124
Orthonik, 133, 869
Orthonol, 133, 869
Output transformers, 540
Oxide magnets, 421
Oxygen in iron, 35, 62, 240

SUBJECT INDEX

Oxygen in silicon-iron, 86

Paramagnetism, 5
  constant with temperature, 462
  free electrons, 467
  gases, 466
  iron group, 464
  Langevin theory, 427, 462
  low temperatures, 465
  measurement, 857
  molecular beams, 468
  quantum theory, 427, 462
  rare earths, 463
  strong, 462, 464
  weak, 462
Paris (Bellevue) magnet, 855, 856
Pearlite, 366
Pendulum magnetometer, 860
Permalloy, 109 ff, 128, 869
  damping, 703, 709
  elastic properties, 687, 694
  loss coefficients, 146, 797
  magnetic, physical properties, 109 ff, 870
  magnetoresistance, 745
  magnetostriction, 664 ff
  stress, 599 ff, 601 ff
  superposed permeability, 543
Permalloy Powder, 131, 146, 870
Permanent magnets, 344 ff
  aging, 377
  demagnetization curve, 344, 349
  design, 359
  energy product, 349
  history, 345
  load line, 363
  magnetic anneal, 348, 389 ff
  magnetoresistance, 758
  magnetostriction, 682 ff, 832
  oxide magnets, 421
  physical basis, 832
  powder magnets, 419
  reversible permeability, 353
  stability, 354
  temperature coefficients, 357
  (See also under materials)
Permanent magnet alloys (see under composition: Iron-carbon alloys, etc.)
Permanite, 869
Permeability, differential, 7
Permeability, incremental, 6, 539 ff
Permeability, initial, 6, 489, 821
  domain rotation, 821
  domain wall movement, 822
  high frequency, 801
  internal strain, 623, 821
  nickel under tension, 612
  upper limit, 626
  (see also under materials)
Permeability maximum, 3, 6
  (see also under materials)
Permeability, normal, 6
Permeability, reversible, 7, 345, 353, 540 ff, 623
Permeability, superposed, 6, 539, 549 ff

Permeameters, 849 ff
Permendur, 197 ff, 261, 869, 870
Permenorm, 127, 130, 133, 869, 870
Permet, 418, 873
Perminvar, 160 ff, 179 ff, 261, 797, 869, 870
PF Magnets, 420, 421
Phase diagram, 15
  cobalt alloys, 283 ff
  iron alloys, 210 ff
  nickel alloys, 299 ff
  (see under alloy compositions)
Physical properties of the elements, 864
Poles, magnetic, 1, 814, 835
Potassium sulfide, 342
Powder magnets, 419
Powder metallurgy, 417
Powder patterns, 532 ff, 559, 834 ff
Precession in field, 450, 807
Precipitation hardening, see Hardening
Preferred orientations, 42, 88, 124
Pressure, see Stress
Properties of the elements, 864
Pyrrhotite, 253, 574

Quadrupole moment, 593
Quality factor, Q, 774
Quantum theory, 428, 430
  Bohr magneton, 430
  Brillouin function, 431
  gyromagnetic ratio, 450
  molecular field, 443
  paramagnetism, 427, 462
  temperature saturation, 428

Radiometal, 869, 870
Rare earths, 463
Rayleigh relations, 489, 688, 708
Recovery, 43, 60
Rectangular loops, 494
Relaxation time, 790, 808
Reluctance, 842
Reluctivity, 484
Remalloy, 39, 261, 346, 347, 383 ff, 873
Remanence, 5, 370, 499 ff, 623, 561
  (see also under materials)
Residual induction, 5, 370, 499 ff, 623
Residual loss, 127, 780
  see also Isoperm, Molybdenum Permalloy powder, Ferroxcube
Resistivity, electrical, alloys, 40, 756 ff
  change with magnetization, 745 ff
  change with temperature, 760 ff
  change with tension, 748
  domain theory, 750 ff
  single crystals, 764 ff
  temperature coefficient (see under specific materials)
Retentivity, 5, 370, 499 ff, 623
Reversible permeability, 7, 353
  (see also under materials)
Rhometal, 869
R-Monel, 312
Rotational hysteresis, 514 ff

Rotation by magnetization, 452
Rotation of domain magnetization, 479, 818

Saturation induction, 3
 see also Saturation magnetization
Saturation magnetization, 7
 approach to saturation, 484
 $T^{3/2}$ law, 448, 719
 temperature variation, 427 ff, 448, 713 ff
 values for Fe, Co, Ni, 54, 264, 269, 867
 (see also under materials)
Saturation moments, see Atomic moments
Saturation permeameter, 850
Search coil, 845
Sendust, 95, 716, 869, 870
Senperm, 186
Sigma phase, see Iron-chromium alloys
Silectron, 88
Silicon alloys, see Iron-silicon alloys, etc.
Silicon-iron or Silicon steel, see Iron-silicon alloys
Silmanal, 39, 415, 873
Silver fluoride, 342
Single crystals, see Crystals
Sinimax, 185, 869, 870
Skin thickness, 771, 777
Slags, 22
S-Monel, 312
Special orientations, 42, 88, 124
Specific heat, 734 ff
Spinels, 244 ff, 726
Spin moment, 430
 magnetic, 435, 450
 mechanical, 450
Spinning electron, 430, 435
Spontaneous magnetization, 12, 429, 477
Spring-back, 353
Stability, 145, 354 ff
Stainless Invar, 207, 644, 675
Stainless steel, 147, 833
Steel, 364, 870, 872
 additions to, 382
 chrome, 346, 347, 374 ff, 872
 cobalt, 346, 347, 379 ff, 872
 magnetostriction, 676 ff
 manganese, 370
 molybdenum, 378
 stainless, see Stainless steel
 tungsten, 346, 347, 371 ff, 872
Steinmetz law, 509, 513
Stellite, 261
Strain, see Stress and Magnetostriction
Stress, effect of, 595 ff
 domain theory, 610
 hydrostatic pressure, 645
 hysteresis, 604
 internal stress, 619, 624
 iron, 598, 602
 iron-nickel alloys, 599, 601
 large stresses, 596
 nickel, 598, 600
 pressure and Curie point, 445, 723 ff
 remanence, 624
 residual stresses, 626

Stress, small stresses, 613
Structural steel, 95, 147
Sucksmith balance, 859
Sulfur as impurity, 36, 62, 65, 86
Superexchange, 475
Superlattice (see under materials)
Supermalloy, 138, 140 ff, 716, 805, 869, 870
Super-nilvar, 160
Super-permalloy, 185, 869
Super-perminvar, 869
Superposed fields and inductions, 539 ff, 552
Superstructures (see under materials)
Susceptibility, 5
 atomic, 456
 crystals, 558 ff
 elements, 456
 gases, 466
 initial, see Initial permeability
 iron group, 464
 measurement, 857
 molecular, 457
 rare earths, 463
 theory, diamagnetism, 458
 theory, paramagnetism, 462
 volume, 455
 see also Permeability
Swedish iron, 51

Temperature, 713 ff
 compensator alloys, 133, 153, 268, 311, 728
 diamagnetism, 461
 high temperatures, 716
 low temperatures, 719
 magnetization, 427 ff, 448, 713 ff
 paramagnetism, 463
 phase change, 16, 715
 resistivity, effect on, 760
 see also Curie point
Textures, see Preferred orientations
Thermal expansion (see under materials)
Thermal expansion, magnetic, 447, 644, 727
Thermalloy, 311, 869, 870
Thermodynamics of magnetic processes, 731
Thermodynamic potential, 732
Thermoperm, 133, 311, 869, 870
Thin films, 45
Ticonal, 389, 394
Time lag, see Lag
$T^{3/2}$ law, 448, 719
Torque curves, 584, 743
Torsion, 636, 675
Torsion balance, 858
Trafoperm, 869
Trancor X, 3X, 88, 869
Transformations, 16
 cobalt, 262, 264
 iron, 53, 54
 magnetic, see Curie point
 nickel, 270, 271
 phase, 16
 (see also under materials)
Transformer "steel," see Iron-silicon alloys
Tromolit, 418, 873
Troostite, 366

Tungsten steel, 39, 346, 347, 371 ff, 872

Ultramicrometer, 628

Vacuum melting, 23
Vanadium, 341
Vanadium Permendur, 204, 545, 869, 870
Vectolite, 421, 873
Vibration, effect of, 700, 798
Vicalloy, 39, 346, 348, 405 ff, 873
Villari reversal, 602
Virgin curve, 6
Virtual Curie point, 190
Viscosity, *see* Lag

Volume magnetostriction, *see* Magnetostriction

Wall displacement, *see* Domain walls
Warburg's law, 508, 520
Water, susceptibility, 461
Weiss domain, *see* Domain theory
Weiss field, *see* Molecular field
Weiss theory, 427 ff
Wiedemann effect, 628

Yokes, 849
Young's modulus, 684 ff
 (see also under materials)